International Conference on Mechanical Engineering, Automation and Control Systems (MEACS 2020)

IOP Conference Series: Materials Science and Engineering
Volume 1064

Novosibirsk, Russia
17 September 2020

Editors:

Vadim Skiba
Nikita Martyushev

ISBN: 978-1-7138-4575-1
ISSN: 1757-8981

Printed from e-media with permission by:

Curran Associates, Inc.
57 Morehouse Lane
Red Hook, NY 12571

Some format issues inherent in the e-media version may also appear in this print version.

Institute of Physics
Dirac House, Temple Back
Bristol BS1 6BE UK

Phone: 44 1 17 929 7481
Fax: 44 1 17 920 0979

techtracking@iop.org

Additional copies of this publication are available from:

Curran Associates, Inc.
57 Morehouse Lane
Red Hook, NY 12571 USA
Phone: 845-758-0400
Fax: 845-758-2633
Email: curran@proceedings.com
Web: www.proceedings.com

TABLE OF CONTENTS

IOP Conf. Series: Materials Science and Engineering 1064 (2021) 011001 doi:10.1088/1757-899X/1064/1/011001

MEACS-2020

International Conference on Mechanical Engineering, Automation and Control Systems 2020 (MEACS-2020) was successfully held on September 17, 2020. Initially, the conference was planned to be held in March, as part of the exhibition of equipment for metalworking and welding "MashExpo Siberia 2020" in Novosibirsk. However, due to the lockdown and all the restrictions associated with the coronavirus pandemic, MEACS-2020 was postponed until autumn. The conference was hosted by Novosibirsk State Technical University (Novosibirsk, Russia). Today, Novosibirsk is one of the largest scientific centers in Russia and in the Commonwealth of Independent States.

MEACS-2020 provided an opportunity for researchers and practitioners to present and discuss up-to-date developments and innovations in the fields of mechanical engineering, automation of production processes, materials science, and industrial engineering. The participants presented papers on the issues of technological machines and equipment, mechanics and mechatronics, materials science in mechanical engineering, manufacturing process management, automation and optimization of production engineering.

After careful reviewing, the most significant and valuable papers were accepted for publication. The volume consists of 79 relevant manuscripts. They present a wide range of technical solutions for automated mechanical engineering, and prospects for using information technologies and modeling methods in engineering.

We would like to express gratitude to the Organizing Committee for hosting MEACS-2020, despite the unfavourable epidemiological situation. We would also like to thank all the participants and the editors for their valuable contribution in preparing the MEACS-2020 proceedings. We hope that readers would find this volume relevant and useful.

On behalf of the editors,
Vadim Skiba,
Nikita Martyushev

Programme Committee

Chairs	
Atapin V.G.	Dr. of Eng., Professor of the department of mechanical properties of airplanes, Novosibirsk State Technical University (Novosibirsk, Russia)
Batayev A.A.	Dr. of Eng., Editor-in-chief "Processing of metals (technology, equipment and tools)", Novosibirsk State Technical University (Novosibirsk, Russia)
Korotkov A.N.	Dr. of Eng., Head of the department of mechanical engineering, Kuzbas State Technical University (Kemerovo, Russia)

Members	
Ivantsivsky V.V.	Dr. of Eng., Deputy editor-in-chief "Processing of metals (technology, equipment and tools)", Novosibirsk State Technical University (Novosibirsk, Russia)
Lobanov D.V.	Dr. of Eng., Head of the department of mechanical properties of airplanes, Chuvash State Technical University (Cheboskary, Russia)
Tregubchak P.V.	Cand. of Eng., Chapter engineer of JSC Novosibisrsk Switch Plant (Novosibirsk, Russia)
Yanpolsky V.V.	PhD, Head of the department of machinery design, Novosibirsk State Technical University (Novosibirsk, Russia)

Organizing Committee

Organizers	
Skiba V.Yu.	Cand. of Eng., co-editor of the volume, Deputy editor-in-chief "Processing of metals (technology, equipment and tools)", Novosibirsk State Technical University (Novosibirsk, Russia)
Mark E.S.	Director of the MashExpo Siberia (Novosibirsk, Russia)
Martushev N.V.	Cand. of Eng., co-editor of the volume, associate professor of department of materials science, Tomsk Polytechnic University (Tomsk, Russia)

Secretaries	
Zvarev E. A.	Cand. of Eng., co-editor of the volume, associate professor of department of machinery design, Novosibirsk State Technical University (Novosibirsk, Russia)
Martynova T.G.	Cand. of Eng., co-editor of the volume, associate professor of department of materials science, Tomsk Polytechnic University (Tomsk, Russia)

IOP Conf. Series: Materials Science and Engineering 1064 (2021) 011002 doi:10.1088/1757-899X/1064/1/011002

Peer review declaration

All papers published in this volume of *IOP Conference Series: Materials Science and Engineering* have been peer reviewed through processes administered by the Editors. Reviews were conducted by expert referees to the professional and scientific standards expected of a proceedings journal published by IOP Publishing.

- **Type of peer review:** Single-blind peer review by two committee members
- **Conference submission management system:** submissions were received and handled via e-mail
- **Number of submissions received:** 129
- **Number of submissions sent for review:** 118
- **Number of submissions accepted:** 84
- **Acceptance Rate (Number of Submissions Accepted / Number of Submissions Received X 100):** 65%
- **Average number of reviews per paper:** 2
- **Total number of reviewers involved:** 26
- **Any additional info on review process:** all the papers were reviewed
  - by the review coordinator or one of his affiliates (co-editors or secretaries if the papers is within the field of expertise);
  - by two independent referees chosen from the members of the programme committee (or their affiliates);
  - the volume comprised of accepted papers vas edited and proofread by the editors.
- **Contact person for queries:** Vadim Skiba, review coordinator, cand. of eng., secretary of the editorial board of the international journal "Processing of metals (technology, equipment and tools)", Novosibirsk State Technical University, skiba@corp.nstu.ru

IOP Conf. Series: Materials Science and Engineering 1064 (2021) 012001 doi:10.1088/1757-899X/1064/1/012001

Modelling of the optimal height of the variable edge of a volumetric anti-filtration geotextile coating for restoring failed water supply structures

M A Bandurin, I P Bandurina, V V Vanzha, P G Pasnichenko and V T Tkachenko

Kuban State Agrarian University named after I.T. Trubilin, 13, Kalinina str., Krasnodar, 350044, Russia

E-mail: chepura@mail.ru

Abstract. This article presents the results of numerical modelling of the optimal height of the variable edge of a volumetric anti-filtration geotextile coating for the restoration of failed water supply structures. After a long period of operation, hydraulic structures continue to operate without scheduled repairs, so it is necessary to perform technical reconstruction of the structure on the basis of innovative, as well as previously proven repair technologies. Most of the long-term operated hydraulic structures in Russia are in unsatisfactory condition, requiring modernization and reconstruction. Volumetric anti-filtration geotextile coating is used for technical restoration of long-term water supply structures that have failed. The volumetric cellular structure is filled with soil and has a variable height of the edge from the ridge to the bottom of the slope. Operational reliability and durability of the volumetric coating are provided by a cellular structure with a variable ridge height, to eliminate the formation of landslides in the underlying soil and in the filling, as well as the loss of water for filtration. A solid-state model of the stress-strain state of a bulk anti-filtration geotextile coating was constructed. The number of elements and the number of units of the ensemble was 127830 and 102740, respectively. The source information was encoded in terms of the increment method, taking into account the fragmentary representation of a three-dimensional anti-filtration geotextile coating with a variable edge height in the form of simple geometric objects-plates. The work of the coating from the load from its own weight and filling the volume cells of the soil at various parameters was numerically modelled. The results obtained are adequate to the calculations and do not exceed the operational parameters of the design. Processing of the obtained simulation results showed the presence of minor displacements, both horizontally and vertically along the structure, which indicates the presence of a large margin of safety of the bulk anti-filtration geotextile coating.

1. Introduction

Most of the long-term operated hydraulic structures in Russia are in a unsatisfactory condition, requiring modernization and reconstruction. After a long period of operation, hydraulic structures continue to operate without scheduled repairs, so it is necessary to perform technical reconstruction of the structure on the basis of innovative, as well as previously proven repair technologies [1].

Bulk anti-filtration geotextile coating is used for technical restoration of long-term water supply structures that have failed.

The bulk cellular structure is filled with soil and has a variable height of the ridge to the rib to the bottom of the slope. Operational reliability and durability of the bulk coating are provided by a cellular structure with a variable ridge height to eliminate the formation of landslides in the underlying soil and in the filling, as well as the loss of water for filtration [2].

During long-term operation of water supply structures, many structural elements fail, this is the complete destruction of the reinforced concrete structure of the water supply structure, the formation of defects that violate the normal operation of the water supply structure; for example, the violation of butt joints, as well as the destruction of individual elements of the structure, etc.

In the water management construction in Russia, waterproof nylon fabrics with rubber and polymer coating are used, produced by Ufa, Kursk, Yaroslavl, Ivanovo and other factories of polymer products and structures. The mass of 1 m^2 of single- and double-layer coated fabrics is from 1.2 to 3.5 kg with a fabric thickness of 0.9 to 3.0 mm. According to the technological parameters of manufacturing at domestic factories, preference is given to single-layer fabrics [3].

Structures made from the mentioned materials are mostly seasonal or temporary. In the USA, Japan, France, Italy and other countries, stronger and heavier fabrics are used in water management. Structures made from them are mainly classified as permanent structures. In the USA, two- and three-layer materials made by DuPont, Goodyear, Firestone and others are used with a mass of 1 m^2 from 2.4 to 7.2 kg.

By the recommendation of the "Committee of the study of large dams made of rubberized fabrics" (Japan), the company "RUBBER DAM" shifted to the use of heavy fabrics that provide a 50-year service life of soft water or air-filled dams [4]. So, the three-layer material on a nylon (nylon) basis of the FABRIDAM company has a thickness from 3.0 to 7.0 mm, and the three-layer material of the RUBBER DAM company – from 10.0 to 25.0 mm. Currently, Japanese companies have built more than 5,000 dams made of fabric materials.

The analysis of the work of various companies around the world shows that a wide range of fabric-based composite materials is used in the water sector. Depending on their purpose, their strength ranges from 100 to 1000 kN/m [5].

Designs of rubberized fabrics differ in the following features:

- textile base – material, method of interweaving threads, number of layers and type of duplication;
- protective coating – material, method of application, quantity (single or double-sided coating).

2. Results

In the course of numerical modelling, it is necessary to determine the possible conditions for further operation of the coating, namely, the formation of various defects. The numerical experiment included the construction of a solid-state model of the stress-strain state of the coating [6].

The use of anti-filtration geotextile coatings to restore the operability of water supply structures is to ensure the operational reliability and durability of anti-filtration coatings of water supply structures, excluding foci of water loss for filtration in the operation of water supply structures [7].

This technical result is achieved in that the applied two-layer coating of geotextile, which ensures no loss of water through seepage from water conveyance channels, the coating is placed around the perimeter of the channel and then into the cavity between the two layers of the coating through the neck, located on the edge of a panel, pumped concrete, which imparts rigidity and structural strength of the coating, both coating layer are connected by a local longitudinal seams staggered (Figure1) [8].

On underlying soil 1 along the entire perimeter of the water supply structure, a two-layer coating is spread out, made of geotextile material 3, formed from two panels connected to each other by longitudinal local seams 5 arranged in a staggered order, for uniform distribution of concrete 2 throughout the two-layer coating, through necks 4 located on the edge of the panel fixed on the edge of the channel.

The operational reliability and durability of the anti-filtration coating of water supply structures is ensured by the joint work of concrete 2, which further solidifies, betrays the rigidity and strength of the coating structure, and geotextile material 3 ensures that there is no loss of water for filtration from water supply channels [9].

The use of the coating will increase the operational reliability of anti-filtration coatings of water supply channels, due to its rigidity and strength, and a double layer of film (geotextile material) will prevent water loss, preventing waterlogging, salinization and flooding of agricultural land. Even if the upper layer of the film facing the water fails during prolonged operation of the coating, the lower layer will continue to exclude water loss for filtration [10].

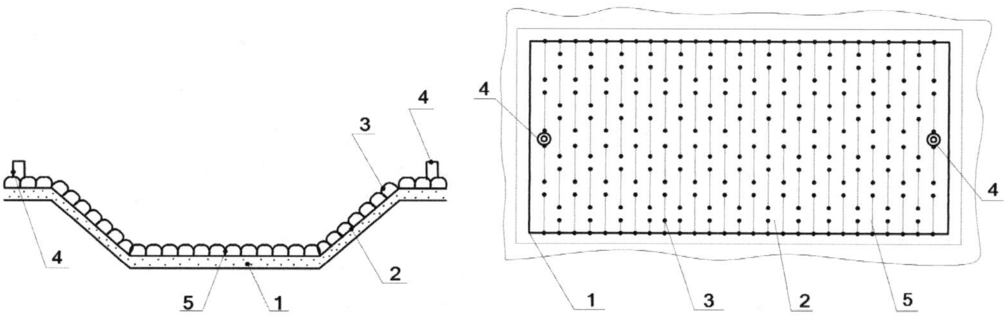

a b

Figure 1. Anti-filtration geotextile coating of water supply structures: a – in cross-section; b – top view.

Also, the use of this coating is possible without concrete, when it can be replaced with a mixture of soil and water, which will later exit through the membranes of the geotextile coating and only the soil will remain, which can also give rigidity to this structure.

Bulk anti-filtration geotextile coating is used for technical restoration of long-term water supply structures that have failed [11].

The bulk cellular structure is filled with soil and has a variable height of the ridge from the rib to the bottom of the slope. Operational reliability and durability of the bulk coating is provided by a cellular structure with a variable ridge height to eliminate the formation of landslides in the underlying soil and in the filling [12], as well as the loss of water for filtration [13].

Figure 2 shows the anti-filtration geotextile coating of a low-pressure earthen dam, which can be used for the construction of an anti-filtration coating of a low-pressure earthen dam, that is, the sides of water supply structures.

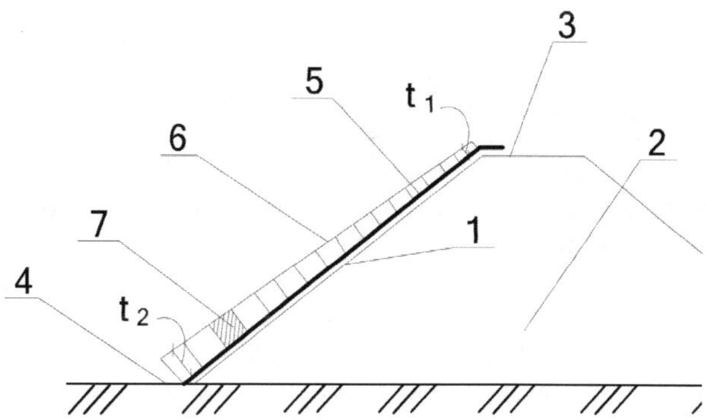

Figure 2. Anti-filtration geotextile coating of water supply structures, in cross-section

The technical result achieved by this anti-filtration geotextile coating of a low-pressure earthen dam is to prevent water losses for filtration and ensure the operational reliability and durability of the anti-filtration geotextile coating of a low-pressure earthen dam.

This technical result is achieved in that the cellular structure is mounted on waterproof geotextile material panel which is laid on the underlying layer at the raised slope of the low-pressure earthen dam, and the cell structures are filled with soil and have a variable depth which gradually increases from the crest to the foot of the dam and $t_2 > t_1$[14].
where t_1 – depth of the rifge of the dam, t_2 – depth of the cell at the foot of the dam [15].

On the underlying layer of soil 1 along the upper slope of low-pressure earthen dam 2, a coating is spread, fixed on ridge 3 and bottom 4 of the dam, and made of a layer of geotextile material 5 in the form of a cellular structure of variable depth 6 filled with soil 7 and attached to coating 5.

Operational reliability and durability of impervious geotextile coating low-pressure earthen dam are provided, cellular structure 5 having a changeable depth of the cell, which gradually increases from ridge 3 to bottom 4 of the dam and $t_2 > t_1$, where t_1 is the depth of the cells on the ridge of the dam, t_2 – the depth of the cell at the bottom of the dam. They prevent the formation of landslide processes in the soil filling it 7, which protects the geotextile material 5 from damage, and the geotextile material 5 prevents water loss for filtration through a low-pressure earthen dam [16].

It is necessary to determine the possible conditions for further use of the coating, namely the formation of various defects. The numerical experiment included the construction of a solid-state model of the stress-strain state of the coating [17].

The work of the coating of the load from its own weight and filling the volume cells of the soil at various parameters was numerically modelled. The results obtained are adequate to the calculations [18] and do not exceed the operational parameters of the structure (Figure 3) [19]

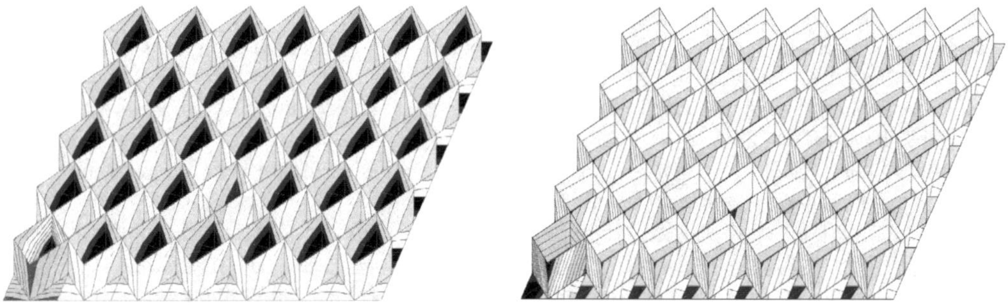

a b

Figure 3. Results of numerical simulation: a – plot of total stresses; b – plot of total displacements.

The result of modelling is characterized by the presence of total displacements that are not dangerous for the coating operation, which indicates the presence of a large margin of safety of the coating [20]. The formation of various defects was also modelled, namely the most dangerous – subsidence of the underlying soil. The formation of a deflection of the coating along the length is shown in the plot of total displacements (Figure 4) from the subsidence of the underlying soil.

Horizontal stresses occur in the support zone, and vertical stresses along the coating occur in the support zone and in the middle of the considered part [21].

In the course of numerical modelling of a volumetric cellular anti-filtration geotextile coating, we obtained diagrams of stresses and displacements from various defects that occur during the operation of the structure, which will increase the reliability of coatings.

The variable height of the edge of the coating cell prevents the formation of landslide processes, and reduces the consumption of geotextile material. One of the most urgent tasks in the field of ensuring the safety of long-term water supply structures is the correct assessment of the remaining resource [22, 23], and this can only be shown when performing numerical modelling of the development of various damages, both individually and in a group.

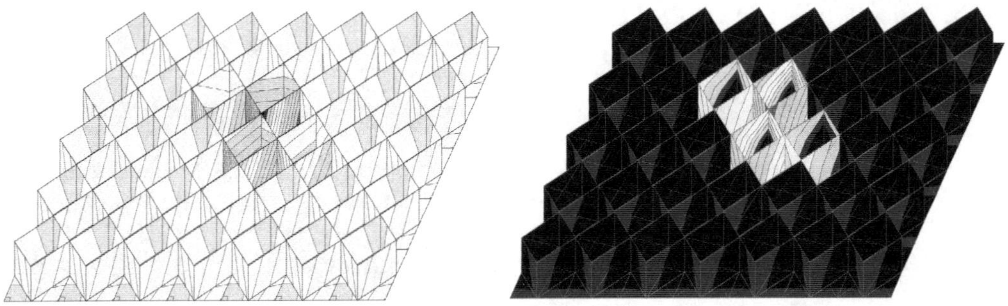

a b

Figure 4. Results of numerical modeling: a – plot of total displacements from subsidence of the underlying soil; b – plot of deformations from subsidence of the underlying soil.

3. Conclusion

The use of impervious geotextile cover will improve the operational reliability of impervious surfaces, low-pressure earth dams due to the cellular structure having a variable depth, filled with soil. This reinforces the upstream slope, and prevents the formation of landslide processes, as well as waterproof coating with geotextile material prevents the loss of water to filter, preventing blooding, salinization, and waterlogging of agricultural land.

During the development of methods of extending the life cycle of water supply structures, anti-filtration geotextile coatings were obtained thanks to which, if the residual resource of the structure is correctly estimated, it is possible to extend its life cycle.

The question arises of the need to assess correctly and timely the remaining resource and load-bearing capacity of the water supply structure, to assess the technical condition and extend its life cycle. To do this, we can apply the proposed anti-filtration geotextile coatings depending on the type of water supply structures of anti-filtration geotextile coating for restoring the operability of water supply channels; anti-filtration geotextile coating for sealing the butt joint of tray channels of irrigation systems and underwater pipelines; anti-filtration cellular geotextile coating of water supply channels and low-pressure embankment dams.

Thanks to the use of modern anti-filtration geotextile coatings, it is possible to extend the life cycle of water supply structures with minimal economic costs.

References

[1] Kireicheva L V and Zakharova O A 2002 The effect of cyclic irrigation with wastewater on the properties of gray forest soils *Eurasian Soil Science* **35(9)** 990-995

[2] Kireicheva L V and Karpenko N P 2015 Evaluation of the efficiency of irrigation in a zonal soil sequence *Eurasian Soil Science* **48(5)** 524-532

[3] Bandurin M A, Yurchenko I F and Bandurina I P 2019 Computer technology to assess the capacity reserve of the irrigation facilities of the agro-industrial complex *Int. Multi-Conf. on Industrial Engineering and Modern Technologies (FarEastCon)* pp 1-5. doi: 10.1109/FarEastCon.2019.8933970

[4] Olgarenko V I, Olgarenko G V and Olgarenko I V 2018 A method of integral efficiency evaluation of water use on irrigation systems *Int. Multidisciplinary Sci. GeoConf. SGEM* **18(3.1)** 3–9

[5] Yurchenko I F 2018 Information support for decision making on dispatching control of water distribution in irrigation *J. of Phys.: Conf. Ser.* **1015(4)** 042063

[6] Ovchinnikov A S, Bocharnikov V S, Skorobogatchenko D A, Borisenko I B, Chernyavsky A N, Abezin V G, Ryadnov A I, Shaprov M N, Kuznetsov N G, Nekhoroshev D A, Sedov A V,

Grigorov S M, Fomin S D and Ol'garenko V I 2018 The optimum geometrical form modeling of the "striegel" type harrow *ARPN J. of Engineering and Appl. Sci.* **13(23)** 9138-9144
[7] Kuznetsov E V, Khadzhidi A E, Poltorak Y A and Kuznetsova M 2019 Operator model to control process of obtaining vermicompost *EurAsian J. of BioSciences* **13(1)** 315-321
[8] Olgarenko I V and Selyukov I V 2011 Software for water use planning in the irrigation systems *Environmental Engineering* **4** 38-40
[9] Kuznetsov E V, Khadzhidi A E, Kilidi K I and Kurtnezirov A N 2018 Management of agro-resource potential for agricultural landscape stability increase *Plant Archives* **18(2)** 2151-2158
[10] Abdrazakov F K, Orlova S S, Pankova T A, Mirkina E N and Mikheeva O V 2018 Risk assessment and the prediction of breakthrough wave during a dam accident *J. of Int. Research* **8(1)** 154-161
[11] Abdrazakov F K, Pankova T A, Zatinatsky S V, Orlova S S and Trushin Yu E 2017 Increasing efficiency of water resources use in forage crops irrigation *Int. J. of Advanced Biotechnology and Research* **8** 283-293
[12] Yurchenko I F 2016 Management information systems of reclamative water utilization system *J. of Agricaltural Sci.* **1** 12-15
[13] Kireicheva L V and Khokhlova O B 2000 Elemental composition of different fractions from the sapropel organic matter *Eurasian Soil Sci.* **33(9)** 947-949
[14] Degtyareva O G, Degtyarev G V, Lavrov N L and Aliev D U 2018 Constructive-technological decisions in regulating the flow of atmospheric precipitation *Magazine of Civil Engineering* **82(6)** 32-48
[15] Bandurin M A, Volosukhin V A and Yurchenko I F 2018 Remote monitoring of reliability for water conveyance hydraulic structures *Mater. Sci. Forum* **931** 209-213
[16] Yurchenko I F 2013 Perfect water use systems as a factor maintaining soil fertility and sustainability of agricultural production in the irrigated agricultural landscapes *Agrochemical J.* **1** 25-27
[17] Gaydzhurov P P, Kravchenko G M and Savelieva N A 2014 Finite element modeling of elastic plastic bending of steel beams with use of rod end elements *Construction mechanics and design of structures* **2** 17-22
[18] Chesnokov B P, Abdrazakov F K, Naumova O V, Krivoschapov D S and Strelnikov V A 2017 The use of ionizing radiation for the tungsten preparation *J. of Industrial Pollution Control* 1-12
[19] Bandurin M A, Yurchenko I F, Volosukhin V A, Vanzha V V and Volosukhin Ya V 2018 Ecological and economic efficiency of diagnostics of technical condition of water supply facilities of irrigation systems *J. Ecology and Industry of Russia* **22(7)** 66-71
[20] Chesnokov B P, Naumova O V, Strelnikov V A, Abdrazakov F K and Tronin B A 2016 Polyethylene production from granules using high voltage *Int. J. of Applied Engineering Research* **11** 2140-2144
[21] Fen N, Kozlov D B and Rumyantsev I S 2016 Hydraulic studies of stepped spillways of various design *Power Technology and Engineering* **49(5)** 337-344
[22] Abdrazakov F K, Ryzhko N F, Ryzhko S N, Horin S A and Botov S V 2018 Electricity consumption decrease at pump stations during watering by multi-support sprinkling units *J. of Fundamental and Appl. Sci.* **10** 1464-1481
[23] Yurchenko I F 2018 Information support system designed for technical operation planning of reclamative facilities *J. of Theoretical and Appl. Information Technology* **96(5)** 1253-1265

Evaluation of the effect of various types of tools on a weld joint during wave strain hardening

A V Kirichek[1], S V Barinov[2], A V Yashin[2], M N Ryzhkova[2], A M Konstantinov[2] and A A Zaitsev[2]

[1]Bryansk State Technical University, 7, 50 let Octyabrya Blvd, Bryansk, 241035, Russia
[2]Murom Institute (branch) of Vladimir State University, 23, Orlovskaya Str., Murom, 602264, Russia

E-mail: box64@rambler.ru

Abstract. The article studies the problem of increasing the durability of weld joints using surface plastic deformation by the method of wave strain hardening (WSH). Earlier, WSH was not used for these purposes. The prospect of using WSH for increasing the durability of weld joints is associated with the possibility of increasing the microhardness and the formation of compressive residual stresses at a depth of more than 10 mm. To identify the most effective conditions for the use of various types of tools for WSH of weld joints and accelerate the search for rational modes and conditions of WSH, finite element modeling was used. As a result of the research, recommendations related to the use of various types of WSH tools, depending on the thickness of weld joints, changes in hardness and stress state of the near-weld zone after welding, were given.

1. Introduction

Increasing the strength of weld joints is an urgent problem in mechanical engineering. The need for additional impact on the weld joint after it is obtained is associated with a decrease in the strength properties of the heat-affected zone (weld and heat-affected zone) due to the burnout of metal chemical elements, rapid local heating and metal cooling, and as a consequence, the appearance of significant tensile stresses [1, 2].

To increase metal durability in the heat-affected zone and reduce the residual stresses in the base metal and in joints, various technological methods such as heat treatment, aging, surface plastic deformation (SPD), etc. are used.

The use of traditional methods to improve the strength properties of the heat-affected zone - heat treatment and aging, is characterized by significant time and energy costs and has certain limitations associated with the overall dimensions of the processed product.

Among the SPD methods, for the treatment of weld joints and near-weld zones, ultrasonic and vibro-shock treatment, rolling, etc. are used.

A sufficiently effective method for processing weld joints and heat-affected zones is the SPD method is wave strain hardening (WSH). Compared to other SPD methods, wave strain hardening has wider technological capabilities for the formation of a hardened surface layer, allows a high degree (up to 150%) and a large depth of hardening (more than 6 ... 10 mm) [3]. A feature of the method is the generation of deformation waves in a shock system with an intermediate link and their communication to the deformation zone, in order to strengthen the critical surfaces of machine parts. The effectiveness of the method is provided due to the almost complete use of the deformation wave energy generated in the shock system for the elastoplastic deformation of the hardened material. Depending on the required parameters of the surface layer hardening, a shock pulse of the required amplitude and duration is formed, the material and geometric parameters of the shock system elements are selected. By controlling the parameters of the deformation wave, it becomes possible to obtain not only a uniformly hardened surface layer, but heterogeneously hardened zones as well. The uniformity of overlapping of plastic prints is estimated through the overlap coefficient K. The range of K variation is from 0 to 1: at K = 0, the prints do not overlap, the edges of the prints border each other; at K = 1, there is a complete overlap of the prints, the strikes are applied to the same place [3].

WSH has a large number of controlled technological parameters, due to the use of various tools it can provide the required effect on the deformation zone. However, at the moment, theoretical and practical experience has been accumulated in the processing of solid materials, the development of the WSH technology of weld joints is at the initial stage, the necessary information is not available. Therefore, studies related to the choice of the type, shape of the tool for WSH of weld joints, parameters that provide the required effect on the deformation zone are relevant.

Carrying out a full-scale experiment is always associated with the difficulties of its implementation, material costs, requirements for the availability of appropriate equipment. Modeling the process using modern engineering analysis tools based on the finite element method significantly speeds up the procedure for obtaining the necessary information.

The purpose of this work is to study the result of the impact of various types of tools in the case of WSH of the weld joint by computer simulation by the finite element method.

2. Experimental technique

The development of a finite element model of the process of strengthening weld joints by the WSH method was carried out in the Ansys program, which is the world leader in the field of engineering analysis [4-7]. The modeling was carried out in four stages.

At the first stage, the geometry of the welded product, weld joint and WSH tool was formed.

At the second stage, a model of the material of the welded product was created. For this, its thermophysical and mechanical properties were set. The data were taken from the reference literature and the results of the corresponding tests of the material used.

At the third stage, the thermal problem of heating the weld was solved. As a result residual stresses in the welded workpiece that arise in the product after welding was obtained.

At the fourth stage, the resulting weld was subjected to WSH with different processing modes with different types of tools. The result of modeling the fourth stage is the determination of the residual stresses and microhardness values obtained in the weld joint after WSH.

To assess the adequacy of the developed finite element model, a preliminary field experiment was carried out. For research, we used plates with dimensions of 600 * 100 * 7 mm, made of 40X steel. With the help of a semi-automatic welding machine, these plates were butt welded and had a final size of 600 * 200 * 7 mm. WSH of weld joints was carried out with an energy of 150 J with a shock frequency of 10 Hz and with modes corresponding to the overlap coefficients K = 0.3 and K = 0.6. A rod roller with a diameter of 10 mm and a width of 40 mm was used as a tool for strengthening. After processing the WSH joints, the welded plate was cut on a band saw along the joint and the microhardness was measured in the resulting section using a KB30S hardness tester. Comparison of the simulation data and the values obtained as a result of a full-scale experiment, under the same initial conditions, was carried out according to the microhardness map in the weld joint section. The comparison results are shown in Figure 1.

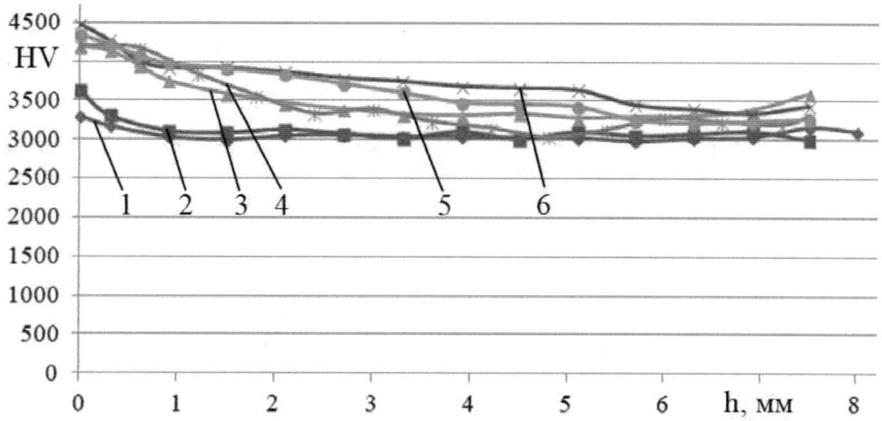

Figure 1. Distribution of microhardness (HV, MPa) over the depth (h, mm) of the weld joint: 1) unreinforced (experimental data); 2) unreinforced (simulation results); 3) hardened with K = 0.3 (experimental data); 4) hardened with K = 0.3 (simulation results); 5) hardened with K = 0.6 (experimental data); 6) hardened with K = 0.6 (simulation results).

The obtained theoretical data on the distribution of microhardness in the surface layer with a satisfactory probability (92%) correspond to the experimental values. The results obtained can be considered significant.

To study the effect of various types of tools for WSH on weld joints of plates made of 40X steel, we simulated the problems in which the following were used as tools: 1) rod rollers 40 mm wide and 10, 18 and 27 mm in diameter; 2) balls with diameters of 18 and 27 mm; 3) rods with a spherical end, 12 mm in diameter and 18 mm high. In the case of obtaining deep imprints of tool impacts in the workpiece being processed and the impossibility of further processing because of this, work was carried out to compose combinations of several simultaneously operating tools according to the multi-contact WSH scheme [8, 9].

3. Assessment of the effect of various types of tools on the weld joint during wave strain hardening

As a result of the conducted studies, it was established that the heat-affected zone in the welded workpiece made of 40X steel is the areas located to the left and right of the axis of the weld at a distance of 18 mm. Thus, in order to influence this heat-affected zone, it is necessary to expose a 40 mm wide section to the WSH. Another condition for conducting WSH is the need to perform machining in one pass of the tool. The fulfillment of this condition is necessary to increase the processing productivity, reduce the risk of edge effects and the overlapping of the edges of the hardened tracks (excluding over-hardening) in the case of WSH of a heat-affected zone for several tool passes.

As a result of studies of the welded seam without hardening, it was found that the average hardness along the thickness was 3000 MPa. On the hardened surface, there are residual tensile stresses of 370 MPa, which at a depth of 6.5 mm begin to transform into compressive ones, and at a depth of 9.5 mm, reach a maximum of -210 MPa (Fig. 2, 4, 5).

As a result, WSH with K = 0.3 rollers with diameters of 10, 18, 27 mm, in comparison with unreinforced weld joints, the microhardness increased by 46, 32, 31%, respectively, and a depth of hardening of 6.3, 5.3 and 5 mm was obtained (Fig. 2, 3). As a result of WSH with K = 0.3, the values of residual stresses on the weld hardening surface changed from tensile +370 MPa to compressive -310 MPa (Fig. 4, 5). The use of rollers of the considered diametrical dimensions did not change the value of the obtained compressive stresses on the hardened joint surface. The transition of compressive stresses to tensile stresses for rollers with diameters of 10 and 27 mm is set at a depth of 7 mm, and for 18 mm rollers - 8.2 mm. The maximum tensile residual stresses + 50..70 MPa, installed on the side opposite to the hardening, is also the same for all rollers of the considered diametrical dimensions.

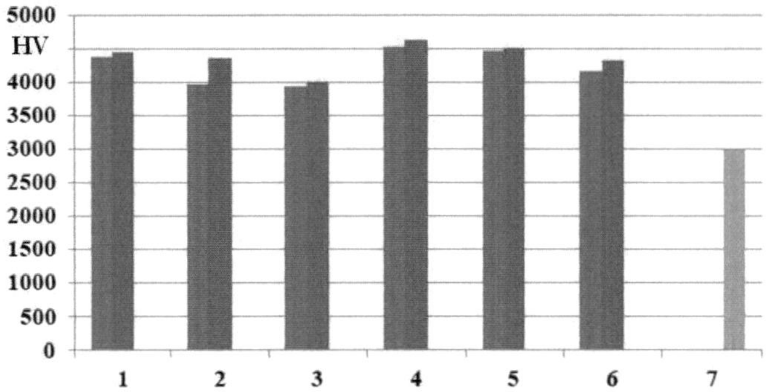

Figure 2. The maximum values of microhardness (HV, MPa) obtained on the surface of the weld joint at its WSH with ■ - K = 0.3; and with ■ - K = 0.6 using: 1) 10 mm roller; 2) 18 mm roller; 3) 27 mm roller; 4) rods with a spherical end of 12 mm; 5) 18 mm balls; 6) 27 mm balls; 7) weld joint without hardening.

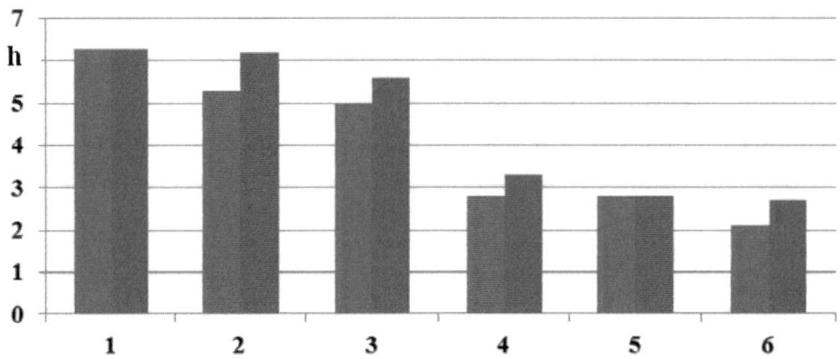

Figure 3. The maximum values of the depths of hardening (HV, MPa) of the welded layer at its WSH with ■ - K = 0.3; and ■ - K = 0.6: using 1) 10 mm roller; 2) 18 mm roller; 3) 27 mm roller; 4) rods with a spherical end of 12 mm; 5) 18 mm balls; 6) 27 mm balls; 7) weld joint without hardening.

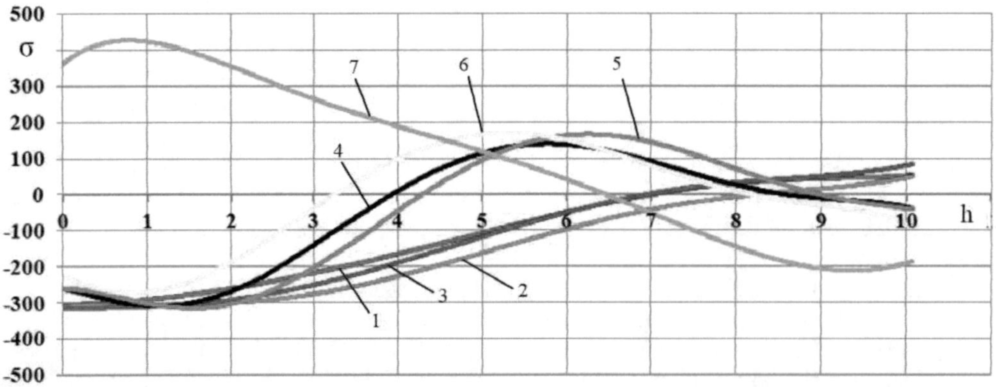

Figure 4. Distribution of residual stresses (σ, MPa) over the thickness of the welded layer (h, mm) at its WSH with K = 0.3 using: 1) 10 mm roller; 2) 18 mm roller; 3) 27 mm roller; 4) rods with a spherical end of 12 mm; 5) 18 mm balls; 6) 27 mm balls; 7) weld joint without hardening.

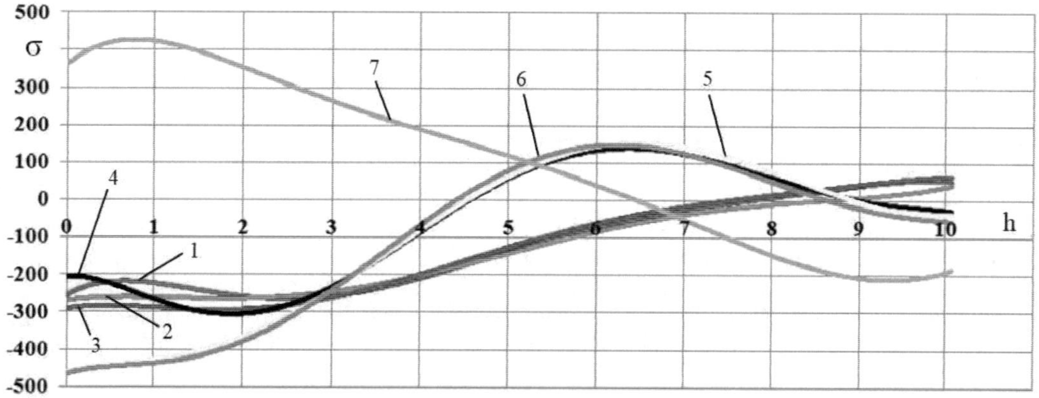

Figure 5. Distribution of residual stresses (σ, MPa) over the thickness of the welded layer (h, mm) at its WSH with K = 0.6 using: 1) 10 mm roller; 2) 18 mm roller; 3) 27 mm roller; 4) rods with a spherical end of 12 mm; 5) 18 mm balls; 6) 27 mm balls; 7) weld joint without hardening.

In case of WSH weld joints with K = 0.6 rollers with diameters of 10, 18, 27 mm, compared with K = 0.3, the microhardness increased by 1.6, 9.5, 2%, respectively, and the depth of hardening for a roller with a diameter of 10 mm remained unchanged, and for other cases it increased by 17 and 12%, respectively. The influence of the considered diametrical dimensions of the rollers with WSH with K = 0.6 on the residual stresses is practically indistinguishable and merged into one line on the graph (Fig. 5). On the hardened side, the value of residual compressive stresses was -250 ... -270 MPa, which is comparable to the values obtained at K = 0.3. The transition to tensile stresses is located at a depth of 7.5 mm, and the maximum of their values + 40..50 MPa is located on the surface opposite to the one being hardened. These values practically coincided with the values of the residual stresses obtained with WSH with K = 0.3 by a roller 18 mm in diameter.

Under the given WSH modes using one tool with a spherical end with a diameter of 12 mm, its deep penetration into the hardened surface is established, which makes it impossible to implement the process. Proceeding from this, to achieve the adopted hardening width of the heat-affected zone equal to 40 mm, tools with a spherical end with a diameter of 12 mm were installed in two rows staggered in a special mandrel, in the amount of 10 pieces. Due to this technical solution, as a result of WSH with specified modes, with K = 0.3, it was possible to increase the microhardness by 51% compared to unreinforced welded joints, and to achieve a hardening depth of 2.8 mm (Fig. 2, 3). The values of the residual stresses on the hardened weld surface changed from tensile +370 MPa to compressive -250 MPa (Fig. 4, 5). The maximum residual stresses of -300 MPa were at a depth of 1 ... 1.5 mm. Unlike rod rollers, tools with a spherical end at a depth of 4 and 8.5 mm show two transitions of compressive stresses into tensile stresses. The value of the residual compressive stresses on the surface opposite to the one being hardened was -40 MPa. The maximum tensile residual stresses are set at a depth of 6 mm and are +130 MPa.

With WSH of weld joints with K = 0.6 (compared with K = 0.3), using tools with a spherical end with a diameter of 12 mm, the microhardness increased by 1.8% respectively and the hardening depth by 18%. The distribution of residual stresses in case of WSH with K = 0.6 in comparison with K = 0.3 differs only near the surface to be hardened. With WSH with K = 0.6 compared with K = 0.3, the residual compressive stresses decreased from -250 MPa to -210 MPa, and their maximum (-300 MPa) shifted from a depth of 1..1.5 mm to a depth of 1.5 ... 2.2 mm.

When using tools for WSH – balls, as in the case of tools with a spherical end, to ensure the required width of hardening of the heat-affected zone of the weld joint, they were installed in a checkerboard pattern in a special mandrel. Balls with diameters of 18 and 27 mm were used in an amount of 10 pieces. So, with WSH with K = 0.3 balls with diameters of 18 and 27 mm, in comparison with unreinforced welded seams, it was possible to increase the microhardness by 49 and 39% and to achieve the hardening depth of 2.7 and 2.1 mm, respectively (Fig. 2, 3).

Thanks to the use of balls with diameters of 18 and 27 mm as WSH tools, it was possible to change the stress state: instead of tensile residual stresses of +300 MPa, compressive stresses of 260 MPa and 220 MPa, respectively, are formed on the hardened surface (Fig. 4, 5). As in the case of the use of tools with a spherical end, the use of balls as tools also led to a double transition of compressive stresses into tensile stresses in the weld section. Such a transition for balls with a diameter of 18 mm occurred at depths of 4.3 and 9 mm, and for balls with a diameter of 27 mm – 3.3 and 8 mm. The value of the residual compressive stresses on the surface opposite to the one to be hardened was -45 and -95 MPa, respectively, for the balls of 18 and 27 mm. The maximum tensile residual stresses of 180 MPa were set for balls of 18 mm and 27 mm at a depth of 6.3 and 5.2 mm, respectively.

With WSH of weld joints with K = 0.6 using balls with diameters of 18 and 27 mm, compared with K = 0.3, the microhardness increased by 1.1 and 3.7%, respectively, and the hardening depth increased by 28.6% only for balls with a diameter of 27 mm. When passing from K = 0.3 to K = 0.6, the character of the curve of the distribution of residual stresses along the depth of the weld became the same for the considered standard sizes of balls. On the hardening surface, the residual compressive stresses amounted to 460 MPa. A double transition of compressive stresses into tensile stresses in the section of the welded seam occurred at a depth of 3.9 and 8.7 mm. The value of the residual compressive stresses on the surface opposite to the hardened surface was 45 and 55 MPa, respectively, for balls of 18 and 27 mm. The maximum tensile residual stresses of + 140 MPa were found for balls of 18 mm and 27 mm at a depth of 6.4 mm.

4. Conclusion

1) The prospects of using WSH for increasing hardness, hardening depth and ensuring the transition of tensile stresses to compressive ones in weld joints and the heat-affected zone have been confirmed.

2) It was found that if it is necessary to increase the hardness to 50% of the heat-affected zone of weld joints with a thickness of up to 5 mm, it is necessary to use balls with a diameter of 18 mm as a tool and carry out WSH according to a multi-contact scheme with K = 0.6.

3) It has been established that to increase the strength of weld joints up to 10 mm thick without a significant decrease in hardness (up to 20%) after welding, it is promising to use rod rollers with a diameter of 10 mm and to carry out WSH with K = 0.3-0.6.

4) It has been established that if the priority is not increasing the hardness, but obtaining residual compressive stresses in the surface layer of the weld with a thickness of up to 10 mm, it is necessary to use a roller with a diameter of 18-27 mm and WSH with K = 0.3.

5. Acknowledgments

The research has been performed with the assistance of the Russian Foundation for Basic Research No. 18-38-20066.

References

[1] Shinkarev A S 2010 *Improvement of the quality of the surface layer of welded seams based on the improvement of ultrasonic shock treatment,* candidate dissertation, p 146

[2] Aborkin A V 2010 *Research and prediction of the durability of machine parts with welded joints,* candidate dissertation, p 173

[3] Kirichek A V, Soloviev D L and Lazutkin A G 2004 *Technology and equipment of static-pulse processing by surface plastic deformation* (Moscow: Engineering)

[4] Pronin A I 2009 *Development of methods for assessing the performance of circular welds of gas pipelines of compressor stations,* candidate dissertation

[5] Porowski J S, O'Donnell W J et al 1990 *Nuclear Engineering and Design* **124** 91-100

[6] Tilenko G A, Morgunov E A and Korobov Yu S 2015 *Welding and diagnostics: collection of reports of the international forum* p 35

[7] Vasiliev K V, Vill V I, Volchenko V N et al 1978-1979 *Welding in mechanical engineering: Handbook in 4 volumes* (Moscow: Mechanical Engineering)
[8] Kirichek A V, Barinov S V and Yashin A V 2017 *Key Engineering Materials* **736** 12-17
[9] Kirichek A V, Barinov S V, Aborkin A V, Yashin A V and Zaicev A A 2018 *IOP Conf. Ser.: Mater. Sci. Eng.* **327** 042011

IOP Conf. Series: Materials Science and Engineering 1064 (2021) 012003 doi:10.1088/1757-899X/1064/1/012003

Investigation of loading speed influence on fatigue wear of - polymer materials

D C S Bataev[1], R U Goytemirov[1,2], M A Gaziyev[1] and M A Dzhamaldinova[2]

[1]Department of Materials Science, Integrated Research Institute named after H. I. Ibragimov, Russian Academy of Sciences, 21a, Staropromyslovskoe highway, Grozny, Chechen Republic, 364051, Russia
[2]Technological and Economic Faculty, Chechen State Pedagogical University, 33, S. Kishieva str., Grozny, Chechen Republic, 364037, Russia

E-mail: kniiran@mail.ru

Abstract. Antifriction and friction materials used in many construction structures, devices and road construction machines simultaneously undergo dynamic and impact loads. The article considers the influence of dynamic and impact loads on the friction and wear of polymer composite materials. The materials of the maslyanit group are established to have high resistance to frictional and bulk fatigue and are capable of long-term operation without failure under cyclic dynamic and impact nature of loading. Test results show that pure kapron, despite its homogeneous structure and greater elasticity than maslyanits, is more prone to bulk fatigue failure under the cyclic impact. This fact is fundamental for confirming the mechanism of fatigue crack blocking by solid particles of the filler. Researches confirm that introduction of the solid particles capable of additional relaxation of stresses in a crack tip zone reduces the speed of distribution of the fatigue crack defined by the speed of polymer loading. The conducted research also confirms high tribotechnical properties of the composite material Maslyanit KSC which has increased hardness of surface layers and high impact viscosity of deep layers as a result of surface "cementation".

1. Introduction

The discreteness of contact of rubbing parts causes high-frequency impact pulses in the areas of microroughness and adjacent areas. However, due to the insignificance of microroughness protrusions and elastic properties of polymers, their impact wave has a small amplitude, and it is impossible to - evaluate its role in fatigue wear separately from the influence of the nominal load in the contact. In addition to the indicated pulses of transverse deformations, the surface layers in the friction contact of parts of construction and road-building, transport machines and technological equipment undergo dynamic loads changing in time, which may manifest as an impact. Correct forecasting of operational fatigue resistance and strength properties of antifriction materials is impossible without studying the influence of their loading dynamics.

2. Problem statement

If the failure in metals results from the formation of "pure" cracks with their clearly defined analytical field stress at the crack tip [1], the failure of polymer composite material is a more complex process. Instead of pure cracking, we encounter a "damage zone", which may include fiber breakage, microblocking of the fiber, fiber pulling, matrix cracking, dislocation, exfoliation, structural irregularities, delamination, or any combination [2, 3].

According to the physical chemistry of polymers [4, 5], in pure polymers put into a force field, there is a process of orienting the parts of chain macromolecules rolled up into globules, or collected in packs, connected with the complex of relaxation phenomena and structural changes of material.

The orientation of the deformed polymer structure leads to the strengthening of the crack tip zone. Therefore, if the orientation speed exceeds the crack growth rate, the strengthening will stop the crack growth for some time. If the speed of the force field is high, the orientation may fail to surpass the crack growth, which will lead to the failure of the surface (wear) or the polymer volume [6]. It follows that critical values of magnitude and speed of load application for homogeneous polymers depend on their relaxation capabilities.

We have found above [7] that the friction of materials causes the orientation of their working surfaces, which gives them high wear resistance. This correlates with the regularity established by G.A. Gorokhovsky and his colleagues for orientation of molecular structure elements of the polymer working layer: PE, PTFE, PCA and FFS and increase of their wear resistance at low loading speeds [6].

3. Materials and methods

Since polymers react differently from metals and low molecular weight solid materials under mechanical loading, it is not enough to use methods of testing these materials for complete description of their behavior under mechanical loading test. Besides, the existing methods of polymer impact failure investigation do not allow to consider the simultaneous effect of impact and friction [8]. For this reason, we have developed a methodology reflecting the special properties of polymers. To compare the effect of dynamic loading and impact loads, we made two types of samples: with eccentricity and with a protrusion (shoulder) on the working surface. We studied the influence of loading speeds (dynamics) on the friction pair operation on the materials previously tested and recommended in our study [9] for the most loaded units of road construction and agricultural machinery (ball and cardan joints, etc.): maslyanits KSC and KS-2 with fiberglass and the most characteristic representative of maslyanits without fibrous filler is maslyanit D(S-1). The comparison included the testing of pure kapron without filler.

We tested on a standard MI-IM friction machine at a linear speed of V – 0.42 m/s using the scheme shown in Figure 1. The height of the shoulder for an impact loaded roller A is 6 mm, the maximum distance of point B from the opposite point (stroke of the loaded roller) is 6 mm.

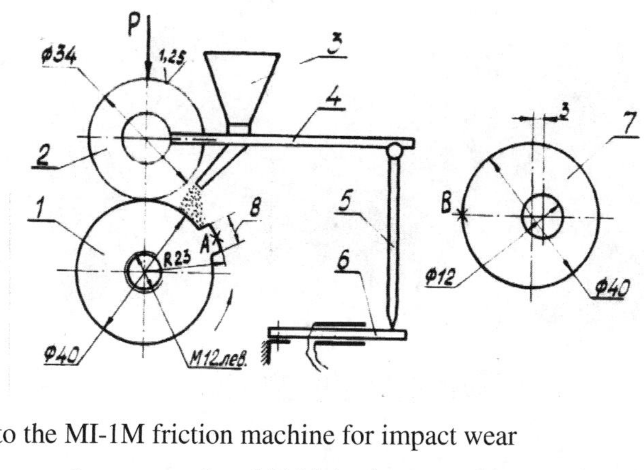

Figure 1. Friction unit to the MI-1M friction machine for impact wear

The load in the impact study was equal to 647 N, in the case of impact-free dynamic loading - 647 N and 1254 N, and the rate of increase of these forces due to eccentricity by 8 ÷ 12 N was 53 ÷ 80 N/s.

We neglected the reduction of speed and load as a result of sample wear, since this factor had the same effect on samples in all experiments.

4. Results and Discussion

Tables 1 and 2 show the results of materials testing.

Table 1. Dynamic loading test results for maslyanits

Materials	Wear speed (mm/h) under loads:		Coefficient of friction under loads:	
	647 H	1254 H	647 H	1254 H
Maslyanit KSC	0.85	1.76	0.06	0.068
Maslyanit KS-2	2.04	3.11	0.08	0.082
Maslyanit D (S-1)	3.73	7.90	0.067	0.075

Table 1 shows that the maslyanit KSC has minimal wear. Material wear tests on other machines showed the same ratios. We have obtained a slightly different pattern of friction coefficients: when they are in the area of lower values, they increase for all materials with an increase in the load on the roller, although the pressure due to an increase in the actual contact area does not increase in - proportion to the load, but by a smaller value. The fact of a general reduction of friction coefficient is possible to explain by rational squeezing out due to "cemented" barrier and consumption of plasticizer material in the friction zone (we used lubricant No. 158), as well as reduction of adhesive component and a smaller value of overlap coefficient, which favourably affects the temperature regime of friction process. The increase in the coefficient of friction with increased load may result from the increased height of the wave of deformed material running in front of the roller.

Figure 2. Samples of maslyanits: D(C-1) (1), KS-2 (2) and KSC (3) tested under dynamic loading (p=1254 N)

Wear on all samples started at maximum load areas. The surfaces of wearing samples (Figure 2) and the flaky shape of the wear particles indicate the fatigue nature of the wear.

It is established [10], that the sizes of the wear products decrease at introduction of fibrous fillers into the material or at reduction of loading. If no filler is available, the wear particle may be proportional to the size of the sample friction path. So, for the case of rolling friction of PA610 under a similar scheme without dynamic loading E. A. Fedorchuk [11] has received that the wear particle with triangular section covers all track of friction and has the thickness to 3 mm depending on pressure.

For dynamic loading, the sample was struck twice in one revolution: when the roller came into contact with the protrusion and jumped off it with a stronger impact. Thus, there are two phases in the impact process: the deformation phase, during which the normal component of the contact point velocity decreases to zero, remaining negative, and the recovery phase, during which the normal component of the contact point velocity increases from zero to the previous value. A crack appeared in this place on the kapron sample, resulting in the failure of the sample after 5 minutes of operation (Figure 3).

Table 2. Impact test results for maslyanits under impact loading of 647 N

Materials	Wear speed (mm/h)
Maslyanit KSC	0,73
Maslyanit KS-2	1,50
Maslyanit D(S-1)	2,62
Capron (control sample)	Sample was destroyed

Figure 3. Samples from maslyanit KSC (1), maslyanit D (S-1) (2) and kapron (3) tested under impact loads.

Thus, the test results show that pure kapron, despite its homogeneity and greater elasticity than maslyanit KSC, is more prone to bulk fatigue failure under the cyclic impact. This fact is fundamental for confirming the mechanism of fatigue crack blocking by solid particles of the filler. As we noted above, in homogeneous polymers, stress peaks arising at low loading speeds are smoothed by relaxation processes seeking to balance the system. The resulting orientations of the molecular chains can stop the growth of fatigue cracks. Research results on the influence of friction impact effects show the low ability of oriented kapron sections to block crack growth.

Studies have shown high wear resistance and contact strength of high-filled maslyanits. The introduction of the solid particles capable of additional relaxation of stresses in a crack tip zone reduces the speed of distribution of the fatigue crack defined by the speed of polymer loading. At the same time, such a mechanism for blocking cracks does not exclude but favours the orientation of molecular chains and packs due to the delay of the crack along the front of its growth. High tribotechnical and strength properties of maslyanite KSC under friction-dynamic loading are also due to the mixture of resins as a binding component: kapron resin with glass fiber KS-30a and phenol-formaldehyde novolac resin SF-010, which gives it a high impact strength.

5. Conclusion

We conclude that the materials of the maslyanit group have high resistance to frictional and bulk fatigue and are capable of long-term operation without failure under cyclic dynamic and impact nature of loading. The conducted research also confirmed high tribotechnical properties of the composite material Maslyanit KSC, which has increased hardness of surface layers and high impact viscosity of deep layers as a result of surface "cementation".

References

[1] Gonzalez K L 2001 *Global Failure Modes in Composites* (California: California Institute of Technology)

[2] Knops M 2008 Analysis of failure in fiber polymer laminates: the theory of Alfred Puck *Springer-Verl.* **XI** 205

[3] Mikitaev A K 2010 New in polymers and polymer composites *Polymers* **1** 3-4

[4] Kargin I A and Slonimsky G L 1967 *Brief essays on the physical chemistry of polymers* (Moscow: Khimiya)

[5] Askadsky A A and Khokhlov A P 2009 *Introduction to the physical chemistry of polymers* (Moscow: Nauchniy Mir)

[6] Gorokhovsky G A, But G P and Bezruk L I 1966 Influence of loading speeds on wear of polymers *Mechanics of polymers* **6** 862-866

[7] Bataev D K-S, Goytemirov R U, Muradov M T and Alarkhanova Z Z 2013 Study of fatigue wear of materials with oscillating alternating character of motion *Proc. of the SRI RAS* (Grozny: SRI RAS) **6** 50-57

[8] National Standard of the Russian Federation GOST R 57734-2017 *Polimer compositions. Determination of the impact energy spent on the destruction of the sample*

[9] Goytemirov R U, Gubarev S M, Kutkov A A and Mamaev N M 1980 *Antifriction polymer compositions in the automotive industry (*Moscow: NIIINavtoprom)

[10] Vettegren V I, Bashkarev A I, Vasiliev K D, Lyashkov A I, Masalimov R I, Savitsky A V and Shcherbakov I P 2012 Investigation of formation of microcracks in a fibrous composite under impact and friction *Questions of materials science* **4** 251-257

[11] Fedorchuk E A 1967 *Investigation of some issues of contact strength of metal-polymer rolling pairs with initial linear contact,* abstract of candidate's dissertation (Novocherkassk)

New nitriding process of high-alloyed maraging steel for cryogenic operation

I S Belashova[1], P S Bibikov[2], L G Petrova[1] and A S Sergeeva[1]

[1] Moscow Automobile and Road Construction State Technical University (MADI), 64, Leningradsky prospect, Moscow, 125319, Russia
[2] Moscow Aviation Institute (National Research University), 4, Volokolamskoe highway, Moscow, 125993, Russia

E-mail: irina455@inbox.ru

Abstract. The paper considers the gas nitriding process with cyclic alternation of atmospheres: ammonia and an ammonia mixture with air for strengthening the corrosion-resistant high-alloyed steel 03Cr11Ni10Mo2Ti, designed to operate at cryogenic temperatures. The formation in steel of a nitrided layer of 130—150μm thickness with increased microhardness (more than 12GPa) is shown experimentally. During the final stage in the ammonia and air environment, iron oxides remain in the compound zone, and a sharp drop in hardness is observed at the border of the core. At the completion of the cyclic process with the stage of saturation in ammonia, a nitrides zone is formed based on the γ'-phase, the layer thickness increases (up to 180...200μm) and the microhardness profile is smoothed, which reduces the risk of embrittlement.

1. Introduction

Corrosion-resistant maraging steels alloyed with chromium and nickel are used for loaded parts operating in aggressive atmospheric conditions (turbomachine disks, gear wheels, spindles, worms, aircraft parts). Steels with a reduced carbon content operate at cryogenic temperatures due to the stabilization of retained austenite: steel 03Cr11Ni10Mo2Ti – at -70ºC/-196ºC, steel 03Cr12Ni10MoTi – at -253ºC. These steels are designed for use as treated by nitriding [1]: nitriding is carried out at temperatures of 540...550°C for 36 hours in ammonia with a dissociation degree 25...40%. In this case, nitrided layers with increased hardness of a thickness of 0.20...0.25 mm are formed. Nitriding maintains the required level of toughness at low temperatures, obtained during preliminary heat treatment, which consists of quenching and tempering. The corrosion-resistant zone is 50...60% of the thickness of the nitrided layer

In recent years, ion and plasma nitriding methods have been widely used as an alternative to furnace and liquid nitriding. Ion-plasma nitriding is reasonably considered a progressive method of surface hardening, the advantage of which is the formation of layers with a given thickness, phase composition and hardness on the outer surfaces of parts [2-4]. Ionic and plasma nitriding is successfully used for hardening of austenitic and martensitic steels [4-6]. Thus, for maraging high-strength steels, methods of ion-plasma nitriding have been developed, which make it possible to reduce the process temperature (up to 400...450ºC) and obtain a nitrided layer with high hardness without a surface zone of chemical compounds in a short time [7]. However, the thickness of the hardened layer does not exceed 20 μm. The disadvantage of the process is also an increase in the roughness of the steel surface.

Gas nitriding technologies are being improved towards processes intensifying, that may be achieved by using multicomponent atmospheres, including techniques of thermal and gas cycling. Thus, cyclic processes with alternating supply of ammonia and air are not inferior to the processes of ion-plasma nitriding in terms of the rate of formation of the nitrided layer; such processes create additional possibilities for regulating the phase composition of the layer and obtaining the desired properties [8, 9].

The aim of this work is to study a new process of gas nitriding with cyclic alternation of atmospheres, consisting of ammonia and air, as applied to high-alloy maraging steel.

2. Materials and methods

The studies were carried out on samples of steel 03Cr11Ni10Mo2Ti; the chemical composition is presented in table 1.

Table 1. Chemical composition of steel 03Cr11Ni10Mo2Ti

Fe	C	Cr	Ni	Mo	Ti	Si	Mn	S	P
Bal.	≤ 0.03	10-11.3	9.0-10.0	1.8-2.3	1.0-1.4	≤ 0.15	≤ 0.1	≤ 0.01	≤ 0.01

The gas nitriding processes were carried out in a two-zone shaft furnace with a gas intake system, which allows one to carry out processes in ammonia-air atmospheres with air concentratons from 10 to 90 vol.%. The temperature control of the heaters is carried out by thermocouples in the upper and lower zones. The presence of two heating zones with separate control ensures uniform temperature distribution over the working space of the furnace (± 3°C).

The process temperature was 580°C. At the stages of the process, the cycles of ammonia supply were alternated with cycles of the supply of a mixture of ammonia and air into the working space. The nitrogen potential of the atmosphere is regulated by changing the ratio of the volumetric concentrations of air (50...70%) and ammonia. The addition of air to the saturating atmosphere causes a sharp temperature rise in the container due to the reaction of hydrogen with oxigen. The temperature increase depends on the concentration of air in the mixture: the overheating was 60°C for 70% air, and it amounts 40°C for 50% of air. Thus, the cyclic air supply causes micro-cycling of the temperature in the oven (Fig. 1).

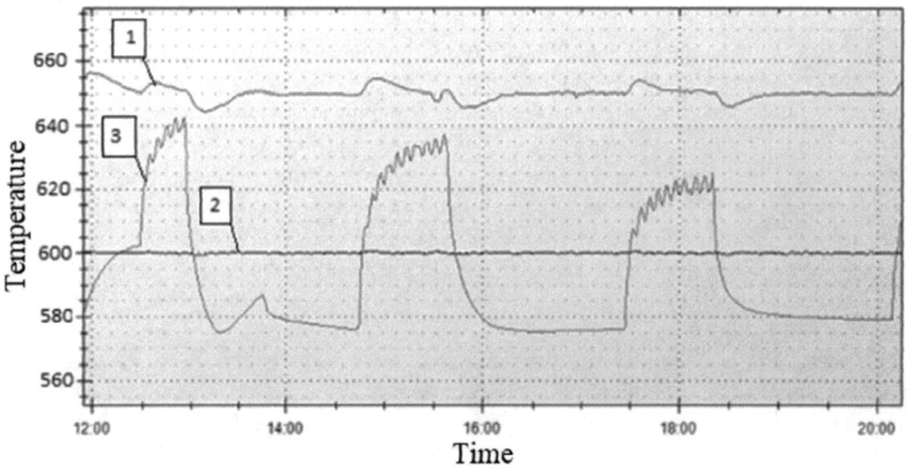

Figure 1. Time-temperature curves of cyclic nitriding ("Need for data" software): temperature splashes at "ammonia+air" stages: 1 — in the upper furnace zone; 2 — in the bottom furnace zone; 3 — in the reactor.

The analyses of the structure of nitrided samples were carried out: the microstructure of the nitrided layer were examined in the optical microscope METAM-LV 34; thickness and hardness of nitrided layers was determined on the PMT-3 device, and phase composition was determined by X-ray diffraction method on DRON-3 difractometer.

3. Results and discussion

Alternation of saturation cycles with high and low nitrogen potential of the atmosphere contributes to the intensification of the nitriding process. At the saturation stage, a mixture of ammonia and air provides a high concentration of nitrogen in the near-surface layer; at the processing stage, nitrogen is moved from the surface to the depth of the layer in ammonia. When such cycles are repeated, nitrogen penetrates deeper, which leads to the formation of diffusion layers of greater thickness. There is an increase in the nitrided layer with a gradual decrease of the nitrogen concentraton.

Experiments provided that during 12 hours nitriding with a high level of activity of the saturating atmosphere, a hardened layer is formed with a thickness of up to 130...150 microns, and with a hardness more than 12 GPA (Fig. 2, curve 1).

The accelerated growth of the diffusion layer is caused by microcyclic temperature fluctuations at the surface. Since the dissociation of ammonia occurs catalytically on the steel surface of the samples, it is in this region that hydrogen is bound by oxygen atoms. As a result, the effect of overheating is mostly appears in the near-surface area, that leads to an increase of the nitrogen concentration, and to the significant acceleration of diffusion processes.

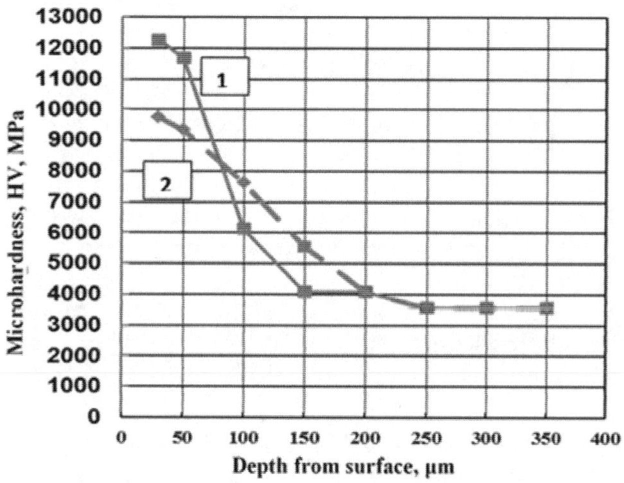

Figure 2. Microhardness profiles of steel after cyclic nitriding: 1 - "NH_3+air" final stage; 2 – "NH_3" final stage.

However, the high concentration of oxygen in the saturating atmosphere results in the formation of iron oxides (Fe_3O_4) in the surface layer (Fig. 3). A compounds zone consists of iron oxides and nitrides (Fig. 4); it has increased brittleness as a consequence of its high microhardness and of the sharp drop in hardness at the layer/core interface (Fig. 2, curve 1). The compound zone is mainly consists of the ε-$Fe_{2.3}N$ phase (table 2).

A process with a final stage in pure ammonia (3 h) provides optimal hardness and reducing the brittleness of the layer due to the "dissolution" of the compound zone. Increase in the degree of ammonia dissociation in the saturating atmosphere leads to the deceleration in nitrogen penetration from the atmosphere into the surface layer. The increase in the nitrided layer (up to 180...200 microns) takes place due to the redistribution of nitrogen in depth. The maximum microhardness of the layer is slightly reduced (about 9.5 GPa), and the microhardness profile becomes smoother (Fig. 2, curve 2), that decreases stresses and risks of microcracks at the layer/core interface.

Another consequence of the dissolution final stage consists in the reduction of oxides that changes the phase composition of the compound zone. A low-nitrogen nitrides zone is formed, consisting mainly of γ'-Fe_4N phase (table 2); besides the X-ray diffraction pattern shows reflexes of a γ-$Fe_{0.7}Ni_{0.3}$ solid solution (Fig. 5). Such structure of the nitrided layer creates preconditions for wear resistance increase and for securing of the required toughness.

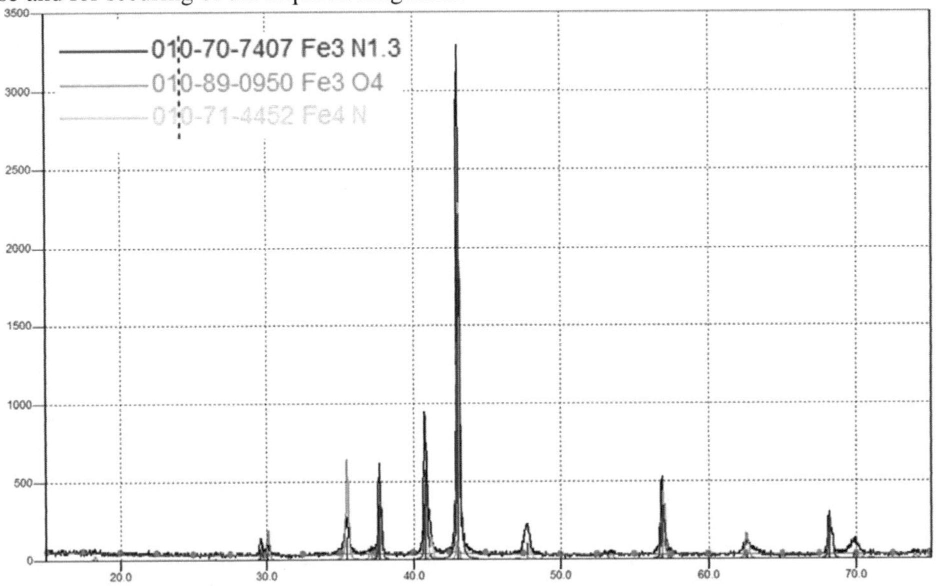

Figure 3. X-ray pattern of steel 03Cr11Ni10Mo2Ti after gas-cyclic nitriding with "NH_3+air" final stage.

Table 2. Phase composition of nitrided steel 03Cr11Ni10Mo2Ti

Gas-cycling mode	Mole fraction of phases			
	ε-Fe_3N	γ'-Fe_4N	Fe_3O_4	Fe_γ(Ni)
"NH_3+air" final stage	0.7969	0.0417	0.1614	-
"NH_3" final stage	0.0290	0.8744	-	0.0966

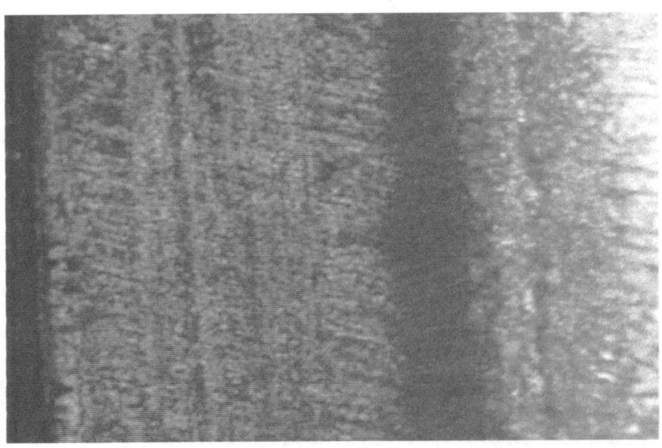

Figure 4. Micrograph of nitrided layer in steel 03Cr11Ni10Mo2Ti (x300).

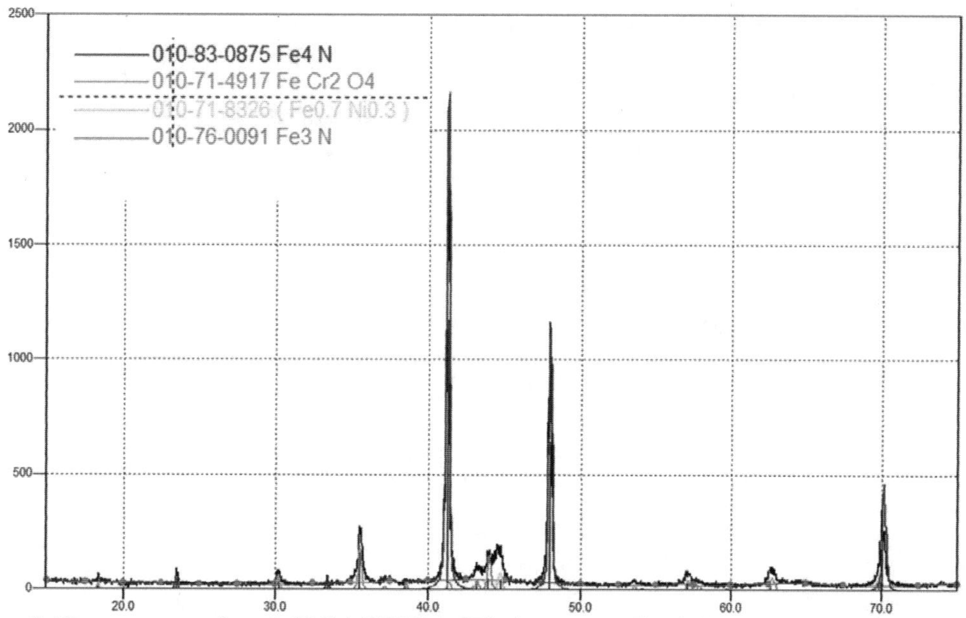

Figure 5. X-ray pattern of steel 03Cr11Ni10Mo2Ti after gas-cyclic nitriding with "NH_3" final stage.

4. Conclusion

The new nitriding process for maraging steel 03Cr11Ni10Mo2Ti, designed for operation at cryogenic temperatures, consists of a cyclic alternation of saturation stages in a mixture of ammonia and air, and the stages of resorption of the formed diffusion layer in ammonia. Gas-cyclic nitriding provides an intensification of the process of saturation of alloy steel with nitrogen due to the effect of microthermocycling. The optimal phase composition of the nitrided layer (nitride zone based on the γ'-phase without oxides) is formed during the final stage of the process with a lower nitrogen potential. During the final stage in pure ammonia, an increase in the thickness of the nitrided layer and a smooth microhardness profile are observed, which contributes to the preservation of plasticity and viscosity at a sufficiently high level of hardening.

5. Acknowledgments

The material was prepared within the framework of scientific research under project no. FSFM-2020-0011 (2019-1342); experimental studies were carried out using the equipment of the MADI Centre of collective usage.

References

[1] Alekseeva G P, Banas I P, Beliakova V I et al 1982 *Aviation industry* **8** 31-32
[2] Budilov V V, Agzamov R D and Ramazanov K N 2007 Ion nitriding in glow discharge with hollow cathode effect *Metal Sci. and Heat Treatment* **7-8(49)** 358-361
[3] Petrova L G, Aleksandrov V A and Demin P E 2012 Thermochemical treatment of steels in the plasma of a hydrostatic glow-spark discharge *Metal Sci. and Heat Treatment* **5-6(54)** 309-314
[4] Semenov M Y, Kai J D, Smirnov A E et al 2019 Use of glow discharge nitriding for raising the surface hardness of bearing parts from precision nickel alloys *Metal Sci. and Heat Treatment* **3-4(61)** 173-177
[5] Zhang Z L and Bell T 1985 Structure and corrosion resistance of plasma-nitrided stainless steel *Surf Eng.* **1** 131-136
[6] Khusainov U G, Ramazanov K N and Esipov R S 2018 Influence of ultrafine-grained structure of martensitic and austenitic steels on diffusion processes at low-temperature ion nitriding

IOP Conf. Series: Materials Science and Engineering 1064 (2021) 012004 doi:10.1088/1757-899X/1064/1/012004

Diffusion and Defect Data. Pt A Defect and Diffusion Forum **383** 167-172
[7] Borisyuk Yu V, Kozlova V V, Mozgrin D V et al 2018 Pulsed abnormal glow discharge with hollow cathode for nitriding of internal cylindrical surfaces *J. Phys.: Conf. Ser.* 012036
[8] Shestopalova L P 2016 Influence of preliminary oxidation on the subsequent nitriding *Russian metallurgy (Metally)* **13** 1241-1248
[9] Belashova I S, Petrova L G, Aleksandrov V D and Demin P E 2018 Improving the properties of low-alloy and carbon steel tools by cyclic nitriding *Russian Engineering Res.* **1(3)** 53-56

Mathematical model of a radial sliding bearing with a porous layer on its operating surface with a low-melting metal coating on shaft surface

K S Akhverdiev, E A Bolgova, M A Mukutadze and V V Vasilenko

Rostov State Transport University, 2, Rostovskogo Strelkovogo Polka Narodnogo Opolcheniya square, Rostov-on-Don, 344038, Russia

E-mail: bolgova_katya6@mail.ru

Abstract. The paper considers the steady flow of a liquid lubricant and a melt of a low-melting metal coating of the shaft surface, which have viscous rheological properties in the operating clearance of an endless radial sliding bearing, as well as in the body of the porous coating of the surface of bearing bush. In order to solve this problem, the authors used the equation of the movement of viscous lubricant for a thin layer, the equation of continuity and also the equation describing the radius of the molten contour of the fusible coating of the shaft surface, taking into account the rate of dissipation of mechanical energy. Having averaged the acceleration of the movement of the liquid lubricant in the operating clearance, the authors obtained the dependence for the contour of the molten coating of the shaft surface. As a result of the integration of the obtained dependence, the expression was found that allowed solving the function characterizing the contour of the molten surface of the coating by the method of successive approximations. Using the well-known method of finding an exact self-modal solution, the velocity field and pressure in the lubricating and porous layer, as well as the load capacity and friction force, were found. On the basis of the performed theoretical studies, the experimental study was carried out taking into account the parameters characterizing the melt of a low-melting metal coating, as well as a porous coating on the surface of the bearing bush and the rheological properties of a viscous lubricant. As a result, the original expression was obtained for the calculation of the load capacity of a radial bearing and the magnitude of the friction force in the range under the following loading conditions v=0.5÷3 m/s, σ=2÷7 MPa. The research results can be used in mechanical engineering, aircraft construction, instrument engineering, where the supply of lubricant is associated with great difficulties.

1. Introduction

Many works are devoted to the hydrodynamic calculation of sliding bearings [1-8] operating in conditions of the melt on one of their operating surfaces. One of the drawbacks of the existing design models [9-15] is that the lubrication process in the considered bearing designs is not self-sustaining.

In order to ensure damping properties and a self-sustaining lubrication process, it is necessary to have a porous layer on the operating surfaces of these bearings [16-20]. In our design model, the presence of a porous layer on the operating surface of the bearing and the presence of melt on the surface of the shaft are taken into account.

2. Results

The authors consider the steady movement of a viscous incompressible fluid between an eccentrically located pin and a bearing. A bearing with a porous layer on the operating surface is stationary, and the pin with the melt on its surface rotates with an angular velocity Ω (Fig. 1). In the polar coordinate system in the center of the bearing, the equations of the contours C_2 and C_3, C_0 and C_1 take the following form: $r'=b$, $r'=b+h$, $O_1M=a-\beta f(\theta)$ - radius equal to the molten shaft contour, a - shaft radius to the melt

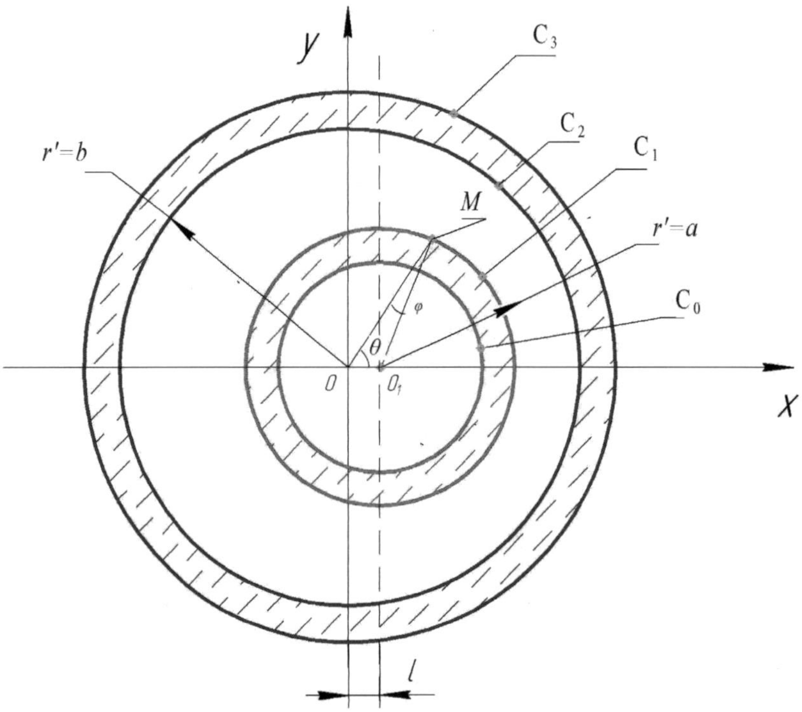

Figure 1. Scheme of the determination of the contours of shaft and bearing

For contour points of pin C_0 we have:

$$OM=OO_1\cos\theta+O_1M\cos\varphi$$

where O_1 – center of pin, O – center of bearing, $OO_1=e$ – eccentricity, θ – polar angle of point M of the contour C_0, φ – angle between OM and O_1M. From the triangle OO_1M, by the sine theorem, we find sin φ, and then cos φ:

$$\sin\varphi=\frac{OO_1}{O_1M}\sin\theta=\frac{e}{a-\beta f(\theta)}\sin\theta=\varepsilon\left(1+\frac{\beta}{a}f(\theta)\right)\sin\theta$$

$$\cos\varphi=\sqrt{1-\xi^2\left(1+\frac{\beta}{a}f(\theta)\right)^2\sin^2\theta}=1+O\left(\varepsilon^2\right) \qquad (1)$$

where $\varepsilon=\frac{e}{a}$ - eccentricity ratio

Taking into account (1) for the C_0 contour in the adopted approximation, we obtain the equation:

$$OM = a\left(1+\varepsilon\cos\theta-\frac{\beta}{a}f(\theta)\right) \tag{2}$$

The initial equations are flow equations of a viscous incompressible fluid for a thin layer, the continuity equation, the equation describing the lubricant flow in the body of the porous bush and the equation describing the radius of the molten contour of the shaft surface.

In the polar coordinate system, the above mentioned system of equations will be written in the following form:

$$\mu\left(\frac{\partial^2 v_\theta}{\partial r'^2}+\frac{1}{r'}\frac{\partial v_\theta}{\partial r'}\right)=\frac{1}{r'}\frac{dp'}{d\theta},\ \frac{\partial v_{r'}}{\partial r'}+\frac{v_{r'}}{r'}+\frac{1}{r'}\frac{\partial v_\theta}{\partial\theta}=0,$$

$$\frac{d}{d\theta}\left[a\left(1+\varepsilon\cos\theta-\frac{\beta}{a}f(\theta)\right)\right]=\frac{2\mu}{\Omega L^*}\int_{H(\theta)}^{b}\left(\frac{\partial v_\theta}{\partial r'}\right)^2 dr',$$

$$H(\theta)=a(1+\varepsilon\cos\theta-\frac{\beta}{a}f(\theta)) \tag{3}$$

For a certain flow of lubricant in the body of the porous bush, we will assume that the bush is homogeneous and isotropic and that the flow in the body follows Darcy's law.

$$\bar{V}'=-\frac{k}{\mu}gradP' \tag{4}$$

Here v_r, v_θ - velocity vector coordinates, p' - hydrodynamic pressure in the lubricating layer, k - permeability coefficient, P' - pore pressure in the bush, μ - dynamic viscosity of the lubricant, a - shaft radius to melt, b - radius of bush, L^* - specific heat of fusion per the unit of volume, $\tilde{H}$ - porous layer thickness.

From the equation (4) and the equation of continuity it follows that the pressure in the pores of the bush is equal to:

$$\frac{\partial^2 P'}{\partial r'^2}+\frac{1}{r'}\frac{\partial P'}{\partial r'}+\frac{1}{r}\frac{\partial P'}{\partial\theta}=0 \tag{5}$$

The boundary conditions for the velocity and pressure in the lubricating and porous layer will be as follows:

$\bar{v}=\bar{\Omega}x\overline{O_1M}$ on C_0;

$v_{r'}=-\frac{k}{\mu}\frac{\partial P'}{\partial r'},\ v_\theta=0$ on C_2, $\frac{\partial P'}{\partial r'}=0$ on C_3, $p'=P'$ on C_2, $p'(0)=p'(2\pi)=p_g$,

$a-\beta f(\theta)=h_0^*$ at $\theta=0$, $\theta=2\pi$ (6)

If we design the velocity vector $\bar{v}=\bar{\Omega}x\overline{O_1M}$ of shaft spindle on the coordinate axis, then we get $v_{r'}=-\Omega(a-\beta f(\theta))\sin\varphi,\ v_\theta=\Omega(a-\beta f(\theta))\cos\varphi,$ on the contour C_0 taking into account (1)

$$v_{r'} = -\varepsilon\Omega a\sin\theta + O\left(\frac{\beta^2}{a^2}\varepsilon\right),\ v_\theta = \Omega a\left(1-\frac{\beta}{a}f(\theta)\right) \text{ on the contour } C_0 \tag{7}$$

Before the formulation of an exact self-similar solution to the system of equations (3) and (5) satisfying the boundary conditions (6) $\frac{\partial v_\theta}{\partial r'}$, we average integrally in the interval $[b, H(\theta)]$, we get:

$$\frac{\partial v_\theta}{\partial r'} = \frac{\Omega a\left(1-\frac{\beta}{a}f(\theta)\right)}{a\left(1+\varepsilon\cos\theta-\frac{\beta}{a}f(\theta)\right)-b}, \text{ then}$$

$$\int_b^H\left(\frac{\partial v_\theta}{\partial r'}\right)^2 dr' = \frac{\Omega^2a^2\left(1-\frac{\beta}{a}f(\theta)\right)^2}{a\left(1+\varepsilon\cos\theta-\frac{\beta}{a}f(\theta)\right)-b} \tag{8}$$

Taking into account (8) for the determination $a-\beta f(\theta)$ we have the following equation:

$$\frac{d\left(a-\beta f(\theta)\right)}{d\theta} = e\sin\theta + \frac{2\mu\Omega}{L^*}\frac{\left(a-\beta f(\theta)\right)^2}{\left(a+e\cos\theta-\beta f(\theta)-b\right)} \tag{9}$$

Integrating this equation, we get

$$\Phi = 1-\frac{\beta}{a}f(\theta) = \frac{h_0^*}{a}+\frac{e}{a}\cos\theta + K\int_0^\theta \frac{\Phi^2(\theta)d\theta}{\Phi(\theta)+\xi\cos\theta-\frac{b}{a}},\quad K = \frac{2\mu\Omega a}{L^*} \tag{10}$$

Solving this equation by the method of successive approximations, we get:

$$\Phi_0 = \frac{h_0^*}{a}+\frac{e}{a},\quad \Phi_1 = \frac{h_0^*}{a}\varepsilon\cos\theta + K\int_0^\theta \frac{\left(\frac{h_0^*}{a}+\varepsilon\right)^2 d\theta}{\frac{h_0^*}{a}+\varepsilon-\frac{b}{a}+\varepsilon\cos\theta}, \tag{11}$$

We integrally average $\Phi_1(\theta)$ in the interval $[0, 2\pi]$. For the averaged radius of a shaft with a molten surface, we get:

$$\alpha^* = \frac{1}{2\pi}\int_0^{2\pi} a\Phi_1(\theta)d\theta \tag{12}$$

Taking into account (12) for *OM*, we finally get the expression in the following form:

$$OM = \alpha^* + e\cos\theta = \alpha^*\left(1+\xi\cos\theta\right), \text{ where } \xi = \frac{e}{\alpha^*} \tag{13}$$

In the lubricating layer, we pass to dimensionless variables by the formulas:

$$v_{r'} = \delta\Omega u,\ v_\theta = \alpha^*\Omega v,\ p' = p^*p,\ r' = b-\delta r,\ \delta = b-\alpha^*,\ p^* = \frac{\mu\Omega\alpha^* b}{\delta^2} \tag{14}$$

In a porous layer, we pass to dimensionless variables by the formulas:

$$P' = p^* P', \quad r' = br^* \tag{15}$$

We substitute (14) and (15) into (3) and (6), respectively, we arrive to the following system of equations and boundary conditions to them (with an accuracy up to the terms $O\left(\frac{\delta}{b}\right)$

$$\frac{\partial^2 v}{\partial r^2} = \frac{\partial p}{\partial \theta}, \quad \frac{\partial u}{\partial r} + \frac{\partial v}{\partial \theta} = 0, \quad \frac{\partial^2 P}{\partial r^{*2}} + \frac{1}{r^*}\frac{\partial P}{\partial r^*} + \frac{1}{r^*}\frac{\partial P}{\partial \theta} = 0, \tag{16}$$

$v = 1, \quad u = -\tilde{\eta}\sin\alpha\theta$ at $r = 1 - \tilde{\eta}\cos\theta, \quad \tilde{\eta} = \frac{e}{\delta},$

$$v = 0, \; u\Big|_{r=0} = -\tilde{K}\frac{\partial P}{\partial r^*}\Big|_{r^*=1}, \; \tilde{K} = \frac{-k\alpha^* b}{h\delta^3}, \; \frac{\partial P}{\partial r^*}\Big|_{r^*=\frac{h}{b}+1} = 0 \tag{17}$$

Exact self-similar solution of problem (16) - (17) up to the terms $O(\tilde{\eta}^2)$ we will find using following form:

$$v = \frac{\partial \psi}{\partial r} + V(r,\theta), \quad u = -\frac{\partial \psi}{\partial \theta} + U(r,\theta), \quad \psi = \tilde{\psi}(\xi), \quad V(r,\theta) = \tilde{v}(\xi), \quad U(r,\theta) = \tilde{u}(\xi)\tilde{\eta}\sin\theta$$

$$\xi = \frac{r}{1-\tilde{\eta}\cos\theta}, \quad \frac{dP}{d\theta} = \frac{\tilde{C}_1}{(1-\eta\cos\theta)^2} + \frac{\tilde{C}_2}{(1-\eta\cos\theta)^3} \tag{18}$$

$$P = R(r^*)\tilde{\eta}\tilde{C}_1\sin\theta$$

Substituting (18) into (16) and (17), we get:

$$\frac{d^3\tilde{\psi}}{d\xi^3} = \tilde{C}_2, \; \frac{d^2\tilde{v}}{d\xi^2} = \tilde{C}_1, \quad \frac{d\tilde{u}}{d\xi} + \xi\frac{d\tilde{v}}{d\xi} = 0, \quad \frac{d^2R(r^*)}{dr^{*2}} + \frac{dR(r^*)}{r^* dr^*} - \frac{R(r^*)}{r^{*2}} = 0 \tag{19}$$

$$\tilde{v} = 0 \text{ at } \xi = 0; \quad \tilde{u}\big|_{\xi=0} = -\tilde{K}\frac{dR}{dr^*}\Big|_{r^*=1}, \quad R(1) = 1, \quad \tilde{v} = 1, \quad \tilde{u} = -1 \text{ at } \xi = 1,$$

$$\frac{dR}{dr^*} = 0 \text{ at } r^* = 1 + \frac{b}{h}, \; \tilde{\psi}' = 0 \text{ at } \xi = 0, \quad \xi = 1, \; \int_0^1 \tilde{v}\,d\xi = \tilde{K}\frac{dR}{dr^*}\Big|_{r^*=1} \tag{20}$$

The solution of the task (19) - (20) in the adopted approximation is found by direct integration. As a result, we have:

$$\psi' = \frac{\tilde{C}_2}{2}(\xi^2 - \xi), \quad \tilde{v} = \tilde{C}_1\frac{\xi^2}{2} + \left(1 - \frac{\tilde{C}_1}{2}\right)\xi - 1, \; R(r^*) = \frac{r^*}{1 + \left(1 + \frac{h}{b}\right)^2} + \frac{\left(1 + \frac{h}{b}\right)^2}{\left(1 + \left(1 + \frac{h}{b}\right)^2\right)r^*}$$

$$p = \tilde{C}_1\tilde{\eta}\sin\theta + \frac{p_g}{p^*}, \quad \tilde{C}_2 = -\tilde{C}_1, \quad ,\tilde{C}_1 = \frac{6}{1 + 12\tilde{K}R'(1)} \tag{21}$$

Using formula (19) for the bearing capacity and friction force, we have:

$$R_y = \Omega\alpha^* \int_0^{2\pi} \tilde{C}_1 \eta \sin^2\theta d\theta = \frac{\mu\Omega\alpha^* b}{\delta^2} \tilde{C}_1 \pi \tilde{\eta}^2,$$

$$L_{TP} = \Omega\alpha^* \int_0^{2\pi} \left(-\frac{\tilde{C}_1}{2(1-\eta\cos\theta)^2} + \frac{1-\frac{\tilde{C}_1}{2}}{1-\eta\cos\theta} \right) d\theta = 2\pi \frac{\mu\Omega\alpha^*}{\delta}\left(1-\tilde{C}_1\right)$$

(22)

Based on the dependences given in the results (20), according to the numerical analysis data, it was proved that a radial bearing with a low-melting metal coating on the shaft surface and a porous coating on the bearing bush surface has a bearing capacity that exceeds standard radial plain bearings by 9 - 11%. In this case, the friction coefficient decreases by 10 - 12%.

In addition, experimental studies confirmed the conclusions of the theoretical research (Table 1).

Table 1. Experimental studies

Coefficient of friction / Parallel experiments	**Radial bearing**		
	Standard	Low-melting metal coating	Low-melting metal and porous coating
1	0,0038	0,0031	0,0021
2	0,0042	0,0032	0,0022
3	0,0043	0,0034	0,0024
4	0,0045	0,0038	0,0028
5	0,0048	0,0041	0,0029
Average	0,0043	0,0035	0,0024

3. Conclusion

According to the complex of theoretical and experimental studies of a mathematical model of a radial sliding bearing with a porous layer on its operating surface with a low-melting metal coating on the shaft surface and taking into account all additional factors, the authors obtained the original expression for the calculation of the load capacity of a radial sliding bearing and the magnitude of the friction force in the range under the following loading conditions v=0.5÷3 m/s, σ=2÷7 MPa.

References

[1] Zakovorotny V L, Shapovalov V V and Kolesnikov V I 2006 *Transport tribotechnics* (Rostov-on-Don: Roszheldor, Rostov State Transport University) vol 2

[2] Shapovalov V V, Chelokhyan A V, Kolesnikov I V, Ozyabkin A L and Kharlamov P V 2010 *Amplitude-phase-frequency analysis of critical states of friction systems* (Moscow)

[3] Zadorozhnaya E, Hudyakov V and Dolgushin I 2020 Evaluation of Thermal Condition of Turbocharger Rotor Bearing *Lecture Notes in Mechanical Engineering* 1183-1193. DOI: 10.1007 / 978-3-030-22041-9_123

[4] Levanov I G, Zadorozhnaya E A, Mukhortov I V and Eschiganov M O 2020 Study of effect of metal oleates on mixed and boundary lubrication *Tribology in Industry* **42(3)** 461-467. DOI: 10.24874 / ti.708.06.19.08

[5] Kandeva M, Rozhdestvensky Y V, Svoboda P, Kalitchin Z and Zadorozhnaya E 2020 Influence of the size of silicon carbide nanoparticles on the abrasive wear of electroless nickel coatings. Part 2 *J. of Environmental Protection and Ecology* **21(1)** 222-233
[6] Zadorozhnaya E, Levanov I and Kandeva M 2019 Tribological research of biodegradable lubricants for friction units of machines and mechanisms: Current state of research *Lecture Notes in Mechanical Engineering* **0 (9783319956299)** 939-947. DOI: 10.1007 / 978-3-319-95630-5_98
[7] Levanov I, Zadorozhnaya E and Vichnyakov D 2019 Influence of friction geo-modifiers on HTHS viscosity of motor oils *Lecture Notes in Mechanical Engineering* **0 (9783319956299)** 967-972. DOI: 10.1007 / 978-3-319-95630-5_101
[8] Mukhortov I, Zadorozhnaya E, Kandeva M and Levanov I 2019 Studying the possibility of using complex esters as AW / EP additives *Tribology in Industry* **41(3)** 355-364. DOI: 10.24874 / ti.2019.41.03.05
[9] Akhverdiev K S, Alexandrova E V, Kruchinina E V and Mukutadze M A 2010 Stratified flow of two-layer lubricant in the clearance of a thrust bearing with increased bearing capacity *Bull. of DSTU* **10(2(44))** 529-536
[10] Akhverdiev K S, Alexandrova E E, Mukutadze M A and Kopotun B E 2009 Stratified flow of two-layer lubrication in the clearance of a radial bearing with increased load-carrying capacity and damping properties *Bull. of RSTU* **4(36)** 133-139
[11] Akhverdiev K S, Alexandrova E E and Mukutadze M A 2010 Stratified flow of a two-layer lubricant in the clearance of a complexly loaded radial bearing of finite length with increased load-carrying capacity *Bulletin of RGUPS* **1(37)** 132-137
[12] Akhverdiev K S, Mukutadze M A, Alexandrova E E and Erkenov A Ch 2011 Mathematical model of the stratified flow of a two-layer lubricant composition in a radial bearing with increased bearing capacity, taking into account heat transfer *Bull. of RSTU* **1(41)** 160-165
[13] Kolesnikov I V, Mukutadze A M and Avilov V V 2018 Ways of Increasing Wear Resistance and Damping Properties of Radial Bearings with Forced Lubricant supply *Proc. of the 4th Int. Conf. on Industrial Engineering, Lecture Notes in Mechanical Engineering (ICIE 2018)* pp 1049-1062
[14] Mukutadze A M 2016 Coefficient of a rolling motion bearing drive *Procedia Engineering* **150** 547–558
[15] Lie K N and Jerry C T Su 2013 A Numerical analysis of optimum air journal bearings *J. of applied mechanical engineering* **2**. DOI: 10.4172 / 2168-9873.1000122.
[16] Mukutadze A M, Zadorozhnaya N S, Pinevich E V and Polyakov E V 2014 Hydrodynamic calculation of a non-uniform porous bearing of finite length operating in a stable non-stationary mode with a combined lubrication supply *Bull. of RGUPS* **2** 139-145
[17] Mukutadze A M 2016 Radial bearing with a porous support sleeve of variable thickness *Science of Science* **8(5).** Retrieved from: http://naukovedenie.ru/PDF/56TVN516.pdf.
[18] Kolesnikov I V, Erkenov A Ch, Boyko N I, Kruchinina E V and Mukutadze A M 2016 Development of a computational model of a radial bearing of finite length with a porous sintered ring of variable thickness for a combined supply of lubricant *Bull. of Rostov State Transport University* **4** 144-151
[19] Kolesnikov I V and Mukutadze A M 2017 Stability of a damper with a porous anisotropic ring *Bull. of Rostov State Transport University* **2** 8-16
[20] Kolesnikov I V and Mukutadze A M 2017 Damper with a porous insert for combined supply of lubricant *Bull. of Rostov State Transport University* **3** 22–31

IOP Conf. Series: Materials Science and Engineering 1064 (2021) 012006 doi:10.1088/1757-899X/1064/1/012006

Automation of the procedure for assessing the quality of transport service for the population in the conditions of digitalization of the railway transportation system

T A Bulokhova[1], V A Olentsevich[1], V Yu Konyukhov[2] and D A Lysenko[2]

[1] Irkutsk State Transport University, 15, Chernyshevskogo str., Irkutsk, 664074, Russia
[2] Irkutsk National Research Technical University, 83, Lermontova str., Irkutsk, 664074, Russia

E-mail: bulokhova@irgups.ru

Abstract. The Strategic Program for the Development of Railway Transport until 2030 focuses on the development of a customer-oriented approach as one of the conditions for the company's competitiveness. The customer focus is one of the elements of anti-crisis management of transport systems, it provides additional opportunities, especially when the volume of traffic decreases, when it is necessary to fight for attracting customers. A customer-oriented approach is closely related to the quality of the organization of transport services for passengers, which is certainly impossible without the use of modern automated technologies. Thus, a modern passenger is characterized by mobility and the habit of being online, in connection with which the passenger infrastructure and rolling stock must be provided with equipment and software that allows the passenger to remain in the familiar information environment.

1. Introduction

The implementation of a customer-oriented approach is possible only under the condition of well-functioning feedback channels, through which they receive an objective and relevant assessment of the quality of services from the passenger. For this purpose, the customer service sector and the marketing group of the Federal Passenger Company regularly analyze the quality of passenger service. The assessment of the quality of the organization of transport services for passengers carried out by the authors revealed a number of problems in this area. So, in marketing research, the assessment is carried out according to six criteria, and the employees of the client services sector, the assessment is carried out according to four, which violates the comparability of the data, in addition, the periods of the assessment do not coincide.

To eliminate the identified shortcomings, the authors proposed an automated procedure for assessing the quality of the organization of transport services for passengers, carried out using the correlation analysis of the pairwise dependence of the indicators of marketing research and the implementation of the parameters of client services. The use of correlation analysis will make it possible to determine the degree of dependence between the estimated indicators in order to further their automated selection and establish a single list of criteria for joint analysis of the quality of transport services [1, 2].

The activity of the railway transport system (RTS) in Russia deserves special attention when creating optimal economic conditions for the development of regions, for modernizing the entire transport sector of the country, during the transition to innovations and ensuring a stable growth of the national economy.

It contributes to creating conditions for ensuring Russia's leadership in the world transport system. In recent years, the holding company Russian Railways has chosen the vector of a customer-oriented approach, which provides additional opportunities in the face of competition in the transport market for a client and a reduction in passenger traffic.

The very model of a customer-centered approach also includes systems that influence the quality of services provided to passengers at stations and on trains. The modern passenger is characterized by mobility and the habit of being online, in connection with which the passenger infrastructure and rolling stock must be provided with equipment and software that allow the passenger to remain in the familiar information environment [3-5].

In the industry, there is an acute problem of increasing the competitiveness of transport systems, including by improving the quality of transport services for the population [6, 7]. It is also important that for the development of the railway passenger complex, there is no need to invent anything superfluous, you just need to modernize services for real time and demand. Modern passenger complexes should be not only a place for serving passengers, but also a territory to attract the urban population.

2. The main tasks of digitalization of the transport system in the field of passenger transportation

The key principle of customer focus is “How a customer orients a company can only be decided by the customer - and never the company manager”. The implementation of this principle in the railway transport system is possible only if there are "feedback" channels through which the client transmits an objective assessment of the quality of the services provided, using, for example, a system of questionnaires and surveys, IT technologies and Internet resources, Figure 1.

Currently, in all structural divisions of the railway transport network, a lot of work is underway to create a client-oriented system for the implementation of services based on:

- creation of unified call-centers, services rendered on the principle of "one window";
- provision of constant "feedback" with passengers;
- development of approaches to the formation of a competitive system of tariff setting for services not regulated by the state;
- formation of marketing and service promotion centers;
- development of the institution of managers for work with clients;
- active promotion of the services of the Russian Railways holding on the Internet;
- development of industry standards for interaction with customers, etc.

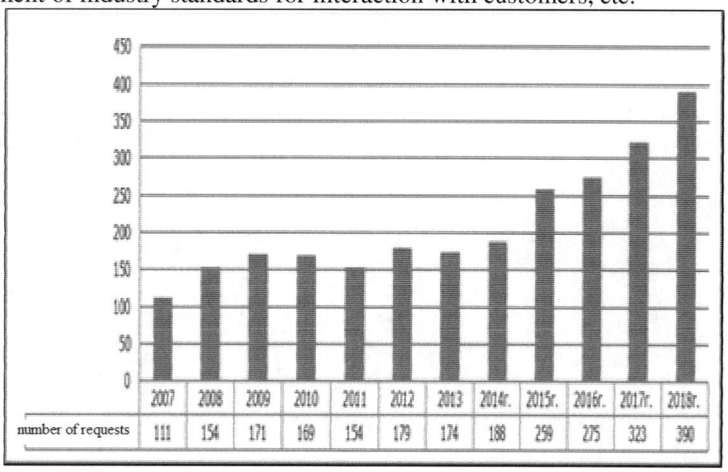

Figure 1. The Statistics of incoming requests from service consumers’ passenger complex.

Over the past eight years (from 2010 to 2018), the number of long-distance passengers transported decreased by 4.3%; the passenger turnover during the period under review decreased by 13.3%. One of the reasons for the reduction in passenger traffic and long-distance passenger traffic is associated with the development of competition with other modes of transport. Therefore, for comparison, over the same period, the passenger turnover in air transport, including international traffic, almost doubled: from 147.1 billion pass-km to 286.3 billion pass-km.

The innovative development of railway transport presupposes not only an increase in the volume of transportation of goods and passengers, but also an improvement in the quality of service provision, ensuring a high level of competitiveness of the industry, due to the digitalization of the railway transport system as a whole [3, 8, 9].

3. Organization and problems of assessing the quality of passenger service

The quality of transport services is understood as the ability of the transport system to satisfy the need for travel, as well as the associated needs, subject to the following conditions must be met [9-11]:

- transport accessibility - everyone who wants to travel can do it;
- customer satisfaction - consumers perceive the services of the transport system positively; they are satisfied with the level of comfort, they do not experience the negative effects of the external environment associated with receiving transport services, etc.;
- compliance with technology - the transport system is fault-tolerant, technologically and economically efficient, safe, the level of rejects does not exceed technologically sound limits.

Assessment of the quality of transport services is carried out according to certain criteria, which are enshrined in a number of regulatory documents [10, 12]. When assessing the quality, it is recommended to single out the following indicators of the quality of transport services for passengers (KTOP) [10, 13]: availability, efficiency, reliability, convenience.

The Federal Passenger Company (hereinafter OJSC FPK) has adopted a two-tier approach to quality assessment. Marketers assess the "perceived" quality of services, the customer service sector - the "provided quality of service" using a questionnaire consisting of 90 questions, with a frequency of once a quarter [10, 14, 15]. The Customer Services Sector conducts a monthly audit evaluating two hundred and fifty questions. Table 1 shows the estimated indicators by areas of responsibility.

Table 1. Estimated indicators by areas of responsibility.

Marketing research	Customer service sector	Area of responsibility
Evaluation of the ticket purchase procedure	Evaluation of fulfillment of requirements for cash services and railway agencies	East Siberian Railway Agency
Assessment of the conductor's work	Service assessment en route	Department of passenger service and provision of services on trains
Ride comfort assessment	–	East - Siberian branch
Assessment of the technical condition of the car fleet	Assessment of the technical condition of cars	Department of repair and operation of passenger cars
Evaluation of catering services provided on trains	Evaluation of food services en route	Food service sector
Evaluation of the convenience of the schedule, the duration of the trip and the ability to travel in the desired direction	–	Department of planning and organization of transportation of passengers, baggage and cargo

It can be seen from the presented table that the assessment of the comfort of the trip and the assessment of the convenience of the schedule fall out of the zone of consideration by the employees of the client services sector. Therefore, these indicators are not analyzed. In addition, the assessment of other "joint" indicators is carried out by researchers using different methods. Marketers, as mentioned above, are guided by a survey of passengers, and the customer service sector is based on personal checks. In this regard, there are often conflicting data on the estimated indicators.

4. Automation of the methodology for researching the quality of passenger service

To solve this problem, the authors propose to conduct a study of the pairwise relationship between the indicators for assessing the quality of services provided by the marketing department and the customer service control department using correlation analysis. The purpose of correlation analysis is to identify an estimate of the strength of the relationship between random variables (features) that characterize some real process. Carrying out this kind of research will allow measuring the degree of closeness and coherence of quality assessment indicators obtained during the study by different structural divisions, as well as establish a single list of criteria for joint analysis of the quality of transport services using an automated assessment system [5, 16, 17].

In addition, it is proposed to use GAP-analysis to compare the current indicators of quality assessment with the target, in order to develop adequate corrective measures to achieve the target state in the field of quality [10, 18, 19].

The essence of the method is to study the problem, which manifests itself as a gap arising in the course of the implementation of the change plan, between the indicators and results, the achievement of which was planned and what happened in reality (Figure 2).

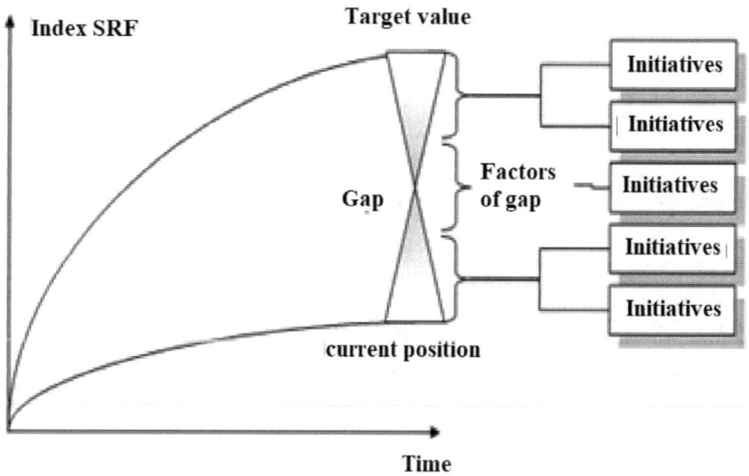

Figure 2. GAP analysis.

When carrying out the correlation analysis, the authors used data on four areas of responsibility:

- cash services;
- on serving passengers along the route;
- on the technical condition of the cars;
- on the provision of food services along the route.

The automated procedure for analyzing the current state of the "perceived" and "supplied" quality of services at JSC FPC is based on a study of the change in time (dynamics) of the average score ($\overline{B}$). It is based on the results of marketing research and changes in time (dynamics) of the average percentage ($\overline{P}$) fulfillment of the controlled quality parameters for a certain period of time from 2015 to 2018 [4, 10, 20]. The implementation of the project will allow forming new principles and models of work of structural units of the railway passenger complex in the conditions of widespread penetration of automated technologies, making it possible to transfer the organization and operation technology to a qualitatively new level, optimize operational work and improving the quality of service for service users.

IOP Conf. Series: Materials Science and Engineering 1064 (2021) 012006 doi:10.1088/1757-899X/1064/1/012006

The article presents an algorithm for automating the procedure for conducting a joint analysis of the "perceived" and "provided" quality of services to passengers, Figure 3.

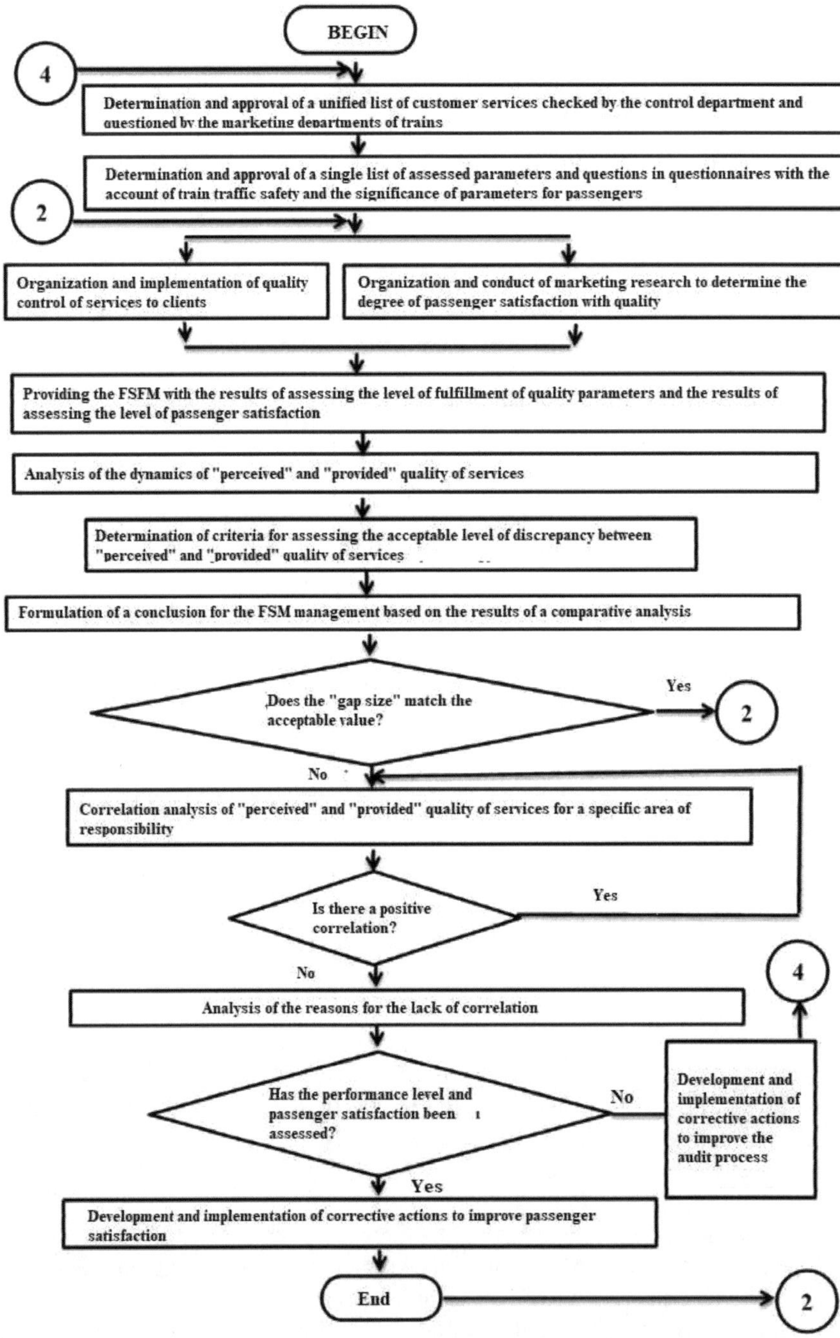

Figure 3. Algorithm for automating the procedure for conducting a joint analysis of the "perceived" and "provided" quality of services to passengers.

The dynamics of the average score of marketing research indicators is presented in table 2.

Table 2. Dynamics of the average score of marketing research indicators.

No	Areas of responsibility	2011	2012	2013	2014
1	Cash service	4.40	4.60	4.45	4.65
2	Service of passengers along the route	4.25	4.50	4.40	4.65
3	Technical condition of wagons	4.40	4.54	4.59	4.70
4	Providing food services along the route	4.13	4.22	4.25	4.65
5	Average score based on marketing research	4.30	4.46	4.42	4.66

Table 3 shows the average percentage of completion of grouped statistics of monitored parameters by the customer services sector over four years.

Table 3. Summary of performance of parameters of client services on average, %.

General controlled parameters	2011	2012	2013	2014
Cash service (average)	80	81	88	90
Technical condition (average)	87	93	94	96
Passenger service (average)	91	92	94	94
Organization of meals along the route (average)	81	82	82	87
Average percentage of parameter completion	88	90	92	93

The data for carrying out the correlation analysis for the East Siberian branch of JSC FPK as a whole for each area of responsibility for all the periods assessed are presented in Table 4.

Table 4. Correlation analysis data.

Name of the responsible subdivision of the East Siberian branch of JSC "FPK"	2011		2012		2013		2014	
	B*	P*	B*	P*	B*	P*	B*	P*
Cash service (Railway agency)	4.4	80	4.6	81	4.45	88	4.65	90
Passenger service en route (Passenger Service Department)	4.25	87	4.5	93	4.4	94	4.65	96
Technical condition of cars (Department of repair and operation)	4.4	91	4.54	92	4.59	94	4.7	94
Providing food services in transit (Food Sector)	4.13	81	4.22	82	4.25	82	4.65	87

* B - level of perceived quality of services, score; P - service quality level, %

Let us build a correlation between the parameters and conduct a gap analysis between the indicators. The calculated data on the correlation coefficient are presented in Table 5.

Table 5. Calculated data on the correlation coefficient.

Indicators	Correlation coefficient
Cash service	0,406
Technical condition	0,894
Passenger service	0,906
Catering on the go	0,997

As you can see from the table 5, according to the indicator "Cash services", the correlation coefficient was 0.4. This value indicates the absence of correlation factors. The reasons are that there are discrepancies in the obtained data for marketing research and the customer services sector, as well as different monitoring periods. In order to eliminate this problem, it is necessary to analyze the reasons for the lack of correlation, as well as develop and implement a single list of quality parameters monitored and evaluated in the questionnaires.

The correlation dependence of the indicator "Technical condition of a passenger car" is shown in Figure 4.

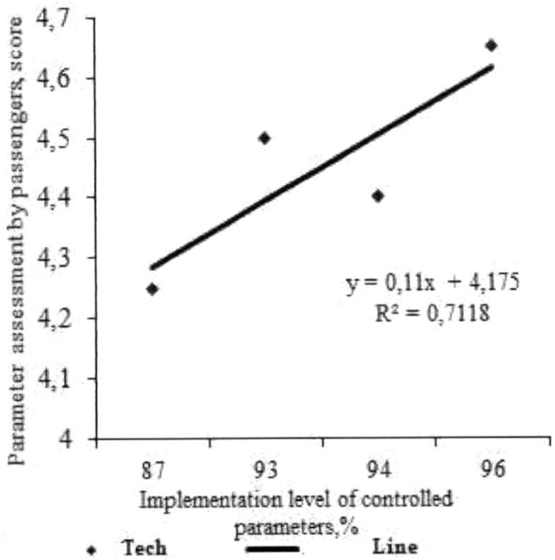

Figure 4. Correlation dependence of the indicator "Technical condition".

As we can see from the constructed dependence of the criterion "Technical condition of the passenger car", the correlation coefficient was R = 0.89, which indicates a high dependence between the data. A "positive" correlation means that an increase in the level of fulfillment of controlled quality parameters on trains leads to an increase in the level of customer focus of the company. And, consequently, this leads to a reduction in the "gap size" between the "provided" to the passenger and the "perceived" passenger quality of services, which in turn leads to improvement of the quality of service for the population as a whole.

Correlation dependence of the indicator "Passenger service» is in Figure 5.

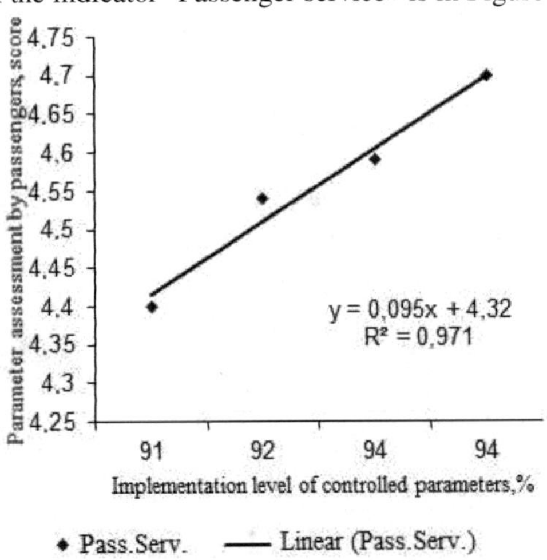

Figure 5. Correlation dependence of the indicator "Passenger service".

According to the constructed dependence of the criterion "Passenger Service", the correlation coefficient was R = 0.9, which indicates a high dependence between the data. The GAP analysis showed that the gap between "perceived" and "provided" quality of services is minimal, which also indicates a high degree of dependence between indicators.

The correlation dependence of the indicator "Organization of meals on the way" is shown in Figure 6.

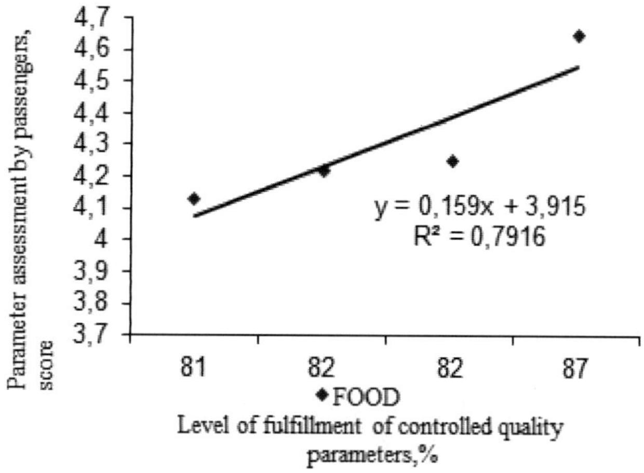

Figure 6. Correlation dependence of the indicator "Organization of meals on the way".

According to the constructed dependence of the criterion "Organization of meals on the way", the correlation coefficient was R = 0.9, which indicates a high dependence between the data.

Below is a general correlation analysis for all indicators considered in the table (Figure 7).

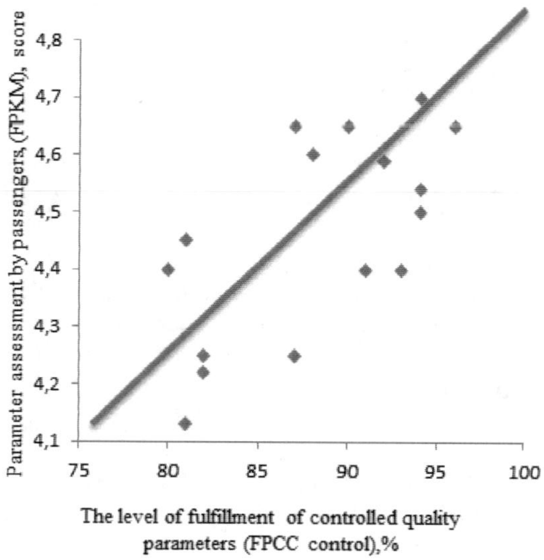

Figure 7. Correlation between indicators.

The calculated value of the criterion was R = 0.65, which is more than the critical value. Thus, there is a relationship between the indicators.

As an example, Figure 8 shows the GAP-analysis according to the criterion "Cash services".

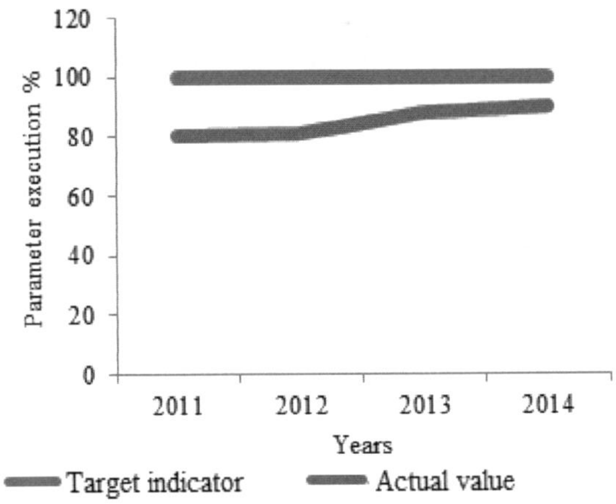

Figure 8. GAP-analysis of the "Cash service" parameter.

As you can see, the gap between the target (100%) indicator of the quality of passenger service and the actual values for the study period is narrowing, which indicates an improvement in the quality of service and the achievement of the target state.

5. Conclusion

Summarizing the above, the following should be noted. The railway transport network operates in conditions of fierce competition from other modes of transport. One of the ways to increase competitiveness is to increase the customer focus of the company, including through the use of modern digitalization tools. To this end, in order to conduct a more complete study of passenger service quality indicators, it is advisable to use an automated correlation analysis procedure to test the hypothesis about the statistical dependence of the values of the estimated indicators obtained from the results of marketing research and the client services sector. In the case of obtaining a correlation between the estimated indicators, it is recommended to supplement the GAP-analysis aimed at determining the degree of discrepancy between the indicators.

The use of the automated procedure presented by the authors in the study of indicators of the quality of passenger service in the railway transport system allows timely identification of discrepancies in the data obtained by the client services sector and the marketing staff. And, therefore, there is a chance to have a more realistic picture of the quality of transport services for the population. Automation of railway services in the field of passenger transportation is important for the development of customer focus of the passenger complex, since of all types of transportation are in this transport system, only a passenger can assess the quality of transportation and, depending on this, make a decision to use the services of railway or another type of transport

References

[1] 2019 *Railway transport in Russia: challenges until 2025* (Institute of Natural Monopolies)
[2] 2018 *Development strategy of Federal Passenger Company JSC until 2030. Reporting materials in accordance with the terms of reference*
[3] Larin A N and Larina I V 2017 Customer-oriented approach in structural divisions of Russian Railways *Economy of railways* **11** 30-37
[4] Tereshina N P, Zhakov V V and Filimonova Z V 2017 Increasing the competitiveness and quality of freight traffic *Economy of railways* **8** 41-49
[5] Elizariev M Yu 2012 Assessment of the quality of transport services for the population *Economy of railways* **5** 92-97
[6] Kargapol'Tsev S K, Gozbenko V E, Kuznetsov B O, Karlina Yu I and Karlina A I 2019 The effect

of the periodic driving force on a system with two degrees of freedom *J. of Phys.: Conf. Ser.* **1333(5)** 052009

[7] Gozbenko V E, Kargapol'Tsev S K, Kuznetsov B O, Karlina A I and Karlina Yu I 2019 Determination of the principal coordinates in solving the problem of the vertical dynamics of the vehicle using the method of mathematical modeling *J. of Phys.: Conf. Ser.* **1333(5)** 052007

[8] Kuznetsov B O, Gozbenko V E, Kargapoltsev S K, Karlina Yu I and Karlina A I 2019 Dynamic vibration protection of the railway carriage *J. of Phys.: Conf. Ser.* **1333(5)** 052018

[9] GOST R51004-96 *Transport services. Passenger Transportation. Nomenclature of quality indicators*

[10] Bulokhova T A 2016 On the issue of improving the quality of transport services *Modern approaches to management in transport and logistics: Coll. of mater. of the All-Russian sci. and pract. conf.* (Moscow: Moscow State Transport University) p 33

[11] Yolkin K S, Yolkin D K, Kolosov A D, Ivanov N A and Shtayger M G 2018 Technologies, which allow to reduce an impact of metal silicon production on the environment *IOP Conf. Ser.: Mater. Sci. Eng.* **411** 012028

[12] Bulokhova T A 2020 Modeling the assessment of the quality of transport services for the population *Modern technologies. System analysis. Modeling* **65** 128-136

[13] Gozbenko V E, Olentsevich V A and Belogolov Yu I 2018 Automation of individual operations of the transportation process in order to ensure sufficient conditions for the optimal functioning of digital transport and logistics *Modern technologies. System analysis. Modeling* **4(60)** 125-132

[14] Konyuhov V Yu, Gladkih A M, Galyautdinov I I and Shchadova E I 2020 Calculations of efficiency in implementing progressive mold forming methods *IOP Conf. Ser.: Mater. Sci. Eng.* **760(1)** 012027

[15] Gozbenko V E, Kargapoltsev S K, Karlina Yu I and Karlina A I 2019 Vibration state of technical facilities *IOP Conf. Ser.: Earth Env. Sci.* **378(1)** 012058

[16] Suslov K, Solonina N and Gerasimov D 2018 Assessment of an impact of power supply participants on power quality *Proc. of Int. Conf. on Harmonics and Quality of Power* pp 1-5

[17] Ilyushin P and Suslov K 2019 Operation of automatic transfer switches in the networks with distributed generation *IEEE Milan PowerTech* 8810450

[18] Voropai N, Ukolova E, Gerasimov D, Suslov K, Lombardi P and Komarnicki P 2019 Simulation approach to integrated energy systems study based on energy hub concept *IEEE Milan PowerTech* 8810666

[19] Kolobnev N I, Ber L B, Khokhlatova L B and Ryabov D K 2012 Structure, properties and application of alloys of the Al - Mg - Si - (Cu) system *Metal Sci. and Heat Treatment* **53(9-10)** 440-444

[20] Olentsevich V A and Gozbenko V E 2019 *Methodological and software support for predicting the values of the safety level of the functioning of the railway transport system,* monograph (Irkutsk)

Development of electric power systems based on the use of intelligent technologies

Y D Burda[1], A V Kosygina[1], I O Volkova[1], M V Gorgisheli[1], A Y Yakovleva[1] and K V Suslov[2]

[1] National Research University, Higher School of Economics, 20, Myasnitskaya str., Moscow, 101000, Russia
[2] Irkutsk National Research Technical University, 83, Lermontova str., Irkutsk, 664074, Russia

E-mail: yburda@hse.ru

Abstract. The development of retail electricity markets was specific in different countries. There is an important condition for further development of an electric power system with active consumers and prosumers' participation and the integration of renewable energy sources (RES), other innovative technologies and smart solutions. This study is aimed at observation of tracks of development and of retail electricity market characteristics in some countries through introducing a uniform framework for comparison. The results can help both experts and authorities to better understand the tasks and the conditions for the processes of energy transition.

1. Introduction

Recent changes in the global power energy sector have been mostly shaped by the transformation of the global economy, energy policy in the developed countries and technological advances in electricity-related technologies. The previously unavailable technologies to the actors of the sector are digital power electronics; technologies for bulk data storage (cheaper drives), its processing (networks and network equipment with higher bandwidth) and security (blockchain algorithms). There are also technologies for real-time demand-response; tools for multivariate modelling of power systems and their components; technologies for automation of commercial relationships in the market (including smart contracts) [1].

These advances, in their turn, re-shape the organizational basis of the industry through changes in the structure of stakeholders (emergence of *prosumers* and *demand aggregators,* dissemination of *demand response practices* and various *complementary services*) and nature of their collaboration. Alongside with above-mentioned new participants of the market, nowadays there is a constellation of new-entrants, providing various *complementary services*. These are digital data operators (who process the data, collected from smart metering devices, installed within the system and allow revealing consumption patterns and optimizing the whole system); providers of forecasting and prognostic services, etc. [2–5].

The described above changes lead to the situation when traditional electric utility market models, despite being considerably reliable and verified, are no longer able to fulfill all the requirements of the participants and to cope with the dynamic environment. This, in turn, brings to the necessity of elaboration of a new model, enabling to take into account mentioned changes, provide value to the actors involved, and stay sustainable in the long-run way.

Meanwhile, such changes require a corresponding external environment, which would allow actors of the power sector to adapt to these trends. This environment should be flexible and incorporate a certain degree of competitiveness. Therefore, it could create adequate stimuli for the participants to invest in new technologies and re-shape their business models. We suggest a uniform framework for the comparison of different models of retail electricity market that would allow for a meaningful analysis of the experience of different countries.

2. Methodology and approach

The practical implementation of the described models varies from country to country. This paper will pay particular attention to the retail markets of the following countries: USA, China, Singapore, Germany, and Russia, which represent both transition and developed economies. The choice was determined by the availability and by cohesiveness of information provided in open sources.

It is worth mentioning that the analysis of each case will start with the retrospective timeline of institutional and industrial changes of the power sector, as they are essential for understanding the backbone of the model applied at a certain country.

This paper has the following structure:

- literature review of theoretical electricity retail models, applied at the moment;
- tracks of retail electricity markets development provided;
- characteristics of retail markets in different countries to be studied: a mechanism of price regulation at the retail market; a level of concentration within the generation segment; a market share of power supply companies; a degree of value chain integration; a choice of power supply company for consumers.

3. Major models for retail electricity markets

For a substantial period of time most of the electricity markets were vertically integrated, when all the elements of the value chain (starting with generation and up to the electric power supply) were controlled by a single firm (often state-owned). Gradual democratization and globalization processes were followed by the implementation of competitive elements into the existing retail market mechanisms, including the changes in the participants and cooperation between market agents.

Conducted analysis of open data on the topic of electricity market liberalization [6–9] allowed one to outline three major models of retail electricity market differentiated by the level of liberalization: Monopoly, Single buyer, Competition in retail.

Each of the models incorporates a unique market structure and channels for the participation of the members. Positive and negative sides of the described models are provided in the following tables (Table 1-3).

Table 1. Advantages and disadvantages of the monopoly model for the electricity retail market

Advantages	Disadvantages
• Greater control of regulators over the whole system • Ability to implement various state policies (absence of resistance) • Simplicity • Greater coordination on the side of supply	• Lack of choice for end-consumers (high dependence on the power supply company functioning in the area) • Lack of incentives for vertically-integrated companies to modernize and increase their efficiency • Possible state interventions in pricing mechanism • All price risks on the side of consumers

Table 2. Advantages and disadvantages of the single buyer model

Advantages	Disadvantages
• Competition among generation companies (incentives for modernization and optimization) • Price averaging through long-term contracts • Regulated tariffs (determined in long-term contracts), providing sufficient investments into capacity	• Artificial long-term price-setting (long-term contracts may inadequately cover arising additional costs of energy suppliers) • Absence of supplier choice for end-consumers • All price risks on the side of consumers

Table 3. Advantages and disadvantages of the competition in the retail model

Advantages	Disadvantages
• Competitive pricing (based on the supply-demand equilibrium in dynamics) • Choice of supply conditions for end-consumers • Possibility for creation of the deregulated market (competition based)	• Necessity for complicated structural and legal transformations for implementation of the model • Necessity for complex infrastructural support of the system • Complication of the metering system

Described models of the retail electricity market are the most common in modern practice. They vary by the level of market liberalization and availability of the choice for consumers. More centralized ones are easier to control and manage, while liberalized models are more flexible and efficient in a stable institution and macroeconomic situation. Nevertheless, it is an open question on which model to choose that will be adjustable and stable under the condition of energy transition and global transformation. However, the choice of the model rests upon certain factors – both of institutional and industrial nature. These factors will be studied in detail in the following sections of the paper.

4. Comparative analysis of retail electricity markets in different countries

Thorough study of the models in every chosen country will be done in this section of the paper. Country tracks of retail electricity markets development and characteristics of retail markets will be observed.

4.1. USA

The development of the USA retail electricity market represents a transition from the local competitive microsystem in the early years to creation of large monopoly and after to a partial competition. Nowadays the market functions in the form of a partially competitive one, as just over 20 states have adopted this model [10] (Figure 1). It is worth mentioning that the specificity of the USA energy sector is determined by the tight bonds between electricity and gas markets, most of which is still exposed to high levels of state intervention.

Power companies are regulated both at the federal and state levels. FERC (Federal Energy Regulatory Commission) is responsible for retail electricity market and interstate transmission, while state commissions are governing the state level market – distribution and sales segments in particular. And one more important fact is that in integrated states (where power companies are vertically integrated) transmission tariff and sales premium are calculated by state commissions basing on the economic rules. At the same time, in deregulated states (where generation is competitive, and sales premium is market-determined) transmission and distribution tariffs are regulated by state commissions [11, 12].

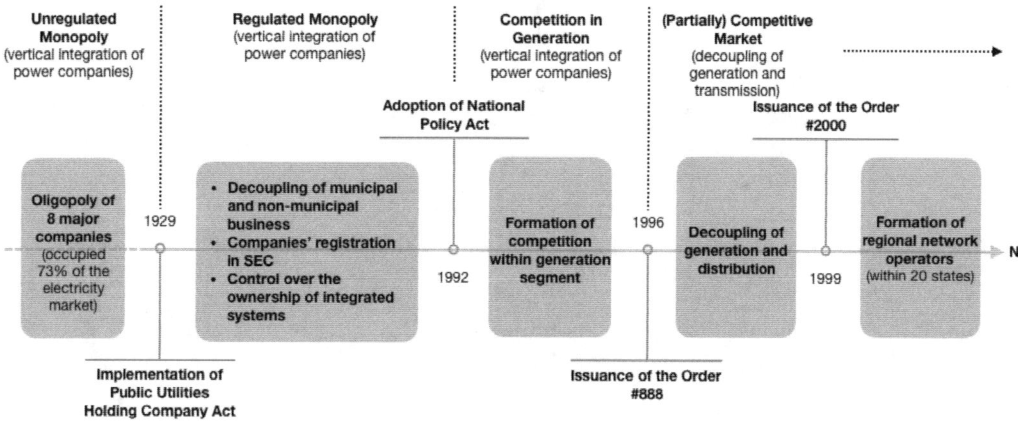

Figure 1. Transformation of the USA retail electricity market (Source: Authors' summary of [10]).

In such manner, by nowadays the USA has adopted two models of the major retail electricity market– centralized market (in two forms) and bilateral contracts (Table 4).

Table 4. USA retail electricity markets [13, 14].

	Centralized Market		*Bilateral Contracts (3)*
	VIC[a] (1)	***Retail choice (2)***	
Mechanism of consumers' price calculation	State-regulated ("costs plus" basis)	Generation tariff is market-determined and transmission tariff is regulated	Determined (either Contract bundle or customized proposition)
Concentration in generation[c]	5 major generating companies occupy 22% of the market $CR_5 = .22$		
Number of power supply companies	5 major power supply companies occupy 56% of the market MS_1 ·19.9%		
Value chain integration	VIC	Generation Transmission Distribution + Sales	VIC
Consumers' choice of power supply company	No	Yes	No
Model	*Monopoly*	*Competition*	*Monopoly*
Implemented in	AR*, CA, IA*, IN, KS, KY*, MI, LA*, MN, MO, MT*, ND*, NE*, NM*, OK, SD*, VA, VT, WI, WV	CT, DE, IL, MA, MD, ME, NH, NJ, NY, OH, OR, PA, RI, TX	AK, AL, AZ, CO, FL, GA, HI, ID, MS, NC, NV, SC, TN, UT, WA, WY
Market share of the model[b]	32%	41%	27%

[a]: VIC – Vertically integrated company.
[b]: market share is calculated basing on the number of consumers living in the mentioned states.
[c]: CR_i – concentration ratio of *i* generating companies with the biggest market share;
*: these states use a hybrid model.

Within the *monopoly* model VICs use "cost plus" pricing, which is criticized by many experts (not only from the power sector) [15] as this mechanism may lead to overinvestments and creates not enough stimuli for increasing efficiency. The *competition* model, on the other hand, provides consumers with the ability to choose the supplier and tariff plan and to develop other active strategies, including demand response, becoming prosumers, implementing energy effective solutions, etc.

Independent power supply companies determine their own competitive rates for production and services, but network tariffs are still regulated by the government. This model was implemented as a reaction to the strongly regulated electricity prices (compared to the relative costs level) in certain states. A lot of consumers find this model to be rather attractive as competitive prices are usually lower compared to those at the regulated markets. However, critics of this model stress the fact that a lot of generators do not consider households as an attractive segment (due to its relatively small share in consumption) and are rarely involved into competition due to a low level of the profit margin [14]. Fast developing new sectors (innovative energy ecosystem, intellectual integrated energy systems, prosumers' generation, energy storage, smart microgrid, etc.) change and complicate the retail electricity market structure, strategies of participants and rules.

4.2. China

China activity for developing of power sector was extensive and ambitious in last thirty years and the country catches up the other leaders except for the competitive retail market model. Starting from the 1980-s key processes of the changes there was decentralization and deregulation with the shift away from a state-regulated vertically-integrated monopoly and creation of a competitive market [16, 17, 18] (Figure 2). However, these shifts are still in progress.

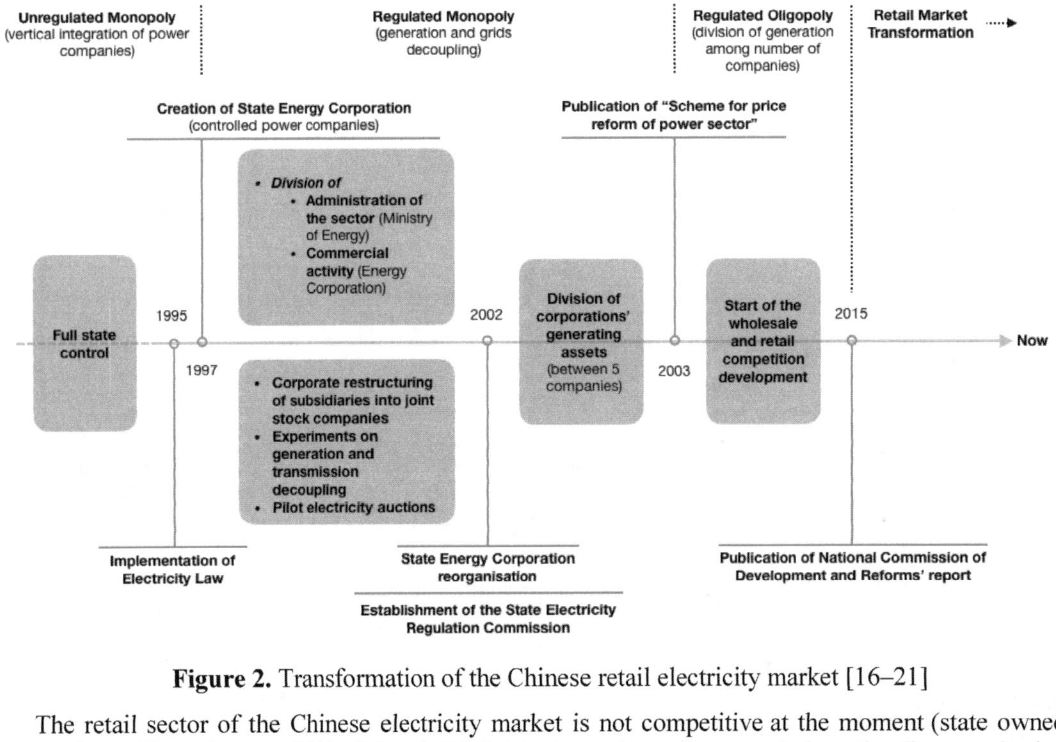

Figure 2. Transformation of the Chinese retail electricity market [16–21]

The retail sector of the Chinese electricity market is not competitive at the moment (state owned enterprises still supply over 80% of the electricity) and consumers have no choice other than purchasing it from local power supply companies (Table 5).

Table 5. The model of the Chinese retail electricity market [19, 21, 22].

Mechanism of consumers' price calculation	Long-term (20 years) contract for power supply company
Concentration in generation[a]	5 major generating companies occupy 100% of the market $CR_5 = 1.00$
Number of power supply companies	5 (affiliated with generators) + 300 local (function nominally) MS_i~20%
Value chain integration	Generation Transmission + Distribution + Sales
Consumers' choice of power supply company	No
Model	*Single buyer*

[a]: CR_i – concentration ratio of *i* generating companies with the biggest market share

However, market transformation is ongoing at the moment and fast development of RES, smart city projects and other innovative sectors in energy support it. In 2016 the National Commission of Development and Reforms (2015a) published the document, which implicates further development of a competitive retail market: shift from direct supplies to competition and implementation of the "Internet of Energy" concept [23].

4.3. Singapore

The creation of a competitive retail electricity market model in Singapore was executed in a rather abrupt time period, caused by massive foreign investment attraction, which required a corresponding infrastructure. For the period between 1995 and 2003 the country was able to shift from regulated monopoly to a competition in retail (Figure 3).

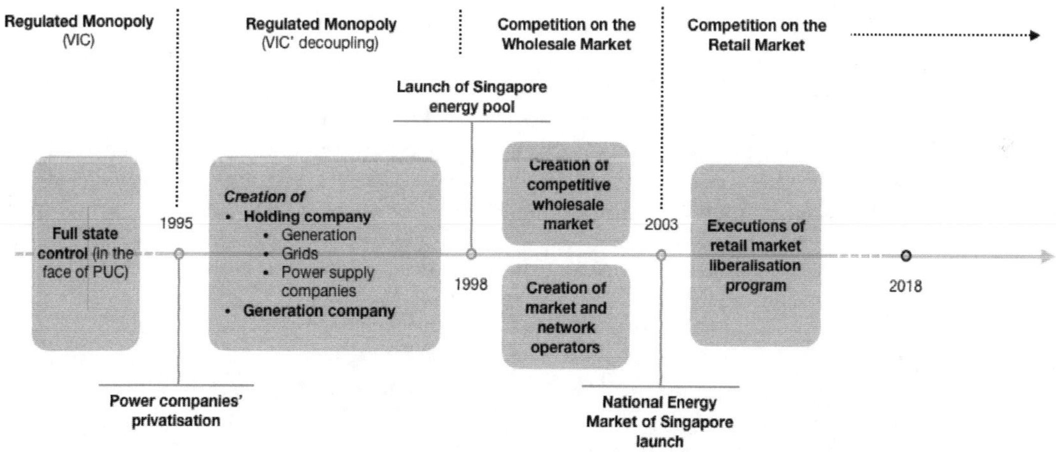

Figure 3. Transformation of the Singapore retail electricity market [24, 25].
Note: PUC – Public Utilities Committee

At the moment, the market model can be described as a partial competition, when only a certain part of the consumers has choice of a supplier (Table 6).

However, even under the following conditions a considerable share of consumers has a choice of a power supplier, what, coupled with the openness of the government, makes the whole system more flexible and attractive for doing business.

Table 6. Singapore retail electricity markets models [25]

	When power supply companies are market participants (1)	**When power supply companies are not market participants (2)**
Mechanism of consumers' price calculation	Competitive (price auction)	State regulation
Concentration in generation [c]	5 major generating companies occupy 78% of the market $CR_5 = .78$	
Number of power supply companies	26 MS_i=3.8%	
Value chain integration	Zero (all activities are decoupled)	
Consumers' choice of power supply company	Yes (for accepted[a]): Via power supply company; Directly on the wholesale market (requires registration); Via SP	No
Model	*Competition*	*Single buyer*
Market share of the model[b]	70%	30%

[a]: "accepted" are the companies with the monthly average consumption over 2.000 kWh (or bill over $400). They have an ability to decline the standard power supply and choose the supplier (data of the Energy Market Authority).

[b]: market share is calculated basing on the volume of consumption, generated by the mentioned segment of the consumers: market and non-market participants [26].

[c]: CR_i – concentration ratio of *i* generating companies with the biggest market share

4.4. Germany

The pre-reform power sector of Germany was different from all other European countries – a state monopoly on the major activities was absent per se. The market included companies with various forms of ownership – both state and private, while vertically integrated structures prevailed. Due to this, the liberalization of the market was executed via two stages. First National Power Law established a competitive model, which in practice was not viable due to a lack of essential legislative bodies. That is why later development can be described as a parallel liberalization of all value chain elements and creation of fully functional competitive market model by 2011.

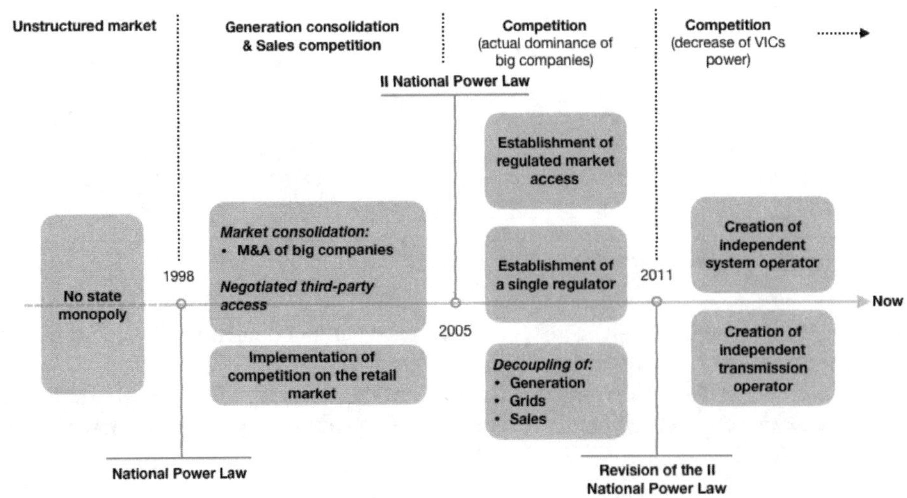

Figure 4. Transformation of the Germany retail electricity market [27–29].

Currently, the retail market is functioning within the competitive model with non-concentrated ownership of the assets and end-consumers have an ability to choose a power supply company (Table 7).

Table 7. Germany retail electricity market model [30]

Mechanism of consumers' price calculation	Price auction
Concentration in generation[a]	5 major generators occupy 45% of the market $CR_5 = .49$
Number of power supply companies	Around 900 $MS_i \sim 1\%$
Value chain integration	Generation + Distribution + Sales Transmission
Consumers' choice of power supply company	Yes
Model	*Competition*

a: CR_i – concentration ratio of *i* generating companies with the biggest market share

The current market includes a sufficient number of participants for supporting the competition: 4 big generating companies (EnBW, E.On, RWE, Vattenfall) and 1000+ small ones; 4 transmission lines operators (Ampirion, EnBW, Tennet, 50 Hertz Transmission); 890 distribution lines operators and 900+ power supply companies.

Still, the German model differs from the traditional competition by the fact that generation, distribution and sales are integrated (via generators' subsidiaries) and transmission is performed by independent operators. Innovative energy sectors (especially RES and green energy solution) have significant promotion for the development and transformation of the retail electricity market model.

4.5. *Russian Federation*

The development of the Russian electricity market represents the transition from regulated monopoly, which constitutes a heritage of the GOELRO plan and later comprehensive plans for energy, economy, and cities in 1950-1990, implemented in the USSR, to a single buyer / competitive model (Figure 5). Now this model need to be transformed due to developed innovative solutions in the energy sector, but this process moves slower than that in many other countries.

The government support the development of RES and other innovative energy niches in both zones with the retail electricity market and grid-off inhabited localities. For example, over the last years rules to support and implement domestic production of equipment for RES, payment for RES capacity and energy on the all retail electricity markets were established. But this activity meets a lot of obstacles, conflict of interests and lack of investment due to high risks and transactions costs.

RAO "ES Vostoka", "Gazprom" and other state and private companies install RES capacities or hybrid diesel-RES capacities in north regions and in grid-off inhabited localities [31]. Some of the national parks (Kenozero and Russian Arctic in Arkhangelsk region, Basegi in Perm Krai, Tigirekskij in Altai Krai and many others) use solar panels for heating and lighting, and some other technologies of alternative energy.

But at the same time majority of urban areas need significant modernization of energy supply systems, and Arctic region, Russia's Far East and some other regions need infrastructure development. This is an opportunity to transform energy systems and retail electricity energy markets on the base of new technologies.

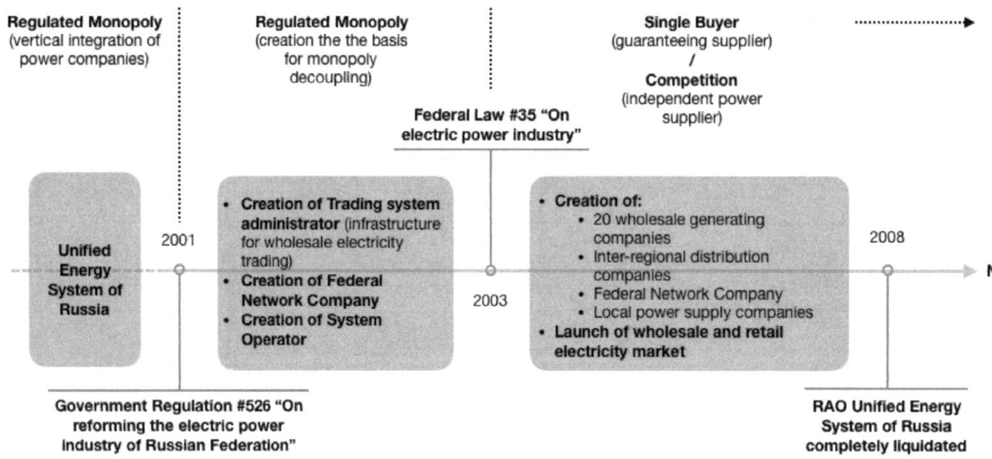

Figure 5. Transformation of the Russian retail electricity market [32, 33].

Currently, the retail market operates in both the single buyer and competitive models with the dominance of the former (Table 8).

Table 8. Russian retail electricity market model [32–34].

	When electricity is bought from a guaranteeing supplier (1)	When the electricity is bought from an independent power supply company[a] (2)
Mechanism of consumers' price calculation	State-regulated price	Agreement-determined
Concentration in generation[d]	8 major generators occupy 70% of the market $CR_5 = .64$	
Number of power supply companies	Local power supplier at each territory (720 in total) $MS_i \sim .1\%$	
Value chain integration	Generation + Sales Transmission + Distribution (UES' subsidiaries)	
Consumers' choice of power supply company	No[b]	Yes
Model	*Single buyer*	*Competition*
Market share of the model	69% of overall consumption[c] 93%	5% of overall consumption[c] 7%

a: this ability is available only to non-household consumers

b: in some cases, households purchase electricity not from the guaranteeing supplier – for instance, when it was liquidated. But these cases are very rare

c: the overall consumption of electricity in Russia is distributed in the following way: bulk consumers – 26%, guaranteeing suppliers – 69%, independent power supply companies – 5%.

d: CR_i – concentration ratio of *i* generating companies with the biggest market share

Consumers, attributed to households, do not have an opportunity to choose the supplier of energy. They have to purchase it from the guaranteeing supplier, although at a regulated price, which is lower comparing to that one for industrial and other corporate consumers (so called "cross-subsidization", when lower price for households is compensated by higher prices for other consumer categories). At the same time, corporate consumers are functioning within the competitive model when they are able to choose their supplier under specific conditions. Large industrial consumers can participate on the wholesale market. In these cases, pricing is purely market-driven.

5. Conclusion

The above-mentioned trends in the power energy industry determine the transformation of retail electricity markets around the globe. Moreover, the cornerstone is reaching flexibility, which has to provide an adequate reaction of the market to the changes in the dynamic environment.

This process of the industry liberalization is heavily dependent on the individual characteristics of national power energy sectors and economy, the specificity of state energy and economic policies, level of development of the national institutional systems, and certain external factors including technological advancement and other institutional pressures.

All considered markets move to the competition. But they have the different tracks and different levels of liberalization and concentration of generation and sales.

A monopoly model is still in use especially in countries with large territories and grid-off locations (USA, Russia).

The government and companies with state participants play an important role for the next transformation of retail electricity markets in Russia and in China.

We may conclude that further transformation of retail electricity markets will have significant specifity in each country. Innovative technologies and solution will change participants and their strategies, structure and rules of retail electricity markets. Probably characteristics and the structure of these markets will be different and the considered comparative model for analysis need further development.

Further research may focus on models of the electricity retail market in other countries, including identifying patterns in the development of these models and factors of influence. In doing so, the future studies can be based on both the framework proposed in the current article and a new expanded approach. It is of particular interest to consider the transformation of models of the electricity retail market in the studied countries under the current energy transition.

References

[1] Kubli M, Loocks M and Wustenhagen R 2018 The flexible prosumer: Measuring the willingness to co-create distributed flexibility *Energy Policy* **114** 540-8
[2] Artac G, Flynn D, Kladnik B, Pantos M and Gubina A F 2013 A new method for determining the demand reserve offer function *Electric Power Systems Research* **100** 55-64
[3] Bryant S, Straker R and Wrigley C 2018 The typologies of power: Energy utility business models in an increasingly renewable sector *J. of Cleaner Production* **195** 1032-46
[4] Hall S and Roelich K 2016 Business model innovation in electricity supply markets: The role of complex value in the United Kingdom *Energy Policy* **92** 286-98
[5] Orvis R and Aggarwal S 2018 Refining competitive electricity market rules to unlock flexibility *The Electricity J.* **31** 31-7
[6] Yassin E G, Volkova I O and Kosareva N B 2017 Some questions of development of retail power markets, urban economy and agglomeration *The Expert Institute of Higher School of Economics*
[7] Kasim N 2014 Study on Single Buyer Model and Pool Trading Model in Deregulated Electricity Market. Retrieved from: http://eprints.uthm.edu.my/5535/1/NURULAIN_NADIAH_BINTI_KASIM.pdf.
[8] Lagendjik V 2011 An experience forgotten today: examining two rounds of European electricity liberalization *History and technology* **27** 291-310
[9] The World Bank Group 2000, The Single-Buyer Model. Retrieved from: http://siteresources.worldbank.org/EXTFINANCIALSECTOR/Resources/282884-1303327122200/225Lovei-1211.pdf.
[10] Electric Choice (b). Timeline and History of Energy Deregulation in the United States. Retrieved from: https://www.electricchoice.com/blog/timeline-history-energy-deregulation/.

[11] Nazarian D 2012 Introduction to U.S. Electricity Markets. Maryland Public Service Commission, Portland. Retrieved from: http://www.camput.org/wp-content/uploads/2013/09/2012-07-21_-Nazarian_US_Electricity_Markets.pdf.
[12] Rudkevich T 2016 White Paper on Developing Competitive Electricity Markets and Pricing Structures. New York State Energy Research and Development Authority. Retrieved from: https://www.bu.edu/pcms/caramanis/NYPSC%20TCR%20Whitepaper April2016.pdf .
[13] Michigan Agency for Energy. What approaches to retail electric market structure have been tried in Michigan and in other states and jurisdictions. Retrieved from: http://www.michigan.gov/energy/0,4580,7-230-72200_68204_54287-293379--,00.html.
[14] Navigant 2013, Evolution of the Electric Industry Structure in the U.S. and Resulting Issues. Retrieved from: http://www.emrf.net/uploads/3/4/4/6/34469793/evolution_of_the_electric_industry_for_emrf_10-8-13.pdf.
[15] Li K and Lin B 2015 How does administrative pricing affect energy consumption and CO2 emissions in China? *Renewable and Sustainable Energy Reviews* **42** 952-62
[16] Andrews-Speed P and Zhenning C 2005 Prospects for Privatization in China's Energy Sector, in Green S and Liu G (Eds), *Exit the Dragon?: Privatization and State Control in China* (Wiley-Blackwell, London) pp 196-213
[17] Ciwei G and Yang L 2010 Evolution of China's power dispatch principle and the new energy saving power dispatch policy *Energy Policy* **38(11)** 7346-57
[18] Speed P 2004 *Energy Policy and Regulation in the People's Republic of China* (Kluwer Law International, Hauger)
[19] Chen L 2010 Playing the Market Reform Card: the Changing Patterns of Political Struggle in China's Electric Power Sector *The China J.* **64** 69-95
[20] National Development and Reform Commission. (2015a) Implementing Suggestions of Promoting Retail Electricity Market Reform. Retrieved from: http://tgs.ndrc.gov.cn/zywj/201601/t20160129_773854.html.
[21] Yeh E and Lewis J 2004 State Power and the Logic of Reform in China's Electricity Sector *Pacific Affairs* **77(3)** 437-465
[22] Pollitt M, Yang C and Chen H 2017 Reforming the Chinese Electricity Supply Sector: Lessons from International Experience. University of Cambride - Energy Policy Research Group. Retrieved from: https://www.eprg.group.cam.ac.uk/wp-content/uploads/2017/03/1704-Text.pdf.
[23] National Development and Reform Commission. (2015b). Relative Policies on Deepening the Reform of Power Industry. Retrieved from: http://tgs.ndrc.gov.cn/zywj/201601/t20160129_773852.html.
[24] Hogan Lovells Lee and Lee 2017 Singapore Energy Market. Hogan Lovells Lee & Lee, Singapore. Retrieved from: https://www.hoganlovells.com/~/media/hogan-lovells/pdf/publication/singapore-energy-marketseptember-2012_pdf.
[25] Wong K and Meng W 2017 Electricity Regulation in Singapore: overview. Thomson Reuters Practical Law. Retrieved from: //uk.practicallaw.thomsonreuters.com/7-637-5107?transitionType=Default&contextData=(sc.Default).
[26] Singapore Energy Statistics 2017 Energy Consumption. Retrieved from: https://www.ema.gov.sg/cmsmedia/Publications_and_Statistics/Publications/ses/2017/downloads/SES2017_Chapter3.pdf.
[27] Bundesnetzagentur. History of liberalisation. Retrieved from: https://www.bundesnetzagentur.de/EN/Areas/Energy/Companies/GeneralInformationOnEnergyRegulation/HistoryOfLiberalisation/historyofliberalisation_node.html.
[28] Deloitte 2015 European energy market reform. Country profle: Germany. Deloitte, Zurich. Retrieved from: https://www2.deloitte.com/content/dam/Deloitte/global/ Documents/Energy-and-Resources/gx-er-market-reform-germany.pdf.
[29] Uwer D, Zimmer D and Mueller H 2017 Electricity regulation in Germany: overview. Thomson Reuters Practical Law. Retrieved from: https://uk.practicallaw.thomsonreuters.com/5-524-

0808?transitionType=Default&contextData=(sc.Default)&firstPage=true&comp=pluk&bhcp=1.
[30] Bayer E 2015 Report on the German power system. Agora Energiewende. Brussels, Belgium. Retrieved from: https://www.agora-energiewende.de/fileadmin/ Projekte/2014/CP-Deutschland/CP_Germany_update_ 1015_web.pdf.
[31] Berdin V, Kokorin A and Yulkin G 2017 Renewable energy in off-grid settlements in the Russian Arctic *WWF Moscow* **45**
[32] Fortum. Market liberalization stages. Retrieved from: http://proekt.fortum.ru/company/energorinok/scheme-levels/.
[33] Ministry of Energy of Russian Federation. Industry condition. Retrieved from: https://minenergo.gov.ru/node/539.
[34] Branan 2013 Power Sector of Russia: Pricing.

Reduction of mine methane emissions for ensuring sustainable development of geotechnologies in the transition to Industry 3.0

A K Dzhioeva[1] and A V Tekhov[2]

[1] North Caucasian Mining and Metallurgical Institute, 44, Nikolaeva str., Vladikavkaz, 362021, Russian Federation
[2] South Ossetian State University named after A. A. Tibilov, 8, Putina str., Tskhinval, 100001, Russian Federation

E-mail: adadak@mail.ru

Abstract. The need to apply innovations in the field of environmental modernization is related to all branches of the domestic industry. Its particular relevance for the enterprises of the mining complex necessitates the neo-industrialization of the integrated development of the subsoil resources. The article defines the conceptual conditions for the development of sustainable geotechnologies. The nonlinear nature of the dynamics of methane release into the methane drainage boreholes has been proved and an explanation of the mechanism for its implementation under the influence of situational geomechanical conditions for developing reserves of high-gas-bearing coal seams has been proposed.

1. Introduction

Innovative development of domestic geotechnical systems is possible only if there is no tangible gap in the technological paradigms of the developed and developing countries. In this regard, the work [1] is one of the first domestic fundamental research (in identifying ways to overcome these gaps), which laid the foundation for the new industrial paradigm – the transition from the raw materials orientation of the Russian economy to the innovative way.

Enhancing the transformation processes of the international labor division in the framework of neo-industrialization (Industry 4.0) poses a threat to the macroeconomic stability, including by reducing the competitiveness of domestic coal mining enterprises [2, 3]. The main uncertainty lies in the difficulties of forecasting the final form of equilibrium in international markets, due to the lagging behind the emerging and more competitive 'technotronic society' (when the equipment begins to operate the equipment). Neo-industrialization should also be considered as a scenario of transition to a sustainable development, which lies in coordination of efforts aimed at synchronizing the processes of industrialization and informatization of coal mines [4]. The compulsory elements of "sustainable development" (in the area of environmentalization of mining production and integrated development of the subsoil resources) include the provision of a productive environment, the transition to a low-carbon economy, and a set of energy saving and recycling measures [5-7].

The need to upgrade geotechnologies for the transformation of the energy sector within the framework of the green and ecological economies concept is indicated by the existence of national programs implemented in developed countries. The United States Department of Energy is implementing this direction in the form of a system of initiatives (for example, the Clean Coal Technology and the Clean Coal Power Initiative). Of the most significant steps in this direction, it is worth noting the message of the European Commission "Sustainable power generation from fossil fuels: aiming for near-zero emissions from coal after 2020".

The mining of coal- and gas-bearing beds by high-loaded longwall faces causes difficulties in ensuring the safety of drainage boreholes, as well as the problems of ensuring the required quality of the bypassing gas-air mixture for cogeneration purposes [8, 9]. The lack of reliability of the circuit drainage network can lead to the impossibility of using coal mine methane as fuel for gas engine generators of cogeneration units [10]. This brings up to date the need to develop a comprehensive program in the field of achieving environmental transformation and improving energy efficiency of mining production.

The environmental orientation of mining production is constrained by the lack of a rational vision of methane, not only as a greenhouse gas, but also a hidden potential able to increase the profitability of coal mining. The use of methane as an alternative energy carrier (in modular gas engine generators (GEG)) allows you to stream the generated electricity to the mine's power system, simultaneously reducing the production cost.

Therefore, the purpose of the paper is to ensure the sustainable development of geotechnologies by identifying the spatial distribution of the circuit drainage network failures, using coal mine methane in cogeneration units.

2. Materials and methods

The work [11] described the existence of a non-linear nature of methane concentration dynamics in worked up degassing boreholes when developing a deep-lying (over 1300m) rich gas-bearing bed (methane content 19-23 m^3/t daf) m_3 at Zasyadko Mine. The authors have found out that the cyclical nature of dynamics changes in the methane concentration in equally oriented boreholes (with taper angle to the bottomhole = 35° and angle of elevation to the horizon = 40°) is evident 50m in front of the longwall face.

The disadvantage of the study was the lack of clarity in the representation of the spatial and temporal nature of the methane emission flow, which restricts localizing the shape of areas with a dangerously low concentration of methane. Therefore, the work's data [11] were used as primary information for constructing the surface (based on the methodology [12]), reflecting the change in methane concentration in this type of boreholes depending on the distance from the coal face line and the slots position relative to the distance to the beginning of the site (the value of which are interrelated with different measurement times).

In contrast to LOESS (local polynomial regression) used in [13] for data smoothing, we applied the method of non-uniform rational B-splines. NURBS is a distance-weighted B-spline. Order in the x and y dimensions were equal – bicubic (order 3/3). NURBS was calculated only for nodes in data, further interpolation was carried out using conventional B-spline methods.

3. Spatial distribution of failures of "drainage holes" as an element of the circuit drainage network (according to the nature of the manifestation)

Due to the fact that the lower margin of the working range of methane concentration required for GEG's basic operation is 25-30%, an increase of the nonproductive (with a concentration of less than 25%) consumption of the mixture in the circuit degassing network poses a significant threat to the normal energy generation. From the theory of reliability it is known that the "failure" of the system is an event, accompanied by the loss of its operational state. From the point of view of reliability engineering, the state of the degassing system as a whole (the network or its elements) in which the concentration of methane in the bypassing mixture exceeds 25% can be considered "efficient". Therefore, it is vital to hunt down the zones in which failures or, on the contrary, the operational state of the drainage boreholes can be traced.

The projection of the resulting surface on the **S-L** axis (the length of the air roadway of the "pickets" of the 18th eastern lava – the distance from the longwall face to the coal face line) is shown in Figure 1.

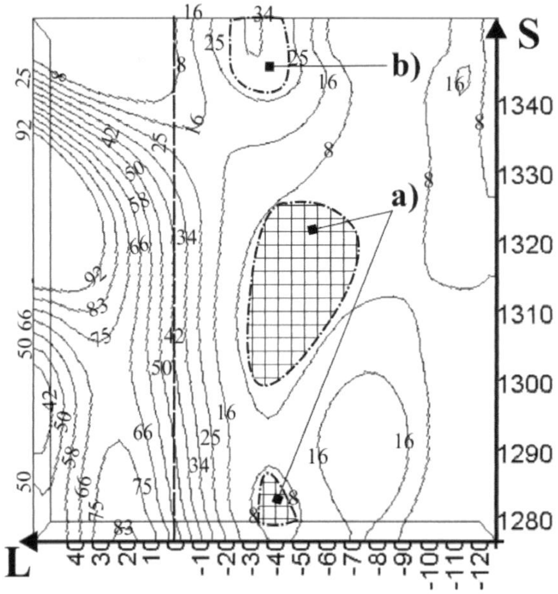

Figure 1. The projection of the time variation of methane concentration depending on **L** (distance relative to the unmined coal) and **S** (distance to the beginning of the mine section) in the boreholes of the circuit drainage network: a) – zones of local minimum of methane release ("borehole failures"); b) – zones of local maximum of methane release ("usable condition of boreholes").

Figure 1 more clearly shows the movement of areas with cyclical changes in the dynamics of methane concentration. The dotted line denotes the line of the "zero" distance to the face (negative values are the distance ahead of the longwall), and the contour line represent the locus of points with the same methane concentration values. From the analysis of contour lines, it is obvious that most of the time when the holes are ahead of the longwall (more than 10m ahead of the lava), their work is characterized by insufficient reliability. Besides, it should be noted that the value and position of local maximum and minimum also non-linearly change their position in the S-L plane. As indicated in [6], apparently, "this process is greatly influenced by the size of the roof span and the distance to the longwall". Moreover, up to approximately 1335m from the beginning of the section (roof span = 25m), the area of the local maximum of methane release at a distance of 30m from the longwall (area b) began to move closer to the longwall, gradually changing to zones of local minimum.

Local maxima on surface of the investigated function were traced at the following distance from the longwall: 1)-th (33% CH_4) – from -30 to -35m; 2)-th (16%) – from -58 to -77m; 3)-th (16%) – from -110 to -120m. The minima of the "zone b)" function are traced: 1)-th – from -30 to -70m; 2)-th – from -35 to -45m. It should be clarified that the movement of the production face line was carried out in reverse. That is, conditionally from the picket 1360m (from the beginning of the panel), therefore, already at the picket 1350 \m, the span of the main roof was 10m.In this regard, L = 40m (the length of goaf) was formed by the time when the face was on S = 1320m. By the same time, we observe the beginning of the formation of the 1)-th minimum at -35m (ahead of longwall) - "zone a)".

Analysis of Figure 1 it follows that for the entire period of increasing the roof span to 80m (with a change in S from 1360 to 1280m), ahead of the face (L from -35 to -120m), a dangerously low concentration of methane in the extracted mixture can be traced. At the same time, operational control (blocking the channels of boreholes at their mouths) of the system's operation, apparently, was not performed.

The next feature is that with the formation of a sufficient roof span (40m and more), due to an increase in the reference pressure, a stable pattern of parallel stripes of isogas up to 24m in front of the lava is traced. Moreover, the parallel lines are directed at a certain angle to the axis L = 0. Isogas 25% with a span of 40m (corresponding to S = 1320m) is located only 10 m ahead of the face (L = -10m). At the same time, at S = 1280m, this distance increases to L = -20m.

Perhaps this is due to the squeezing out of the edge zone of the reservoir and the redistribution of part of the rock pressure deep into the massif. Indirectly, the revealed features confirm the results of foreign studies in the conditions of mines in China [14] (surveys of the state of underground wells), the data of laboratory experiments [15], as well as the conclusions of domestic studies [6].

4. Conclusion

Based on the obtained results, we can conclude:

- that the nonlinear nature of the dynamics of methane release is realized in the form of cyclical changes (alternation of minimum and maximum) of methane concentration in local areas of the rock mass that move relative to the coal face under the influence of changes in the geomechanical conditions of the excavation site;
- an increase in the span of the main roof from 11m to 66m is accompanied by an increase in the area of distribution of the zone of maximum methane emission at a distance of 70-90m in front of the face, but it is still insufficient to achieve a safe level of methane concentration in the degassing network;
- when mining high-methane-bearing seams of Donbass (at depths over 1000m), the inclusion of separate wells in the operation of the degassing network does not ensure the reliability of their operation until the face of the face approaches □ 40m.

Based on the theory of sustainable development, in our opinion, the essence of the environmental transformation of geotechnologies should lie in developing a set of measures based on a balanced interaction of society and natural biosystems, while reducing the emission of harmful substances (primarily mine methane) from air workings.

In order to significantly reduce mine methane emissions, it is necessary to monitor (in real time) the state of the degassing system. Based on the studies carried out, in most of the operation time (L from -120 to -20m), in front of the working face, an air-gas mixture with a methane concentration of □ 25% is discharged. This makes it necessary to create a network of sensors combined with shut-off valves at wellhead of each underground borehole. Obtaining the ability to quickly shut off ineffective wells will allow implementing the concepts of neo-industrialization of mining.

References

[1] Glazyev S Yu and Fetisov G 2013 *Economist* **1** 3–13

[2] Tolkachev S A 2017 *Humanities. Financial University Bulletin* **1** 86–91

[3] Litvinenko V S 2020 *Nat. Resour. Res.* **29(3)** 1521–1541. DOI: 10.1007/s11053-019-09568-4

[4] Safiullin A R 2017 *Sci. and Tech. J. of Peter the Great St. Petersburg Polytechnic University. Economics* **10(6)** 40–49

[5] Rylnikova M V 2020 *Sustain. Dev. Mt. Territ.* **12(1)** 154–161. DOI: 10.21177/1998-4502-2020-12-1-154-161

[6] Kachurin N M, Ermakov A Y and Sencus V V 2018 *Mining Informational and Analytical Bull.* **7** 61–68. DOI: 10.25018/0236-1493-2018-7-0-61-68

[7] Golik V I, Razorenov Yu I and Karginov K G 2017 *Sustain. Dev. Mt. Territ.* **9(2)** 163–171. DOI: 10.21177/1998-4502-2017-9-2-163-171

[8] Yu E A and Kachurin N M 2018 *Mining Informational and Analytical Bull.* **6** 193–206. DOI: 10.25018/0236-1493-2018-6-0-193-206

[9] Rylnikova M V, Ainbinder I and Radchenko D 2018 *3rd Int. Innovative Mining Symp. (IIMS 2018) E3S Web of Conf.* **41** 01033. DOI: 10.1051/e3sconf/20184101033

[10] Brigida V S, Dmitrak, Yu V, Gabaraev O Z and Golik V I 2019 *Work safety in industry* **747(3)** 7–11. DOI: 10.24000/0409-2961-2019-3-7-11
[11] Brigida V S and Zinchenko N N 2014 *J. of Min. Sci.* **50(1)** 60–64. DOI: 10.1134/S1062739114010098
[12] Golik V I, Razorenov Yu I, Brigida V S and Burdzieva O G 2020 *Bull. of Tomsk Polytechnic University. Geo Assets Engineering* **331(6)** 175–183. DOI: 10.18799/24131830/2020/6/2687
[13] Brigida V S, Kozhiev Kh Kh, Saryan A A and Dzhioeva A K 2020 *Mining Informational and Analytical Bull.* **4** 20–32. DOI: 10.25018/0236-1493-2020-4-0-20-32
[14] Xue F, Zhang N, Feng X, Zheng X and Kan J 2015 *PLoS One* **10(1)** e0115874. DOI: 10.1007/s11053-019-09568-4
[15] Niu Y, Zhang X, Wang E, Li Z, Cheng Z, Duan X, Li H, Wei Y, Qian J, Cai G, Lin S and Li B 2020 *Measurement* **154** 107521. DOI: 10.1016/j.measurement.2020.107521

Photo-, cathodic- and electroluminescence-band models in solid $(SiC)_{1-x}(AlN)_x$ luminescence centers and SiC/SiC-AlN LEDs

R Kh Dadashev[1,2], A V Sankin[3], V I Altukhov[3], G K Safaraliev[4], D Z Elimkhanov[1,2] and V F Antonov[4]

[1] Faculty of Physics and Mathematics, Chechen State University, 32, Sheripova ave., Grozny, 364907, Russia
[2]Department of Physical, Mathematical and Chemical Research, Academy of Sciences of the Chechen Republic, 13, Esambaeva ave., Grozny, 364024, Russia
[3]North Caucasus Federal University (branch), 1, 40 years of October ave., Pyatigorsk, 357500, Russia
[4]Department of Experimental Physics, Dagestan State Technical University, 70, Imama Shamilya ave., Makhachkala, 367026, Russia

E-mail: edzhabrail@mail.ru

Abstract. The paper presents models of bands (levels) in solid $(SiC)_{1-x}(AlN)_x$ luminescence centers and **n- SiC/p-$(SiC)_{1-x}(AlN)_x$** heterostructures (light-emitting diodes). The diagram of **$(SiC)_{1-x}(AlN)_x$** energy gaps shows the positions of luminescence levels, subject to x. A **SiC/SiC–AlN** series of electroluminescence bands, for the first time, is found to have a continuous relationship between the positions of short-wave and long-wave band maxima in a K-minimum conduction band, respectively, as continuous functions of contact current density.

1. Introduction

Wide-band semiconductors and, in particular, solid $(SiC)_{1-x}(AlN)_x$ solutions with the energy gap (E_g) from 3.3 to 6.0 (eV) are of considerable interest for creating high-temperature, stable, high-voltage instruments, optoelectronic devices in a short-wave region of electro-optic emission. With the advent of new technology, it became possible to obtain heterostructures and almost ideal Schottky barriers, thereby enabling to design power electronic components and devices with extreme characteristics. This is related to an enormous interest from researchers in the physics and wide-band heterostructures. These materials are currently used to develop LEDs, sensors, microwave signal mixers, low-inertia modulated photodetectors and field-effect transistors [1-3]. Their characteristics are attributed to the height of a potential barrier and to surface states and mechanisms of thermionic, thermo-field and field emissions [4-6].

SiC solid solutions are currently being massively studied. In particular, in [7-10], the regions of stabilization of a number of polytypes (3C, 2H, 4H, 6H, 15R) of $(SiC)_{1-x}(AlN)_x$ solid solutions at T = 2370-2190 °C for compositions $0.04 \cdot x \cdot 0.95$. The data on photo-, cathodic- and electroluminescence of $(SiC)_{1-x}(AlN)_x$ solid solutions and resulting structures are presented. The technology is implemented enabling variband heterostructures of $SiC/(SiC)_{1-x}(AlN)_x$ LEDs for ultraviolet (UV) radiation sources.

The authors, based on experimental data available [7], proposed models enabling to determine the positions of levels of a series of photo- and cathodoluminescence (PL/CL) bands as continuous functions of the composition 0.05□ x □0.95 of the target $(SiC)_{1-x}(AlN)_x$ radiation spectrum from 1.8 to 3.8 (eV). It is shown that, subject to a slight variance (K_x=1), a similar model is also suitable for the position of the maxima of a series of electroluminescence emission bands of n-SiC/p-$(SiC)_{1-x}(AlN)_x$ LEDs (heterostructures). A continuous relationship is obtained between the positions of the maxima of a series of long-wave and short-wave transitions in K-, M-minima of the band structure, respectively, and the contact current density. It is shown that as the composition x increases, the energy levels of the luminescence centers shift towards the corresponding energy gaps of $(SiC)_{1-x}(AlN)_x$ LEDs towards a conduction band. This is due to the relocation of luminescence centers in SiC, which are even likely to emerge on the surface of an AlN crystal at x □1.

2. Results

Photo- and cathodoluminescence (PL/CL) (Figure 1). Based on the experimental data [7] for intense maxima positions of long-wave transitions to the K-minimum conduction band in the case of SiC-AlN luminescence, the following semiempirical formula can be proposed:

$$E_0^x = E_g^x - (1 - K_x\, x)\, \Delta E_x \qquad (1)$$

Here E_0^x is the position of a transition band maximum with a Gaussian distribution:

$$I(E) = A\, G_0 / [(E - E_0)^2 + G_0^2], \qquad (2)$$

where A is a relative normal distribution at the luminescence band maximum. In case [7] $A=65G_0$ and max I (E_0)=65 RU; G_0 is the width of the distribution of 0.2 eV order or the width of a possible effective impurity band formed by the levels of luminescence centers with different local nearest neighbours and different depths of their levels.

It is known [7] that the energy gap E_g^x in $(SiC)_{1-x}(AlN)_x$ solid solutions can be represented as:

$$E_g^x = E_g - ax + bx^2\,; \qquad E_g = 3{,}3\ \text{eV};\ a=0.56 \text{ and } b=3.86\ \text{(eV)}. \qquad (3)$$

Thus, for the compositions used in the experiment x_1=0.12; x_2=0.24 and x_3=0.54 (Figure 1), as well as x_1=0.05; x_2=0.24 (Fig. 2.) we get:

1, 1′ x=0.05; 1, 2 – J = 3 μA; 1′, 2′ – J=20 μA (direct current). 2, 2′, 3 – x = 0.24; 3 – J = 30 μA (reverse current); E^1 ,E^2 are transitions to the K-minimum and E^1_i, $E_i^{\,2}$ – to the M-minimum; E^3, E^4 – J=20 and E^5 – J= 30 (μA), respectively.

E_g^1 =3.28 (x_1=0,05); $E`_g^1$=3.29 (x_1=0.12); E_g^2=3.39 (x_1=0.24) and E_g^3=4.115 (eV). The parameters $K_1=K_3$=0.25; K^2=0.02 in expression $(1-K_x x)$ are related to the degree of approximation of the E_0^x level to the corresponding energy gap (Figure 3).

The energy ΔE_x is an average depth of impurity luminescence centers: donors (for example, V is vanadium and O is oxygen) and acceptors (Al-aluminum, N-nitrogen). The depth of donor levels refers to the E_c conduction band, and of the acceptor levels to the E_v valence band, respectively. According to [11, 12, 13], for the depths of impurity centers (V, O and Al, N) we have:

V, E_c - 0.97 ÷ 0.70; G'_0 = 0.27 ; Al, E_v ~ 0.27÷0.10; G'_0 =0.17;
O, E_c - 0.90 ÷ 0.95; G'_0 = 0.05; N, E_v ~ 0.142÷0.081; G'_0 =0,061 (eV).(4)

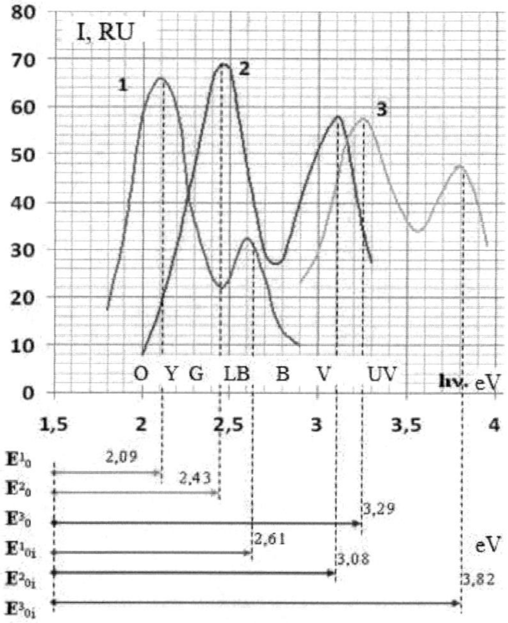

Figure 1. Spectra and level diagram of photoluminescence 1,2 and cathodoluminescence 3, at x: (1) – 0 .12, (2) - 0.24, (3) – 0.54; T=77 °K for $(SiC)_{1-x}(AlN)_x$ LEDs [7]

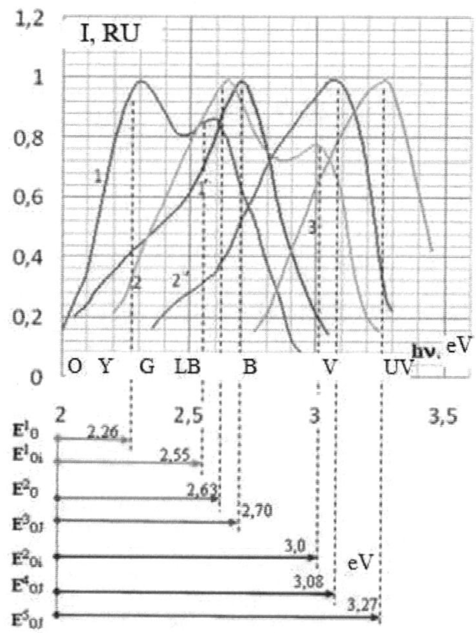

Figure 2. Spectra and diagram of electroluminescence levels of $SiC/(SiC)_{1-x}(AlN)_x$ LEDs at 300 °K [7]

The average value of G_0 in the experiment is 0.2 eV.

Certainly, other luminescence centers, defects and impurities found to be present in SiC can also contribute to the emission. In this case, as can be seen from the data presented in (1) for probable emission centers (V, O and Al, N), the distribution of depths or widths of arising impurity zones is generally quite significant. Therefore, assessing the positions of emission band maxima, subject to the experimental data, is based on the following average values:

For V and O: $\Delta E_1 = \Delta E_3 = 1.24$; $(K_1 = K_3 = 0.25)$,

Whereas, for Al and N: $\Delta E_2 = 1.05$, $(K_2 = 0.02 = K)$. (5)

Then, for the positions of SiC-AlN luminescence maxima in the K-minimum conduction band for intense long-wave transitions, according to (1), we find:

$$E_0(x_1,x_2,x_3) = \begin{cases} 2.29 - (1 - 0.25 \cdot 0.12)1.24 = 2.09 = E_0^1, \\ 2.39 - (1 - 0.2 \cdot 0.24)1.05 = 2.41 = E_0^2, \\ 4.115 - (1 - 0.25 \cdot 0.54)1.24 = 3.24 = E_0^3\,(\text{eV}). \end{cases} \quad (6)$$

The output values of energy maxima for the corresponding emission bands are in line with the experimental data and are given in Table 1.

Table 1. Calculated (C) and experimental (E) positions of PL/CL emission band maxima of SiC/AlN at 77 °K

Uel		PL (eV)			CL (eV)	
E_0	E^1_0	E^2_0	E^1_{0i}	E^2_{0i}	E^3_0	E^3_{0i}
x	0.12	0,24	0.12	0.24	0.54	0.54
E_0^E	2.10	2.44	2.61	3.08	3.25	3.82
E_0^C	2.09	2.41	2.62	3.10	3.24	3.81

For less dense short-wave emission bands, which are most likely due to transitions to the K-minimum conduction band, based on the relevant parameters of the occurrence of levels (states) ξ_i and experimental outputs, we obtain:

$$E_{0i}^x = E_g^x(\xi_i + x). \quad (7)$$

Where $\xi_1=0.68\approx0.7=\xi_2=\xi_3$ parameters are meant to indicate that the bands corresponding to this emission are likely to correspond to impurity centers located near the heterointerface, the surface of the $(SiC)_{1-x}(AlN)_x$ epitaxial layer or surface states [11]. In this case, for E_{0i}^{x1} and E_{0i}^{x2} emission maxima, we find:

With cathodoluminescence (CL) of $(SiC)_{1-x}(AlN)_x$ epitaxial layers (EL) at x=0.54, the value of E_g^x=4.115 eV is too high (> 3.9 eV) and, in accordance with the experimental data, it is necessary to reduce E_g^x by $\Delta E_2 \equiv \Delta E$=1.05. Then, we find:

$$E_{0i}(x_1,x_2) = \begin{cases} 3.29 \cdot (0.681+0.12) = 2.62 = E_{0i}^1, \\ 3.39 \cdot (0.7+0.24) = 3.10 = E_{0i}^2 \text{ (eV)}. \end{cases} \quad (8)$$

$$E_{0i}^3 = (E_g^x - \Delta E)(\xi_3 + x) = (4.115-1.05)(0.7+0.54) = 3.81 \text{(eV)}. \quad (9)$$

The output values by (7) and (9) agree with the experimental data and are shown in Table 1. As a result, formulas (1), (7), and (9) ensure good agreement with the data on the $(SiC)_{1-x}(AlN)_x$ EL luminescence in a wide energy range $1.80 \le h\nu \le 3.90$ (eV).

Electroluminescence of SiC/$(SiC)_{1-x}(AlN)_x$ LEDs (Figure 2). For a series of n-SiC/p-$(SiC)_{1-x}(AlN)_x$ LED bands, similarly to (1) at K_x=1, in accordance with the experimental data, we find:

$$E_0^x = E_g^x - (1-x)\,\Delta E. \quad (10)$$

Here, $\Delta E \equiv \Delta E_2 \equiv 1.05$, and the factor (1-x) determines the degree of approximation of the levels (bands) E_0^x to the corresponding gap E_g^x as the composition x enhances (see Fig. 3). In this case, for the maxima of E_0^{x1} and E_0^{x2} long-wave luminescence bands at x_1=0.05; E_g^1=3.28 and x_2=0.24; E_g^2=3.39, according to (10), we obtain:

$$E_0^{1,2} = \begin{cases} 3.28-(1-0.05)1.05 = 2.26 = E_0^1, \\ 3.39-(1-0.24)1.05 = 2.63 = E_0^2 \text{ (eV)}. \end{cases} \quad (11)$$

For the maxima of less intense short-wave emission bands of E_0^1 and E_0^2 LEDs at the corresponding values x_1=0.05; x_2=0.24 and ξ_1=0.73; ξ_2=0.65, according to (7) we find:

$$E_{0i}^{1,2} = \begin{cases} 3.28\,(0.73+0.05) = 2.56 = E_{0i}^1, \\ 3.39\,(0.65+0.24) = 3.01 = E_{0i}^2 \text{ (eV)}. \end{cases} \quad (12)$$

The presence of J_x current density through the LED contact results in a deviation of emission bands to a short-wave region of the spectrum. In this case, if the derivative $dE_{0i}^{x}/dJ_x=tg\alpha$ is denoted by β, then, similarly to (10), with an increment of the current ΔJ_x for the maxima of emission bands, we obtain:

$$E_J^{x} \cong E_{0J}^{x,Э} = E_{0i}^{x,Л} + \beta(1-x)\Delta J_x \,. \qquad (13)$$

Whereas the experimental data, $\beta \approx 0.010 \div 0.012$. If we assume that at x=0.05 the maximum of the luminescence band is $E_{0i}^{x,L}$=2,53 and ΔJ_x=3, then for the initial E_{Jx}^{1} band (i.e., for the maximum of the first electroluminescence band E_{Jx}^{1}) according to formula (13) we find:

$$E_J^{x} = 2.53+0.01(1-0.05)3 = 2.56 \text{ (eV)}. \qquad (14)$$

This corresponds to an experimental value of 2.55 eV (Table 2).

Table 2. Calculated vs experimental positions of SiC/AlN emission maxima

$E_0=E^i_{0E}$ (eV)	E^1_0	E^1_{0i}	E^2_0	E^2_{0i}	E^3_{0J}	E^4_{0J}	E^5_{0J}
J, μA		3				20	30
x	0.05		0.24			0.24	
E_0^E	2.25	2.51	2.62	3.00	2.69	3.10	3.26
E_0^C	2.26	2.56	2.63	3.01	2.70	3.12	3.27

For the maxima of electroluminescence bands in the presence of contact currents, similarly to (13), we obtain:

$$E_{0J}^{i} = E_{0J}^{x} + \beta(1-x_i)\,\Delta J_i,\ \Delta J_i = Jxi - Jx(i-1) \qquad (15)$$

Then for the values x_1=0.05; x_2=x_3=0.24 and contact currents J_1=3, J_2=20 and J_3=30 (μA), given that ΔJ_1=17, ΔJ_2=17 and ΔJ_3=27, according to formula (15), we find:

$$E_{0I}(J_1, J_2, J_3) = \begin{cases} 2.55+0.01(1-0.05)\,17=2.70= E_{0J}^{3} \equiv E_0^{3}, \\ 3.00+0.01(1-0.24)\,17=3.12= E_{0J}^{4} = E_0^{4}, \\ 3.00+0.01(1-0.24)\,27=3.27= E_{0J}^{5} \equiv E_0^{5} \text{ (eV)}. \end{cases} \qquad (16)$$

The output values of the maxima of the emission bands in the presence of contact currents agree with the experimental data [7] for $SiC/(SiC)_{1-x}(AlN)_x$ and are given in Table 2.

Level positions of luminescence centers as continuous functions of composition x.

Let us estimate the relationship between the composition x and the level deviations of the $(SiC)_{1-x}(AlN)_x$ luminescence centers for the maxima of the long-wave ∇^x_L and short-wave ∇^x_S transitions, respectively. These deviations of the levels of the luminescence centers with increasing x are estimated with respect to the energy gaps of the band structure and appear when x deviate from 0 to 1 (Figure 3).

The calculated values of maxima positions of E_0^x and E_{0i}^x photo- and cathodoluminescence bands can be plotted on the diagram of the main energy gaps ($(SiC)_{1-x}AlN_x$) ([7] Figure 4). The positions of these maxima enable to determine the corresponding variance of the luminescence centers for the compositions x_1, x_2, x_3, as well as for arbitrary x using formulas (1), (7), (10), (15). The corresponding variance of the luminescence centers are shown in Fig. 3. It can be seen that along lines 5, 5′ and 6, 6′, (dashed line), the emission maxima approximate the corresponding gaps of $(SiC)_{1-x}AlN_x$) band as the composition enhances from x=0.05 to x=0.65. Such deviations of luminescence levels are apparently caused by the relevant relocation of luminescence centers, defects and variance in their immediate environment. It is possible that at $x \to 1$ some defects could appear on the surface of AlN crystal. It would be interesting to try to verify the latter assumption in the course of special experiments. This assumption is associated with both the displacement of the corresponding luminescence centers and the rearrangement of the band structure. With a variance in x from 0 to 1, the energy gap increases according to (6) from 3.3 to 6.6 eV during the transition from SiC to AlN. The depths of impurity levels and ∇^x_L reduce, which can be seen from Table 3. The values ∇^x_L=0.90 and ∇^x_S=0.31 (eV), (Table 3, line 3), within the limits 0.05÷0.10 eV, escape the general scheme of band gaps ([7] Figure 4), since the corresponding cathodoluminescence transitions occur in deeper levels of luminescence centers.

Table 3. Parameters of luminescence depths from x_1, x_2, x_3 compositions

x	∇^x_L (eV)	∇^x_S (eV)
0.12	1.20	0.69
0.24	1.01	0.31 0,34
0.54	0.90	0.31

The derived curves are shown in Figure 3.

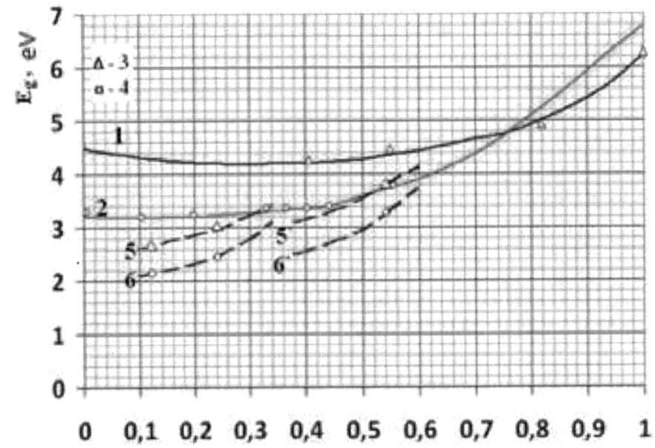

Figure 3. Composition x vs $(SiC)_{1-x}(AlN)_x$ energy gaps ([7] Figure 4) and luminescence depths for ∇^x_S and ∇^x_L transitions

Curves 5, 5^1 and 6, 6^1 constructed according to formulas (1), (7), (9); points (Δ – 5 , 5^1) and (O – 6, 6^1) were calculated for a series of compositions x_i, $(SiC)_{1-x}(AlN)_x$, respectively.

3. Conclusion

For the first time, models were proposed, enabling to determine the positions of emission band maxima in the region 1.80≤ hν≤3.90 (eV) as continuous functions of the composition x in the region of stable polytype structures $(SiC)_{1-x}(AlN)_x$ at x≥0.05 and T≥2100 K. On the energy diagram of the SiC – AlN gaps, the positions of the luminescence levels are determined as continuous functions of the composition. It is shown that as the composition x (0.05≤x≤1.0) increases, the luminescence depths deviate towards the conduction band. This is apparently due to the fact that the photoluminescence centers are relocated or possibly even emerge on the surface of the AlN crystal at x →1. The authors defined the relationship between the position of the maxima of the bands of direct and indirect transitions of photo-, cathode -, electroluminescence $(SiC)_{1-x}(AlN)_x$ and SiC/SiC – AlN LEDs, the composition x and the value of the contact current density of the LED transition.

Long-wavel luminescence bands characteristic of SiC appear to arise due to the transition of electrons from the $(SiC)_{1-x}(AlN)_x$ epitaxial layers to a narrower-gap SiC substrate. Short-wave bands are attributed to the emission of impurities from epitaxial layers located near heterointerfaces or surface states in the contact region. Formulas (1), (7), (10), (15) make it possible to continuously trace the rearrangement of emission bands and predict the position of the maxima of the emission spectra bands of $(SiC)_{1-x}(AlN)_x$ luminescence centers and $SiC/(SiC)_{1-x}(AlN)_x$ LEDs over the entire region of the visible spectrum when the composition of x changes from 0 to 1.

References

[1] Safaraliev G K 2011 *SiC-based solid solutions* (Moscow) 296 p
[2] Lebedev A I 2008 *Physics of semiconductor devices* (Moscow: Fizmatlit) 488 p
[3] Schubert F E 2008 *LEDs* (Moscow: Fizmatlit) 496 p
[4] Safaraliev G K, Bilalov B A, Kurbanov M K, Kasyanenko I S, Altukhov V I and Sankin A V 2015 Calculation of the height of the Schottky barrier at the contact of a metal with a semiconducting solid solution $(SiC)^{1-x}(AlN)^x$ *Microelectronics, Elements of Integrated Circuits* **44(6)** 453-458
[5] Altukhov V I, Sankin A V, Sigov A S, Sysoev D K, Yanukyan E G and Filipova S V 2018 Schottky barrier model non-linear with respect to the concentration of surface states and calculation of current-voltage characteristics of SiC-based diodes and its solid solutions in a composite model of current transfer *Physics and Technology of Semiconductors* **52(3)** 366-369
[6] Altukhov V I, Kasyanenko I S, Sankin A V, Bilalov B A and Sigov A S 1991 Calculation of the Schottky barrier and current-voltage characteristics of metal-solid SiC-based solutions *Physics and Technology of Semiconductors* **50(9)** 1190-1194
[7] Safaraliev G K, Tairov Yu M and Tsvetkov V F 2016 Wide-gap $(SiC)_{1-x}(AlN)_x$ solid solutions *Physics and Technology of Semiconductors* **25(8)**
[8] Safaraliev G K, Bilalov B A, Kurbanov M K, Kargin N I, Ramazanov Sh M and Gusev A S 2013 Features of the formation of a potential barrier in *Me*/$SiC_{1-x}(AlN)_x$/SiC structures *Bull. of NRNU MEPhI* **2(2)** 273
[9] Safaraliev G K, Kargin N K, Kurbanov M K, Bilalov B A, Ramazanov Sh M and Gusev A S 2014 Investigation of the influence of processing mode parameters on the height of the potential barrier of $Me/(SiC)_{1-x}(AlN)_x$ structure *Bull. of NRNU MEPhI* **3(1)** 63–67
[10] Kargin N I, Safaraliev G K, Kharlamov N A, Kuznetsov G D and Ryndya S M 2013 Kinetic features of the preparation of $(SiC)_{1-x}(AlN)_x$ solid solution films by ion sputtering *Proc. of Universities. North Caucasian region. Technical Sci.* **6**
[11] Altukhov V I, Sankin A V, Dadashev R Kh, Sigov A S, Kargin N I and Kardashova G D 2019 *Technologies for obtaining wide-gap SiC-based materials, heterostructures, diodes and calculation of their characteristics,* monograph (Grozny: State Unitary Enterprise Book Publishing House) 102 p
[12] Siklitsky V 2003 *Electronic arhive "New semiconductor materials, Characteristics and Properties": - SiC, AlN, GaN*
[13] Lebedev A A 1999 Parameters of deep centers in silicon carbide *FTP* **33(2)** 129-125

Designing mining machinery screw modules

V A Evstratov[1], E Yu Voronova[1], Yu N Linnik[2], V Yu Linnik[2], A S Apachanov[1], V I Grigoryev[1] and V A Suxarnikova[1]

[1] Shakhty Automobile and Road Construction Institute (Branch) of Platov SRSPU(NPI), 1, Lenina Square, Shakhty, Rostov region, 346500, Russia
[2] State University of Management, 99, Ryazanskiy ave., Moscow, 109542, Russia

E-mail: vae602@yandex.ru

Abstract. Screw modules are widely used in technological and transporting machines. Their main disadvantage is low productivity due to the fact that the material moves partly in the direction of the feed and partly rotationally as a result of the torque transmitted by the screw blade to the material. The rotational component of the material movement significantly reduces the performance of the screw. The work considers the possibility of mining machines with screw modules efficiency improving by determining the rational screw shaft configuration, depending on the properties of the feed material and the operating conditions of the module, realized at the designing stage. The aim is achieved by means of a mathematical description of the screw feed process that takes into account processes occurring both on the surfaces of the material contacts with the working bodies of the machine and inside the material mass, including the occurrence of a sliding surface in the material. Furthermore, some part of the material sticks to the screw and moves in concentric circles, with no axial movement in the feed direction, which reduces the performance of the screw. The screw shaft configuration is determined by means of design procedures based on the mathematical description of the screw feeding process, given in the work. The description takes into account the processes occurring in the mass of the feed material. The resulting mathematical model of the screw feeding process allows designing screw shafts with shape identical to the sliding surface in the material, which will increase the performance of the screw module by 20-23%, by reducing the rotational component of the material movement and increasing the translational component in the feed direction.

1. Introduction

Mining production and transporting machines perform their main functions via physical impact of their working bodies on the materials being processed or moved. The performance of mining equipment depends largely on the qualitative and quantitative indicators of the processes occurring in the materials [1]. Therefore, the properties of the original material and the final product specify the machine structure and the parameters of its working bodies. The complexity of the selection is due to the wide range of materials and the variety of their physical and mechanical properties. Moreover, the properties of the same material may change significantly if temperature, humidity, pressure, etc., change.

At present, there is theoretical interpretation and mathematical description of the machinery working bodies' interaction processes with loose, moist, plastic, dust-like and fine-grained materials. But the phenomena occurring on the working bodies contact surfaces and in the material are not always adequately described, which makes it very difficult to create high-performance machines [2, 3]. The lack of an accurate mathematical description of the processes occurring in the materials when interacting with the mining machinery working bodies often results in the working bodies parameters selection on the basis of experimental data or on the basis of experience gained in other industries processing materials with similar properties [4]. This approach to design and construction often leads to building low-performance mining and transporting machines, whose working bodies have structures and operating parameters which significantly differ from the rational ones [5].

It can be totally applied to processes in different materials interacting with the screw working bodies, which are quite common in mining machines [6].

The majority of cutter-loaders for mechanized coal extraction in the mining face have working bodies in the form of two screws equipped with cutters. Cutter-loaders do not have any special loading devices. The broken-down coal is loaded onto the conveyor by means of screws together with the loading shields.

For laying prepared concrete mixture behind the formwork to make an excavation support with monolithic concrete, concreting machines with screw feeders are used. For a lateral monolithic concrete lining erection, a mobile machine with a screw mixing and transporting working body is widely used [7]. Augers are used to excavate very thin and thin coal seams with soft walls.

Drill-rods with a screw blade are widely used to remove drill-cutting waste from wells. For dry backfill materials transportation and the mechanized preparation of solutions for plugging operations, a movable cement mixing unit is widely used, in which the supply of dry components to the mixing chamber is provided by a screw module. In drill and blast tunneling method, the drilling machine, which has a screw feeder, is widely used.

2. Materials and methods

Performance improvement of mountain machinery screw modules through the structural and parametric synthesis of their working bodies can be realized only on the basis of modern computer technologies for the selection of technical solutions and design features [8]. This enables to take into account the processes occurring both on the surfaces of the contacts of the material with the module working bodies and in the material massive under different module service conditions and different properties of the materials supplied [9].

Let us consider the balance of the material volume element cut from the channel formed by the internal cylindrical surface of the module tube, the shaft and the screw blade (figure 1).

The forces exerted upon the material volume element as follows.

Pressure force:

$$F_4 = PS_{ch} \tag{1}$$

where P - pressure of the material in the conveyor, Pa; $S_{ch} = (R-r)t$ - screw channel section area, formed by the internal surface of the tube, the shaft and the screw blade, m^2; t - screw pitch, $t = 2\pi Rtg\alpha$, *m*; α - pitch angle of the screw blade; R - radius of the screw blade, *m; r* - radius of the shaft, *m*.

Counterforce:

$$F_5 = (P + \Delta P)S_{ch} \tag{2}$$

where ΔP - pressure difference on the front and back surfaces of the material to be considered, *Pa*.

Since the material being fed cannot be considered a Newtonian fluid [10], the values of the material pressure in the planes perpendicular to the plane of action of the pressure force and the counterforce can be determined using the formula:

$$P_x = P_z = \mu P \tag{3}$$

where μ - side pressure coefficient.

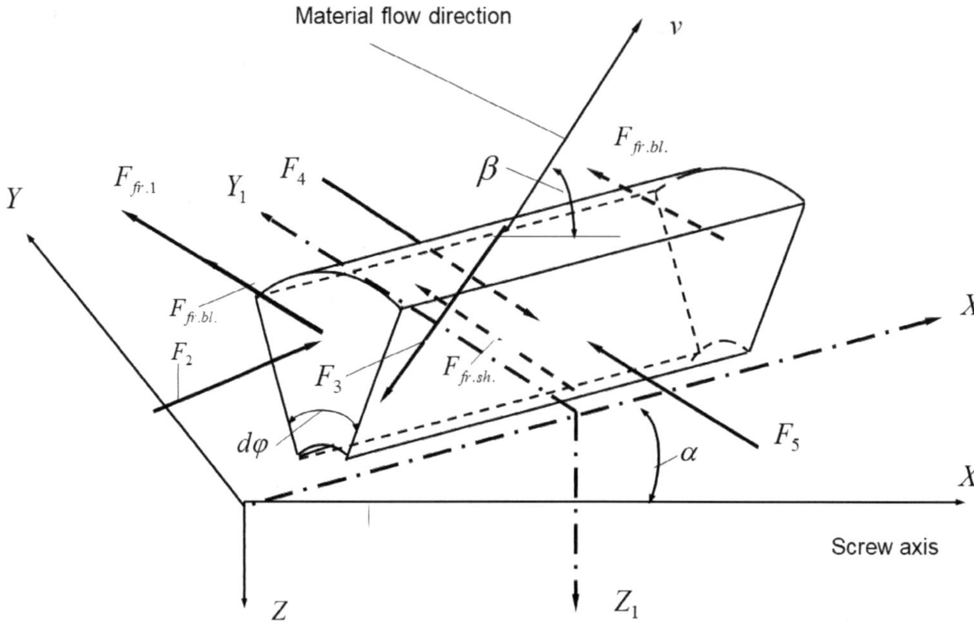

Figure 1. Diagram of the forces acting on the material volume element in the screw channel of the screw module.

Therefore, the force of the material friction on the internal surface of the module tube:

$$F_3 = \mu P S_h f_h \tag{4}$$

where $S_h = tRd\varphi = 2\pi R^2 d\varphi tg\alpha$ - contact area of the volume considered and the inner surface of the tube; f_h - coefficient of the material friction on the internal surface of the module tube; $d\varphi$ - angle of cut material sector.

Normal pressure on the screw blade dependence on force F_3:

$$F_2 = F_3 \cos(\beta - \alpha) \tag{5}$$

where β - angle between the material movement direction and the screw axis.

Friction force of the material friction on the screw:

$$F_{fr.aug.} = F_{fr1} + F_{fr2} \tag{6}$$

where $F_{fr1} = F_3 \cos(\beta - \alpha) f_{aug}$ - component of the material friction force on the blade of the screw due to the effect of force F_3; $F_{fr2} = F_{fr.sh.} + 2F_{fr.bl.}$ - component of the material friction force on the screw due to the effect of pressure P; $F_{fr.sh.} = \mu P S_{sh} f_{aug}$ - material friction force on the screw; $S_{sh} = 2\pi Rrd\varphi tg\alpha$ - area of the considered material volume contact with the screw shaft surface; $F_{fr.bl.} = \mu P S_{bl} f_{aug}$ - component of the material friction force on the screw blade due to the effect of pressure P; $S_{bl} = (R^2 - r^2) d\varphi / (2\cos\alpha)$ - area of the considered material volume element contact with the screw blade surface.

Equilibrium of the considered material volume element follows the equation system:

$$F_2 - F_3\cos(\beta-\alpha) = 0$$
$$F_{fr.sh.} + 2F_{fr.bl.} + F_{fr.1} + \Delta PS_{ch} - F_3\sin(\beta-\alpha) = 0 \quad (7)$$
$$N_ц - N_{sh} = 0.$$

The solution of the equations of statics (7) gives the condition for determining the angle β between the direction of the material movement and the screw axis. The result of substitutions and transformations is:

$$f_h f_{aug}\cos(\beta-\alpha) + \frac{r}{R} f_{aug} + \frac{f_{aug}(R^2-r^2)}{2\pi R^2\sin\alpha} + \frac{(R-r)\Delta P}{R\mu P d\varphi} -$$
$$- f_h\sin(\beta-\alpha) = 0 \quad (8)$$

Pressure in the screw is a function of φ, it transforms linearly:

$$P(\varphi) = P_0 + a\varphi \quad (9)$$

where P_0 - screw feed pressure; a - proportionality coefficient, depending on the properties of the feed material and the screw geometrics. P from (9) and $dP = ad\varphi$ were substituted into (8) to integrate it with respect to $d\varphi$ within the interval from 0 to $2\pi n$ (n - number of threads):

$$\int_0^{2\pi n}[-R(R-r)\frac{a}{\mu(P_0+a\varphi)} + f_h R^2\sin(\beta-\alpha) - f_{aug}Rr -$$
$$-R^2 f_h f_{aug}\cos(\beta-\alpha) - \frac{f_{aug}}{2\pi\sin\alpha}(R^2-r^2)]d\varphi = \quad (10)$$
$$= -\frac{R(R-r)}{\mu}\ln\left|\frac{2\pi na+P_0}{P_0}\right| + 2\pi nf_h R^2\sin(\beta-\alpha) -$$
$$-2\pi nR^2 f_h f_{aug}\cos(\beta-\alpha) - 2\pi nf_{aug}Rr - \frac{nf_{aug}(R^2-r^2)}{\sin\alpha} = 0$$

where $2\pi na + P_0 = P_{out}$ - screw discharge pressure.

Analysis of the material equilibrium condition (10) shows that its movement direction depends on the following parameters:

- coefficient of the material friction on the screw;
- coefficient of the material friction on the inner surface of the tube;
- screw feed pressure and screw discharge pressure ratio;
- pitch angle of the screw blade;
- number of threads of the screw blade;
- area of feed material contact with the screw blade surface and the shaft;
- area of feed material contact with the screw module inner surface.

The first two parameters depend on the feed material properties and the quality of the screw module working bodies' surfaces. Together with other properties of the material, they determine rational values of the structural parameters of the screw module.

The third parameter considers the operating conditions of the module - a feeding pressure at the input of the screw (the height of the material in the loading bunker, the use of feed rolls, etc.) and a locking pressure at the output of the screw (the resistance of the forming, grinding or other working bodies of the machine).

Consider the material volume element cut from the screw channel, taking into account the coordinates of the forces and loads application points. The load application type of the material volume element, which is a tapered thickness plate, as well as the slight difference between the coefficients of external and internal friction of the material, suggests the possibility of a curved sliding surface connecting the edges of the screw blade (figure 2). The maximum lateral force is applied to the cylindrical surface of the contact between the material and the module tube. However, there are a number of hollow rotational surfaces connecting the edges of the screw blade, whose area is smaller than the area of the corresponding cylindrical surface. If the shear stresses in the material in the direction of the axis $Y1$ (figure 2) exceed maximum permissible limits on any of these surfaces, the material layers will slide relatively to one another and the material will be split into two parts. Furthermore, the part of the material located closer to the screw will gyrate in concentric circles, which will adversely affect the screw module performance. The material shear will take place on a surface where the tangent stresses are maximal:

$$\tau = Q/S_m = \max \tag{11}$$

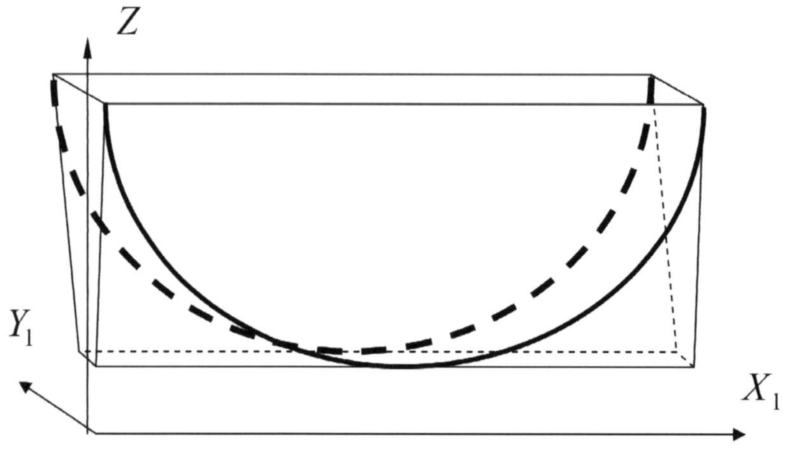

Figure 2. Curved sliding surface in material being fed.

The value of shear stresses on any curvilinear lateral surface of a material element depends on the area of that surface S_m and the lateral force Q applied to that surface in the direction of the axis $Y1$ (figure 2).

The value of the lateral force Q at the material cross-section depends on the shape of the surface (figure 2) and can be determined from the equation:

$$Q = F_{fr.sh.} + 2F_{fr.bl.} + \Delta P S'_{ch},$$

where S'_{ch} - area of the part of the screw channel limited by the intersection line of the material cross-section with the plane $X1Z1$ (curves 1, 2 and 3, figure 3) and line segments $Z1 = r$, $X1 = -t/2$ and $X1 = t/2$; ΔP - pressure difference on the front and back surfaces of the material to be considered.

The screw channel section area S_{ch} (figure 3) is:

$$S_{ch} = \int_{-t\backslash 2}^{t/2} (R - r)dx, \text{ so } S'_{ch} = \int_{-t\backslash 2}^{t/2} (f(x) - r)dx,$$

where $f(x)$ - function whose graph is the intersection of the surface being considered with the plane $X1Z1$ (curves 1, 2 and 3, figure 2); t - screw pitch; R - radius of the screw blade; r - radius of the shaft.

The surface being considered is formed by the movement of a certain curve $z(x)$ in a spiral direction in such a way that, in one revolution around the axis X it is displaced one pitch distance t along the axis X. The surface area can be determined by the formula:

$$S_m = 2\pi \int_{-t/2}^{t/2} \sqrt{f^2(x)+T^2}\sqrt{1+f'^2(x)}dx \qquad (12)$$

where $f'^2(x)$ - square of the first derivative of function $f(x)$; $T = t/2\pi$.

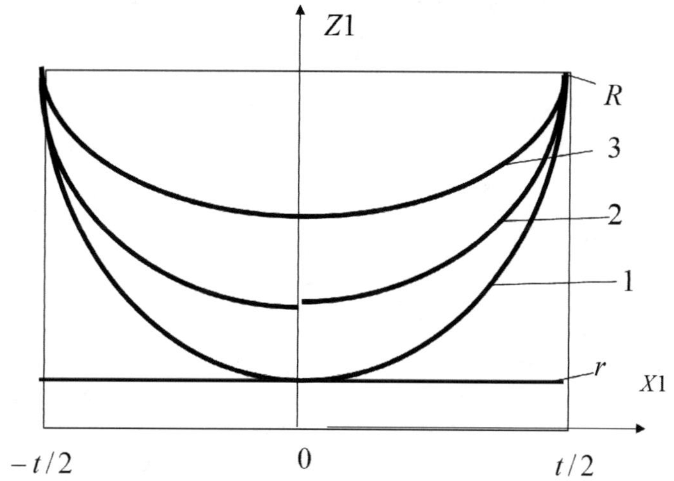

Figure 3. Possible projections of the sliding surfaces in the material on the X1Z1 plane.

Taking into account symmetry of the considered material element about Z1 and equilibrium equations (7), condition (11) can be written as follows:

$$\frac{\tau}{P} = \frac{2\pi n f_{aug} Rr + \dfrac{n f_{aug}(R^2-r^2)}{\sin\alpha} + 2\pi n R^2 f_h f_{aug}\cos(\beta-\alpha)}{4\pi n \int_0^{t/2} \sqrt{f^2(x)+T^2}\sqrt{1+f'^2(x)}dx} +$$

$$+\frac{\dfrac{2}{\mu}\ln\left|\dfrac{2\pi n a + P_0}{P_0}\right| \int_0^{t/2}(f(x)-r)dx}{4\pi n \int_0^{t/2}\sqrt{f^2(x)+T^2}\sqrt{1+f'^2(x)}dx} = \max \cdot \qquad (13)$$

The resulting equation determines the shape of the sliding surface in the material - $f(x)$. The analysis of the values in (13) shows that $f(x)$ generally depends on the friction coefficients of the material on the screw and on the internal surface of the tube, the geometrical parameters of the screw, and the ratio of the screw feed pressure and the screw discharge pressure.

When solving the equations (13), function $z = f(x)$ was predetermined as power function $z = c + mx^n$ with different coefficients. The study of the dependence (13) enables to determine effects of the screw discharge pressure, the properties of the feed material and the geometrical parameters of the module working bodies on the shape of the shear surface in the material in the screw module channel.

Figure 4 shows the profiles of the material shear surfaces in the screw channel at different pressure values at the outlet of the module and at the following values of the geometric parameters of the module working bodies: the radius of the screw blade $R = 0{,}2m$; the radius of the shaft $r = 0{,}05m$; the pitch angle of the screw blade $\alpha = 20^{\circ}$; the number of threads $n = 4$. Analysis of the obtained results shows that the shape of the material shear surface in the screw channel of the module depends on the ratio of the screw feed pressure and the screw discharge pressure.

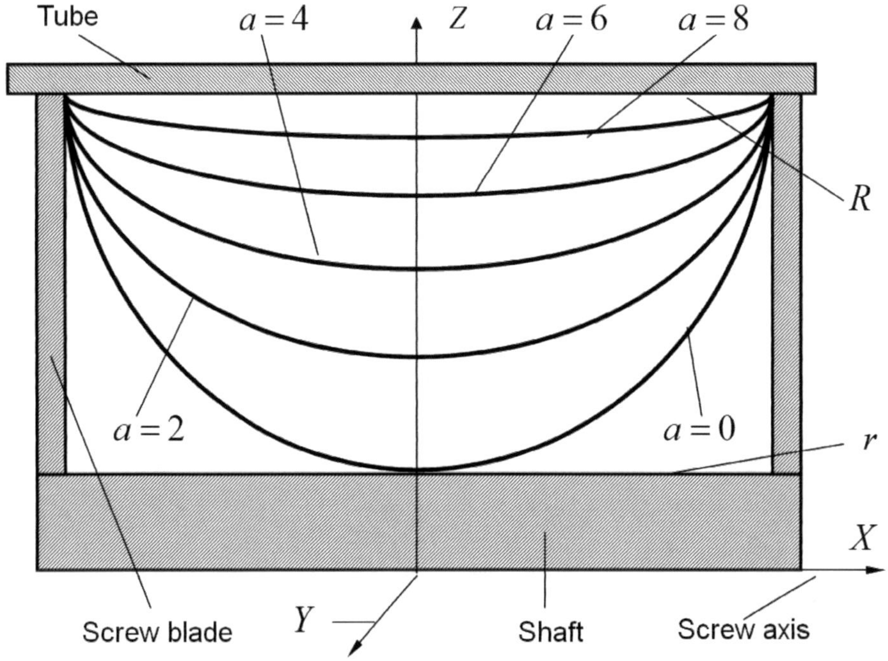

Figure 4. Dependence of the configuration of the material shear surface in the screw channel on the locking pressure value at the module output.

When $P_0 = P_{out}$ (curve $a = 0$, figure 4), the shear surface in the material is the same as the surface with minimum area, whose line of intersection with plane XZ connects the blade edges, and is copunctal for the line $Z = r$. In this case, the area of the part of the screw channel in which the material has a progressive movement is maximum. It is shown in figure 4 as the area of the figure bounded by lines $a = 0$ and $Z = R$.

With the increase in the locking pressure, the intersection lines of the material sliding surfaces with the plane XZ are lower sloped, and the passive region in the screw channel filled with material having no axial movement in the feed direction increases. The higher the locking pressure at the screw discharge, the more the material moves in concentric circles without moving in the direction of the feed. At a certain pressure, the movement of the material in the screw channel of the module ceases.

Similar results can be obtained by studying the dependence of the sliding curve on the number of threads, the friction coefficients of the material (external and internal), and the screw pitch.

3. Conclusion

The determination of the material flow patterns in the screw module, taking into account that a sliding surface can occur therein, is of great practical importance. It allows for more rational selection of geometric parameters of working bodies, when designing screw modules, and makes it possible to use various design and structural solutions that ensure maximum performance of the screw module, taking into account the properties of the feed materials and the technological regulation requirements [11]. The resulting mathematical model of the screw feeding process allows designing screw shafts with shape identical to the sliding surface in the material, which will increase the performance of the screw module by 20-23% by reducing the rotational component of the material movement and increasing the translational component in the feed direction.

References

[1] Bogdanov V S, Fadin Y M, Vasilenko O S, Demchenko C E and Trubaeva V A 2018 *IOP Conf. Ser.: Mater. Sci. Eng.* **327(4)** 042113
[2] Zareiforoush S, Komarizadeh M H and Alizadeh M R 2010 *J. of Sci. Review* **2(1)**
[3] Bogdanov V S, Semikopenko I A and Vavilov D V 2018 *IOP Conf. Ser.: Mater. Sci. Eng.* **327(5)** 052006
[4] Asli-Ardeh E A and Mohsenimanesh A 2012 *Sci. World J.* DOI: 10.1100/2012/136218
[5] Khakhalev P A, Bogdanov V S and Kovshechenko V M 2018 *IOP Conf. Ser.: Mater. Sci. Eng.* **327(4)** 042049
[6] Owen P J and Cleary P W 2009 *Proc. of the 7th Int. Conf. on CFD in the Minerals and Process Industries* pp 1- 7
[7] Bogdanov V S, Eltsov M Y, Shirokova L Y and Khakhalev P A 2017 *ZKG Int.* **70(5)** 64-66
[8] Zareiforoush S, Komarizadeh M H and Alizadeh M R 2010 Oriza Sativa L. *Grains. Nature and Science* **8(6)**
[9] Voronova E Yu and Khazanovich G Sh 2019 *IOP Conf. Ser.: Mater. Sci. Eng.* **560(1)** 012048
[10] Rud A V, Evstratova N N, Evstratov V A, Bogdanov D V, Lozovaya S Y and Lunev A S 2014 *ARPN J. of Engineering and Appl. Sci.* **9** 2376–2382
[11] Latyshev S S, Voronov V V, Bogdanov V S, Bazhanova O I and Maslovskaya A N 2018 *IOP Conf. Ser.: Mater. Sci. Eng.* **327(2)** 022046

Prospects for power supply infrastructure enhancement

E V Gavrikova[1], M V Gorgisheli[1], A V Kosygina[1], I O Volkova[1], A Y Yakovleva[1], I A Dolmatov[1] and K V Suslov[2]

[1] National Research University, Higher School of Economics, 20, Myasnitskaya Street, Moscow, 101000, Russia
[2] Irkutsk National Research Technical University, 83, Lermontova Street, Irkutsk, 664074, Russia

E-mail: egavrikova@hse.ru

Abstract. Traditional retail electricity market models transform under the condition of new technologies development and institutional changes in countries and worldwide. New models of the retail electricity market need to be adopted to recent requirements made by different stakeholders. The purpose of this study is to advance the understanding of transformation of the electricity retail market in different countries through introducing a uniform framework for comparison. The findings provide insights for experts and authorities on how the changes in the dynamic environment led to the emergence of new market models in six surveyed countries. Specific features of different models are highlighted to ensure optimal working conditions and incorporation with managerial practices in the industry as well as to help policy makers to understand what additional measures may boost the competition in retail electricity markets.

1. Introduction

The challenges of increasing the energy infrastructure efficiency, fighting energy poverty and transitioning to low carbon energy systems are tightly coupled with energy policies on national levels. There is evidence that polycentric governance and decentralization of energy solutions allow one to reach higher levels of experimentation and innovation required for providing reliable electricity supply [1]. Liberalization of retail electricity markets has a potential to produce benefits for different stakeholders, i.e. consumers, who can gain economically from switching service providers. Competition may also boost the enhancement of different electric services as well as accelerate the employment of distributed generation capacities and renewables, developing smart solutions and, thus, contribute to the decarbonization of the energy system. At the same time the reformation and liberalization of existing market models imply establishing an open access to technologies and information that would allow one to maximize efficiency of the energy system, while information security and privacy issues have to be addressed [2]. This paper will provide an overview of the retail electricity market models applied in different countries. Furthermore, beside the differentiation between countries, different electricity market models might appear within the same country. It may happen due to historical, geographic and some economic prerequisites. For example, in Russia, USA, China liberalization activity is complicated due to some factors. It is worth adding that the particular attention to the liberalization aspect, as an explicit indicator of the flexibility of the external environment and environment readiness for incorporating new technologies and practices, need to be observed. The authors will propose a unified framework for the comparison of different retail electricity markets in order to assess the level of their liberalization and subsequent readiness for the described changes. Three main models of retail electricity market will be considered: monopoly, competition, single buyer.

2. Methodology and approach

The level and speed of the dissemination of new market models vary from country to country; thus, the presented paper is based on a descriptive case study consisting of multiple holistic cases in order to establish sequences of events during the period of liberalization and unravel underlying mechanisms.

2.1. Proposition of research

There is a noticeable transition towards more liberalized electricity markets with a broader set of actors and various methods of collaboration among them. This requires a corresponding environment – more flexible and ready for this type of changes. Regarding this, the central research question is:

RQ. Is the competitive model currently dominating in retail electricity markets?

From this research question, the following central propositions are derived:

P1. The higher generation capacity concentration within the generating segment the less competitive would be the retail electricity market.

P2. The higher generation capacity concentration within the sales segment, the less competitive would be the retail electricity market.

2.2. Countries for the analysis

There is a noticeable transition towards more liberalized electricity markets with a broader set of actors and various methods of collaboration among them. This requires a corresponding environment – more flexible and ready for this type of changes. The authors will pay particular attention to the retail markets of the following countries, as those being examples of similar processes ongoing in both developed and transition economies (Table 1) [3].

Table 1. Countries for the analysis

Country	Reason for inclusion into analysis
USA	• A bright example of the initially non-united energy sector with a considerably high level of decentralization of electricity value-chain elements • Highly distributed governance of the sector (determined by the federal and regional structure) • High interdependence of 2 major commodities: natural gas and electricity • Innovative development of the electric power sector
China	• A high level of state control and intervention in the electric power sector coupled with the strong trend towards the liberalization and renewables development • Fast development of new trends in the electric power sector
Singapore	• An interesting example of extraordinary influence of the government coupled with the extremely high level of market activity
Germany	• Direction towards the development to a low carbon energy system • A high level of electricity consumers' fragmentation
Russian Federation	• One of the most representative cases of the development and transformation of the electricity power sector under the condition of post-soviet and transition economics • A strong trend towards competition within a highly regulated environment in the energy sector • Existence of zones with different conditions for the organization of the retail electricity market due to the features of the territory • Domestic carbon fuel and low carbon resources for electricity production • Significant concentration of production in all energy industries and high interconnection among them

2.3. Research design

All the six cases were thoroughly studied through content and narrative analyses. Particular attention was paid to the examination of differences and similarities of the liberalization processes as well as to the underlying triggers of this transition.

At the first stage of the research, through the literature review, the authors achieved two major outcomes. They found that retail electricity markets may be assessed by the level of their liberalization (through a set of measurable parameters). The stage also allowed one to outline models for the classification of retail electricity markets with the corresponding set of indicators. It is worth noting that during the research, the authors paid specific attention to the market share attributed to a particular model within a given country (based upon a premise that several models may be functioning on one market at the same time).

Stages of the research determine the structure of the paper:

- comparative model for research of retail electricity markets in different countries is presented;
- comparative characteristics of retail electricity markets in different countries on the base of the comparative model will be provided.

This study concludes about the discussion of implications and directions for future research.

2.4. Comparative model

There is a noticeable transition towards more liberalized electricity markets with a broader set of actors and various methods of collaboration among them. This requires a corresponding environment – more flexible and ready for this type of changes.

As for the creation of the comparative model, the conducted cross-case synthesis allowed us to uncover a set of indicators that reflect the liberalization processes within the studied countries:

- Mechanism of the price regulation at the retail market (whether it is formed by market laws or regulated by the government. In the former case it may be more beneficial for the producers, but the latter protects consumers as the most vulnerable actor in the electricity market).
- Level of the concentration within the generation segment – will be measured with the generation capacity, Concentration Ratio index, CR5, which estimates a market share of the 5 largest companies in a certain market in 2017. (This allows one to assess the influence of the companies).
- Market share of power supply companies will be assessed through the calculation of a market share of a given company, MSi. (This helps to assess the level of market power of a given company).
- Degree of the value chain integration (this demonstrates which of the major stages are integrated: generation; transmission; distribution; sales. It indicates the level of liberalization of a given retail electricity market).
- Choice of the power supply company by consumers (whether it is possible to make a switch: yes / no).

2.5. Classification framework

The framework can be applicable to various cases. It should incorporate indicators which are (1) unified across the examining countries, (2) can be calculated in a representative way and (3) reflect the transition towards more liberalized electricity markets.

Considering indicators, provided in section 2.4 of the paper, only two of them may correspond to the stated above guidelines: "Value chain integration" (might be called a "Liberalization level") and "Choice of power supplier", i.e. choice available for the consumers on a given market (Figure 1).

To author's mind, the examined countries could be classified by the given indicators into six quadrants of the matrix, depending on the level of their value chain integration and availability of choice for consumers. The formulated framework might help us to answer the research question as it will visualize positions of the surveyed countries on the liberalization scale.

In order to support the propositions of the study and to reveal the possible link between the scale of the competitive model adoption and concentration within generation and sales segments, respectively (P1 and P2), the authors of the paper will apply the following matrix (Figure 2).

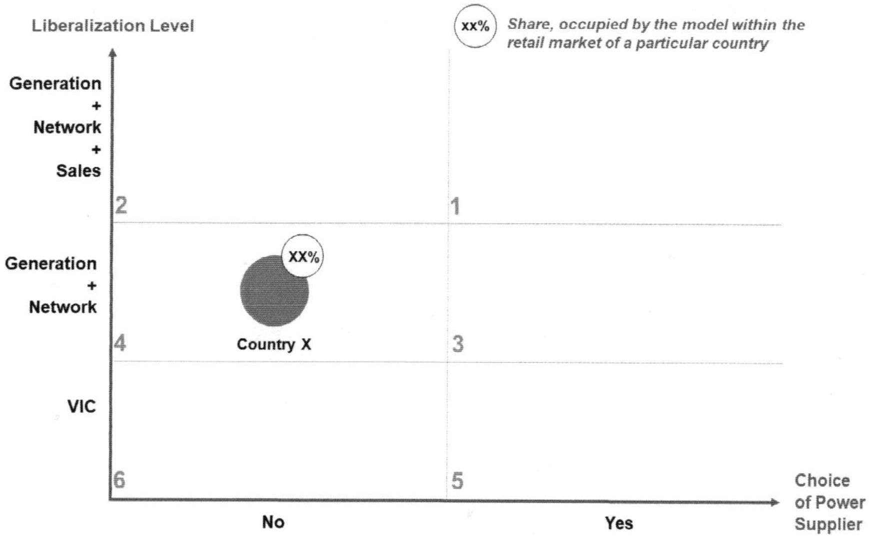

Figure 1. Classification framework [1]

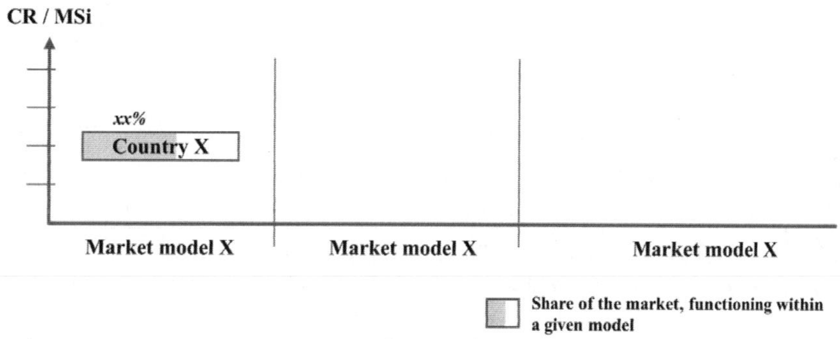

Figure 2. Classification framework [2, 3]

The matrices mentioned above are similar except from the value of OY. The Y-axis exhibits the concentration ratio in the generation segment for supporting P1 and exhibits the market share of a single power supply company for supporting P2.

3. Results

The conducted case study could bring us to the following results. We can see that among the countries surveyed, the competitive model in the retail electricity market is most widely used (5 out of 6 surveyed countries have adopted this model). This fact might be considered as a positive answer to the research question. However, there are cases when a single country adopts few models at the same time: Russia (single buyer and competitive models); Singapore (single buyer and competitive models), USA (monopoly and competition). It is worth mentioning that surveyed countries vary by the level of liberalization and availability of the choice of the power supplier – two major factors that all these countries have promoted during the development of their power sectors. Considering these indicators, we conclude that the observed electricity markets' model might be plausibly grouped upon these indicators (Figure 3).

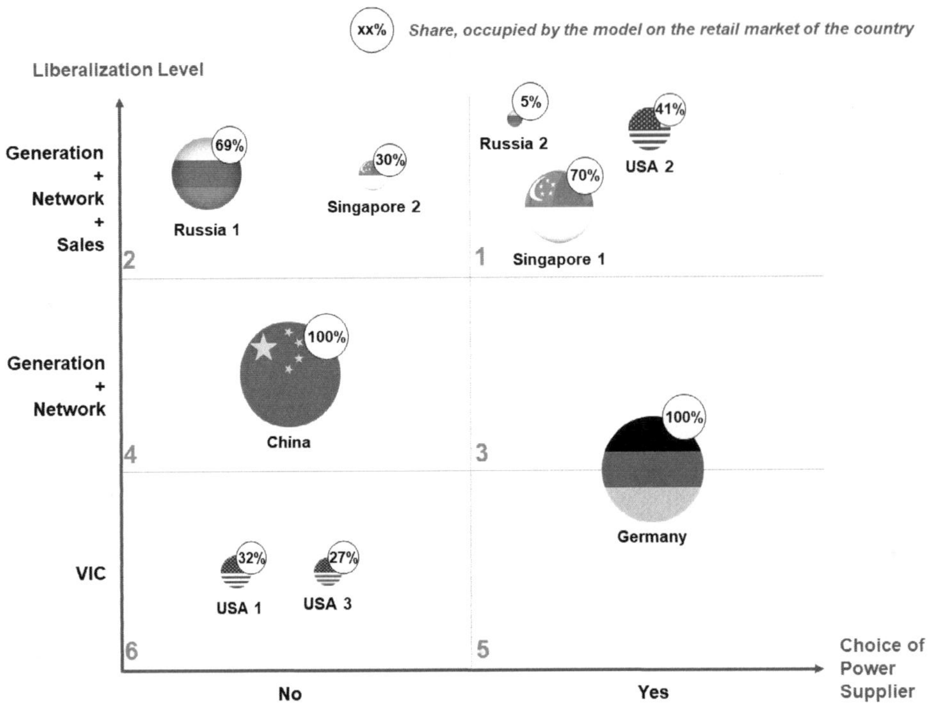

Figure 3. Classification of currently used retail electricity markets' models [4–9][a]

[a]Note 1: USA 1 – VIC; USA 2 – Retail choice; USA 3 – Bilateral contracts; Singapore 1 – when power supply companies are market participants; Singapore 2 – when power supply companies are not market participants; Russia 1 – when electricity is acquired from guaranteeing supplier; Russia 2 – when the electricity is bought from independent power supply company

Note 2: the size of the flag represents the share, occupied by the model within the retail market of a given country (this share is also doubled in the circle in the upper right corner of the flag).

The given classification incorporates 6 categories of the models (quadrants of the matrix): (1) USA 2, Singapore 1 and Russia 2 – the most liberalized markets; (2) Russia 1 and Singapore 2 – these markets include activities decoupling, but consumers lack choice of a supplier; (3/5) Germany – a hybrid of 3 and 5 categories as generation, distribution and sales are integrated, and transmission is decoupled; (4) China – average level of activities decoupling and lack of consumers' choice; (6) USA 1 and USA 3 – the least liberalized retail electricity markets.

After study and assessment of the parameters, indicated in the comparative model (see paragraph 2.4 for details), we might conclude that the levels of concentration in both generation and sales segment significantly vary (Table 2).

As can be seen in some countries (USA, Russia) there are two models of the retail electricity market. This is due to the complex structure of the market itself. Large territories of these countries are characterized by heterogeneous density - in places of high population density there is an opportunity to create a certain unified energy system, in places of low population density there are isolated territories, energy supply of which is particularly difficult.

The brightest example in this case is Russia. The Russian electric power market is a mosaic one in terms of distribution and sales of electricity. Pricing depends on the entry of a region into a particular zone: there are price zones, non-price zones and isolated areas. In last ones, power is usually supplied by local heat and electricity systems. Consequently, in isolated areas power services and production are expensive due to significant transportation costs for fuel and equipment. And what is more, tariffs for generators are set by regional authorities based on their actual cost. Such conditions make it almost impossible to organize market relations in the energy systems of these areas. This feature of the Russian market structure leads to the formation of completely different models of the retail power market. A main model for the most part of the territory will not be relevant for isolated areas. But at the same time, taking into account the development of technological base and distributed generation [10], available low carbon (nuclear and local RES) energy sources, it is possible to assume the appearance of hybrid models with wide implementation of integrated intellectual energy solutions and innovative energy ecosystems which are different from the existing ones in such areas. Study of these transformations and their phenomenon may become one of the further directions of research.

Table 2. Comparative analysis of market models of surveyed countries

Country	Degree of concentration within generation segment (concentration ratio)[a]	Degree of concentration within sales segment (market share)[b]	Current market model	Market share of the model[c]
USA	$CR_5 = .22$	5 major power supply companies occupy 56% of the market MS_i~19.9%	*Monopoly*[d]	32%
			Competition	41%
			Monopoly[e]	27%
China	$CR_5 = 1.00$	5 (affiliated with generators) + 300 local (function nominally) power supply companies MS_i~20%	*Single buyer*	100%
Singapore	$CR_5 = .78$	26 power supply companies MS_i=3.8%	*Competition*	70%
			Single buyer	30%
Germany	$CR_5 = .49$	900 power supply companies MS_i~1%	*Competition*	100%
Russia	$CR_5 = .64$	Local power supplier at each territory (720 in total) MS_i~.1%	*Single buyer*	93%
			Competition	7%

[a]: CR – concentration ratio (calculated upon the share of generation by 5 biggest generating companies in the total volume of generation of a particular country – in 2017). Note: calculations include only that electricity generation, which was performed by a particular company on the territory of a particular country;
[b]: MS_i – market share in generation capacity of a single power supply company in the total market;
[c]: data for calculations is retrieved from previous sections of the paper and web-sites of Ministries of Energy of the given countries;
[d]: in centralized market;
[e]: in two-sided contract.

In order to support or reject the propositions, it is worth studying two next tables. There is no substantial connection between the degree of generation capacity concentration and predominance of a particular market model. For instance, in China, with the highest value of CR we may perceive an absolute dominance of the single buyer model; in USA, where CR has the lowest values, the monopoly model dominates; in Germany, where CR has average values, we can see the prevalence of the competitive market model (Figure 4).

As for the link between the degree of the concentration of power supply companies and predominance of a particular model, there is a bidirectional linkage: there are cases when a high concentration is observed within the dominance of a monopoly model (USA) while there are also cases when such connection is less evident (China). Meanwhile, considering the case of Singapore, relatively a small number of power supply companies (12 and 26 respectively) exist within a competitive model (Figure 5).

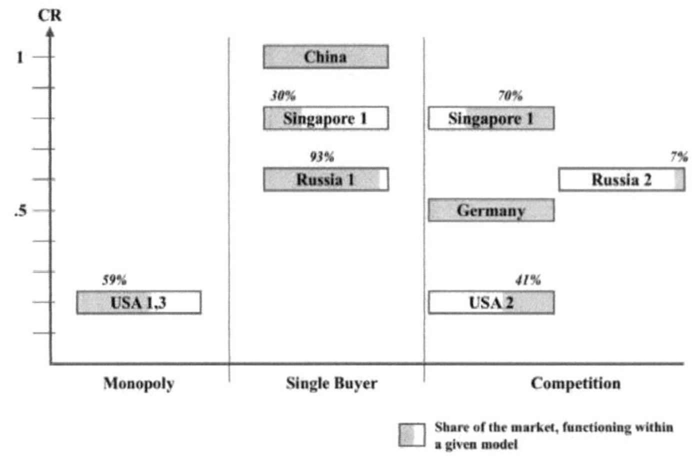

Figure 4. Link between the concentration in the generation and applied model of the retail electricity model

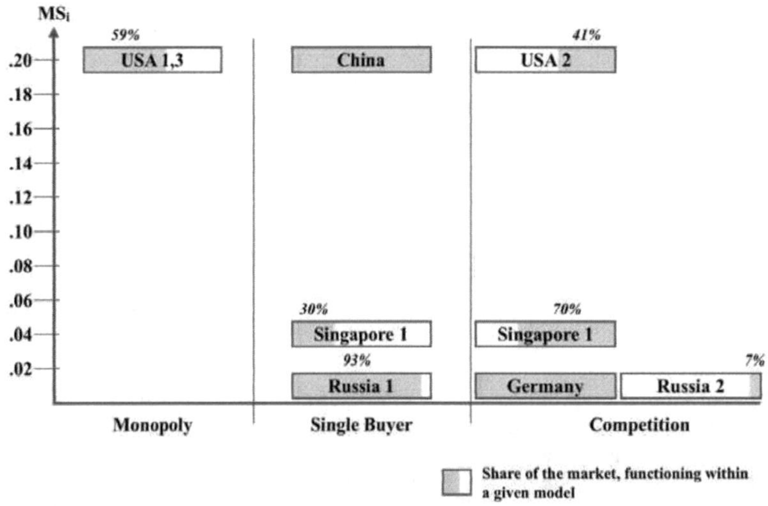

Figure 5. Link between the concentration in sales and the applied model of the retail electricity model

Altogether, the facts mentioned above lead to a conclusion that more evidence is needed in order to support or discard the proposition of the existence of a link between the concentration within the sales segment and predominance of less competitive market models. We might sum up the results in the following table (Table 3).

Table 3. Support of the proposition

Research question / Propositions	Result
RS	Yes. The competitive model is currently dominating (among the countries surveyed)
P1	Not supported
P2	Partially supported – more evidence is needed

4. Conclusion

The conducted case study analysis allows one to draw some conclusions. As it was shown by means of the proposed framework for the comparison and retrospective timeline analysis, the energy policy in the examined countries pursues the establishment of open and competitive retail electricity markets. in this way, the results show that currently competitive models dominate among those that were analyzed, while in some countries the complete transition to purely competitive markets may take more time and efforts both from the government and the business. At the same time, levels of the concentration in certain segments (generation or sales) do not necessarily directly determine the level of liberalization on the market.

The energy policy and liberalization reforms differ from country to country in terms of institutional constraints, scope, geographic areas, number of players, consumer disparity etc., while regulatory requirements tend to evolve over time in all discussed countries and this evolutionary process is continuous over time. The results of the study may be applicable to energy policy reformation processes taking place in developing countries, where liberalization in public utility sectors may be still at an incipient stage. Also, it seems to be useful for authorities in terms of developing additional regulatory rules or introducing subsidies that would provide incentives for market entry and boost competition, integration of markets and of distributed generation, etc.

Despite the evident trend towards retail electricity markets liberalization there are still certain obstacles slowing down this process. The liberalization of retail electricity markets might require a high switching rate between different suppliers among consumers as far as it is one of the main outcomes of the deregulatory reform. Yet consumer activism and initiative to create a competitive market seem to be considerably low when it comes to switching between electricity retailers [11]. This may be due to the long-term institutional constraint of default supply, thereby, the practice of switching between electricity retailers seems to be new or even unknown for the majority of consumers. Another source of low consumer involvement can be associated with the inertial decision-making or budget constraint. Consumer inertia arises both from the incurred costs of searching alternatives and the incumbent's brand advantage [12].

Electricity is an instantaneous commodity and a virtually intangible product, both its production and consumption take place simultaneously; storage is still expensive, and the demand is very inelastic and follows certain seasonal patterns and fluctuations. Modern electric grids have to be balanced with certain technologies, such as high capacity storage systems, demand response programs, smart grid solutions and non-dispatchable renewable generation [13].

Considering all these barriers, more empiric evidence is needed in order to understand whether switching to purely competitive market models is the only alternative. On the one hand, deregulatory reforms are aimed at increasing economic efficiency and provide price reductions for end-users of electricity. On the other hand, it can be noted that there is no direct evidence in favor of a competitive electricity market instead of a regulated monopoly [11]. Also it should be noted that new ecology requests and development of intellectual integrated energy systems and innovative energy ecosystems may lead to next deep transform of retail energy markets, roles and strategies of their participants, price and capacity reserves mechanisms, regulation and policy of the energy sector and economy.

A noticeable limitation of this study is that its results are anchored to the cases of the 5 countries and their specific energy policy reforms. Hence, we cannot infer directly that the same conclusions would be representative for other cases.

Nevertheless, the proposed framework allows doing further comparative analysis and, thus, can be applied to other cases in future research in order to validate (or not) that certain patterns of deregulatory processes coincide in various retail electricity markets.

Further research may involve comparative analyses of liberalization processes in other utilities (landline telecom, rail, natural gas, etc.) under the creation of different intelligent power systems and change the consumer' role. This could help to assess the effects of deregulatory reforms and technology development and improve our understanding of the consumer behavior as well as incumbent responses to liberalization and new entrants' incentives to join the market.

References

[1] Goldthau A 2014 Rethinking the governance of energy infrastructure: Scale, decentralization and polycentrism *Energy Research & Social Science* **1** 134-40

[2] Bae M, Kim H, Kim E, Chung A, Kim H and Roh J 2014 Toward electricity retail competition: Survey and case study on technical infrastructure for advanced electricity market system *Applied Energy* **133** 252-73

[3] Yassin E G, Volkova I O and Kosareva N B 2017 Some questions of development of retail power markets, urban economy and agglomeration *The Expert Institute of Higher School of Economics*

[4] Nazarian D 2012 Introduction to U.S. Electricity Markets. Maryland Public Service Commission, Portland. Retrieved from: http://www.camput.org/wp-content/uploads/2013/09/2012-07-21_-Nazarian_US_Electricity_Markets.pdf.

[5] Michigan Agency for Energy. What approaches to retail electric market structure have been tried in Michigan and in other states and jurisdictions. Retrieved from: http://www.michigan.gov/energy/0,4580,7-230-72200_68204_54287-293379--,00.html.

[6] Speed P 2004 *Energy Policy and Regulation in the People's Republic of China* (Kluwer Law International, Hauger)

[7] Hogan Lovells Lee and Lee 2017 Singapore Energy Market. Hogan Lovells Lee & Lee, Singapore. Retrieved from: https://www.hoganlovells.com/~/media/hogan-lovells/pdf/publication/singapore-energy-marketseptember-2012_pdf.

[8] Uwer D, Zimmer D and Mueller H 2017 Electricity regulation in Germany: overview. Thomson Reuters Practical Law. Retrieved from: https://uk.practicallaw.thomsonreuters.com/5-524-0808?transitionType=Default&contextData=(sc.Default)&firstPage=true&comp=pluk&bhcp=1.

[9] Ministry of Energy of Russian Federation. Industry condition. Retrieved from: https://minenergo.gov.ru/node/539.

[10] Lombardi P, Sokolnikova T, Suslov K, Voropai N and Styczynski Z A 2016 Isolated power system in Russia: a chance for renewable energies? *Renewable Energy* **90** 532-41

[11] Shin K and Managi S 2017 Liberalization of a retail electricity market: Consumer satisfaction and household switching behavior in Japan *Energy Policy* **110** 675-85

[12] Hortacsu A, Madanizadeh S and Puller S 2017 Power to choose? An analysis of consumer inertia in the residential electricity market *American Economic J.: Economic Policy* **9** 192-226

[13] Muratori M, Schuelke-Leech B and Rizzoni G 2014 Role of residential demand response in modern electricity markets *Renewable and Sustainable Energy Reviews* **33** 546-53

Description of automated functions of high-tech production of nanostructures based on silicon dioxide to improve the properties of building and structural materials

R S Zott[1], **A M Gladkih**[2], **V Yu Konyuhov**[2]

[1] LLC "Tarasovsky coal", 1, Dzerzhinskogo Str., Irkutsk, 664003, Russia
[2] Irkutsk National Research Technical University, 83, Lermontova Str., Irkutsk, 664074, Russia

E-mail: gladkikh@ex.istu.edu

Abstract. Within the framework of the project "Development of an integrated resource-saving technology and organization of high-tech production of nanostructures based on carbon and silicon dioxide to improve the properties of building and structural materials", an automated control system for technological thermal vortex enrichment has been created, designed to obtain a modifying additive based on dioxide silicon. The main parameters of the operation of technological lines for thermal vortex enrichment of raw materials, automatic control when threshold values are reached, as well as signaling for the use of control action from automated technological lines are presented. The functions of collecting and primary processing of data, creating control actions, creating and maintaining requests and reporting, providing reporting and data analysis tools are subject to automation. Automation of these functions in the form of a graphical human-machine interface allows for continuous technological control of equipment operation and technological process parameters; ensuring the safety of the technological process in the production of a modifying additive based on silicon dioxide. The collection of data on technological processes and equipment operation, their processing, display and documentation will ensure the optimization of the technological process using advanced visualization tools, modern control algorithms and analysis of the accumulated technological information. Automation of control of the technological process of thermal vortex enrichment will ensure the minimization of the influence of the human factor on the processes of collecting and processing information about the technological process; automatic prevention of the development of emergency situations.

1. Introduction

Known methods for the production of nanostructures in the form of spherical silicon dioxide, characterized by the requirement to use special capacities, increased consumption of raw materials and energy [1-4]. Microsilica is an integral by-product of the production of silicon and ferrosilicon. It is obtained by purifying gaseous emissions from the metallurgical industry, which improves the environmental situation at factories, but creates a significant disposal problem [5-6]. In Norway, after the official adoption of technological solutions in the form of standards and relevant specifications, since 1980, microsilica has been used as a modifying additive in concrete [7]. At present, "nanosilica" is of the greatest interest to researchers as a modifier of building and structural materials [8-11]. The development of technological solutions for the creation of high-tech production for the mass use of nanomaterials is urgent. In this case, the technology should be based on the selection of nanostructures rather than microparticles. The analysis shows that it is advisable to use the technology of incidental production of nanostructure concentrates in the production of aluminum by trapping dust from ore-thermal electric arc furnaces in the production of silicon [12-15]. In this paper, the automated functions of the control system for the technological process of obtaining a modifying additive based on silicon dioxide (APCS) are considered.

2. The goals of the development of the APCS

The APCS is designed to monitor and control the technological process of thermal vortex enrichment to obtain a modifying additive based on silicon dioxide.

The purpose of creating an automated process control system is:

- collection of data on technological processes and equipment operation, their processing, display and documentation;
- ensuring the safety of the technological process by preventing the development of emergency situations;
- optimization of the technological process through the use of advanced visualization tools, modern control algorithms and analysis of the accumulated technological information;
- minimizing the influence of the human factor on the collection and processing of information about the technological process.

To achieve the goals of creating an automated process control system, the automation of the following processes has been implemented:

- automation of the process of collecting and primary processing of analog and discrete information from sensors, actuators, thyristor converters (analog and discrete signals);
- processing of primary information according to the specified control algorithms and the formation of analog and discrete control actions on the output modules of the USO in order to regulate and maintain technological indicators at a given level;
- backup of data on technological parameters, on equipment operation;
- visualization of pre-configured reporting in tabular and graphical forms.

3. Description of the automation object and parameters to be measured and adjusted

The object is five technological lines for thermal vortex enrichment of raw materials to obtain a modifying additive based on silicon dioxide.

The production line capacity is 200 kg / h. Each line includes the following nodes:

- feedstock bunker;
- dosing device;
- combustion chamber;
- pipe for thermal vortex enrichment of raw materials;
- cooling gas duct;
- bag filter.

The principle of operation of the technological line is as follows. An air flow of a given temperature enters the vortex enrichment tube from the combustion chamber and feedstock is supplied through the metering device. When the dust-gas mixture passes through the vortex enrichment pipe, carbon is reburned from the feedstock. Further, in the cooling gas duct, the temperature of the dust-gas mixture drops below 200 °C and the finished product is captured by a bag filter.

The system provides for the measurement and adjustment of the following technological parameters.

Parameters to be measured:

- temperature of the air entering the combustion chamber;
- temperature of the air flow at the exit from the combustion chamber;
- air flow temperature at the beginning of the vortex enrichment tube;
- air flow temperature at the end of the vortex enrichment tube;

- temperature of the air flow in the cooling gas duct;
- temperature of the air entering the cooling gas duct;
- temperature in the bag filter hopper;
- gas pressure entering the burner;
- compressed air pressure;
- pressure at the beginning of the vortex enrichment pipe;
- pressure in the cooling gas duct;
- pressure in the dirty chamber of the bag filter;
- pressure in the clean chamber of the bag filter;
- gas consumption into the combustion chamber;
- rotational speed of the blower;
- rotation frequency of the dosing device for feeding the starting material;
- position of the air supply gate;
- the presence of material in the vortex enrichment pipe;
- the upper level in the feedstock bin;
- the lower level in the feedstock bin;
- upper level in the bag filter hopper;
- the lower level in the baghouse filter.

Parameters to be regulated:

- the temperature of the air flow in the combustion chamber;
- temperature of the dust-gas mixture in the vortex enrichment pipe;
- temperature in the cooling gas duct;
- vacuum in the vortex enrichment pipe;
- consumption of starting material;
- air consumption in the combustion chamber;
- pressure drop across the baghouse filter.

Parameters to be signaled:

- the lower value of the vacuum in the vortex enrichment pipe;
- the lower value of the temperature of the air flow in the combustion chamber;
- the upper value of the pressure drop in the baghouse filter;
- shutdown of the blower;
- turning off the smoke exhauster;
- shutdown of the dosing device for feeding the starting material;
- the lower level in the feedstock bin;
- upper level in the bag filter hopper;
- the temperature in the cooling gas duct is close to the upper limit;
- the temperature in the vortex enrichment pipe is close to the lower value;
- upper temperature level in the cooling gas duct;
- the lower level in the bag filter hopper;
- the upper level in the feed bin.

4. Description of Automated Functions

The automated control system for the technological process of thermal vortex enrichment has a modular structure with a clear division of functions between its subsystems. The performance of the functions of the APCS, as well as access to the data and control structures contained in it, is through specialized components that have mechanisms based on a human-machine interface for interacting with users. In the process of functioning of this system, the procedures for exchanging data between subsystems are carried out using software [10, 16-18].

APCS is a set of automated systems of separate departments, which can function both separately, while ensuring the functioning of individual subsystems at each level in full within a separate automated system, and jointly ensuring the functioning of subsystems within the framework of general information interaction [10, 19-20].

The APCS includes the following subsystems:

- subsystem for collecting primary information, processing, generating control actions and data transmission;
- data storage subsystem;
- subsystem for generating and visualizing reports.

The general diagram of the functional structure of the APCS is shown in Fig. 1.

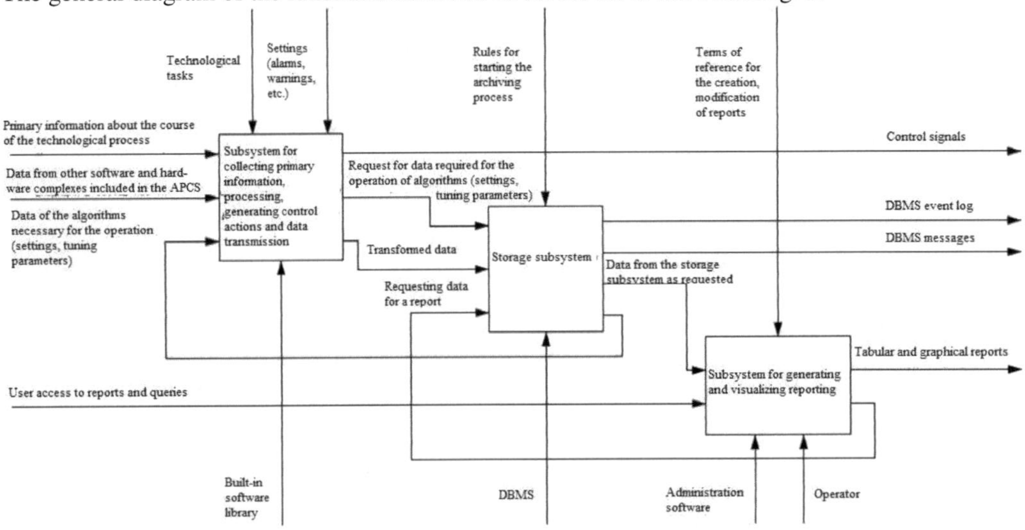

Figure 1. General diagram of the functional structure of the APCS.

The subsystem for collecting primary information, processing, generating control actions and transmitting data is designed to implement the processes of collecting primary information about the course of the technological process, bringing this information to a form convenient for software processing in accordance with algorithms, generating control actions, and data exchange with other automated process control systems via digital communication lines.

The function of collecting and primary processing of information includes the following tasks:

- reception and primary processing of analog and discrete information from sensors, executive mechanisms, thyristor converters (analog and discrete signals);
- reception and primary processing of information and commands (settings, settings) coming from operators through the SCADA system.

The function of forming control actions includes the following tasks:

- processing of primary information in accordance with the specified control algorithms and technological settings;
- formation of analog and discrete control actions on the output modules of the USO.

The subsystem for the formation and visualization of reporting, designed to generate information about the course of the technological process, archived data in a tabular and graphical form.

The function of creating and maintaining a logical presentation of information includes the following tasks:

- creation of a logical representation of information in the form of a description of the stored data;
- modification of the logical presentation of information.

The function of creating and maintaining queries and reporting includes the following tasks:

- creation of data request templates;
- setting up tabular forms and graphs of data analysis.

The reporting function and data analysis tools include the following tasks:

- providing the ability to perform mathematical operations on indicators;
- providing the ability to perform group operations on data in real time;
- visualization of pre-configured reporting in tabular and graphical form.

The table contains a list of subsystems with a description of the process of performing functions.

Table 1. List of subsystems with a description of the process of performing functions.

Subsystem	Process description
The subsystem for collecting primary information, processing, generating control actions and data transmission	The processes of collection and primary processing of analog and discrete information from sensors, actuators, thyristor converters (analog and discrete signals). These processes are automatic, processing is by a microprocessor controller. The process of receiving and primary processing of information and commands (settings, settings) coming from operators through the human-machine interface. Data entry is human.
The subsystem for generating and visualizing reports	The processes of presenting information, conducting mathematical operations on indicators; visualization of pre-configured reporting in tabular and graphical form. All processes are automatic.

Work on the development of APCS was taking into account the following requirements:

- support for generally accepted technologies for processing and presenting information;
- use of standard sets of tools within the framework of the APCS;
- openness of data structures;
- focus on reliable equipment;
- construction taking into account ease of use for the end user;
- construction of an automated process control system taking into account safety requirements.

The relationship between the upper and lower layers is carried out at the driver level via the TCP / IP protocol. The SCADA system has full access to the read and write memory cells of the programmable logic controller. The interconnection of the APCS parts is via the Ethernet network.

5. Conclusion

When organizing the production of nanostructures based on silicon dioxide to improve the properties of building and structural materials, methods of automation of production and management processes were used. Automation of the functions of collecting and primary processing of information allows the processing of primary information in accordance with the specified control algorithms and technological installations. The function of providing reporting and data analysis tools makes it possible to carry out mathematical operations on indicators; providing the ability to perform group operations on data in real time; visualization of pre-configured reporting in tabular and graphical form. The implementation of these functions in the form of a graphical human-machine interface allows achieving the goals of creating an automated process control system.

References

[1] Vasin V A, Fatyanova N G, Troshin B A, Vasichev B N and Stepanchikov S V 2012 Problems of

creation of nanoproducts using nanotechnology *Electrotechnical and informtion complexes and systems* **8(4)** 37-43
[2] Gerasimenko N, Volokhovsky A and Zaporozhan O 2017 Features of control of technology of silicon nanostructures *Nanoindustry* **5(76)** 36-51
[3] Kodolov V I, Bondar A Yu and Kuznetsov A P 2007 Technology of production of carbon-herodometal-containing nanostructures in polymeric matrix nanoreactors (by the example of pva) *Nanotechnics* **1(9)** 38-41
[4] Khokhlun A R 2008 Equipment and technologies for production of mems, micro and nanostructures *Electronics and electrical equipment of transport* **6** 36-38
[5] Ershov V A, Kondratiev V V, Karlina A I, Kolosov A D and Sysoev I A 2018 Selection of control system parameters for production of nanostructures concentrates *J. of Phys.: Conf. Ser.* **1118(1)** 012014
[6] Kuzmin M P, Shestakov S S, Kuzmina M Yu and Zhuravleva A S 2015 Innovative development of the metallurgical complex of the Irkutsk region *Bulletin of ISTU* **5(100)** 236-240
[7] KondratEv V V, Nemchinova N V, Ivanov N A, Ershov V A and Sysoev I A 2013 New production solutions for processing silicon and aluminum pro-duction waste *Metallurgist* **57(5-6)** 455-459
[8] Yolkin K S, Yolkin D K, Kolosov A D, Ivanov N A and Shtayger M G 2018 Technologies, which allow to reduce an impact of metal silicon production on the environment *IOP Conf. Ser.: Mater. Sci. Eng.* **411** 012028
[9] Munkhtuvshin D, Balabanov V B and Putsenko K N 2017 Experience of application of additional-wok micro- and nanosilic from waste of silicon production in concrete technologies *Izvestiya vuzov. Investments. Building. The property* **7(3(22))** 107-115
[10] Kondratiev V V, Petrovskiy A A, Ershov V A, Sysoeva T I and Karlina A I 2017 Results of research with revealing of technological parameters of processes of recycling and neutralization of the first and second cut of the spent lining of electrolyzers for reception of aluminum fluoride by pyrolytic and hydro chemical method *Int. J. of Applied Engineering Research* **12(23)** 13898-13904
[11] Yurovsky V S, Glukhatkina L G, Pankratova G M, Sorokina I E and Maksimova T V 2016 Development of heat-frost-resistant rubbers for sleeves on the basis of siloxane rubber and domestic silica fillers *Industrial production and use of elastomers* **1** 28-31
[12] Guseva E A, Konstantinova M V, Kargapoltsev S K, Gozbenko V E, Sivtsov A V, Karlina A I and Shtayger M G 2019 Methods of graphitized steels obtaining *J. of Phys.: Conf. Ser.* **1353(1)** 012063
[13] Ivanchik N N, Balanovsky A E, Shtayger M G, Sysoev I A and Karlina A I 2018 Capability enhancement of production of activating fluxes for arc welding using ultradispersed products of silicon waste processing *IOP Conf. Ser.: Mater. Sci. Eng.* 012035
[14] Kuzmin M P, Larionov L M, Kondratyev V V, Kuzmina M Yu, Grigoriev V G, Knizhnik A V and Kuzmina A S 2019 Obtaining silumins with the use of silicon waste *Proc. of higher educational institutions. Non-ferrous metallurgy* **4** 4-15
[15] Karlina A I, Kondratev V V, Kolosov A D, Balanovskiy A E and Ivanov N A 2019 Production of new nanostructures for modification of steels and cast irons *IOP Conf. Ser.: Mater. Sci. Eng.* 012183
[16] Ershov V A, Kondratiev V V, Sysoev I A and Mekhnin A O 2013 Extraction of car-bon nanoparticles from fluorinated alumina during aluminum production *Metallurgist* **56(11-12)** 952-956
[17] Buryanina N, Korolyuk Y, Koryakina M, Lesnykh E and Suslov K 2019 Algoritm of Current Protection Based on Three Instantaneous-Value Samples *Proc. of 2019 IEEE PES Innovative Smart Grid Technologies Europe, ISGT-Europe* 8905542
[18] Suslov K, Shushpanov I, Buryanina N and Ilyushin P 2020 Flexible power distribution networks: New opportunities and applications *SMARTGREENS 2020 – Proc. of the 9th Int. Conf. on Smart Cities and Green ICT Systems*

[19] Suslov K, Solonina N and Gerasimov D 2018 Assessment of an impact of power supply participants on power quality *Proc. of Int. Conf. on Harmonics and Quality of Power* pp 1-5
[20] Petrovskaya V N and Kondratiev V V 2012 Gas-hydrodynamic nature of the anode effect *Metallurgist* **56(3-4)** 215-221

The main technical solutions for the creation of an automated control system for the technological process of thermal vortex enrichment

V O Gorovoy, A S Govorkov, V V Kondrat'ev, R V Kononenko and A M Pestova

Irkutsk National Research Technical University, 83, Lermontova Street, Irkutsk, 664074, Russia

E-mail: vog@ex.istu.edu

Abstract. The main technical solutions for the creation of an automated control system for the technological process of thermal vortex enrichment were adopted for the creation of an integrated resource-saving technology and the organization of high-tech production of nanostructures based on carbon and silicon dioxide to improve the properties of building and structural materials using control digitalization techniques technological processes. The decisions were made taking into account the dependence of the process of obtaining a modifying additive containing nanostructures in the form of spherical silicon dioxide on a large number of interrelated variables, a variety of operating modes of the technological line, and instantaneous changes in parameters. Technical solutions have been developed for the structure, means and methods of communication for information exchange; interconnection with adjacent systems, ensuring its compatibility; modes of operation, diagnostics of system operation; number, qualifications and functions of personnel, modes of their work, the order of interaction. This includes also requirements for ensuring the given consumer characteristics of the system, which determine its quality; the composition of functions, complexes of tasks implemented by the system; a set of technical means, its placement at the facility. Besides, it concerns the composition of information, volume, methods of its organization, types of machine media, input and output documents and messages, information processing sequence and other components; the composition of software, languages of activity.

1. Introduction

The main stream of the science in the future is resource and energy saving technologies [1-4]. A characteristic feature of the process of obtaining a modifying additive containing nanostructures in the form of spherical silicon dioxide is the dependence of its flow on a large number of interrelated variables, a variety of operating modes of the technological line, and instantaneous changes in parameters [1-4].

Ensuring the stability of the technology is achieved by collecting and processing data on the state of the technological process every second. To ensure constant technological control of equipment operation and parameters of the technological process, to ensure its safety, an automated control system for the technological process of obtaining a modifying additive based on silicon dioxide has been developed [5-8]. When creating the system, the following technical solutions were adopted [9-12].

IOP Conf. Series: Materials Science and Engineering 1064 (2021) 012013 doi:10.1088/1757-899X/1064/1/012013

2. Decisions on the structure, means and methods of communication for information exchange

Step The automated control system for the technological process of thermal vortex enrichment has a modular structure with a clear division of functions between its subsystems. The performance of the functions of an automated control system for the technological process of thermal vortex enrichment, as well as access to the data and control structures contained in it, is carried out through specialized components that have mechanisms based on a human-machine interface for interacting with users. In the process of functioning of the automated control system for the technological process of thermal vortex enrichment, the procedures for exchanging data between the subsystems are carried out using software [13-15].

The automated control system for the technological process of thermal vortex enrichment is a set of automated systems of separate departments, which can function both separately, while ensuring at each level the functioning of individual subsystems in full within a separate automated system, and jointly ensuring the functioning of subsystems within the framework of general information interaction [16-19].

The automated control system for the technological process of thermal vortex enrichment includes the following subsystems:

- a subsystem for collecting primary information, processing, generating control actions and data transmission;
- data storage subsystem;
- a subsystem for generating and visualizing reports.

Work on the development of an automated control system for the technological process of thermal vortex enrichment was carried out taking into account the following requirements:

- support of generally accepted technologies for information processing and presentation;
- use of standard sets of tools within the automated control system for the technological process of thermal vortex enrichment;
- openness of data structures;
- focus on reliable equipment;
- building taking into account ease of use for the end user;
- construction of an automated control system for the technological process of thermal vortex enrichment, taking into account safety requirements.

3. Decisions on modes of operation, diagnostics of system operation

The automated control system for the technological process of thermal vortex enrichment is designed to function during the working time set by the Customer, except for cases caused by the occurrence of force majeure circumstances [20-21].

The following operating modes are provided:

1) Normal mode of production operation:

- execution of automated functions of technological process control;
- displaying the parameters of the technological process, technological equipment;
- archiving and backup of technological information;
- administration, management of basic settings.

2) Operation in emergency situations:

- execution of automated functions of technological process control according to the algorithm.

In the event of force majeure circumstances leading to the downtime of the APCS 2, the interaction between the subjects is temporarily suspended until the moment of full restoration of working capacity, while the control systems of individual departments can continue to function in normal modes.

4. Decisions on the number, qualifications and functions of personnel, their modes of operation, the order of interaction

For the technical and system support of the system, to perform the functions of system administration, the creation of a technical support group is required. The recommended number of technical service personnel required to maintain the system is given below.

The following is the requirements for the qualifications and positions of the system technical support group(s).

The automation engineer performs the following tasks:

- ensuring the operability of technical means and software of automated information and measuring systems (AIIS) and automated control systems (ACS) in technological shops and departments, operation of systems for the protection of technological equipment;
- ensuring the uninterrupted functioning and maintenance of systems transferred to industrial operation in accordance with the requirements and rules set out in the technical documentation;
- metrological support of the automated control system for the technological process of thermal vortex enrichment;
- implementing production control over compliance with industrial safety requirements in the sector;
- coordinating and directing the work of the engineering and technical personnel of the sector on maintenance and modernization of software, on the development and correction of working documentation for software. Control the timeliness of making all necessary changes and additions to the programs and work instructions;
- coordinating the work of the department of the automated control system for the technological process of thermal vortex enrichment with the technological department;
- constantly improving their qualifications and qualifications of personnel, organize the study and implementation of advanced methods in work. Organize in the sector and conduct work on training and preparation for independent work of young specialists newly entering the sector, as well as those transferred from other workshops in accordance with the rules and regulations in force in the association.

Persons with a higher technical education in the specialty "Automation of technological processes and production", work experience of at least five years are appointed to the position of an automation engineer.

The instrumentation and control locksmith on duty performs the following tasks:

- repair, adjustment, installation of automatic and electronic equipment;
- check of electrical measuring instruments of all types and systems;
- check and adjustment of automation units, control units on automatic machines with programmed control;
- identification and elimination of equipment malfunctions;
- determination of the degree of wear of parts and assemblies;
- commissioning and comprehensive testing of control schemes and automation of technological equipment;
- assembly of circuits for testing automation devices;
- mathematical processing of measurement results and registration of the necessary materials.

Persons with higher professional engineering and technical or secondary vocational education are appointed to the position of the duty mechanic of instrumentation and automation.

The required number of personnel for the technical support group is shown in Table 1.

Table 1. Required number of technical support staff.

Personnel name	Quantity
Automation engineer	1
Locksmith on duty for instrumentation and automation	3

The functions of personnel (operator, shift supervisor) working with an automated control system for the technological process of thermal vortex enrichment include:

- remote control and monitoring of the technological process, presented on mnemonic diagrams;
- prompt notification of service personnel about pre-emergency process situations and malfunctions of technical automation equipment.

5. Information on ensuring the specified consumer characteristics of the system that determine its quality

Ensuring the compliance of the developed automated control system for the technological process of thermal vortex enrichment with the technical requirements is carried out through the following measures:

- automatic interrogation of sensors and devices, processing of input signals is achieved by using a distributed data collection system, built on the USO of Siemens, which interact with each other via Ethernet;
- control of technological processes and equipment is carried out by the control of the burner, smoke exhauster, screw feeder motor and air supply damper to the cooling zone.

Requirements for the organization of input data are provided:

- using sensors and devices with an output analog signal (4-20 mA);
- using sensors, instruments and devices in a discrete format (dry contact, 24 V);
- using a digital communication channel Eternet for communication between distributed input-output systems, controllers, AWPs and operator panels, frequency converters;
- using automated workstations, as well as operator panels to ensure the input of user commands.

The requirements for the organization of the output are provided by:

- using discrete signal output modules (24V);
- using a digital communication channel (Ethernet, etc.) for communication between distributed input-output systems, controllers, workstation and operator panels, frequency converters;
- using automated workstations, as well as operator panels for displaying graphic information.

Timing requirements are provided by using I / O modules with a polling time of less than 2 s;

Requirements for the composition and parameters of technical means are provided by the use of an appropriate set of technical means.

Siemens programmable logic controllers, Simatic S7-1214C with parameters:
operation time is no more than 2.3 μs;
the number of channels in the local input-output system is not more than:

- discrete input -146;
- discrete output-142;
- analog input - 67;
- analog output -33.
- built-in interfaces - Ethernet;
- loadable memory (MMC) not less than 4MB.

Personal computer must be with the following parameters:

- processor with a clock speed of at least 3.0 GHz;
- RAM of at least 4 GB;

- hard drive with at least 1 TB of free space;
- monitor capable of displaying at a resolution of 1280x1024;
- DVD-ROM;
- 101 key or Windows-compatible keyboard;
- Windows compatible mouse.

6. The composition of functions, complexes of tasks (tasks) implemented by the system

The system consists of the following subsystems and tasks.

The system includes the following functional subsystems:

- subsystem for collecting primary information, processing, generating control actions and data transmission, designed to implement the processes of collecting primary information about the course of a technological process, bringing this information to a form convenient for program processing in accordance with algorithms, generating control actions, and so the same for data exchange via digital communication lines;
- subsystem for generating and visualizing reports, designed to generate information about the progress of the technological process, archived data in tabular and graphical form.

7. Solutions for a set of technical means, its placement at the facility

A distributed control system was chosen as the structure of the automated control system for the technological process of thermal vortex enrichment. This structure provides independent control functions for each of the 5 thermal vortex enrichment lines. If the equipment of the adjacent departments fails, the rest continue to function normally. The same situation occurs when the integrity of the network cable is violated.

When using the control system, a line of Siemens controllers of the Simatic S7 1200 series was chosen.

Rittal modular cabinetry was chosen to accommodate the control systems.

The following primary technical means were selected to control the technological process:

- measuring the temperature of the air entering the combustion chamber, the temperature of the air at the exit from the combustion chamber, the temperature of the air flow at the beginning and end of the vortex enrichment tube, the temperature in the cooling gas duct, the temperature of the air entering the cooling gas duct, the temperature in the bag filter hopper - thermal converter with unified output signal Metran 2700;
- signaling the presence of material in the pipeline after the feeder - microwave sensor for the presence / absence of flow of solid bulk materials MWS-DP-3;
- speed control for blower M1-1, smoke exhauster M2-1, screw feeder M3-1 - Schneider Electric Altivar Process ATV600 frequency converters;
- vibration level switch VEGAWAVE 61 for signaling the upper and lower levels in the raw material hopper and the bag filter hopper;
- controlling the gas flow rate in the combustion chamber - vortex flowmeter Rosemount 8800;
- measuring the pressure of compressed air in the main, gas pressure in the main, pressure at the beginning of the vortex enrichment pipe, pressure at the end of the vortex enrichment pipe, pressure in the cooling gas duct, pressure in the clean and dirty chambers of the bag filter - pressure sensors of the Metran 55 series;
- valves for supplying compressed air for bag filter regeneration - SMC solenoid valves for cleaning bag filters of the VXF2 series.

All primary converters are located on process pipelines, equipment in such a way as to ensure ease of maintenance, as well as to fulfill the requirements for the installation of devices specified in the operating manuals.

The power supply for the control cabinets envisaged in this project is carried out from the existing power supply boards. Power supply for the lower level of automation is provided from control cabinets.

8. Decisions on the composition of information, volume, methods of its organization, types of machine media, input and output documents and messages, information processing sequence and other components

The information consists of data coming from the following sources:

- sensors and primary converters;
- buttons and switches;
- automated workstations (AWP);
- operator panels;
- personal computers (PC).

By volume, the information is divided into:

- small (sensors and primary converters);
- medium (operator panels, remote data collection stations);
- large (AWP, PC).

Organization method:

- analog (two-three-wire single-point: current loop, resistance);
- digital (two-wire, multipoint: RS-485);
- digital (multi-wire, multipoint: Ethernet).

Types of media:

- flash memory of the controller, operator panel;
- hard drive AWP, PC.

Output documents and messages:

- process trends;
- logs of emergency messages.

The sequence of information processing:

- raw information from sensors, instruments and devices enters the PLC via analog or digital communication channels;
- brought to real values, check for validation of raw information from sensors, bringing it to real values with processing of pre-alarm and emergency limits. Saving data in the PLC memory;
- processed data are used in control algorithms and form data for controlling devices and mechanisms. And they also enter the upper levels via digital communication channels: operator panels, AWP and PC;
- information received from the lower level is displayed on mnemonic diagrams of AWP.

9. Decisions on the composition of software, languages of activity

The delivery set of the automated control system for the technological process of thermal vortex enrichment includes the developed software (hereinafter referred to as the program) for the PLC and PC.

The PLC program is a project developed in Siemens Simatic Step7 environment.

The PC program (SCADA) is a project developed in the Siemens TIA Portal environment.

To develop an automated control system for the technological process of thermal vortex enrichment, the following programming languages, queries, representations, visual modeling are used:

- standard IEC 61131-3 (LD, FBD);
- STL (IL IEC 61131-3 modified by Siemens Simatic Step7);
- ANSI C.

For the functioning of the automated control system for the technological process of thermal vortex enrichment at the AWP, Microsoft Windows 8.1 (OS) software is required:

10. Conclusion

The main technical solutions for the creation of an automated control system for the technological process of thermal vortex enrichment were adopted according to:

- structure, means and methods of communication for information exchange;
- interconnection with related systems, ensuring its compatibility;
- modes of operation, diagnostics of system operation;
- number, qualifications and functions of personnel, modes of their work, the order of interaction;
- requirements for ensuring the specified consumer characteristics of the system, which determine its quality;
- composition of functions, complexes of tasks implemented by the system;
- set of technical means, its placement at the facility;
- composition of information, volume, methods of its organization, types of machine media, input and output documents and messages, sequence of information processing and other components;
- composition of software, languages of activity.

The main technical solutions for creating an automated control system for the technological process of thermal vortex enrichment are necessary and sufficient to achieve the goal of creating an automated control system for the technological process of thermal vortex enrichment.

References

[1] Kondratiev V V, Petrovskiy A A, Ershov V A, Sysoeva T I and Karlina A I 2017 Results of researches with revealing of technological parameters of processes of recycling and neutralization of the first and second cut of the spent lining of electrolyzers for reception of aluminum fluoride by pyrolytic and hydro chemical method *Int. J. of Applied Engineering Research* **12(23)** 13898-13904

[2] Yurovsky V S, Glukhatkina L G, Pankratova G M, Sorokina I E and Maksimova T V 2016 Development of heat-frost-resistant rubbers for sleeves on the basis of siloxane rubber and domestic silica fillers *Industrial production and use of elastomers* **1** 28-31

[3] Guseva E A, Konstantinova M V, Kargapoltsev S K, Gozbenko V E, Sivtsov A V, Karlina A I and Shtayger M G 2019 Methods of graphitized steels obtaining *J. of Phys.: Conf. Ser.* **1353(1)** 012063

[4] Ivanchik N N, Balanovsky A E, Shtayger M G, Sysoev I A and Karlina A I 2018 Capability enhancement of production of activating fluxes for arc welding using ultradispersed products of silicon waste processing *IOP Conf. Ser.: Mater. Sci. Eng.* 012035

[5] Vasin V A, Fatyanova N G, Troshin B A, Vasichev B N and Stepanchikov S V 2012 Problems of creation of nanoproducts using nanotechnology *Electrotechnical and information complexes and systems* **8(4)** 37-43

[6] Gerasimenko N, Volokhovsky A and Zaporozhan O 2017 Features of control of technology of silicon nanostructures *Nanoindustry* **5(76)** 36-51

[7] Kodolov V I, Bondar A Yu and Kuznetsov A P 2007 Technology of production of carbon-herodometal-containing nanostructures in polymeric matrix nanoreactors (by the example of pva) *Nanotechnics* **1(9)** 38-41

[8] Yolkin K S, Yolkin D K, Kolosov A D, Ivanov N A and Shtayger M G 2018 Technologies, which allow to reduce an impact of metal silicon production on the environment *IOP Conf. Ser.: Mater. Sci. Eng.* **411** 012028

[9] Ilyushin P and Suslov K 2019 Operation of automatic transfer switches in the networks with distributed generation *IEEE Milan PowerTech, PowerTech* 8810450

[10] Buryanina N, Korolyuk Y, Koryakina M, Lesnykh E and Suslov K 2019 Algoritm of Current Protection Based on Three Instantaneous-Value Samples *Proc. of 2019 IEEE PES Innovative Smart Grid Technologies Europe, ISGT-Europe* 8905542

[11] Suslov K, Shushpanov I, Buryanina N and Ilyushin P 2020 Flexible power distribution networks: New opportunities and applications *SMARTGREENS 2020 – Proc. of the 9th Int. Conf. on Smart Cities and Green ICT Systems*
[12] Voropai N, Ukolova E, Gerasimov D, Suslov K, Lombardi P and Komarnicki P 2019 Simulation approach to integrated energy systems study based on energy hub concept *IEEE Milan PowerTech, PowerTech* 8810666
[13] Munkhtuvshin D, Balabanov V B and Putsenko K N 2017 Experience of application of additional-wok micro- and nanosilic from waste of silicon production in concrete technologies *Izvestiya vuzov. Investments. Building. The property* **7(3(22))** 107-115
[14] Khokhlun A R 2008 Equipment and technologies for production of mems, micro and nanostructures *Electronics and electrical equipment of transport* **6** 36-38
[15] Ershov V A, Kondratiev V V, Karlina A I, Kolosov A D and Sysoev I A 2018 Selection of control system parameters for production of nanostructures concentrates *J. of Phys.: Conf. Ser.* **1118(1)** 012014
[16] KondratEv V V, Nemchinova N V, Ivanov N A, Ershov V A and Sysoev I A 2013 New production solutions for processing silicon and aluminum pro-duction waste *Metallurgist* **57(5-6)** 455-459
[17] Konyuhov V Yu, Konstantinova M V and Gladkih A M 2019 Determination of restored units spectrum of equipment and development of the assembly unit repair method at industrial enterprises *J. of Phys.: Conf. Ser.* **1353** 012047
[18] Petrovskaya V N and Kondratiev V V 2012 Gas-hydrodynamic nature of the anode effect *Metallurgist* **56(3-4)** 215-221
[19] Karlina A I, Kondratev V V, Kolosov A D, Balanovskiy A E and Ivanov N A 2019 Production of new nanostructures for modification of steels and cast irons *IOP Conf. Ser.: Mater. Sci. Eng.* 012183
[20] Gorovoy V O, Kondrat'ev V V, Kolosov A D, Kononenko R V and Konyukhov V Yu 2020 Description of software for an automated control system for the technological process of thermal vortex enrichment *J. of Phys.: Conf. Ser.* **1661** 012096
[21] Kondrat'ev V V, Gorovoy V O, Kolosov A D, Kononenko R V and Konyukhov V Yu 2020 Description of the complex of technical means of an automated control system for the technological process of thermal vortex enrichment *J. of Phys.: Conf. Ser.* **1661** 012101

IOP Conf. Series: Materials Science and Engineering 1064 (2021) 012014 doi:10.1088/1757-899X/1064/1/012014

Automation of individual operations of the transport process to create sufficient conditions for the efficient functioning of digital transport and logistics

E A Guseva[1], **M V Konstantinova**[1], **V A Olentsevich**[2], **V Yu Konyukhov**[1] **and A A Olentsevich**[2]

[1] Irkutsk National Research Technical University, 83, Lermontova Street, Irkutsk, 664074, Russia

[2] Irkutsk State Transport University, 15, Chernyshevskogo St., Irkutsk, 664074, Russia

E-mail: guseva@istu.edu

Abstract. The current evolution in information technology will enable the railway industry to bring the concept of a customer-centric marketing approach to life. At present, the main factor for ensuring the competitiveness of the transport company, which emphasizes the position in the transport services market, is to provide a service with a qualitatively new modern level of services, based on the opinion of consumers of these services. The digitalization of railway transport and logistics, the use of modern technical means and programs allows one to improve the efficiency of the transport complex in general and on separate subsystems, as well as to ensure high quality of services. The introduction of an automated procedure for the optimal placement and fastening of the cargo in the production process will allow transport and logistics market operators to predict the occurrence of risk cases at each stage of the transport process. Algorithms for selecting the type of rolling stock to transport the cargo of the relevant category and the typical scheme of placing cargo in it, as well as the method of fixing cargo, have been developed and presented. Since the digitization of supply chains depends more on the type of cargo and the type of rolling stock chosen for transport, the construction of algorithms is based on the example of a flat-base cargo. The development of the automated procedure is based on the construction of a model scheme for placing and fastening a particular type of cargo. The results of using the developed automated procedure make it possible to determine the magnitude of the load shift and tension in the fasteners under the influence of various factors.

1. Introduction

Logistics is inherently a unique area of the transport business, permeating all sectors of the railway transport industry, infrastructure (exactly rails [1-5]) and users of transport services, product manufacturers, dealers and of course consumers of services. Large companies such as Russian Railways must be constantly prepared to meet the challenges of time while being able to separate fashionable and useless innovations from those that bring real benefits and highlight those associated with maintaining a high level of safety in both freight and passenger transport. This is usually achieved through regular testing of the effectiveness of a particular innovation.

The development of modern automation systems in the railway transport system (RTS) has recently become relevant and requiring development again. If the first developments in automation dealt with cybernetics and have been evolved in aviation, road and rail transport, we are returning to the issue of automation of control systems at a qualitatively new level. The modern stage of the development of automated technologies in the transport sector involves the development of elements of artificial intelligence - digitalization in most sectors of the economic sector [6-9].

Today, the public policy of the prosperous countries in the world is aimed at developing the digital economy in the sectors of the transport sector of the economy. The digital economy is not a single sphere of use; the foundation allows one to create qualitatively new models of logistics, production, business, and trade. It changes the format of education, health care, public administration, human communications and sets a qualitatively new stage in the development of the state, its economy and society as a whole [10].

2. The digitization of logistics as a single customer-oriented mechanism for the development of the transport industry

Today, the transport and logistics services market already uses technologies that reduce monetary costs and out-of-schedule detention of rolling stock units, optimize business processes, help to automate and mechanize many production processes in the industry. Therefore, of course, the digitization of logistics will give a boost to the creation of the "United Digital Platform of the Russian Transport Complex" [11].

"Digital transport and logistics", in our opinion, implies a single customer-oriented mechanism for the development of transport systems and economy, based on digital interaction, value flows of specific data, necessary to obtain a network effect in solving operational and strategic tasks of the railway transport system. The digitalization of the logistics sector is the main element of competitiveness, without which the railway transport system simply cannot survive in the modern transport market.

According to the results of the market research of freight transport conducted in the OAO Russian Railways, it was revealed that representatives of large cargo shippers note the imperfections of existing electronic platforms. They are:

- providing a railcar fleet, which does not allow cargo shippers to order not only the amount of rolling stock they need, but also the corresponding kind;
- pointing to a number of legal problems in the service of non-public tracks;
- the relevance of the introduction of electronic document management is increasing more and more when passing through customs procedures that take a lot of time;
- expressing dissatisfaction with the procedure of accepting the cargo for transportation and fixing it on rolling stock, which is labor-intensive for them and contains a large amount of complex regulatory documentation.

All of this is only a small list of issues requiring urgent solutions.

The modern evolution of information technology will allow railway transport system to bring the concept of a marketing customer-centric approach into effect. Currently, the main factor in ensuring the competitiveness of the railway transport system emphasizes its position in the transport services market, that is, provides a qualitatively new modern level of services, based on the opinion of consumers of these services.

The main prerequisite for the formation of a system of "digital" transport and logistics in the railway transport system is to implement a single customer-oriented mechanism for the development of the industry's economic systems, based on digital interaction, the value streams of data consumed, necessary to obtain a network-wide effect in solving operational and strategic tasks (Figure 1).

IOP Conf. Series: Materials Science and Engineering 1064 (2021) 012014 doi:10.1088/1757-899X/1064/1/012014

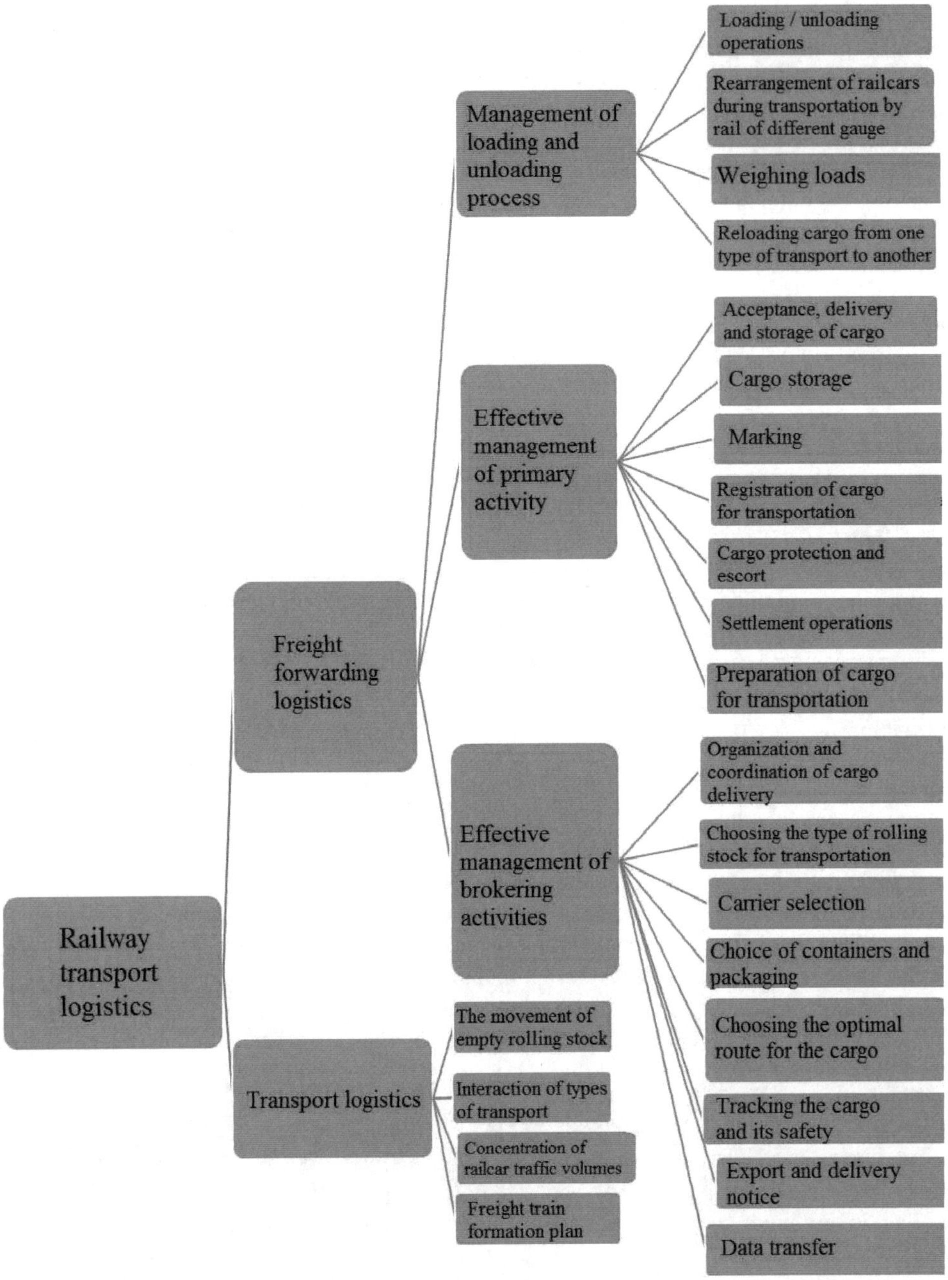

Figure 1. Transport logistics structure of the railway transport system.

There are several levels of "digital" transport and logistics of the railway transport system, and a prerequisite for the operation of this system is the development of a technical and technological framework embracing all transportation vehicles, technical devices, linear communications, automation of a number of operations (Figure 2).

Examining the customer level, it should be noted that the process of interaction of technical and technological elements should clearly fit into the existing economic system of service users, to help meet the basic needs of the client. The presence of a variety of control devices and mechanisms connected by a single global information network will be used in the digital transport and logistics system, which will lead to significant changes in the transport service procedure. Some of the tasks of this interaction are:

- automation of individual operations (loading, unloading, packing, cargo fastening, weighing, cargo release, etc.);
- automation of business processes (storage, transportation);
- interaction of the industry's fixed assets;
- collecting and processing the necessary data;
- providing information services;
- cooperation with already used application software products and systems, etc.

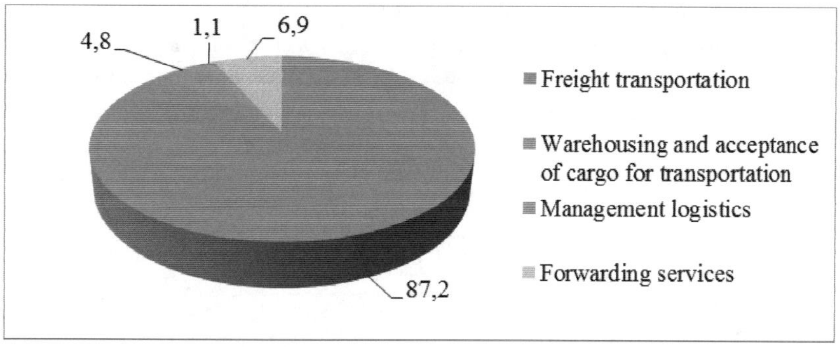

Figure 2. The structure of transport and logistic services of the railway transport system.

Digitalization of transport and logistics in the railway transport system, as one of the main social sectors of the country, the use of modern technical means, programs, allows one to increase the efficiency of the transport complex as a whole and in individual subsystems, as well as to ensure the high quality of services provided.

The benefits of "digital logistics" for transportation process participants should be considered from the perspective of:

1) Transportation process participants:

- ability to obtain cargo transportation services under more favorable conditions;
- reducing the cost of finding better operators;
- reducing the corruption factor;
- understanding the market situation and forecasting.

2) From the perspective of carriers - creation of an orderly and open system of sales of rail services.

3) From the perspective of operators - creation of new mechanisms for reaching potential customers. Improving one's own activities through professionalism and staff development.

4) From the perspective of owners of non-public tracks - creation of a new mechanism for the implementation of their own services.

5) From the perspective of transport market regulators - emergence of levers of influence on the freight market, transparency of transport, preventing the creation of monopolization of market positions leading to speculative price increases.

3. The digitization of the supply chains depending on the type of cargo

In the described conditions, the task of the transport and logistics market operator is to select a given algorithm of action from the library of standardized solutions. And the ability to operate solutions is created by the digitalization of supply chains and depends on the type of cargo, as well as the type of rolling stock for transportation, their adaptation to the parameters of the customer subsystem. All of this will make it possible, based on the study of customer behavior in the past period, to build an accurate forecast model of possible needs for services in the future period [12]. Predictive needs can transform the way transport products are consumed before shared use and can reduce the necessary resources for all involved in the transportation process.

The automated decision-making procedure (developed and tested by the authors) on the optimal placement and fastening of cargo on an open rolling stock introduced into the production process of railway transport system operation, by placing a flat-support cargo in it, will allow operators of the transport and logistics market involved in the preparation of cargo for transportation to make a forecast of the occurrence of risk cases at each stage of the transportation process. The automated procedure will ensure the safety of the structural units and subsystems involved in the organization of the transportation process, as well as the railway transport system itself. This will minimize or eliminate the number of mistakes made by employees in the development and implementation of technical conditions, improve the quality of the industry and its attractiveness, minimize the operating costs of the participants in the transportation process, increase productivity by automating the processes and reducing the time for preparatory and related operations of the transportation cycle [13].

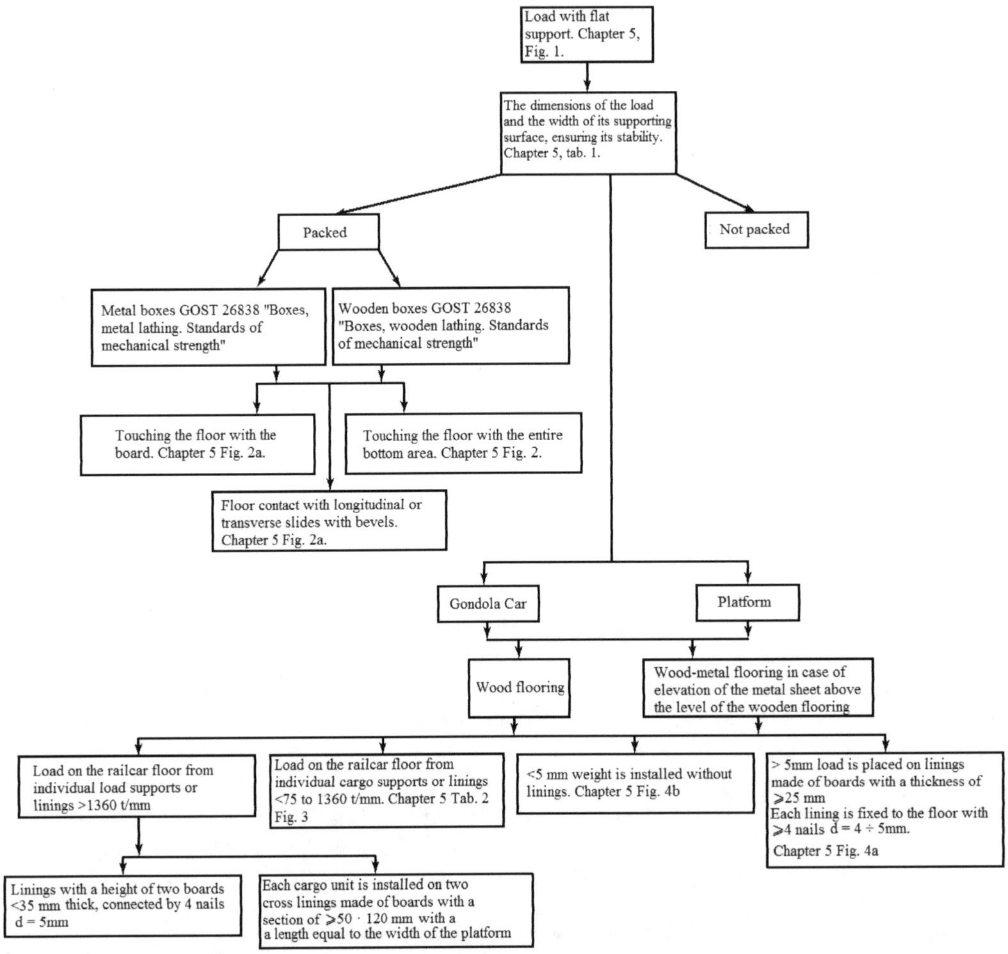

Figure 3. The algorithm for selecting the type of rolling stock for transporting and placing cargo in it (by the example of a flat-support cargo).

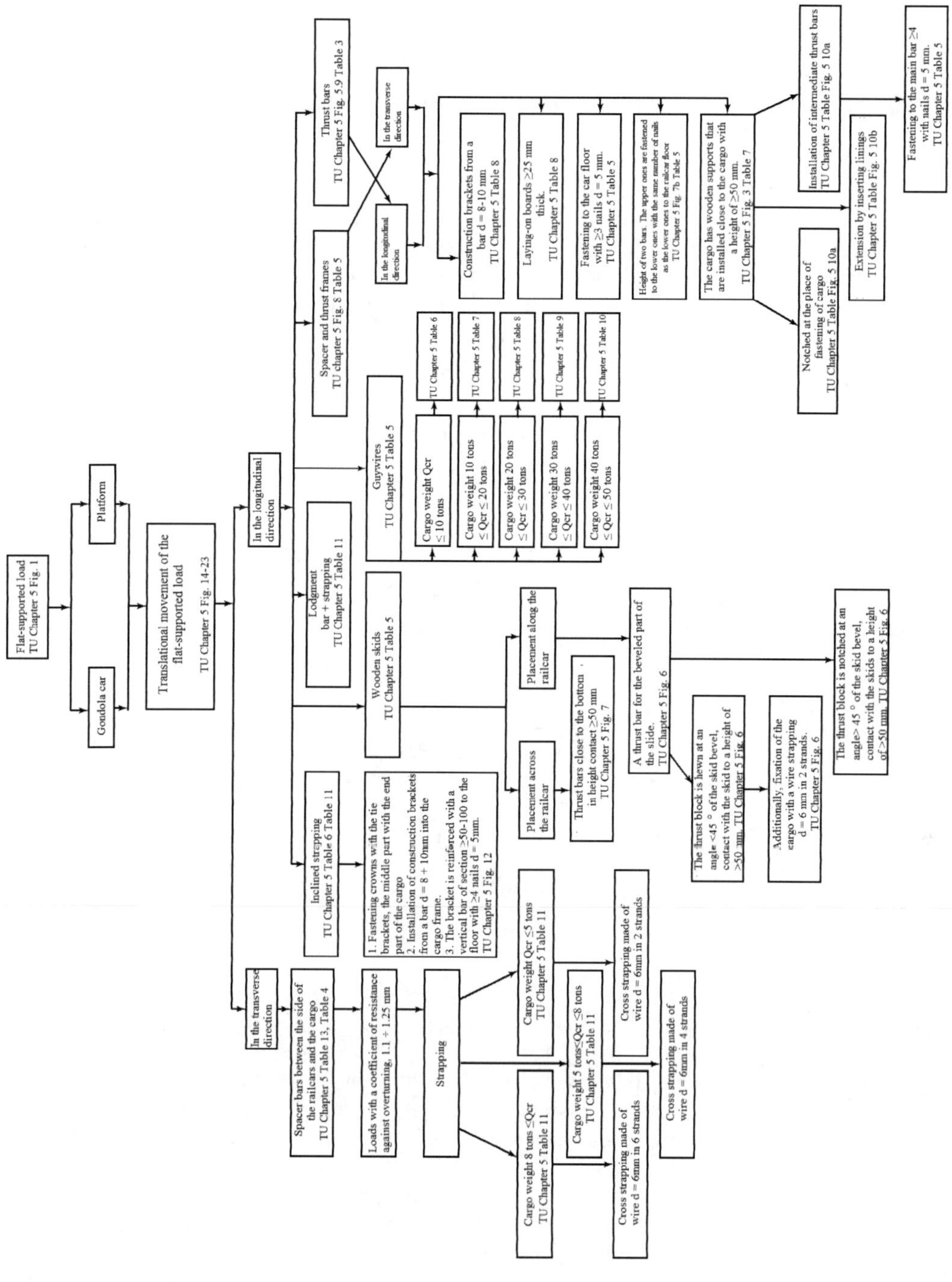

Figure 4. Cargo fastening method algorithm (by the example of a flat-support cargo).

Algorithms for choosing the type of rolling stock for the transportation of cargo of the corresponding category and a typical scheme for placing cargo in it, as well as the method of fastening cargo, have been developed and presented, see Fig. 1 and 2. Since the digitization of supply chains depends more on the type of cargo and the type of rolling stock chosen for transport, the construction of algorithms is based on the example of a cargo with a flat base. The development of an automated procedure is based on the construction of a typical layout and fastening of a specific type of cargo, based on the nomenclature of freight effective in the railway transport system [14].

Most of the existing regulations do not provide for an assessment of the safety of the operation of the railway transport system, including the technical conditions of placement and fastening of cargo in railcars and containers, do not forecast the occurrence of risk cases in the transportation of cargo depending on different natural-climatic and geographical conditions. In our opinion, in the development of an automated procedure for preparing the technical conditions for the placement and fastening of cargo in railcars and containers, it is necessary to take into account the parameters of the assessment of the train safety level by the criterion of the cargo shift across the railcar.

The results of the developed automated procedure allow one to determine the amount of weight and tension in the fastenings from the impact of spatial and, as a case, flat force system, depending on the variation in the longitudinal dynamics ratio of the railcar k_{dx} = 1.2 (k_{dy} = 0.46 and k_{dz} = 0.66) are presented in Table 1.

Table 1. Generalizing computation results.

Railcar dynamics ratio	Resulting force, kN	The equivalent of stiffness, KN/m	Cargo shift, mm	Tension in the fastenings, KN				
k_{dx}, k_{dy}, k_{dz}	ΔF $\frac{\Delta F_x}{\Delta F_y}$	c_{equ} $\frac{c_{eq\,x}}{c_{eq\,y}}$	Δs $\frac{\Delta x}{\Delta y}$ $\frac{\Delta x_1}{\Delta y_1}$	R_1 $\frac{R_{np1}}{R_{n1}}$	R_2 $\frac{R_{np2}}{R_{n2}}$	R_3 $\frac{R_{np3}}{R_{n3}}$	R_4 $\frac{R_{np4}}{R_{n4}}$	R_5 $\frac{R_{np5}}{R_{n5}}$
1.2 0.46 0.66	270.6 $\frac{266.1}{49.2}$	2365 $\frac{2266}{1327}$	114 $\frac{113}{21}$ $\frac{117}{37}$	12.36 $\frac{12.7}{7.0}$	16.34 $\frac{16.6}{6.96}$	19.0 $\frac{19.3}{6.92}$	33.7 $\frac{34.2}{6.5}$	73.2 $\frac{74.3}{10.5}$

Based on the data obtained, the graphical dependence $R_i = f(l_i)$ is constructed, which is presented in Figure 5.

Analysis of the data of Figure 5 shows that in the given initial data, if the length of the fastening l_i exceeds 2.284 meters, then there is a possibility of rupture of this fastening and shift of the cargo relative to the axis of the floor of the freight car.

The advantages of automating this process are the reduction of the preparation time of technical conditions for the placement and fastening of the cargo, the elimination of errors or human interference, the improvement of the cost-effectiveness of work and, as a result, increased safety of the operation of the railway transport system.

The disadvantages associated with the automation of this process include software errors, errors of workers performing the placement and fastening of the cargo directly on the freight car, ignoring the requirements of the Technical conditions by the operating workers.

Evaluating the statistics on train safety and shunting operations requires identifying the main problems of the industry and developing directions to stabilize them. Among them, in our opinion, significant is the automation of the procedure of receiving goods for transportation, because the safety of the operation of the railway transport system begins with a high-quality operation to receive goods for transportation, as well as control of its implementation by the responsible persons of the freight sector.

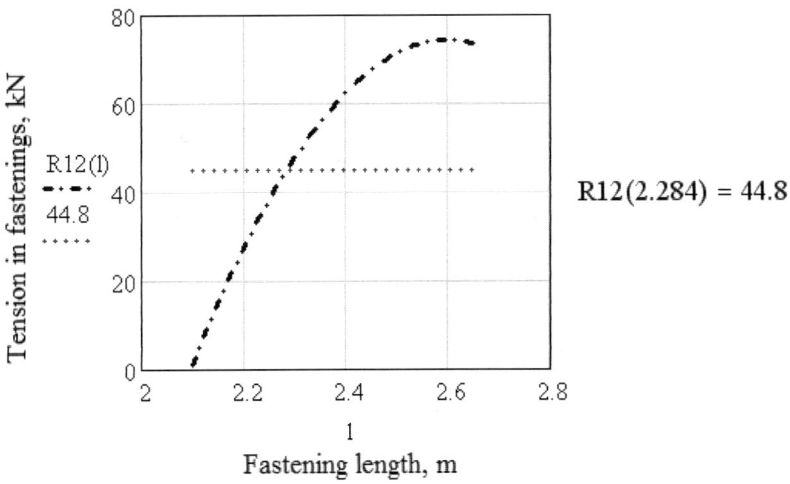

Figure 5. Graphical dependence $R_i = f(l_i)$ at $k_{dx} = 1.2$.

In order to ensure the protected state of the railway transport system, it is necessary to minimize the impact of all factors, which implies the following:

- presence of solid knowledge of the employees involved;
- unconditional compliance with the standards of maintenance of technical means in good condition;
- strict maintenance of the order and rules of all parts of the transportation process;
- regular changes to existing specifications and normative standards, requirements, procedures and rules of operation;
- introducing better technology;
- other conditions requiring revision and adjustment of the existing specifications and regulations of the freight sector.

4. Conclusion

The developed and presented automated procedure provides the operator of the transport and logistic market with the basis of reference and regulatory information on the used freight cars of the common fleet of the Russian Railways, allowable loads on the body elements of the railcars and elements of fastening of goods installed in accordance with the effective technical conditions, the necessary materials and their required quantity used for fastening the cargo.

Exploring the customer railway transport system, we note that the above increase in the level of digital service will also represent the need to simplify user access to services, visualize user's needs, train users, provide customer support, develop mobile customer services, move from a fixed service model to a "service on demand" and improve the operation quality of systems. This is due to the fact that a new user mentality, formed in the digital age, is becoming economically active [15-17].

In turn, the task of creating a sufficient environment for the efficient functioning of "digital" transport and logistics is to create for users the necessary conditions and opportunities to ensure interaction between the customer and the operator of logistic services. Such work is premised on the interaction of platforms based on the existing transport-territorial framework, as well as taking into account the topographic features of a specific area and region. Thus, the use of "digital" transport and logistics in the railway transport system will help to increase the speed of transportation of goods, improve the level of customer service, the quality of life of the country's population, reduce the transport components in the structure of goods and services produced by the country's national economy.

References

[1] Konstantinova M V, Balanovskiy A E, Gozbenko V E, Kargapoltsev S K, Karlina A I, Shtayger M G, Guseva E A and Kuznetsov B O 2019 Application of plasma surface quenching to reduce rail side wear *IOP Conf. Ser.: Mater. Sci. Eng.* **560** 012146

[2] Guseva E A, Kargapoltsev S K, Balanovskiy A E, Karlina A I, Shtayger M G, Gozbenko V E, Konstantinova M V and Sivtsov A V 2019 Comparative evaluation of corrosion resistance of wheel and rail steels in various media *IOP Conf. Ser.: Mater. Sci. Eng.* **560** 012181

[3] Shtayger M G, Balanovskiy A E, Kargapoltsev S K, Gozbenko V E, Karlina A I, Karlina Yu I, Govorkov A S and Kuznetsov B O 2019 Investigation of macro and micro structures of compounds of high-strength rails implemented by contact butt welding using burning-off *IOP Conf. Ser.: Mater. Sci. Eng.* **560** 012190

[4] Kolosov A D, Gozbenko V E, Shtayger M G, Kargapoltsev S K, Balanovskiy A E, Karlina A I, Sivtsov A V and Nebogin S A 2019 Comparative evaluation of austenite grain in high-strength rail steel during welding, thermal processing and plasma surface hardening *IOP Conf. Ser.: Mater. Sci. Eng.* **560** 012185

[5] Balanovskiy A E, Shtayger M G, Karlina A I, Kargapoltsev S K, Gozbenko V E, Karlina Yu I, Govorkov A S and Kuznetsov B O 2019 Surface hardening of structural steel by cathode spot of welding arc *IOP Conf. Ser.: Mater. Sci. Eng.* **560** 012138

[6] Voropai N, Ukolova E, Gerasimov D, Suslov K, Lombardi P and Komarnicki P A 2019 Study on Cost-Effectiveness of Energy Supply Based on the Energy Hub Concept *Proc. of 2019 IEEE PES Innovative Smart Grid Technologies Europe, ISGT-Europe 2019* 8905736

[7] Buryanina N, Korolyuk Y, Koryakina M, Lesnykh E and Suslov K 2019 Algoritm of Current Protection Based on Three Instantaneous-Value Samples *Proc. of 2019 IEEE PES Innovative Smart Grid Technologies Europe, ISGT-Europe* 8905542

[8] Suslov K, Shushpanov I, Buryanina N and Ilyushin P 2020 Flexible power distribution networks: New opportunities and applications *SMARTGREENS 2020 – Proc. of the 9th Int. Conf. on Smart Cities and Green ICT Systems*

[9] Suslov K V, Solonina N N and Smirnov A S 2014 Distributed power quality monitoring *Proc. of Int. Conf. on Harmonics and Quality of Power, ICHQP*

[10] Munkhtuvshin D, Balabanov V B, Putsenko

[11] Olentsevich V A and Gozbenko V E 2013 Analysis of the reasons for the disruption of the safety of the railway transport system *Modern technologies. System Analysis. Modeling. Irkutsk* **1(37)** 180-183

[12] Retrieved from: http://www.rzd.ru

[13] 2015 *Technical conditions for storing and securing cargo in railcars and containers* (Moscow: Yurtrans) p 544

[14] Balalaev S V 2008 *Traffic safety on railways: a textbook. Part 1: Safety Fundamentals* (Khabarovsk: DVGUPS) p 125

[15] Ivankov A N and Ivankova L N 1997 *Design of a new marshalling yard with an automated slide: a resource book* (Irkutsk: IRGUPS) 50 p

[16] Konyuhov V Yu, Konstantinova M V and Gladkih A M 2019 Determination of restored units spectrum of equipment and development of the assembly unit repair method at industrial enterprises *J. of Phys.: Conf. Ser.* **1353** 012047

[17] Gozbenko V E, Kargapoltsev S K and Olentsevich V A 2017 Automated system of placing and fastening cargo on the open rolling stock of railway transport *Herald of Irkutsk State Technical University* **21(4(123))** 157-165

Identification of generalized Cobb-Douglass production functions with multiplicative errors in variables

I L Sandler[1], D V Ivanov[2], M Yu Lifshits[1] and A N Diligenskaya[1]

[1]Samara State Technical University, 244, Molodogvardejskaya St., Samara, 443100, Russia
[2]Samara State University of Economics, 141, Sovetskoi Armii St., Samara, 443090, Russia

E-mail: dvi85@list.ru

Abstract. The article discusses the identification of a class of generalized Cobb-Douglas production functions with multiplicative errors in all variables. The article proposes a generalization of Cobb-Douglas production functions in the presence of memory for input and output variables. Parametrization of noise in the form of multiplicative noise in all observable variables is proposed. The logarithmic transformation of such production functions leads to the need to solve the problem of estimating the parameters of a linear difference equation in the presence of additive errors in all variables. To identify the parameters of production functions, a modification of the method of total least squares was used. Computational experiments have shown high accuracy of parameter estimation based on the proposed algorithm

1. Introduction

Currently, production is being actively developed methods of modeling socio-economic systems. Production functions are the main tools used to model socio-economic systems. The parameters of production functions are not known in advance, so they have to be estimated from statistical data. Socio-economic models usually contain a stochastic component that makes it difficult to estimate the parameters of such models.

The Cobb-Douglas function and their modifications have found the widest application among production functions [1]. There are known methods for estimating the parameters of Cobb-Douglas production functions in the presence of multiplicative and additive noise [2-4].

In [5] nonparametric methods of identification of dynamic production functions are considered. In the article [5], the production function is described by a non-linear homoscedastic ARX (autoregressive with exogenous input) process.

This article proposes a Cobb-Douglas function with memory. It should be noted that the authors are not aware of any work on the estimation of production functions in which the stochastic component is interpreted as an error in variables. This article discusses the estimation of the parameters of a production function in the presence of multiplicative noise in variables.

2. Problem statement

Economic and mathematical modeling widely uses the method of constructing mathematical models in the form of a non-uniform production function (PF) of Cobb-Douglas:

$$Y(\mathrm{t}) = \mathrm{A} \cdot \mathrm{K}(\mathrm{t})^{\alpha} L(\mathrm{t})^{\beta} . \quad (1)$$

The production function of Cobb-Douglas can take into account the influence of scientific and technological progress:

$$Y(\mathrm{t}) = \mathrm{A} \cdot \mathrm{K}(\mathrm{t})^{\alpha} L(\mathrm{t})^{\beta} e^{\tau t} , \quad (2)$$

where Y is the release of the final product; $\mathrm{K(t)}$ - capital resources, $L(\mathrm{t})$ - labor resources, α, β - characteristic of resource use efficiency - elasticity indicator, A - scale factor of transformation technology, τ - factor of influence of scientific and technological progress.

The Cobb-Douglas production function assumes that the output of the final product does not depend on the previous values of capital and labor resources. This article proposes to use the generalized Cobb-Douglas function using past values of capital and labor resources. Let us write the equation for the dynamic Cobb-Douglas function:

$$Y(t)\prod_{m=1}^{r_2} Y(t-m\cdot\Delta t)^{\gamma_m} = \mathrm{A}\cdot\prod_{m=0}^{r} \mathrm{K}(\mathrm{t}-m\cdot\Delta \mathrm{t})^{\alpha_m} \prod_{m=0}^{r_1} L(\mathrm{t}-m\cdot\Delta \mathrm{t})^{\beta_m} e^{\tau t}. \quad (3)$$

For discrete values, function (3) will take the form:

$$Y_i \prod_{m=1}^{r_2} \left(Y_{i-m}\right)^{\gamma_m} = \mathrm{A}\cdot\prod_{m=0}^{r} \mathrm{K}_{i-m}^{\alpha_m} \prod_{m=0}^{r_1} L_{i-m}^{\beta_m} e^{\tau i}. \quad (4)$$

The values of capital and labor resources are usually known with errors; we will assume that measurements with multiplicative noise are available:

$$\tilde{Y}_i = Y_i \cdot \exp\left(\xi_i^{(y)}\right),\ \tilde{K}_i = K_i \cdot \exp\left(\xi_i^{(K)}\right),\ \tilde{L}_i = L_i \cdot \exp\left(\xi_i^{(L)}\right),$$

where sequences of independent random variables $\{\tilde{Y}_i\}, \{\tilde{K}_i\}, \{\tilde{L}_i\}$ with

$$\mathrm{E}\left(\xi_i^{(y)}\right) = 0,\ \mathrm{E}\left(\xi_i^{(K)}\right) = 0,\ \mathrm{E}\left(\xi_i^{(L)}\right) = 0,$$

$$\mathrm{E}\left(\left(\xi_i^{(Y)}\right)^2\right) = \sigma_Y^2,\ \mathrm{E}\left(\left(\xi_i^{(K)}\right)^2\right) = \sigma_K^2,\ \mathrm{E}\left(\xi_i^{(L)}\right) = \sigma_L^2.$$

Let us logarithm the equation (4):

$$\ln\left(\tilde{Y}_i\right) + \sum_{m=1}^{r_2} \gamma_m \ln\left(\tilde{Y}_{i-m}\right) = \ln(\mathrm{A}) + \sum_{m=0}^{r} \alpha_m \ln\left(\tilde{\mathrm{K}}_{i-m}\right) + \sum_{m=0}^{r_1} \beta_m \ln\left(\tilde{L}_{i-m}\right) + \tau i. \quad (5)$$

Let us write the equation for the output of the final product at a given time:

$$\ln\left(\tilde{Y}_i\right) = -\sum_{m=1}^{r_2} \gamma_m \ln\left(\tilde{Y}_{i-m}\right) + \ln(\mathrm{A}) + \sum_{m=0}^{r} \alpha_m \ln\left(\tilde{\mathrm{K}}_{i-m}\right) + \sum_{m=0}^{r_1} \beta_m \ln\left(\tilde{L}_{i-m}\right) + \tau i. \quad (6)$$

Applying the property of the logarithm to observations with noise, we get:

$$\ln\left(\tilde{Y}_i\right) = \ln\left(Y_i\right) + \xi_i^{(Y)},\ \ln\left(\tilde{K}_i\right) = \ln\left(K_i\right) + \xi_i^{(k)},\ \ln\left(\tilde{L}_i\right) = \ln\left(L_i\right) + \xi_i^{(L)}. \quad (7)$$

It is required to estimate the parameters of equation (6) using noisy sequences of observations $\{\tilde{Y}_i\}$, $\{\tilde{K}_i\}, \{\tilde{L}_i\}$.

3. Identification algorithm based on total least square

We will define estimates in the form of minimizing the criterion:

$$\min_{\theta} \frac{\sum_{i=1}^{N}\left(\ln\left(Y_i\right)-\varphi_i^T\theta\right)^2}{1+\theta^T D\theta}, \quad (8)$$

where $\theta=\left(\gamma_1 \ \dots \ \gamma_{r_2} \mid \alpha_1 \ \dots \ \alpha_r \mid \beta_1 \ \dots \ \beta_{r_1} \mid \tau \ \ln(A)\right)^T$,

$\varphi_i^T=\left(\ln\left(\tilde{Y}_i\right) \ \dots \ \ln\left(\tilde{Y}_{i-r_2}\right) \mid \ln\left(\tilde{K}_i\right) \ \dots \ \ln\left(\tilde{K}_{i-r}\right) \mid \ln\left(\tilde{L}_i\right) \ \dots \ \ln\left(\tilde{L}_{i-r_2}\right) \mid i \ \ 1\right)^T$,

$$D=\begin{pmatrix} I_{r_2} & 0 & 0 & 0 \\ 0 & I_r & 0 & 0 \\ 0 & 0 & I_{r_1} & 0 \\ 0 & 0 & 0 & 0_2 \end{pmatrix}.$$

The minimum criterion (8) can be found as a solution to the biased normal system:

$$\hat{\theta}=\left(\Phi^T\Phi-\sigma^2 I\right)^{-1}\Phi^T\tilde{Y}, \quad (9)$$

where

$$\Phi=\begin{pmatrix} \ln\left(\tilde{Y}_1\right) & \dots & 0 & \ln\left(\tilde{K}_1\right) & \dots & 0 & \ln\left(\tilde{Y}_1\right) & \dots & 0 & 1 & 1 \\ \ln\left(\tilde{Y}_2\right) & \dots & 0 & \ln\left(\tilde{K}_2\right) & \dots & 0 & \ln\left(\tilde{Y}_1\right) & \dots & 0 & 2 & 1 \\ \vdots & \ddots & \vdots & \vdots & \ddots & \vdots & \vdots & \ddots & \vdots & \vdots & \vdots \\ \ln\left(\tilde{Y}_N\right) & \dots & \ln\left(\tilde{Y}_{N-r_2}\right) & \ln\left(\tilde{K}_N\right) & \dots & \ln\left(\tilde{K}_{N-r}\right) & \ln\left(\tilde{L}_N\right) & \dots & \ln\left(\tilde{L}_{N-r_1}\right) & N & 1 \end{pmatrix},$$

$Y=\left(\ln\left(\tilde{Y}_{i-1}\right) \ \dots \ \ln\left(\tilde{Y}_{i-N}\right)\right)^T$,

$$\sigma^2=\frac{1}{\lambda_{\max}\left(\left(\bar{\Phi}^T\bar{\Phi}\right)^{-1}\bar{D}\right)},$$

$\lambda_{\max}\left(\left(\bar{\Phi}^T\bar{\Phi}\right)^{-1}\bar{D}\right)$ is maximum eigenvalue of the matrix $\left(\bar{\Phi}^T\bar{\Phi}\right)^{-1}\bar{D}$.

$\bar{\Phi}=\left(Y,\bar{\Phi}\right)$,

$$\bar{D}=\begin{pmatrix} 1 & 0 \\ 0 & D \end{pmatrix}.$$

4. Test Example

The generalized Cobb-Douglas production function is described by the equation:

$$\tilde{Y}_i\left(\tilde{Y}_{i-1}\right)^{-0.9}=e\cdot \tilde{K}_i^{0.3}\tilde{K}_{i-1}^{0.7}\tilde{L}_i^{0.4}\tilde{L}_{i-1}^{0.5}e^{0.1i}. \quad (10)$$

After the logarithmic transformation, the generalized Cobb-Douglas production function (10) is described by the equation:

$$\ln\left(\tilde{Y}_i\right)=1+0.9\ln\left(\tilde{Y}_{i-m}\right)+0.3\ln\left(\tilde{K}_i\right)+0.7\ln\left(\tilde{K}_{i-1}\right)+0.4\ln\left(\tilde{L}_i\right)+0.5\ln\left(\tilde{L}_{i-1}\right)+0.1i. \quad (11)$$

The vector of true parameters is $\theta=\left(0.9 \quad 0.3 \quad 0.7 \quad 0.4 \quad 0.5 \quad 0.1 \quad 1\right)^T$.

Sequences $\left\{K_i\right\}$, $\left\{L_i\right\}$ are white noise. The number of observations is $N=48$.

Signal-to-Noise Ratio for function (11) is defined by:

$$SNR_Y = \frac{\sigma_\xi^{(Y)}}{\sigma_Y} = 0.02,\ SNR_K = \frac{\sigma_\xi^{(K)}}{\sigma_K} = 0.02,\ SNR_L = \frac{\sigma_\xi^{(L)}}{\sigma_L} = 0.02.$$

The algorithms were compared by root-mean-square error (RMSE) of the coefficient estimation vector:

$$\delta\theta = \sqrt{\frac{\left\|\hat{\theta} - \theta\right\|^2}{\left\|\theta\right\|^2}} \cdot 100\%.$$

Table 1 shows errors in estimating the parameters according to least-square (LS) estimate and the estimate based on total least square (TLS) .

Table 1. The results of the identification of the model (11).

True Parameters	LS	TLS
$\gamma_1 = 0.9$	0.8816	0.8883
$\alpha_1 = 0.4$	0.3624	0.3532
$\alpha_2 = 0.5$	0.7677	0.7573
$\beta_1 = 0.3$	0.3442	0.3457
$\beta_2 = 0.7$	0.5547	0.5529
$\tau = 0.1$	0.1390	0.1238
$\delta\theta$	0.1009	0.0675

5. Conclusion

The results of the test example show that the proposed modification of total least square based for estimation production function parametrs can significantly improve the accuracy of estimating system parameters.

The direction of further research is the development of a recursive version of the algorithm for estimating the parameters of generalized Cobb-Douglas production functions [7].

When estimating parameters based on real data, the condition of interference independence is often violated. Methods based on instrumental variables for models with errors in variables can be developed to estimate interference with autocorrelated interference [8].

References

[1] Cobb C W and Douglas P H 1928 A theory of production *American Econ. Rev.* **8** 139–1965

[2] Felipe J and Adams F G 2005 The estimation of the Cobb-Douglas function: A retrospective review *Eastern Econ. J.* **31** 427–445

[3] Stephen M G and Richard E Q 1970 The Estimation of Cobb-Douglas Type Functions with Multiplicative and Additive *Int. Economic Review* **11(2)** 251-257

[4] Kelejian H H 1972 The Estimation of Cobb-Douglas Type Functions with Multiplicative and Additive Errors: A Further Analysis *Int. Economic Review* **13** 179-182

[5] Hossain M M, Majumder A I and Basak T 2012 An Application of Non-Linear Cobb-Douglas Production Function to Selected Manufacturing Industries in Bangladesh *Open J. of Statistics* **2** 460-468. DOI: http://dx.doi.org/10.4236/ojs.2012.24058

[6] Koshkin G and Kitayeva A 2011 Nonparametric Identification of Static and Dynamic Production Functions *Int. J. of Applied Mathematics* 276-280

[7] Ivanov D V, Sandler I L and Kozlov E V 2018 Identification of fractional linear dynamical systems with autocorrelated errors in variables by generalized instrumental variables *IFAC PapersOnLine* **51(32)** 580-584. DOI: https://doi.org/10.1016/j.ifacol.2018.11.485

[8] Ivanov D V and Katsyuba O A 2010 Recursive identification dynamic systems with errors-in-variables using nonlinear least square *Proc. of the IASTED Int. Conf. on Automation, Control, and Information Technology-Control, Diagnostics, and Automation (ACITCDA 2010)* pp 267-271

Selection of tools and cutting modes for turning small-sized high-precision parts of micro-wave electronics from beryllium bronze

Yu I Karlina[1], S K Kargapoltsev[2], V E Gozbenko[2, 3], A I Karlina[1] and D S Leonovich[1]

[1] Irkutsk National Research Technical University, 83, Lermontova str., Irkutsk, 664074, Russia
[2] Irkutsk State Transport University, 15, Chernyshevskogo str., Irkutsk, 664074, Russia
[3] Angarsk State Technical University, 60, Chaykovskogo str., Angarsk, 665835, Russia

E-mail: gozbenko@irgups.ru, karlinat@mail.ru

Abstract. Beryllium bronze has the property of plasticity after heat treatment, which is a decisive factor when choosing the material of conductors in unpressurized coaxial connectors. The presence of burrs on these parts according to technical requirements is not allowed. When turning details of beryllium bronze, difficult to remove burrs appear. The aim of the study is to select a tool and cutting modes for turning small-sized high-precision parts of microwave electronics from beryllium bronze, in which there are no burrs or their size and quantity are minimal. In this study, empirical methods are used in the form of experimental selection of cutting modes and tools for the purpose of combating burrs, and analytical methods in the form of analysis of the character of burrs using the provisions of the theory of metal cutting, analysis of factors affecting the formation of burrs during machining surface parts. On CNC lathes CITIZEN CINCOM studied typical of this type of production tools ten leading manufacturers of precision cutting tools for turning. As a result of the study, it was not possible to completely exclude the appearance of burrs during machining. The smallest size and the number of burrs were obtained when machining with carbide tools manufactured by Swiss companies Fraisa, Utilis, IFANGER, Applitec. TiAlN coating is well suited for beryllium bronze treatment. When turning with a new tool at low speeds, the number of burrs decreases, their character changes, they become small drop-shaped. The processing modes recommended by the tool manufacturer differ from those selected experimentally. For further use of the selection results, it is necessary to create a database of regulatory and reference information on the types of processing for specific materials, tools, machine tools according to the accuracy criteria, taking into account the minimization of the number and size of burrs.

1. Introduction

Parts for coaxial radio components, which are external and internal conductors in coaxial transmission lines, are made of certain materials and have strict requirements for surface quality. So, the inner and outer conductors of leaky connectors are made mainly of BrB2 beryllium bronze, since this material has the property of plasticity after heat treatment, which is necessary to obtain a tight connection of parts during assembly. Beryllium bronze case lamellas are easily crimped to connector assemblies. The signal is transmitted in coaxial radio components over the surface of the parts, in order to create a highly conductive surface layer, a galvanic coating of 2-3 "skin layers" or about 2-3 microns is applied to the parts [1]. The coating can be two-layer or even three-layer, depending on the possibility of direct application of the coating material to the material of the part. The highly conductive layer should be uniform, and the surface of the parts should not have defects that would impair the conductivity of the surface layer during assembly. Therefore, according to technical requirements, sinks, burrs, sharp edges, and the presence of contaminants are not allowed on the surface of these.

When choosing a machining strategy, selecting a tool, and choosing machining modes, the most important factor is minimizing the number and size of burrs., Since removing them is a difficult task and requires considerable time and labor. This study is devoted to the selection of tools and cutting modes for the manufacturing of small-sized high-precision parts of electronic equipment made of beryllium bronze and the creation of the efference guide on the recommended processing modes, tools depending on the material and size of details to meet the requirements of minimizing the number and size of burrs.

2. Materials and research methods

When studying the problem of burrs on parts for microwave electronics, processed on CNC machines, in the framework of this study, from the entire nomenclature of parts processed on CNC machines, we selected parts manufactured on lathes of longitudinal turning with CNC CITI-ZEN CINCOM K16E-VII, CITIZEN CINCOM K16-VI, CITIZEN CINCOM R07-VI (body of rotation) from materials beryllium bronze Wire Brb2 GOST 15834-77, Bar Brb2 GOST 15835-2013.

In this study, analytical and experimental research methods are applied. The experimental methods were carried out in the form of full-scale experiments - this is an experimental selection of cutting modes and tools in which no burrs appear, or the number of burrs and their size are minimal. Analytical studies were performed in the form of analysis of the character of burrs using the provisions of the theory of metal cutting, analysis of factors affecting the formation of burrs during machining of the surface of parts.

The selection of cutting modes and tools in which the number of burrs was minimally carried out on cutting tools from Fraisa (Switzerland), IFANGER (Switzerland), Utilis (Switzer-land), Applitec (Switzerland), Seco (Sweden), Sandvik (Sweden), Schunk (Germany), Guhring (Germany), Mitsubishi (Japan), WhizCut (Taiwan), Sharp Tool (China).

The list of tools used in machining on longitudinal turning machines for the manufacture of parts having hard-to-reach internal surfaces of the type Housing made of BrB2 beryllium bronze material for which processing modes were selected:

1. Cutting plate 731R (L) -1.5 H1.5 (right and left direction).
2. Plate for cross turning 744x-1.5-R.
3. Frontal passage plate CCGT 09T301FN-F23.
4. Plate for reverse turning Back turn 743x-1.2.
5. Cutter boring frontal SDG 435 192-R Ø1,92.
6. Grooving boring tool SDS 440 292-R Ø2, 92.
7. Drill centering DRP 442 090 R-C Ø4.
8. Twist drill B570140220 Ø2,2.
9. Mill disk cutting 1101-0.35 40z (80z).
10. End mill E124F-Dc5 0 Ø5.
11. End mill E102F-Dc1 0 Ø1.
12. End mill E102F-Dc1 5 Ø1.5.
13. Twist drill B570140160 Ø1.6.
14. Twist drill 2023-1.05 Ø1,05.

Quality control of parts was carried out under the Instrument Microscope IMC-150x50 and the Microscope stereoscopic pancratic Altami SM0745, and using a measuring tool:

1. Micrometer MK25-1 GOST 6507-90.

2. The indicator ICh-10 GOST 577-68.
3. A support for the PA9370570 indicator.
4. Tip indicator PA9372294.
5. Gauge plug Ø2.67 H10 (+0.04) RA9600745.
6. Gauge plug Ø2.52 N9 (+0.025) RA9600745.
7. Gauge plug Ø1.7 (+ 0.02 / + 0.07) RA9600745.
8. Gauge plug Ø2.25 ((+ 0.02 / + 0.07) RA9600745.
9. Gauge plug Ø1.6 (+ 0.02 / + 0.07) RA9600745.
10. Gauge plug Ø1.1 (+ 0.02 / + 0.07) RA9600745.
11. Gauge plug Ø1.25 (+ 0.02 / + 0.07) RA9600745.
12. Gauge plug Ø2.8 (+ 0.02 / + 0.04) RA9600745.
13. Gauge plug Ø2.37 H9 RA9600745.
14. Caliber bracket Ø2.9 h9 (-0.025) RA9600751.
15. Caliber bracket Ø0.35 (+0.06) RA9600750.
16. Watch GOST 23350-98.
17. Samples of roughness parameters for ISO TESA turning.
18. Magnifier LPZ-6x TUZ-3 1809-84.

An analysis of the nature and size of the burrs was carried out under the scanning electron microscope SEM Multi Beam System JEOL JIB-Z4500.

3. Research results and discussion

Beryllium bronze BrB2 according to GOST 18175-78 has characteristic properties - high strength and wear resistance, high spring properties, good anti-friction properties, average electrical conductivity and thermal conductivity, very good deformability in the hardened state, and is intended for the manufacture of springs, spring parts for critical purposes, wear-resistant parts of all kinds, non-sparking tools [2]. It is the property of very good deformability in the hardened state that is decisive when choosing a material for the manufacture of external conductors, since during assembly it is necessary to ensure tight compression of the lamellas of the case.

Thermally hardened BrB2 copper-beryllium alloy becomes more durable, elastic and ductile. Initially, it is brought into a soft state by heating to 760-780 ° C, and then subjected to aging in water at a temperature of 310-330 ° C for 3 hours. Upon heating and subsequent cooling of the alloy to room temperature, beryllium dissolves in copper to form a saturated solid solution. Subsequent hardening leads to its deposition, as a result of which BrB2 bronze acquires high hard-ness up to 350 - 400 HB.

Due to their high strength and hardness, BrB2 beryllium bronze is a hard-to-work material: when cutting along the entire surface, there are solid burrs that are firmly mated to the surface, and when removed by a scraper, there is a high probability of gouging, and other mechanical removal methods are also more complicated and take longer. This leads to an increase in the complexity and production time of parts. In this regard, the selection of tools and cutting modes in order to minimize the size and number of burrs is relevant.

CAM-system CAMWorks allows you to produce:

- selection of processing strategies and setting parameters of selected strategies;
- selection of machined and bounding surfaces on the model of the product (details) [3];
- formation of the tool path taking into account machining strategies, selected surfaces, cut-ting tool, automatic control of notches and optimization of the tool path taking into account the current state of the workpiece;
- automatic tracking of changes made to the model of the product (details) [3];
- the use of standard technological templates for processing and forming the tool path with a partial change of parameters to obtain similar parts [3-6].

So, at the stage of creating a control program for processing a part on CNC longitudinal turning machines, the problem arises of the need to select the types and parameters of processing depending on the material being processed, the tool used, the required surface quality, of the volume of material being removed. When choosing a sequence and the number of passes, choosing the selection of cutting modes, such as feed, the number of spindle revolutions, depth of cut must be considered:

- shavings should not be clogged into the tool;
- it is necessary as quickly as possible to ensure exit to the specified design dimensions and to control the size of the allowances obtained at each stage of processing;
- the quality of the treated surface must meet the requirements of the drawing;
- it is necessary to ensure the planned wear resistance of the tool.

If chips quickly become clogged, then it is necessary to reduce the feed, or increase the number of passes, or use intermittent turning. The new CITIZEN machines have a wave turning function, due to which chips are less clogged and less burrs occur.

The surface roughness is provided by the selection of cutting speed and allowances. At low speeds and small allowances, the surface quality is higher, the number and size of burrs are reduced. However, at low loads, swift and crushing occur, which negatively affects the dimensional stability and wear resistance of the tool.

Tool manufacturers attach tables to it with recommended ranges or specific values of processing modes for various materials on different equipment, however they are focused only on ensuring tool wear resistance and the ranges of recommended modes can be quite wide, and the selection of real processing modes for a particular part is multifactorial, requires the establishment of specific parameter values that differ from those recommended. In this regard, the information from these tables is not applicable in practice and is refined by conducting field experiments. Figure 1 shows a fragment of the Cutting specification table from the Utilis Multidec Swiss Type Tools catalog with recommended processing parameters for various materials, in particular for bronze when using a tool with UHM 20 HX cutting material. The recommended cutting speed range Vc is from 5000 to 18000 mm / s. Figure 2 shows that for machining bronze with a tool for drilling and longitudinal turning SDG in diameter of 2.9 mm, the recommended feed values f is in the range from 0.015 mm / rev to 0.035 mm / rev. It has been experimentally established that when setting the cutting speed Vc = 904 mm / s, feed f = 0.012 mm / rev, using the material of the cutting part UHM 20 HX of the tool for drilling and longitudinal turning SDG, the requirements or stability of dimension and ensuring tool wear resistance, and surface quality requirements for subsequent processing are fulfilled. As you can see, the feed value is below the minimum recommended, and the cutting speed is in the middle of the recommended range.

	Rostfreier Stahl Acier inoxydable Stainless steel			Rostfreier Stahl Acier inoxydable Stainless steel			Aluminium Aluminium Aluminium			Messing Laiton Brass		
Härte (HB) Dureté (HB) Hardness value (HB)	180–220			220–330			60–130			–		
Kategorie Catégrie Category	V			VI			VII			VIII		
Bearbeitung Usinage Machining method	▼	▼▼	▼▼▼	▼	▼▼	▼▼▼	▼	▼▼	▼▼▼	▼	▼▼	▼▼▼
Schnittgeschwindigkeiten Vitesses de coupe Cutting speeds	V_c (m/min)											
Schneidstoff Hartmetall Matériaux de coupe carbure Cutting material carbide												
UHM 20	–	–	20–80	–	–	20–60	–	–	50–220	–	–	30–110
UHM 20 HX	–	–	30–120	–	–	30–100	–	–	60–350	–	–	50–180
UHM 20 SX	–	–	30–100	–	–	30–80	–	–	60–300	–	–	50–150

Figure 1. Recommended processing parameters for various materials from the Utilis Multidec Swiss Type Tools catalog, Cutting specification table [7].

FEED AND DEPTHS OF CUT MULTIDEC®-BORE MICRO

SDG – SXG – SDH – SDI – SXI – SDY – SDZ

	Stahl unlegiert Acier non allié Steel unalloyed		Stahl niedriglegiert Acier faibl. allié Steel low alloyed		Stahl hochlegiert Acier fortem. allié Steel high alloyed		Rostfreier Stahl Acier inoxydable Stainless steel		Titan Titane Titanium		Aluminium Aluminium Aluminium		Messing Laiton Brass	
D (mm)	f (mm)	a_p (mm)	f (mm)	a_p (mm)	f (mm)	a_p (mm)	f (mm)	a_p (mm)	f (mm)	a_p (mm)	f (mm)	a_p (mm)	f (mm)	a_p (mm)
≤1	0.01–0.02	0.1–0.2	0.01–0.017	0.1–0.17	0.007–0.017	0.07–0.17	0.007–0.017	0.07–0.17	0.006–0.02	0.06–0.2	0.01–0.025	0.1–0.25	0.01–0.025	0.1–0.25
2	0.012–0.022	0.12–0.22	0.012–0.02	0.12–0.2	0.008–0.018	0.08–0.18	0.008–0.018	0.08–0.18	0.008–0.02	0.08–0.2	0.015–0.03	0.15–0.3	0.015–0.03	0.15–0.3
3	0.015–0.025	0.15–0.25	0.014–0.024	0.14–0.24	0.009–0.019	0.09–0.19	0.009–0.019	0.09–0.19	0.01–0.02	0.1–0.2	0.015–0.035	0.15–0.35	0.015–0.035	0.15–0.35
4	0.015–0.027	0.15–0.27	0.015–0.025	0.15–0.25	0.01–0.02	0.1–0.2	0.01–0.02	0.1–0.2	0.01–0.02	0.1–0.2	0.015–0.035	0.15–0.35	0.015–0.035	0.15–0.35
6	0.015–0.03	0.15–0.3	0.015–0.025	0.15–0.25	0.01–0.02	0.1–0.2	0.01–0.02	0.1–0.2	0.01–0.025	0.1–0.25	0.015–0.04	0.15–0.4	0.015–0.04	0.15–0.4
8	0.015–0.03	0.15–0.3	0.015–0.025	0.15–0.25	0.01–0.02	0.1–0.2	0.01–0.02	0.1–0.2	0.01–0.025	0.1–0.25	0.015–0.05	0.15–0.5	0.015–0.04	0.15–0.4

Figure 2. Recommended processing parameters for various materials from Utilis Multidec Swiss Type Tools catalog, Feed and depth of cut table for SDG tools for drilling and longitudinal turning [7].

Figure 3 shows that for bronze processing with a 2.9 mm diameter SDS tool, the recommended feed values f is in the range from 0.007 mm / rev to 0.020 mm / rev. It was experimentally established that when setting the cutting speed Vc = 5501 mm / s, feed f = 0.010 mm / rev using the material of the cutting part UHM 20 HX of the tool for boring the SDS groove, the requirements for dimensional stability, ensuring tool wear and surface quality requirements are fulfilled. The number, size and size of burrs are minimal. As you can see, the feed and cutting speeds are in the middle of the recommended range.

SDR – SDS

Stahl unlegiert Acier non allié Steel unalloyed	Stahl niedriglegiert Acier faibl. allié Steel low alloyed	Stahl hochlegiert Acier fortem. allié Steel high alloyed	Rostfreier Stahl Acier inoxydable Stainless steel	Titan Titane Titanium	Aluminium Aluminium Aluminium	Messing Laiton Brass
f (mm)	f (mm)	f (mm)	f (mm)	f (mm)	f (mm)	f (mm)
0.007–0.020	0.005–0.015	0.005–0.015	0.005–0.015	0.005–0.015	0.007–0.020	0.007–0.020

Figure 3. Recommended processing parameters for various materials from Utilis Multidec Swiss Type Tools catalog, Feed and depth of cut table for SDS tools [7].

Figure 4 shows that for processing bronze with a B57014 spiral drill with a diameter of 1.6 mm, the recommended feed f is 0.025 mm / rev, the cutting speed Vc is 3000 mm / min, the number of revolutions n is 5970 min-1. And for processing with a drill with a diameter of 2.2 mm, recommended feed f = 0.034 mm / rev with a cutting speed Vc is equal to 4000 mm / min. The number of revolutions n is equal to 5785 min-1. It was established experimentally that when setting the cut-ting speed Vc = 3044 mm / min, feed f = 0.020 mm / rev. The number of revolutions is n = 6000 min-1 for drilling with a drill bit B57014 with a diameter of 1.6 mm. The requirements for dimensional stability, ensuring tool wear and surface quality requirements are fulfilled, the number and the size of burrs are minimal. For drilling with a twist drill B57014 with a diameter of 2.2 mm, feed values f = 0.027 mm / rev were selected with a cutting speed Vc equal to 3454 mm / min, the number of revolutions n equal to 5000 min-1. As can be seen, the feed and speed values are close to recommended, but small reducing these parameters reduces the size and number of burrs while maintaining the remaining requirements.

d1 [mm]	v_c [m/min]	f [mm]	n [min^{-1}]	v_f [mm/min]	Q [cm^3/min]	T [sek]
1.10	30	0.017	8680	150	0.0	2.2
1.20	30	0.018	7960	145	0.0	2.5
1.30	30	0.020	7345	145	0.0	2.7
1.40	30	0.022	6820	150	0.0	2.8
1.50	30	0.023	6365	145	0.5	3.1
1.60	30	0.025	5970	150	0.5	3.2
1.70	30	0.026	5615	145	0.5	3.6
1.80	30	0.028	5305	150	0.5	3.6
1.90	30	0.029	5025	145	0.5	4.0
2.00	40	0.031	6365	195	0.5	3.1
2.10	40	0.032	6065	195	0.5	3.2
2.20	40	0.034	5785	195	0.5	3.4
2.35	40	0.036	5420	195	1.0	3.6
2.50	40	0.038	5095	195	1.0	3.9
2.60	40	0.040	4895	195	1.0	4.0

Figure 4. Recommended processing parameters for solid materials from Fraisa catalog for twist drills B57014 [8].

When selecting cutting tools, the best quality of surface treatment, proceeding from the requirements of ensuring dimensional stability, tool wear and surface quality, minimum number and size of burrs, showed carbide tools manufactured by Swiss companies Fraisa, Utilis, IFANGER, Applitec. The TiAlN titanium-aluminum-nitrite coating, which is characterized by high heat resistance (up to 900 degrees) and corrosion resistance, is well suited for drilling durable BrB2 materials with or without cooling. The results of selecting a cutting tool and cutting conditions for a small-sized part from BrB2 material with the aim of combating burrs are presented in table 1.

In a number of cases, when establishing the boundary values of the ranges recommended by the manufacturer, it was noted that when setting the minimum feed and maximum cutting speed, the cutters overheat. When setting the minimum feed and speed, the necessary chip-breaking space and chip removal are not provided, and at maximum feed and speed large shavings form and the tool burns. Combustion occurs due to overheating, which can be caused by in-creased friction due to tool wear, poor chip heat dissipation due to thinning, operation with low feeds or with the feed turned off [9].

An analysis of the change in the nature and the number of burrs on beryllium bronze parts during the selection of cutting tools showed that when the speed and depth of cut are reduced to values acceptable to satisfy the above requirements and the use of a new cutting tool manufactured by Swiss companies Utilis, Fraisa, IFANGER, Applitec the number and the size of burrs are minimal. At higher cutting speeds with a small feed, burrs randomly occur over the entire surface and become long, stretching, with a reinforced base. Their number increases and the base area reaches 0.1 mm. This indicates that the material heats up, its fluidity increases and it splashes on-to the surface in the form of burrs. Tool wear increases friction, and as a result, material heating increases. When turning with a new tool at low speeds, the number of burrs decreases, their character changes, they become small drop-shaped. It was not possible to completely eliminate the appearance of burrs on parts made of BrB2 material by selecting a cutting tool and cutting modes.

To solve the problem of selecting the processing regimes by experimental methods, it is proposed to use standard technological templates for processing and forming a tool path with a partial change in parameters to obtain similar details. To create standard technological templates for processing and forming the tool path with a partial change in the parameters for obtaining such parts, it is necessary to create a database of normative and reference information on the types of processing for specific materials, tools, machines for accuracy standards taking into account the minimization of the number and size of burrs.

IOP Conf. Series: Materials Science and Engineering 1064 (2021) 012016 doi:10.1088/1757-899X/1064/1/012016

Table 1. Cutting tools and machining modes for KAPD.724212.030 housing (-01, -02)

	Name	F - feed, mm/min	n – number of spindle revolution	Cutting depth, mm	Number of moves	Comment
1	Cutting plate 731R-1.5 TIALN	0.01	8000	1.5	1	Cut Ø5
2	Frontal passage plate CCGT 097301FIN-F23 CTR2120	0.02 0.023	8000 5000	0.25 0.1	2	Butt trimming Ø5. Groove Ø5 → Ø4.9
3	Plate for cross turning 744-1,0-R TIALN	0.03 0.03	6000 6000	0.5 0.2	2	Ø5 → Ø3
4	Cutter boring frontal SDG 435 192-R Ø1,92 UHM 20 HX	0.012	1500	0.3	1	Groove width Ø1.92
5	Grooving boring tool SDS 440 292-R Ø2, 92 UHM 20 HX	0.01	6000	0.5 0.3	2	0.5 to Ø3.5 0.3 to Ø2.2 Ø2.92
6	Drill centering DRP 442 090 R-C Ø4 UHM 20 HX	0.017	5500	1.25	1	Deep centering 1.25 Ø4
7	Twist drill B570140220 Ø2, 2 TIALN	0.027	5000	0.5	12 / 2 Intermittent turning	Ø2.2 The number of interruptions 12 is due to the need for chip removal
8	Mill disk cutting 1101-0,35 x 40z	0.4	5000	0.4	1	Lamella cross cutting
9	End mill E124F-Dc5 0 Ø5 TIALN	0.083	6000	0.3	1	Ø5 → Ø4.7
10	End mill E102F-Dc1 0 Ø1 TIALN	0.012	5000	0.1	1	Ø1 → 1.1 ÷ Ø 1.7
11	End mill E102F-Dc1 5 Ø1,5 TIALN	0.003	6000	0.2	1 3	Ø1 → Ø1.25 Ø1.6 → Ø2.25
12	Twist drill B570140160 Ø1,6 MG10	0.003	6000		1	Ø1.6
13	Twist drill 2023-1.05 Ø1,05 TIALN	0.003	6000		1	Ø1.05
14	Plate for reverse turning Back turn 743x-1,2 TIALN	0.005	5000	0.1	1	Ø2.9 ÷ Ø3.25

4. Conclusion

The selection of tools and cutting modes for turning small-sized high-precision parts of radio electronic equipment made of beryllium bronze, manufactured on CNC lathes, showed that beryllium bronze is a difficult to process material. And it was not possible to completely exclude the appearance of burrs during machining.

The best quality of surface treatment, based on the requirements of ensuring dimensional stability, tool wear and surface quality, minimum number and size of burrs, was shown by carbide tools manufactured by Swiss companies Fraisa, Utilis, IFANGER, Applitec. TiAlN titanium-aluminum-nitrite coating, characterized by high heat resistance (up to 900 degrees) and corrosion resistance, is well suited for drilling durable materials like BrB2. When turning with a new tool at low speeds, the number of burrs decreases, their character changes, they become small drop-shaped. The processing modes recommended by the tool manufacturer differ from those selected experimentally.

For further use of the selection results, it is necessary to create a database of normative and reference information on the types of processing for specific materials, tools, machines for accuracy standards, taking into account minimizing the number and size of burrs.

Acknowledgments

The reported study was funded by RFBR, project number 19-38-90184.

References

[1] Dzhurinsky K B 2006 *Miniature coaxial radio components for SHF microelectronics. Connectors, coaxial microstrip transducers, adapters, SHF-inputs, glands, low-frequency inpits, insulating stands, interference filters* (Moscow: Tekhnosfera) p 216

[2] GOST 18175-78. 2004 *Interstate standard. Tinless bronzes, pressure treated. Non-ferrous metals. Bronze* (Technical conditions Stamps, Moscow: IPK Publishing House of Standards) p 148

[3] Averin V V and Guselnikov V S 2010 Automation of the design of control programs *Bull. of Universities. Instrument Engineering* **6** 67-71

[4] Voronenko V P, Sedykh M I and Shashin A D 2017 Design and efficient operation of production sites of multi-machine-building production *Bull. of Russian State Pedagogical University named after P.A. Solovyov* **1** 182-189

[5] Voronenko V P and Shashin A D 2017 Pilot production planning taking into account its current state *Computer-aided design in mechanical engineering: Mater. of the Vth int. correspondence sci.-pract. conf.* vol 5 (Novokuznetsk: SIC MS) p 134

[6] Madanov A V 2015 Analysis of problems in the processing of parts of complex geometry and ways to solve them at the stage of technological preparation of production *Problems of Sci.* **1** 14-16

[7] *Unilis Multidec 2018 Tools Catalog.* Retrieved from: https://drive.google.com/file/d/16RklcSx8hpMf1kysP6b0e5IJNu9172Oc/view?usp=sharing.

[8] *Fraisa tool catalog.* Retrieved from: https://drive.google.com/file/d/18-Zug4pifm1PE5D6INBhCWsPSBoOlL5Z/view?usp=sharing.

[9] Kuznetsov A V and Kuznetsova E N 2018 Features of the machining of parts made of magnesium alloys *University complex as a regional center of education, science and culture: Mater. of the All-Russian sci. and methodological conf.* pp 718-722. Retrieved from: http://elib.osu.ru/bitstream/123456789/5754/1/718-722.pdf

[10] Karlina Yu I, Kargapoltsev S K, Gozbenko V E and Karlina A I 2019 Removal of burrs from small-size high-precise parts for SHF electronics *IOP Conf. Ser.: Earth Env. Sci.* **378(1)** 012015

Automation of preproduction processes for high-precision small-sized parts on CNC machines

Yu I Karlina[1], **S K Kargapoltsev**[2], **V E Gozbenko**[2, 3], **D S Leonovich**[1] **and A I Karlina**[1]

[1] Irkutsk National Research Technical University, 83, Lermontova str., Irkutsk, 664074, Russia
[2] Irkutsk State Transport University, 15, Chernyshevskogo str., Irkutsk, 664074, Russia
[3] Angarsk State Technical University, 60, Chaykovskogo str., Angarsk, 665835, Russia

E-mail: kck@irgups.ru, karlinat@mail.ru

Abstract. The theoretical basis of the processes of preparation for the production of high-precision small-sized parts of coaxial radio components on CNC machines allows you to supplement the existing CAD / CAE / CAM with fragments of algorithmizing. To integrate the stages of technological preparation for the production of high-precision small-sized parts of coaxial radio components on CNC machines, a study of the features of the production of parts and the standard functionality of CAM systems was carried out. A classification of these parts by the type of equipment, types, serial production, variable parameters within the type, materials, and tools has been developed. The possibilities of the extensibility of the functions of CAM systems, the existing methods of creating and verifying control programs are considered. An example of a successful expansion of the functionality of a standard CAM system in the radio-electronic industry is given - a module for increasing machining accuracy with a constant cutting mode by optimizing the mutual positioning of the tool and the workpiece. Requirements have been developed for the creation of a specialized combined method for verification of control programs using all existing verification capabilities and minimal use of debugging and verification directly on the machine with the maintenance of reference books of worked processing modes in which the number and size of burrs are minimal, as well as a library of control programs equipped with a search system for the parameters of the difference in the execution of parts.

1. Introduction

The theoretical basis of the processes for preparing the production of high-precision small-sized parts of coaxial radio components on CNC machines allows you to supplement the existing CAD / CAE / CAM with fragments of algorithms built on the basis of the main provisions of symbolic logic, which allows you to describe the designer's course of action within the framework of the concept of exact formalism and to draw up a model on its basis designer behavior within the constraints. Coaxial radio components are small, high-precision and are made of difficult-to-machine materials. The manufacturing precision reaches 7 quality standards, the overall dimensions are from 0.39 mm to 6 mm in diameter and up to 15 mm in length. The choice of materials for manufacturing is limited by technical requirements [1, 2]. For the production of internal and external conductors of sealed coaxial radio components, alloy 29 NK is used to obtain coordinated thermal expansion coefficients of joints with S52-1 glass [3]. Conductors of leaking coaxial radio components are produced from beryllium bronze, less often brass, which is due to the springy properties of these materials, which are necessary for tight compression of the lamellas during assembly. During blade machining, burrs are formed on the surfaces of parts made of materials, which cannot be prevented. However, control of the parameters of burrs can significantly reduce the cost of removing them. To minimize the number and size of burrs, it is necessary to select the tool and modes of blade processing, and during processing, periodically measure both the overall dimensions and the burrs. This pre-production stage requires studying the recommended ranges for processing modes, setting specific values and testing on CNC machines. The test is carried out until the permissible number and size of burrs are obtained over a period of time that allows assessing the fulfillment of the requirement to ensure tool wear resistance and permissible parameters of burrs. The costs of creating control programs and full-scale tests can be reduced by using a number of measures to integrate the stages of pre-production and processing on CNC machines.

To carry out the integration of the stages of preparation for the production of parts on CNC machines, it is necessary to research the following issues [4-7]:

- classification of parts by type of equipment, types, serial production, variable parameters within the type, materials, tool;
- extensibility of the functionality of CAM systems;
- methods of verification of control programs;
- methods of creating control programs taking into account the classification of parts.

2. Classification of parts by type of equipment, serial production, standard designs with variable parameters, materials, tool

According to the inaccessibility of surfaces for deburring, the parts of coaxial radio components can be divided into two groups: internal conductors - parts that do not have internal hard-to-reach surfaces, external conductors - parts with internal hard-to-reach surfaces (grooves, pockets, grooves, etc.). Deburring of internal conductors (axles, bushings, etc.) is easily done in a tumbler, and a large number of larger burrs are acceptable. Deburring parts with hard-to-reach internal surfaces is challenging. Thermopulse deburring can be used to eliminate manual plumbing deburring [8, 9]. For high-quality thermal impulse deburring, it is necessary to control their size. The maximum size of the burr to remove depends on the type of material of the part, as well as on the presence of thin-walled elements. The ratio of the burr size to the minimum part thickness should not exceed 1: 4, otherwise thin-walled structural elements will be deformed due to annealing [10, 11].

By seriality, parts must be distinguished into serial and single orders of small volumes. One-time orders with the prospect of further serial production should be classified as serial. Verification of serial parts must be carried out taking into account ensuring the planned wear resistance of the tool; for one-time orders, it is enough to make sure that the tool retains its durability, the design dimensions and surface quality correspond to the drawing over several cycles of the control program.

By typical designs with the presence of various designs, parts can be grouped by the name of the part and the parameters of the designs, which are indicated in the drawing. The execution of parts can have different lengths, diameters, threads, end geometry, additional structural elements. For example, a part of the current lead of a radio frequency connector has 52 versions, which differ in the length of the current lead L, mm, the length of the overhang l, mm, and the presence of a radius at the end of the current lead. A fragment of a drawing of a part that has a multiple design, showing the difference in size, is shown in Figure 1.

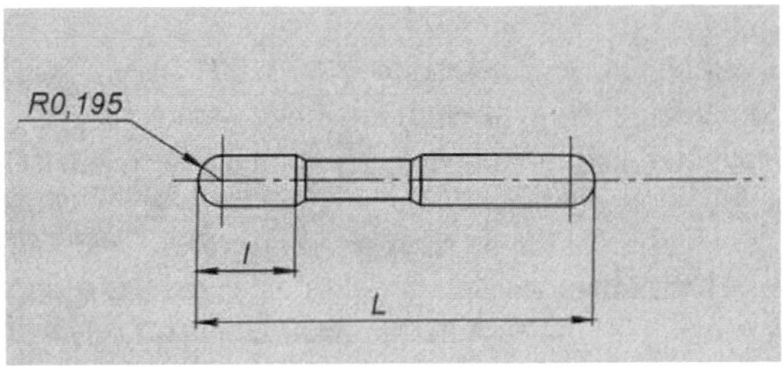

Figure 1. Fragment of a drawing of a part that has a multiple design with a difference in the presence of a radius at the end of the current lead

A fragment of a drawing of a part having a multiple design, demonstrating the difference in geometric design elements, is shown in Figure 2.

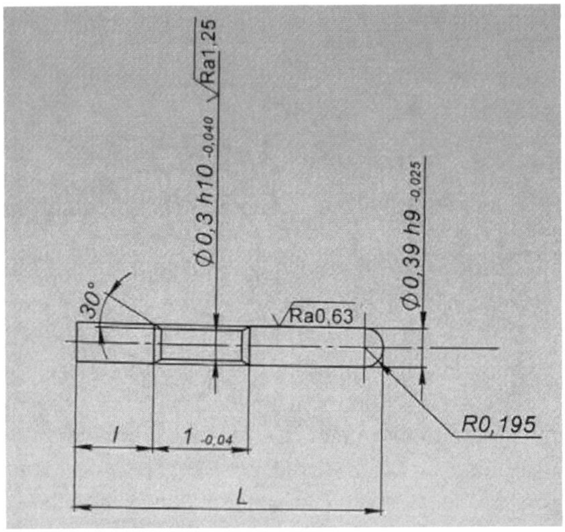

Figure 2. Fragment of a drawing of a part having multiple execution with a difference in lengths L and l

The use of materials and tools from the list of acceptable substitutes also leads to the need to adjust the control program with its subsequent verification. For example, changing the diameter of the wire for the manufacture of the part shown in Figure 1 will change the depth of cut, the volume of material removed and the rate of consumption. Replacing a tool with an analogue acceptable according to the technological documentation, or even replacing a tool manufacturer with the same catalog number leads to the need to change the planned wear resistance and adjust the cutting conditions. Cutting data and tooling can also depend on the characteristics of specific equipment.

3. A set of functional capabilities of CAM systems and the possibility of their expansion

The variety of existing CAM systems has led to their specialization in areas of application and set of functionalities. Currently, the following classification of CAM systems is being considered: by the level of automation; by type of equipment; by the number of processing coordinates; by the method of forming the processing technology.

It is advisable to divide the level of automation into 4 classes:

- manual development, in which the writing of a program is carried out manually by a programmer-technologist using general-purpose text editors;

- automation of the function of calculating the coordinates of control points and constructing the tool path (for example - KOMPAS CNC);
- automation of functions for calculating cutting conditions and forming a processing technology with a preliminary manual formulation by a programmer-technologist of a processing scheme, equipment, tool, detailing an operation (for example - Gemma);
- automation of the function of creating a processing technology with the participation of a programmer-technologist (for example, SprutSAM).

By types of equipment, CAM systems are specialized for turning, milling, electrical discharge machining, machining centers, woodworking, engraving.

By the number of coordinates and the complexity of the operation performed, there are: 2, 2.5. 3, 4. 5-coordinate CAM systems.

By the method of forming the processing technology:
- Operation-based CAM systems require the user to follow numerous steps to select a machining strategy for each surface of a workpiece:
- Structural CAM systems use a set of processing elements to describe the finished part:
- Process-based CAM systems are typically used for programming similar processes and operations [12].

The choice of a CAM system is significantly limited by the compatibility with the CAD system used in the enterprise. Electronic models created with a CAD system must be readable by the CAM system. Modern CAM systems are modular [13, 14]. The composition of modules in terms of functionality and extensibility can be selected for the needs of a specific type of production. The presence of built-in programming tools makes it easier to customize the CAM system for the needs of production: describe postprocessors, connect additional modules created using high-level programming languages.

The presence of requirements for high accuracy, small size, and the absence of burrs on the parts of coaxial radio components leads to the need for a special setting of processing parameters and verification of control programs for their production using a full-scale experiment. When such tasks arise that cannot be formalized by typical means of general-purpose CAM systems, it becomes necessary to develop additional software products to automate and reduce the cost of compiling and verifying control programs.

4. Verification techniques for control programs

An important stage in the preparation of the control program is the verification and debugging procedures. Various options are also possible here.

The built-in ability to check the syntax and correct syntax errors of the control program by means of coding the control program with further debugging of the control program on the machine by processing test parts and correcting the identified inconsistencies is the simplest version of the verification method for control programs. This debugging technique is ineffective, time consuming, time consuming, and distracting equipment from the production process.

Modern CAM systems usually have built-in verifiers, which provide the ability to graphically build and check the tool path according to the control program [15]. At a higher level, the processing is simulated using the control program with an indication of the errors that occur. In the extended verification mode, not only debugging is performed, including analysis and optimization of the control program. The use of such simulators to reproduce the results of the control program operation significantly increases the efficiency of the control program preparation.

However, the estimation of dimensions, the number of burrs throughout the entire life cycle of the tool, and their control are not covered by the existing functionality of CAM systems. Therefore, it is necessary to develop a specialized combined method for verification of control programs using all existing verification capabilities and minimal use of debugging and verification directly on the machine using reference books of worked out processing modes in which the number and size of burrs are minimal, as well as a library of control programs parameterized by classification parameters according to standard designs with the presence of various designs.

Possible strategies for developing control programs.

The following strategies of the enterprise in the field of automated development of control programs in the technological preparation of production can be distinguished:

Manual creation of control programs directly on the machine.

Automated preparation of UE. for the implementation of which it may be necessary to purchase computer equipment, a general-purpose CAM system or a specialized CAM system: purchase or develop modules for expanding functionality; postprocessors for existing equipment.

Extended verification without performing.

Using simulators for verification:

- with the involvement of means of verification of the CAM system;
- with the involvement of specialized verifiers:
- using integrated analysis and optimization tools [16, 17] (for example, Tecnomatix).

When choosing the appropriate strategy, the following production conditions must be considered:

The serial production.

Urgency of the order.

Characteristics of the available equipment.

Available software.

The complexity of the product being developed.

Requirements for products being developed.

Personnel qualifications.

5. Examples of expanding the functionality of CAM systems.

Improving the machining accuracy with a constant cutting mode is possible by optimizing the mutual positioning of the tool and workpiece. However, in the process of developing control programs for CNC machine tools, the programmer-technologist is faced with a significant amount of calculations, among which the calculation of the optimal positioning parameters plays a significant role [18]. Unfortunately, such calculations, as a rule, are not automated by standard CAM systems [19, 20].

6. Conclusion

Investigation of the peculiarities of the production of parts for coaxial radio components and the functionality of CAM systems made it possible to classify parts by the type of equipment, types, serial production, variable parameters within the type, materials, tools. This allows one to develop requirements for the creation of a specialized combined methodology for verifying control programs using all existing verification capabilities and minimal use of debugging and verification directly on the machine with maintaining reference books of worked processing modes in which the number and the size of burrs are minimal, including control program libraries.

Acknowledgments

The reported study was funded by RFBR, project number 19-38-90184.

References

[1] STD-01. WBTC. 1996 *draft. – Deburring Technology International* (Inc., KansasCity, MO)

[2] OST 4GO.070.014-79 1992 *Details of electronic equipment* (Moscow: LLC Radio Standard-TsNIIRES) p 15

[3] Dzhurinsky K B 2006 *Miniature coaxial radio components for SHF microelectronics. Connectors, coaxial microstrip transducers, adapters, SHF-inputs, glands, low-frequency inpits, insulating stands, interference filters* (Moscow: Tekhnosfera) p 216

[4] Kulikov D D, Padun B S and Yablochnikov E I 2014 Prospects for automation of technological preparation of production *Bull. of Universitites. Instrument making* **8(57)** 7-12

[5] Zilberburg L I, Milkman V I and Yablochnikov E I 2008 *Information technologies in design and production* (Saint Petersburg: Polytechnic) 304 p

[6] Gavrilova T A and Khoroshevsky V F 2000 *Databases of intelligent systems* (Saint Petersburg: Peter) 384 p
[7] Afanasyev M Ya, Salomatina A A, Alyoshina E E and Yablochnikov E I 2011 Application of multi-agent technologies for the implementation of the management system of a virtual enterprise *Sci. and Tech. Bull. Information technology, mechanics and optics* **5(75)** 105-111
[8] *Modern technologies for deburring holes in small diameters in the manufacture of parts for hydraulic equipment.* Retrieved from: https://stanko-arena.ru/article/sovremennye-tekhnologii-udaleniya-zausentsev-v-otverstiyakh-malykh-diametrov.html
[9] Halladay J 2016 *The practical application of thermal and electrochemical deburring methods.* Retrieved from: http://stankoforward.ru/texnologii-udaleniya-zausencev
[10] Suslova A G, ed 2008 *Engineering of the surface of parts* (Moscow: Mechanical engineering) 320 p
[11] Titov S N 2016 Deburring with a Thermal Polar Unit Pulsar TI-576 *Bull. of Sci. Conf.* (Tambov: Consulting Company LLC) **12-4** 176-177
[12] Kulikov D D and Yablochnikov E I 2011 Application of evaluative metrics for the analysis of technological preparation of production *Sci. and Tech. Bull. Information Technology, Mechanics and Optics* **6(76)** 109-112
[13] Yablochnikov E I, Fomina Yu N and Salomatina A A 2010 Organization of technological preparation of production in a distributed environment *Bull. of Universities. Instrument Making* **53(6)** 12-15
[14] Averin V V and Guselnikov V S 2010 Design automation of control programs *Bull. of Universities. Instrument Making* **6(53)** 67-71
[15] Voronenko V P, Sedykh M I and Shashin A D 2017 Design and efficient operation of production areas of multi-product machine-building production *Bull. of P.A. Solovyov* **1(40)** 182-189
[16] Voronenko V P and Shashin A D 2017 Pilot production planning taking into account its current state *Computer-aided design in mechanical engineering: Proc. of the V Int. correspondence sci. and pract. conf.* (Novokuznetsk: SRC MS) No 5 p 134
[17] Madanov A V 2015 Analysis of problems in the processing of parts of complex geometry and ways to solve them at the stage of technological preparation of production *Problems of Sci.* **1(1)** 14-16
[18] 2008 The future of mechanical engineering in Russia *Proc. of the All-Russian conf. of young scientists and specialists* (Moscow: MSTU named after N. E. Bauman) p 48
[19] Lieberman Ya L and Chernogolova S A 2008 Improving the accuracy of CNC *STIN systems* **L** 6-9
[20] Lovygin A A, Vasiliev A V and Krivtsov S Yu 2006 *Modern machine with CNC and CAD. CAM-system* (Moscow: Elf PPR) 286 p

Algorithm for calculating the time of thermal pulse processing of small-sized high-precision parts

A I Karlina[1], V A Ershov[1], S K Kargapoltsev[2], V E Gozbenko[2, 3], and Yu I Karlina[1]

[1] Irkutsk National Research Technical University, 83, Lermontova str., Irkutsk, 664074, Russia
[2] Irkutsk State Transport University, 15, Chernyshevskogo str., Irkutsk, 664074, Russia
[3] Angarsk State Technical University, 60, Chaykovskogo str., Angarsk, 665835, Russia

E-mail: karlinat@mail.ru, gozbenko@irgups.ru

Abstract. The thermal impulse method of deburring is one of the most promising methods for removing unwanted liquids formed on the surface of small-sized high-precision parts of electronic equipment at the stage of blade machining. The thermal impulse method of deburring and rounding of sharp edges is characterized by constant processing time and quality, subject to the normalized size of liquids. Such qualities can significantly increase the stability of the release of parts in time, and, thanks to this, successfully implement a system of operational scheduling of production. This article describes an algorithm for calculating the time for thermal pulse processing of small-sized parts of coaxial radio components. An example of a basic generalized calculation of the time spent on thermal impulse processing of typical parts using a Pulsar-VKF thermal impulse installation of various configurations is given. Taking into account the peculiarities of thermal pulse processing of parts of coaxial radio components, a method for calculating the batch size and thermal pulse processing time, subject to algorithmization, is developed, conditions and restrictions, input and output information for the algorithm are described. Formulas for calculation are given.

1. Introduction

The thermal pulse method of deburring is one of the promising methods for removing unwanted liquids formed on the surface of parts at the stage of blade machining. Undesirable liquids include burrs and sharp edges, the presence of which is unacceptable for technical requirements. Parts of electronic equipment are small-sized and high-precision. Parts manufactured by blade processing on automatic lathes have overall dimensions of Ø0.4 ÷ 16mm × 80mm, on milling machining centers - dimensions of up to 400mm × 400mm. They require the execution of sizes according to 7 ÷ 10 accuracy grades. One of the main reasons for the failure of electronic devices is defects in the manufacture of parts resulting from the shaping of their surfaces. Burrs make it difficult to assemble products, lead to damage caused to mating surfaces and wires, and ultimately to a violation of the performance of the device. Sharp edges of parts of electronic equipment are stress concentrators and contribute to the destruction of the parts themselves and contacting surfaces [1]. According to international and industry standards [2, 3], all workpieces must be descaled, burrs removed, sharp edges blunted. Insufficient preparation of the surface

IOP Conf. Series: Materials Science and Engineering 1064 (2021) 012018 doi:10.1088/1757-899X/1064/1/012018

of parts before coating (burrs, sharp edges, the presence of contamination) leads to deterioration of adhesion, the formation of build-ups on the surface of parts.

The thermal pulse method is based on thermophysical processes in which burrs and other unwanted liquids are removed by reflow, followed by the discharge of combustion products from the working chamber, preventing the condensation of oxides. The advantages of the method include flexibility, high productivity, combination of cleaning and finishing of surfaces and edges, the ability to process high-precision parts, including parts of electronic equipment. The disadvantage is the limitation of the size of liquids depending on the material and design features of the parts [4]. So, the maximum size of burrs for parts made of metals with low thermal conductivity is 0.3 mm, for parts made of highly heat-conducting metals - 0.1 mm. Such design features of parts as the presence of thin-walled elements also limit the maximum size of burrs, it should be no more than 1/6 of the minimum thickness of the part.

The use of the thermal impulse deburring method eliminates manual metalworking of parts, which reduces the time and cost of manufacturing parts, and eliminates the element of uncertainty associated with the human factor inherent in manual processing of parts. The thermo-pulse method of deburring and rounding of sharp edges is characterized by constant processing time and quality, provided that the liquid size is normalized [5, 6]. Such qualities make it possible to significantly increase the stability of the production of parts in time, and, thanks to this, successfully implement a system of operational scheduling of production [7-9]. This article describes an algorithm for calculating the time for thermal pulse processing of small-sized parts of coaxial radio components.

2. An example of calculating the time spent on thermal pulse processing of typical parts when using a thermal pulse Pulsar-VKF installation of various configurations

For thermal pulse processing of small-sized parts of radio electronic equipment, it is advisable to use thermal pulse Pulsar-VKF installations of Russian production by Alfa Steel in St. Petersburg. The design of these units is well developed, and the quality of processing and reliability are confirmed by the experience of practical use at industrial enterprises. At the same time, the cost of these installations is several times lower than the cost of imported analogues [5, 6].

As a typical part for calculating the time of thermal impulse treatment in Pulsar-VKF installations, a part has been adopted that has the overall dimensions of a typical part of electronic equipment - the body of a coaxial radio component. Parameters of this part: cylindrical part Ø22 mm, H = 18 mm, $V_d = 6840$ mm (Figure 1).

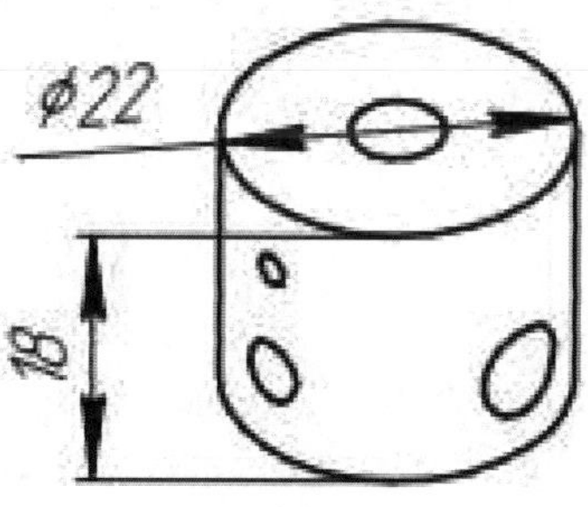

Figure 1. Schematic representation of a typical small-sized part - the housing of a coaxial radio component

According to the operating conditions of the Pulsar VKF thermal impulse installation, the effective working volume of the chamber is 0.65 of the actual volume of the chamber (taking into account the fixing devices and containers, the free required volume for filling the space between the parts with gas, the geometry of a typical part).

The processing cycle of one batch of parts is no more than 30 minutes and it consists of:

- 5-10 minutes to start from cold off state;

- 5-10 minutes placing typical parts in containers and fixing them in the working area;
- 1,5 minutes automatic processing cycle;
- 5-10 minutes to cool down and remove containers with parts from the working area.

Table 1 shows the parameters for calculating the processing time of a batch of typical parts in Pulsar VKF thermal pulse units of various configurations.

Table 1. The parameters for calculating the processing time of a batch of typical parts in Pulsar VKF thermal pulse units of various configurations.

Model	Working chamber size, mm	Actual volume / Working volume, mm^3	The number of processed typical parts for 1 processing cycle	The number of processed typical parts per shift
VKF 3.150	Ø120 x H170	Vf = 1 922 000 Vw = 1 249 300	182	4368
VKF 3.250	Ø220 x H250	Vf = 9 500 500 Vw = 6 180 300	903	21672
VKF 3.350	Ø310 x H340	Vf = 25 650 000 Vw = 16 672 500	2500	60000
VKF 3.450	Ø400 x H350	Vf = 44 000 000 Vw = 28 600 000	4180	100320

The calculations in Table 1 are made according to the following formulas:

The actual volume of the chamber of the thermal impulse installation, mm^3:

$$V_f = \pi r^2 \cdot H \quad (1)$$

where r, H – radius, part height, mm.

The working volume of the chamber of the thermal impulse installation, mm^3:

$$V_w = V_f \cdot k \quad (2)$$

where k – coefficient of the effective working volume of the chamber, k = 0,65.

The number of processed standard parts per 1 processing cycle, pieces:

$$N_d = \frac{V_w}{V_d}. \quad (3)$$

The number of processed standard parts per shift, pieces:

$$N_{ws} = N_d \cdot N_{ch} \cdot T_{ws} \quad (4)$$

where N_{ch} – number of machining cycles per hour, N_ch = 3, T_{ws} – number of hours per shift, T_{ws}= 8.

In this example, the time of thermal pulse treatment was calculated based on the averaged data on the effective working volume of the chamber, taking into account the fixing devices and containers, the free required volume to fill the space between the parts with gas, and the geometry of a typical part. For a more accurate calculation of the batch processing time for thermal pulse processing, which is applicable when calculating the schedule of work centers in production, it is necessary to carry out additional formalization and supplement the calculation with updated data.

3. Algorithm for calculating the size of a batch of parts to be loaded into a thermal impulse installation and the time of thermal impulse processing of parts for a production program

The algorithm for calculating the batch size for loading parts into a thermal impulse unit and the time for thermal impulse processing of parts for a release program is designed to establish the size of the batch for loading parts into a thermal impulse unit in the operating chart, taking into account the permissible deviation, establishing the actual batch size for loading parts into a thermal impulse unit depending on the batch size parts coming from the previous operation, as well as determining the time of thermal impulse processing of parts to determine the output, the shift factor for this operation, as well as to calculate the schedule within the program for coordinating the production of products [10-14]. The calculation of the batch size for loading parts into a thermal impulse installation and the time for thermal impulse processing of parts for a release program can be performed using the formulas below. The calculation is based on the following conditions and restrictions:

- parts are high-precision, easily damaged, small-sized, so they will be placed in a tooling with a lid in one layer so as to provide gas access to all surfaces;
- placement of different parts in one load with separation by a partition is not provided, as this can increase the downtime due to waiting for the next parts for processing, reducing the size of the last batch of loading by more than 25% after evenly dividing the total amount into batches of acceptable size. It allows one to ensure the quality of processing; it is compensated by the placement of ballast, separated by a partition.

Input information is:

- overall dimensions of parts obtained from the CAD system, mm;
- diameter of the bottom of the chamber of the thermal impulse installation, obtained from the technical documentation of the thermal impulse installation, mm;
- wall thickness of the tooling for placing parts, obtained from the CAD system, mm;
- time for processing one batch of parts, obtained by the timing of the operation, minutes.

The output information is:

- maximum number of parts in a batch of loading into a thermal impulse installation, pieces;
- number of parts in the batch of loading into the thermal impulse installation, pieces;
- processing time of parts in a thermal impulse installation, minutes.

Since the parts are placed in one layer, it is necessary to calculate the area occupied by one part and the number of parts to be placed on the bottom area of the tooling, so as to provide access of the gas-air mixture to all surfaces of the parts.

The areas of parts occupied when placed on different surfaces are calculated using the formulas:

$$S_{wl} = W \cdot L, \tag{5}$$

where S_{wl} – area of the part occupied when placed in width and length, mm^2; W – part width, mm; L – part length, mm.

$$S_{hl} = H \cdot L, \tag{6}$$

where S_{hl}– area of the part occupied when placed in height and length, mm^2; H – part height, mm; L – part length, mm.

$$S_{wh} = W \cdot H, \tag{7}$$

where S_{wh}h – area of the part occupied when positioned in width and height, mm^2; W – part width, mm; H – part height, mm.

The area occupied by one part in a thermal impulse installation, mm^2, is calculated by the formula:

$$S_d = \max(S_{wl};\ S_{hl};\ S_{wh}) \cdot (1 + k_{fs}), \tag{8}$$

where k_{fs} – coefficient of the inter-part space to ensure free access of the gas-air mixture to the surfaces of the part and loose fit; let us take k_{fs}= 0.05.

The maximum number of parts in a batch placed in a thermal impulse installation, pieces, is calculated by the formula:

$$N_{dm} = \frac{\pi \cdot (D_{tu} - T_{sn}) \cdot (1 - k_{fs})}{S_d}, \tag{9}$$

where D_{tu} – diameter of the bottom of the chamber of the thermo-impulse installation, mm; T_{sn} – wall thickness of the tooling for placing parts, mm.

The number of parts in a batch placed in a thermal impulse installation, pieces, depends on the size of the batch of parts arriving for processing, and is calculated by averaging the size of the batch of loading parts, taking into account the permissible deviation of the size of the batch of loading by the formula:

$$N_d = \frac{N_{di}}{max\left(\frac{N_{di}}{N_{dm}};N_{dm} \cdot 0{,}75\right)}, \tag{10}$$

where N_{di} – number of parts arriving for thermal pulse processing from the previous operation, pcs.

The time of thermal impulse processing of parts for the release program is calculated by the formula:

$$t_{tp} = \sum_{i=1}^{n} \left(\frac{N_{pl_i}}{N_{d_i}}\right) \cdot t_1, \tag{11}$$

where N_{pl_i} – planned number of parts of the i -type of the nomenclature, pieces; N_{d_i} – number of parts of the i -type of the nomenclature in the batch placed in the thermal impulse installation, pieces; t_1 – processing time of one batch of parts in a thermal pulse installation, according to the results of the timing of the thermal pulse processing operation – t_1= 20 minutes.

4. Conclusion

To calculate the number of parts placed in the thermal impulse installation and the time of thermal impulse processing of parts for the release program, a calculation method has been developed that is subject to algorithmization. Calculation of the batch size for loading parts into a thermal impulse unit and the time for thermal impulse processing of parts for the release program is intended to establish the size of the batch for loading parts into a thermal impulse unit in the operating chart, taking into account the permissible deviation, establishing the actual size of the batch for loading parts into a thermal impulse unit depending on the size of the batch of parts coming from the previous operation, as well as determining the time of thermal impulse processing of parts to determine the output, the shift ratio for this operation, as well as to calculate the schedule as part of the product coordination program.

The calculation is based on the following conditions and restrictions:

- parts are high-precision, easily damaged, small-sized, so they will be placed in a tooling with a lid in one layer so as to provide gas access to all surfaces;
- placement of different parts in one load with separation by a partition is not provided, as this can increase the downtime due to waiting for the next parts for processing, reducing the size of the last batch of loading by more than 25% after evenly dividing the total amount into batches of acceptable size. This allows one to ensure the quality of processing; it is compensated by the placement of ballast, separated by a partition.

Input information is:

- overall dimensions of parts obtained from the CAD system, mm;
- diameter of the bottom of the chamber of the thermal impulse installation, obtained from the technical documentation of the thermal impulse installation, mm;
- wall thickness of the tooling for placing parts, obtained from the CAD system, mm;
- time for processing one batch of parts, obtained by the timing of the operation, minutes.

The output information is:

- maximum number of parts in a batch of loading into a thermal impulse installation, pieces;
- number of parts in the batch of loading into the thermal impulse installation, pieces;
- processing time of parts in a thermal impulse installation, minutes.

Acknowledgments
The reported study was funded by RFBR, project number 19-38-90184.

References

[1] Tamarkin M A, Smolentsev E V and Kolganova E N 2019 Analysis of the current state of finishing methods of processing in the environment of free abrasives of parts with small grooves and holes *Bull. of Voronezh State Technical University* **1(15)** 122-129

[2] STD-01. WBTC. 1996 *draft. –Deburring Technology International* (Inc., KansasCity, MO)

[3] OST 4GO.070.014-79 1992 *Details of electronic equipment* (Moscow: LLC Radio Standard-TsNIIRES) p 15

[4] Loseva O A and Losev A V 2009 Treatment of the edges of parts by the thermal impulse method *Open information and computer integrated technologies: Coll. of sci. papers of the National Aerospace University* (Zhukovsky KhAI) iss 42 pp 120-126

[5] Titov S N 2016 Deburring with the Pulsar TI-576 thermal pulse installation *Bull. of sci. conf.* **12-4** 176-177

[6] Karlina Yu I, Kargapoltsev S K, Gozbenko V E, Karlina A I and Leonovich D S 2020 Overview of electro physicochemical methods for deburring small-sized high-precision details of coaxial radio components *J. of Phys.: Conf. Ser.* **1582(1)** 012041

[7] Kulikov D D, Padun B S and Yablochnikov E I 2014 Prospects for automation of technological preparation of production *Bull. of universities. Instrument making* **8(57)** 7-12

[8] Zilberburg L I, Molochnik V I and Yablochnikov E I 2008 *Information technologies in design and production* (SPb: Polytechnic) p 304

[9] Gavrilova T A and Khoroshevsky V F 2000 *Databases of intelligent systems* (SPb: Peter) p 384

[10] Afanasyev M Ya, Salomatina A A, Alyoshina E E and Yablochnikov E I 2011 Application of multi-agent technologies for the implementation of a virtual enterprise management system *Sci.-Tech. Bull. Information Technology, Mechanics and Optics* **5(75)** 105-111

[11] Kulikov D D and Yablochnikov E I 2011 Application of evaluative metrics for the analysis of technological preparation of production *Sci. and Tech. Bull. Information Technology, Mechanics and Optics* **6(76)** 109-112

[12] Yablochnikov E I, Fomina Yu N and Salomatina A A 2010 Organization of technological preparation of production in a distributed environment *Bull. of Universities. Instrumentation* **53(6)** 12-15

[13] Voronenko V P, Sedykh M I and Shashin A D 2017 Design and efficient operation of production areas of multi-product machine-building production *Bull. of the P.A. Solovyov* **1(40)** 182-189

[14] Voronenko V P and Shashin A D 2017 Planning a pilot production taking into account its current state *Computer-aided design in mechanical engineering: Proc. of the V int. correspondence sci. and pract. conf.* (Novokuznetsk: SRC MS) p 134

Interaction algorithm of the mechanisms of relative manipulation in a 6-DOF robot control system

S Y Khalapyan[1], L A Rybak[2], E V Gaponenko[2] and L R Kholoshevskaya[2]

[1]Stary Oskol Technological Institute named after A A Ugarov NUST MISiS, 42, Makarenko mkr., 309516, Stary Oskol, Russia
[2]Belgorod State Technological University named after V G Shukhov, 46, Kostyukova str., Belgorod, 308012, Russia

E-mail: khalapyan@yandex.ru, rl_bgtu@intbel.ru

Abstract. The paper considers the features of management of a 6-DOF robot, developed on the basis of two interacting mechanisms of a parallel structure. Each mechanism provides movement in three independent degrees of freedom: a flat 3-RPR mechanism for installing a part and a 3-RPS mechanism (tripod) for installing a tool. It is shown that when these mechanisms are jointly controlled, it becomes necessary to correct the movement of one of them in connection with the parasitic movements of the other. The authors propose the method for the description of the position of the movable platform of the tripod, which allows using only five output coordinates. According to the authors, the platform rotation representation used in this case is more convenient to describe the interaction of mechanisms than the traditional use of Tait – Bryan angles. The algorithm was developed to correct the task by the position of the planar mechanism in order to compensate parasitic movements of the movable platform of the tripod in the horizontal plane during the control of a 6-DOF robot. On the basis of a specially constructed 6-DOF robot model with mechanisms of relative manipulation, the simulation in the MATLAB package was carried out, which confirmed the effectiveness of the proposed algorithm.

1. Introduction

Modern machine tools and systems of complex processing require the organization of tool access to the surface of the part from different sides and at different angles, i.e. the movement in 6 degrees of freedom. On the other hand, parallel robots with 6 degrees of freedom have complex kinematics and therefore are difficult to control [1-6].

Under these conditions a promising system is a system of 2 parallel robots. Each of them has 3 independent degrees of freedom, and as a result, the system provides movement along 6 degrees of freedom (movement in three-dimensional space and rotations around three coordinate axes).

For example such a system can move the workpiece in the horizontal plane (x, y) and rotate it around the vertical z-axis using the 3-RPR mechanism [7] and move the tool intended for machining the workpiece along the z-axis and rotate it around the x and y axes using the 3-RPS [8].

Each of these mechanisms separately was studied well enough, the dependences describing their kinematics were obtained and approaches to control were proposed. However, when these mechanisms are jointly controlled, it becomes necessary to correct the movement of one of them in connection with the parasitic movements of the other. In addition, the use of the Tait – Bryan angles to describe the rotations of the 3-RPS mechanism leads to the use of three variables for two rotational degrees of freedom and is generally not very convenient for users to represent the tilt and orientation of a tool.

At the first stage the method of two-stage control of a parallel robot proposed in the work [7] provides the construction of a safe trajectory of movement of its output link, located at a sufficient distance from the boundaries of the working space. In the case of joint control of two parallel robots at this stage, it is necessary to take into account the effect of the movement of one mechanism during the formation of a task for the second one.

2. Description of 6-DOF Robot with Relative Manipulation Mechanisms

6-DOF robot with relative manipulation mechanisms contains a module for the installation of a tool 1 and a module the installation of a workpiece 2, made on the basis of mechanisms of a parallel structure (Fig. 1).

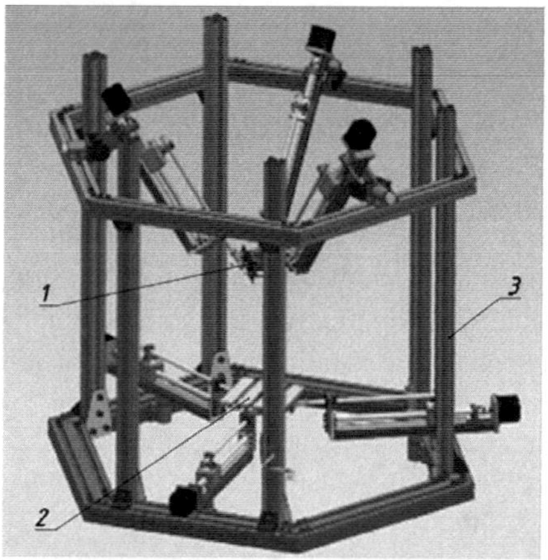

Figure 1. 6-DOF robot with relative manipulation mechanisms: 1 - module for tool installation; 2 - module for workpiece installation; 3 - frame.

The lower module is developed on the basis of a flat 3-RPR mechanism. It is a mechanism consisting of a movable platform in the form of an equilateral triangle, to the vertices of which three rods of variable length are attached using rotators. With the other end of the rod, using the same rotators, they are attached to the fixed base at the points corresponding to the vertices of the equilateral triangle. These rotators are directed in such a way that when the lengths of the rods change, the movable platform moves in the horizontal plane and rotates around the vertical axis.

The input coordinates of the lower module are the lengths of its three rods, which are changed under the action of the drive motors. The output coordinates are the position of the center of the movable platform relative to the center of the stationary one, the movement along the x and y axes and rotation of φ around the z axis. In order to describe the kinematics of the mechanism, the formulas are used that allow determining the lengths of the drive links by the known output coordinates presented in the work [7].

The upper module is developed on the basis of a 3-RPS-mechanism or tripod and also has a movable platform in the form of an equilateral triangle. However, it is attached to the bars of variable length using spherical rotators.

At the other end the rods are attached to a fixed base at points corresponding to the vertices of an equilateral triangle, using pivot joints directed in the way that the rods move in vertical planes passing through the center of the base. As well as in the flat mechanism, the input coordinates of the tripod are the lengths of the rods equipped with an electric drive. The output coordinates are the position of the center of the movable platform relative to the center of the stationary platform, the movement along the z-axis and rotations ψ and θ around the x and y axes, respectively. The kinematic analysis of the

IOP Conf. Series: Materials Science and Engineering 1064 (2021) 012019 doi:10.1088/1757-899X/1064/1/012019

mechanism allows determining the relationship between the output and input coordinates presented in the work [8].

The difficulty of the use of a tripod in the system of interacting robots is that, as it can be seen from formulas (5) - (7) presented in the work [8], in addition to useful movement along the vertical z axis and rotations around the horizontal x and y axes, during the movement of the mechanism, parasitic movements along the x and y axes also occur: it is clear that the second robot of the studied system will have to correct such displacements. For this, a corresponding correction signal will be added to its own task.

3. Description of the rotation of tripod platform. Horizontal displacement and its compensation

The value of the horizontal displacement of the tripod platform at the same angles of rotation is determined by its geometric dimensions. It is easy to show it by the example of the platform rotation from the initial horizontal position to the inclined one by the extension of one bar (Fig. 2). The spherical rotators of the other two rods do not change their position in space. Therefore, the value of the horizontal displacement of the center of the movable platform at the same angles of inclination to the horizontal is proportional to the radius of the circle described around the triangle, at the vertices of which there are rotators attached to the movable platform.

Therefore, in order to minimize parasitic movements and reduce the corresponding additional load on the drive of the planar mechanism, which can deteriorate the speed characteristics of the system as a whole, it is useful to reduce the geometric dimensions of the movable platform of the tripod. Such a decrease will lead to the acceleration of angular displacements, although it will reduce their accuracy to some extent. It also will lead to some deceleration of the useful linear movement of the movable platform along the vertical axis.

As for the sixth degree of freedom of the tripod - rotation around the vertical z-axis it is absent because of the location of the pivot joints. At the same time, the traditional use of the Tait-Bryan angles to describe the rotation of the movable platform leads to the fact that, as it can be seen from formulas (5) - (7) presented in the work [7], two angles are not enough to describe it: the angle ψ of rotation about the axis x and the angle θ of rotation about the y' axis resulting from the first rotation of the original coordinate system. On the other hand, the use of the Tait-Bryan angles, which is usually quite acceptable for the description of the state of the output link of robots of a parallel structure, is not very convenient for a system of interacting robots. For example, in such a system it may be necessary to feed the tool perpendicular to the surface of the part, while turning it with the specified side.

Under such conditions, the rotation of the platform can be described as a single rotation through a certain angle around a certain axis w passing through the geometric center of the movable platform. The impossibility of its rotation around the vertical axis shows that the rotation axis can only lie in the horizontal plane. Then, in order to describe this axis, it is enough to indicate the angle ζ between its direction and the positive direction, for example, the x axis. For the description of the rotational motion of the platform, it is only necessary to add the angle ξ of its rotation around the selected axis (Fig. 3).

As a result, the angle ζ determines the direction of the working end of the tool, the angle ξ determines the direction of its feed relative to the vertical axis (or horizontal plane), and the angle φ of rotation of the movable platform of the planar mechanism around the vertical axis z allows the workpiece to be turned with the necessary side to the tool.

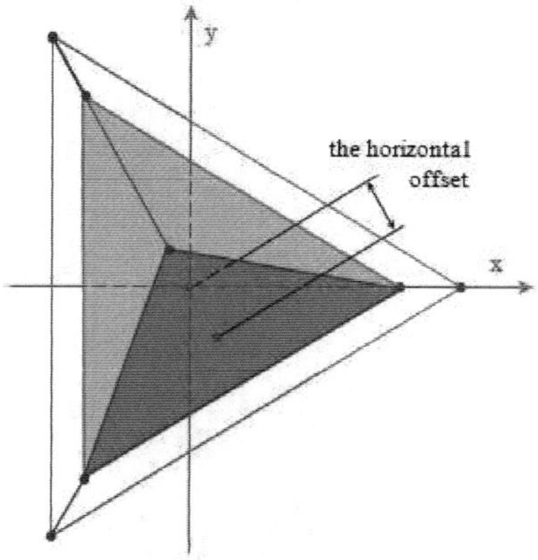

Figure 2. Horizontal displacement of the platform during the change of the length of one of the rods: blue triangle - initial; blue - new platform position

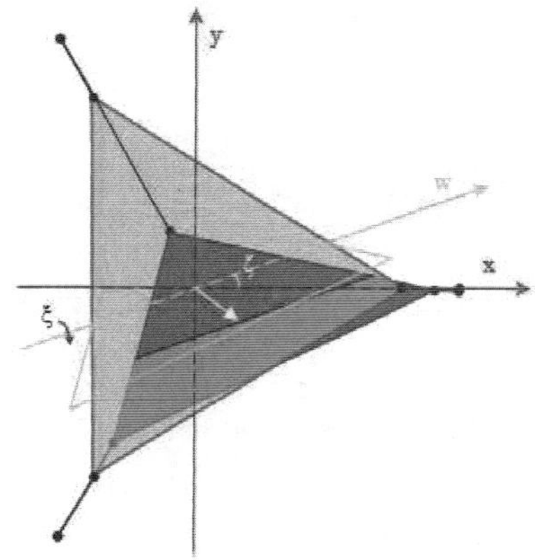

Figure 3. Rotation around the w axis ($\zeta = 19^{o}$, $\xi = 60^{o}$) and parallel transfer of the tripod platform: blue triangle - initial state; orange - rotation without considering horizontal displacement; blue - new platform position

Thus, the movement of the platform from the initial position in the general case can be described by its parallel translation and rotation around axis w passing through the center of the platform.

Let us suppose that initially the platform was in a certain horizontal position and a parallel transfer was performed, in which its center was at the point P (x_0, y_0, z_0). Let us consider the displacement of platform points as a result of turning more thoroughly (Fig. 4). The arbitrary point A (x_0, y_0, z_0) of the mobile platform, which is in a horizontal position, as a result of rotation (ζ, ξ) will move to some position B (x_2, y_2, z_2).

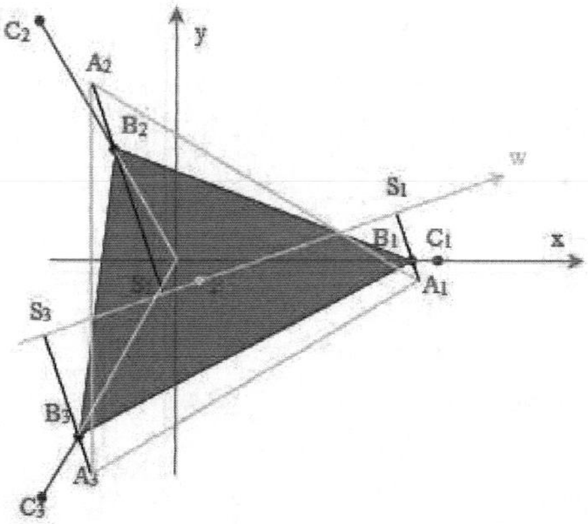

Figure 4. Determination of the coordinates of the vertices of the movable platform of the tripod: orange triangle - parallel transfer is performed without rotation; blue - new platform position

In order to determine it, we draw a perpendicular AS to the w axis. The equations of lines PS and AS in the plane PSA || (x, y) respectively:

$$y = k\,(x - x_0) + y_0, \quad (1)$$

$$y = -\frac{1}{k}\,(x - x_1) + y_1, \quad (2)$$

where $k = tg\,\zeta$.

The coordinates of the point $S(x_3, y_3, z_0)$ of their intersection are determined by joint solution of equations (1) and (2):

$$k(x_3 - x_0) + y_0 = -\frac{1}{k}(x_3 - x_1) + y_1$$

Here we get

$$x_3 = \frac{\frac{1}{k}x_1 + y_1 + k x_0 - y_0}{k + \frac{1}{k}} = \frac{x_1 + k(y_1 - y_0) + k^2 x_0}{k^2 + 1}$$

Substituting this expression in (1), we get

$$y_3 = k\left(\frac{x_1 + k(y_1 - y_0) + k^2 x_0}{k^2 + 1} - x_0\right) + y_0 = \frac{k(x_1 - x_0) + k^2 y_1 + y_0}{k^2 + 1}$$

Taking into account the value k:

$$x_3 = \frac{x_1 + (y_1 - y_0)\, tg\,\zeta + x_0 tg^2\zeta}{tg^2\zeta + 1}$$

$$y_3 = \frac{(x_1 - x_0)\, tg\,\zeta + y_1\, tg^2\zeta + y_0}{tg^2\zeta + 1}$$

Since $tg^2\zeta + 1 = 1/\cos^2\zeta$

$$x_3 = x_0 + (x_1 - x_0)\cos^2\zeta + (y_1 - y_0)\sin\zeta\cos\zeta \quad (3)$$

$$y_3 = y_0 + (x_1 - x_0)\sin\zeta\cos\zeta + (y_1 - y_0)\sin^2\zeta \quad (4)$$

it allows determining x and y for point B:

$$x_2 = x_3 + (x_1 - x_3)\cos\xi = x_3(1 - \cos\xi) + x_1\cos\xi =$$
$$((x_1 - x_0)\cos^2\zeta + (y_1 - y_0)\sin\zeta\cos\zeta + x_0)(1 - \cos\xi) + x_1\cos\xi \quad (5)$$

$$y_2 = y_3 + (y_1 - y_3)\cos\xi = y_3(1 - \cos\xi) + y_1\cos\xi =$$
$$((x_1 - x_0)\sin\zeta\cos\zeta + (y_1 - y_0)\sin^2\zeta + y_0)(1 - \cos\xi) + y_1\cos\xi \quad (6)$$

To calculate z_2, we determine the length AS (radius of the rotation arc):

$$|AS| = \sqrt{(x_3 - x_1)^2 + (y_3 - y_1)^2} = (((y_1 - y_0)\sin\zeta\cos\zeta - (x_1 - x_0)\sin^2\zeta)^2 + ((x_1 - x_0)\sin\zeta\cos\zeta - (y_1 - y_0)\cos^2\zeta)^2)^{1/2} = |(y_1 - y_0)\cos\zeta - (x_1 - x_0)\sin\zeta|$$

The indicator of the change in z during rotation depends on the position of point A relative to the w axis. This dependence is determined by the indicator of the vector product $e_w \times \overrightarrow{PA}$, where $e_w(\cos\zeta, \sin\zeta, 0)$ is the unit vector corresponding to the w axis:

$$(\cos\zeta, \sin\zeta, 0) \times (x_1 - x_0, y_1 - y_0, 0) = (0, 0, (y_1 - y_0)\cos\zeta - (x_1 - x_0)\sin\zeta)$$

Consequently

$$z_2 = z_0 + |(y_1 - y_0)\cos\zeta - (x_1 - x_0)\sin\zeta|\, sign((y_1 - y_0)\cos\zeta -$$
$$(x_1 - x_0)\sin\zeta)\sin\xi = z_0 + ((y_1 - y_0)\cos\zeta - (x_1 - x_0)\sin\zeta)\sin\xi \quad (7)$$

The resulting formulas allow, in particular, determining the coordinates $B1$, $B2$, $B3$, the vertices of the mobile platform $A_1(x_0 + r, y_0, z_0)$, $A_2(x_0 - r/2, y_0 + r\sqrt{3}/2, z_0)$, $A_3(x_0 - r/2, y_0 - r\sqrt{3}/2, z_0)$ after its rotation.

B_1:

$$x_2 = x_0 + ((x_1 - x_0)\cos^2\zeta + (y_1 - y_0)\sin\zeta\cos\zeta)(1 - \cos\xi) + (x_1 - x_0)\cos\xi = x_0 + r(\cos^2\zeta(1 - \cos\xi) + \cos\xi)$$

$$y_2 = y_0 + ((x_1 - x_0)\sin\zeta\cos\zeta + (y_1 - y_0)\sin^2\zeta)(1 - \cos\xi) + (y_1 - y_0)\cos\xi = y_0 + r\sin\zeta\cos\zeta(1 - \cos\xi)$$

$$z_2 = z_0 + ((y_1 - y_0)\cos\zeta - (x_1 - x_0)\sin\zeta)\sin\xi = z_0 - r\sin\zeta\sin\xi$$

B_2:

$$x_2 = x_0 + ((-\cos\zeta + \sqrt{3}\sin\zeta)\cos\zeta(1 - \cos\xi) - \cos\xi)\frac{r}{2}$$

$$y_2 = y_0 + ((-\cos\zeta + \sqrt{3}\sin\zeta)\sin\zeta(1 - \cos\xi) + \sqrt{3}\cos\xi)\frac{r}{2}$$

$$z_2 = z_0 + (\sin\zeta + \sqrt{3}\cos\zeta)\frac{r}{2}\sin\xi$$

B_3:

$$x_2 = x_0 + ((-\cos\zeta - \sqrt{3}\sin\zeta)\cos\zeta(1 - \cos\xi) - \cos\xi)\frac{r}{2}$$

$$y_2 = y_0 + ((-\cos\zeta - \sqrt{3}\sin\zeta)\sin\zeta(1 - \cos\xi) - \sqrt{3}\cos\xi)\frac{r}{2}$$

$$z_2 = z_0 + (\sin\zeta - \sqrt{3}\cos\zeta)\frac{r}{2}\sin\xi$$

Since for point B1, due to the corresponding rotator $y_2 = 0$

$$y_0 = -r\sin\zeta\cos\zeta(1 - \cos\xi) = -\frac{r}{2}\sin 2\zeta(1 - \cos\xi), \quad (8)$$

and for point B2 due to the rotator $y_2 = -\sqrt{3}x_2$

$$y_0 + ((-\cos\zeta + \sqrt{3}\sin\zeta)\sin\zeta(1 - \cos\xi) + \sqrt{3}\cos\xi)\frac{r}{2} = -\sqrt{3}(x_0 + ((-\cos\zeta + \sqrt{3}\sin\zeta)\cos\zeta(1 - \cos\xi) - \cos\xi)\frac{r}{2}),$$

which, taking into account (8), allows determining x_0

$$x_0 = (\cos^2\zeta - \sin^2\zeta)(1 - \cos\xi)\frac{r}{2} = \frac{r}{2}\cos 2\zeta(1 - \cos\xi) \quad (9)$$

Then for B_1:

$$x_2 = (\cos^2\zeta - \sin^2\zeta)(1 - \cos\xi)\frac{r}{2} + r(\cos^2\zeta(1 - \cos\xi) + \cos\xi) = ((2\cos^2\zeta - \frac{1}{2})(1 - \cos\xi) + \cos\xi)r$$

$$y_2 = 0$$

$$z_2 = z_0 - r\sin\zeta\sin\xi$$

for B_2:

$$x_2 = (\cos^2\zeta - \sin^2\zeta)(1 - \cos\xi)\frac{r}{2} + ((-\cos\zeta + \sqrt{3}\sin\zeta)\cos\zeta(1 - \cos\xi) - \cos\xi)\frac{r}{2} = \frac{r}{2}((-\sin^2\zeta + \sqrt{3}\sin\zeta\cos\zeta)(1 - \cos\xi) - \cos\xi)$$

$$y_2 = -r\sin\zeta\cos\zeta(1 - \cos\xi) + ((-\cos\zeta + \sqrt{3}\sin\zeta)\sin\zeta(1 - \cos\xi) + \sqrt{3}\cos\xi)\frac{r}{2} = \frac{r}{2}\sqrt{3}((-\sqrt{3}\sin\zeta\cos\zeta + \sin^2\zeta)(1 - \cos\xi) + \cos\xi)$$

$$z_2 = z_0 + \frac{r}{2}(\sin\zeta + \sqrt{3}\cos\zeta)\sin\xi$$

for B_3:

$$x_2 = (\cos^2\zeta - \sin^2\zeta)(1 - \cos\xi)\frac{r}{2} + ((-\cos\zeta - \sqrt{3}\sin\zeta)\cos\zeta(1 - \cos\xi) - \cos\xi)\frac{r}{2} = \frac{r}{2}((-\sin^2\zeta - \sqrt{3}\sin\zeta\cos\zeta)(1 - \cos\xi) - \cos\xi)$$

$$y_2 = -r\, sin\, \zeta\, cos\, \zeta\, (1 - cos\, \xi) + ((-cos\, \zeta - \sqrt{3}\, sin\, \zeta)\, sin\, \zeta\, (1 - cos\, \xi) - \sqrt{3}\, cos\, \xi)\, \frac{r}{2}$$
$$= \frac{r}{2}\sqrt{3}\, ((-\sqrt{3}\, sin\, \zeta\, cos\, \zeta - sin^2\, \zeta)(1 - cos\, \xi) - cos\, \xi)$$
$$z_2 = z_0 + \frac{r}{2}\, (sin\, \zeta - \sqrt{3}\, cos\, \zeta)\, sin\, \xi$$

Then, taking into account the coordinates of the vertices of the fixed base $C_1(R, 0, 0)$, $C_2(-\frac{R}{2}, \frac{R}{2}\sqrt{3}, 0)$, $C_3(-\frac{R}{2}, -\frac{R}{2}\sqrt{3}, 0)$ it is possible to determine the lengths of the drive links $L_i = |C_iB_i|$:

$$L_1 = \sqrt{(R - r\, ((2cos^2\, \zeta - \tfrac{1}{2})(1 - cos\, \xi) + cos\, \xi))^2 + (z_0 - r\, sin\, \zeta\, sin\, \xi)^2} \qquad (10)$$

$$L_2 = \left((R - r\, ((sin^2\, \zeta - \frac{\sqrt{3}}{2}\, sin\, 2\zeta)(1 - cos\, \xi) + cos\, \xi))^2 + (z_0 + \frac{r}{2}\, (sin\, \zeta + \sqrt{3}\, cos\, \zeta)\, sin\, \xi)^2\right)^{\frac{1}{2}} \qquad (11)$$

$$L_3 = \left((R - r\, ((sin^2\, \zeta + \frac{\sqrt{3}}{2}\, sin\, 2\zeta)(1 - cos\, \xi) + cos\, \xi))^2 + (z_0 + \frac{r}{2}\, (sin\, \zeta - \sqrt{3}\, cos\, \zeta)\, sin\, \xi)^2\right)^{\frac{1}{2}} \qquad (12)$$

The comparison of relations (10) - (12) with the formulas (5) - (7) from the work [8] shows the additional advantage of the proposed method to describe the rotation - the compactness of expressions for the solution of the inverse problem of kinematic analysis, i.e. fewer calculations.

Let us imagine the sequence of activities to compensate the parasitic movements of the tripod during control of a 6-DOF robot by correcting the task for the planar mechanism in the form of an algorithm:

1. Loading the trajectory (of x [n], y [n], z [n], ξ [n], ζ [n], φ [n] arrays, where *n* is the number of trajectory steps).
2. Organization of cyclic enumeration of the steps of the trajectory i = 1 ... *n*.
3. Capability checking of the implementation of the current step by a tripod. If a point with coordinates *z[i], ξ[i], ζ[i]* is outside its working space, the algorithm ends with a message about the impossibility to realize the trajectory.
4. Calculation of the required lengths of the tripod rods $L_1[i]$, $L_2[i]$, $L_3[i]$ from the known values of *z[i], ξ[i], ζ[i]* using the formulas (10) - (12).
5. Calculation of parasitic displacements of the tripod x_0 and y_0 from the values *ξ[i], ζ[i]* using the formulas (8), (9).
6. Correction of the task: calculation $x1 = x[i] + x0$, $y1 = y[i] + y0$.
7. Recalculation of the task from the global coordinates x_1, y_1, $\varphi[i]$ into the coordinates of the planar mechanism x', y', φ' according to the formulas determined by the mutual arrangement of the modules.
8. Capability checking of the implementation of the current step by a planar mechanism. If a point with coordinates x', y', φ' is outside its working space, the algorithm ends with a message about the impossibility to implement the trajectory.
9. Calculation of the required lengths of rods of the lower module $l_1[i]$, $l_2[i]$, $l_3[i]$ by the known values of x', y', φ' by the formulas from the work [7].
10. Switch to the next step.

As a result of the algorithm, the arrays $L_1[n]$, $L_2[n]$, $L_3[n]$, $l_1[n]$, $l_2[n]$, $l_3[n]$ of tasks for control loops [7] of input coordinates of the 6-DOF robot will be generated.

4. Discussion

The proposed algorithm was tested by computational experiment based on a specially constructed model of the system of interacting robots in the MATLAB package. Upper module parameters are as follows: $R = r = 30$ cm, $L_{max} = 1$ m, $L_{min} = 50$ cm; lower module are: $R = 1$ m, $r = 25$ cm, $L_{max} = 1$ m, $L_{min} = 50$ cm; fixed bases of robots are located in parallel planes at a distance of 1 m from each other., The workpiece is fixed on the movable platform of the planar robot. It is a sphere with a radius of 10 cm on a stand with an effective height of 5 cm, and there is a machining tool 10 cm long in the center of the movable platform of the tripod.

The task of the experiment was to draw a horizontal circular line on the surface of the part with the working tip of the tool. The tripod rotates the tool so that it always remains perpendicular to this surface and the planar robot moves the part horizontally.

It is necessary to note that in order to implement this trajectory of the tripod movement using the method proposed in this paper for the description of the rotation, it is enough to fix the variables z and ξ at the required level, and vary ζ within the range from 0 to 2π.

Let us compare the simulation results with the correction of tasks of the planar robot proposed in the work and without it. Figure 5 shows the trajectories of its movement in each of these cases. The figure shows that the correction of the task leads to the complication of the trajectory. Figure 6 shows the trajectories of movement of the working tip of the tool relative to the center of the processed sphere.

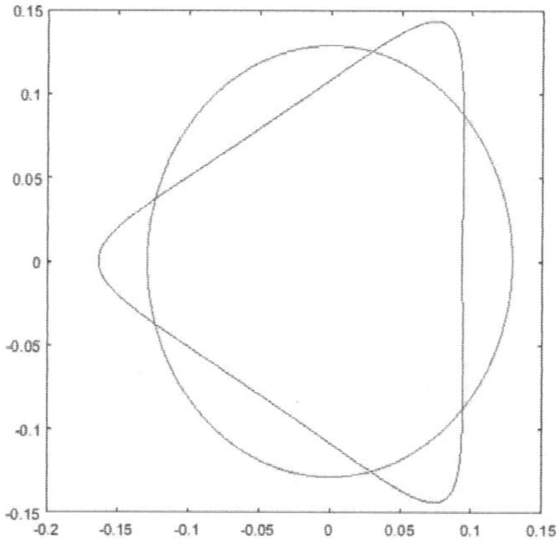

Figure 5. Trajectories of movement of a planar robot (x, y in meters): red line - no task correction; blue - with correction

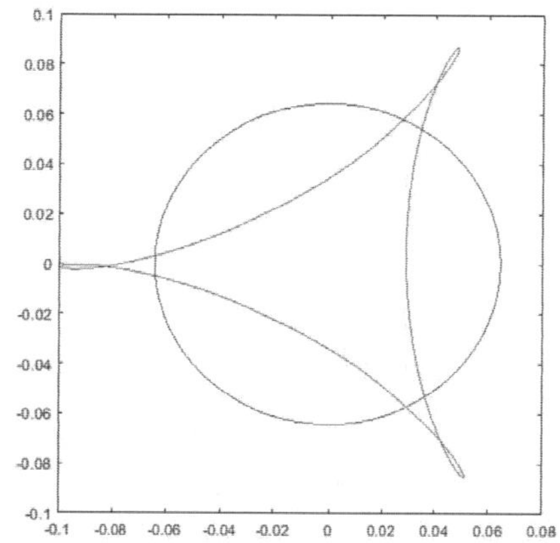

Figure 6. Trajectories of the movement of tool tip relative to the center of the processed sphere: red line - without correction; blue - with correction

We can see from the figure that without correction of the task, the tool is expected to move along an unacceptable trajectory, cutting deeply into the workpiece or breaking away from its surface. The correction of the reference allows ensuring the correct movement of the tool around a circle with a radius of $0.1 \sin 40^{\circ} \approx 0.064$ (m).

5. Conclusion

The paper studies the features of the management of a 6-DOF robot, created on the basis of two interacting modules. Each of them provides movement in three independent degrees of freedom. The authors proposed the method to describe the state of the movable platform of a tripod. According to the authors this method is more convenient for the case of the interaction of mechanisms than Tait – Bryan angles. The authors developed the algorithm to compensate parasitic movements of a tripod during control of a 6-DOF robot. The simulation of a 6-DOF robot with mechanisms of relative manipulation was carried out and the effectiveness of the developed algorithm was proved.

Acknowledgments

The reported study was funded by the Russian Science Foundation, according to the research project No. 16-19-00148, using equipment of High Technology Center at BSTU named after V G Shukhov

References

[1] Hao R, Wang J, Zhao J and Wang S 2016 *IEEE Transactions on Automation Science and*

Engineering **13(3)** 1399-1408
[2] Tang J, Cao D and Yu T 2019 *Proceedings of the Institution of Mechanical Engineers* **233(1)** 132-145
[3] Pi Y and Wang X 2010 *Mechatronics* **20(6)** 648-655
[4] Zhao D, Li S and Zhu Q 2010 *Int. J. Control Autom. Syst.* **8** 1189–1197
[5] Zheng S, Gao C, Niu X, Cong D and Han J 2019 *IEEE Access* **7** 90094-90105
[6] Virabyan L G, Khalapyan S Y and Kuzmina V S 2018 *Bulletin of BSTU named after V G Shukhov* **9** 106-113
[7] Khalapyan S, Rybak L and Malyshev D 2020 *Advances in Artificial Systems for Medicine and Education III* 368-379
[8] Posypkin M, Rybak L, Malyshev D and Gaponenko E 2018 *Proceedings at the 9th ICCM2018* 949–960

Automation of failure forecasting on the subsystems of the railway transport complex in order to optimize the transportation process as a whole

M V Konstantinova[1], **A A Olentsevich**[2], **V Yu Konyukhov**[1], **E A Guseva**[1] **and V A Olentsevich**[2]

[1] Irkutsk National Research Technical University, 83, Lermontova Street, Irkutsk, 664074, Russia
[2] Irkutsk State Transport University, 15, Chernyshevskogo St., Irkutsk, 664074, Russia

E-mail: konstantinova_mv@istu.edu

Abstract. The existing technology of working with railway rolling stock belonging to various operators should improve the safety of the transportation process. This is one of the priority issues of the operation of the railway transport complex and is aimed at improving the safety of the goods transported, reducing the cost of repairing losses associated with loss or damage to cargo, reducing the non-productive costs of eliminating various types of traffic safety violations. At the same time, the majority of "new" operators providing services do not set out to ensure the safety of the transportation process. The main condition for them is to get the maximum amount of income and profit. These reasons have led to significant changes in the pattern of relations between cargo owners, carrier, rolling stock operator companies and require the speedy automation of most production cycles in order to ensure full control of the situation on the smooth organization of the transportation process. The article assesses the safety of the railway transport complex and constructs a mathematical model of the impact of specific indicators of the violation types on the overall safety level. The described system of mathematical predictive regression-type models can be used in the future to predict the level of security in the selected areas of activity. The factors that have the most significant impact on the safety of technical means by using the developed program complex of failure forecasting have been identified. The proposed program will enable structural managers to make reasoned decisions to improve the safety of the transportation process in both freight and passenger transport.

1. Introduction

Throughout the modern history of Russia's development as a state, as well as in the foreseeable future, the railways of the Russian Federation retain the role of the system-forming dominant element of the general transport complex. They provide about 80% of the freight turnover, bring service to about 80 large and small regions, more than 70,000 enterprises of various industries, while providing more than 50% of export-import freight transportation. In addition to the role of the national transport integrator, railway transport is providing employment for more than 1.5 million people throughout the country. It plays a huge social role, while at the same time being an important component for ensuring Russia's defense capability and mobilization readiness [1, 2]. Biggest role belongs to infrastructure works [3-7] and to arranging the optimization transport process.

The third stage of the reforms being carried out in the railway transport complex was characterized by a reduction in the share of the inventory fleet railcars in connection with the creation of a series of separate freight transport companies and the transfer to operational management of part of the rolling stock fixed assets – the inventory fleet railcars. After that, these companies moved into the category of "private" that formed a competitive sector of the transport market in the industry. The existing technology for working with rolling stock belonging to various operators in fact should ensure an increase in safety of the transportation process. This is one of the priority issues of the functioning of the railway transport complex in any conditions and is aimed at: increasing the safety of transported goods, reducing costs for compensation for losses associated with loss or damage to cargo, as well as reducing the non-productive costs of the industry aimed at eliminating various kinds of traffic safety violations.

At the same time, the long period of railway transport in these conditions shows that a significant part of the «new» operators providing services does not set their task to ensure the safety of the transportation process. The main condition for them is to get the maximum amount of income and profit. These reasons have led to significant changes in the relationship patterns between cargo owners, carrier, and rolling stock operator companies.

The safety of the functioning of the railway transport complex includes the following main elements: safety of the transportation process; environmental safety; safety of transported goods, rolling stock and infrastructure; personal safety of staff and users. Ensuring safety and minimizing risks in the operation of all components of the complex depends on a number of factors:

- studying and obtaining timely, adequate information about the conditions of operation;
- working with the staff of operator companies;
- implementation of the transport process only in the conditions of serviceable vehicles and infrastructure;
- careful planning of the rolling stock operation.

The role of the above factors is multiplied when the cargo is transported, weighting down the consequences of a possible failure. Implementation of these areas of development, aimed primarily at retaining existing customer bases and creating new ones, is impossible without the speedy automation of most production cycles in order to ensure full control of the situation on the smooth organization of the transportation process. This is particularly the case with regard to the acceptance of cargo in transport, since it is this production cycle, with the existing technology of handling rolling stock that belongs to different operators, is the most labour-intensive and has a greater impact on the safety of the transportation process at all its stages [8].

2. Use of a system of mathematical predictive models to forecast the level of security in the designated areas of operation

At the current stage of development, the industry regularly increases the volume of traffic in all areas of activity, as well as the quality of services provided, on the premise of the safety of trains and the production of maneuvering, the preservation of material values, nature and other components of the system and its subsystems. Since rail transport is a multi-industry economy, representing a transfer conveyor of huge length, its uninterrupted and accident-free work depends on the operation of each individual part of its components - subsystems. The problem is to identify the factors that affect the safety level of the railway transport complex and its subsystems, taking into account the degree of this influence [9-12]. The solution to the problem involves the decomposition of the object of research, considered as a complex system.

In accordance with the accepted structure of the organization of the railway transport complex and the degrees of influence on the transportation process safety indicators, it is advisable to distinguish the following subsystems:

- traffic department;
- traction directorate;
- service of railcar facilities;

- track service;
- automation service;
- passenger service;
- freight and commercial operation service;
- electrification service.

A generalized characteristic of the level of safety of the railway transport complex at the federal and regional levels will be the vector of $y = (y_1, y_8)$, components of which are the number of defects in the activities of the allocated subsystems, occurred for a variety of reasons ($x_1 - x_{42}$). Let us introduce designations that characterize defects in work that occurred for the following reasons:

x_1 – departure for an occupied station-to-station block;

x_2 – derailment of railcars;

x_3 – defects of the railcar fleet;

x_4 – damage to the locomotive fleet;

x_5 – break of the automatic coupling

x_6 – train delay of more than one hour due to locomotive malfunctions;

x_7 – uncoupling of railcars in transit due to malfunction of roller axle-boxes;

x_8 – uncoupling of railcars in transit due to other technical faults;

x_9 – train delay of more than one hour due to technical faults of the railcars;

x_{10} – train derailment;

x_{11} – rail breakage;

x_{12} – train delay of more than one hour due to technical failures of the track;

x_{13} – train departs for an unprepared route.

x_{14} – blocking of the signal, which caused the running a red light at the station;

x_{15} – malfunction of alarm, centralization and locking devices with train delays of more than one hour;

x_{16} – uncoupling of the passenger railcars for technical faults;

x_{17} – malfunction of the wheel sets;

x_{18} – wreck of the cargo in transit;

x_{19} – uncoupling of railcars due to the violation of the technical conditions for the placement and fastening of cargo and cargo loading;

x_{20} – contact network malfunction;

x_{21} – auto-locking malfunction;

x_{22} – malfunction of the rolling stock inspection systems;

x_{23} – low quality and violation of the rolling stock depot maintenance and overhaul technology;

x_{24} – decrease in the quality of trains inspection;

x_{25} – high physical wear and ageing of fixed assets;

x_{26} – lack of and low quality of spare parts and necessary materials;

x_{27} – reduction in the number of employees.

x_{28} – increase in the guaranteed haul distances of mileage of loaded and empty railcars;

x_{29} – imperfection of the technical means and technologies of preventing defects cases;

x_{30} – violation of the splitting-up and braking mode of the cuts on the gravity sorting yards;

x_{31} – non-compliance with the requirements of instructions on the maintenance of inclined tracks and sorting tracks, arrangement of gravity sorting yards;

x_{32} – descent and collision in shunting operations;

x_{33} – drop of the parts of the rolling stock and cargo on the track;

x_{34} – non-fencing of the places of trackwork by stop signals.

x_{35} – technical condition of the body elements of freight railcars;

x_{36} – inconsistency in the actions of third-party organizations and transport companies;

x_{37} – low labor and technological discipline, insufficient staff level, poor knowledge and non-compliance with regulatory documentation;

x_{38} – theft of fixed assets;

x_{39} – violation of the established work and rest schedule of the workers;
x_{40} – loading of railcars in excess of the established standards;
x_{41} – lack of proper control by rail workers in the reception of cargo for transportation.
x_{42} – other reasons for the unspecified nature.

The level of security of each subsystem is estimated by one indicator - the number of defects within the range of the service.

Based on the results of the East Siberian Railways, the Russian Railways branch has built a mathematical model of the impact of specific indicators of the types of violations x_i, $i=\overline{1.42}$, on the general state of the level of safety of the regional railway transport complex with its decomposition on subsystems y_j, $i=\overline{1.8}$ [13-15].

The following criteria are used to assess the adequacy of models:
R – of multiple determination;
t – Student's;
F – Fischer's;
DW – Darbin-Watson's;
E – average relative approximation error.

To build each regression model for y_j, $j=\overline{1.8}$, several hundred (depending on the nature of the data) of its alternatives, followed by a choice of the best of them, were built based on their vector criterion of adequacy. With that, the algorithm described in work [2, 16-17], was used.

Traffic department $y_1 = 2.47 + 1.3x_1 + 1.48x_2 + 1.48x_3$,
(3.16) (11.66) (3.16)
$R = 0.975$, $F = 59.26$, $DW = 2.83$, $E = 7.85\%$,
Traction Directorate $y_2 = -77.78 + 0.98x_4 + 5.67\sqrt{x_5} + 17.24\sqrt{x_6}$,
(6.55) (3.86) (24.18)
$R = 0.993$, $F = 214.87$, $DW = 1.63$, $E = 0.76\%$,
Service of Railcar Facilities $y_3 = 9.4 + 7.54 \cdot 10^{-5} \cdot x_7^3 + 8.42\sqrt{x_8} + 10.57\sqrt{x_9}$,
(13.78) (7.46) (14.6)
$R = 0.991$, $F = 168.34$, $DW = 1.93$, $E = 0.69\ \%$,
Track Service $y_4 = 12.87 + 4.6\sqrt{x_{10}} + 0.28x_{11}^2 + 12.87\sqrt{x_{12}}$,
(4.92) (19.49) (10.40)
$R = 0.986$, $F = 106.8$, $DW = 3.02$, $E = 1.20\ \%$,
Automation Service $y_5 = 0.93 + 0.68x_{13} + 0.45x_{14}^{1.5} + 0.68x_{15}^{1.5}$,
(1.68) (7.71) (3.51)
$R = 0.926$, $F = 18.76$, $DW = 1.12$, $E = 5.08\ \%$,
Passenger Service $y_6 = 0.56 + 0.75x_{16} + 0.95x_{17}$,
(2.42) (3.21)
$R = 0.981$, $F = 92.51$, $DW = 1.74$, $E = 8.19\ \%$,
Freight and Commercial Service $y_7 = 1.52 + 0.64x_{18}^{1.5} + 0.31x_{19}^{1.5}$,
(10.2) (14.25)
$R = 0.983$, $F = 133.18$, $DW = 1.59$, $E = 6.71\ \%$,
Electrification Service $y_8 = 1.28 + 0.9x_{20} + 1.15x_{21} + 0.92x_{22}$,
(22.04) (17.59) (4.47)
$R = 0.993$, $F = 212.35$, $DW = 1.71$, $E = 1.26\ \%$.

Here, under the parameters of the models in brackets, there are the values of t - the Student's criterion.

An analysis of these adequacy criteria for each of the eight models shows that they all have high values that far exceed critical boundaries. Exception, in some cases, is Darbin-Watson's criterion (DW), indicating the presence of autocorrelation in some models (y_1, y_2, y_4, y_5). This only means that model data should only be used to address short-term forecasting tasks.

This system of mathematical predictive models of regression type is later used for medium-term forecasting of the level of safety of the East Siberian Railways of the Russian Railways OAO branch in the designated areas of operation. The results of the analysis are presented in Table 1.

Table 1. Failure rating for subsystems.

No.	Service	Failures
1	Traffic department	1. Car derailments 2. Other defects 3. Departure to an occupied station-to-station block
2	Traction directorate	1. Locomotive malfunction 2. Damage to locomotives 3. The break of the automatic coupling
3	Service of railcar facilities	1. Train delays of more than 1 hour due to technical faults of railcars 2. The uncouplings of railcars in transit due to the malfunction of roller axle boxes 3. The uncouplings of railcars in transit due to other technical faults
4	Track service	1. Rail breakage 2. Train delays of more than 1 hour due to technical track malfunctions 3. Train derailments
5	Automation service	1. Signal blocking, which caused the red light running at the station 2. Signaling devices malfunction with train delay of more than 1 hour 3. Departure of the train for an unprepared route
6	Passenger service	1. Faulty wheel sets 2. The uncoupling of the railcars for other technical reasons
7	Freight and commercial operation service	1. Uncoupling of railcars from the train due to violation of the Technical conditions for loading cargo 2. The collapse of the cargo in transit
8	Electrification service	1. Overhead system malfunction 2. Automatic blocking malfunction 3. Malfunction of special self-propelled railway equipment

3. Automation of the procedure for forecasting failures for the subsystems of the railway transport complex

In order to automate the procedure for predicting failures on the subsystems of the railway transport complex, a corresponding software package has been developed in the Delphi programming environment, using the example of the East Siberian Railways of the Russian Railways OAO branch. From the start, it includes the initial failure statistics for 2010-2019, as well as a regression model for each subsystem that allows you to make a prediction of the occurrence of a risk in any of the subsystems. The feature of the package is that when new data for the reporting period, the user in automatic mode can easily and conveniently rebuild any model and get new forecast values in all areas of the transportation process.

Forecasting stages:

- the first stage is to choose a service;
- the second is to set a forecast period;
- the third is to enter the upper and lower boundary for each factor;
- the fourth is to get a forecast.

The software package allows you to get three types of forecasts: optimistic, pessimistic and neutral. The optimistic forecast shows the dynamics of failures in a better situation, i.e. all negative factors are considered to be minimal. A pessimistic forecast is when all negative factors are considered to be maximal. A neutral forecast assumes that negative factors take on average values between the maximum and the minimum ones. The results of failure forecasting are presented in Table 2.

Based on the data of the mathematical model of the software package, a failure forecasting was compiled by the example of the East Siberian Railways branch of the Russian Railways for the period of 2018-2022. [13, 14]. The results of the failure prediction are presented in Table 2. The failure forecasting based on the complexes is graphically shown in Figure 1.

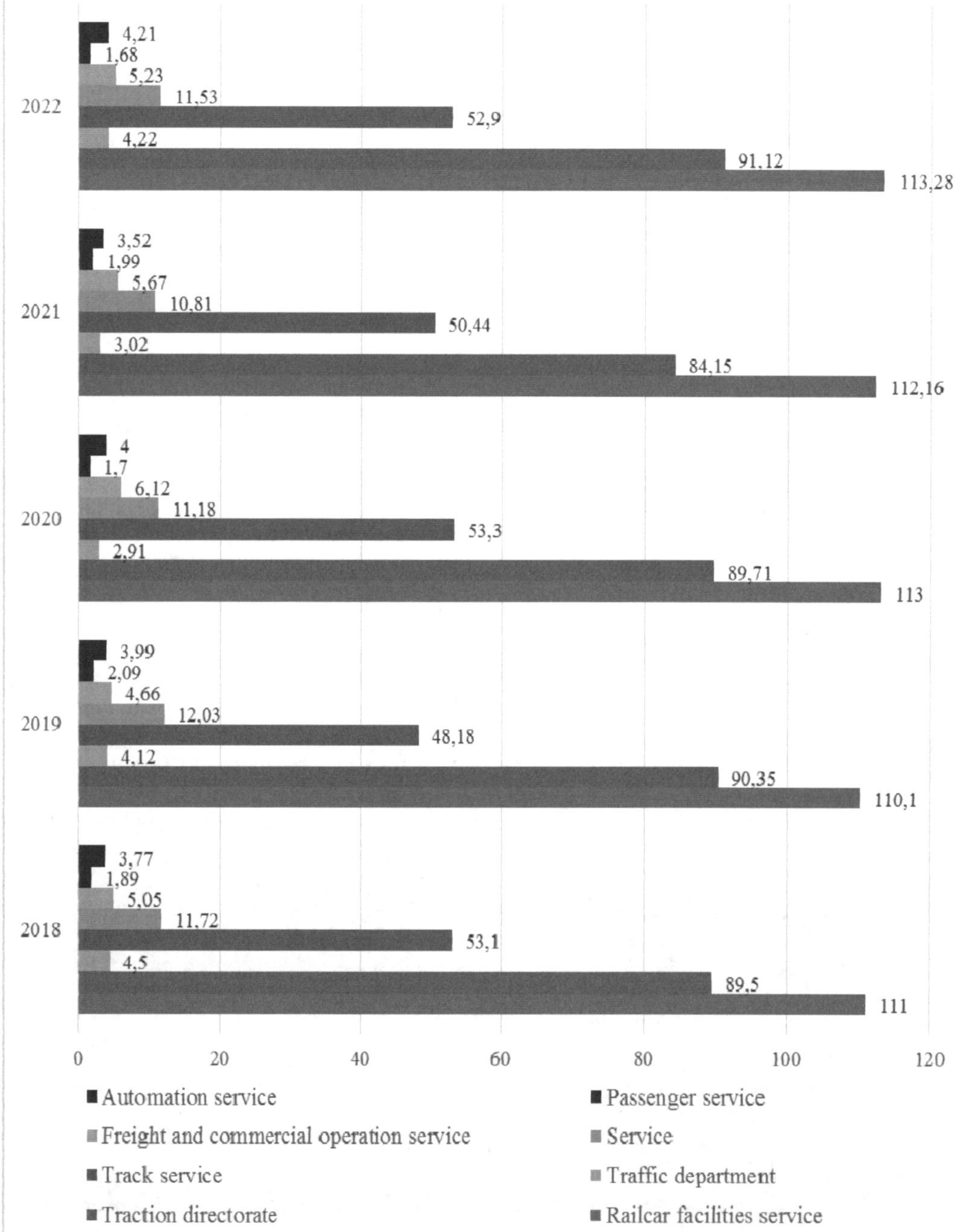

Figure 1. Failure forecasting for the period from 2018 to 2022.

Table 2. Failure forecasting results with rating.

No.	Service	Year	Failure forecasting	No.	Service	Year	Failure forecasting
1	Traffic department	2018	4.50	5	Automation service	2018	3.77
		2019	4.12			2019	3.99
		2020	2.91			2020	4.00
		2021	3.02			2021	3.52
		2022	4.22			2022	4.21
2	Traction directorate	2018	89.50	6	Passenger service	2018	1.89
		2019	90.35			2019	2.09
		2020	89.71			2020	1.70
		2021	84.15			2021	1.99
		2022	91.12			2022	1.68
3	Service of railcar facilities	2018	111.00	7	Freight and commercial operation service	2018	5.05
		2019	110.10			2019	4.66
		2020	113.90			2020	6.12
		2021	112.16			2021	5.67
		2022	113.28			2022	5.23
4	Track service	2018	53.10	8	Electrification service	2018	11.72
		2019	49.18			2019	12.03
		2020	53.30			2020	11.18
		2021	50.44			2021	10.81
		2022	52.90			2022	11.53

4. Conclusion

As the analysis of the rating of failures showed, for the transportation complex the second most important reason for failures and violations of normal operation is the lack of capacity of railway stations, i.e. the inconsistency of the level of development of the network of federal railways, the legislative framework for the organization of the process of transportation and fleets of cars moving on the network. The situation is aggravated by the lack of an effective system of centralized transportation planning, which leads to errors in the management of railcar fleets.

The transition to work with a private railcar fleet has had a negative impact on the efficiency of the entire transportation management system, the lack of proper infrastructure on the approaches to sea and river ports, the marshalling capacity of railway stations, focused on the handling of empty railcars traffic volume. It virtually eliminates the possibility of forming routes from empty railcars, creating additional load on the technical stations of the most heavy traffic areas.

As can be seen from the analysis of failures, the priority of the railway transport complex in the context of the constant increase in the volume of work is to ensure the safety of the transportation process organization. It is necessary to multilaterally consider various failures and their causes affecting the safe functioning of all subsystems. The development of probabilistic scenarios of regional growth involves continuous expansion and improvement of predictive tools.

References

[1] Olentsevich V A and Gozbenko V E 2013 Analysis of the reasons for the disruption of the safety of the railway transport system *Modern technologies. System Analysis. Modeling. Irkutsk* **1(37)** 180-183
[2] Retrieved from: http://www.rzd.ru
[3] Gozbenko V E, Korchevin N A, Kargapoltsev S K, Strakhov V M, Karlina A I and Karlina Y I 2020 Replacement of Graphite by Petroleum Coke in Rail Lubricants *Coke and Chemistry* **63(4)**

183-187

[4] Gozbenko V E, Kargapoltsev S K and Karlina A I 2019 Synthesis and structure of sulfur-containing polymers based on polymer industrial waste applied for rail lubrication *IOP Conf. Ser.: Earth Env. Sci.*

[5] Gozbenko V E, Kargapoltsev S K and Karlina A I 2019 Environmental benefits of new industrial waste-based lubricant compositions *IOP Conf. Ser.: Earth Env. Sci.*

[6] Kolosov A D, Gozbenko V E, Shtayger M G, Kargapoltsev S K, Balanovskiy A E, Karlina A I, Sivtsov A V and Nebogin S A 2019 Comparative evaluation of austenite grain in high-strength rail steel during welding, thermal processing and plasma surface hardening *IOP Conf. Ser.: Mater. Sci. Eng.* **560** 012185

[7] Konstantinova M V, Balanovskiy A E, Gozbenko V E, Kargapoltsev S K, Karlina A I, Shtayger M G, Guseva E A and Kuznetsov B O 2019 Application of plasma surface quenching to reduce rail side wear *IOP Conf. Ser.: Mater. Sci. Eng.* **560** 012146

[8] 2015 *Technical conditions for storing and securing cargo in railcars and containers* (Moscow: Yurtrans) p 544

[9] Suslov K V, Solonina N N and Smirnov A S 2014 Distributed power quality monitoring *Proc. of Int. Conf. on Harmonics and Quality of Power, ICHQP*

[10] Voropai N, Ukolova E, Gerasimov D, Suslov K, Lombardi P and Komarnicki P A 2019 Study on Cost-Effectiveness of Energy Supply Based on the Energy Hub Concept *Proc. of 2019 IEEE PES Innovative Smart Grid Technologies Europe, ISGT-Europe* 8905736

[11] Suslov K, Shushpanov I, Buryanina N and Ilyushin P 2020 Flexible power distribution networks: New opportunities and applications *SMARTGREENS 2020 – Proc. of the 9th Int. Conf. on Smart Cities and Green ICT Systems*

[12] Buryanina N, Korolyuk Y, Koryakina M, Lesnykh E and Suslov K 2019 Algoritm of Current Protection Based on Three Instantaneous-Value Samples *Proc. of 2019 IEEE PES Innovative Smart Grid Technologies Europe, ISGT-Europe* 8905542

[13] Balalaev S V 2008 *Traffic safety on railways: a textbook. Part 1: Safety Fundamentals* (Khabarovsk: DVGUPS) p 125

[14] Konyuhov V Yu, Konstantinova M V and Gladkih A M 2019 Determination of restored units spectrum of equipment and development of the assembly unit repair method at industrial enterprises *J. of Phys.: Conf. Ser.* **1353** 012047

[15] Konyuhov V Yu, Gladkih A M and Semenov V V 2019 Measures to the improvement of efficiency of a repair enterprise *J. of Phys.: Conf. Ser.* **1353** 012046

[16] Ivankov A N and Ivankova L N 1997 *Design of a new marshalling yard with an automated slide: a resource book* (Irkutsk: IRGUPS Publ.) 50 p

[17] Gozbenko V E, Kargapoltsev S K and Olentsevich V A 2017 Automated system of placing and fastening cargo on the open rolling stock of railway transport *Herald of Irkutsk State Technical University* **21(4(123))** 157-165

Machine-building enterprise performance and quality improvement tools

V Yu Konyuhov[1], A M Gladkih[1] and R S Zott[2]

[1] Irkutsk National Research Technical University, 83, Lermontova str., Irkutsk, 664074, Russia
[2] LLC "Tarasovsky coal", 1, Dzerzhinskogo str., Irkutsk, 664003, Russia

E-mail: konyuhovvy@ex.istu.edu

Abstract. Mechanical engineering is the leading sphere in the industrial countries. The engineering enterprises products take the top place in a realization of the scientific and technological progress achievements in the all economy areas. The development of mechanical engineering is impossible without quality management system introduction at enterprises (hereinafter QMS), which aim is to improve technological processes, develop human resources, reduce production costs and improve competitiveness. This paper reviews the influence of quality management system on the technological efficiency of a machine-building enterprise. To effectively use the concept of total quality management (TQM), a model of a quality management system at a machine-building enterprise is proposed in four areas: quality, process, personnel and resource management. Defining the quality management concept of engineering products, current trends in the sphere of engineering enterprise development were classified according to the products life cycle stages. Developing the total quality management concept, machine-building enterprises should form their own quality concept and its continuous improvement, as well as product competitiveness. The product quality improvement concept should include an adaptation of the enterprise system model leading to the concept of total quality management.

1. Introduction

The products quality is the most important criterion of a machine-building enterprise activity, which determines its success indicators in the market. Trends of the product quality level increase are dictated by the survival tendencies in a competitive market, where the enterprise sets the pace of a production technological development, promotes innovation, accelerates the process of improving production efficiency, an appropriate policy of enterprise resource consumption is being implemented [1-4]. In the modern market relations, the competition between enterprises is growing at the level of product quality.

The concept of product quality is based on the state standard GOST R 53603-2009 "Conformity assessment. Product certification schemes of the Russian Federation".

The products quality and services of an organization are determined by their ability to satisfy consumers and by their deliberate or unintended influence on relevant stakeholders. A quality-oriented organization encourages a culture that is reflected in behavior, attitudes, actions, and processes that determine the value by meeting the needs and expectations of consumers and other relevant stakeholders [2, 5]. The quality concept is closely related to the concept of "a production technological level" - a relative product quality characteristic based on comparing the indicators values that characterize the evaluated product technological perfection with the corresponding basic indicators and their values [6-8]. A product quality improvement in the sphere of the machine-building contributes to a significant economic renovation effect of all national economy sectors with new equipment.

When new machines are produced or previously manufactured ones are improved, when their quality system is being planned, it is necessary to use the machines quality indicators.

A quality standard indicator is a quantitative expression of one or more product properties relation to the conditions of the product manufacturing, operation and consumption [9].

There are two types of product quality indicators: individual and complex.

Individual quality indicators are related only to one of the product properties. Let's consider these indicators in relation to machine-building products. So, the internal combustion engine is characterized by the following individual indicators – power, speed, specific fuel consumption per power unit, etc. Individual quality indicators are used in the engineer's products, when studying the characteristics of a new part or modernization of a previously mastered one, as well as during technical and technological control of their production.

Complex indicators are created for a complete assessment of the machine quality only by their most important properties, which are used in technical and economic planning of machine-building production to assess the quality dynamics for certain periods of time [10-12].

When the quality of machine-building products is considered, it should be noted that almost the entire organization takes part in the process of "the product quality creation". Thus, the concept of total quality management (TQM) was born. TQM is a quality-oriented approach to management, based on the participation of all organization members and aimed at achieving long-term success by meeting the needs of the consumer and benefits for all organization and society members.

This idea is considered as the most progressive and implies the active participation of each employee in the products quality improvement of machine-building production processes.

2. Engineering products certification

Certification of the machine-building enterprise products is a variety of these products quality assessment, which determines their compliance with the established requirements of specific standards or other regulatory documents. The product certification process does not show the quantitative level of quality or technical characteristics of an industrial product, but only confirms that produced goods meet the established quality requirements [8, 13, 14].

Products certification is divided into mandatory and voluntary. Mandatory certification is a state control over the manufactured products quality and safety. The certification boundaries are determined by mandatory standards.

Any product is a subject of the voluntary certification if its manufacturer intends to obtain certain economic benefits from this product type manufacturing, as well as to guarantee its export and competitiveness. A certificate of conformity is a document issued according to the quality management system certification rules to confirm that the certified product meets the established requirements. Issuance of this document type completes the product certification process.

In addition to the certificate of conformity, an industrial enterprise or company can receive a conformity mark confirming that the product meets the requirements established in the standards. National and foreign products certification is carried out according to the same rules and principles.

Certification in mechanical engineering is a very complex procedure that includes several stages. The main purpose of such control is to verify that a product of any category comply with norms and standards.

The basis of the Russian mechanical engineering products standardization system consists of technical regulations. So, at the head of products conformity verification in this industry there are the technical regulations that formulate safety requirements for machines and all equipment types [1, 15].

Quality and safety verification process of the manufactured products study such conditions that are created specifically for engineering equipment manufacturing, installation, transportation, also there are recycling requirements. At the same time, for some kind of machines is issued a certificate or a declaration of conformity [5, 16].

In addition, mechanical engineering certification is also checked by the emergency situations ministry. This confirmation of the safety compliance is mandatory only for equipment used in fire extinguishing. The fire safety certification is issued at the first stage of standardization by the certification departments since the permit document of the technical regulations system is issued only with the permission of an expert accredited by the Ministry of emergency situations [8, 17].

If the fire safety certification is not mandatory, the manufacturer has right to obtain a permit on a voluntary basis.

The product certification process includes: establishment of products range to be certificated, specific certification requirements determination of products types, introducing them in regulatory and technical documentation, development of documents establishing the rules for products certification, conclusion of product certification agreements, certification of exporting enterprises production, accreditation of testing organizations that conduct certification tests of products from the list, certificate registration and issuance, supervision and control over the certification correctness and certified products quality, as well as information about the results of certification.

In the practice of foreign quality assessment system are used several certification models types. The most popular type of them is self-certification, when the product manufacturer itself, without involving relevant certification experts, declares that its products meet all the requirements and norms of the standard. Such product certification system can be used only by a manufacturer with a perfect reputation.

However, an independent third-party certification system is the most popular among the all certification systems. It has eight types.

The first type of certification system is based on the conducting standard experiments on product samples to meet the requirements of standards. Results of the experiments may prove or may not prove the submitted product sample conformity.

The second third-party type of the certification system is based on conducting experiments on product samples in specially approved testing organizations through periodic inspections of the product quality control.

The third type of certification system is based on conducting the same experiments as the first and second types, but with subsequent quality control of manufactured products, by periodic control tests of samples taken before sending them to the retail network [5, 18].

The fourth type of certification system includes conducting experiments on the product samples with subsequent products quality control by conducting periodic tests of samples that are taken from their production and trade spheres.

The fifth type of certification system involves conducting experiments on the product samples in specially approved organizations using evaluating systems to ensure product quality at the enterprise, followed by conducting periodic tests for samples quality control taken from the trade and production spheres. This system is the most complex, costly and expensive, but it receives a high level of distribution in industrialized countries and in international certification systems, because it provides to the consumer confidence of the product high quality level.

The sixth type of certification system implies only a preliminary assessment of the product quality assurance system at the enterprise. This type is used when the standard does not specifically regulate the requirements of final product quality, but only sets requirements for production process.

The seventh type of certification system is based on testing samples from each finished product batch. Their formation is carried out by special organizations, processing of test results using the mathematical statistics methods.

The last eighth type of certification system is based on testing each product for compliance with quality standards [7, 19].

The product certification process is preceded by the quality system certification process. The certificate of conformity of quality system is a document that verify quality system conformity to the ISO 9001 standard, which indicates the potential ability of the company to consistently produce high-quality products.

The main certification rules are defined by the international standard ISO 10011. The standard pays special attention to the qualification requirements for quality system inspectors (education, competence, work experience in a specific field of technology and in the field of quality assurance, personal qualities characteristics).

3. Implementation of a quality system concept at machine-building enterprises

Considering the problem of machine-building products quality management, it is necessary to take into account its impact on the production technical level in consumer industries. Also, it should be noted that machine-building products are always purchased by the customer to achieve their economic or social goals, so the customer achieves the planned results using machine-building products [9, 19, 20].

For effective use of the total quality management concept, based on the management concept developed by professor G. D. Segezzi at the University of Saint Gallen (Switzerland), the quality management system model at a machine-building enterprise in the quality management areas is proposed (table 1).

Table 1. Universal quality management elements and quality management levels.

Quality management levels	TQM elements	
	Aspects	Components
1. Normative level	Resource standards: - material; - labour; - financial; - temporary.	Resource costs + quality concept.
2. Strategic level	Structure: - organization; - main production; -auxiliary production; - service farms; - structural divisions of the quality management system.	Time + strategy qualities.
3. Operational level	- activity; - staff behavior.	Universal quality management + regulations + cost, quality and demand ratio.

The quality management levels, aspects and components shown in table 1 are recommended for the use in TQM principles implementing at the enterprise that already have been tested abroad and have a certain positive result. At the same time, large machine-building enterprises use all TQM levels at the same time. Defining the quality management concept in the machine-building industry, it is necessary to classify current trends (table 2).

Thus, as an example of this concept application can be UZTM "URALMASHZAVOD", which produces and processes iron ore, steel casting, oil, natural gas products and coal mining. UZTM "URALMASHZAVOD" machines and equipment are used in 42 European countries, Asia and Latin America. Now the company is aimed at the welding production quality improvement. Introduction of new innovative processes will improve the metal structures quality, thereby it will significantly reduce the work processes complexity. Standardization company is aimed not only at modernization of the welding production technologies but also at the welding production itself, which determines the quality improvement need of produced machines and parts.

Table 2. Theoretical content of the concept of the engineering product quality improvement.

Products lifecycle stage	Modern tendencies in machine-building industry
1. Design stage	- equipment aggregation; - high-capacity aggregates creation; - microprocessor- based systems management use; - resource saving; - ecological safety; - new materials use.
2. Production stage	- cooperation with other companies; - complete equipment production; - rotary and rotary-conveyor lines production; - new technologies use; - innovations dispersion.
3. Exploitation stage	- personnel training and retraining; - follow the operating instructions for machines and equipment.

To increase the efficiency of the plant's production, there was made a decision to expand investment program by allocating additional funds for the technical welding production re-equipment and for the new quality management technologies and implementation of standardization. The implementation of new standardization processes and automation increase of the welding production has significantly reduced machining share in the field of metal structures production, which means overall decrease of work laboriousness (on average, from 10% to 15% for different technological processes) [7, 20].

Another one investment program that was aimed at standardization and product quality management, made it possible to purchase a spectrum analyzer for metals and alloys, which aim is to improve quality and to reduce production time of the final product. This device (price - 6.5 million rubles) was delivered to "URALMASHZAVOD" in September 2018. The unit is made in Germany. With the use of this unit you can easily check metal workpieces for quality standards compliance directly at the production stage, as well as increase the speed of their analysis. Before innovations were made, due to the lack of own chemical laboratory, verification of metals that were received from suppliers had to be ordered from third-party companies. This process was 2 weeks long, which was slowing the equipment production time. With a use of own testing device, the most complex analyses are performed for a maximum of 1 hour long time.

In order to achieve the total quality management concept, engineering companies must create their own unique quality management concept. In general, the proposed concept of improving the quality management of engineering products is based on the seven main standardization and quality management principles:

- principle of resources and transport services;
- principle of process technologies;
- principle of process equipment;
- principle of personnel training;
- principle of enterprise structure organization
- principle of enterprise structure management;

- principle of organization's culture creation and improvement;
- principle of relationships with suppliers and consumers.

4. Conclusion

Taking into account these principles, relationships between the supplier and the consumer should be based on compromises, while all logistics chain participants should have a limited profitability and efficiency level. This concept implementation should be considered by all participants of the technological process [9].

Influence sphere of the enterprise economic and technological interests covers the strategic, organizational and operational decision-making levels in the field of product quality improvement.

In the field of industry standardization the strategic decision-making level means the choice of activities, consumers, processes, suppliers. Therefore, when enterprise chooses suppliers, there are important such moments as their parts quality, flexibility and logistics chains reliability. The organizational level of standardization includes decision-making in the field of production and market organization, shipment warehousing methods. The operational decision-making level consists of the continuous production and products sales organization.

In our opinion, the concept of product quality improvement should include system modeling of a company that adapts total quality management concept [10]. This quality management concept definition for engineering products is most capacious, specific and takes into account all the features of it.

References

[1] Bondaruk A M 2017 *Automated quality management systems in technological processes* (Ufa: Monograph) p 144

[2] Efimov V V 2018 *Quality management means and methods. Study Guide* (Moscow: KnoRus) p 670

[3] Kibirev Yu V, Potemkina M V, Lontsikh P A, Livshits I I and Kunakov E P 2017 Process efficiency and safety assessment of engineering-construction organizations based on lean manufacturing *Quality. Innovation. Education* **8(147)** 36-45

[4] Kunakov E P and Lontsikh P A 2017 The role of problematic personnel and increase of motivation in information security terms. Basic methods of working with personnel within the organization's information security system *School of Graduate Students: Collection of Articles of the All-Russian Scientific Conference* (Irkutsk) pp 44-50

[5] Livshits I I and Lontsikh P A 2015 Current certification analysis trends in the information security management systems according to the ISO 27001 requirements *Bull. of Irkutsk State Technical University* **3(98)** 268-273

[6] Marquardt M 2017 *The right questions - an effective management method. How leaders find optimal solutions by asking questions* (Moscow: Omega- L, SmartBook) p 240

[7] *Legal and regulatory technical documentation official website.* Electronic Fund. Retrieved from: http://docs.cntd.ru/document/1200124393

[8] Serenkov P S 2018 *Quality management methods. Risk management standardization methodology* (Moscow: INFRA-M) p 221

[9] Khokhlova G I, Kretova N V and Sergeev V A 2018 Problems and factors affecting the development of innovation activities in Russia *Management of economic systems* **3** 33

[10] Shishkin I F 2018 *Metrology, standardization and quality management* (Moscow: Standards) p 342

[11] Leonovich D S, Zhuravlev D A, Karlina Yu I, Govorkov A S and Karlina A I 2020 Automated assessment of the low-rigid composite parts influence on the product assemblability in the GePARD system *IOP Conf. Ser.: Mater. Sci. Eng.* **760(1)** 012038

[12] Karlina A I, Gozbenko V E, Kargapol'Tsev S K and Karlina Y I 2019 Methods for controlling the vibration state of technical facilities *J. of Phys.: Conf. Ser.* **1384(1)** 012019

[13] Gozbenko V E, Kargapol'Tsev S K, Karlina Yu I and Karlina A I 2019 Approximation of amplitude-frequency characteristics using equidistants *J. of Phys.: Conf. Ser.* **1384(1)** 012014
[14] Gozbenko V E, Kargapol'Tsev S K, Kuznetsov B O, Karlina A I and Karlina Yu I 2019 Determination of the principal coordinates in solving the problem of the vertical dynamics of the vehicle using the method of mathematical modeling *J. of Phys.: Conf. Ser.* **1333(5)** 052007
[15] Kuznetsov B O, Gozbenko V E, Kargapol'Tsev S K, Karlina Yu I and Karlina A I 2019 Dynamic vibration protection of the railway carriage *J. of Phys.: Conf. Ser.* **1333(5)** 052018
[16] Voropai N, Ukolova E, Gerasimov D, Suslov K, Lombardi P and Komarnicki P 2019 Simulation approach to integrated energy systems study based on energy hub concept *IEEE Milan PowerTech* 8810666
[17] Ilyushin P and Suslov K 2019 Operation of automatic transfer switches in the networks with distributed generation *IEEE Milan PowerTech* 8810450
[18] Suslov K, Solonina N and Gerasimov D 2018 Assessment of an impact of power supply participants on power quality *Proc. of Int. Conf. on Harmonics and Quality of Power* pp 1-5
[19] Konyuhov V Yu, Gladkih A M, Galyautdinov I I and Shchadova E I 2020 Calculations of efficiency in implementing progressive mold forming methods *IOP Conf. Ser.: Mater. Sci. Eng.* **760(1)** 012027
[20] Konyuhov V Y, Gorban A V and Gladkih A M 2019 Investment in the improvement of maintenance service efficiency of processing equipment of an industrial enterprise *J. of Phys.: Conf. Ser.* **1353(1)** 012102

IOP Conf. Series: Materials Science and Engineering 1064 (2021) 012022 doi:10.1088/1757-899X/1064/1/012022

Assessment of the adhesion strength of composite curved elements

S I Koryagin, O V Sharkov and N L Velikanov

Immanuel Kant Baltic Federal University, 14, A. Nevskogo str., Kaliningrad, 236016, Russia

E-mail: skoryagin@kantiana.ru

Abstract. Modern composites have good strength properties. But insufficient adhesive strength often leads to the delamination of a coating. The Poisson's ratio, modulus of elasticity, and thermal coefficient of linear expansion are of particular importance among the characteristics of materials that determine different properties of the coating and metal. The ratio of metal and coating thicknesses is also important. The work suggests that the temperature deformations associated with the polymerization of a coating influence the process of delamination of the reinforced polymer coating. Evaluation of the potential of the proposed calculated dependencies was carried out according to the results of experiments. A polymer coating consisting of «Sprut-Plus» adhesive composition and a filler (T-11-GVS-9 glass fabric) was applied to a surface of St.3 steel. For experimental studies, the reinforced polymer coating was kept in water and machine oil for 100 days. The thickness of the coating was taken to be 10 mm, and the thickness of the metal - from 3 to 6 mm. Calculated and experimental dependences of radial stresses on the relative thickness of the polymer coating are obtained. An increase in radial stresses at the interface between the layers of the metal and the reinforced polymer coating after holding the composite material in water has been established; it does not exceed 10%. Analysis of the results showed that the destruction of the adhesive metal-polymer joint is more likely than that of the polymer-polymer adhesive joint. In addition, the radial stresses are at their maximum at the interfaces. Prospects for further research in this direction are associated with express methods of repairing metal structures of complex configuration. It is necessary to increase the efficiency and reliability of calculations of curved elements consisting of metal with a reinforced polymer coating. This can be achieved by a combination of experimental and theoretical research.

1. Introduction

Composite materials combine the protective properties of the coating with the mechanical strength of the metal. Therefore, their application in real operating conditions of machine-building structures is promising [1–3]. In modern mechanical engineering, composite materials are increasingly used for the manufacture of structural elements with various functional purposes [4–8].

The bearing capacity of composite elements largely depends on the adhesive joint between the metal and the reinforced polymer coating [9–14], which is determined by the magnitude of the radial stresses σ_r at their interface.

It should be noted that some of the engineering structural elements have a curved shape [15–17]. Information on stresses in polymer coatings on a curved metal surface is limited. The existing calculation methods are applicable when the metal-coating composition is deformed without bending during the formation process. But in real structural elements, areas with a large curvature are, as a rule, in a moment stress state.

2. Materials and Methods

The authors use theoretical and experimental methods to perform a comparative assessment of the radial stresses σ_r at the interface between metal layers and reinforced polymer coatings that occur during the manufacture of composite elements with a large curvature of the working surfaces.

The effect of aggressive environmental factors (water, air, and engine oil) on the adhesive strength of composite elements is taken into account.

Theoretical study is based on the theory of the stress-strain state of thin-walled shells [18] and the general theory of elasticity [19]. At experimental studies were used the methods of turning and cutting rings [20, 21], together with a method of strain measurement [22].

The test specimens were composite elements with a curved surface with a radius R. St.3 steel was used as a base layer, on which a polymer coating consisting of «Sprut-Plus» adhesive composition and a filler (T-11-GVS-9 glass fabric) was applied. The application of a reinforced polymer coating on the metal surface of the element was carried out in air at a temperature of 18...20 °C.

In the process of manufacturing curved composite elements, such effects as a disruption of the uniformity in the coating layers, a delamination of the coating, and a cracking of the adhesive composition are possible. These defects may be caused by the following factors:

1. Internal residual stresses in the coating caused by differences in the physical and chemical properties of the adhesive composition and filler.
2. Formation of gaseous products during polymerization.
3. Thermal stresses due to the difference in the coefficients of linear expansion of the metal and the coating.
4. Thermal stresses due to temperature gradients in the coating during polymerization.

In addition, the curvature of the surface also changes the temperature field.

The above factors significantly affect the mechanical properties of the composite and can cause delamination and porosity of the material. It is known that the mechanical interaction between the metal and the coating is mainly influenced by the difference in their Poisson`s ratios and, to a lesser extent, by the elastic moduli.

3. Theoretical estimation of radial stresses

The stress state of a curved composite element (Figure 1) can be considered as a set of plane stress state on the coating surface and three-dimensional stress state at the metal-coating interface. The plane stress state is estimated by the circumferential σ_c and longitudinal (meridian) σ_m stresses, and the three-dimensional stress – by a combination of three stresses (σ_x, σ_y and σ_z), of which only radial $\sigma_y = \sigma_r$ stresses are necessary for subsequent calculations.

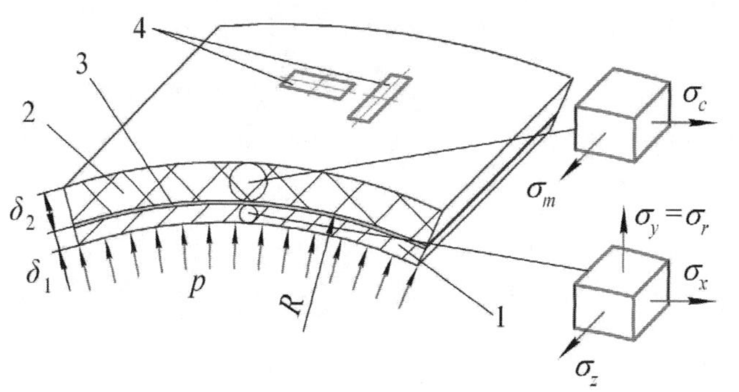

Figure 1. Calculation diagram of a curved element: 1 – metal layer (base); 2 – polymer coating; 3 – adhesive layer; 4 – strain gages.

Assuming that the main factors of delamination of the reinforced polymer coating are temperature deformations arising in the process of polymerization of the coating applied to the metal surface, the values of radial stresses can be estimated by the dependence:

$$\sigma_r = \frac{E_1\delta_1 E_2\delta_2 \Delta t(\alpha_1 - \alpha_2)}{R(E_1\delta_1 + E_2\delta_2)}, \tag{1}$$

where E_1, E_2 are the moduli of elasticity of the metal and coating; δ_1, δ_2 are the thicknesses of the metal layer and coating; α_1, α_2 are the thermal expansions of metal and coating, respectively; Δt is the temperature difference, $\Delta t = t_2 - t_1$; R is the radius of curvature.

Dependence (1) was obtained by solving the problem for a thin-walled curved element [18, 19] from the strain compatibility condition of the metal and coating layers.

4. Computational and experimental evaluation of radial stresses

To confirm the correctness of the calculated dependence (1), experimental studies of composite curved elements were carried out.

The study considered a case unfavorable for an adhesive joint when the resulting stresses of the base material were 0.3...0.5 of the yield point.

The approaches described in the works [20, 21] was used for determining the radial stresses. The stresses were determined by a ring method with variations – with the removal of polymer coating layers, without removing the layers, and cutting the ring in the radial direction.

Without removing the polymer coating, the radial stresses, assuming the correspondence $p = \sigma_r$, were found using the well-known Barlow's formula [22, 23]:

$$\sigma_r = \frac{\sigma_{c2}(\delta_1 + \delta_2)}{R}, \tag{2}$$

where σ_{c2} is the circumferential (shear) stresses on the surface of the composite element.

The value of stresses σ_{c2} was determined using strain measurement [24]. The strain gauges were glued to the convex surface of the composite element from the side of the coating.

The transformed Barlow's formula was used to determine the stresses in the case of successive removal of polymer coating layers from a metal surface:

$$\sigma_r = \frac{\varepsilon_{\tilde{n}1} E_1 \delta_1}{R}, \tag{3}$$

where $\varepsilon_{\tilde{n}1}$ is the deformation in the metal layer.

In this and subsequent cases, the strain gauges were glued to a metal base.

The stresses were also determined through the volume fraction of the polymer coating and the mechanical characteristics of the composite:

$$\sigma_r = \frac{\varepsilon_{max} k_1 (1-2\nu_1)(\nu_1 - \nu_2)\nu_2(1-\nu_2)}{\nu_2(1-2\nu_1)\left(1-\dfrac{k_1}{k_2}\right)+1+(1-2\nu_1)\dfrac{k_1}{k_2}}, \tag{4}$$

where ν_1, ν_2 are Poisson's ratio of the base and coating, respectively; $k_1 = E_1/(1+\nu_1)(1-2\nu_1)$; $k_2 = E_2/(1+\nu_2)(1-\nu_2)$ are the constants of the base and coating material, respectively; ε_{max} is the ultimate deformation.

While simultaneously cutting the rings and removing the coating, the stresses were determined by the following dependence:

$$\sigma_r = \frac{D}{\pi\delta_1 R^3 \Delta_R}, \tag{5}$$

here Δ_R is the change in the ring radius; D is the reduced rigidity of the ring in bending

$$D = \frac{E_2 \delta_2^3 (1 + 4a_0 a_1^3 + 6a_0 a_1^2 + 4a_0 a_1 + a_0^2 a_1^4)}{12(1 + a_0 a_1)},$$

where $a_0 = \dfrac{E_1}{E_2}$; $a_1 = \dfrac{\delta_1}{\delta_2}$.

5. Comparative analysis of the results

The effect of the thickness of the bearing metal layer at a constant thickness of the reinforced polymer coating on the value of radial stresses σ_r under various environmental factors has been investigated.

Calculations of radial stresses were carried out according to dependencies (1)–(5), in which the following magnitudes were taken: Δt =100 °C; E_1=2·10^5 MPa, E_2=1.9 10^4 MPa; α_1=12·10^{-6} 1/deg, α_2=9·10^{-5} 1/deg; R=300 mm; ν_1=0.30, ν_2=0.15.

When carrying out calculations and experimental studies, the thickness of the reinforced polymer coating δ_2 was taken equal to 10 mm, and the metal thickness δ_1 was taken in the range from 3 to 16 mm.

In addition, the composite elements made in air at a temperature of 18...20 °C were kept in water and machine oil for 100 days.

Tables 1–2 show the behavior of radial stresses obtained by theoretical and calculation-experimental methods, depending on the thickness of the polymer coating.

Table 1. Calculated and experimental radial stresses.

δ_2, mm	Value of σ_r, MPa				
	By dependencies			Ring turning method	Ring cutting method
	(1)	(2)	(4)		
3	80.0	74.4	71.3	68.1	76.3
5	88.8	82.4	76.9	75.5	83.9
8	94.8	88.2	84.8	80.6	89.8
10	96.9	90.1	86.7	81.4	92.0
12	98.4	91.5	88.4	83.7	92.9
14	99.6	92.6	89.2	84.7	93.7
16	100.3	93.2	89.9	85.3	94.4

A slight increase (by 5…9%) in radial stresses at the interface between the layers of the metal and the reinforced polymer coating after keeping the composite material in water for 100 days can apparently be explained by the interaction of the coating with water (Table 2).

Table 2. Radial stresses after exposure to water and oil.

δ_2, mm	Value of σ_r, MPa			
	Ring turning method		Ring cutting method	
	In oil	In water	In oil	In water
3	68.4	71.8	76.5	80.3
5	75.3	79.1	84.0	88.2
8	80.7	84.7	89.9	93.6
10	81.2	85.3	92.2	96.6
12	83.9	88.0	93.2	97.7
14	84.5	88.8	93.9	98.2
16	85.4	89.7	94.5	99.7

The mechanism of the effect of water on the adhesive strength of bond between a metal and a reinforced polymer coating is far from clear.

Observations have shown that polymer-coated metal adhesive joints are often destroyed in an interfacial manner, i.e. the interface is particularly sensitive to a decrease in adhesion strength under the influence of water.

In this case, water is accumulated near the interface, and its further penetration into the adhesive joint of the metal and the coating and between the layers of the polymer coating is difficult. The observations made it possible to establish that the destruction of the adhesive metal-polymer joint is more likely than the polymer-polymer adhesive joint destruction.

The density of the reinforced polymer coating decreases slightly (by about 1%) due to the destruction of macromolecules of the adhesive composition at the interface under the influence of water.

After exposure for 100 days in engine oil, the values of radial stresses were almost equal to those in air at a temperature of 18...20 °C (Table 2).

Studies have shown that the value of the radial stress at the interfaces reaches its maximum and decreases as it approaches the surface, i.e. with an increase in the thickness of the composite element, the magnitude of this stress increases.

6. Conclusion

A good qualitative and quantitative agreement was obtained between theoretical and experimental values of radial stresses at the interface between the metal and polymer coating layers. The greatest discrepancies between the maximum and minimum stress values do not exceed 17...20%.

It is shown that the main reason for the appearance of radial stresses in composite elements is the difference in Poisson's ratios and linear expansion of materials. In addition, as a result of plastic flow of the metal, the difference between the Poisson's ratios of the base and the coating increases, since the material of the base (metal) has a higher Poisson's ratio, i.e. the magnitude of stresses increases significantly with the development of plastic flow.

The correctness of using dependencies (1)–(5) for engineering estimation of radial stresses at the interface between metal and polymer coating layers has been experimentally confirmed.

References

[1] Zhang K, Zhang W and Ding X 2019 *Procedia CIRP* **85** 114–120
[2] Kheruvimov A, Nikonov A and Aliukov S 2018 *SAE Technical Papers* **2019-April (April)**
[3] Gay D, Hoa S V and Tsai S W 2002 *Composite materials: Design and applications* (Boca Raton: CRC Press) p 552
[4] Somireddy M, Singh C V and Czekanski A 2020 *Engineering Failure Analysis* **107** 104232
[5] Sharma S, Sudhakara P, Nijjar S, Saini S and Singh G 2018 *Materials Today: Proceedings* **5(14)** 28195–28202
[6] Sun H, Tao Y and Zhang J 2020 *J. of Alloys and Compounds* **847** 156527
[7] Koryagin S I, Sharkov O V and Velikanov N L 2018 *Materials Science Forum* **938** 46–53
[8] Xu S, Chena L, Gong M, Hu X, Zhang X and Zhou Z 2017 *Composites Part B: Engineering* **111** 143-147
[9] Nurprasetio I P, Budiman B A and Aziz M 2018 *Int. J. of Adhesion and Adhesives* **85** 193–201
[10] Sun S, Yang R, Han J, Guo H and Xing L 2016 *Surface and Coatings Technology* **297** 19–26
[11] On S Y, Kim M S and Kim S S 2017 *Composite Structures* **159** 636–645
[12] Ekrem M and Avcı A 2018 *Composites Part B: Engineering* **138** 256–264
[13] Bartkowiak M, Czech Z, Mozelewska K and Nowak M 2020 *Polymer Testing* **90** 106603
[14] Zhai L L, Ling G P and Wang Y W 2008 *Int. J. of Adhesion and Adhesives* **28(1–2)** 23–28
[15] Zappino E, Zobeiry N, Petrolo M, Vaziri R, Carrera E and Poursartip A 2020 *Composite Structures* **241** 112057
[16] Li B, Li H, Hu X, Feng G, Yao X and Wang P 2020 *J. of the European Ceramic Society* **40(8)**

2768–2780
[17] De Miguel A G, De Pietro G, Carrera E, Giunta G and Pagania A 2018 *Computer Methods in Applied Mechanics and Engineering* **337** 481–500
[18] Ventsel E and Krauthammer T 2001 *Thin Plates and Shells: Theory: Analysis, and Applications* (Boca Raton: CRC Press) p 688
[19] Lurie A I 2010 *Theory of Elasticity* (Berlin: Springer Science & Business Media) 1050 p
[20] Totten G E, Howes M and Inoue T 2002 *Handbook of Residual Stress and Deformation of Steel* (Ohio: ASM International) p 499
[21] Niku-Lari A 2002 *Advances in surface treatments: Technology-Applications-Effects. Vol. 4. Residual Stress* (Amsterdam: Elsevier) p 586
[22] Sinclair G B and Helms J E 2015 *Int. J. of Pressure Vessels and Piping* **128** 1–7
[23] Salehi M M, Khalkhali T and Davoodi A A 2016 *Polymer Science Series A* **58** 567–577
[24] Keil S 2017 *Technology and Practical Use of Strain Gages: With Particular Consideration of Stress Analysis Using Strain Gages* (Berlin: Ernst & Sons) p 512

A method for monitoring the violation of the integrity of pipeline communications using a radio wave vibration sensor

A E Pilipenko[1], E V Rudyaga[1], V V Kukartsev[1,2], V S Tynchenko[1,2], S O Kurashkin[1] and D V Rogova[1]

[1]Reshetnev Siberian State University of Science and Technology, 31, Krasnoyarsky Rabochy av., Krasnoyarsk, 660037, Russian Federation
[2] Siberian Federal University, 79, Svobodny pr., Krasnoyarsk, 660041, Russian Federation

E-mail: pilipenko.alesya@mail.ru

Abstract. The article discusses the implementation of a continuous vibroacoustic control system, in order to detect leaks and quickly prevent emergencies, describes the principles of their functioning. The existing method of recognizing leaks is considered. The analysis of non-contact methods for measuring vibration parameters is carried out, radio wave methods are highlighted as the most promising. A non-contact radio wave vibration sensor is considered, which is introduced into the pipeline monitoring system. The proposed systems of vibroacoustic control allow one to effectively diagnose the state of the pipeline at low time costs and with a low error in detecting a useful signal. The physical and mathematical meaning of the method being introduced was also considered. The scope of application of the research results relates to industries (heat power, gas and oil fields), in which complex dynamic objects are used with the need to use a non-contact method for monitoring vibration and displacement parameters.

1. Introduction

In the modern world, very little attention is paid to diagnostics of the technical condition of pipelines. The service life of pipeline systems is sometimes more than 30 years. Over the course of this time, they are affected by many different factors, for example, natural and climatic influences, changes in pressure inside the pipes. In addition, the geographical features of the place of laying the pipes have a great influence.

All of the above factors lead to the fact that the service life of the systems is reduced and leads to a large number of leaks, the subsequent development of which can cause an emergency. This is the reason for the need for prompt and accurate identification of possible problem areas of leaks, localization of earthworks and fast reconstruction of water supply and heating networks. Since the costs of a complete replacement of pipeline systems are extremely high, long-term and not always possible, for example, pipelines under highways, it is necessary to implement pipeline monitoring systems. The main challenge is to optimize, improve the quality, reliability and safety of pipelines. To achieve this goal, it is necessary to reduce the amount of damage to the water supply network, as well as to reduce the cost of repair work. Thus, the solution to the problem of monitoring the condition and rapid response to emerging pipeline failures is important and relevant today. We consider a system for diagnosing the state of pipeline systems using a radio wave vibration sensor and describe their effectiveness.

2. Existing method for diagnosing leaks

The most widespread defects are thinning and destruction of the integrity of the walls, which leads to metal corrosion, cracking, all of these factors lead to leakage. Detecting a defect as quickly as possible is of paramount importance, as early identification and elimination of the problem will prevent further major accidents.

These defects can be detected using a radio-wave vibration sensor, which is based on the excitation of vibroacoustic signals in the pipe shell with their subsequent diagnosis at a given area. The result of the operation of the investigated object is signaled by a sensor that provides a small amount of information about the state of the pipeline section. But the combination of such signals allows you to identify defects and assess the technical condition of the product pipeline.

Today, the principle of detecting leaks is as follows. A pair of copper conductors is located around the perimeter of the pipeline, under a thermal insulating coating. To diagnose the occurrence of leaks and determine the distance to it, the change in resistance value is measured. The low efficiency of the current method is manifested in the fact that, firstly, if a leak occurs from the reverse side of the pipe, then the short circuit of the conductors may not occur immediately after the leak appears. Secondly, shorting of conductors can occur without malfunctions. Defects on the walls of the hot water supply pipeline lead to a significant decrease in energy efficiency in case of violation of thermal insulation, and if damage is not detected for a long time, accidents are possible. At the moment, the diagnosis of such damage is not performed, because of this, their presence can be judged only by how much the ground surface over the pipeline is warming up [1].

3. Comparative analysis of sensors

To begin with, we will consider the existing types of sensors and evaluate their effectiveness for subsequent implementation into the system (Table 1).

Table 1. Sensor comparison.

Duration of implementation	Optical	Ultrasonic	Radio waves
	Depends on excavation	Depends on excavation	Small
Is excavation necessary	yes	yes	no
Influence of natural conditions	Smoke, water vapor and dust prevent	Does not affect	Does not affect
Frequencies	0-20 MHz	0-3000 Hz	0-250 kHz
Working temperature	-60 to + 60°C	-40 – 85	-60 °C to +80 °C
Does it register interference	no	no	no
Signal reading range	0.1-10 m	No more than 1.5-2 m	0.2-10 m
Is there a possibility of replacement without destroying the integrity of the system	no	no	yes
Disadvantages	Complexity, high cost of equipment; high energy consumption.	Dependence on pressure, temperature, error is possible. Low resolution, low dynamic range.	Calibration complexity (for amplitude methods)
Advantages	Fire safety high precision, small size, corrosion resistance.	Suitable for pipes of any diameter. Cheapness and compactness of equipment.	Wide dynamic range, measuring poor under conditions

When choosing sensors, we consider the following characteristics:

- Compact and sturdy mechanism that will allow for installation in certain operating conditions.
- The design should guarantee the maximum sensitivity of the fiber to various influences, depending on the conditions of use.
- The use of various types of structures in terms of materials and properties, based on the impact of environmental conditions (aggressive chemicals, high or low temperatures, crushing loads, etc.).
- The structure should be light, facilitating the installation work [2].

As shown by a comparative analysis, the most advantageous sensor is a radio wave.

4. Introduction of a radio wave vibration sensor

The leak detection method used today is not the most reliable and effective, so we will try to introduce a radio wave vibration sensor into the system. After considering existing sensors, we concluded that the radio wave vibration sensor has a low cost, while having many advantages. Among them, the possibility of measuring vibration occurring at distances inaccessible for ultrasonic measurements, at distances of more than 2 m. This makes it possible to install the sensor not too close to the pipeline, thereby not carrying out excavations, which reduces the cost of implementation (time and money for workers and equipment) (Figure 1).

Also, a significant advantage is the ability to replace it with a new observation point without dismantling or stopping the controlled object, which cannot be realized by contact sensors, since they require excavation of the pipeline to directly install or replace the sensor with them, thereby increasing the cost of additional excavations. The superiority of a radar vibration sensor over laser ones is the ability to measure vibration on an object, even if there is water vapor or other optical interference. Also, the advantage of a radar vibration sensor is the accuracy of determining the frequency and amplitude of vibration using a sensor operating in the millimeter wavelength range, in contrast to sensors that use traditional radio bands (3-30 GHz). This is explained as follows. In order for the phase of the reflected signal at the receiver to change 360 degrees, the distance must change by half the wavelength. One and a half millimeters for a 3-mm 94 GHz sensor gives 360 degrees, and for a 3-cm 9 GHz it is only a tenth. In other words, the 94 GHz sensor turns out to be more sensitive to vibration compared to lower frequency radar wavelength ratio sensors [3].

Frequency and amplitude measurements using 1D, 2D or 3D can be used to analyze vibration parameters by taking data from vibration sensors. For this, it is necessary to introduce, respectively, 1, 2, or 3 radar vibration sensors [4].

Since vibration measurement can take place at a distance of more than two meters, the installation of the sensor does not take place too close to the pipeline, which makes it possible to reduce the cost of implementation, since you will not need to carry out large-scale excavations to install it.

5. The physical meaning of the radio wave method

Radio wave measurement methods are based on the use of dependencies on the controlled value of various parameters of electromagnetic systems used as primary measuring transducers. These parameters include: the amplitude and phase of the reflected signal; frequency of electromagnetic oscillations of the system, its quality factor; the number of excited types of oscillations; transit time of an electromagnetic wave from a radiation source to a controlled object, etc. In radio wave methods, 2 groups are distinguished.

Resonator. To work with indicators, the device, the degree of vibration of which must be determined, is placed in the region of the microwave resonator, which ensures their small error. The disadvantage here is a complex design, without the possibility of diagnosing at long distances, a rather complex presentation of the study of the data obtained, which, as a result, does not allow them to be used in many areas of industry [5].

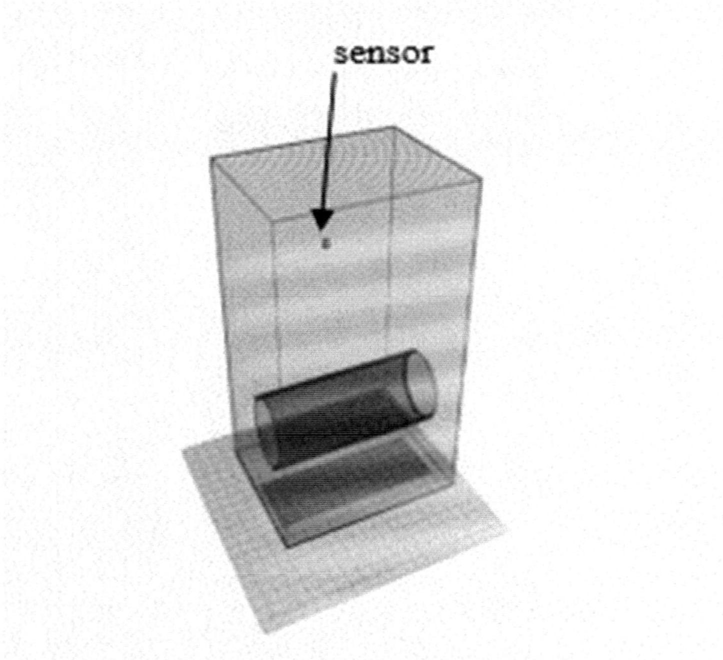

Figure 1. Diagram of a pipeline with a radio wave vibration sensor.

Interference is using microwave sounding, in their subsequent consideration in reflection from an obstacle. Due to interference and electromagnetic influences, a standing wave appears, which changes its amplitude under the influence of vibrations.

In the head of the radio wave vibration sensor we have considered, there is an interference radio wave method for assessing the phase of the reflected signal. Vibration of the sensor leads to phase and amplitude modulation of the reflected wave, and subsequently to the formation of a beat signal. As a consequence, the phase of the reflected signal includes all information about the vibration characteristics of the object under study.

One of the main advantages of the phase method is that the amplitude of the reflected signal is not used in the calculation of vibration parameters. This made it possible to invent an intelligent sensor that can operate at a long distance to the pipeline, with high metrological characteristics and unified analog and digital interfaces [6].

The economic efficiency of such a sensor is that there is no need to carry out large-scale excavations for its implementation or replacement in the pipeline system, which significantly reduces the cost of implementing this method.

The sensor measures such parameter as vibration velocity. This characteristic shows the power of the oscillatory (vibration) process [7]:

$$V = 2\pi \mathrm{f} S, \tag{1}$$

where f is the vibration frequency, which helps to identify the vibration source; S – value of vibration displacement.

The structure of the radio-wave vibration sensor is considered below (Figure 2).

The vibrometer includes a transceiver module (TCM), an analog signal input-output module (ASIOM) and a personal computer (PC).

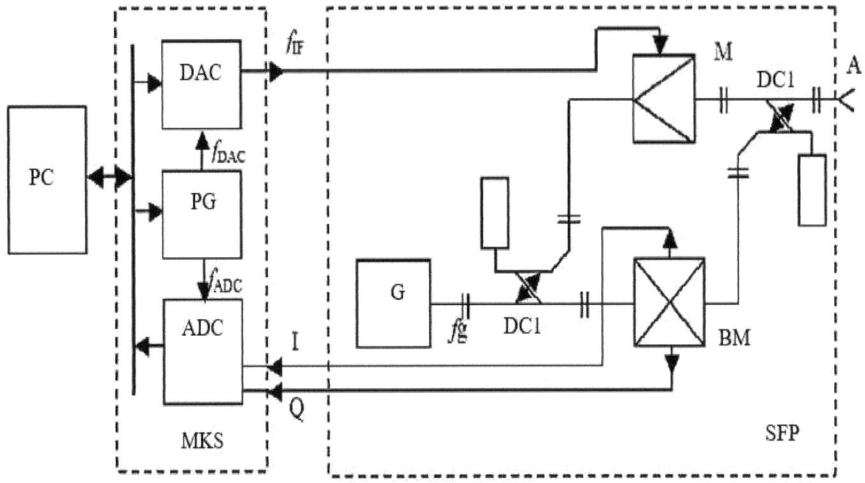

Figure 2. Vibrometer structure.

TCM contains a generator of EHF range, directional coupler HO1, with the help of which the measuring and reference channels are formed, phase manipulator M, which provides phase modulation of the probing signal with a signal of intermediate frequency f_{IF}. The horn antenna A is both receiving and transmitting; directional coupler HO2, which selects EHF waves reflected from the measurement object. A balanced BS mixer, in which the interference of the modulated reflected signals with the unmodulated reference signals is carried out and at its outputs the Q and I quadrature signals of the intermediate frequency, carrying information about the phase of the signal and, therefore, about the vibration parameters, are extracted (Figure 2).

A two-channel analog-to-digital converter is part of the MVVAS structure, which provides the ability to change the Q and I quadrature signals of the intermediate frequency into digital form and transfer them to a PC, a digital-to-analog converter that processes the baseband f_{IF} signal, a pulse generator f_{ADC}, f_{DAC} that clocks the operation of the ADC and DAC. In the PC, processes are performed using mathematical processing of incoming signals, calculation and determination of the deviation of the measured vibration parameters [8].

6. The mathematical meaning of the radio wave method

The following mathematical model of quadrature signals is used to determine vibration parameters by the radio-wave phase method (Figure 3).

From the phase value and the known wavelength of the probing signal λ, it is easy to calculate (1) the instantaneous value of the displacement D (t) and determine the vibration parameters:

$$Q = X(t) = Acos(\varphi(t)), \quad (2)$$

$$I = Y(t) = Asin(\varphi(t)), \quad (3)$$

where φ (t) is the instantaneous value of the phase of the reflected signal associated with the displacement D (t) of the object surface by the ratio:

$$\varphi(t) = (D(t)4\pi)/\lambda, \quad (4)$$

where λ is the wavelength of the probe signal.

We calculate the phase from the values of the quadratures as follows:

$$\varphi(t) = arctg(Y(t)/X(t)). \quad (5)$$

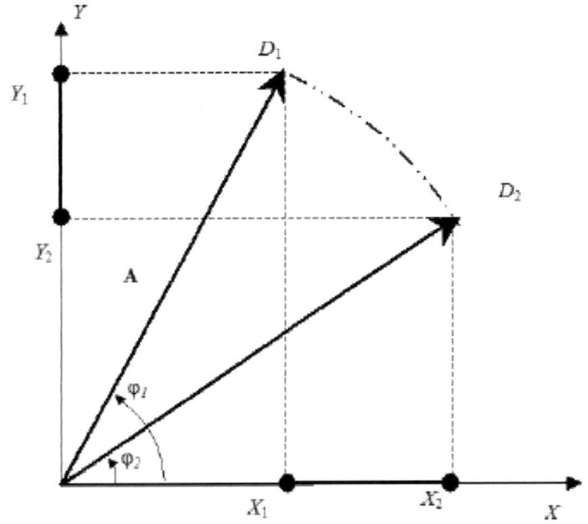

Figure 3. Mathematical model of quadrature signals: where D1, D2 are the end points of the vector (the required vibration range).

Based on the presented phase method with the introduction of modern digital signal processing, a non-contact radio wave vibration sensor has been created that operates at a long distance to the pipeline, vibration amplitudes, frequencies and displacements, characterized by high metrological properties, unified analog and digital interfaces. The radio wave vibration sensor belongs to a new generation of sensors designed for continuous remote monitoring of dynamic and static characteristics of structural elements of any source.

Based on the foregoing, we can say that an intelligent sensor is a sensor that performs the functions that are mandatory for the implementation of an accurate description and provision of the measured value. In addition to self-testing, the sensor performs such functions as measuring, ordering and correcting the signal we receive. An individual quality of a contactless intelligent vibration sensor is digital signal processing immediately from the output of the initial transducer. This ensures the correctness and stability of its properties in the permissible measurement range, as well as low sensitivity to external interference.

To ensure the implementation of various functions of changing the controlled values, which can further improve the properties of the realized processes in the sensor, digital signal processing is used, including the possibility of improving the software.

For each value we determine vibration displacement, vibration velocity and vibration acceleration, statistical and spectral analysis is performed, based on which the average value, root mean square value, minimum and maximum values, swing and amplitude spectrum are determined. The characteristic property of the sensor is to provide monitoring of vibration changes from 0 Hz, which provides the function of a non-contact micrometer, as well as to perform low-error assessment of low-frequency vibrations.

7. Conclusion

As a result of the study, the authors substantiated the need for the introduction of a radio wave vibration sensor into the monitoring system of pipeline communications. A method for determining the vibration velocity of a radio wave signal reflected from an object is considered, and its mathematical model is described. The most effective sensor for the existing system has been identified, which gives an idea of the capabilities of radio wave devices and its effective application in practice.

References

[1] Vorobiev A E, Khabarov D N, Mussa M T and Yankevsky A V 2013 An innovative system for detecting leaks and monitoring the activity of pipelines *Bull. of Peoples' Friendship University of Russia, Series: Engineering Research* **3** 211-9

[2] Sadler M and Shostakovskiy P 2019 Remote dispatching of heating networks in a smart city *Control Engineering Russia* **1** 175-91

[3] Volkovets A I, Rudenko D F, Gusinsky A V and Kostrikin A M 2007 Non-contact radio wave method for measuring motion and vibration parameters *Reports of the Belarusian State University of Informatics and Radioelectronics* **4(20)** 58-64

[4] Islamov R R, Agineyi R V and Isupova E V 2017 Analysis of means and methods for monitoring the stress state of underground oil and gas pipelines operating in difficult engineering and geological conditions *Transportation and storage of petroleum products and hydrocarbon raw materials* **6** 83-94

[5] Suslov Yin T and Zhu H P Selection of masters in dynamic reduction-based structural health monitoring using Bayesian experimental design *Mechanical Systems and Signal Processing* **150(1)** 107294

[6] Quy T B and Kim J M Crack detection and localization in a fluid pipeline based on acoustic emission signals *Mechanical Systems and Signal Processing* **150(1)** 107254

[7] Foorginezhad S, Mohseni-Dargah M, Firoozirad K, Aryai V, Razmjou A, Abbass, R and Asadnia M 2020 Recent advances in sensing and assessment of corrosion in sewage pipelines *Process Safety and Environmental Protection* **147** 192-213

[8] Zheng S, Yu Y, Qiu M, Wang L and Tan D A modal analysis of vibration response of a cracked fluid-filled cylindrical shell *Applied Mathematical Modelling* **91(1)** 934-58

The use of reliability indicators in the design of die tooling of technological processes of hot die forging

A V Kutyshkin[1], A V Shashok[2] and T B Brylova[3]

[1]Khanty-Mansiysk Yugra State University, 16, Chehova str., Khanty-Mansiysk, 628012, Russia
[2]Rubtsovsk Industrial Institute (branch) Altai State Technical University named after I. I. Polzunov, 2/6, Tractornaya str., Rubtsovsk, 658207, Russia
[3]Omsk State Transport University, 35, Marksa pr., Omsk, 644046, Russia

E-mail: avk_200761@mail.ru

Abstract. The article discusses the use of a number of indicators of the reliability of the functioning of the hot forming process to substantiate the choice of the number of die inserts (die streams) and selection of grades of tool steels for their manufacture. There are such used reliability indicators as the probability of the absence of typical defects in the shape of the stamped forgings, the probability of predominant types of exploitative refusals of the stamp, and the absence of failures in the implementation of the technological process as a whole. The rationale for the expediency of choosing a pre-stream of the die for stamping an axisymmetric forging with a high degree of complexity taking into account the possibility of occurrence of such shape defects as clamping, shooting, and incomplete forging radii during stamping, is considered. When the choice of the grade of tool steel for die inserts was taken into account, there was the occurrence of such typical types of their exploitative refusals as adhesion wear of die streams, the formation of hot cracks and plastic crushing of the forming elements of the stamp streams. The appropriate organizational and technical decisions are adopted on the basis of a comparison of cost indicators characterizing the production costs due to scrap forgings. They are caused by the formation of shape defects during stamping and the occurrence of exploitative refusals of dies. In addition, there are the costs for the manufacture of more complex die tooling and its exploitation, i.e. production quality costs.

1. Introduction

At present, a great deal of positive experience has been gained in designing die tools for hot die forging. At the same time, it should be noted that when designing this tool, reliability indicators are extremely rarely used, including its operation and the corresponding technological process as a whole [1-3]. Direct design of die tooling consists of a number of complex steps and procedures, which include the determination of the number of die streams and the choice of tool steel for tools manufacture. The solution of this kind of problems, despite the use of modern design tools, is often carried out by the designers on the basis of their own accumulated experience in this field, as well as using their well-known solutions - analogues provided by the databases of design tools. As a consequence, it is quite relevant to solve these problems when designing the tool in question, taking into account the estimates of the reliability indicators of its operation when stamping forgings.

2. Materials and methods

Often, decision makers in their activity use a comparative analysis of cost indicators characterizing a particular option for a possible solution to a task or problem. A similar approach is proposed to be used to select the number of forming die streams for hot stamping, as well as to determine the grade of die steel for their manufacture.

As part of the first task, which involves choosing between die with one insert and a die with two inserts, it is advisable to verify the fulfillment of the following cost ratio (hereinafter, the subscript "1" of variables corresponds to a die with one insert, the lower index "2" variables - a die with two inserts):

$$C_{1.1}\left[\left(\gamma_{sh.1}-\gamma_{sh.2}+P_{sh.2}\gamma_{sh.2}-P_{sh.1}\gamma_{sh.1}\right)+\left(\gamma_{st.1}-\gamma_{st.2}+P_{st.2}\gamma_{st.2}-P_{st.1}\gamma_{st.1}\right)\right]>\dots$$
$$\dots>\gamma_{st.2}C_{1.2}\left(1-P_{tp.2}\right)+\left(N_{d.2}\right)^{-1}\left[C_3+C_4\left\{\left(N_{d.2}\right)\left(N_{d.1}\right)^{-1}-1\right\}\right]. \quad (1)$$

Here $C_{1.1}$, $C_{1.2}$ - cost of producing one defective forging during stamping; $\gamma_{sh.1}$,$\gamma_{sh.2}$ - proportions of rejects associated with the formation of mold defects, stamped forgings [2]; $\gamma_{st.1}$,$\gamma_{st.2}$ - proportions of rejects of the forgings caused by the occurrence of predominant types of operational die failures [1]; $P_{sh.1}$,$P_{sh.2}$ - probabilities of failures associated with the formation of defects in the shape of the stamped forging [2]; $P_{st.1}$,$P_{st.2}$ - probabilities of failures that associated with the primary types of operational failure of the die [1]; $N_{d.1}$,$N_{d.2}$ - estimated average die life [1]; C_3 - cost of ensuring the durability of both dies at a given level $N_{d.1}$; C_4 is the cost of ensuring the durability of die with two inserts at a given level $N_{d.2}$; $P_{tp.2}$ - probability of failure of the technological process of stamping forgings in a die with two inserts as a whole [2].

The expression on the left side of inequality (1) characterizes the cost savings at the expense of reducing forgings during stamping in die with two inserts with respect to marriage of forgings stamped in a die with one insert caused by the appearance of shape defects in the forgings. This includes non-completion of the rounding radii of the protruding forgings elements, the formation of the defects type of "clamp", "chamber" and "extrusion defect" [2, 4–7]) and the formation of the predominant types of failures of the die tool during its operation [1, 4, 8-12]. The first term in the right-hand side of inequality (1) characterizes the costs associated with the production of defective forgings, which is caused by the failure of the die with two inserts due to the occurrence of predominant types of failures in it. The second term (1) estimates the costs associated with measures to increase the average resistance of the stamp from the level of $N_{d.1}$ to the desired level of $N_{d.2}$.

The probability of the absence of predominant types of failures P_{st} and during operation of the inserts die for hot stamping is estimated by the following expression [1, 2]:

$$P_{st}=\left(B\right)^{-1}\left[1+\left(B-1\right)\exp\left(-B\right)\right];B=\sum_{j=1}^{J}\left(P_{st.j}\right)^{-1}-\left(J-1\right). \quad (2)$$

$$P_{st.j}=1-\left\{1+\beta_j\left[1+\frac{\beta_j^2}{1+\beta_j}+\frac{\beta_j^4}{\left(1+\beta_j\right)\left(2+\beta_j\right)}+\frac{\beta_j^6}{\left(1+\beta_j\right)\left(2+\beta_j\right)\left(3+\beta_j\right)}+\dots\right]\right\}^{-1},\beta_j=\frac{N_{0.j}}{l_jN_b}. \quad (3)$$

Here J is the total number of predominant types of operational failures of the die inserts for hot stamping; l_j is the number of die insertion elements most critical to the occurrence of the j-th type of primary failure; $N_{0.j}$ - number of forgings, stamping of which is possible before the occurrence of the j-th type of primary failure of the die insert; N_b - batch of stamped forgings.

The predominant types of failures of the die insert of the preliminary stream are plastic deformation of the elements of the die impression (j = 1) and the formation of high-pressure cracks in the die impression (j = 2) [3, 6-8, 10-12]. For the final die impression, in addition to those already indicated, adhesive wear of the adhesive wear of the die bridge of the final creek of the stamp (j = 3) is also possible [1, 3, 6-11].

The values $N_{d.1}$, $N_{d.2}$ are proposed to be estimated by the following dependence [1, 2]:

$$N_{d.(1,2)} = \min_{j=1}^{J_{(1,2)}} \left\{ N^2_{0.j(1,2)} \left(l^2_{j(1,2)} N_b \right)^{-1} - \left(N_{0.j(1,2)} \left(N_b \right)^{-1} - 1 \right) P_{st.j(1,2)} \right\}, \quad (4)$$

where $J_{(1,2)}$ is the number of preferred types of failures of the corresponding type of stamp, the occurrence of which is possible during stamping; $N_{0.j(1,2)}$ - number of forgings, stamping of which is possible before the occurrence of the *j*-th type of preemptive failure in the corresponding type of stamp; $l_{j(1,2)}$ is the number of elements of the corresponding type of stamp that are most critical to the occurrence of the *j*-th type of primary failures; $P_{u.н.j(1,2)}$ - probability of the absence of the *j*-th type of primary failures during stamping of forgings in a die with one insert and in a die with two inserts (4).

The probabilities $P_{sh.2}$, $P_{sh.1}$, characterizing the absence of failures associated with the formation of the defects of the forging form listed above, stamped in a die with two inserts and a die with one insert, are evaluated by a similar structure (3) [2], in which:

$$B = \sum_{f=1}^{F_{(1,2)}} \left(P_{sh.f(1,2)} \right)^{-1} - \left(F_{(1,2)} - 1 \right), \quad (5)$$

where $P_{sh.f(1,2)}$ - probabilities of the absence of forging defects in the corresponding type of die during forging: non-completion of the rounding radii of the protruding forgings elements (f = 1), the formation of defects of the "clamp" type (f = 2), "chamber" (f = 3) and "extrusion defect" (f = 4), determined according to the technique given in [2, 3]; $F_{(1,2)}$ - total number of form defects in the forgings, stamped in a die with one insert and die with two inserts.

The proportions of rejects of the forgings $\gamma_{sh.1}$, $\gamma_{sh.2}$, $\gamma_{st.1}$, $\gamma_{st.2}$, caused by the formation of shape defects during stamping of the forgings and the occurrence of primary types failures of the die tool, are determined by the dependencies:

$$\gamma_{sh.(1,2)} = \left(1 - P_{sh.(1,2)}\right)\left(P_{sh.(1,2)}\right)^{-1}; \gamma_{st.(1,2)} = \left(1 - P_{st.(1,2)}\right)\left(P_{st.(1,2)}\right)^{-1}. \quad (6)$$

If inequality (1) holds, then the inclusion of a preliminary stream in the stamp design seems rational.

When solving the second task associated with the selection or replacement of die steel used in the manufacture of die inserts, it is proposed to use the following dependence:

$$Q\gamma_{st}\left(1 - P_{tp}\right)C_5\left(P_{new} - P_{old}\right) > \sum_{i=1}^{m}\left(C_i \bar{x}_i \left(\sigma_{x_i}\right)^{-1}\right) + C_6 Q \left(N_b\right)^{-1}\left(N_{0.new}\left(N_{0.norm}\right)^{-1} - 1\right). \quad (7)$$

Here Q - forgings release program; C_5 - cost of producing one defective forgings; $N_{0.norm}$ - standard resistance of die tooling in the considered technological operation; $N_{0.new}$ - average resistance of die tooling from the stamp steel grade proposed for replacement; C_6 - cost required to increase the durability of die tooling: $N_{0.new} > N_{0.norm}$; $\bar{x}_i$ - average values of technological parameters that affect die tool failures; σ_{x_i} - standard deviations of x_i; C_i - expenses for providing a unit value $\bar{x}_i \left(\sigma_{x_i}\right)^{-1}$ for the i-th technological parameter of the stamping operation; γ_{st} is the proportion of forgings defective due to failures in die tooling, which is determined similarly to (6); P_{tp} - probability of failure of the technological process of stamping forgings as a whole [2]; P_{new}, P_{old} are the probabilities of failure-free operation of die tooling from the steel proposed for replacement and the replaceable initial tool steel (2,3).

The left-hand side of inequality (7) characterizes the cost effect of reducing the marriage of manufactured forgings, which is caused by failures of die tooling. The expression on the right-hand side of inequality (7), in turn, characterizes the costs associated with replacing the grade of tool steel used for the manufacture of die inserts.

The average durability of die tooling $N_{0.new}$ is the proposed for the replacement of stamped steel grades is determined by the dependence [1]:

$$N_{0.new}=\frac{1}{3}\sum_{j=1}^{J}\left\{\left[N_{0.j.new}\right]^2\left(l_jN_b\right)^{-1}-\left[N_{0.j.new}\left(l_j\right)^{-1}-N_b\right]P_{st.j.new}\right\} \quad (8)$$

where $P_{st.j.new}$ is the probability of the absence of the *j*-th type of primary failures of die inserts from the proposed die steel replacement; $N_{0.j.new}$ - number of forgings, stamping of which is possible before the occurrence of the *j*-th type of primary failure of the die insert from the stamp steel proposed for replacement.

If condition (7) is met, the decision to replace the stamped steel grade is justified.

3. The rationale for the choice of the number of die forge inserts for die of hot forging and tool steel for their manufacture

Let us consider now the feasibility of making a decision on the use of the blocking impression of the die for stamping forging, a sketch of which is shown in Figure 1.

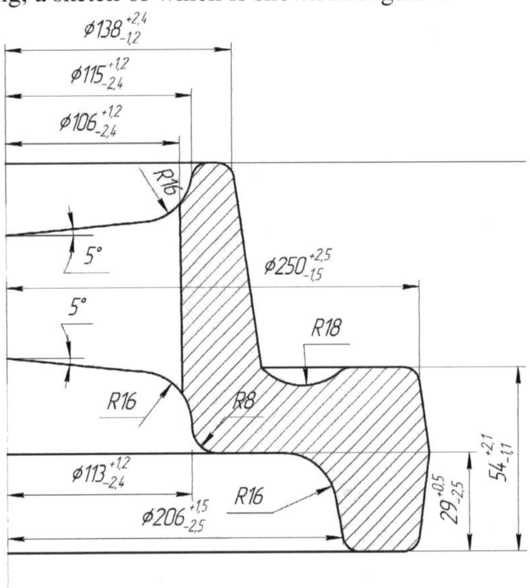

Figure 1. Sketch of the forging

The cost of producing one defective forging is equal to: $C_{1.1}$ = 20 thsd rubles (RUB), $C_{2.2}$ = 24 thsd rubles. C2 costs for ensuring the durability of both types of dies at the level of $N_{d.1}$ are C_2 = 210 thsd rubles; the cost of increasing the durability of a two-strand stamp to a value of $N_{d.2}$ is C_3 = 150 thsd rubles. The forgings release program is Q = 19200 pcs. Forgings are stamped in batches N_b = 2000 pcs. Forging material is steel 50G (steel 1050 USA), mechanical properties at t = 20 ^{0}C are $\sigma_в$ = 700 MPa, ψ = 0,55, δ = 0,18; initial blank is ∅120×241±2,5 mm; hot-rolled bars of normal accuracy. Pre-heating of the workpieces is carried out in a semi-methodical furnace to a temperature of 1210 ± 10 ^{0}C.

The average temperature of the workpieces before stamping is T_{forg} =1150 ± 8 ^{0}C. The average temperature of the pre-heating die is T_{st} =200 ± 8 ^{0}C. The material of die inserts steel is 4KhMFS (C 0,37…0,45; Si 0,5…0,8; Mn 0,5…0,8; Ni up to 0,4; S up to 0,03; P up to 0,03; Cr 1,5…1.8; Mo 0,9…1,2; V 0,3…0,5; Cu up to 0,3), 41...46 HRC$_э$. The tensile strength at a temperature of 20 ^{0}C is $\sigma_в$ = 1520 MPa; the temperature of the beginning of intense softening – T_s = 430 ^{0}C; σ_{T_s} = 1100 MPa [4, 5]. Lubrication is a solution of sulfite-alcohol stillage (KBZH) (KBZH 10...15%, graphite 1...30%, borax 2...3%, alkali 0.5%).

The calculated values of the parameters included in the dependence (1 - 6) are given in tables 1 - 2.

Table 1. Probabilities of the absence of shape defects in the forgings (table 1), stamped in dies with one insert and two inserts [2].

f	Types of failures associated with defects form of the forgings	Probabilities absence of the failures	
		Forging in the die with one insert	Forging in the die with two inserts
1	Non-completion of the rounding radii of the protruding forgings elements	$P_{sh.1.1}$ = 0,987	$P_{sh.1.2}$ = 0,993
2	The formation in a forging of the defects of the clamp	$P_{sh.2.1}$ = 1*	$P_{sh.2.2}$ = 1*
3	The formation in a forging of the defects of the type "chamber".	$P_{sh.3.1}$ = 0,996	$P_{sh.3.2}$ = 0,999
4	The formation in a forging of the defects of the type "extrusion defect"	$P_{sh.4.1}$ = 1*	$P_{sh.4.2}$ = 1*

* - The probability value of the absence of this defect during stamping of the considered forging is 1, because there are no elements in the forging configuration where the formation of this defect is possible.

The probability of failure-free operation of the technological process of stamping forgings in a die with two insert according to [2] is equal to $P_{tp.2}$ = 0,988. Values of the quantities $\gamma_{sh.(1,2)}$ and $\gamma_{st.(1,2)}$ for dies with one insert and two inserts according (6) respectively equal $\gamma_{sh.1}$ = 0,0132; $\gamma_{sh.2}$ = 0,004; $\gamma_{st.1}$ = 0,1135; $\gamma_{st.2}$ = 0,004. In turn, the average durability $N_{d.(1,2)}$ of each type of dies in accordance with (4) is equal to: $N_{d.1}$ = 1800 die, $N_{d.2}$ = 2400 pcs. Substituting the calculated and accepted values of the corresponding variables into the dependence (1) we obtain: 232,9 rubles > 101,4 rubles. When taking into account the size of the production program Q, it is obvious that the use of a preliminary brook is economically profitable, and the costs associated with its manufacture and operation of more complex die tooling are compensated by reducing costs due to failures during stamping of forgings.

Table 2. Probabilities of the absence of primary failure modes in the die with one insert and in the die with two inserts [1, 2]

j	Preferred types of die insert failures	Forging in the die with one insert		Forging in the die with two inserts	
		$N_{d.j.1}$, pcs	$P_{st.j.1}$	$N_{d.j.2}$, pcs	$P_{st.j.2}$
1	Plastic deformation of the elements of the die impression	4000	0,924	5700	0,991
2	Adhesive wear of the die bridge	6000	0,994	7400	0,999
3	Formation of high-pressure cracks in the die impression	3400	0,88	4700	0,974

Now we consider the feasibility of replacing 4KhMFS steel used in the manufacture of die inserts with 4Kh5V2FS steel (C 0,35…0,45; Si 0,8…1,2; Mn 0,15…0,45; Ni up to 0,4; S up to 0,03; P up to 0,03; Cr 4,5…5,5; W 1,6…2,2; V 0,6…0,9; Cu up to 0,3), which has higher wear resistance. Forgings are stamped in a die with one insert. Physico - mechanical properties of the steel 4Kh5V2FS 41...46 HRC$_э$ [4, 8, 12, 13] are as follows: the tensile strength at a temperature of 20 ^{0}C is $\sigma_в$ = 1780 MPa; the temperature of the beginning of intense softening is T_S = 500 ^{0}C; σ_{T_S} = 1200 MPa. In this case, the predominant types of failure of die inserts during stamping of forgings are also plastic deformation of the elements of the die impression (j = 1), adhesive wear of the die bridge (j = 2) [3, 6-8, 10-12] and the formation of high-pressure cracks in the die impression (j = 3) [1, 3, 6-11]. The predominant types of failures of die inserts (Table 3) are similar to those considered earlier (Table 2).

Table 3. The calculated values of durability and the probability of the absence of primary types of failures of die tooling from replaceable and proposed for replacement die steel [1, 2]

j	Preferred types of die insert failures	The calculated values of the die tooling life (8), pcs.		Estimated probability of failure's absence of the die (2)	
		4HMFS	4Kh5V2FS	4HMFS	4Kh5V2FS
1	Plastic deformation of the elements of the die impression	$N_{0.1.old}$ = 4000	$N_{0.1.new}$ = 5000	$P_{st.1.old}$ = 0,924	$P_{st.1.new}$ = 0,995
2	Adhesive wear of the die bridge	$N_{0.2.old}$ = 6000	$N_{0.2.new}$ = 7000	$P_{st.2.old}$ = 0,994	$P_{st.2.new}$ = 0,999
3	The formation of high-pressure cracks in the die impression	$N_{0.3.old}$ = 3400	$N_{0.3.new}$ = 4200	$P_{st.3.old}$ = 0,88	$P_{st.3.new}$ = 0,974
	The average resistance of the die tooling	$N_{0.old}$ = 3500	$N_{0.new}$ = 4500		
	The probability of functioning die tooling without failure			$P_{st.old}$ = 0,898	$P_{st.new}$ = 0,978

We take the following values of a number of variables of dependence (7): $\gamma_{st.1}$ = 0,1135 (6); C_5 = 20000 RUB; C_6 = 60000 RUB; $N_{0.н}$ = $N_{0.norm}$ = 2500 pcs; $P_{st.old}$ = 0,898; $P_{st.new}$ = 0,978 (Table 3); P_{tp} = 0, 982 [2]; $N_{0.old}$ = 3500 pcs; $N_{0.new}$ = 4500 pcs (Table 3).

Since the organizational and technological infrastructure of production does not change with respect to the use of both steels for the production of die inserts, we can assume that the value of the term $\sum_{i=1}^{m} C_i \bar{x}_i \left(\sigma_{x_i}\right)^{-1}$ expression (7) for both steel grades is the same and may not be taken into account in the calculations. Given there are values of the variables, we obtain the estimate of the left side of expression (7): 2.73 105 rubles. The value of the right side of this expression is 3.42 105 rubles. Thus, the cost effect of reducing forgery rejects due to failures in die tooling does not exceed the costs associated with replacing the grade of tool steel used to manufacture this tooling. Consequently, the decision to replace the tool (die) steel for the manufacture of die inserts is not advisable.

4. Conclusion

The results allow us to conclude that it is advisable to use such indicators of the reliability of the hot forming process as the probability of failure in terms of quality of manufactured forgings and the absence of primary types of operational failures of die tooling at the design stage. Along with this, the use of these reliability indicators is also relevant at the stage of supporting the technological process, for example, in developing organizational and technical solutions for replacing die steel for the manufacture of die inserts.

References

[1] Stepansky L G 2015 *Prediction of the reliability of technological processes, tools and metal forming machines* (Moscow: Moderat)
[2] Kutyshkin A V 2006 *Reliability of the technological systems of a hot forging processes* (Barnaul: Altai State University)
[3] Kutyshkin A V 2003 *KShP.OMD.* **4** 11–18
[4] Aliev I S, Alieva L I, Lobanov A I and Savchinsky I G 2007 *Metalworking* **5(41)** 22–28
[5] Altan T, Deshpande M and Groseclose A 2011 *FIA Tech. Conf.* p 644
[6] Hawryluk M and Jakubik J 2016 *Engineering Failure Analysis* **59** 396–409

[7] Choi C, Groseclose A and Altan T 2012 *J. of Materials Processing Technology* **212** 1742–1752
[8] Abachi S, Akkok M and Gokler M I 2010 *Tribology Int.* **43** 467–473
[9] Santosh B D, Ronge B P and Misal N D 2014 *Int. J. of Application or Innovation in Engineering & Management (IJAIEM)* **11(3)** 152–158
[10] Wilkus M, Polak S, Gronostajski Z, Kaszuba M, Rauch Ł and Pietrzyk M 2015 *Comp. Meth. Mater. Sci.* **15** 311–321
[11] Banaszek G, Berski S, Dyja H and Kawałek A 2013 *J. of Iron and Steel Res. Int.* **20** 111–116
[12] Vlasov A V 2016 *Steel* **5** 49–53
[13] Tylkin M A, Vasiliev D I et al 1977 *Dies for hot deformation of metals* (Moscow: Higher school)

Formation of new principles and models of operation of structural units of the industry under the conditions of implementation of digital technologies

D A Lysenko[1], V A Olentsevich[2], N V Vlasova[3] and V Yu Konyukhov[1]

[1] Irkutsk National Research Technical University, 83, Lermontova str., Irkutsk, 664074, Russia
[2] Irkutsk State Transport University, 15, Chernyshevskogo str., Irkutsk, 664074, Russia
[3] East Siberian Branch of the JSC "Russian Railways", 12/1, 2nd Zheleznodorozhnaya, Irkutsk, 664005, Russia

E-mail: lysenko@istu.edu

Abstract. The construction of a high-tech modern model "Digital Railroad" is currently a key element, the implementation of which is planned for the near future. The project will create new principles and models for the work of all structural divisions of the industry in the context of the widespread penetration of digital technologies. The emergence of digital technologies will make it possible to transfer the organization and technology of functioning of transport and other systems to a qualitatively new level, new information technologies will make it possible to optimize the operation of rolling stock. According to the results of the analysis of the performance indicators of the service locomotive depot Irkutskoye, according to the model of "full service" of the locomotive fleet, it was revealed that the indicators negatively affects the financial results of the depot and negatively affects the quality of the repair performed in general. To solve this problem, it is proposed to use an automated locomotive fleet reliability management system in the Irkutsk service locomotive depot, the implementation of which will increase the availability factor, reduce downtime in repair depots, and minimize resource consumption for all types of repair and maintenance. The main consumer of the services of the locomotive fleet reliability management system is, first of all, the Russian Railways holding company and the traction directorate. They are interested in the reliable operation of the rolling stock, since during the operation of locomotives there are risks: comments on the operation of the locomotive fleet, failures and malfunctions of various degrees of complexity, cases violations of train traffic safety.

1. Introduction

According to the current system of reforming the holding, JSC Russian Railways (hereinafter referred to as JSC Russian Railways), repair and maintenance of the locomotive fleet of the industry are carried out in full by service locomotive depots, which were created on the basis of previously existing repair locomotive depots. At the same time, they are not included in the system of Russian Railways, but are included in the structure of TMH-Service LLC (Transmashholding), established in June 2010 according to the concept of "Development of a service system and creation of a single center of responsibility for the technical condition of locomotives", approved by Russian Railways [1-4]. Despite the reforms, the holding continues to be the owner of these repair depots, locomotive fleet maintenance points (PTOL), as well as the repair equipment and other fixed assets available in structural divisions [5-7].

The system of work is in accordance with the established model of "full service" for the locomotive fleet belonging to JSC Russian Railways, which provides for a procedure for full responsibility for the technical condition of the mobile units by the service company during their full-service life. At the same time, the service company produces both direct maintenances, planned and unscheduled types of repair (HP), and also carries out high-quality training of personnel, investing in the technological equipment of repair enterprises [4, 8].

2. Evaluation of the efficiency of the "full service" model of the locomotive fleet owned by Russian Railways by the example of the service locomotive depot Irkutskoye

Taking into account the established operating technology, service depots should focus on performing high quality repairs of the locomotive fleet, subject to minimizing financial and time costs, which should ensure high-quality performance of the assigned types of work.

The service locomotive depot Irkutskoye, in accordance with the established work technology, performs the following types of work: repair and maintenance of electric locomotives for passenger traffic EP1, EP1P, freight electric locomotives for pushing traffic 3ES5K (Ermak) in the volume of TR-1 and TR-2 and shunting locomotives in the volume of TO-3 and TR -1. The depot includes a PTOL of network significance, located at station Irkutsk-Sortirovochny and PTOL at station Slyudyanka [5, 9-10].

To assess the labor intensity of the production of certain types of repairs and series of locomotives in the given repair units, special regulatory coefficients are used, which are separately set for each series of a mobile unit. Table 1 shows the given locomotive repair program in the given units.

Table 1. Indicators of the repair program of the service locomotive depot Irkutskoye.

2016 fact	2017 fact	Change by 2016, %	2018				
			Plan	Fact	Implementation of a plan, %	Change, %	
						by 2016	by 2017
1621.70	1580.57	97.46	1523.78	1539.66	101.04	94.94	97.41

The main reasons for the reduction in the repair program of the service locomotive depot Irkutskoye are reduction in the repair program for electric locomotives; uneven loading of stalls of repair shops due to stopping locomotives for a complaint, unscheduled repairs, waiting for work. They also include untimely statement for repair due to excess downtime for repairs owing to a large number of additional work; lack of timely provision of the depot by OOO TMH-Service with the necessary spare parts and equipment [2, 11-13].

For a more detailed description of the work and in order to develop measures to optimize the activity of the service locomotive depot Irkutskoye, it is necessary to analyze the quality indicators. They include the coefficient of technical readiness of locomotives (CTR), idle time of locomotives for scheduled and unplanned types of repair, the volume of failures of the used technical means, the number of calls of mobile units for unscheduled types of repairs, the number of events related to the violation of train traffic safety and the occurrence of risk cases.

CTR of locomotives directly depends on the level and quality of maintenance and repair of this rolling stock. Timely detection and elimination of recorded faults, compliance with the rules of technical operation contribute to an increase in CTR.

According to the values presented in Figure 1, it can be seen that the KTG both for electric locomotives and for diesel locomotives has a tendency to decrease, while it should be noted that the value of the indicator depends on other qualitative indicators of the enterprise, the dynamics of which is presented in Table 2.

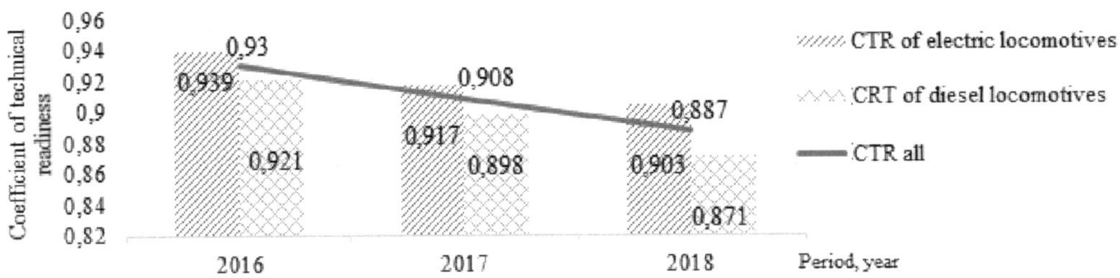

Figure 1. The coefficient of technical readiness of locomotives.

Table 2. Quality indicators of the service locomotive depot Irkutskoye.

Index		2016 fact	2017 fact	Change by 2016, %	2018 Fact	2018 Change, % by 2016	2018 Change, % by 2017
Electric locomotives							
Down-time for scheduled repairs	Norm Maintenance-2, hour	72.00	72.00	100.00	72.00	100.00	100.00
	Maintenance-2, hour	114.09	96.00	84.14	112.80	98.87	117.50
	Norm Maintenance-1, hour	31.00	31.83	102.68	31.00	100.00	97.39
	Maintenance-1, hour	102.46	91.95	89.74	76.80	74.96	83.52
	Norm Services-2u, hour	30.00	30.00	100.00	30.00	100.00	100.00
	Services-2u, hour	160.70	113.10	70.38	109.70	68.26	96.99
Unsche-duled repairs	Cases	352.00	371.00	105.40	403.00	114.49	108.63
	Downtime, hour	18685.23	19654.85	105.19	21506.01	115.10	109.42
Diesel locomotives							
Down-time for scheduled repairs	Norm Maintenance-1, hour	45.15	38.00	84.16	38.00	84.16	100.00
	Maintenance-1, hour	143.12	130.60	91.25	117.20	81.89	89.74
	Norm Services-3, hour	27.00	27.00	100.00	27.00	100.00	100.00
	Services-3, hour	58.56	54.60	93.24	66.40	113.39	121.61
Unsche-duled repairs	Cases	69.00	50.00	72.46	66.00	95.65	132.00
	Downtime, hour	10705.21	9013.65	84.20	10401.25	97.16	115.9

The idle time of electric locomotives for all types of scheduled repairs exceeds the established standard. So, on Maintenance-2, the average downtime for 2018 is 112.8 hours, with the established rate not exceeding 72 hours, on Maintenance-1, the downtime is 76.8 hours with the established rate of 31 hours and on Services-2u - 109.7 hours when set to 30 hours. Such dynamics have such reasons as: exceeding the idle rates of electric locomotives for repairs associated with waiting for a free stall for repairs, lack of spare parts, untimely delivery of electric locomotives for scheduled repairs (out of schedule), absence or employment of repair personnel, performance of overcycle work, etc.

The downtime of diesel locomotives on planned types of repairs has an uneven trend: a decrease is observed in the repair cycle Maintenance-1, and an increase in indicators is noted in the Services-3 cycle. The average idle time of diesel locomotives for repair in the Maintenance-1 cycle in 2018 amounted to 117.2 hours, which is 25.92 hours less than it was in 2016, or 18.11%. According to the Services-3 cycle, in 2018 compared to 2016, there is an increase in the idle time of diesel locomotives by 7.84 hours, or by 13.39%. The idle rates of diesel locomotives for repairs, as well as for electric locomotives, are not met, the idle times for planned types of repairs exceed the standard indicator at least twice.

The main reasons for exceeding the standards of idle time for diesel locomotives in repair are similar to those associated with electric locomotives, but there are a number of other reasons. Thus, failure to comply with the idle time standards for diesel locomotives during scheduled repairs of Maintenance -1 is associated with additional repairs due to violation of operating modes, including interference in the proper operation of locomotive systems and additional work. The list of additional work to be carried out contains oil change, wheelset turning, elimination of water leakage through the cylinder covers of a diesel engine, elimination of oil leakage through the connection of valve boxes with cylinder covers, etc. The main reason for the increase in the downtime of diesel locomotives at scheduled Services -3 is additional work. These are replacement and repair of the air cooler, revision of the compressor with removal from the locomotive, replacement and repair of the water pump, elimination of oil and water leaks through the cylinder covers of the diesel engine. These also include repair of the traction motors, cooling fan, repair of the clutch compressor, repair of the synchronous single-phase exciter drive (replacement of bearings), etc. [7, 14].

In case of failure of units and assemblies of mobile units, they are sent for unscheduled repair (UR), which leads to the need to divert the workforce from the planned work, requires additional costs of labor resources, spare parts, materials. These financial costs are non-productive, lead to an increase in the cost of current types of repairs and maintenance, since the costs of unscheduled repairs of mobile units are related to the type of repair and maintenance that led to the failure.

The increase in the number of cases of UR of electric locomotives in 2018 compared to 2016 is 51 cases, and compared to 2017, 32 cases, or an increase occurred by 14.49% and 8.63%, respectively. In 2018, the number of UR diesel locomotives was 66 cases, which is 32% more than in the previous year. An increase in the volume of calls of electric locomotives to UR occurs due to the failure of traction motors, wheelsets with axle boxes, non-observance of the mode of driving freight trains, excess of weight standards and poor-quality repairs.

The greatest downtime at UR was 551 hours of the EP1P # 023 electric locomotive due to the presence of current surges in operation. There are 221 hours of electric locomotive EP1P # 057 due to the presence of current surges due to failure of the UVO2P cassettes. 451.38 hours of electric locomotive EP1P # 021 due to the restoration of the electric locomotive after the alarm and fire extinguishing system have been triggered. 179.74 hours of the electric locomotive 1.5VL80tk No. 1014/2016 are due to the failure of the traction transformer. The overestimated downtime is due to a long decision on the part of DT to organize work to put the electric locomotive into operation with its subsequent transfer to the balance of the Krasnoyarsk railway.

During operation, there are failures in the operation of individual units of the locomotive, which are caused by wear and tear, design flaws of mobile units, unsatisfactory quality of repair and maintenance, violation of operation technology, which is presented in Table 3.

In 2018, 88 cases of failures were recorded in the service locomotive depot Irkutskoye, which is 3.53% higher than this indicator for the same period in 2017. In 2017, compared to 2016, there was a decrease of 2.3%. For the entire analyzed period, the largest number of failures was recorded for electrical equipment, 45 cases in 2016, which is 51.72% of the total number of failures, and during the period under review, failures for electrical equipment are increasing.

According to the analysis of the causes of malfunction in the electrical equipment of mobile units, the following were revealed: short circuit in the wire circuits, malfunctions of the main controller and control circuits, damage to the insulation of electrical equipment, breakage of the current collector. So, in 2018, five cases of failures were recorded for auto brake equipment. Faults: breakage of the connecting sleeve of the end valve of the feed line, malfunction of the driver's valve, breakage of the brake line pipe elbow. On safety devices 8 cases. Fault: short circuit of the armature winding, interturn short circuit of the traction motor poles.

Table 3. Distribution of equipment failures at the service locomotive depot Irkutskoye.

Equipment type	2016 fact	2017 fact	Change by 2016, %	2018		
				Fact	Change, %	
					by 2016	by 2017
Electrical equipment	45	48	106.67	51	113.33	106.25
Auto brake equipment	8	6	75.00	5	62.50	83.33
Safety devices	12	5	41.67	8	66.67	160.00
Mechanical equipment	4	6	150.00	7	175.00	116.67
Electronic equipment	5	8	160.00	4	80.00	50.00
Traction motor	4	5	125.00	3	75.00	60.00
Auxiliary machines	2	1	50.00	3	150.00	300.00
Locomotive equipment	2	2	100.00	4	200.00	200.00
Other	5	4	80.00	3	60.00	75.00
Total:	87	85	97.70	88	101.15	103.53

Further, it seems appropriate to analyze the distribution of failures for organizational reasons, table 4.

Table 4. Distribution of failures due to organizational reasons of the service locomotive depot Irkutskoye.

Reasons for malfunction	2016 fact	2017 fact	Change by 2016, %	2018		
				Fact	Change, %	
					by 2016	by 2016
Total rejections	87	85	97.70	88	101.15	103.53
Through the fault of the Service Locomotive Depot, including:	43	38	88.37	41	95.35	107.89
- locomotive maintenance point	21	16	76.19	7	33.33	43.75
- Electrical machine shop	1	2	200.00	4	400.00	200.00
- Electrical apparatus shop	1	0	0.00	2	200.00	-
- Electronics	0	2	-	1	-	50.00
- Hitchhiking shop	5	3	60.00	5	100.00	166.67
- Maintenance workshop	9	9	100.00	16	177.78	177.78
- Automatic shop	6	4	66.67	5	83.33	125.00
- Hardware workshop	0	2	-	1	-	50.00
Wines of repair of other Service locomotive depots and traction part repair	11	17	154.55	15	136.36	88.24
Wines of the locomotive brigade Traction part operational	12	5	41.67	2	16.67	40.00
Wines Traction part operational-5	0	1	-	0	-	0.00
Factory wines	0	1	-	0	-	0.00
Constructive flaw	1	2	200.00	3	300.00	150.00
Degradation wear	2	8	400.00	6	300.00	75.00
Track part	0	1	-	0	-	0.00
Power supply distance	0	3	-	1	-	33.33
Center manager	3	1	33.33	1	33.33	100.00
other	15	8	53.33	19	126.67	237.50

Of the total number of marriages in 2018 in the number of 88 cases, the distribution according to the perpetrators occurred: due to the fault of the Service Locomotive Depots, 41 cases were recorded (unsatisfactory quality of maintenance and repairs), including 16 cases due to the fault of the maintenance shop. The reasons for the marriage were the violation of the requirements of the instructions for maintenance and repair of electric locomotives of alternating current, poor-quality repair of electric locomotives, lack of control by the master of the complex team over the quality of repair of electrical equipment.

The analysis of the qualitative indicators of the service depot Irkutskoye shows the level of service at the depot, characterizes production activities and their effectiveness. It should be noted that, in addition to analyzing qualitative indicators, it also seems appropriate to analyze costs and their dynamics. It is necessary to consider the costs both for the main types of repairs (maintenance and repair), and for unscheduled repairs, which is necessary in order to determine in which areas the greatest expenditure of funds is going. In general, the cost analysis should show whether the operation of the service locomotive depot is effective from a financial point of view, Table 5.

Table 5. Distribution of expenses for planned and unplanned types of repairs.

Costs	2016 fact	2017 fact	Change by 2016, %	2018		
				Fact	Change, % by 2016	Change, % by 2017
Total for scheduled maintenance and repair of locomotives, million rubles:	92.658	90.588	97.766	84.519	91.216	93.300
1. For scheduled maintenance and repair of electric locomotives, million rubles:	80.679	78.871	97.759	73.341	90.905	92.989
-Maintenance-1	64.307	63.154	98.207	59.582	92.652	94.344
-Maintenance-2	5.858	5.586	95.357	5.586	95.357	100.000
-Services-2u	10.514	10.131	96.357	8.173	77.734	80.673
2. For scheduled maintenance and repair of diesel locomotives, million rubles:	11.979	11.717	97.813	11.178	93.313	95.400
-Maintenance-1	2.265	2.181	96.291	1.846	81.501	84.640
-Services-3	9.714	9.536	98.168	9.332	96.068	97.861
For unscheduled maintenance and repair of locomotives, million rubles	16.572	16.587	100.091	18.324	110.572	110.472
Total, million rubles	109.23	107.175	98.119	102.843	94.153	95.958

In general, the costs of all types of repairs have a steady downward trend. Expenses for scheduled types of repairs of electric locomotives are steadily decreasing, this decrease for two years amounted to 7.338 million rubles or 9.1%, for diesel locomotives there is also a decrease in costs, which amounted to 0.801 million rubles or 6.69%. The reasons for the cost savings lie not in optimizing material cost management or improving the repair process, but in reducing the locomotive repair program. While the number of unscheduled repairs is increasing both for diesel locomotives and electric locomotives, which in turn leads to an increase in the cost of unscheduled repairs. So in 2018, the costs associated with the implementation of unscheduled repairs amounted to 18.324 million rubles, which is 1.752 million rubles more than in 2016 or 10.57%.

Thus, despite the stable decrease in repair costs, the quality indicators have a negative trend: the number of failures is growing, the number of unscheduled repairs is increasing, the standards of locomotive downtime for repairs are not observed, and the CTR of locomotives decreases [1, 15-17].

Based on the results of the analysis of the qualitative indicators of the operation of the service locomotive depot Irkutskoye according to the "full service" model of the locomotive fleet owned by Russian Railways, it should be noted that the deterioration of these indicators negatively affects the financial results of the depot and the quality of the repairs performed in general. Thus, quality indicators, being the key factors in increasing the efficiency of the depot, should be minimized. So, ideally, the number of equipment failures and the number of unscheduled repairs (UR) should tend to zero, while the downtime should also decrease on planned and unplanned types of repair. At the same time, the downtime for scheduled repairs should not exceed the established standard values.

3. Automated locomotive fleet reliability management system

To solve this problem, it is proposed to use the automated locomotive fleet reliability management system (ASCTR) in the service locomotive depot Irkutsk.

ASCTR is aimed at the following key performance indicators of the enterprise:

- increase in the availability of locomotives;
- reduction of downtime of locomotives in repair depots;
- reduction of resource consumption for maintenance and repair of locomotives.

The rest of the ASCTR performance indicators are of a secondary nature.

During the operation of locomotives, failures occur in the operation of individual units of the locomotive. Failures are caused by wear, structural deficiencies of the RTS, unsatisfactory quality of repair and maintenance, etc. In order to identify pre-failure conditions, eliminate emerging failures as soon as possible and generally increase the reliability of locomotives, it seems appropriate to use ASCTR.

ASCTR is a set of technological, technical, mathematical, labor and other types of resources, which are combined into an integral system in order to increase the level of reliability of using the locomotive fleet, reduce the downtime of locomotives for repairs, and reduce the cost of maintenance and repair. The main attention is paid to the technology for carrying out such types of repairs as maintenance and repair of locomotives, the structure of enterprise management, the implementation of the principle of control and optimization of performance indicators.

ASCTR consists of three main blocks: software, technological and resource support. The construction of a high-tech modern model "Digital Railway" at JSC "Russian Railways" is currently a key element in ensuring the digital economy of the Russian Federation, the implementation of which is planned for the near future. The project will create new principles and models for the work of all structural divisions of the industry in the context of the widespread penetration of digital technologies. The emergence of digital technologies will make it possible to transfer the organization and technology of functioning of transport and other systems to a qualitatively new level. New information technologies will allow one to optimize the operation of rolling stock and infrastructure: high-speed data transmission networks, big data processing technologies, intelligent systems, mobile applications, the Internet of Things [3, 18]. In connection with this strategy, we will pay most attention to the consideration of the ASCTR software block, which includes a unified system for monitoring the technical condition of locomotives (USMT), data decoding systems for on-board microprocessor systems of a locomotive (MSUT), an automated system for technical diagnostics (ASTD), information systems of railway transport (ASCRT).

USMT is essentially an automated process control system (APCS), which should combine in a single information space all the data on the reliability of locomotives, thereby creating the prerequisites for integrated management of locomotive reliability. USMT is used to identify incidents (failures, pre-failure states, violations of operating modes, remarks of drivers, station attendants and dispatchers, other situations other than normal) entering the locomotive system using remote control of data transmission from various sources - MSUT, ASTD, ASCRT. Of all the listed information systems, information about possible incidents in the form of diagnostic messages is transmitted to the USMT, where an incident registration sheet is automatically created. The next step is incident lifecycle management, problem management and service level management.

In the process of locomotive control, MSUT interrogates sensors, collects and accumulates diagnostic data. The MSUT data are read and analyzed at specialized workplaces (AWS MSU), where the incident is found using special algorithms and the decryption data are attached to confirm the malfunction. Information from the MSU AWS is sent to the USMT in a unified format: date and time of the incident, series, number, locomotive postscript, incident code, brief description, place and circumstances of the incident.

The use of diagnostic data from the MSUT significantly increases the amount of diagnostic information and, as a consequence, a more complete understanding of the technical condition of the locomotive, it becomes possible to fix the pre-failure state in real operating conditions, hence it becomes possible to control the operating modes of locomotives. The operating experience of modern traction rolling stock testifies to the growing role of built-in and on-board diagnostic systems [6].

Another diagnostic tool for locomotives is ASTD - portable or stationary diagnostic systems, as a rule, are located in the depots themselves and at the locomotive maintenance points and are designed to monitor and predict the technical condition of the locomotive and its individual components and assemblies. Information from ASTD is also sent to the USMT in a unified format.

All initial information is recorded in various information systems of the ASCRT - integrated automated accounting system, monitoring the elimination of failures of technical means (KASANT), integrated automated subsystem for accounting and analysis of technological failures (KASAT).

Management of these risks is carried out on the basis of three sources of information: AWS MSU, ASTD and ASCRT. In the process of constantly receiving information, a database is formed, which is the basis of a single information space for ASCNT. Based on the above, it seems important to consider the structure of the functioning of the system, which is shown in Figure 2.

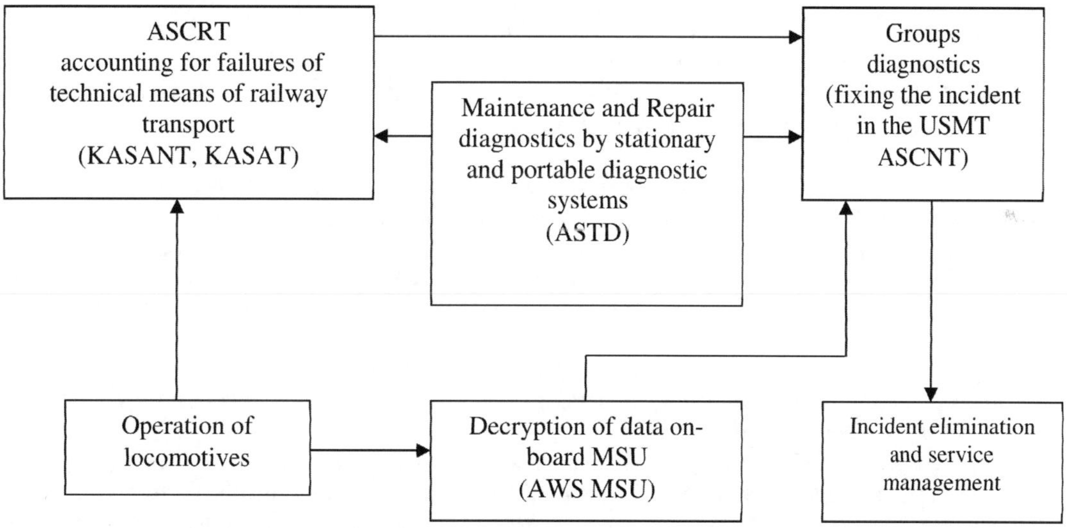

Figure 2. Structure of information flows of ASCTR.

The proposed ASCTR makes it possible to consider pre-failure states, identify failures, as well as violations of operating modes, identify trends in parameter changes, the intensity of failures, use previous statistical data on the rate of development of a pre-failure state into failure. In addition to these parameters, the data of the entire life cycle of the incident are analyzed: the time of setting up for repair, the time of repair, the time to search for a spare part, etc., the quality of the repair performed, the number of participants in the elimination of the incident, the consumption of other types of resources.

4. Conclusion

The introduction of the proposed ASCTR will allow you to get an effect due to:

- reducing the volume of malfunctions of the equipment of the locomotive fleet by eliminating or minimizing the number of violations of their operating modes;
- minimization of failures of the fleet of locomotives on the line by identifying pre-failure conditions and taking preventive measures;
- reducing the downtime of locomotives in the depot for repair and maintenance operations by reducing the number of failures and timely planning the supply of required spare parts of appropriate quality;
- reducing the cost of maintenance and repair due to the elimination of high-cost failures (due to their prevention);
- optimal use of the cash resources of the service depot aimed at innovative development through factor analysis and identification of the weakest points in production.

References

[1] Lebedev D V and Grigorieva N N 2018 Transaction costs in transport *Transport infrastructure of the Siberian region: Proc. of the Ninth Int. Sci. and Pract. Conf.* (Irkutsk: IrGUPS) vol 2 pp 135-139

[2] Olentsevich V A, Belogolov Yu I and Grigoryeva N N 2020 Analysis of reliability and sustainability of organizational and technical systems of railway transportation process *IOP Conf. Ser.: Mater. Sci. Eng.* 012061

[3] Urusov A V 2018 Digital railway *Automation, communication, informatics* **1** 6-8

[4] 2013 *Development strategy of the Russian Railways Holding for the period up to 2030* (Moscow: JSC Russian Railways)

[5] *Official website of Russian Railways:* http://rzd.ru

[6] Ryabov D K, Shestov V V, Buznik V M and Antipov V V 2016 Design concept of multifunctional layered composite materials for operation in arctic regions *Inorganic Mater.: Appl. Res.* **7(4)** 547-553

[7] Shestov V V, Antipov V V and Ryabov D K 2017 Corrosion Resistance and Mechanical Properties of Layered Structural Material Based on Aluminum Alloy and Fiberglass Thin Sheets *Metallurgist* **60(11-12)** 1191-1196

[8] Yolkin K S, Yolkin D K, Kolosov A D, Ivanov N A and Shtayger M G 2018 Technologies, which allow one to reduce an impact of metal silicon production on the environment *IOP Conf. Ser.: Mater. Sci. Eng.* **411** 012028

[9] Konyuhov V Y, Gorban A V and Gladkih A M 2019 Investment in the improvement of maintenance service efficiency of processing equipment of an industrial enterprise *J. of Phys.: Conf. Ser.* **1353(1)** 012102

[10] Olentsevich V A, Gozbenko V E, Kargapoltsev S K and Kramynina G N 2019 Complex of organizational, technical and reconstructive measures aimed at improving the performance of the site based on the study of system connections and patterns of functioning of the railway transport system *Modern technologies. System analysis. Modeling* **3(63)** 171-179

[11] Ilyushin P and Suslov K 2019 Operation of automatic transfer switches in the networks with distributed generation *IEEE Milan PowerTech* 8810450

[12] Voropai N, Ukolova E, Gerasimov D, Suslov K, Lombardi P and Komarnicki P 2019 Simulation approach to integrated energy systems study based on energy hub concept *IEEE Milan PowerTech* 8810666

[13] Buryanina N, Korolyuk Y, Koryakina M, Lesnykh E and Suslov K 2019 Algoritm of Current Protection Based on Three Instantaneous-Value Samples *Proc. of 2019 IEEE PES Innovative Smart Grid Technologies Europe, ISGT-Europe* 8905542

[14] Leonovich D S, Zhuravlev D A, Karlina Yu I, Govorkov A S and Karlina A I 2020 Automated assessment of the low-rigid composite parts influence on the product assemblability in the

GePARD system *IOP Conf. Ser.: Mater. Sci. Eng.* **760(1)** 012038
[15] Karlina A I, Gozbenko V E, Kargapoltsev S K and Karlina Y I 2019 Methods for controlling the vibration state of technical facilities *J. of Phys.: Conf. Ser.* **1384(1)** 012019
[16] Gozbenko V E, Kargapol'Tsev S K, Karlina Yu I and Karlina A I 2019 Approximation of amplitude-frequency characteristics using equidistants *J. of Phys.: Conf. Ser.* **1384(1)** 012014
[17] Kargapoltsev S K, Gozbenko V E, Karlina Yu I and Karlina A I 2019 Introduction of additional inertial couplings in mathematical models of problems of vibration protection and vibration isolation *IOP Conf. Ser.: Earth Env. Sci.* **378(1)** 012059
[18] Belogolov Yu I, Olentsevich V A and Astashkov N P 2018 Improving the operational management of transport processes in railway transport *Proc. of 6th Int. Symp. on Innovation and Sustainability of Modern Railway ISMR 2018* (Irkutsk: ISTU) pp 602-609

IOP Conf. Series: Materials Science and Engineering 1064 (2021) 012026 doi:10.1088/1757-899X/1064/1/012026

Analytical method in the study of kinematics of process machinery mechanisms

B N Orlov, M A Karapetyan and A S Matveev

Russian State Agrarian University-Moscow Timiryazev Agricultural Academy, 49, Timiryazevskaya str., Moscow, 127550, Russia

E-mail: matveev@rgau-msha.ru

Abstract. Among all technological machines and equipment, the object of the study was a drainage machine. The machine operates in particularly difficult conditions. The quality of operation depends on the maintenance of process parameters. The kinematics of a working element of the drainage machine and its suspension on the base machine significantly affect the accuracy of holding the given slope of a drain trench and slot. In recent years this issue has received considerable attention. However, the existing analytical methods for kinematic studies of drainage machine mechanisms are complex and therefore not always acceptable. The geometrical methods are used more often. The developed analytical method based on the use of complex numbers conveniently combines the analytical study of the kinematics of mechanisms with geometrical representation.

1. Introduction

The developed analytical method based on the use of complex numbers conveniently combines the analytical study of the kinematics of mechanisms with the geometrical representation (1).

Each complex number $z=x+iy$ can be represented by a point on a plane (2). The real part is placed on the horizontal line, and the imaginary part of the number z – on the vertical line [1, 2]. Let us suppose a complex number as vector $\overline{z}$, the origin of which coincides with the origin of coordinates, and the end – with the point z on the complex plane. The vector $\overline{z}$ is characterized by the distance l of the z point to the origin and the angle φ with abscissa axis.

Thus,

$$\begin{aligned} &x = l\cos\phi;\ y = l\sin\phi, a \\ &\overline{z} = x + iy = l(\cos\phi + i\sin\phi)\cdot \end{aligned} \quad (1)$$

The vector module $|\overline{z}|$ is based on the formula

$$|\overline{z}| = l = (x^2 + y^2)^{1/2}, \quad (2)$$

and its inclination according to formulas

$$tg\phi = \frac{y}{x}; \sin\phi = y\cdot(x^2 + y^2)^{1/2}; \cos\phi = x\cdot(x^2 + y^2)^{1/2}. \quad (3)$$

It should be noted that formula (1) can be represented by a set of points of a curve with coordinates

$$x = l\cos\phi;\ y = l\sin\phi. \tag{4}$$

At $l = const$, the path described by a point is a circle with radius l when φ is changed from 0 to 2π.

Formulas (1...4) make it possible to determine the trajectory of movement of any point of a link or a working tool at a variable l in the time function, given that $\phi = \omega t$ (here ω – angular velocity of a mechanism link, t – time) [3].

The speed of the link point $\overline{V}$, its modulus, the coordinates of the velocity vector are found as follows.

By differentiating the equation (1), we get:

$$\overline{V} = \overline{z}' = l'(\cos\phi + \sin\phi) + \phi' li(\cos\phi + \sin\phi) = \frac{\overline{z}}{l}(l' + \phi' l), \tag{5}$$

where $\phi' = \omega$ – angular velocity of a link; l' – linear velocity of a link if its length is variable over time.

It follows from equation (5) that the vector $\overline{z}'$ is shorter or longer than the vector $\overline{z}$ by an amount $\frac{l' + i\phi' l}{l}$.

Dividing the real R_e and imaginary parts J_m into equation (5), we get

$$\overline{V} = \overline{z}' = (l'\cos\phi + \phi' l\sin\phi) + i(\phi' l\cos\phi + l'\sin\phi) = x' + iy'. \tag{6}$$

The derivative $\overline{z}'$, i.e. the speed of a link point, is depicted by the vector with the beginning at the point with the coordinates

$$\begin{aligned} V_x &= x^1 = l'\cos\phi - \phi' l\sin\phi \\ V_y &= l'\sin\phi + \phi' l\cos\phi = y^1 \end{aligned} \tag{7}$$

The values x' and y' can also be obtained by differentiating the equations (4).

The module of the vector $|\overline{z}'|$, i.e. the absolute velocity V of the point, is

$$|\overline{z}'| = V = (x'^2 + y'^2)^{1/2} = (l'^2 + l^2\phi'^2)^{1/2}. \tag{8}$$

From the equation (8) we see that at $l = const$ the velocity of the point equals $l\phi'$.

To find the acceleration $\overline{W}$ of a link point by differentiating $\overline{z}'$ (see equation (5)) we get

$$\overline{W} = \overline{z}'' = l''\frac{\overline{z}}{l} + 2l'\phi' i\frac{\overline{z}}{l} + i^2 y'^2\,\overline{z} + \phi'^2\, i\overline{z} \tag{9}$$

or in trigonometric form

$$\begin{aligned} \overline{W} = \overline{z}'' = l''(\cos\phi + i\sin\phi) + 2l\phi' i(\cos\phi + i\sin\phi) + \\ + i^2 l y'^2(\cos\phi + i\sin\phi) + l\phi'' i(\cos\phi + i\sin\phi) \end{aligned} \tag{10}$$

or by separating the real and imaginary parts, $\overline{W} = \overline{z}'' = x'' + iy''$.

The second derivative $\overline{z}''$ corresponding to the acceleration vector is represented by a vector with the beginning at a point with coordinates

$$W_x = x'' = l''\cos\phi - 2l'\phi'\sin\phi - l\phi''\sin\phi - l\phi'^2\cos\phi$$
$$W_y = y'' = l''\sin\phi + 2l'\phi'\cos\phi - ly'^2\sin\phi + l\phi''\cos\phi$$

The module of the vector $\left|\overline{z}''\right|$ equal to the full acceleration W of a link point is determined from the formula

$$W = \left|\overline{z}''\right| = \left(x''^2 + y''^2\right)^{1/2} = \left(l''^2 + 4l'^2\phi'^2 + l^2\phi''^2 + l^2\phi'^4\right)^{1/2}. \quad (11)$$

2. Results

It should be noted that regardless of the complexity of the mechanism, the determination of speeds and accelerations of link points according to the developed method is reduced to the compilation and solution of linear equations. This facilitates the task [4, 5]. For example, let us conduct a kinematic study of the suspension mechanism of the excavating attachment of the drainage machine on the base machine.

The kinematic scheme of this mechanism is shown in Figure 1, and Figure 2 illustrates an analogue in the form of vector polygons [5, 6].

In this case, the closedness equations of vector polygons should be differentiated by a generalized coordinate, and not by time, since the base machine is influenced by variable bond forces.

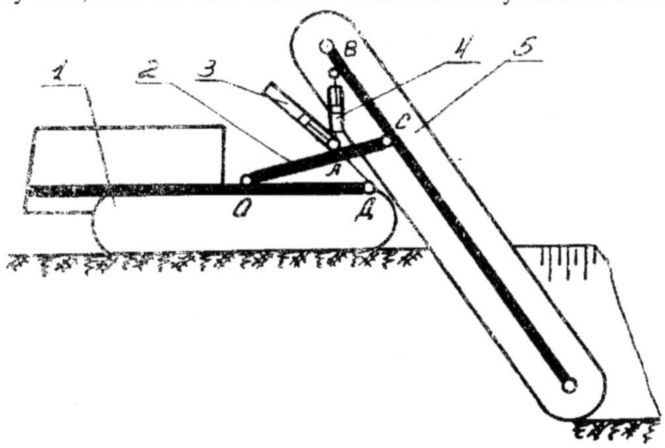

Figure 1. Kinematic diagram of an excavator – drainage machine: 1 – base machine; 2 – drive unit; 3, 4 – hydraulic drives; 5 – working element

If the frame of the base machine is stationary or in rectilinear and uniform movement, then $\varphi = t$ (φ-radians).

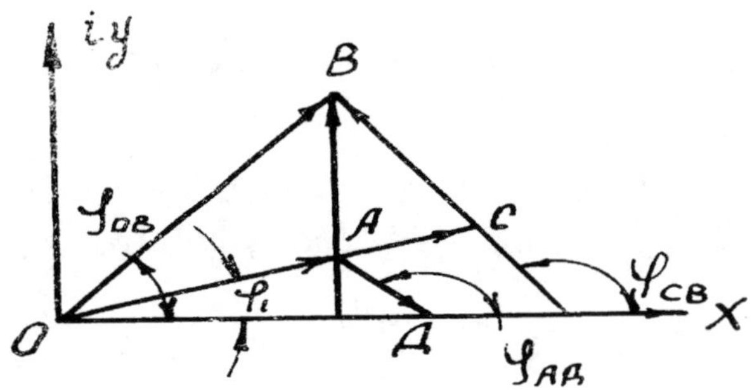

Figure 2. Vector analogue of an excavator – drainage machine kinematic scheme

The dependence of the generalized coordinate φ_1 on t is found from the dynamics equation of a drive link (2), which moves under the action of a hydraulic cylinder (3) (Figure 1), which overcomes the resistance of forces from the drive link. Angle φ_1 is chosen as the generalized coordinate.

So we get the velocity v*A* from the equation.

$$\bar{z}'_{OA} = \bar{v}A = -\phi_1' OA\sin\phi_1 + i\phi_1' OA\cos\phi_1. \tag{12}$$

Considering that $i^2 = -1$, and $\phi_1' = 1$, we get

$$\bar{v}A = \bar{z}'OA = iOA(\cos\phi_1 + i\sin\phi_1) = i\bar{z}OA \ . \tag{13}$$

From the equation we see that the vector $\bar{v}_A$ is perpendicular to the vector $\bar{z}_{OA}$, the beginning of the vector has coordinates $y'_A = OA\cos\phi_1; x'_A = -OA\sin\phi_1$, i.e. coincides with the end of the vector $\bar{z}_{OA}$.

Module $\left|\bar{z}'OA\right| = v_A = \left(x_A'^2 + y_A'^2\right)^{1/2} = OA$.

To find the acceleration of point *A*, let us differentiate the equation:

$$\overline{W}_A = \bar{z}''_{OA} = i^2 OA(\cos\phi_1 + i\sin\phi_1) = i^2 \bar{z}_{OA}. \tag{14}$$

From equation (14) we see that the vector $\overline{W}_A$ is turned in relation to the vector $\bar{z}_{OA}$ by 180° and directed to the center *O*. The beginning of the vector $\overline{W}_A$ at the point with coordinates $x''_A = -OA\cos\varphi_1; y''_A = -OA\sin\varphi_1$, and the modulus

$$W_A = \left|\bar{z}''_{OA}\right| = \left(x_A''^2 + y_A''^2\right)^{1/2} = OA.$$

Let us find the change in the length of a hydraulic cylinder rod (3), its inclination angle, angular velocity and angular acceleration at φ_1 change [7].

From the vector triangle *OAD* we have (when compiling the vector closedness equation you should bypass the perimeter in the set direction, if the vector is directed against the bypass, it is taken with minus):

$\bar{z}_{AO} = \bar{z}_{AD} - \bar{z}_{OA}$,

or, by substituting the values of the vectors, we get

$$\bar{z}_{AD} = OD(\cos\varphi_{OD} + i\sin\varphi_{OD}) - OA(\cos\varphi_1 + i\sin\varphi_1).$$

However, given that φ_{OD}=0, we finally have

$$\bar{z}_{AD} = OD - OA(\cos\phi_1 + i\sin\phi_1). \tag{15}$$

According to equation (15): $x_{AD} = OD - OA\cos\varphi_1; y_{AD} = OA\sin\varphi_1$.

The length *A D* changes by law

$$AD=\left(x_{AD}^{2}+y_{AD}^{2}\right)^{1/2}=\left(OD^{2}-2OD\cdot OA\cos\phi_{1}+OA^{2}\right)^{1/2}.$$

The inclination of a hydraulic cylinder (3) is found from the formula

$$\cos\phi_{AD}=\frac{x_{AD}}{\left(x_{AD}^{2}+y_{AD}^{2}\right)^{1/2}}=\frac{OA\cos\phi_{1}}{\left(OD^{2}-2OD\cdot OA\cos\phi_{1}+OA^{2}\right)^{1/2}} \qquad (16)$$

or

$$tg\phi_{AD}=\frac{y_{AD}}{x_{AD}}=\frac{OA\sin\phi_{1}}{OD-OA\cos\phi_{1}}. \qquad (16a)$$

To find the angular velocity of the hydraulic cylinder (3), let us differentiate the equation (16a)

$$\frac{\varphi'_{AD}}{\cos^{2}\varphi_{AD}}=\frac{(OD-OA\cos\varphi_{1})OA\varphi'_{1}\cos\varphi_{1}-OA\sin\varphi_{1}\varphi_{1}\varphi'_{1}OA\sin\varphi_{1}}{(OD-OA\cos\varphi_{1})^{2}}=\frac{\varphi'_{1}\left(OD\cdot OA\cos\varphi_{1}-OA^{2}\right)}{(OD-OA\cos\varphi_{1})^{2}}.$$

From this equation, given that $\varphi'_{1}=1$, we find

$$\omega_{AD}=\phi'_{AD}=\frac{OA\cos^{2}\phi_{AD}\left(OD\cos\phi_{1}-OA\right)}{\left(OD-OA\cos\phi_{1}\right)^{2}}. \qquad (17)$$

By substituting the value $\cos^{2}\varphi_{AD}$ from the equation (16), we finally get the angular velocity *AD*

$$\omega_{AD}=\varphi'_{AD}=\frac{OA(OD\cos\varphi_{1}-OA)(OA\cdot\cos\varphi_{1})^{2}}{\left(OD^{2}-2OD\cdot OA\cos\varphi_{1}+OA^{2}\right)(OD-OA\cdot\cos\varphi_{1})^{2}}. \qquad (17^{1})$$

To find the angular acceleration of link *AD* (hydraulic cylinder (3)), it is necessary to differentiate the equation (17).

By grouping the real and imaginary parts, we have

$$\bar{z}_{OB}=\left(OC\cos\phi_{1}+CB\cos\phi_{CB}\right)+i\left(OC\sin\phi_{1}+CB\sin\phi_{CB}\right). \qquad (18)$$

From the equation (18) we find the coordinates of point *B*

$$x_{B}=OC\cos\phi_{1}+CB\cos\phi_{CB};$$
$$y_{B}=OC\sin\phi_{1}+CB\sin\phi_{CB}. \qquad (19)$$

After grouping, we get

$$\bar{z}_{OB}=(OA\cos\varphi_{1}+AB\cos\varphi_{AB})+i(OA\sin\varphi_{1}+AB\sin\varphi_{AB}). \qquad (20)$$

From the equation (20) we have the coordinates of point *B*

$$x_{B}=OA\cos\phi_{1}+AB\cos\phi_{AB};$$
$$y_{B}=OA\sin\phi_{1}+AB\sin\phi_{AB}. \qquad (21)$$

The module of vectors $\bar{z}_{OB}$ equals

$$OB=\left|\bar{z}_{OB}\right|=\left(OA^{2}+2OA\cdot AB\cos\left(\phi_{1}-\phi_{AB}\right)+AB^{2'}\right)^{1/2}=.$$

By equating the coordinate values of point *B* found from the equation (19) and (21), we get

$$\sin\phi_{CB}=\frac{-AC\sin\phi_{1}+AB\sin\phi_{AB}}{CB} \qquad (21^{1})$$

where – *OA*+*OC*=*AC*.

From the last equation we find φ_{CB}.

After differentiating φ_{CB} we get the angular velocity ω_{CB} of the *CB* link, and after secondary differentiation – its angular acceleration ε_{CB}.

The velocity and acceleration of point *B* can be found by differentiating the vector $\bar{z}_{OB}$ (see equation (20), or equations $\bar{z}_{AB} = \bar{z}_{AC} + \bar{z}_{CB}$, or

$$\bar{z}_{AB} = AC\left(\cos\phi_1 + i\sin\phi_1\right) + CB\left(\cos\phi_{CB} + i\sin\phi_{CB}\right). \tag{22}$$

Then

$$x_B = AC\cos\phi_1 + CB\cos\phi_{CB};$$
$$y_B = AC\sin\phi_1 + CB\sin\phi_{CB}$$

– derived from the closedness equation of the *ACB* vector triangle

$$AB = \left(x_B^2 + y_B^2\right)^{1/2} = \left(AC^2 + 2AC\cdot CB\cos\left(\phi_1 - \phi_{CB}\right) + CB^2\right)^{1/2}$$

By differentiating the vector $\bar{z}_{AB}$ (according to equation (22)), we get

$$\begin{aligned}\bar{V}_B = \bar{z}'_{AB} = i^2\left(AC\sin\phi_1 + CB\phi'_{CB}\sin\phi_{CB}\right) + \\ + i\left(AC\cos\phi_1 + CB\phi'_{CB}\cos\phi_{CB}\right),\end{aligned} \tag{23}$$

or after conversion

$$\begin{aligned}\bar{V}_B = i\left[AC\left(\cos\phi_1 + i\sin\phi_1\right) + \phi'_{CB}CB\left(\cos\phi_{CB} + i\sin\phi_{CB}\right)\right] = \\ = i\bar{z}_{AC} + i\phi'_{CB}\bar{z}_{CB}.\end{aligned} \tag{23}$$

The inclination angle of the velocity vector $\bar{V}_B$ equals

$$tg\phi_{\bar{V}_B} = \frac{y'_B}{x'_B} = \frac{AC\cos\phi_1 + CB\phi'_{CB}\cos\phi}{-AC\sin\phi_1 - CB\phi'_{CB}\sin\phi_{CB}}. \tag{24}$$

It shall be noted that according to formulae (23) and (24), at φ_{CB} =const, i.e. while maintaining a given angle φ_{CB}, during adjustment with the help of hydraulic cylinders (3) and (4) $\phi'_{CB} = 0$, and $\phi_{OB} = \phi_1 + 270^\circ$, i.e. the velocity vector of point *B* should always be directed to the vector $\bar{z}_{OA}$ at an angle $\left(270^\circ + \phi_1\right)$ [8, 9, 10].

After differentiating the vector $\bar{z}'_{AB}$, we get the acceleration vector $\bar{W}_B$ of point *B*

$$\begin{aligned}\bar{W}_B = \bar{z}''_{AB} = \left(-AC\cos\phi_1 - CB\phi'^2_{CB}\cos\phi_{CB} - CB\phi''_{CB}\sin\phi_{CB}\right) - \\ - i\left(AC\sin\phi_1 + CB\phi'^2_{CB}\sin\phi_{CB} - CB\phi''_{CB}\cos\phi_{CB}\right)\end{aligned} \tag{25}$$

or after grouping we get

$$\overline{W}_B = i\left[\left(CB\varphi''_{CB}\cos\varphi_{CB} + iCB\varphi''_{CB}\sin\varphi_{CB}\right)\right] + \\ + i^2\left[\left(AC\cos\varphi_1 + iAC\sin\varphi_1\right) + \left(CB\varphi'^2_{CB}\cos\varphi_{CB} + CB\varphi'^2_{CB}\sin\varphi_{CB}\right)\right] = \\ = i\varphi''_{CB}\bar{z}_{CB} + i^2\bar{z}_{AC} + i\varphi'^2_{CB}\bar{z}_{CB}. \tag{25^1}$$

We will find the inclination angle $\overline{W}_B$ by the formula

$$tg\varphi_{\overline{W}_B} = \frac{\varphi''_B}{x''_B} = \frac{AC\sin\varphi_1 - CB\varphi^{11}_{CB}\cos\varphi_{CB} + CB\varphi^{12}_{CB}\sin\varphi_{CB}}{AC\cos\varphi_1 + CB\varphi^{11}_{CB}\sin\varphi_{CB} + CB\varphi^{12}_{CB}\cos\varphi_{CB}}$$

(x''_B and y''_B – defined from equation (25)).

Let us study the mechanism of suspension of the working tool shown in Figure 3 (the vector analogue of this mechanism is shown in Figure 4).

The mechanism is actuated by a hydraulic cylinder (1). The *OA* link is made in the form of a hydraulic cylinder (2). Additional condition is the perpendicularity of the *AB* link to X axis regardless of the hydraulic cylinder position (1).

Vector $\bar{z}_{OA}$ can be written as

$$\bar{z}_{OA} = OA\left(\cos\varphi_{OA} + i\sin\varphi_{OA}\right), \tag{26}$$

where *OA* – length of the *OA* link (variable);

φ_{OA} – inclination angle of the *OA* link to X axis.

We get

$$\left(\ln\bar{z}_{OA}\right) = \left(\ln OA\right)' + i\phi'_{OA}$$

or

$$\ln\bar{z}_{OA} + i\phi_{OA} \tag{27}$$

Next, the equation (27) can be widely used in the analysis of mechanisms, but the function $\ln\bar{z}_{OA}$ is multi-valued, i.e. $\ln\bar{z}_{OA} = \ln OA + i\varphi_{OA} + 2K\pi i$, but the main value is obtained at $K = 0$.

Turning to the mechanism analysis (Figure 3, 4), based on the closedness of a triangle *OAB*, we can write

$$\bar{z}_{AB} = \bar{z}_{OA} - \bar{z}_{OB} \tag{28}$$

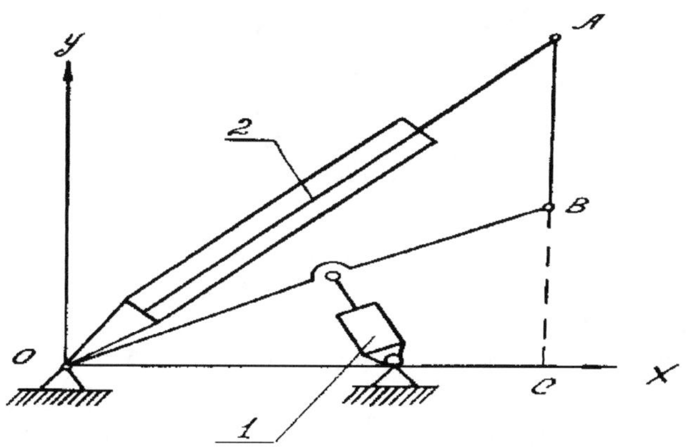

Figure 3. Pendulum motion of the working tool suspension on the base machine

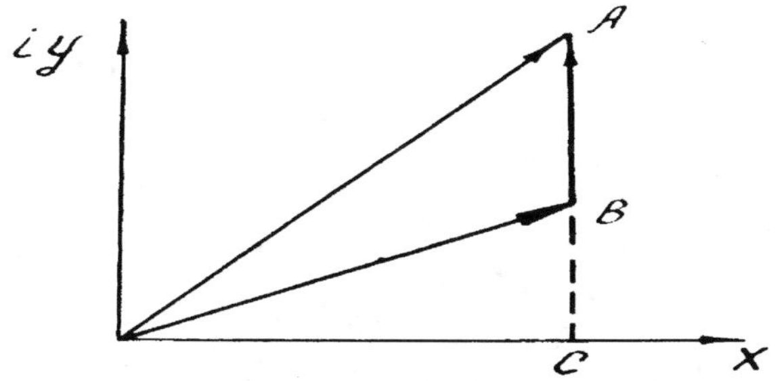

Figure 4. Vector analogue of pendulum suspension mechanism

Let us differentiate the vector $\bar{z}_{AB}$ and set the result to zero.

Vector $\bar{z}'_{OA}$ equals

$$\bar{z}'_{OA} = (\ln OA)'\bar{z}_{OA} + i\bar{z}_{OA}\varphi'_{OA} \tag{28a}$$

Vector $\bar{z}'_{OB}$ equals

$$\bar{z}'_{OB} = i\bar{z}_{OB}\varphi'_{OB}.$$

(there is no valid part since OB=const).

Based on the equality of vectors $\bar{z}'_{OA}$ and $\bar{z}'_{OB}$ we get

$$\bar{z}'_{OA} - \bar{z}'_{OB} = (\ln OA)'\bar{z}_{OA} + i\bar{z}_{OA}\varphi'_{OA} - i\bar{z}_{OB}\varphi'_{OB} = 0.$$

If the vector equals zero, then $R_e=J_m=0$, i.e.

$$(\ln OA)'\bar{z}_{OA} = \bar{z}_{OA}\varphi'_{OA} - \bar{z}_{OB}\varphi'_{OB} \tag{29}$$

(this equality can also be obtained after moving J_m to the right and squaring the left and right parts).

The ratio of vectors from the equation (29) is

$$\frac{\bar{z}_{OA}}{\bar{z}_{OB}} = -\frac{\varphi'_{AB}}{(\ln OA)' - \varphi'_{OA}} = -\frac{\omega_{OB}}{(\ln OA)' - \omega_{OA}},$$

where ω_{OB} and ω_{OA} – angular speeds of OA and OB links.

Considering the obtained expression, the equation (28) can be written as

$$\bar{z}_{AB} = \bar{z}_{OA} + \bar{z}_{OA}\frac{(\ln OA)' - \omega_{OA}}{\omega_{OB}}. \tag{30}$$

By differentiating the equation (30) and equating the result to zero, we get

$$\bar{z}'_{OA} + \bar{z}'_{OA}\left[\frac{(\ln OA)' - \omega_{OA}}{\omega_{OB}}\right] + \bar{z}_{OA}\left[\frac{(\ln OA)' - \omega_{OA}}{\omega_{OB}}\right]' = 0.$$

Let us put $\bar{z}'_{OA}$ outside the brackets and, taking into account the equation (28a), we get

$$\left[(\ln OA)'\bar{z}_{OA} + i\bar{z}_{OA}\varphi'_{OA}\right]\left[1 + \frac{(\ln OA)' - \dot{\omega}_{OA}}{\omega_{OB}}\right] + \bar{z}_{OA}\left[\frac{(\ln OA)' - \omega_{OA}}{\omega_{OB}}\right]' = 0.$$

By dividing R_l and J_m considering that $\bar{z}OA$, we get

$$(\ln OA)'\left[1 + \frac{(\ln OA)' - \omega_{OA}}{\omega_{OB}}\right] + \left[1 + \frac{(\ln OA)' - \omega_{OA}}{\omega_{OB}}\right]' + i\omega_{OA}\left[1 + \frac{(\ln OA)' - \omega_{OA}}{\omega_{OB}}\right] = 0.$$

If the vector is zero, then $R_l = J_m = 0$.

Therefore, we have two equations of complex trajectories of the working tool points of the drainage machine and its suspension.

3. Conclusion

The developed analytical method of analyzing the mechanisms of a drainage machine makes it possible to study their kinematics by analytical calculation, to determine quite complex movement trajectories of the mechanism points, their velocity and acceleration.

The velocity equation:

$$\omega_{OA}\left[1 + \frac{(\ln OA)' - \omega_{OA}}{\omega_{OB}}\right] = 0$$

$$(\ln OA)'\left[1 + \frac{(\ln OA)' - \omega_{OA}}{\omega_{OB}}\right] + \left[1 + \frac{(\ln OA)' - \omega_{OA}}{\omega_{OB}}\right]' = 0$$

Based on the first equation, we get (taking into account that $\omega_{OA} \neq 0$) quite interesting kinematic relations:

$$\begin{aligned} \omega_{OA} - \omega_{OB} &= (\ln OA)'; \\ \phi'_{OA} - \phi'_{OB} &= (\ln OA)'; \\ \phi_{OA} - \phi_{OB} &= \ln OA. \end{aligned} \tag{31}$$

As a result, $\omega'_{OA} - \omega'_{OB} = (\ln OA)''$, where ω_{OA} and ω_{OB} – angular velocities of *OA* and *OB* links.

The second equation gives more complex dependencies, which should be solved in relation to $(\ln OA)''$ taking into account the equations (31).

After transformations, it is equated to following

$$y''B + y'(A + y'B) + C = 0,$$

where $$\begin{aligned} y &= \ln OA; A = \left(\omega_{OB}^2 - \omega_{OB} \cdot \omega_{OA} - \varepsilon_{OB}\right); \\ B &= \omega_{OB}; C = \left(-\omega_{OB}\varepsilon_{OA} + \omega_{OA}\varepsilon_{OB}\right); \end{aligned}$$ +

ε_{OB} and ε_{OA} – angular accelerations of links *OB* and *OA* .

After solution, we get $OA = f(\omega_{OB})$.

This makes it possible to improve the accuracy and quality of laying drainage trench shelters in especially harsh operating conditions of process machines and equipment.

References

[1] Orlov B N, Karapetyan M A and Matveev A S 2018 The impact of the industrialization of agriculture on the structural reliability of agri-business machines *Int. tech.-economic j.* **3** 72-77

[2] Orlov B N and Matveev A S 2015 Software problems FEM structural strength *Ground transport and technological complexes and facilities: Mater. of the Int. sci. and tech. conf.* (Tyumen Industrial University) pp 194-198

[3] Fedorenko V F, Buklagin D S, Golubev I G and Nemenushchaya L A 2015 Review of Russian nanoagents for crops treatment *Nanotechnologies in Russia* **10(3-4)** 318-324

[4] Melekhov V I, Bykov B B and Pyrikov P G 2016 The anniversary of professor Evgeniy A. Pamfilov *News of higher educational institutions. Forest magazine* **6(354)** 175-177

[5] Golubev I G, Spitsyn I A, Bykov V V and Golubev M I 2018 Prospects of application of additive technologies in the repair of agricultural machinery *Works of GOSNITI* **130** 214-219

[6] Prokhorov V Yu and Bykov V V 2017 Increase of durability and wear resistance of sliding bearings of mounted processing equipment *Machinery and equipment for rural area* **8** 36-39

[7] Starovoytov V, Starovoytova O, Aldoshin N and Manohina A 2017 Jerusalem artichoke as a means of fields conservation *Acta Technologica Agriculturae* **20(1)** 7-10

[8] Aldoshin N and Didmanidze O 2018 Harvesting lupinus albus axial rotary combine harvesters *Res. in Agricult. Eng.* **64(4)** 209-214

[9] Aldoshin N V, Kravchenko I N, Kuznetsov Yu A, Kalashnikova L V and Korneev V M 2018 New working element of stripper header "Ozon" *INMATEH – Agricult. Eng.* **55(2)** 69-74

[10] Golubev I G, Fadeev A Yu and Makuev V A 2010 Tractors Servicing Quality Assessment *Machinery and equipment for rural area* **7** 40-41

Description of information support of the automated control system for the technological process of thermal vortex enrichment

S A Nebogin, V V Kondrat'ev, V O Gorovoy, R V Kononenko and T A Oparina

Irkutsk National Research Technical University, 83, Lermontova Str., Irkutsk, 664074, Russia

E-mail: nebogin@ex.istu.edu

Abstract. The information support of the automated control system for the technological process of thermal vortex enrichment is described. Solutions for the organization of information support, collection and transmission of information have been developed. A classification and coding system has been built. The intra-machine information base is organized according to the principle of dividing into high-level and low-level databases. It is possible to maintain an off-machine information base by creating reports for outputting information in the form of logs on paper. The composition of the information base in terms of databases is given.

1. Introduction

The automated control system for the technological process of thermal vortex enrichment is part of a project to create an integrated resource-saving technology and organize high-tech production of nanostructures based on carbon and silicon dioxide to improve the properties of building and structural materials. Information support for the automated control system for the technological process of thermal vortex enrichment consists of high-level and low-level databases. All databases are provided complete and ready to go [1-3]. The database is filled only by software specialists. Changing database fields is provided by the user interface. All DBMS maintenance procedures, including health diagnostics and data backup, must be performed in accordance with the general principles of DBMS administration [4, 5].

The organization of the information base does not provide for integration with external information resources. For the subsystem of integration with external information resources, databases of other subsystems are used [6-8].

Information from the databases is displayed on the screen of the operator's workstation and operator panels in varying degrees of detail. Through visualization systems, changes can be made to a limited number of database fields [9]. High-level databases can be accessed by external information systems.

2. Organization of information support

Low-level data from monitoring and control devices are stored in the energy-dependent memory of the controller on an electronic medium (NFlash micro memory card). Access to data from other AS is closed with the exception of trusted sources and receivers [10-13].

Top-level data to maintain continuous long-term archives of all process parameters, a hard-magnetic disk (HDD) and Microsoft SQL databases are used. Data can be accessed by means of OPC, ODBC, SQL queries.

3. Organization of collection and transmission of information

The information consists of data coming from the following sources:

- sensors and primary converters;
- frequency converters;
- executive mechanisms;
- operator panels;
- automated workstations (AWP);
- personal computers (PC).

According to the intensity of the incoming information, they are divided into:

- continuous (operator panels, sensors and primary converters, frequency converters, PC, AWP);
- frequent (executive mechanisms);
- rare (switches).

By volume, the information is divided into:

- small (sensors and primary converters, frequency converters, etc.);
- medium (operator panels, remote data collection stations);
- large (AWP, PC).

The collection of information is carried out in automatic mode with diagnostics about the state of information quality.

4. Building a classification and coding system

Data classification:

- raw process parameters;
- processed process parameters;
- DB of regulators;
- DB of regulatory bodies (RO);
- discrete variables;
- others.

Class encoding:

- Raw process parameters

RAW_*_**

where * – measured parameter, ** – short name of the position

- Processed process parameters

* _ **

where * - measured parameter, ** - short name of the position

- DB of regulators

FROM_*_**

where * - adjustable parameter, ** - short name of the position

- DB of regulatory bodies (RO)

V _ * _ **

where * - location of RO installation, ** - short name of the position

- DB of archive parameters

* _ ** _ SUM

where * - measured parameter, ** - short name of the position

- Discrete variables

* _ALARM / HI / LO / ON / OFF

where * is the measured parameter or the short name of the position, HI / LO / ON / OFF is selected depending on the type of the sensor in the triggered state (logical unit).

- Others.

The rest of the parameters are coded logically for reasons of a clear concept of what information the variable carries [14-16].

5. Formation of database numbers DB

DB "Reading and processing analog signals" is as follows.

Primary measuring devices: the position of the primary converter is taken from the automation circuit ABSP.441199.001 C3 (the number before the separator in the form "-"); this value is multiplied by 10 and the line number is put at the end (from 1 to 5).

The rotational speed is of the blower, smoke exhauster, feeder auger motors and the position of the actuator. The position of the primary measuring transducer in the circuit of which the regulating device is involved is taken. This value is multiplied by 10, the zero at the end changes to 7 and the line number is put at the end (from 1 to 5) [17-18].

DB "Control of the regulatory body" and DB "PID-regulator" includes the following.

The position of the primary measuring transducer is taken in the loop of which this regulator is located, this value is multiplied by 10, the zero at the end changes to the number 1 and at the end the line number is put (from 1 to 5).

DB "Archive values of process parameters" consists of the following.

Primary measuring devices: the DB number of the DB “Reading and processing analog signals” is taken and the zero before the line number is changed to 8.

The rotational speed of the blower motors, the smoke exhauster, the feeder auger and the position of the actuator: the DB number of the database “Reading and processing analog signals” is taken and the number 5 before the line number is changed to 8.

6. Organization of the in-machine information base

The information base is organized according to the following principle:

- Low-level databases:
  - DB "Reading and processing of analog signals";
  - DB "Management of the regulatory body";
  - DB "PID-controller";
  - DB "Archive values of process parameters".
- High-level databases:
  - DB "Long-term archive of process parameters";
  - DB "Alarms and warnings".

The structure of the in-machine information base is shown in Fig. 1.

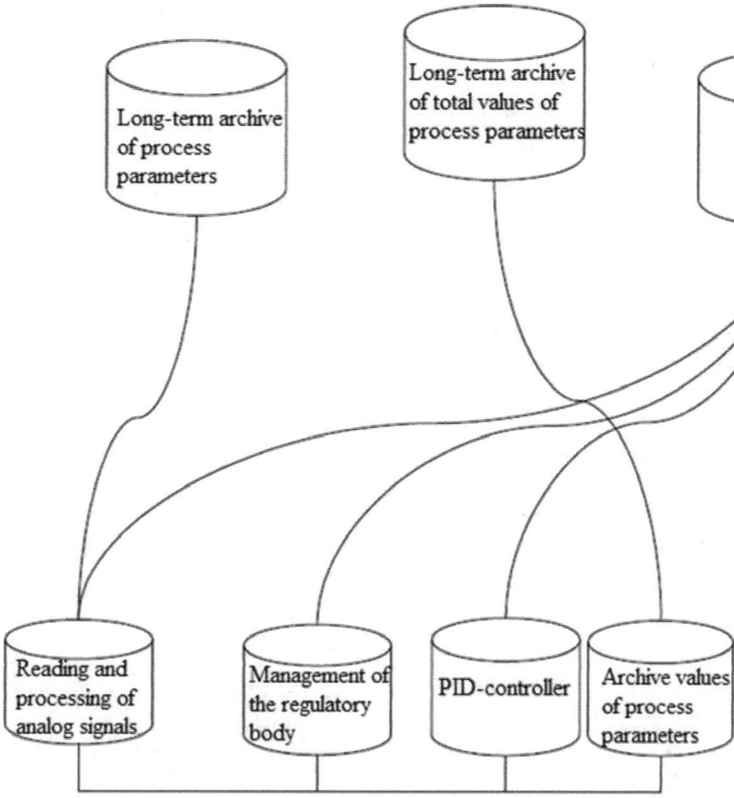

Figure 1. Algorithm for reading and processing analog signals.

7. Organization of the off-machine information base

The out-of-machine information base is built on the regulatory process indicators, ore and concentrate sampling logs, as well as on the basis of technical data of the applied control, management and regulation devices. In accordance with which information is entered into the in-machine databases through the user man-machine interface of the used SCADA system [19-20].

Keeping logs of process indicators is not provided, since the system, at the user's request, can issue the necessary information on paper. But the process data can be recorded in the journals of shift, daily, monthly or annual work [21-23].

8. Composition of information support

The databases of the lower level of the automated system are described in tables 1 - 4.

Table 1. DB "Reading and processing of analog signals".

No	Designation	Code	Name
1	T_vozd_in	DB 101	Temperature of the air entering the combustion chamber
2	T_vozd_out	DB 201	Air temperature at the exit of the combustion chamber
3	T_beg_react	DB 301	Air flow temperature at the beginning of the vortex enrichment tube
4	T_end_react	DB 401	Air flow temperature at the end of the vortex enrichment tube
5	T_cool	DB 501	Air flow temperature in the cooling gas duct
6	T_in_cool	DB 601	Air temperature entering the cooling gas duct
7	T_prod_bunk	DB 701	Bag filter bin temperature
8	P_gas	DB 801	Gas pressure entering the burner
9	P_air	DB 901	Compressed air pressure
10	P_out_topk	DB 1001	Pressure at the beginning of the vortex enrichment tube
11	P_cool	DB 1101	Cooling duct pressure
12	P_in_filtr	DB 1201	Pressure in the dirty bag filter chamber
13	P_out_filtr	DB 1301	Pressure in the clean chamber of the bag filter
14	F_gas	DB 1401	Gas consumption entering the combustion chamber
17	FREQ_M1_1	DB 257	Blower M1-1. Rotation frequency.
18	FREQ_M2_1	DB 1157	Smoke exhauster M2-1. Rotation frequency.
19	FREQ_M3_1	DB 1557	Material feed screw drive M3-1. Rotation frequency.
20	Poloz_5_1	DB 557	Gate 5-1s position

Table 2. DB "Management of the regulatory body".

No	Designation	Code	Name
1	V_ F_air_cool	DB 551	Regulator of air supply to the cooling flue

Table 3. DB "PID-controller".

No	Designation	Code	Name
1	C_ T_vozd_out	DB 251	Combustion chamber temperature controller
2	C_ P_cool	DB 1151	Vacuum regulator in the cooling gas duct

Table 4. DB "Reading and processing of analog signals".

No	Designation	Code	Name
1	T_vozd_in_Sum	DB 181	Temperature of the air entering the combustion chamber
2	T_vozd_out_Sum	DB 281	Air temperature at the exit of the combustion chamber
3	T_beg_react_Sum	DB 381	Air flow temperature at the beginning of the vortex enrichment tube
4	T_end_react_Sum	DB 481	Air flow temperature at the end of the vortex enrichment tube
5	T_cool_Sum	DB 581	Air flow temperature in the cooling gas duct
6	T_in_cool_Sum	DB 681	Air temperature entering the cooling gas duct
7	T_prod_bunk_Sum	DB 781	Bag filter bin temperature
8	P_gas_Sum	DB 881	Gas pressure entering the burner
9	P_air_Sum	DB 981	Compressed air pressure
10	P_out_topk_Sum	DB 1081	Pressure at the beginning of the vortex enrichment tube
11	P_cool_Sum	DB 1181	Cooling duct pressure
12	P_in_filtr_Sum	DB 1281	Pressure in the dirty bag filter chamber
13	P_out_filtr_Sum	DB 1381	Pressure in the clean chamber of the bag filter
14	F_gas_Sum	DB 1481	Gas consumption entering the combustion chamber
17	FREQ_M1_1_Sum	DB 2581	Blower M1-1. Rotation frequency.
18	FREQ_M2_1_Sum	DB 1181	Smoke exhauster M2-1. Rotation frequency.

The databases of the lower level of the automated system are described in tables 5, 6.

Table 5. DB "Long-term archive of process parameters".

No	Designation	Name
1	T_vozd_in.PV	Temperature of the air entering the combustion chamber
2	T_vozd_out.PV	Air temperature at the exit of the combustion chamber
3	T_beg_react.PV	Air flow temperature at the beginning of the vortex enrichment tube
4	T_end_react.PV	Air flow temperature at the end of the vortex enrichment tube
5	T_cool.PV	Air flow temperature in the cooling gas duct
6	T_in_cool.PV	Air temperature entering the cooling gas duct
7	T_prod_bunk.PV	Bag filter bin temperature
8	P_gas.PV	Burner gas pressure
9	P_air.PV	Compressed air pressure
10	P_out_topk.PV	Pressure at the beginning of the vortex enrichment tube
11	P_cool.PV	Cooling duct pressure
12	P_in_filtr.PV	Pressure in the clean chamber of the bag filter
13	P_out_filtr.PV	Pressure in the dirty bag filter chamber
14	F_gas.PV	Gas consumption entering the combustion chamber
15	F_air.PV	Combustion air consumption
16	F_air_cool.PV	Air flow entering the cooling gas duct
17	FREQ_M1_1.PV	Blower M1-1. Rotation frequency.
18	FREQ_M2_1.PV	Smoke exhauster M2-1. Rotation frequency.
19	FREQ_M3_1.PV	Material feed screw drive M3-1. Rotation frequency.
20	Poloz_5_1.PV	Gate 5-1s position
21	C_ T_vozd_out.PV	Temperature control from the combustion chamber
22	C_ P_cool.PV	Regulation of vacuum in the cooling gas duct
23	C_ F_air.PV	Combustion air flow control
24	C_ F_air_cool.PV	Regulation of air supply to the cooling flue
25	C_FREQ_M3_1	Material feed screw drive M3-1. Speed control.

Table 6. DB "Alarms and warnings".

No	Designation	Name
1	Burner_alarm	Burner. Crash.
2	bunk_1_HL	Feedstock bin. Upper level.
3	bunk_1_LL	Feedstock bin. Lower level.
4	mater_on	Material availability
5	bunk_2_HL	Product hopper. Upper level.
6	bunk_2_LL	Product hopper. Lower level.
7	M1_1_on	Blower M1-1. Work.
8	M1_1_alarm	Blower M1-1. Frequency drive failure
9	M2_1_on	Smoke exhauster M2-1. Work.
10	M2_1_alarm	Smoke exhauster M2-1. Frequency drive failure
11	M3_1_on	Material feed screw drive M3-1. Work
12	M3_1_alarm	Material feed screw drive M3-1. Frequency drive failure
13	P_cool	The lower value of the vacuum in the tube of vortex enrichment
14	P_delta_filtr_	The upper value of the pressure drop in the baghouse filter
15	T_cool	The temperature in the cooling flue is close to the upper limit
16	T_out	Upper combustion chamber temperature
17	T_out	Lower combustion chamber temperature
18	T_out	The temperature in the combustion chamber is close to the upper value
19	T_out	The temperature in the combustion chamber is close to the lower value
20	T_cool	Upper temperature level in the cooling flue

9. Conclusion

Information support of the automated control system for the technological process of thermal vortex enrichment consists of high-level and low-level databases. The collection of information is carried out automatically with diagnostics about the state of the quality of information from sensors, transducers, buttons and switches, automated workstations, operator panels, personal computers.

The developed classification of data and coding of classes makes it possible to fully structure information, which is clearly reflected in the presented composition of the information base. Generation of reports for maintaining an off-machine database is based on the regulatory process indicators, ore and concentrate sampling logs, as well as on the basis of the technical data of the control, management and regulation devices used.

References

[1] Munkhtuvshin D, Balabanov V B and Putsenko K N 2017 Experience of application of additional-wok micro- and nanosilic from waste of silicon production in concrete technologies *Izvestiya vuzov. Investments. Building. The property* **7(3(22))** 107-115

[2] Kondratiev V V, Petrovskiy A A, Ershov V A, Sysoeva T I and Karlina A I 2017 Results of research with revealing technological parameters of processes of recycling and neutralization of the first and second cut of the spent lining of electrolyzers for reception of aluminum fluoride by pyrolytic and hydro chemical method *Int. J. of Applied Engineering Research* **12(23)** 13898-13904

[3] Yurovsky V S, Glukhatkina L G, Pankratova G M, Sorokina I E and Maksimova T V 2016 Development of heat-frost-resistant rubbers for sleeves on the basis of siloxane rubber and domestic silica fillers *Industrial production and use of elastomers* **1** 28-31

[4] Guseva E A, Konstantinova M V, Kargapoltsev S K, Gozbenko V E, Sivtsov A V, Karlina A I and Shtayger M G 2019 Methods of graphitized steels obtaining *J. of Phys.: Conf. Ser.* **1353(1)** 012063

[5] Ivanchik N N, Balanovsky A E, Shtayger M G, Sysoev I A and Karlina A I 2018 Capability enhancement of production of activating fluxes for arc welding using ultradispersed products of

silicon waste processing *IOP Conf. Ser.: Mater. Sci. Eng.* 012035
[6] Vasin V A, Fatyanova N G, Troshin B A, Vasichev B N and Stepanchikov S V 2012 Problems of creation of nanoproducts using nanotechnology *Electrotechnical and information complexes and systems* **8(4)** 37-43
[7] Gerasimenko N, Volokhovsky A and Zaporozhan O 2017 Features of control of technology of silicon nanostructures *Nanoindustry* **5(76)** 36-51
[8] Kodolov V I, Bondar A Yu and Kuznetsov A P 2007 Technology of production of carbon-herodometal-containing nanostructures in polymeric matrix nanoreactors (on example pva) *Nanotechnics* **1(9)** 38-41
[9] Khokhlun A R 2008 Equipment and technologies for production of mems, micro and nanostructures *Electronics and electrical equipment of transport* **6** 36-38
[10] Buryanina N, Korolyuk Y, Koryakina M, Lesnykh E and Suslov K 2019 Algoritm of Current Protection Based on Three Instantaneous-Value Samples *Proc. of 2019 IEEE PES Innovative Smart Grid Technologies Europe, ISGT-Europe* 8905542
[11] Suslov K, Shushpanov I, Buryanina N and Ilyushin P 2020 Flexible power distribution networks: New opportunities and applications *SMARTGREENS 2020 – Proc. of the 9th Int. Conf. on Smart Cities and Green ICT Systems*
[12] Voropai N, Ukolova E, Gerasimov D, Suslov K, Lombardi P and Komarnicki P A 2019 Study on Cost-Effectiveness of Energy Supply Based on the Energy Hub Concept *Proc. of 2019 IEEE PES Innovative Smart Grid Technologies Europe, ISGT-Europe 2019* 8905736
[13] Suslov K, Solonina N and Gerasimov D 2018 Assessment of an impact of power supply participants on power quality *Proc. of Int. Conf. on Harmonics and Quality of Power, ICHQP* pp 1-5
[14] Ershov V A, Kondratiev V V, Karlina A I, Kolosov A D and Sysoev I A 2018 Selection of control system parameters for production of nanostructures concentrates *J. of Phys.: Conf. Ser.* **1118(1)** 012014
[15] Kuzmin M P, Shestakov S S, Kuzmina M Yu and Zhuravleva A S 2015 Innovative development of the metallurgical complex of the Irkutsk region *Bull. of ISTU* **5(100)** 236-240
[16] KondratEv V V, Nemchinova N V, Ivanov N A, Ershov V A and Sysoev I A 2013 New production solutions for processing silicon and aluminum pro-duction waste *Metallurgist* **57(5-6)** 455-459
[17] Yolkin K S, Yolkin D K, Kolosov A D, Ivanov N A and Shtayger M G 2018 Technologies, which allow to reduce an impact of metal silicon production on the environment *IOP Conf. Ser.: Mater. Sci. Eng.* **411** 012028
[18] Kuzmin M P, Larionov L M, Kondratyev V V, Kuzmina M Yu, Grigoriev V G, Knizhnik A V and Kuzmina A S 2019 Obtaining silumins with the use of silicon waste *Proc. of higher educational institutions. Non-ferrous metallurgy* **4** 4-15
[19] Karlina A I, Kondratev V V, Kolosov A D, Balanovskiy A E and Ivanov N A 2019 Production of new nanostructures for modification of steels and cast irons *IOP Conf. Ser.: Mater. Sci. Eng.* 012183
[20] Petrovskaya V N and Kondratiev V V 2012 Gas-hydrodynamic nature of the anode effect *Metallurgist* **56(3-4)** 215-221
[21] Konyuhov V Yu, Konstantinova M V and Gladkih A M 2019 Determination of restored units spectrum of equipment and development of the assembly unit repair method at industrial enterprises *J. of Phys.: Conf. Ser.* **1353** 012047
[22] Kondrat'ev V V, Gorovoy V O, Kolosov A D, Kononenko R V and Konyukhov V Yu 2020 Description of the complex of technical means of an automated control system for the technological process of thermal vortex enrichment *J. of Phys.: Conf. Ser.* **1661** 012101
[23] Gorovoy V O, Kondrat'ev V V, Kolosov A D, Kononenko R V and Konyukhov V Yu 2020 Description of software for an automated control system for the technological process of thermal vortex enrichment *J. of Phys.: Conf. Ser.* **1661** 012096

Development and justification of parameters of a microwave installation for pre-planting treatment of vegetable seeds

O V Mikhailova[1], G V Novikova[2], M V Belova[3], A I Kotin[3], A V Kazakov[4] and A A Tikhonov[4]

[1]Chair «Technology info-communications and connection systems», Nizhny Novgorod state engineering and economic university, 22a, Oktyabrskaya Str., Knyaginino, 606340, Russia
[2]Nizhny Novgorod state engineering and economic university, 22a, Oktyabrskaya Str., Knyaginino, 606340, Russia
[3]Chair «Electrification and automation», Nizhny Novgorod state engineering and economic university, 22a, Oktyabrskaya Str., Knyaginino, 606340, Russia
[4]Chair «Metal technology and car fixing», Nizhny Novgorod state agricultural academy, 97, Gagarina Prospect, Nizhny Novgorod, 603107, Russia

E-mail: NovikovaGalinaV@yandex.ru

Abstract. The article is devoted to the analysis of the dielectric parameters of potato and onion tubers for the development of a microwave installation that provides pre-planting treatment by means of a complex effect of electrophysical factors. The sources of energy were magnetrons, Ultratone or Darsonval. The originality of the research includes analysis of the volume of planting material for determination of the required plant performance; analysis of the dielectric parameters of potatoes depending on the frequency of EMF, humidity, and temperature for calculating the power of dielectric losses per unit volume. Theoretical studies of the dielectric parameters of onions presented their structure as a "capillary model". The development of the design of a microwave installation for provides a complex effect of electrophysical factors when calibrating onions. The justification of the corona discharge intensity of electric-gas-discharge lamps is in UHFEMF. There is the study of the dynamics of onion-sowing heating, justification of effective treatment modes. The dynamics of onion heating shows that with the power of the microwave generators of 3.4 kW and Darsonval of 80 W, the drum drive power of 0.25 kW, it is possible to ensure the installation capacity of 150-200 kg/h at an energy cost of 0.025 kWh/kg.

1. Introduction

In medium-sized farms in Russia, pre-planting processing of 125 tons of potato tubers is carried out. 20-30 days before planting, the onion is calibrated. For planting onion-sowing on 1 hectare, a small fraction (10-14 mm) of 960-1120 kg is required; an average fraction (14-17 mm) is 1714-2000 kg; a sample of onions (21-24 mm) is 4616-5385 kg [1, 2]. In order to prevent bolting, the sowing is warmed up 3-4 days before planting. In order to prevent diseases, pests, to accelerate germination and obtain a good harvest, they are treated with growth stimulants, i.e. various agricultural products. They protect the crop from diseases at all stages of growth, but do not exclude side effects on humans.

The main problem is to analyze the dielectric parameters of potato and onion tubers of the developed microwave installation (patent no. 2703062) [3, p. 1–10], which provides pre-planting treatment of seeds by a complex effect of electrophysical factors.

The ultra-high-frequency (microwave) installation with a toroidal resonator for pre-planting processing of potato tubers provides for a complex effect of such electrophysical factors as: ultra-high-frequency electromagnetic field, corona discharge, ultraviolet rays, ozone. The sources are a microwave generator, an Ultraton or Darsonval [4, p. 1], and an induction cooker. Our research shows that the effect of their influence can be increased by placing electric-gas-discharge lamps connected to kilohertz frequency sources in an ultrahigh frequency electromagnetic field (UHF) [5, p. 89].

This unit can be used for calibrated planting material, including large-fraction onion-sowing, otherwise small fractions together with ectoparasites are subjected to induction heating. For effective complex influence of electrophysical factors, it is necessary to exclude induction heating and use pre-calibrated planting material, this will allow you to identify the exact dose of UHFEMF exposure for each fraction, including for onion-sowing. Therefore, we are developing a new installation that provides a complex effect of electrophysical factors on calibrated onions.

2. Materials and methods

When analyzing humidity, temperature, onion density, and the size of each fraction, GOST 30088-93 "Onion-sowing and onion-samples. Seed quality. General technical conditions was used [1]. The sources of the kilohertz frequency were the Ultraton "AMP-2 INT" (frequency 22 kHz) or the Darsonval "AMD-Iskra-4", with a capacity of 80 VA with a scallop electrode. The frequency of pulse-modeled high-frequency oscillations is 110 kHz [4, p. 110-112]. The sources of microwave energy were magnetrons. powerful generators of electromagnetic waves in the centimeter range (frequency 2450 MHz, wavelength 12.24 cm), the principle of operation of which is based on the braking of electrons in crossed magnetic and electric fields. We used magnetrons with forced air cooling of the Galanz model (M24FA-410A, 800 W), JM002 (700 W), 2M210-M1 (900 W), 2M213 (700 W). The concentration of aeroions near the electro-gas discharge lamp was measured by light ion counters (MAS-01).

The algorithm of the research includes:

* analysis of the amount of planting material (potato tubers, onion-sowing) to determine the required productivity of the plant for farms;

* analysis of changes in the dielectric parameters of potatoes depending on the UHF frequency, humidity, and heating temperature to calculate the power of dielectric losses per unit volume;

* theoretical studies of the dielectric parameters of the onion-sowing by presenting its structure as a "capillary model";

* development of a design version of a microwave installation with volumetric resonators that provide a complex effect of electrophysical factors in the process of onion-sowing calibration;

* theoretical justification of the corona discharge intensity when scallop electro-gas discharge lamps are located in the UHF;

* experimental studies of the heating dynamics of calibrated onion seeds;

* preliminary justification of effective modes of pre-planting processing of onion-sowing.

3. Results

In order to disinfect and activate the cells of the planting material in order to increase their sowing and productive indicators, we propose to perform pre-planting treatment of onion sowing and potato tubers in an installation with ultra-high frequency electromagnetic field sources (UHF) and electro-gas discharge lamps powered by Darsonval or Ultraton. The effect of ultra-ionizing or darsonvalization is manifested by the action of ionized and ozonized air on the raw material.

The paper analyzes the dielectric characteristics of early-grade potatoes "Priekulsky" depending on the frequency of EMF, temperature and humidity at a frequency of 2400 MHz (Fig. 1) according to I. A. Rogov [6, p. 92-93]. The analysis shows that with an increase in the UHF frequency from 433 MHz to 3000 MHz, the dielectric loss factor decreases from 33 to 15.5 at a temperature of 20°C and a humidity of 81.3 %. The dielectric parameters of the EMF frequency are described by empirical expressions:

$$\varepsilon = 66.788e^{-2E\text{-}05 \cdot f},\ k = 34.17e^{-3E\text{-}04 \cdot f},\ tg\delta = 0.5114e^{-3E\text{-}04 \cdot f}. \quad (1)$$

When the temperature changes from 20 °C to 95 °C, the dielectric constant (ε) decreases from 63.1 to 47.1, the coefficient of dielectric loss (k) increases from 16.9 to 18, that is, at 19.5 %, the tangent of the dielectric loss angle ($tg\delta$) increases from 0.268 to 0.38. These characteristics are described by empirical expressions:

$$\varepsilon = 69.78e^{-0{,}004x},\ k = 17.66e^{0.0008x},\ tg\delta = 0.254e^{0.0046x}. \quad (2)$$

The increase in the dielectric loss factor in the specified temperature range is explained by the fact that at a temperature above 30 °C, denaturation of proteins begins, accompanied by the release of moisture. Over 60°C, the dielectric loss factor decreases due to moisture evaporation and thermal movements of polar molecules that prevent their dipole orientation in EMF [6, p. 92-93].

Changes in the dielectric parameters of potatoes from humidity in the range of 70-80 % at a temperature of 20 °C are described by empirical expressions:

$$\varepsilon = 9.415e^{0.023 \cdot w}, k = 0.408e^{0.046 \cdot w},\ tg\delta = 0.0432e^{0.023 \cdot w}. \quad (3)$$

In the specified humidity range, the dielectric loss factor increases from 10.8 to 20.4. One of the main factors affecting the dielectric parameters of potatoes is humidity, which must be taken into account when justifying heat treatment modes.

a)

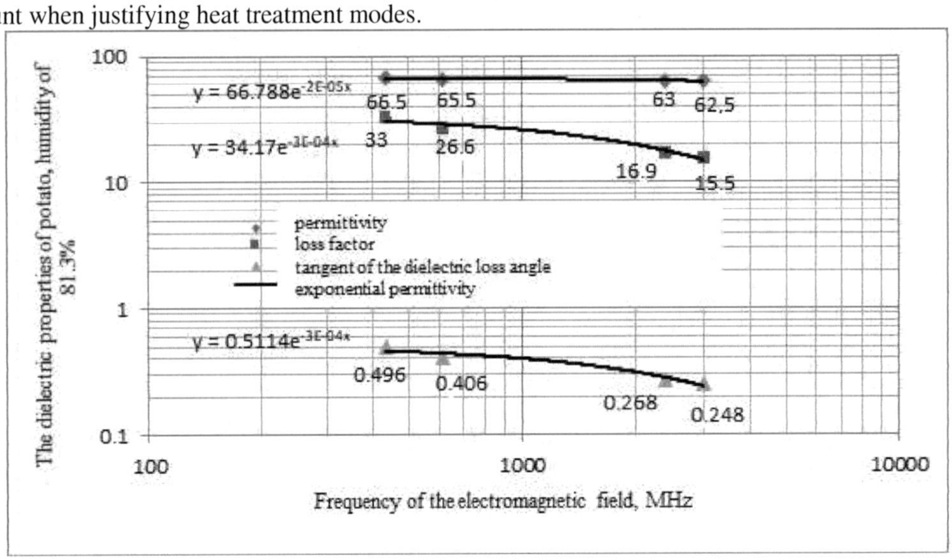

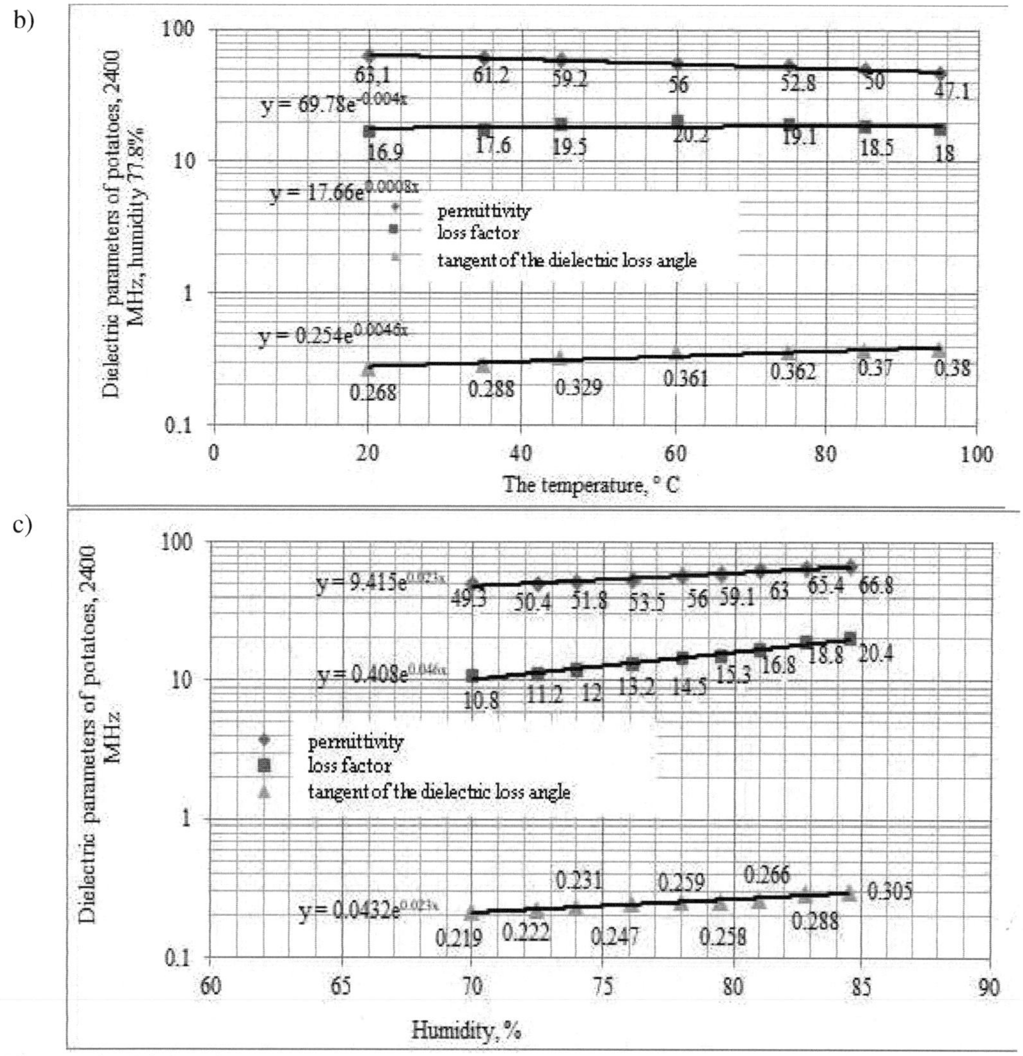

Figure 1. Dielectric parameters of potato tubers depending on EMF frequency (a), heating temperature (b), humidity (b)

The permittivity and the coefficient of dielectric loss depending on humidity are described by expressions [6, p. 64]:

$$\varepsilon = 81 \cdot \left(\frac{\rho_s}{\rho_w}\right) \cdot (1-m) \cdot \frac{W}{(1-W)}, \quad k = 2 \cdot \left(\frac{\rho_s}{\rho_w}\right) \cdot (1-m) \cdot \frac{\sigma}{\omega} \cdot \left[\frac{W}{(1-W)}\right], \tag{4}$$

where ρ_s – density of dry substance, kg/m^3; ρ_w – density of water, kg/m^3; m – porosity of the onion; W – humidity of the onion, %; σ – specific conductivity of the liquid in capillaries.

With $\rho_w = 10^3$ kg/m^3, $\rho_s = 1.5 \cdot 10^3$ kg/m^3, $m = 0.85$ and with the frequency of 2450 MHz these expressions look like this:

$$\varepsilon = 81 \cdot \left(\frac{1.5 \cdot 10^3}{10^3}\right) \cdot (1-0.75) \cdot \frac{W}{(1-W)} = 30.375 \cdot \frac{W}{(1-W)},$$

$$k = 2 \cdot \left(\frac{1.5 \cdot 10^3}{10^3}\right) \cdot (1-0.75) \cdot \frac{\sigma}{2 \cdot \pi \cdot f} \cdot \left[\frac{W}{(1-W)}\right] = 0.75 \cdot \frac{\sigma}{6.28 \cdot 2450 \cdot 10^6} \cdot \left[\frac{W}{(1-W)}\right]. \tag{5}$$

If we take into account the specific conductivity of the liquid in the capillaries, equal to $31.91\cdot10^{10}$ sm/m, the following is obtained:

$$k = 0.485\cdot10^{-10}\cdot31.91\cdot10^{10}\cdot\left[\frac{W}{(1-W)}\right] = 15.48\cdot\left[\frac{W}{(1-W)}\right],\ k = 15.48\cdot\left[\frac{W}{(1-W)}\right]. \quad (6)$$

The expression describing the change in the onion-sowing dielectric loss factor as a function of frequency is:

$$k = 0.12\cdot\frac{\sigma}{f}\cdot\left[\frac{W}{(1-W)}\right]. \quad (7)$$

If we take into account, the equation describing the change in the dielectric constant of water from the frequency of EMF [6, p. 30; 7, p. 318] $\varepsilon = 2\cdot10^{-9}\cdot f + 79.53$ (fig. 2a), then you can get an approximate expression of the change in the dielectric constant of the onion from the frequency and humidity:

$$\varepsilon = \varepsilon(f)\cdot\left(\frac{1.5\cdot10^3}{10^3}\right)\cdot(1-0.75)\cdot\frac{W}{(1-W)},\quad \varepsilon = \left(2\cdot10^{-9}\cdot f + 79.53\right)\cdot0.375\cdot\frac{W}{(1-W)}. \quad (8)$$

a)

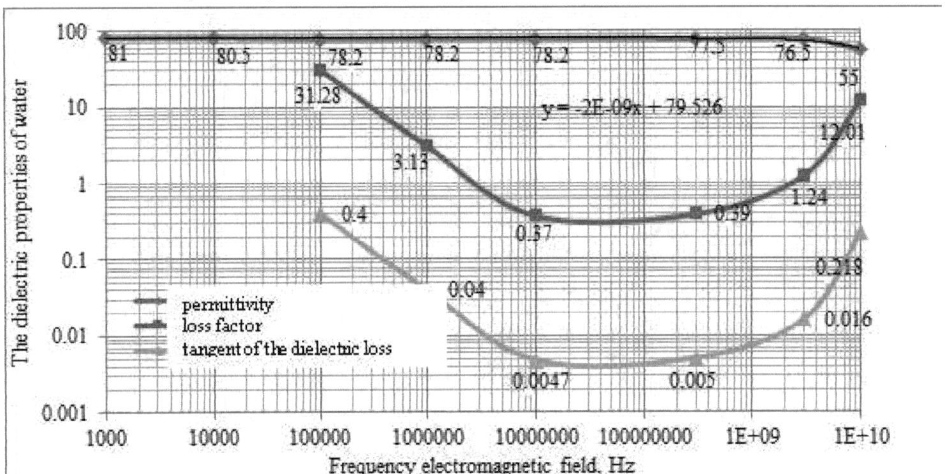

b)

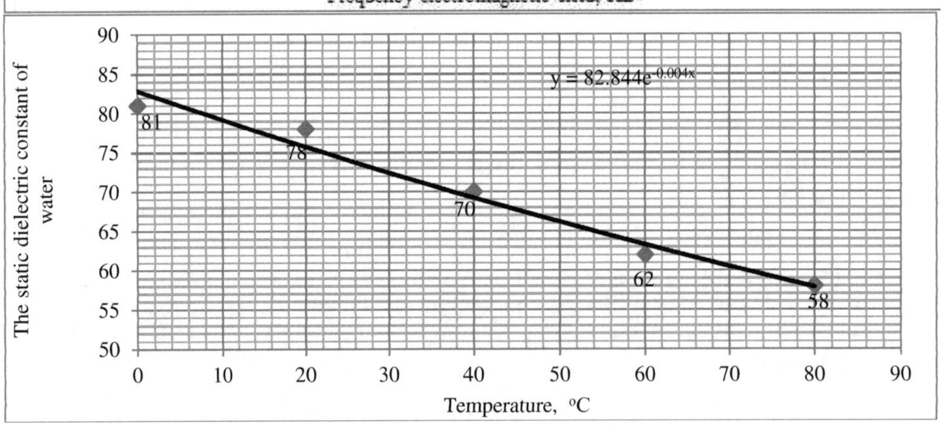

c)

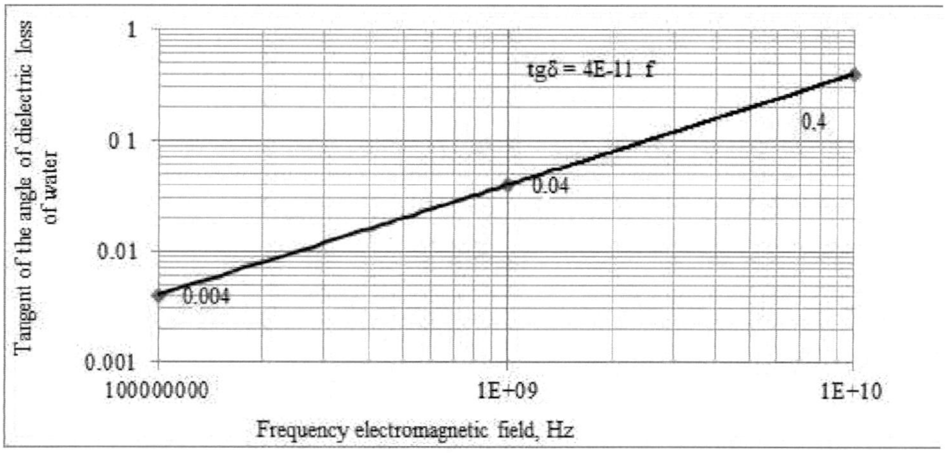

Figure 2. Dielectric parameters of water depending on the EMF frequency (a, b) and heating temperature (b)

Since the permittivity of water depends on temperature, it means that for the studied frequency of 2450 MHz, the equation can be written as follows:

$$\varepsilon = \varepsilon(T)\cdot 0.375\cdot\frac{W}{(1-W)} = \left(82.844\cdot e^{-0.004\cdot T}\right)\cdot 0.375\cdot\frac{W}{(1-W)}.\ \varepsilon = 31.066\cdot e^{-0.004\cdot T}\cdot\frac{W}{(1-W)}. \quad (9)$$

So, the behavior in the ultrahigh frequency electromagnetic field of the onion sowing becomes more clear as a result of the analysis of the dielectric parameters, their frequency, temperature and humidity dependence.

The technical task is to develop an installation for pre-setting processing of onion sowing by the complex influence of an ultrahigh frequency electromagnetic field and ionized air during calibration in a continuous mode. Taking into account the features of already developed microwave installations with different working chambers for heat treatment of raw materials [8, p. 57; 9, p. 64; 10, p. 55] another design is proposed.

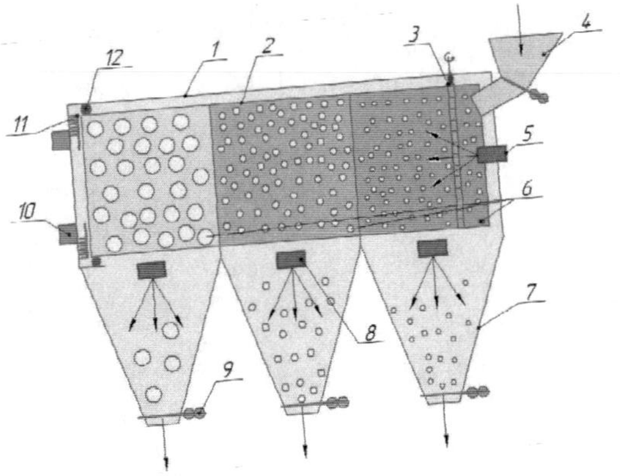

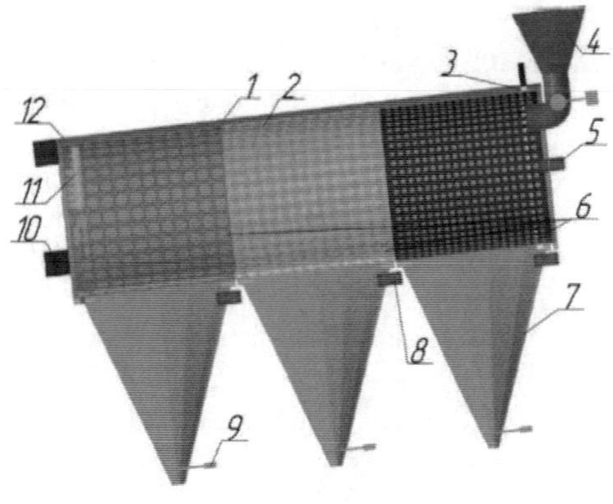

Figure 3. Installation for calibration of onions: 1 – cylindrical shielding body; 2 – section perforated cylindrical resonator; 3 – crown from the drive sprocket on the motor shaft; 4 – collecting bottle with a valve; 5 – microwave generator; 6 – sections with different diameters of perforation holes of the cylindrical cavity; 7 – inverted truncated pyramidal cavities with dampers 9; 8 – magnetron emitters which are directed inside the pyramidal cavities; 10 – generators kilohertz frequency; 11 – scallop electric gas discharge lamps; 12 – support rollers

Installation for calibration and pre-planting processing of onion sowing under the influence of electrophysical factors (Fig. 3) contains a coaxially mounted sectional cylindrical perforated resonator 2 made of non-ferromagnetic material without bases in a cylindrical shielding housing 1 located at an angle to the horizontal plane.

The technological process of calibration and pre-planting processing of onions under the influence of electrophysical factors is as follows. Close the flap of the receiving container 4 and fill in the UN-calibrated onion. Turn on the electric motor to rotate the sectional cylindrical perforated non-ferromagnetic resonator 2 by engaging the leading sprocket with the crown 3. Turn on the generators of the kilohertz frequency 10, after which the electro-gas discharge lamps 11 begin to corona between the base of the shielding housing and the scallops. The air was ionized and ozonized. Open the flap of the receiving vessel 4 and turn on the microwave generators 5 to excite the ultra-high frequency electromagnetic field.

It is known that when placing electro-gas discharge lamps powered by a kilohertz frequency generator in an ultra-high frequency electromagnetic field, the corona discharge increases, and consequently, the concentration of ionized and ozonized air and the radiation of a bactericidal stream of ultraviolet rays increases. The onion of different fractions, falling into a sectional cylindrical perforated resonator when rotating at a speed less than the critical one, is calibrated. The critical speed of rotation is when the onion, pressed against the resonator shell by centrifugal forces, does not break away from it and begins to make a full revolution. The smallest onion-sowing passes through the perforation holes of the first section 6 of the resonator into the corresponding pyramidal resonator 7. Further, the raw material is moved to the second section due to the tilt and rotation of the cylindrical resonator, where the diameter of the perforation holes is larger than in the first section. Therefore, the onion is calibrated into fractions of the corresponding diameter in the next pyramidal resonator 7.

The calibration process continues until the last section, where the largest bulbs pass through holes of the corresponding diameter of the resonator 2 perforation. During the calibration process, the onion is endogenously warmed up, disinfected, and stored in ionized air of an effective concentration controlled by the gap between the base of the scallop electric-gas discharge lamp and the power of the kilohertz frequency generators. Ionized air is also distributed in pyramidal resonators, where it also acts effectively on a biological object. Radiation through the perforation holes of the cylindrical resonator sections towards the pyramidal resonators and vice versa occurs, while the power of the radiation flow is higher through the perforation holes of the last section of the resonator. Radiation outside the installation is limited by a cylindrical shielding housing made of non-ferromagnetic material (aluminum, copper, etc.) and the use of dampers on the receiving vessel and on small bases of pyramidal resonators. Calibrated, pre-heated, ionized and ozonated onion in pyramidal resonators is subjected to additional exposure to the electromagnetic field of ultrahigh frequency, different doses in each resonator 7, so the size of the onion in them are different. Thus, the implementation of a complex effect of electrophysical factors on raw materials in the calibration process is possible in an ultra-high-frequency installation with a sectional cylindrical perforated rotating resonator that provides calibration of the onion into fractions, disinfection due to ozonated air and treatment with ionized air of high concentration.

Experimental studies of the dynamics of heating of onion tubes of different calibrations were carried out (Fig. 4). The control of the heating temperature of the onion with the Testo 925 pyrometer after exposure to the EMP SHOWS that at the same specific power, the calibrated onion of different diameters is heated in the same way. If you do not calibrate, the onion tubes of different diameters is heated at different speeds, for example, for 10 seconds, the onion tubes with a diameter of 1.3-1.5 cm is heated 5-6°C higher.

Based on the known data of the positive effect of negative aeroions on a biological object, studies of the complex effect of EMF and ionized air by placing an electro-gas discharge lamp in it are conducted. The research was based on the scientific works of leading scientists [11, p. 49-50; 12, p. 154; 13, p. 93; 14, p. 321-322; 15, p. 103-104]. According to the results of experiments, the operating modes of the microwave installation with a working chamber that provides heat treatment of onions in ionized air were corrected.

The corona discharge intensity is justified when placing a scallop electro-gas discharge lamp, when pulse-modeled high-frequency oscillations (110 kHz) are amplified due to an ultra-high frequency electromagnetic field of certain intensity.

An inhomogeneous electric field is created between the scallop electro-gas discharge lamps and the base of the cylindrical resonator, the intensity of which is maximum on the scallops. If you reduce the interelectrode gap, the electric field intensity reaches a value at which air ionization begins. Corona discharge is accompanied by the formation of ozone and electromagnetic radiation in the range from radio waves to UV rays. Let us consider the calculation of the main parameters of the corona discharge for a scallop electro-gas discharge lamp powered by a kilohertz frequency generator.

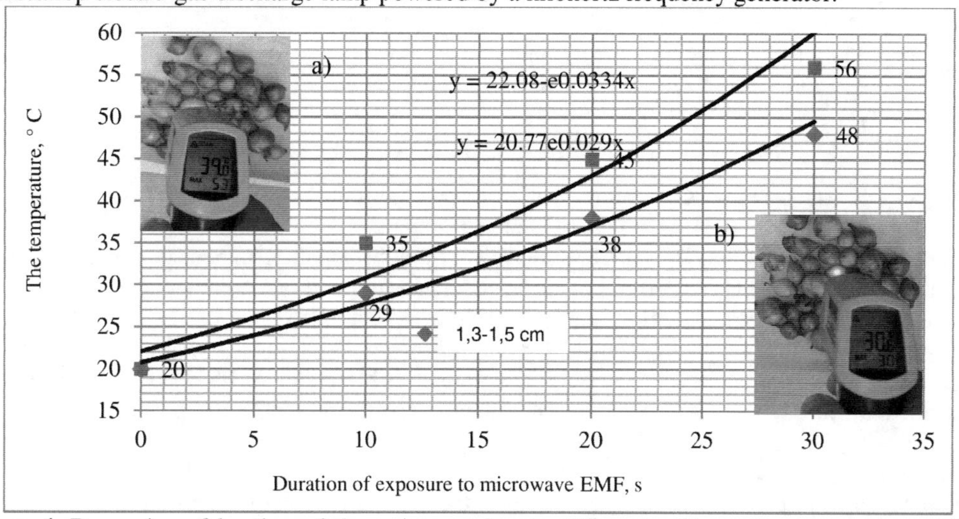

Figure 4. Dynamics of heating of the onion-sowing in different-caliber microwave oven with an equal specific power of 8.5 W/g: a) temperature on the surface of a large fraction; b) temperature on the surface of a small fraction

We determine the initial tension by the empirical Peak formula (E_o) corona discharge through atmospheric pressure (p, PA), air temperature (T, K), and air density (δ) [16, p. 204]:

$$E_0 = 30.3 \cdot 10^4 \cdot \delta \cdot \left(1 + \frac{0.0298}{\sqrt{\delta \cdot r_0}}\right), \quad \delta = 289 \cdot 10^{-5} \cdot \frac{p}{T}, \tag{10}$$

If $p = 1.013 \cdot 10^5$ Pa, $T = 293$ K, so $\delta = 1$.

Then the initial intensity of the corona discharge at the radius of the corona electrode $r_o = 5$ mm is $E_0 = 30.3 \cdot 10^4 \cdot \left(1 + \frac{0.0298}{\sqrt{5 \cdot 10^{-3}}}\right) = 4.3\ kV/cm$, which is much lower than 30 kV/cm, i.e. the breakdown strength of the electric field at an atmospheric air pressure of 100 kPa [17. p. 62].

On the other hand, the corona discharge intensity depends on the geometric parameters of the corona electrodes (scallop electro-gas discharge lamps as high-potential electrodes) and the system of their location relative to the low-potential electrode (the base of the cylindrical resonator). Using the well-known expression [16, p. 206; 20, p. 37], describing the relationship of the initial electric field strength, geometric parameters of this system of electrodes, it is possible to determine the initial corona discharge voltage. For example, if the distance between high-potential electrodes and low-potential electrodes is 20 mm:

$$U_0 = E_0 \cdot r_0 \cdot \left[\frac{2 \cdot \pi \cdot h}{d} - \ln\left(\frac{2 \cdot \pi \cdot r_0}{d}\right)\right]. \tag{11}$$

$$U_0 = 43 \cdot 10^4 \cdot 10 \cdot 10^{-3} \left[\frac{2 \cdot \pi \cdot 20 \cdot 10^{-3}}{10 \cdot 10^{-3}} - \ln\left(\frac{2 \cdot \pi \cdot 10 \cdot 10^{-3}}{20 \cdot 10^{-3}}\right)\right] = 4.91\ kV. \tag{12}$$

The mobility of negative ions at an electric field strength of 430 kV/m and a distance between the corona electrodes of 20 mm is $2 \cdot 10^{-4}$ (m^2/V·s) [16, p. 207]. The normalized parameters of ionized air are the concentration of light ions, which for a biological object is in the range of 10-25 thousand ions/cm^3.

Further, the known formula was used to calculate the thermal energy released per unit volume of raw materials because of heating in an ultra-high frequency electromagnetic field [14, p. 321-322; 19, p. 228].

Power:

$$\mathrm{P} = 0.555 \cdot 10^{-10} \cdot k \cdot f \cdot \mathrm{E}^2 = 0.555 \cdot 10^{-10} \cdot 17 \cdot 2450 \cdot 10^6 \cdot 2 \cdot 10^4 = 46231.5\ W/m^3 \quad (13)$$

where P – power, W/m^3; k – dielectric loss factor equal to the product of the relative permittivity of the onion by the tangent of the loss angle; E – electric field strength in the volume of raw materials, W/m.

Let us determine the speed of heating the onion at this specific power generated per unit volume [20, p. 536]:

$$\frac{\Delta T}{\Delta \tau} = \frac{P_{y\partial} \cdot \eta}{\rho \cdot C} = \frac{46232 \cdot 0.7}{500 \cdot 1382} = 0.047 \frac{{}^{o}C}{s}, \quad (14)$$

where ρ – onion density, kg/m^3, C – onion heat volume, kilo joule /kg·°C, η – thermal efficiency of heating, taking into account the loss of heat to the environment.

According to GOST 10832-2009, the bulk density of onions was determined. It, depending on the fraction, is 490-520 kg/m^3. The specific heat capacity of the onion is 1382 kJ/kg·s.

Effective heating rate of the pearl onion in APPSVC of 0.05 °C/s is possible at specific power of the microwave generator of 0.42 In/yr. If the total useful power of the microwave generators is 3400 watts at a time up to 8 kg onion sets will be in the resonators effects of APPSVC power density 0.42 W/g the exposure duration will be 3-3.3 minutes, the unit performance 145-150 kg/h.

4. Conclusion

The efficiency and productivity of the installation for the complex effect of electrophysical parameters on potato and onion tubers are provided by a high level of microwave energy transfer to the raw material and the choice of the design of interconnected resonators, such as a sectional perforated cylindrical resonator combined with pyramidal resonators. Due to the fact that the specific heat capacity, conductivity and dielectric properties of materials depend on moisture, temperature, to ensure stable operation of the microwave generator and the electric discharge lamp is recommended to maintain the ratio of raw material to the volume of the resonator, i.e., fill the cavity with raw materials 10-20 %.

The technological process of pre-planting treatment of onions can be successfully carried out in a multigenerator installation with the appropriate choice of effective design, technological and operating parameters. Studies of the dielectric constant and the onion loss factor from frequency, temperature and humidity confirmed the possibility of transmitting microwave energy to raw materials with an efficiency of at least 0.7.

Experimental studies of the dynamics of heat onion sets show that when power microwave generators of 3.4 kW and Darsonval of 80 watts of drive power of the drum is 0.25 kW to ensure the plant capacity of 150-200 kg/h at energy costs of 0.025 kWh/kg.

5. Acknowledgments

We express our gratitude to the colleagues who participated in the publication of the article and note the fact that the work was carried out with the financial support of the grant-giver.

References

[1] GOST 30088-93 *onion sowing and onion-samples. Seed quality. General specifications*

[2] Semenov D A and Gugin M S 2019 *Theory of electromagnetic compatibility of radio-electronic devices and systems* (Knyaginino: NGIEU) p 278

[3] Kotin A I, Novikova G V, Shamin E A, Mikhailova O V and Belova M V, RF Patent No. 2703062 IPC AO1C1/08 *Installations for pre-planting processing of potato tubers under the influence of electrophysical factors* (15 October 2019)
[4] Osokin V L and Kirillov N K 2013 *Light engineering* (Knyaginino: NGIEU) p 380
[5] Kotin A I and Zaitsev P V 2019 Research and development of an installation for pre-planting processing of potato tubers under the influence of electrophysical factors *Bull. of Kazan state agrarian University* **1(52)** 89-93
[6] Rogov I A 1981 *Electrophysical, optical and acoustic characteristics of food products* (Moscow: Light and food industry) p 288
[7] Lykov A V 1968 *Theory of drying* (Moscow: Energia) p 470
[8] Belova M V and Yershova I G 2016 Volumetric resonators in ultrahigh frequency technology *Innovation in agriculture* **2(17)** 115-119
[9] Belova M V, Yershova I G and Mikhailova O V 2016 In innovations in technologies of agricultural raw materials processing *ARPN J. of Engineering and Applied Sciences* **11(6)** 1269
[10] Zhdankin G V, Zaitsev P V and Storchevoy V F 2017 Development and justification of parameters of a microwave installation for heat treatment of raw materials in the process of grinding *Scientific life* **11** 25-36
[11] Didenko A N 2003 *Microwave energy: Theory and practice* (Moscow: Nauka) p 446
[12] Ilchenko M E, Vzyatyshev V F et al 1989 *Dielectric resonators* (Moscow: Radio communication) p 320
[13] Ivashov V I 2001 *Technological equipment of enterprises of the meat industry* (Moscow: Kolos) p 551
[14] Azarov B M 1988 *Technological equipment of food production* (Moscow: Agropromizdat) p 467
[15] Levitskiy S M 1974 *Collection of problems and calculations on physical electronics* (Kiev: University) p 210
[16] Basov A I, Bykov V G, Laptev A V and Fain V B 1970 *Electrotechnology* (Moscow: Kolos) p 255
[17] Pchelnikov Yu N 1981 *Electronics of ultrahigh frequencies* (Moscow: Radio communication) p 96
[18] Ginzburg A S 1985 *Calculation and design of drying plants for the food industry* (Moscow: Agropromizdat) p 336
[19] Strekalov A V and Strekalov Yu A 2014 *Electromagnetic fields and waves* (Moscow: RIORINFRA-M) p 375
[20] Listov P N 1974 *Application of electric energy in agricultural production* (Moscow: Kolos) p 623

Automation of the splitting-up processes of freight trains on the gravity sorting yards in the railway transport system

A A Olentsevich[1], V Yu Konyukhov[2], E A Guseva[2], M V Konstantinova[2] and V A Olentsevich[1]

[1] Irkutsk State Transport University, 15, Chernyshevskogo St., Irkutsk, 664074, Russia
[2] Irkutsk National Research Technical University, 83, Lermontova Street, Irkutsk, 664074, Russia

E-mail: olentsevich_aa@irgups.ru

Abstract. The basic reasons for insufficient reduction of a number of safety violations in a railway transport network are systemic and are related to the current condition of the technical and technological base of transport complexes. Aging, poor conditions of fixed assets and errors in the maintenance of the transport system are in most cases the main causes of accidents and casualties. The basis for the automation of most technological railway processes is the redistribution of material, energy and information flows in accordance with the accepted criterion of optimality. According to the analysis of the statistics, this article reveals a problem that has a direct impact on the safety of train service and shunting operations in the railway transport system - that is, the number of railcars that has been damaged on public and non-public approach lines. The dependence of the number of damaged freight cars on the speed of their collision the splitting-up of the railcars on the gravity-sorting yard of the network railway stations has been determined. Automation of the technological sorting process in a single production cycle will provide the basis for the introduction of production management systems and railway management systems as a whole, while improving the safety of the transportation process itself and the production of shunting operations. In the conditions of high-speed traffic development and the introduction of innovative technologies, these requirements need to be tightened, including the widespread use of automated processes for the splitting-up of freight trains on the gravity sorting yards. Improving the operation of the freight classification yards is a major factor in accelerating and cheapening the transportation process. In recent years much attention has been paid to the automation of station operation processes, the use of advanced labor methods, the introduction of new equipment, without which the development of the station cannot ensure the efficiency of the industry as a whole. In order to eliminate these shortcomings, it is proposed to automate the process of splitting-up freight trains on the gravity-sorting yard. For this purpose, it is necessary to equip the gravity sorting yards of railway stations with a comprehensive system of the sorting process control automation.

1. Introduction

The efficient operation of the railway transport system (RTS) is based on the available rolling stock resources, the estimated capacity of transport infrastructure, the availability of track resources of railway lines and sections. In order to ensure the smooth operation of the railway transport system, it is necessary to determine its reserves. They are often expressed in some redundancy - additional means (locomotives, railcars, railway track, rails, fuel, spare parts, etc.) or additional features (throughput capacity, carrying capacity, estimated capacity of the railway transport system facilities) above the minimum necessary for optimal working condition and solving the goals and objectives in due course [1-3].

The goals are achieved by implementing the following basic tasks of the transportation process automation of the process of transportation:

- improving the quality of transportation regulation;
- increasing the ratio of readiness of fixed assets to the change of traffic volume;
- improving the ergonomics of the work of transportation process operators;
- ensuring the reliability of information on material components used in enterprises and in the structural units of the railway transport system, as well as related organizations;
- storing information about the process, emergency situations, and risk cases in production activities.

Automation of technological processes in the railway transport system is useful to consider as a set of methods and means designed to implement a system that makes it possible to the manage the process itself without the direct involvement of the human factor, or giving a person the right to make the most responsible decisions. The basis for the automation of most technological railway processes is the redistribution of material, energy and information flows in accordance with the accepted criterion of optimality [4-7].

The solution to the problems of technological process automation in OAO "Russian Railways" (OAO "RZD") is mainly carried out with the help of the introduction of modern automation methods and the introduction of modern automation tools. Automation of the technological sorting processes in a single production cycle will provide the basis for the introduction of production management systems and railway management systems as a whole, while improving the level of safety of the transportation process itself and the production of shunting operations.

2. Analysis of the low safety factors of the rail transport process

Aging, poor conditions of fixed assets and errors in the maintenance of the transport system are in most cases the main causes of accidents and casualties. Why is that so? In the recent past, the railway transport system had been functioning clearly, smoothly, there was a regular addition of new equipment, equipment and highly skilled personnel, but over the years of market transformations, there was a set of unresolved issues. These include, first of all, the aging of vehicles, poor logistics and maintenance, the cost of training and retraining, the lack of a clear regulatory framework for human security, incompetence, negligence and irresponsibility of officials and transport system workers. The most difficult problem remains the notorious human factor - the failure to comply with prescribed instructions, regulatory documentation, and technologies. In one way or another, all these shortcomings of operation are connected with the presence of many different operators, owners of rolling stock, whose activity basis is to make a profit in the shortest possible time and with minimal costs.

Recently, OAO Russian Railways has noted a sufficiently low level of safety in the production of train service and shunting operations at the approach lines adjacent to the railway stations of public and and non-public use. The main reason for this fact is the reluctance of the owners of these lines to adhere to the requirements of the Charter of Rail Transport and other regulations and documents, regulating the activities of the railway transport system [8]. This problem is quite serious and has a direct impact on the safety of train service and shunting operations in the railway transport system, as the statistics confirm. The number of freight railcars that were damaged on the approach lines reached 336 railcars in 2019. Repeated cases of unauthorized train traffic confirm the seriousness of the situation.

According to statistics, 4.105 violations of the requirements of the legislation of the Russian Federation have been identified on the largest railways for the two quarters of this year, of which 3,713 cases have been eliminated and 1,335 are related to the threat to traffic safety in the railway transport system and require the application of prohibitive measures [9]. In the total volume of freight railcars with commercial faults, requiring an uncoupling to eliminate this type of defects, on average, 25% accounts for railcars loaded in violation of the requirements of technical conditions for the placement and fastening of cargo in railcars and containers (TU). 48.5% is the disorder of the established scheme of loading cargo on the railcar on the cargo travel route from the loading point to the destination point. 28.7% are other types of commercial faults. Specifically, the violation of loading and fastening conditions in the railcar are accounted for by more than 75% of all types of commercial faults [10]. Other reasons include:

- violation and breakage of the existing locking devices;
- leakage flow from tanks;
- spontaneous opening of gondola car hatches;
- broken loops of the railcar hatches;
- malfunctions of the floor of the railcar, etc.

Statistics show that a large percentage of the causes of violations are accounted for by the excess of the collision speed of the railcars during the splitting-up from the gravity sorting yards of railway stations. This is what, largely, leads to a violation of the Technical conditions for the placement and fastening of goods during transportation.

During the period under review, there has been an increase in the volume of damaged freight railcars at railway stations by an average of 8.5%. At the same time, up to 300 railcars out of the total volume on the station gravity sorting yards are damaged, which is on average 60% of the total number of sorted railcars. This volume includes:

- violation of the regime of splitting-up and braking of railcar cuts;
- non-observance of the requirements of the instructions for the maintenance of inclined railroad tracks and classification railroad tracks;
- malfunction of gravity sorting yard devices;
- human factor.

Taking into account the existing methodology [11-12], the number of freight railcars damaged on gravity sorting yards, on average per 1000 recycled ones, due to malfunctions of gravity sorting devices, is determined by the formula (1):

$$n_{\text{of the overtaking}} = 0.00143 V_{\text{coll}} \tag{1}$$

where V_{coll} is the average speed of collision of freight railcars during the splitting-up, m / s;

$n_{\text{of the overtaking}} = 0.00143 \cdot 2.93 = 0.00417$ railcars.

The volume of damaged freight railcars when splitting-up at the gravity sorting yard of a railway station with daily rehandling of 2485 railcars per year can be calculated using the formula (2):

$$N_{\text{damage}} = \frac{N \cdot 365}{1000} \cdot n_{\text{damage}} \tag{2}$$

where N is the average daily number of freight railcars rehandled on the downhill section of the gravity sorting yard.

$N_{\text{of the overtaking}} = \dfrac{2485 \cdot 365}{1000} \cdot 0{,}00417 = 3{,}782$ railcars.

If the permissible speed of unification of cuts is exceeded, various degrees of damage to freight cars occur, requiring one or another type of repair. According to statistical data, the average cost of repairing one such railcar is about 10-15% of the market value of a new freight railcar [5]. Taking into account the average cost of a freight railcar, the approximate average statistical cost of repairing a traction vehicle unit will be 967 thousand rubles, while the total cost of repairing the total amount of damaged railcars that were damaged when being split-up from a sorting gravity yard will amount to 3.15 million rubles per year.

In order to determine the dependence of the number of damaged freight railcars (N_{damage}) on the speed of their collision (V_{coll}) during the splitting-up, we calculate the data according to Table 1 with a collision speed interval of 0.5 m/s and taking into account the data calculated according to formulas 1 and 2.

Table 1. Number of damaged railcars during the splitting-up on the gravity-sorting yard.

Railcar collision speed, m/s	The number of damaged railcars per 1000 rehandled ones during the splitting-up of the cuts, railc.	The number of damaged railcars per year during the daily rehandling of 2.485 railcars, railc.
0.5	0.00072	0.649
1	0.00143	1.297
1.5	0.00215	1.946
2	0.00286	2.594
2.5	0.00358	3.243
2.93	0.00418	3.787
3	0.00429	3.891
3.5	0.00501	4.540
4	0.00572	5.188
4.5	0.00644	5.837
5	0.00715	6.485
5.5	0.00787	7.134
6	0.00858	7.782
6.5	0.00930	8.431
7	0.01001	9.079
7.5	0.01073	9.728
8	0.01144	10.376
8.5	0.01216	11.025
9	0.01287	11.673
9.09	0.01311	11.790
9.5	0.01359	12.322
10	0.01430	12.970

Based on the calculation data of table 1, we will build a histogram of the number of damaged railcars per year depending on the collision rate of the cuts (Figure 1).

The submitted dependency determines that the values of the collision rate of the cuts during the splitting-up on the railway station gravity-sorting yard and the number of damaged freight cars are linearly dependent. That is, the greater the collision rate of the cuts, the more railcars will be damaged in the period under review, and accordingly there will be more costs for the restoration and repair of rolling stock.

An analysis of the statistics [5] of failures shows that most of them are related to technical equipment. Technical equipment failures, as the main causes of operational security breaches, dominate in the main technical complexes of railway transport systems. Figure 2 presents the diagram of the distribution of failures related to the complexes that are part of the railway transport system in 2019.

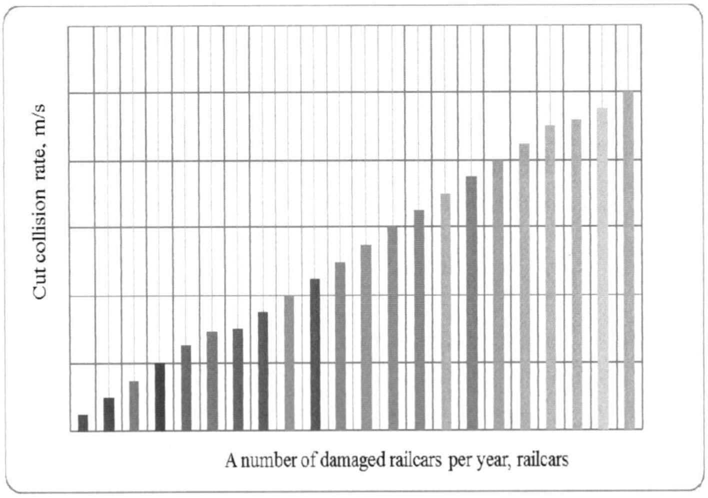

Figure 1. The histogram of the number of damaged freight railcars per year depending on the collision rate of the cuts during the splitting-up on the railway station gravity-sorting yard.

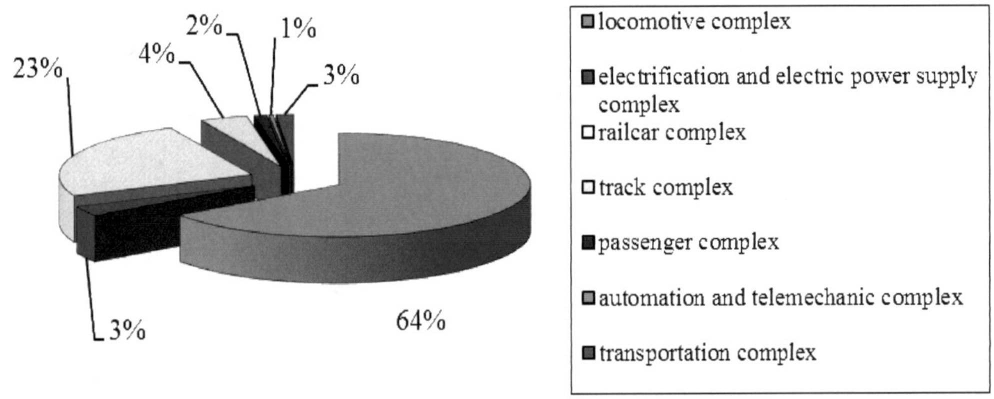

Figure 2. Distribution of traffic safety violations related to complexes in 2019.

Analysis of the results of the studies allows one to identify bottlenecks and to outline reserves in the use of sorting devices at railway stations, as well as to develop timely organizational and technical measures to phase out the limiting elements of their throughput and estimated capacities of the railway-transport system facilities.

To date, according to the data [8], the main reason for the lack of throughput and estimated capacity of the most freight classification yards of the OAO Russian Railways is the inconsistency of the level of development of the federal railway network, the legislative framework for the transportation process and the railcar fleets moving along the network. The situation is aggravated by the lack of an effective system of centralized transportation planning, which leads to errors in the management of railcar fleets.

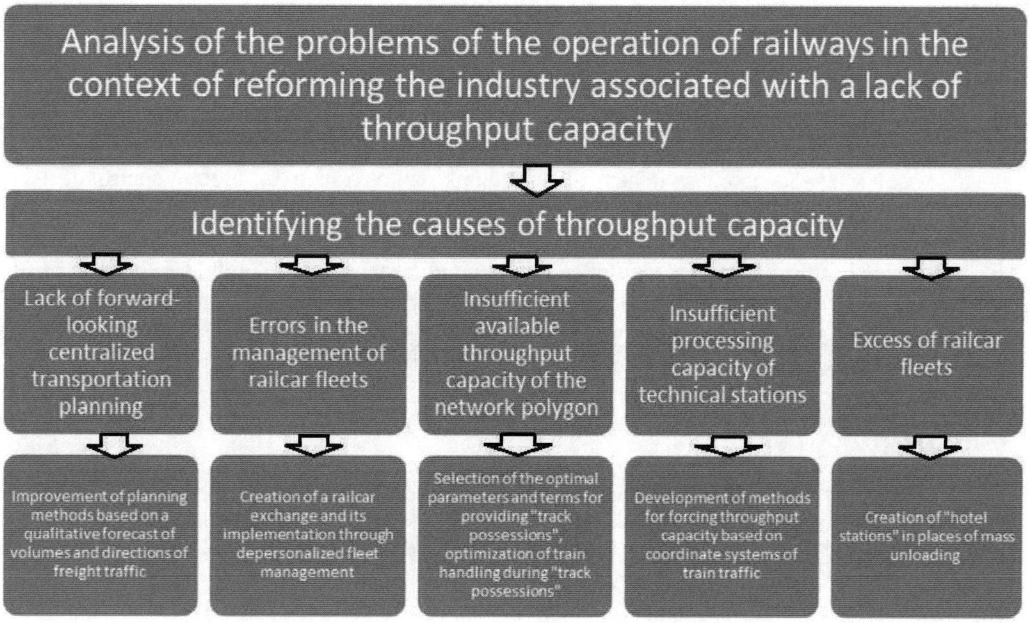

Figure 3. The structure of the study of the reasons for the lack of throughput capacity.

3. Automation of the splitting-up processes of freight trains on the railway transport system gravity sorting yards

An analysis of the statistics [5, 6] in recent years has shown that the reasons for the insufficient reduction in the number of security violations are systemic and related to the current state of the technical and technological base of the complexes. In these circumstances, the formation of requirements for the railway equipment and processes developed, including reliability, safety and overall quality indicators, plays a major role in reducing safety violations. In the conditions of high-speed traffic development and the introduction of innovative technologies, these requirements need to be tightened, as well as the widespread use of automated processes of control of their execution [13-15].

In order to eliminate these shortcomings, it is proposed to automate the process of splitting-up freight trains on the gravity sorting yards of the whole railway transport system [14-18]. For this purpose, it is necessary to equip the railway station gravity sorting yards with a comprehensive system of the sorting process control automation. For the efficient and uninterrupted operation of the railway stations of the railway transport system, the implementation of the specified amount of work of the sorting system should include the following subsystems of the gravity sorting yard complex:

- automatic switching with the introduction of the accumulation of railcars in the marshaling yard (AS-MA), which will allow one to ensure the automatic formation of the rolling-down routes of cuts in automatic and manual modes, automatic tracking of movement on the gravity sorting yard;
- subsystem of automated control of the speed of the rolling-down and control of target braking (ACS-CTB), which will provide automated control of the speed of the rolling-down of the cuts through interval-target braking on the gravity and yard braking positions.

To achieve the optimal level of automation of the studied processes of the sorting slide, as well as to improve the safety level, it will be necessary to install radar indicators of the speed of rolling-down of the cuts "RISVZ-M" on the braking positions of the descent part of the gravity sorting yard and the first yard braking position. The AS-MA measuring section should be equipped with devices that control the track occupancy along the entire tracks of the station's marshalling yard in order to form a cut describer that includes:

- four axle count devices of a CPRD-21-type to count railcars in the cuts;
- radiotechnical switching sensors to record the splitting-up of the train into cuts;
- weight meter to determine the load on each wheel set of the railcar;
- axles count sensors on the switches and hump braking positions – for precise control of the cut position in these sections and for organizing track sections on the axle count devices without track circuits.

The calculations and predicted values showed that the proposed solutions will lead to an increase in the labor productivity of workers of freight classification yards by an average of 11%, ensure an increase in the safety level of the splitting-up of freight trains on gravity sorting yards by 12%. They will also increase the safety of the car fleet and the volume of transported goods of the railway system by 7%, allow bringing the processing capacity of the gravity sorting yard of the railway station to the permissible design level, increasing the throughput and estimated capacity of the station as a whole by 22%.

The effectiveness of the implementation of the technology of automated control of technological processes for splitting-up trains on gravity sorting yards is achieved by reducing the hump interval, eliminating additional shunting operations, increasing the operating resources of floor equipment and shunting locomotives, reducing operating costs associated with repairing damaged railcars after collisions. It is also reached by eliminating cases of damage to goods, repair, with a decrease in the shunting work of pushing railcars back in the marshalling centre and rearrangement of "outsiders" (a railcar cut deviated from the splitting-up program), with a decrease in the time for rehandling railcars, a reduction in fuel consumption for shunting movements.

The main items of savings in operating costs when introducing automated systems in the new operating conditions of railway transport systems are:

- reduction in the volume of expenses associated with the presence of railcars at the railway station in the process of breaking-up;
- reduction in the number of required operating personnel and the associated reduction in the payroll;
- reduction of cases of damage to goods, rolling stock and related costs for repairs, detention of railcars and compensation for losses to the consumer of transport products;
- reduction in the required number of locomotives employed in shunting operations and, as a consequence, in locomotive crews and shunting masters.

When introducing an automation system, damage cases are foreseen in the amount of 15% of the base case due to failures of automation devices. The savings from reducing the time spent by railcars at the railway station in the process of breaking-up will amount to 6.109.06 thousand rubles. The annual savings on wages for maintenance personnel, taking into account mandatory contributions, will amount to 4018.62 thousand rubles. The amount of fuel saved for the year is 1722.8 thousand rubles. The total costs associated with the elimination of one commercial malfunction are 16397.76 rubles / railcar (Table 2).

The annual savings due to a reduction in the number of damaged railcars and damage to cargo during the introduction of the hump complex will amount to 2.2 million rubles. The total savings in operating costs per year will amount to 7.3 million rubles. The payback period of the proposal is 3.14 years. Annual economic effect is 722.213 thousand rubles.

Table 2. Summary table for calculating unproductive operating costs of the industry.

Type of charges	Amount of costs, rub/railcar
Expenses associated with train detention while waiting for uncoupling of railcars with malfunctions	998
Costs of additional shunting operations for uncoupling and attaching railcars in violation of technical specifications;	938.4
Costs for railcar-hours of detention time of railcars under elimination of violations	370.8
Remuneration for the work of commercial inspection stations engaged in correcting violations	91133.56
Operating costs of loading and unloading machines	2457
The cost of materials for the restoration of the technical conditions for the placement and securing of cargo	2500
The total costs associated with the elimination of a risk event	16397.76

4. Conclusion

However, in modern practice, interruptions in the regularity of operation pace of railway-freight classification yards of OAO RZD often lead to stops in train traffic volumes. This leads to a decrease in local speed and locomotive productivity, while the fleet of locomotives sharply increases and, coincidently, there is a shortage of them at the locomotive shift points. The transition to the operation of the locomotive fleet in long circulation journeys seriously increases the price of errors in adjusting the locomotive fleet. The use of technology for handling empty trains with a length of 100 conditional railcars leads to the emergence of oddness.

As a result, on the Trans-Siberian Railway there is traditionally an excess of circulation on the eastern borders and, vice versa, a lack of circulation on the western ones. All these factors mean that at present, and even more so in the near future, many directions and objects of the railway network will operate in load modes that are close to the limit, and sometimes in overload modes, which requires timely technical decisions of varying complexity. Improving the operation of the freight classification yards is a major factor in accelerating and cheapening the transportation process. In recent years, much attention has been paid to the automation of station operation processes, the use of advanced labor methods, the introduction of new equipment, without which the development of the station cannot ensure their operation efficiency.

Automation devices are the most important elements of the technical equipment of railway transport and make it possible to effectively solve the problems of the transportation process, contributing to an increase in the throughput capacity of railway facilities, ensuring the safety of train traffic and uninterrupted communication between all divisions of railway transport.

References

[1] Shtayger M G, Balanovskiy A E, Kargapoltsev S K, Gozbenko V E, Karlina A I, Karlina Yu I, Govorkov A S and Kuznetsov B O 2019 Investigation of macro and micro structures of compounds of high-strength rails implemented by contact butt welding using burning-off *IOP Conf. Ser.: Mater. Sci. Eng.* **560** 012190

[2] Konstantinova M V, Balanovskiy A E, Gozbenko V E, Kargapoltsev S K, Karlina A I, Shtayger M G, Guseva E A and Kuznetsov B O 2019 Application of plasma surface quenching to reduce rail side wear *IOP Conf. Ser.: Mater. Sci. Eng.* **560** 012146

[3] Guseva E A, Kargapoltsev S K, Balanovskiy A E, Karlina A I, Shtayger M G, Gozbenko V E, Konstantinova M V and Sivtsov A V 2019 Comparative evaluation of corrosion resistance of wheel and rail steels in various media *IOP Conf. Ser.: Mater. Sci. Eng.* **560** 012181

[4] Ilyushin P and Suslov K 2019 Operation of automatic transfer switches in the networks with

distributed generation *IEEE Milan PowerTech, PowerTech* 8810450

[5] Buryanina N, Korolyuk Y, Koryakina M, Lesnykh E and Suslov K 2019 Algoritm of Current Protection Based on Three Instantaneous-Value Samples *Proc. of 2019 IEEE PES Innovative Smart Grid Technologies Europe, ISGT-Europe* 8905542

[6] Suslov K, Shushpanov I, Buryanina N and Ilyushin P 2020 Flexible power distribution networks: New opportunities and applications *SMARTGREENS 2020 – Proc. of the 9th Int. Conf. on Smart Cities and Green ICT Systems*

[7] Voropai N, Ukolova E, Gerasimov D, Suslov K, Lombardi P and Komarnicki P 2019 Simulation approach to integrated energy systems study based on energy hub concept *IEEE Milan PowerTech, PowerTech* 8810666

[8] Munkhtuvshin D, Balabanov V B, Putsenko

[9] Olentsevich V A and Gozbenko V E 2013 Analysis of the reasons for the disruption of the safety of the railway transport system *Modern technologies. System Analysis. Modeling. Irkutsk* **1(37)** 180-183

[10] Retrieved from: http://www.rzd.ru

[11] 2015 *Technical conditions for storing and securing cargo in railcars and containers* (Moscow: Yurtrans) p 544

[12] Balalaev S V 2008 *Traffic safety on railways: a textbook. Part 1: Safety Fundamentals* (Khabarovsk: DVGUPS) p 125

[13] Ivankov A N and Ivankova L N 1997 *Design of a new marshalling yard with an automated slide: a resource book* (Irkutsk: IRGUPS) 50 p

[14] Gozbenko V E, Kargapoltsev S K and Olentsevich V A 2017 Automated system of placing and fastening cargo on the open rolling stock of railway transport *Herald of Irkutsk State Technical University* **21(4(123))** 157-165

[15] Kargapol'Tsev S K, Gozbenko V E, Kuznetsov B O, Karlina Yu I and Karlina A I 2019 The effect of the periodic driving force on a system with two degrees of freedom *J. of Phys.: Conf. Ser.* **1333(5)** 052009

[16] Kuznetsov B O, Gozbenko V E, Kargapol'Tsev S K, Karlina Yu I and Karlina A I 2019 Dynamic vibration protection of the railway carriage *J. of Phys.: Conf. Ser.* **1333(5)** 052018

[17] Konyuhov V Yu, Konstantinova M V and Gladkih A M 2019 Determination of restored units spectrum of equipment and development of the assembly unit repair method at industrial enterprises *J. of Phys.: Conf. Ser.* **1353** 012047

[18] Konyuhov V Yu, Gorban A V and Gladkih A M 2019 Investment in the improvement of maintenance service efficiency of processing equipment of an industrial enterprise *J. of Phys.: Conf. Ser.* **1353** 012102

IOP Conf. Series: Materials Science and Engineering 1064 (2021) 012030 doi:10.1088/1757-899X/1064/1/012030

Mathematical model of a journal bearing with low-melting and porous coating in its structure on different contacting surfaces with incomplete filling of operating clearance

M A Mukutadze, A V Morozova, A N Opatskikh and N S Zadorozhnaya

Rostov State Transport University, 2, Rostovskogo Strelkovogo Polka Norodnogo Opolcheniya Sq., Rostov-on-Don, 344038, Russia

E-mail: Opatskih@yandex.ru

Abstract. The authors consider the development of a mathematical design model of an infinite journal bearing lubricated with a lubricant with Newtonian rheological properties and a melt of a low-melting metal alloy covering the surface of the bearing bush, as well as a porous coating on the surface of the shaft, taking into consideration the dependence of the viscosity characteristics of the lubricant, the melt and the permeability of the porous coatings on the pressure with incomplete filling of the operating clearance. The authors found asymptotic solution of the system of differential equations with respect to the parameter K due to the melt and the rate of dissipation of mechanical energy and the exact self-similar solution for the zero approximation without regard for the melt and the first approximation taking into consideration the melt. This solution was found according to the equation of motion of a viscous incompressible fluid for a thin layer, the equations of continuity, Darcy equation and the equation determining, taking into account the expression for the rate of dissipation of mechanical energy, the profile of the molten contour of the bearing bush with regard to the dependence of the permeability of the porous coating, the viscosity of the lubricant depending on pressure. As a result, the fields of velocities and pressures in the lubricating and porous layers were determined taking into consideration the dependence of the lubricant viscosity and the permeability of the porous coating on pressure, as well as the function $\Phi_1(\theta)$, reasoned by the melt of the bearing bush surface. In addition, the main performance characteristics were determined: load-bearing capacity and friction force.

1. Introduction

It is well-known that modern machines are designed in view of the increase in static and shock loads, which are determined by latest engineering research. Furthermore, it is known that one of the most important elements of slide bearings is the lubricant. In our opinion, one of the methods to solve operational problems is the use a low-melting metal alloy as a lubricant that covers the surface of the bearing bush and a porous coating on the surface of the shaft, taking into consideration the dependence of the rheological properties of the lubricant, the melt and the permeability of the porous coating on pressure. Using a melt as a lubricant has been considered in many publications [1-5].

There are also a lot of works [6-10] that discuss the hydrodynamic calculation of sliding bearings taking into consideration the damping properties and the dependence of viscosity on pressure, as well as the non-Newtonian properties of lubricants. However, in these works there is no simultaneous consideration of the dependence of the viscosity of the applied lubricant and the permeability of the porous layer on pressure. A distinctive feature of this work is the development of a mathematical design model of a journal bearing with a porous coating of the journal shaft and a low-melting metal coating of the bearing bush surface, taking into consideration the dependence of the viscosity of the lubricant, the melt and the permeability of the porous layer on pressure, which provides a hydrodynamic friction mode with incomplete filling of the operating clearance.

2. Problem statement

In this article the authors consider the movement of a lubricant in the clearance of a journal bearing, the surface of the bearing bush of which is coated with a low-melting metal alloy, and the neck shaft – with a porous layer.

The rotational speed of the neck journal is Ω, and the bearing bush is stationary. All the heat during rotation goes to melting the surface of the bearing bush coating material covered with a fusible metal coating.

The dependence of the rheological properties of a liquid lubricant and the material permeability is set in the form:

$$\mu' = \mu_0 e^{\alpha' p'}, \quad \tilde{k}' = \tilde{k}_0 e^{\alpha' p'} \tag{1}$$

In the coordinate system that we have set, the equations of the contours of the neck journal, the neck journal with a porous layer, the bearing bush with a fusible metal coating are written in the form:

$$C_0 : r' = r_0 - \tilde{H}; \; C_1 : r' = r_0; \; C_2 : r' = r_1(1+H) + \lambda' f(\theta), \tag{2}$$

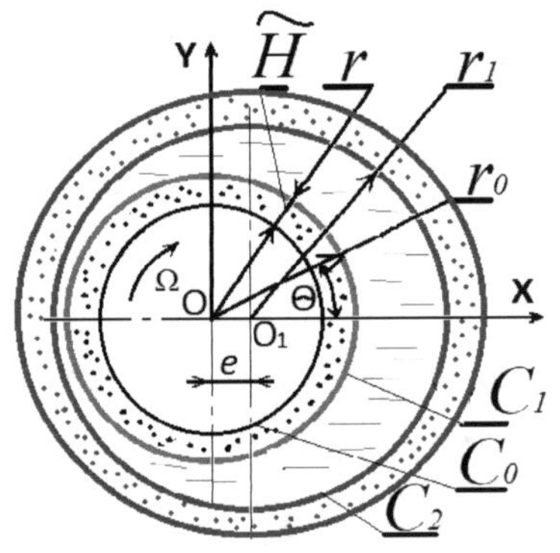

Figure 1. Functional diagram

The initial equation is a system of dimensionless equations of motion of the lubricant, Darcy's law, the equation of continuity and the equation that determines the profile of the molten contour of the bearing bush coated with a low-melting metal alloy with an accuracy of terms $O(\xi^2)$. This equation takes into account the expression for the rate of dissipation of mechanical energy. It is presented in the following form:

$$\frac{\partial p}{\partial r} = 0; \quad \frac{\partial^2 v}{\partial r^2} = e^{-\alpha p}\frac{dp}{d\theta}; \quad \frac{\partial u}{\partial r} + \frac{\partial v}{\partial \theta} = 0; \quad \frac{\partial^2 P}{\partial r^{*2}} + \frac{1}{r^*}\frac{\partial P}{\partial r^*} + \frac{1}{r^{*2}}\frac{\partial^2 P}{\partial \theta^2} = 0, \; \frac{d\Phi(\theta)}{d\theta} = K \int_{r_0}^{1+\eta\cos\theta+\Phi(\theta)} \left(\frac{\partial v}{\partial r}\right)^2 dr; \tag{3}$$

where $K = \dfrac{2\mu\Omega r_0}{L'\delta}$, $\eta = \dfrac{e}{\delta}$, $\eta_1 = \dfrac{\lambda'}{\delta}$, $\Phi(\theta) = \eta_1 f(\theta)$.

The boundary conditions for equation (3) are as follows:

$$u = 0,\ v = 0 \text{ at } r = 1 + \eta\cos\theta + \Phi(\theta),\ u\big|_{r=0} = \tilde{M}\frac{\partial P}{\partial r^*}\bigg|_{r^* = \frac{r_0}{\tilde{H}}},\ v(0) = 1,\ p = P \text{ at } r^* = \frac{r_0}{\tilde{H}},\ \frac{\partial P}{\partial r^*}\bigg|_{r^* = \frac{r_0}{\tilde{H}} - 1} = 0,$$

$$P(\theta_1) = P(\theta_2)=0 \tag{4}$$

where $\tilde{M} = -\frac{k^* r_0^2}{\tilde{H}\delta^3}$, θ_1 and θ_2 –respectively, the slope ratios of the beginning and end of the free surface of the lubricant.

The relationship between dimensionless and dimensional quantities is as follows:

- in the lubricating layer:

$$v'_{r'} = \Omega\delta u,\ v'_0 = \Omega r_0 v,\ r' = r_0 + \delta r,\ \delta = r_1 - r_0,\ p' = \tilde{p}p,\ \ \tilde{p} = \frac{\mu\Omega r_0^2}{\delta^2},\ \mu'=\mu_0\mu,\ \alpha'=\frac{\alpha}{\tilde{p}}\ . \tag{5}$$

- in the porous layer:

$$P' = p^* P,\ r' = \tilde{H}r^*,\ \tilde{k}'=\tilde{k}_0^*\tilde{k}^* \tag{6}$$

Let us introduce the notation, let $z = e^{-\alpha p}$. Differentiating both sides of the equality, we get:

$$e^{-\alpha p}\frac{dP}{d\theta} = -\frac{1}{\alpha}\frac{dz}{d\theta} \tag{7}$$

Then equation (3) will take the following form:

$$\frac{\partial P}{\partial r} = 0,\ \frac{\partial^2 v}{\partial r^2} = -\frac{1}{\alpha}\frac{dz}{d\theta},\ \frac{\partial u}{\partial r} + \frac{\partial v}{\partial \theta} = 0,\ \frac{\partial^2 P}{\partial r^{*2}} + \frac{1}{r^*}\frac{\partial P}{\partial r^*} + \frac{1}{r^{*2}}\frac{\partial^2 P}{\partial \theta^2} = 0\ \ z\frac{d\Phi(\theta)}{d\theta} = K\int\limits_{r_0}^{1+\eta\cos\theta+\Phi(\theta)}\left(\frac{\partial v}{\partial r}\right)^2 dr \tag{8}$$

With the corresponding boundary conditions:

$$u = 0,\ v = 0 \text{ at } r = 1 + \eta\cos\theta + \Phi(\theta),\ u\big|_{r=0} = \tilde{M}\left.\frac{\partial P}{\partial r^*}\right|_{r^*=\frac{r_0}{\tilde{H}}}\ \ v(0) = 1,\ p = P \text{ at } r^* = \frac{r_0}{\tilde{H}},\ \left.\frac{\partial P}{\partial r^*}\right|_{r^*=\frac{r_0}{\tilde{H}}-1} = 0,$$

$$z(\theta_1) = z(\theta_2)=1 \tag{9}$$

Taking the function Φ (θ) as a small parameter K due to the melt, we find:

$$\Phi(\theta)=K\Phi_1(\theta)+K^2\Phi_2(\theta)+K^3\Phi_3(\theta)+...=H(\theta)\ . \tag{10}$$

Boundary conditions on the contour are:

$$v(1+\eta\cos\theta+H(\theta)) = v(1+\eta\cos\theta) + \left.\left(\frac{\partial v}{\partial r}\right)\right|_{r=1+\eta\cos\theta}\cdot H(\theta) + \left.\left(\frac{\partial^2 v}{\partial r^2}\right)\right|_{r=1+\eta\cos\theta}\cdot H^2(\theta) - ... = 0;$$

$$u(1+\eta\cos\theta+H(\theta)) = u(1+\eta\cos\theta) + \left.\left(\frac{\partial u}{\partial r}\right)\right|_{r=1+\eta\cos\theta}\cdot H(\theta) + \left.\left(\frac{\partial^2 u}{\partial r^2}\right)\right|_{r=1+\eta\cos\theta}\cdot H^2(\theta) - ... = 0 \tag{11}$$

Asymptotic transformation of equations (8):

$$v(r,\theta)=v_0(r,\theta)+Kv_1(r,\theta)+K^2v_2(r,\theta)+...;$$
$$u(r,\theta)=u_0(r,\theta)+Ku_1(r,\theta)+K^2u_2(r,\theta)+...;$$
$$\Phi(\theta)=-K\Phi_1(\theta)-K^2\Phi_2(\theta)-K^3\Phi_3(\theta)-...;$$
$$z(\theta)=z_0(\theta) + Kz_1(\theta)+K^2z_2(\theta)+K^3z_3(\theta)+... \tag{12}$$

Substituting (12) into (8) taking into consideration (9), we get:
excluding melting:

$$\frac{\partial^2 v_0}{\partial r^2} = -\frac{1}{\alpha}\frac{dz_0}{d\theta},\ \frac{\partial u_0}{\partial r} + \frac{\partial v_0}{\partial \theta} = 0,\ \frac{\partial^2 P_0}{\partial r^{*2}} + \frac{1}{r^*}\frac{\partial P_0}{\partial r^*} + \frac{1}{r^{*2}}\frac{\partial^2 P_0}{\partial \theta^2} = 0 \tag{13}$$

with boundary conditions:

$$u_0 = 1,\ v_0 = 1 \text{ at } r = 0,\ v_0 = 0,\ u_0 = 0 \text{ at } r = 1+\eta\cos\theta,\ u_0\big|_{r=0} = \tilde{M}\left.\frac{\partial P}{\partial r^*}\right|_{r^*=\frac{r_0}{\tilde{H}}},\ p_0 = P_0 \text{ at } r^* = \frac{r_0}{\tilde{H}},\ \left.\frac{\partial P_0}{\partial r^*}\right|_{r^*=\frac{r_0}{\tilde{H}}-1} = 0,$$

$$z_0(\theta_1) = z_0(\theta_2)=1 \tag{14}$$

including melting:

$$\frac{\partial^2 v_1}{\partial r^2}=-\frac{1}{\alpha}\frac{dz_1}{d\theta},\ \frac{\partial u_1}{\partial r}+\frac{\partial v_1}{\partial \theta}=0,\ \frac{\partial^2 P_1}{\partial r^{*2}}+\frac{1}{r^*}\frac{\partial P_1}{\partial r^*}+\frac{1}{r^{*2}}\frac{\partial^2 P_1}{\partial \theta^2}=0,\ z_0\frac{d\Phi_1(\theta)}{d\theta}=K\int_0^{1+\eta\cos\theta}\left(\frac{\partial v}{\partial r}\right)^2 dr. \tag{15}$$

with boundary conditions:

$$v_1=\left(\frac{\partial v_0}{\partial r}\right)_{r=0}\cdot\Phi_1(0);\ u_1=\left(\frac{\partial u_0}{\partial r}\right)_{r=0}\cdot\Phi_1(0);\ v_1=0,\ u_1=0 \text{ at } r=1+\eta\cos\theta,$$

$$u_1\big|_{r=0}=\tilde{M}\left.\frac{\partial P}{\partial r^*}\right|_{r^*=\frac{r_0}{\tilde{H}}},\ p_1=P_1 \text{ at } r^*=\frac{r_0}{\tilde{H}},\ \left.\frac{\partial P_1}{\partial r^*}\right|_{r^*=\frac{r_0}{\tilde{H}}-1}=0,$$

$$z_1(\theta_1)=z_1(\theta_2)=0,\ K\Phi_1(0)=K\tilde{\alpha},\ \Phi(\theta_1)=\Phi(\theta_2)=\tilde{\alpha} \tag{16}$$

3. Exact solution

We are looking for the exact solution for the zero approximation:

$$v_0=\frac{\partial\psi_0}{\partial r}+V_0(r,\theta),\ u_0=-\frac{\partial\psi_0}{\partial r}+U_0(r,\theta),\ \psi_0(r,\theta)=\psi_0(\xi),\ \xi=\frac{r}{h(\theta)}, \tag{17}$$

$$V_0(r,\theta)=\tilde{v}(\xi),\ U_0(r,\theta)=-\tilde{u}_0(\xi)\cdot h'(\theta),\ h(\theta)=1+\eta\cos\theta.$$

Substituting (17) into (13) - (14), we get:

$$\tilde{\psi}_0'''=\tilde{C}_2,\ \tilde{v}_0''=\tilde{C}_1,\ \tilde{u}_0'(\xi)+\xi\tilde{v}_0'(\xi)=0,\ \frac{dz_0}{d\theta}=-\alpha\left(\frac{\tilde{C}_1}{h^2(\theta)}+\frac{\tilde{C}_2}{h^3(\theta)}\right). \tag{18}$$

and boundary conditions:

$$u_0\big|_{\xi=0}=\tilde{M}\left.\frac{\partial P}{\partial r^*}\right|_{r^*=\frac{r_0}{\tilde{H}}}\ \psi_0'(0)=0,\ \psi_0'(1)=0,\ \tilde{u}_0(0)=0,\ \tilde{v}_0(0)=1,\ \int_0^1\tilde{v}_0(\xi)d\xi=0,$$

$$\left.\frac{\partial P}{\partial r^*}\right|_{r^*=\frac{r_0}{\tilde{H}}-1}=0,\ p_0=P_0 \text{ at } r^*=\frac{r_0}{\tilde{H}}. \tag{19}$$

Integrating (18), we have:

$$\tilde{\psi}_0'(\xi)=\frac{\tilde{C}_2}{2}(\xi^2-\xi),\ \tilde{v}_0(\xi)=\tilde{C}_1\frac{\xi^2}{2}-\left(1+\frac{\tilde{C}_1}{2}\right)\xi+1,\ z_0=-\alpha\left[\tilde{C}_1(\theta-2\eta\sin\theta)+\tilde{C}_2(\theta-3\eta\sin\theta)\right]+1 \tag{20}$$

From $z_0(0)=z_0(2\pi)=1$ the following expression is obtained:

$$\tilde{C}_2=-\tilde{C}_1\left(1+\frac{\eta}{\theta_2-\theta_1}(\sin\theta_2-\sin\theta_1)\right) \tag{21}$$

Substituting the value $\tilde{C}_2$ into equation (20) for z_0, we get:

$$z_0=1-\alpha\eta\tilde{C}_1\left[\sin\theta-\sin\theta_1-\frac{\theta-\theta_1}{\theta_2-\theta_1}(\sin\theta_2-\sin\theta_1)\right] \tag{22}$$

Using the asymptotic expansion to determine *P*, we approach to the following approximate equation:

$$1-\alpha p+\frac{\alpha^2p^2}{2}-1+\alpha\eta\tilde{C}_1\left[\sin\theta-\sin\theta_1-\frac{\theta-\theta_1}{\theta_2-\theta_1}(\sin\theta_2-\sin\theta_1)\right]=0 \tag{23}$$

In order to find the hydrodynamic pressure, we solve equations (23) accurate to $O(\eta^2)$ as a result, we get:

$$P_0=\frac{1}{2}\eta\tilde{C}_1\left[\sin\theta-\sin\theta_1-\frac{\theta-\theta_1}{\theta_2-\theta_1}(\sin\theta_2-\sin\theta_1)\right] \tag{24}$$

Considering expression (24), the Darcy equation will have the form:

$$P(r^*,\theta)=R(r^*)\tilde{C}_1\frac{\eta}{2}\left[\sin\theta-\sin\theta_1-\frac{\theta-\theta_1}{\theta_2-\theta_1}(\sin\theta_2-\sin\theta_1)\right] \tag{25}$$

Substituting (25) into the Darcy equation to determine the function, let us pass to the following differential equation:

$$R''(r^*)+\frac{R'}{r^*}-\frac{R'}{r^{*2}}=0 \tag{26}$$

and boundary conditions:

$$\left.\frac{dR}{dr^*}\right|_{r^*=\frac{r_0}{\tilde{H}}-1}=0,\ R\left(\frac{r_0}{\tilde{H}}\right)=1 \tag{27}$$

Direct integration (26) taking into consideration (27) for the function $R(r^*)$ allows us to get the equation:

$$R(r^*)=\frac{r_0\tilde{H}r^*}{2r_0^2-2\tilde{H}r_0+\tilde{H}^2}+\frac{r_0\left(r_0^2-2\tilde{H}r_0+\tilde{H}^2\right)}{\tilde{H}\left(2r_0^2-2\tilde{H}r_0+\tilde{H}^2\right)r^*} \tag{28}$$

Taking into consideration the expression

$$\tilde{M}\left.\frac{\partial P}{\partial r^*}\right|_{r^*=\frac{r_0}{\tilde{H}}}=\int_0^1\tilde{v}(\xi)d\xi \tag{29}$$

For $\tilde{C}_1$ we use:

$$\tilde{C}_1=6\left[1-6\tilde{M}\frac{\tilde{H}^2(2r_0-\tilde{H})}{r_0\left(2r_0^2-2\tilde{H}r_0+\tilde{H}^2\right)}\left(1-\frac{\sin\theta_2-\sin\theta_1}{(\theta_2-\theta_1)\cos\theta}\right)\right] \tag{30}$$

Then for P_0 we get:

$$P_0=3\eta\left[1-6\tilde{M}\frac{\tilde{H}^2(2r_0-\tilde{H})}{r_0\left(2r_0^2-2\tilde{H}r_0+\tilde{H}^2\right)}\left(1-\frac{\sin\theta_2-\sin\theta_1}{(\theta_2-\theta_1)\cos\theta}\right)\right]\left[\sin\theta-\sin\theta_1-\frac{\theta-\theta_1}{\theta_2-\theta_1}(\sin\theta_2-\sin\theta_1)\right] \tag{31}$$

For $\Phi_1(\theta)$ we have:

$$\frac{d\Phi_1(\theta)}{d\theta}=\frac{1}{z_0}h(\theta)\int_0^1\left(\frac{\tilde{\psi}_0'(\xi)}{h^2(\theta)}+\frac{\tilde{v}_0(\xi)}{h(\theta)}\right)^2 d\xi. \tag{32}$$

Taking into account $K\Phi(\theta_1)=K\Phi(\theta_2)=K\tilde{\alpha}$, we get:

$$\Phi_1(\theta)=\frac{1}{\sup\limits_{[\theta_1;\theta_2]}z_0}\left[\frac{\tilde{C}_2^2}{12}(\theta-3\eta\sin\theta)+\frac{1}{6}\tilde{C}_1\tilde{C}_2(\theta-2\eta\sin\theta)+4(\theta-\eta\sin\theta)\right]_{\theta_1}^{\theta}=\frac{1}{\sup\limits_{[\theta_1;\theta_2]}z_0}\left[(\theta-\theta_1)\left(\frac{\tilde{C}_2^2}{12}+\frac{1}{6}\tilde{C}_1\tilde{C}_2+4\right)-\eta(\sin\theta-\sin\theta_1)\left(\frac{\tilde{C}_2^2}{4}+\frac{1}{3}\tilde{C}_1\tilde{C}_2+4\right)\right]+\alpha \tag{33}$$

We find the solution for the first approximation using:

$$v_1=\frac{\partial\psi_1}{\partial r}+V_1(r,\theta);\ u_1=-\frac{\partial\psi_1}{\partial\theta}+U_1(r,\theta);\psi_1(r,\theta)=\tilde{\psi}_1(\xi);\ \xi=\frac{r}{h(\theta)};V_1(r,\theta)=\tilde{v}(\xi);\quad U_1(r,\theta)=-\tilde{u}_1(\xi)\cdot h'(\theta);$$

(34)

Substituting (34) in (15) - (16), we get:

$$\tilde{\psi}_1'''=\tilde{\tilde{C}}_2;\ \tilde{v}_1''=\tilde{\tilde{C}}_1;\ \tilde{u}_1'(\xi)+\xi\tilde{v}_1'(\xi)=0;\ \frac{dz_1}{d\theta}=\frac{\tilde{\tilde{C}}_1}{h^2(\theta)}+\frac{\tilde{\tilde{C}}_2}{h^3(\theta)}. \tag{35}$$

and boundary conditions:

$$\tilde{\psi}_1'(0)=0,\ \tilde{\psi}_1'(1)=0,\ \tilde{u}_1(1)=0,\ \tilde{v}_1(1)=0;\ \left.\frac{\partial P_1}{\partial r^*}\right|_{r^*=\frac{r_0}{\tilde{H}}-1}=0,$$

$$u_1\Big|_{\xi=0} = \tilde{M}\frac{\partial P_1}{\partial r^*}\Big|_{r^*=\frac{r_0}{\tilde{H}}} \qquad p_1 = P_1 \text{ at } r^* = \frac{r_0}{\tilde{H}}, \qquad \tilde{u}_1(0)=0,\ \tilde{v}_1(0)=M,\ \int_0^1 \tilde{v}_1(\xi)d\xi = 0. \tag{36}$$

Integrating (35), we have:

$$\tilde{\psi}_1'(\xi) = \frac{\tilde{\tilde{C}}_2}{2}(\xi^2-\xi), \quad \tilde{v}_1(\xi) = \tilde{\tilde{C}}_1\frac{\xi^2}{2} - \left(\frac{\tilde{\tilde{C}}_1}{2}+M\right)\xi + M, \tag{37}$$

From $z_1(\theta_1) = z_1(\theta_2) = 0$, , we get

$$\tilde{\tilde{C}}_2 = -M\tilde{\tilde{C}}_1\left(1+\frac{\eta}{\theta_2-\theta_1}(\sin\theta_2-\sin\theta_1)\right), \tag{38}$$

where

$$M = \sup_{\theta\in[\theta_1;\theta_2]}\frac{\partial v_0}{\partial r}\Big|_{r=0} \cdot \Phi_1(\theta) = \sup_{\theta\in[\theta_1;\theta_2]}\left|(-2+2\eta\cos\theta+\eta\tilde{C}_1\cos\theta)\cdot\frac{1}{z_0}\left[(\theta-\theta_1)\left(\frac{\tilde{C}_2^2}{12}+\frac{1}{6}\tilde{C}_1\tilde{C}_2+4\right)-\eta(\sin\theta-\sin\theta_1)\left(\frac{\tilde{C}_2^2}{4}+\frac{1}{3}\tilde{C}_1\tilde{C}_2+4\right)+\tilde{\alpha}\right]\right|$$

Similarly, as well as for P_0, finally for P_1 we get:

$$P_1 = 3M\eta\left[1-6\tilde{M}\frac{\tilde{H}^2(2r_0-\tilde{H})}{r_0(2r_0^2-2\tilde{H}r_0+\tilde{H}^2)}\left(1-\frac{\sin\theta_2-\sin\theta_1}{(\theta_2-\theta_1)\cos\theta}\right)\right]\left[\sin\theta-\sin\theta_1-\frac{\theta-\theta_1}{\theta_2-\theta_1}(\sin\theta_2-\sin\theta_1)\right] \tag{39}$$

4. Research results

Based on the research results, we get:

$$R_x = \frac{\mu_0\Omega r_0^3}{\delta^2}3\eta(1+MK)\int_{\theta_1}^{\theta_2}\left[1-N\left(1-\frac{S}{\cos\theta}\right)\right]\left[\sin\theta-\sin\theta_1-(\theta-\theta_1)S\right]\cos\theta d\theta =$$

$$\left(1-6\tilde{M}\frac{\tilde{H}^2(2r_0-\tilde{H})}{r_0(2r_0^2-2\tilde{H}r_0+\tilde{H}^2)}\right)\left(\frac{\sin(\theta_2+\theta_1)\cdot\sin(\theta_2-\theta_1)}{2}-\sin\theta_1\sin\theta_2+\sin^2\theta_1\right)-(\sin\theta_2-\sin\theta_1)\times$$

$$\times\left(\sin\theta_2+\frac{\cos\theta_2-\cos\theta_1}{\theta_2-\theta_1}\right)-6\tilde{M}\frac{\tilde{H}^2(2r_0-\tilde{H})}{r_0(2r_0^2-2\tilde{H}r_0+\tilde{H}^2)}\cdot\frac{(\sin\theta_2-\sin\theta_1)^2}{2}$$

$$R_y = \frac{\mu_0\Omega r_0^3}{\delta^2}3\eta(1+MK) = \left(1-6\tilde{M}\frac{\tilde{H}^2(2r_0-\tilde{H})}{r_0(2r_0^2-2\tilde{H}r_0+\tilde{H}^2)}\right)\times\left[\frac{\theta_2-\theta_1}{2}-\frac{\sin(\theta_2-\theta_1)\cos(\theta_2+\theta_1)}{2}+\sin\theta_1(\cos\theta_2-\cos\theta_1)\right]+$$

$$+\frac{\sin\theta_2-\sin\theta_1}{\theta_2-\theta_1}\left[\theta_2\cos\theta_2-\theta_1\cos\theta_1-(\sin\theta_2-\sin\theta_1)\right]+\frac{\sin\theta_2-\sin\theta_1}{\theta_2-\theta_1}\times\left(1-6\tilde{M}\frac{\tilde{H}^2(2r_0-\tilde{H})}{r_0(2r_0^2-2\tilde{H}r_0+\tilde{H}^2)}\right)\times$$

$$\times\left[\ln\left|\frac{(1+\cos\theta_2+\sin\theta_2)(1+\cos\theta_1-\sin\theta_1)}{(1+\cos\theta_2-\sin\theta_2)(1+\cos\theta_1+\sin\theta_1)}\right|+\sin\theta_1\cdot\ln\left|\frac{\cos\theta_2}{\cos\theta_1}\right|-\cos\theta_2(\theta_2-\theta_1)-\left(\frac{\theta_2^3-\theta_1^3}{3}+\frac{\theta_2^5-\theta_1^5}{15}+\theta_1\cdot\ln\left|\frac{\cos\theta_2}{\cos\theta_1}\right|\right)\times\frac{\sin\theta_2-\sin\theta_1}{\theta_2-\theta_1}\right]$$

$$L_{\text{тр}} = \mu\int_{\theta_1}^{\theta_2}\left[\frac{\partial v_0}{\partial r}\Big|_{r=0}+K\frac{\partial v_1}{\partial r}\Big|_{r=0}\right]d\theta = \frac{\mu_0\Omega r_0^3}{\delta}\{-2(\theta_2-\theta_1)+\eta(2+\tilde{C}_1+2K\tilde{C}_1)(\sin\theta_2-\sin\theta_1)+KM \tag{40}$$

As a result of the numerical analysis, graphs were built:

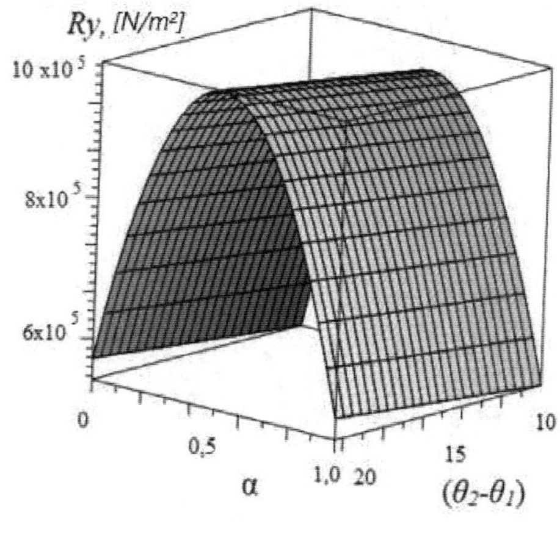

Figure 2. Dependence of R_y on the parameters α and $(\theta_2 - \theta_1)$

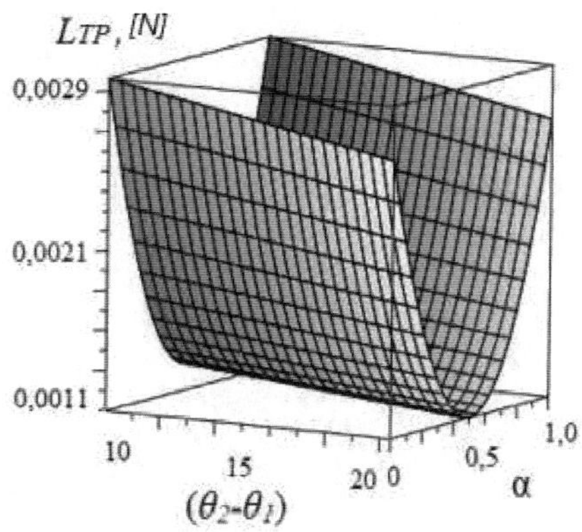

Figure 3. Dependence of L_{TP} on the parameters α and $(\theta_2 - \theta_1)$

5. Conclusion

The theoretical design models and their numerical analysis obtained as a result of the research showed a significant contribution of the parameter (α) characterizing the dependence of the viscosity of the lubricant and the porous layer on pressure as well as parameters (K) that is determined by the melting of an easily fusible metal coating, $(\tilde{H})$ which characterizes the thickness of the porous coating on the surface of the neck journal on the performance characteristics of the plain bearing. A large reduction in frictional force and an increase in bearing capacity have been proven. Tribotechnical design models were approximately refined in terms of bearing capacity by 22%, in terms of friction force by 25%.

References

[1] Kolesnikov V I, Shapovalov V V and Kokhanovsky V A 2006 *Transport tribotechnics*, in: *Friction and wear of materials* vol 1 (Rostov-on-Don: ROSZELDOR, Rostov State Transport University)
[2] Shapovalov V V, Chelokhyan A V and Kolesnikov I V 2010 *Amplitude-phase-frequency analysis of critical states of friction systems*, monograph (Moscow)
[3] Shapovalov V V, Kiselev I V and Fendrikov A I, RU Patent 2011586 C1. *Lubricator* (30 April 1994)

[4] Levanov I, Doykin A, Zadorozhnaya E and Novikov R 2017 Investigation of Antiwear Properties of Lubricants with the Geo-Modifiers of Friction *Tribology in Industry* **39(3)** 302-306
[5] Zadorozhnaya E 2013 The Research of Non-Newtonian Properties and Rheology of Thin Lubricant Layers in Hydrodynamic Journal Bearings *Society of Tribologists and Lubrication Engineers Annual Meeting and Exhibition 2013* pp 95-97
[6] Opatskikh A N, Mukutadze M A and Mukutadze A M 2019 V-Shaped Journal bearings Using Micropolar Lubricants Caused by a Melt Accounting for the Dependence of Lubricant Viscosity and Porous Layer Permeability on Pressure *J. of Phys.: Conf. Ser.* **1353** 012025. DOI: 10.1088 / 1742-6596 / 1353/1/012025
[7] Mukutadze M A, Vasilenko V V, Mukutadze A M and Opatskikh A N 2019 Mathematical Model of a Plain Bearer Lubricated with Molten Metal *IOP Conf. Series: Earth Env. Sci.* **378** 012021. DOI: 10.1088 / 1755-1315 / 378/1/012021
[8] Kropachev D Yu, Grishin A A and Maslo A D 2012 Methods of operational measurement of the temperature of the melt of metals for the needs of machine-building enterprises *Casting and metallurgy* **3(66)** 126-127
[9] Biryukov V P, Tatarkin D Yu, Khriptovich E V and Fishkov A A 2017 Determination of the influence of laser surfacing modes and the composition of powder materials on the wear resistance of coatings *Problems of mechanical engineering and reliability of machines* **1** 63–66
[10] Beretta G P, Niro A and Silvestri M 1992 Plain bearings lubricated by their own melt or sublimation product *Proc. of the American Society of Mechanical Engineers* **1** 86–90

Using refrigeration machines and heat pumps in the cycle of oil preheating using electricity

K V Osintsev[1] and A N Shishkov[2]

[1] South Ural State University, 76, Lenina ave., Chelyabinsk, 454080, Russia
[2] Moscow Polytechnic University, 38, Bolshaya Semenovskaya str., Moscow, 107023, Russia

E-mail: pte2017pte@mail.ru

Abstract. Tank electric heating systems are designed to compensate for heat losses to the environment. The main way of heating technological machines in the negative ambient temperatures is electric heating. Also, electric heating serves to prevent phase changes in the energy carrier. Furthermore, it is needed to control the required viscosity and maintain the set temperatures. The problem of limited resources has always faced humanity. In the field of energy this problem is now more urgent than ever before, because every year, the planet's hydrocarbon reserves are becoming less and less. Hydrocarbon cost is increasing, which leads to an increase in the cost of the final product. Therefore, it is necessary to use resources more efficiently, increase the efficiency of equipment, and introduce new technologies. In addition to limited resources, the problem of environmental pollution is no less acute.

1. Introduction

The use of electrical heating systems varies across the energy industry. If a liquid heat carrier is stored in the tank in winter, then its temperature can be maintained by heating and thermal insulation. An optimal system should have these two components. Heating by one of the components in the absence of the other is ineffective due to large heat losses [1, 2]. Thermal insulation provides a reduction in heat loss. Problems that can arise in cold weather are crystallization of the medium, coagulation of particles with precipitation, condensation, changes in the physical and chemical properties of the product. Heating is not used to heat the medium to a higher temperature [3]. It is needed to provide the desired temperature, compensating for heat loss.

2. Scientific novelty

The regenerating system of the low-power steam generator is less effective without low temperature utilization machines. However, when applying methods of additional energy generation in the form of heat, it is possible to increase the thermal efficiency of a thermal power plant. The authors propose the use of refrigeration machines and heat pumps in the regenerating system of the steam generator [4, 5]. The most effective at the moment are absorption chillers, as well as heat pumps in the ORC cycle. One of the first methods of heating and the most common is heating with steam supplied directly to the environment. Steam heating has a large number of disadvantages, namely the presence of technology for generating steam, the need for large human and technological resources. In addition, the disadvantage should be attributed to the complexity of control of the temperature of the environment, the formation of water condensate when using steam, low efficiency, increased costs to obtain more steam [6, 7].

The supply of steam to the inter-wall space in double-walled tanks, along with conventional steam heating, has disadvantages. First of all these are the following disadvantages - the elimination of condensate from the inter-wall space, uneven heating of the heated medium. In addition, these disadvantages are an increase in the heating time. The heating of the medium is slower or does not occur entirely, the possibility of effective use in small tanks.

Heating of tanks with a heating cable occurs by heating the wall of the tank itself. This system is the most effective and widespread due to its simplicity and reliability. When using electric heating, the design of the system is greatly simplified, its service life is increased, the service and the repair period of this system are reduced due to its independence from the technological cycles of the boiler room.

Typically, a cable heating system includes such elements as a cable and the necessary fasteners, a control system (thermostats, temperature sensors, controller), a power supply system (power cables, junction boxes, control cables).

This heating method is used in the absence of steam, or in case of insufficient production. The heating cable is laid out on the surface of the tank in a spiral or serpentine manner [8]. Fastening is carried out using special fastening materials: aluminum tape, professional mounting tape, mesh. On top of the cable, the container is covered with a layer of thermal insulation and a protective casing. A casing is required to prevent the insulation from getting wet [9, 10].

The preheating scheme using ORC and an absorption refrigeration machine is working by pump 1 in figure 1. We use water line 2 and water output 3, absorption machine 4. The main machine is oil-water electric heat exchanger 5. The cold-water input 6 is going to water tank 7. The pump has an electric drive 8 and electric cable input 9, electric cable output 10. Furthermore, CHP or boiler unit has a low temperature source 11, compressor 12 and high temperature source 13, ORC turbine 14. It is realized in an ORC with low temperature source, electric drive 15 and electric cable 16.

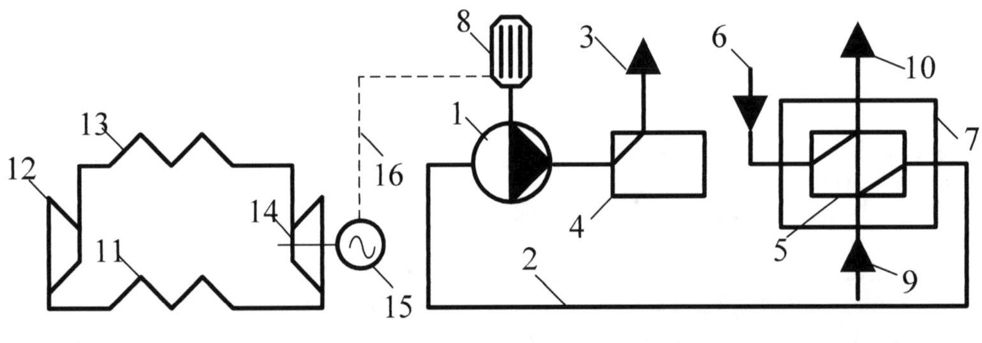

Figure 1. The preheating scheme using ORC and an absorption refrigeration machine: 1 – pump; 2 – water line; 3 – water output; 4 – absorption machine; 5 – oil-water electric heat exchanger; 6 – cold water input; 7 – water tank; 8 – electric drive; 9 – electric cable input; 10 – electric cable output; 11 – low temperature source; 12 – compressor; 13 – high temperature source; 14 – ORC turbine; 15 – electric drive; 16 – electric cable

3. Results

Research and development have shown that it is possible to significantly increase the efficiency of utilizing heat and mass of blow down water by using the high energy potential of this water in individual technological processes of boiler plants. At the same time, one should strive to use blow down water as a working medium directly in the apparatus where these technological processes take place, in order to partially or completely exclude special utilization equipment from the boiler room scheme. Naturally, when using new methods of utilizing blow down water, the rules for the design and safe operation of boilers and pipelines provided for by the guidelines for boiler supervision must be strictly observed.

One of the possible ways to implement the above idea is the use of blow down water as a heating agent for the jet-bubbling vacuum degasser. This method makes it possible to fully use the energy potential of the purge water in the process of desorption of dissolved gases from the boiler feed water or the make-up water of the heating system. This solution reduces or eliminates the need for steam or direct mains water commonly used as a heating agent in the vacuum degasser. In low-pressure boilers with closed heating systems, the blow down water can be supplied directly to the evaporator compartment of the vacuum degasser [11, 12].

In high-pressure boilers, it is advisable to use unevaporated water as a heating agent in a vacuum degasser after the first stage continuous blow down separator, excluding the second stage separator from the boiler room scheme. The use of blow down water as a heating agent for vacuum degassing of the make-up water of the heating network also contributes to the stabilization of the temperature regime of operation of the vacuum degasser. To ensure normal degassing, the temperature of the heating agent should not be less than 90 - 100 ° C in installations with horizontal vacuum degassers and 120 - 150 ° C - with vertical vacuum degassers [13, 14].

Typically, in vacuum degassers for replenishing the heating network, the heating agent is direct network water, the temperature of which has significant seasonal fluctuations and does not reach the above minimum values in the warm season. Since the temperature of the blow down water is constant, the use of this water instead of the mains water makes it possible to stabilize the temperature regime and improve the quality of vacuum degassing of the make-up water. It should be noted that when using blow down water for the technological needs of the boiler room instead of other, more valuable heat carriers and with a high degree of utilization of heat and mass of blow down water, it is not necessary to strive for the minimum course of continuous blow down of boilers [15, 16].

The purge flow rate may slightly exceed the norm and is determined by the energy requirement of the process carried out with the help of purge water. This consideration is also valid when implementing another technical solution, according to which the blow down water is used to regulate the supply of fuel oil to the boiler furnace. This method is implemented in steam-mechanical nozzles, where, instead of steam, blow down water is used, taken from the boiler drum.

Heaters made in the form of perforated rings can be installed in fuel oil tanks in the area where the suction pipes of fuel pumps are connected to them. Calculations and accumulated operational experience show that with the correct application of the technology under consideration, the possibility of excessive watering of fuel oil or accumulation of water in fuel oil tanks is excluded. This solution is much easier to implement in boiler plants, however, the economic and environmental effect of its use is practically the same as when using the previous method.

The most effective technology provides for the use of blow down water simultaneously for heating and transporting fuel oil from the fuel oil facility to the boiler using jet pumps. When supplying steam to some industrial plants. That do not require a high purity of the process steam. It is recommended to use blow down water, which ensures complete utilization of its heat and mass. This invention provides for the complete mixing of blow down water with steam used in production processes, which include steaming of reinforced concrete products, contact heating of raw materials, fuel, that is, processes for which the presence of a certain amount of mineral salts in the steam is not essential. In low pressure boilers with steam superheating, blow down water can be supplied to the live steam pipeline supplied to the consumer. In boilers of medium and high pressure, it is advisable to introduce the blow down water into the bleed-off steam pipeline of the turbines or use it as a cooling medium in reduction-cooling plants.

4. Conclusion

The process in the heat exchanger is very beneficial, since with a decrease in the temperature of a weak solution, the absorption proceeds more vigorously, and with an increase in the temperature of a strong solution, savings are achieved in the heat consumption for evaporating ammonia in the boiler generator.

Since the pressure in the boiler-generator is high, and in the absorber is low, a weak solution enters the throttle valve on the way from the first to the second. Reserve fuel heating currently has a lot of solutions. Considering the methods and technology of systems for ensuring the required temperature value of the fuel in a gas boiler house, based on the availability of media used for these purposes, two possible and effective methods are considered. Summarizing the above, each of the methods has both advantages and disadvantages. Based on the available technological cycles of the boiler room, the use of electric heating is the most effective due to its simplicity, reliability and maintainability.

References

[1] Rudig W 1986 Combined heat and power for district heating *Physics in Technology* **17** 125
[2] Hammond G 1977 Combining district heating and power generation *Physics in Technology* **8** 163
[3] Danilov O L, Garyaev A B and Yakovlev I V 2011 *Energy Saving in Heat Power Engineering and Heat Technologies,* ed A V Klimenko (Moscow: MPEI Publishing house) p 424
[4] Gershkovich V F 2000 About a possibility of implementation of regulation of heat consumption of buildings by the method of periodic interruption of a stream of the heat carrier *News of heat supplying* **2** 63-71
[4] Osintsev K V, Prikhodko I S and Zavyalova M I 2018 Methods for improving energy efficiency of air handling unit using factor analysis of data *IOP Conf. Ser.: Earth Env. Sci.* **194** 052019 7-12
[5] Osintsev K V, Prikhodko I S and Psenitsyna T A 2018 Choice of Chelyabinsk and Kuznetsk coals as main fuel for steam generator PK-14 using elements of cluster analysis *IOP Conf. Ser.: Earth Env. Sci.* **194** 052018 6-10
[6] Batrakov P A, Mikhailov A G and Ignatov V Yu 2018 Fire-tube boiler optimization criteria and efficiency indicators rational values defining *J. of Phys.: Conf. Ser.* **944** 012009
[7] Bukowska M, Nowak K, Proszak-Miąsik D and Rabczak S 2017 Concept of Heat Recovery from Exhaust Gases *IOP Conf. Ser.: Mater. Sci. Eng.* **245** 052057 23-25
[8] Connor S 2018 Total cost of ownership of a high-pressure steam boiler: Six factors to consider before making this long-term investment *HPAC Heating, Piping, AirConditioning Engineering* **90(6)** 13-18
[9] Shaposhnikov V and Biryukov B 2018 On the efficiency of heat and electric power plants based on combined-cycle plants with overexpansion of the working fluid in the gas turbine and injection of steam into the gas path *Chemical and Petroleum Eng.* **54(1-2)** 94-99
[10] Ghosh D, Ray S, Mandal J, Mandal N and Shukla A 2018 Failure analysis of PRDS pipe in a thermal power plant boiler *J. of The Institution of Engineers (India): Series C* **99(2)** 233-238
[11] Zhan L and Zhao J 2007 Implement of increment-model PID control of PLC in constant-pressure water system *Electronic Measurement and Instruments* 4336-4339
[12] Askari B, Sadegh M and Ameri M 2015 Energy management and economics of a trigeneration system considering the effect of solar PV, solar collector and fuel price *Energy for Sustainable Development* **26** 43-55
[13] Cocco D and Cau G 2015 Energy and economic analysis of concentrating solar power plants based on parabolic trough and linear Fresnel collectors *J. of Power and Energy* **229(6)** 677-688
[14] Pugsley A, Zacharopoulos A, Mondol J and Smyth M 2016 Global applicability of solar desalination *Renewable Energy* **88** 200-219
[15] Wu J, Hou H, Hu E and Yang Y 2018 Performance improvement of coal-fired power generation system integrating solar to preheat feedwater and reheated steam *Solar Energy* **163** 461-470
[16] Wang Y, Jin J, Liu D, Yang H and Kou X 2018 Understanding ash deposition for Zhundong coal combustion in 330 MW utility boiler: Focusing on surface temperature effects *Fuel* **216** 697-706

Modernization of technological equipment in the waste water purification process behind the coke oven using the organic Rankine cycle

A N Shishkov[1] and K V Osintsev[2]

[1] Moscow Polytechnic University, 38, Bolshaya Semenovskaya str., Moscow, 107023, Russia
[2] South Ural State University, 76, Lenina Ave., Chelyabinsk, 454080, Russia

E-mail: shan1982@mail.ru

Abstract. The authors analyzed the state of the process scheme, which is a structural part of the water supply treatment facilities. The research results can be applied in oil and gas, chemical, metallurgical and energy industries. In particular, the authors for the first time propose an efficient scheme for the reuse of heat to obtain purified water. Currently, a way to reduce energy consumption by introducing a plant based on an organic Rankine cycle has been considered. The metallurgical production has such equipment as converters, coke ovens, scrubbers, gas ducts, cooling boilers, waste heat boilers. The heat recovery schemes developed by the authors increase the efficiency of the process equipment in the metallurgical industry. In the work of enterprises, low-potential heat of soil, water, air is increasingly used. One of the main directions of energy saving to increase the energy efficiency of the boiler house is to increase the heat utilization rate of the fuel at power plants. Boiler houses that run on gas lose at least 13-18% of heat with flue gases. In the article, the authors propose to combine the methods of heat utilization. Namely, a contact heat exchanger and an organic Rankine cycle are used. The water temperature is significantly reduced due to the fact that the collectors are designed to transport water. In this regard, experts consider it expedient to utilize the heat of wastewater not only at treatment plants, but also directly in the building itself. Prospects for the use of the authors' developments are associated with their further implementation at industrial enterprises of the metallurgical industry.

1. Introduction

The amount of sewage produced in huge quantities by large cities remains virtually unchanged throughout the year. The wastewater temperature is lower than the outdoor temperature in summer and higher in winter. This makes them an ideal source of low grade heat for use in heat pumps. Various devices for the utilization of waste water heat have been developed and used for about 30 years [1, 2]. The most common system is the use of heat pumps installed in waste water treatment plants. Such systems centrally collect waste water heat, which saves a lot of energy. At the same time, energy efficiency experts say that a significant amount of wastewater heat energy literally goes into the ground. Thus, the authors will need to develop methodological approaches to carry out the exergy method of thermodynamic analysis based on new methods for approximating piecewise linear functions. In addition, it is proposed to use this method not only to optimize exergy losses, but also to analyze the flow energy, since any type of energy, according to the fundamental foundations of physics, consists of exergy and anergy.

2. Materials and methods

The top-level program with elements of artificial intelligence incorporated into the control system to optimize the operation of the power plant is adaptive and trainable. When first switched on, the program must test the sensor system of the power plant. After stabilization of the power plant operating mode, the self-learning system records the sensor readings. This procedure must subsequently be repeated under different operating conditions to build the power curve. Moreover, between these measurements, the installation can already operate in a certain safe test mode, but not with full power. During operation, the parameters of the power plant may change slightly, therefore, the described iterations for adjusting the characteristics can be repeated periodically. As a result of constructing the desired graph, the control system has all the data and, taking this into account, builds the load characteristics of the generator. These dependencies will be studied in detail during the project. An alternative to calculating the graph in this way can be the manufacturer's standard load curves. In order to reduce the cost of electricity and heat energy, much attention is paid to inexpensive, low-cost technologies that can be put into operation in a short time. Basically that means technologies for increasing the efficiency of boiler plants using secondary energy resources [3, 4]. The overwhelming majority of treatment facilities are in need of reconstruction or repair. At the same time, it is impossible to use standard solutions that are implemented during design. Reconstruction is a process that requires atypical tasks to improve the energy and economic performance of sewerage systems [5, 6]. The balance of the enterprise may include about 400 pumping stations, more than 3000 km of water supply networks. In addition, control over water quality is carried out by specialists in laboratories. Several thousand people work around the clock so that the inhabitants of megacities receive high-quality water. Mechanical treatment facilities do not meet the requirements for such facilities at the present time [7, 8]. Due to the wear of the mechanical cleaning grate and the sand catcher, the quality of water treatment is not ensured. It should be noted that it is often necessary to replace the system for removing sediment accumulating at the bottom of the plant. Compaction and flushing of the waste can seldom be missing. Biological treatment facilities may not have equipment to remove phosphorus and nitrogen compounds from wastewater. Thus, the goal of the modernization of treatment facilities is not only to improve the quality of water purification to standardized requirements, but also to reduce energy consumption [9, 10].

3. Results

The tasks of efficient use of energy resources, reduction of heat losses in industrial and domestic economy are becoming more and more urgent. The rise in prices for energy resources is forcing the use of low-potential energy sources, the heat of which is emitted into the environment. Domestic waste water is one such source. After passing through the purification system, the water is discharged back into the purified water channel and the cycle is closed. Waste heat from water treatment plants is low-grade. Utilization of this heat into useful work is necessary to improve the energy efficiency of the treatment facilities. The solution to this problem contributes to the development of decentralized energy. There are various thermodynamic cycles for converting low grade thermal energy. Increasingly, the organic Rankine cycle is used using an alternative working fluid, the boiling point of which is significantly lower than the boiling point of water. In the installation scheme, based on the organic Rankine cycle, sewage purified water is used as a low-grade heat source (figure 2). In a heat exchanger, water transfers its heat to the working fluid. Reaching the saturation temperature, the refrigerant boils and in a state of vapor enters the piston expander, where in the cylinder when the piston moves, it expands, and the thermal energy goes into the mechanical work of the rotation of the rotor of the electric generator. After the expander, the working fluid is fed into the condenser, where it is cooled by the flow water and condenses. In the liquefied state, the refrigerant is fed back to the heat exchanger by pump, driven by the engine. The electricity generated by the rotation of the rotor of an electric generator can be used for the company's own needs. Heaters made in the form of perforated rings can be installed in fuel oil tanks in the area where the suction pipes of fuel pumps are connected to them. Calculations and accumulated operational experience show that with the correct application of the technology under consideration, the possibility of excessive watering of fuel oil or accumulation of water in fuel oil tanks is excluded. This solution is much easier to implement in boiler plants, however, the economic and environmental effect of its use is practically the same as when using the previous method. The most effective technology provides for the use of blow down water simultaneously for heating and transporting fuel oil from the fuel oil facility to the boiler using jet pumps. It is recommended to use blow down water, which ensures complete utilization of its heat and mass. This invention provides for the complete mixing of blow down water with steam used in production processes, which include steaming of reinforced concrete products, contact heating of raw materials, fuel, that is, processes for which the presence of a certain amount of mineral salts in the steam is not essential. In low pressure boilers with steam superheating, blow down water can be supplied to the live steam pipeline supplied to the consumer. In boilers of medium and high pressure, it is advisable to introduce the blow down water into the bleed-off steam pipeline of the turbines or use it as a cooling medium in reduction-cooling plants.

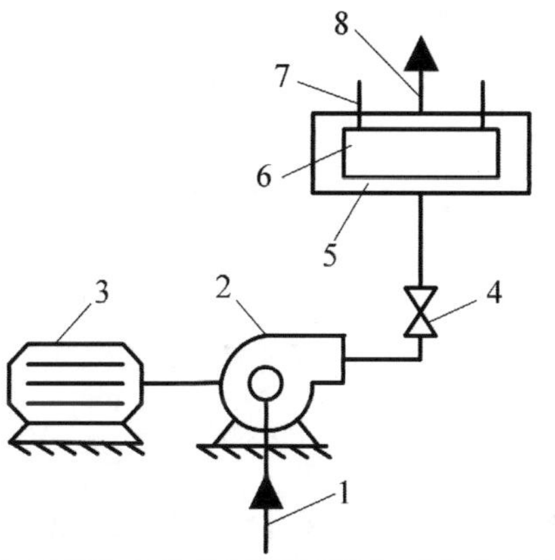

Figure 1. The process scheme using the conacting heat exchanger: 1 – gases; 2 – gas fan; 3 – electric drive; 4 – valve; 5 – gas duct; 6 – water tank; 7 – water input; 8 – gas out

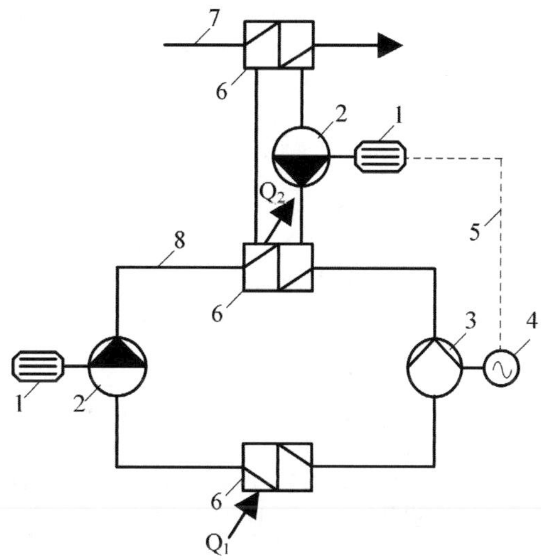

Figure 2. The process scheme using ORC: 1 – electric drive; 2 – pump; 3 – detander; 4 – electric generator; 5 – electric cable; 6 – heat exchanger; 7 – water line

The advantage of the Rankine cycle over the Sterling, Brighton and Kalina cycles is the ability to adapt to various sources of thermal energy, low cost equipment and ease of implementation. In addition, this system can operate in a binary cycle in both high-temperature and low-temperature areas. Industrial plants release a large amount of thermal energy into the environment along with the flow of combustion products and heat carrier at a relatively low temperature. This circumstance often has an extremely adverse effect on the environment, polluting it and leading to climate change. In addition, the cost of the energy released is ultimately determined by the cost of production. Accordingly, the proposed method can be adapted for any control system of a power plant of any capacity, with any generator, with any control system. Moreover, the algorithm can also work with the corresponding power take-off control, which also has to be simulated in the project and tested on real power plants. Thus, the proposed methods and approaches make it possible to successfully solve the assigned tasks and carry out the project at a level significantly exceeding the world level.

4. Discussion

Similar methods of heat recovery were used for the prospect of further increasing the energy efficiency of a thermal power plant based on a contact heat exchanger with an active packing. This heat exchanger is planned to be installed behind the KVG-58.2-150 hot water boiler. This will allow reducing the cost of heat production. The use of contact heat exchangers is one of the most effective ways to save metals and other scarce materials, reduce capital and operating costs, and improve equipment reliability [11, 12]. Water heated to 40 degrees Celsius in a contact heat exchanger can be used for hot water supply in a private school. Here, with the help of heating elements, the water will be heated to 65-70 degrees Celsius. The school also plans to build a swimming pool. The heated water in heat exchanger can be used in this case as well. It should be noted that the school is located at a short distance from the boiler room, which is one of the conditions for using a contact heat exchanger. A contact heat exchanger with an active nozzle is a recuperative-mixing type device, it is designed to utilize the heat of flue gases and to heat water for various needs in the temperature range from 5 to 50 degrees Celsius [13, 14]. It consists of a body made of sheet steel, forming a chamber with an irrigation system, an active nozzle made in the form of a bundle of pipes with a coolant circulating in them, and a separation device. Simultaneously with the heat exchange processes in heat exchanger, figure 1, the utilized gases are cleaned from mechanical impurities of incomplete fuel combustion. Mechanical impurities are captured by the scrubbing liquid [15, 16]. Selective purification from gaseous components can be carried out depending on the composition of the scrubbing liquid [17, 18].

5. Conclusion

As noted in [12], if a CHP plant uses only 37% of the fuel energy, this means that 63% of this energy is lost forever. Therefore, in recent years, great efforts have been made to utilize thermal waste from industrial enterprises with the possibility of generating electricity. Numerous studies show that there are quite a few industrial enterprises with unused reserves of thermal energy. For example, cement production is very energy intensive. This sector of the economy accounts for 12-15% of the total energy consumed by industry, while up to 15% of fuel energy is dissipated into the environment. The temperature of the combustion products at the outlet of a gas turbine usually exceeds 400 ° C, the petrochemical industry emits gases with a temperature of 150-300 ° C. In this case, the ORC unit also provides the opportunity to use the remaining thermal energy, however, disposal of these reserves, as a rule, is challenging task [19, 20]. The planned result is new knowledge about the formation of an exergy method for analyzing economic analysis in order to optimize the operation of a power plant for desalination of water based on the study of applicable hydrodynamic, mechanical, heat and mass transfer and management laws. Scientific significance lies in obtaining new knowledge that is essential and important for world science [21, 22]. The applied significance lies in the use of this knowledge to create a computer model of a power plant.

References

[1] Ma W, Liu T, Min R and Li M 2018 Effects of physical and chemical properties of working fluids on thermodynamic performances of medium-low temperature organic Rankine cycles (ORCs) *Energy Conversion and Management* **171** 742-749
[2] Zhu Y, Li W, Sun G and Li H 2018 Thermo-economic analysis based on objective functions of an organic Rankine cycle for waste heat recovery from marine diesel engine *Energy* **158** 343-356
[3] Bo Z, Sang Z, Lu X and Weng Y 2018 Effects of different working fluids on the performance of a radial turbine in an organic Rankine cycle power system *J. of Mechanical Sci. and Technology* **32(9)** 4503-4515

[4] Toropov E V, Osintsev K V and Alyukov S V 2018 Analysis of the calculated and experimental dependencies of the combustion of coal dust on the basis of a new methodological base of theoretical studies of heat exchange processes *Int. J. of Heat and Technology* **36(4)** 1240-1248
[5] Urazmetov S, Kraev V, Verevkin A, Kiryushin O and Skatov K 2013 Method for determining the viscosity of nonlinear viscous liquids and the device for its implementation *Federal Intellectual Property Service Official Bull.* **16** 21-24
[6] Baniasad A, Oukati S and Ameri M 2015 Energy management and economics of a trigeneration system Considering the effect of solar PV, solar collector and fuel price *Energy for Sustainable Development* **26** 43-55
[7] Cocco D and Cau G 2015 Energy and economic analysis of concentrating solar power plants based on parabolic trough and linear Fresnel collectors *J. of Power and Energy* **229(6)** 677-688
[8] Pugsley A, Zacharopoulos A, Mondol J and Smyth M 2016 Global applicability of solar desalination *Renewable Energy* **88** 200-219
[9] Alabugin A A, Alyukov S V and Osintsev K V 2020 Approximation Methods for Analysis and Formation of Mechanisms for Regulating Heat and Mass Transfer Processes in Heat Equipment Systems *Int. J. of Heat and Technology* **38(1)** 45-58
[10] Aidarova S, Bekturganova N, Kerimkulova M and Sharipova A 2014 The influence of surfactants to the stability of coal water suspension *Periodica Polytechnica Chemical Engineering* **58** 21-26
[12] Danilov O L, Garyaev A B, Yakovlev I V et al 2011 *Energy Saving in Heat Power Engineering and Heat Technologies: studies. for high schools* (ed A V Klimenko) (Moscow: MPEI Publishing house) p 424
[13] Sadhu S 2011 *Handbook of Mechanical Engineering* (S Chand & Company Ltd) p 2080
[14] Emami M, Shahbazian H and Sunden B 2019 Effect of operational parameters on combustion and emissions in an industrial gas turbine combustor *J. of Energy Resources Technology, Transactions of the ASME* **141(1)** 012202
[15] Villagran-Villegas L, Hernandez-Gomez L, Patiño-Ortiz M, Cuellar-Orozco M and Toledo-Velazquez M 2019 Analysis of the wear damage on offshore gas turbine blades *Advanced Structured Materials* **92** 221-236
[16] Saha A, Saikia L, Rajbongshi R, Saha D and Tasnin W 2019 AGC of thermal-split shaft gas turbine system integrating IPFC and ultra-capacitor *Advances in Intelligent Systems and Computing* **757** 105-115
[17] Bukowska M, Nowak K, Proszak-Miąsik D and Rabczak S 2017 Concept of heat recovery from exhaust gases *IOP Conf. Ser.: Mater. Sci. Eng.* **245** 052057 23-25
[18] Shatskikh Y V, Sharapov A I and Byankin I G 2017 Analysis of deep heat recovery from flue gases *J. of Phys.: Conf. Ser.* **891** 012188 1-5
[19] Pashchenko D I, Semenov B A, Shchelokov A I, Naplekov I S and Mustafin R M 2018 Thermodynamic analysis of thermochemical recuperation of high-temperature flue gas heat *J. of Phys.: Conf. Ser.* **1111** 012009 6-10
[20] Batrakov P A, Mikhailov A G and Ignatov V Yu 2018 Fire-tube boiler optimization criteria and efficiency indicators rational values defining *J. of Phys.: Conf. Ser.* **944** 012009
[21] Khavanov P and Chulenyov A 2017 Boiler water regime *IOP Conf. Ser.: Earth Env.* **90** 012199 8-13
[22] Vershinina K Y, Glushkov D O and Strizhak P A 2017 Characteristics of the ignition of the drops of organic coal–water fuels based on waste oils and industrial oils *Solid Fuel Chemistry* **51(3)** 188-194

IOP Conf. Series: Materials Science and Engineering 1064 (2021) 012033 doi:10.1088/1757-899X/1064/1/012033

Increasing the energy efficiency of the industrial enterprise technological and mechanical equipment due to the use of converter steam

K V Osintsev[1] and A N Shishkov[2]

[1] South Ural State University, 76, Lenina Ave., Chelyabinsk, 454080, Russia
[2] Moscow Polytechnic University, 38, Bolshaya Semenovskaya str., Moscow, 107023, Russia

E-mail: pte2017pte@mail.ru

Abstract. The article discusses the possibility of using converter steam to improve energy efficiency. The development of the authors will improve the performance of technological equipment in mechanical engineering. The problem of inefficient use of heat and loss of energy exists at mechanical engineering enterprises. In addition, environmental safety must be ensured at a high technological level. The above scientific problems are solved in this article by the authors. To solve the set tasks, scientific methods are used related to the generalization of data sets, their processing and the construction of an optimization model. In addition, variants of technical solutions at the enterprises of the energy industry are shown. To determine the long-term direction of increasing the energy efficiency of the facility, it is necessary to conduct a comparative analysis of the options. We have analyzed the use of converter steam for heating process equipment. It can be concluded that the method has mainly strengths, but since steam production has an almost constant cycle, the use of this method will not have a positive effect on eliminating problems. Prospects for using the technological solutions developed by the authors are associated with their practical application at metallurgy and mechanical engineering enterprises. It should be noted that the authors were the first to propose such a method for utilizing heat.

1. Introduction

The relevance of the study of methods for increasing the energy efficiency of an enterprise using secondary energy resources is determined by economic and technological indicators, when conducting a study on the utilization of converter steam at an industrial enterprise. The problems should include several issues, namely the utilization of converter steam. Utilization will save natural gas; besides, the cost of steelmaking will decrease. The authors would like to emphasize that the utilization of converter steam reduces the burden on the environment. Technological, political and economic factors are the main environmental factors that determine the increase in energy efficiency. The technological factor of the external environment manifests itself in the need to increase energy efficiency in case of use of secondary energy resources [1, 2]. The economic factor of the external environment is manifested in the saving of natural gas due to the use of converter steam. The use of a disposal facility may be influenced by political factors [3, 4]. They are expressed in economic sanctions, due to which problems arise with the supply of imported equipment for the utilization of converter steam, which is often more efficient than domestic products. Therefore, the consumer is forced to purchase only available units on the power equipment market, which may not correspond to the required parameters of energy consumption, power and quality of the output product. At the same time, a recovery plant requires large capital investments [5, 6]. This is reflected in the high cost of design work, automation work, and equipment costs. Consequently, a contradiction arises between the need to improve energy efficiency through the utilization of converter steam and the high costs of developing and purchasing equipment [7, 8]. This contradiction determines the following research problems. First of all these problems include the loss of thermal energy due to the discharge of converter steam into the environment. In addition, the disadvantages are damage to the environment and noise impact on the population. The unsolved problems of these problems determine the following goal of the research work [9, 10]. It is necessary to identify an energetically efficient way of using the thermal energy of the converter steam. To achieve the goal, it is necessary to solve the following tasks. The most necessary thing is to study the basic concepts of the theory and problems of improving the use of secondary energy resources. In addition, it is necessary to analyze the existing methods of using secondary resources, to choose the most efficient and rational method of using the converter steam. The objects of research in the research work will be existing installations for the utilization of secondary energy resources. The research subjects will be methods of improving the use of secondary energy resources [11, 12]. It is necessary to carry out research work to improve the methods of using secondary energy resources.

2. Materials and methods

The scientific methods developed by the authors are the formulation of the concept of technology, scientific substantiation of the concept, the search for technological approaches to the implementation of the concept, identification of advantages over alternative approaches, determination of the feasibility of further development of the technological concept and assessment of the risks of its implementation. The planned result is the ways and methods of applying the previously studied phenomena and the knowledge gained to solve practical problems of maximizing the operation of a power plant, the presentation of the results in a scientific and technical report on research work, the results of a computer model study. Scientific significance lies in the creation of new knowledge important for world science in the form of universal algorithms. The applied significance lies in the use of this knowledge to create software that implements the algorithm for the operation of a power plant. Scientific and technical development and simulation analysis with analytical definition of the objective function and experimental confirmation of the concept based on testing algorithms as part of the software for the control system of scaled components in laboratory conditions is planned to be carried out using the equipment of a mechanical engineering enterprise. These are proof-of-concept, software development, laboratory and field studies of individual technology elements, modeling and testing. The planned final of the research is checking the operability of software on operating power plants, the final scientific and technical report on the results of research work, patenting of technical solutions and software, publications in accordance with the assumed obligations. Scientific significance also lies in the universalization of the created solutions, which is important for the global energy industry and creates the preconditions for a significant economic effect. The applied significance lies in the replication of the created software on a global scale.

3. Results

To increase the energy efficiency, it is necessary to select and install a device for utilizing converter steam to transfer the maximum amount of heat from converter steam generated by waste heat exchangers OKG-160 to the return heating water, taking into account the observance of the temperature schedule for the heating water. Secondary energy resources are energy potential of products, waste, by-products and intermediate products generated in technological units, which is not used in the unit itself, but can be partially or fully used to supply power to other units. Recovery steam is generated in a recovery waste heat boiler. In many enterprises, steam is used as a heat carrier for technological purposes. Often steam is used for following purposes. Firstly, these are heating return water supply, heating raw water and heating of technological equipment. The problem ranking model is used. With its help, the main problem is determined. To rank problems, one should assess the degree of importance and urgency of problem solving. The problem of heat energy losses has a high degree of importance of the solution and a high degree of urgency of the solution due to the fact that today the tendency of the Russian energy sector is associated with energy conservation [13, 14]. To identify the factors for solving the main problem and determine its genesis, a causal diagram of heat energy losses has been developed [15, 16]. The use of steam with such parameters is possible for the following needs. These are heating return network water and heating process equipment. The network water parameters are 1500 m^3/h of the flow rate and 0.18 MPa of pressure; temperature was 70°C. Let us base on the parameters of the converter steam and the return network water. The utilization unit must provide heating of the return network water to a temperature of 115°C with steam at a temperature of 250°C, water flow rate of 1500 m^3/h and steam consumption 100 t/h. To ensure heating of the return heating water to the required temperature in the recycling unit of converter steam and safe operation of the unit, excluding hydraulic shocks, a system for regulating the pressure and flow rate of converter steam and return network water is required [17, 18]. To maintain the pressure and circulate the return supply water, supply water pumps with frequency drives are required, and condensate pumps with frequency drives are needed to drain condensate from the converter steam recovery unit [19, 20]. The solutions described above lead to high costs for the design and installation of a utilization plant for heating return network water [21]. The schematic diagram of the recycling plant is shown in figure 1.

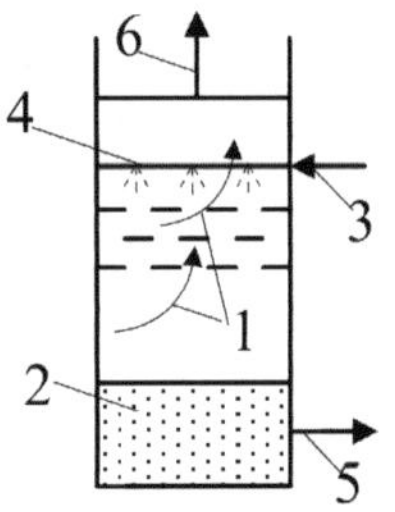

Figure 1. The process scheme using hot gases: 1 – gases; 2 – hot water; 3 – cold water; 4 – nozzeles; 5 – water output; 6 – gas output

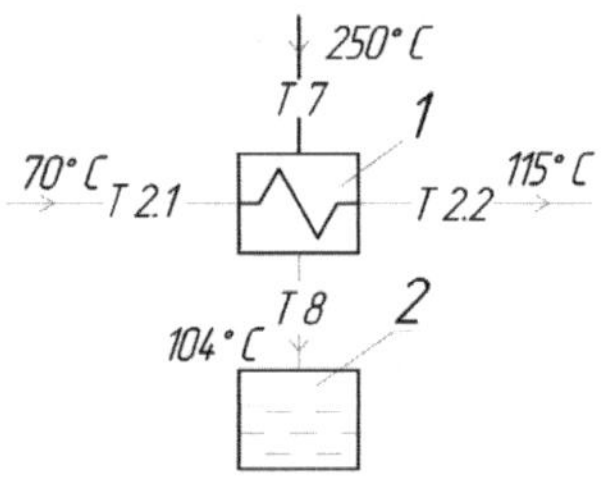

Figure 2. The new process scheme: 1 – heat exchanger; 2 – water tank

4. Discussion

The introduction of a waste heat boiler will significantly reduce the cost of purchasing natural gas, which is used in the production of steam and water. Also with the installation of this utilization system, the degree of purification of process gases from harmful substances will increase due to the low temperature at the gas-cleaning unit itself. Recycling of waste gas heat was not used due to the low temperature of the gases. At present, the problems of energy conservation are the main challenges in industrial production. At the enterprises of this or that industry, factories, combines there are large losses due to insufficient use of energy in technological processes. For example, the heat of the process gas generated during the production process is either not used efficiently enough, or is simply not simply emitted into the atmosphere. Such an example leads, firstly, to environmental pollution and environmental degradation in industrial and urban areas, and secondly, to extensive and unnecessary energy losses in industrial production (figure 2). A huge amount of metallurgical equipment, coupled with a large consumption of fuel and energy resources, has a rather low efficiency, which in turn leads to losses of thermal energy into the atmosphere and large economic losses of the enterprise. For example, in high-temperature metallurgical units during the production of main types of products, high-temperature process gases are formed, which can be used as fuel in heating furnaces of the plant or enterprise itself to generate electricity or hot water, superheated steam and dry air. These types of fuel resources are called secondary energy resources, the energy potential of which is in products, secondary and intermediate elements that are not used in the furnace, but can be partially or completely used in the energy supply of secondary consumers. Secondary energy resources are divided into groups. In metallurgy, in ferroalloy production, one of the most important processes in technology is the production of alloys in high-temperature furnaces, subsequently from which technologically gases of medium or low temperature are released. For the use of waste gases from ferroalloy furnaces, take into account the fact that the gas temperatures are relatively low – 180 - 500°C.

We note the new approaches of the team of authors to the scientific problem being solved. The team forms a theoretical justification for the application of the combined desalination technology in close cooperation, which conducts modeling and begins to develop a hybrid plant for desalination of water based on the fundamental theoretical basis of the way of operation of a new technological cycle, represented by the collective in the form of a mathematical model. The partners are jointly developing a draft heat recovery and recovery scheme using the Rankine cycle, Kalina cycle and solar collectors in combination with a groundwater desalination system. Cooperation of partners is carried out in terms of registration of intellectual property, coverage of the results of the project in leading peer-reviewed international journals of the corresponding profile.

5. Conclusion

The advantage of the introduction of waste heat boilers is a significant decrease in the temperature of flue gases and, accordingly, improved filtration in the gas cleaning plant, which will lead to a decrease in the emission of fine dust into the atmosphere by 99.9%. Waste heat boilers are classified according to the following criteria. These features include the type of heat exchange in the boiler; boiler design; layout and design features of the gas-tube boiler; the presence of circulation of feed and boiler water; arrangement of the execution of gas ducts. In the near future, the installation of a new closed-type ferroalloy furnace will take place, and under which a project for the introduction of a smoke-fired waste heat boiler will be considered. The concept of the project was previously tested in the Matlab Simulink software package. The proposed control scheme in a simplified form was pre-tested with hardware. Both tests showed that it is impossible to increase the efficiency of a power plant without applying the exergy method of economic analysis. However, in order to prove the concept and further increase the efficiency, it is necessary to conduct more in-depth research and accurate modeling in combination with an increase in the accuracy and reliability of calculations with practical testing of the software to be developed in the control system of a real power plant. Over time, authors generate proof of concept proofs, fabricate individual scaled technology elements, and conduct laboratory tests of basic technology elements. In addition, scientists retain the planned final of fundamental research – the use of the developed foundations of the exergo-energetic approach to the exergo-economic method of thermodynamic analysis of low-temperature power-engineering installations. Authors present results of life cycle analysis and economic analysis of desalination system transients and optimal management strategies. At the end of the project, the partners begin to commercialize the technology in countries with the creation of prerequisites for a significant economic effect.

References

[1] Kaczmarczyk T, Zywica G and Ihnatowicz E 2017 The impact of changes in the geometry of a radial microturbine stage on the efficiency of the micro CHP plant based on ORC *Energy* **137** 530-543

[2] Frigo S, Gabbrielli R, Puccini M, Seggiani M and Vitolo S 2014 Small-scale wood-fuelled CHP plants: a comparative evaluation of the available technologies *Chem. Eng. Trans.* **37** 847-852

[3] Klonowicz P, Witanowski L, Suchocki T and Lampart P 2017 A turbine based domestic micro ORC system *Energy Procedia* **129** 930-993

[4] Wang M, Jing R, Zhang H, Li N and Zhao Y 2018 An innovative Organic Rankine Cycle (ORC) based Ocean Thermal Energy Conversion (OTEC) system with performance simulation and multi-objective optimization *Appl. Thermal Engineering* **145** 743-754

[5] Koukou M, Vrachopoulos M, Tachos N and Tsolakoglou N 2018 Design and development of a 10-kW ORC installation working with low-temperature sources *Int. J. of Sustainable Energy* **37(9)** 857-872
[6] Gotelip Correa Veloso T, Sotomonte C, Coronado C and Nascimento M 2018 Multi-objective optimization and exergetic analysis of a low-grade waste heat recovery ORC application on a Brazilian FPSO *Energy Conversion and Management* **174** 537-551
[7] Behzadi A, Gholamian E, Houshfar E and Habibollahzade A 2018 Multi-objective optimization and exergoeconomic analysis of waste heat recovery from Tehran's waste-to-energy plant integrated with an ORC unit *Energy* **160** 1055-1068
[8] Majumdar R, Singh S and Saha S 2018 Quasi-steady state moving boundary reduced order model of two-phase flow for ORC refrigerant in solar-thermal heat exchanger *Renewable Energy* **126** 830-843
[9] Rosset K, Mounier V, Guenat E and Schiffmann J 2018 Multi-objective optimization of turbo-ORC systems for waste heat recovery on passenger car engines *Energy* **159** 751-765
[10] Talluri L, Fiaschi D, Neri G and Ciappi L 2018 Design and optimization of a Tesla turbine for ORC applications *Appl. Energy* **226** 300-319
[11] Kaczmarczyk T, Zywica G and Inhatowicz E 2018 Experimental research on the domestic ORC micro power plant with a commercial biomass boiler *Energy and Environmental Protection* **46** 1-8
[12] Babatunde A and Sunday O 2018 A review of working fluids for Organic Rankine Cycle (ORC) applications *Mater. Sci. Eng.* **413** 138
[13] Pantaleo A, Fordham J, Oyewunmi O, De Palma P and Markides C 2018 Integrating cogeneration and intermittent waste-heat recovery in food processing: microturbines vs. ORC systems in the coffee roasting industry *Appl. Energy* **225** 782-796
[14] Toropov E V, Osintsev K V and Alyukov S V 2018 Analysis of the calculated and experimental dependencies of the combustion of coal dust on the basis of a new methodological base of theoretical studies of heat exchange processes *Int. J. of Heat and Technology* **36(4)** 1240-1248
[15] Alabugin A A, Alyukov S V and Osintsev K V 2020 Approximation Methods for Analysis and Formation of Mechanisms for Regulating Heat and Mass Transfer Processes in Heat Equipment Systems *Int. J. of Heat and Technology* **38(1)** 45-58
[16] Connor S 2018 Total cost of ownership of a high-pressure steam boiler: Six factors to consider before making this long-term investment *HPAC Heating, Piping, Air Conditioning Engineering* **90(6)** 13-18
[17] Danilov O L, Garyaev A B, Yakovlev I V et al 2011 *Energy Saving in Heat Power Engineering and Heat Technologies: studies. for high schools* (ed A V Klimenko) (Moscow: MPEI Publishing house) p 424
[18] Sadhu S 2011 *Handbook of Mechanical Engineering* (S Chand & Company Ltd) p 2080
[19] Emami M, Shahbazian H and Sunden B 2019 Effect of operational parameters on combustion and emissions in an industrial gas turbine combustor *J. of Energy Resources Technology, Transactions of the ASME* **141(1)** 012202 1-5
[20] Villagran-Villegas L, Hernandez-Gomez L, Patiño-Ortiz M, Cuellar-Orozco M and Toledo-Velazquez M 2019 Analysis of the wear damage on offshore gas turbine blades *Advanced Structured Materials* **92** 221-236
[21] Saha A, Saikia L, Rajbongshi R, Saha D and Tasnin W 2019 AGC of thermal-split shaft gas turbine system integrating IPFC and ultra-capacitor *Advances in Intelligent Systems and Computing* **757** 105-115

Combined thermochemical treatment of steel construction parts to increase their corrosion resistance

L G Petrova, P E Demin and A S Sergeeva

Dept. Technology of Structural Materials, Moscow Automobile and Road Construction State Technical University (MADI), 64, Leningradskiy ave., Moscow, 125319, Russia

E-mail: petrova_madi@mail.ru

Abstract. Electrochemical corrosion of construction parts made of structural steels appears due to different electrochemical potentials (ECP) of contacting materials. The values of ECP of carbon steels may be more or less compensated by forming of alloyed layers or coatings by methods of thermochemical treatment (TCT). The following combinations of TCT were examined for surface modification of steels: (a) application of zinc coatings by a method of cold zinc-plating with subsequent classical gas nitriding in ammonia; (b) joint diffusion saturation by zinc and nitrogen; (c) joint diffusion saturation with chromium and nickel. Metallographic and microhardness methods were used for the study of modified layers and coatings. Strengthened layers with a thickness up to 140…150 microns were formed in low-carbon steels after nitriding of zinc coatings at 540°C for 24 hours, and after combined TCT with multicomponent saturation by Cr+Ni at 1000°C for 7 hours. Both methods of TCT increase the corrosion resistance of low-carbon steels and low-alloyed steels. Corrosion products examination in 5% NaCl salt fog shows that the dissolution of nitrided Zn coating and formation of corrosion products slow down with the increase of zinc concentration in the coating.

1. Introduction

Elements of metal constructions such as steel bridges carcass, parts of movement joints, fixings (clinchers, bolts or bolt-rivets) work in the corrosive environment of atmosphere, precipitations, humidity, and condensed moisture that leads to the development of corrosion processes. Corrosion of bridgework parts is the main reason for their lifetime decrease: it reduces mechanical properties of steels, causes cracks in concrete, and its exfoliation from a metal carcass.

In many cases, the main reason for steel parts corrosion consists in the difference of electrochemical potentials of contacting metals in a couple "bolt – girder". This effect is a consequence of steels' different chemical composition. As a rule, fixing elements are made of low-carbon steels (0.1…0.2%C), whereas constructive elements of bridges are made of alloyed steels containing chromium, silicon, nickel, and copper [1].

Two ways are applicable for corrosion resistance increase of carbon steels working in contact with alloyed steels: deposition of protective coatings or forming of alloyed layers with electrochemical potentials close to ECP of contacting steel.

Zinc-plating is one of the most popular techniques for fixing elements protection against corrosion [2, 3]. Zinc coatings provide two mechanisms of corrosion protection: cathodic (electrochemical)

mechanism and barrier mechanism similar to the protection mechanism of painted films [4, 5]. The application of protective coatings leads to the formation of a sharp boundary with a steel substrate. Low adhesion of coatings decreases the resistance of bolt-rivets to mechanical stresses during its mounting. Usage of such coatings for fixing elements may cause their exfoliation during gripping or press-fitting.

Techniques of thermochemical treatment (TCT) using the principles of diffusion saturation form modified layers in steels with gradual structural changes give smooth microhardness depth profiles [6]. To a certain extent, the microhardness gradient represents the residual stress differential through the layer. Therefore, the smooth microhardness profile of a layer is a factor of its good adhesion. It was shown [7] that using TCT for steels with coatings gives the opportunity of their adhesion strength increase due to the transition diffusion layer forming at the interface of a coating with steel substrate.

Diffusion methods of surface alloying by metals are alternative to the coatings' application. TCT techniques allow providing joint saturation of steels by several metals or by metals in combination with non-metal elements, such as nitrogen [8].

The objective of this paper consists in the study of processes of thermochemical treatment for forming the strengthened corrosion-resistant coatings and layers in steels traditional for bridge metal constructions.

2. Experimental Methods

For the experimental study, specimens from three types of steel were prepared (table 1): 10x10x1 mm plates for metallographic examinations, 20x40x5 mm plates for electrochemical tests, and cylinders with the diameter of 4 mm for corrosion tests in a salt fog. All specimens were annealed at 850ºC, mechanically grinded, and the surface was degreased.

Table 1. Chemical compositions of investigated steels

Ref. No.	%C	%Mn	%Si	%Cr	%Ni	%Cu	S_{max}	P_{max}
A	0.14–0.22	0.4–0.65	0.15–0.3	<0.3	< 0.3	<0.3	<0.05	<0.4
B	0.07–0.14	0.35–0.65	0.17–0.37	<0.15	<0.25	<0.25	<0.04	<0.035
C	0.12–0.18	0.40–0.70	0.40–0.70	0.60–0.90	0.30–0.60	0.20–0.40	<0.035	<0.035

Three experimental TCT processes were studied:

Nitriding of zinc coatings. Zinc coatings were applied to the steel substrate by painting with a primer containing finely dispersed Zn-powder and film-forming substance: organic binder (modified polyurethane), or silicate binder (potassium water glass with the density 1.2 g/sm^3). Zinc concentration varied up to 85% in organic-based primer and up to 75% in silicate-based primer. The coatings were dried for 48 hours. Nitriding of specimens with Zn coatings was carried out in ammonia atmosphere at a temperature of 540ºC for 24 hours.

Joint diffusion saturation by zinc and nitrogen. Specimens were covered by slip suspension containing metal zinc powder (25%), quartz sand (70%) as a filler, ammonium chloride NH_4Cl (5%) as a transportation agent, and PVA-glue as a binder. Diffusion saturation was carried out in ammonia at temperatures of 600…1000°C during 1…4 hours.

Multicomponent diffusion surface alloying. Specimens were covered by slip suspension containing 50% of alloying metal powder (chromium, nickel, or their mixtures), 50% of ammonium chloride NH_4Cl, and zapon-varnish as a binder. Diffusion saturation was carried out in the induction furnace blown by CO_2 protective gas at temperatures of 900°C and 1000°C during 1…7 hours.

Metallographic examinations of coatings and diffusion layers were made in the AXIOVERT 25CA microscope. Microhardness tests were carried out by the PMT-3 measurer.

Corrosion tests of specimens were conducted in a salt fog chamber Ascott S120iP; the 5%-solution of sodium chloride (NaCl with pH=6.5) was sprayed inside the chamber by compressed air. The testing temperature was 35ºC, and humidity was 95…100%. Specimens were placed in the chamber below the spraying level that excluded the direct action of salt solution. The salt fog was condensed at specimens' surface with the constant rate; testing time was from 24 to 240 hours. The results of the tests were estimated visually by examination of corrosion products.

Electrochemical corrosion tests were made consisting of tracing the anodic and potentiodynamic (PD) curves in the direct passage and the reverse passage. The solution of sulfuric acid (0.5M H_2SO_4) was used as a testing environment. The following parameters of the corrosion process were determined: electrochemical potentials of the start and the end of passivation (E_p^s and E_p^e), and the passivating interval (E_p^e- E_p^s). Parameters for evaluation of local corrosion protection properties of coatings were measured: the stationary potential of free corrosion E_c, i.e. the potential of the beginning of the anodic dissolution of metal and maximum value of anodic current density i_{max}. The higher the negative value of E_c is and the lower the value of i_{max} is, the better the corrosion resistance of a tested metal is.

3. Results and discussion

3.1.Consequent Process: Nitriding of Zinc Coatings

Visible coating 50 microns thick with maximum microhardness is formed in low-carbon steel after nitriding of previously applied zinc coating (figure 1). Gas nitriding of specimens of both carbon steel and low-alloyed steel with zinc coatings forms modified layers with gradually decreasing microhardness profile. The diffusion sub-layer of about 100 microns thickness was identified as the internal nitriding zone containing elevated concentrations of zinc and nitrogen in ferrite. Increased microhardness of this zone compared to the substrate's microhardness is the consequence of solid solution strengthening.

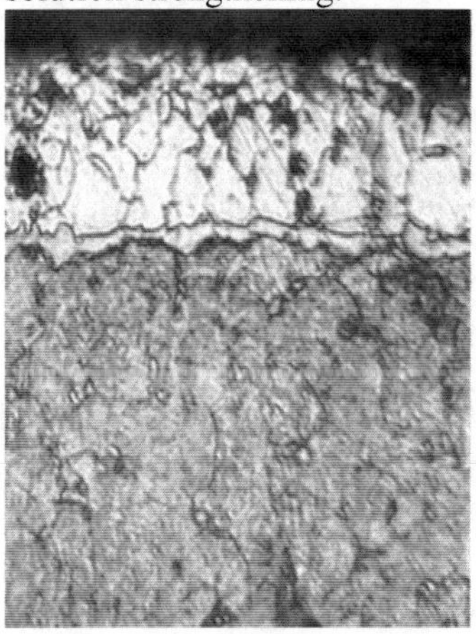

(a)

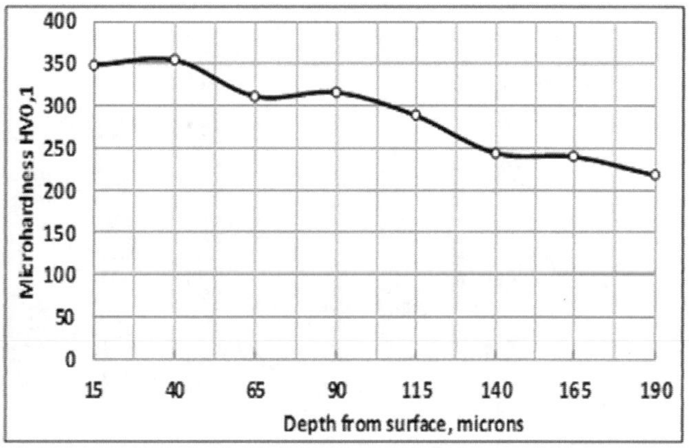

(b)

Figure 1. Micrograph: x500 (a) and microhardness profile (b) of nitrided steel "A" with Zn coating.

Both alloyed steel and carbon steel with nitrided zinc coatings have increased corrosion resistance in the low-salt environment. It was experimentally shown that the corrosion velocity of nitrided steel "A" with a zinc coating (>80%Zn in the primer) in the 5%-NaCl solution is about 0.25 mm per year that is several times less compared to the annual corrosion rate of untreated steel.

The corrosion resistance of nitrided zinc coatings depends on the initial thickness of a coating, on zinc concentration in the primer, and the type of a binder. Nitrided coatings with different zinc concentrations demonstrate different progress of the corrosion process during the tests in the salt fog chamber. Dissolving the coating with 45%Zn was observed after about 100 hours of tests. After 240 hours of tests, the square of dissolved coating reached 10% of the surface: centers of "red corrosion" of iron appeared (figure 2). The coating with 65%Zn remained intact after 240 hours of tests; in this case, the products of the so-called "white corrosion" of Zn plays as the additional protector.

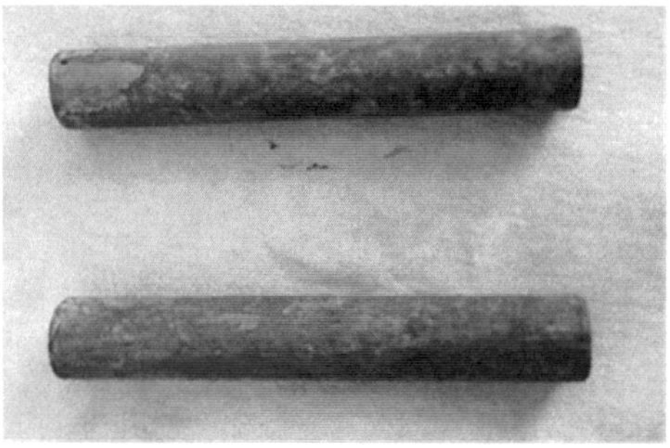

Figure 2. View of nitrided specimens of steel "A" with coatings containing 45%Zn (a), and 65%Zn (b) after 240 hours of tests in the salt fog chamber.

Nitriding increases the resistance of zinc coatings against electrochemical corrosion in 0.5M H_2SO_4 solution (figure 3). Increase of the E_c negative value and decrease of the i_{max} value are observed compared to the parameters for un-nitrided organic-based coating. Minimum values of i_{max} are measured for nitrided double-layer zinc coating. For nitrided silicate-based zinc coating the value of i_{max} decreases 4 times, whereas some reduction of E_c parameter is observed. Corrosion parameters of low-alloyed steel "C" after combined treatment have similar values for both types of film-former. After the most effective combination of cold zinc-plating with nitriding, the 25-time decrease of the i_{max} parameter is achieved compared to the value for the as-received specimen without treatment. The optimum process provides the decrease of the current density value i_{max} more than 6 times compared to this parameter for the steel after hot zinc-plating.

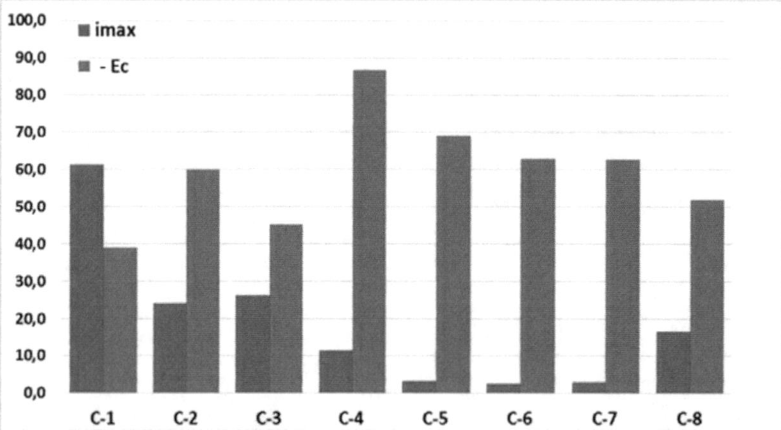

Figure 3. Corrosion parameters E_c (x10^{-1}, mV) and i_{max} (mA/sm^2) measured from PD-curves for specimens of steel "C": C-1 – as-received; C-2 – nitriding 540ºC/24 h; C-3 – single-layer organic-based Zn coating; C-4 – single-layer silicate-based Zn coating; C-5 – single-layer organic-based Zn coating and nitriding 540ºC/24 h; C-6 – double-layer organic-based Zn coating and nitriding 540ºC/24 h; C-7 – single-layer silicate-based Zn coating and nitriding 540ºC/24 h; C-8 – polyurethane-based hot-plated Zn coating.

3.2. Joint Diffusion Saturation with Zinc and Nitrogen

The processes of joint saturation of steels with zinc and nitrogen form modified layers with diffusion zones. Thus, after saturation during 4 hours at 600^0C, a modified layer was received in low-alloyed steel "C" with a thickness of about 140 microns. The layer consists of the external zone (40 μm) enriched by zinc at the surface up to 86%Zn, of the intermediate zone of nitrides (30…40 μm), and of the internal nitriding zone – solid solution of Zn and N in ferrite with nitrides precipitations (figure 4). Such structural changes in the modified layer provide the flowing microhardness depth profile with a slight maximum corresponding to the nitrides interlayer. Achieving strengthening is higher than the hardening of nitrided zinc coatings (figure 1).

The increase of the diffusion process' temperature leads to some significant changes in the structure of modified layers. After the treatment at 1000°C the surface layer of 50 microns is formed without any transition zone (figure 5). The absence of the transition zone explains sharp microhardness decline under the layer. The increased volume fraction of pearlite is observed at the boundary with the layer; this may be caused by carbon migration out of the layer into the steel core. High temperatures of saturation activate zinc diffusion; this decreases the diffusion ability of nitrogen and reduces the depth of its penetration.

Electrochemical tests show that steel "C" after high temperature diffusion saturation by zinc and nitrogen is not passivized in the H_2SO_4 solution (figure 6).

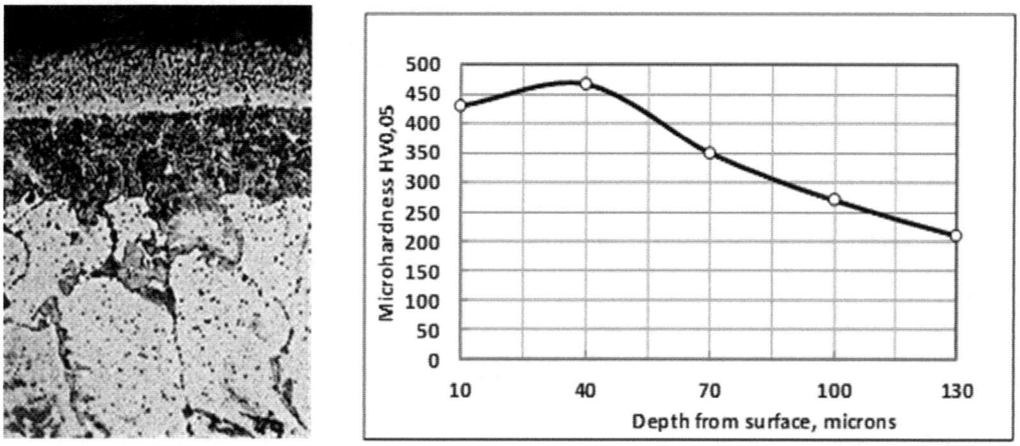

(a) (b)

Figure 4. Micrograph: x500 (a), and microharndess profile (b) of steel "C" after joint diffusion saturation with zinc and nitriding (600ºC, 4 h).

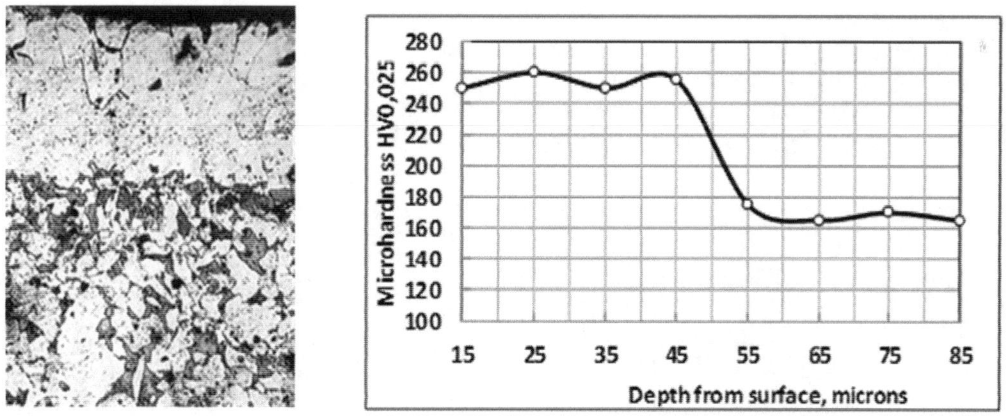

(a) (b)

Figure 5. Micrograph, x500 (a), and microhardness profile (b) of steel "C" after Zn+N diffusion saturation (1000ºC, 1 h).

3.3. Multicomponent Diffusion Surface Alloying

For the forming of coatings on carbon steels with protective properties comparable to the corrosion resistance of alloyed steels, the processes of diffusion saturation by chromium and nickel were applied. Taking into account the presence of a nitrogen-containing component (ammonium chloride) in slip suspensions, we can suppose the parallel nitrogen diffusion into the ferrous matrix at high temperatures of such processes.

The temperature of 900ºC was determined as the lowest temperature for effective diffusion metallization. It was experimentally found the possibility of suitable diffusion layers forming at the temperature 1000ºC and the process duration up to 7 hours. Under these conditions, after joint diffusion saturation by chromium and nickel of steel "B", the modified layer has a thickness of 150 microns. Two zones are observed in the layer (figure 7): surface compound zone (70…80 μm) with maximum strengthening due to the high volume fraction of uniaxial nitride particles and the internal sub-layer of austenite containing lamellar precipitations typical of ferrous nitrides (γ'-phase). Microhardness increase is the consequence of joint strengthening mechanisms: solid solution hardening and dispersion strengthening by nitrides. Microhardness falls gradually corresponding to the reduction of saturating elements concentrations with the depth.

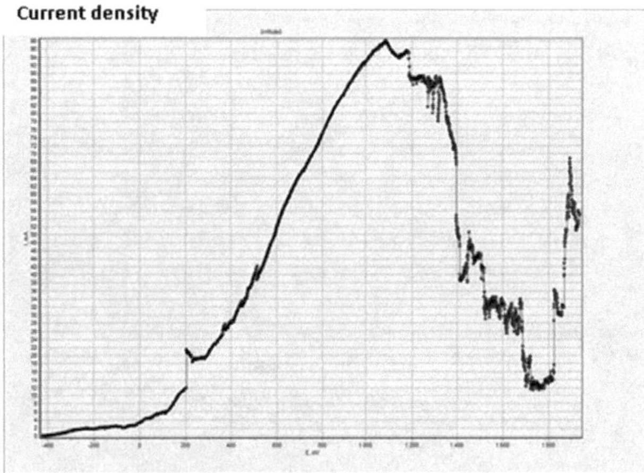

Figure 6. PD-curve of steel "C" after Zn+N diffusion saturation in ammonia (900ºC/4 h) tested in 0.5M solution of H_2SO_4.

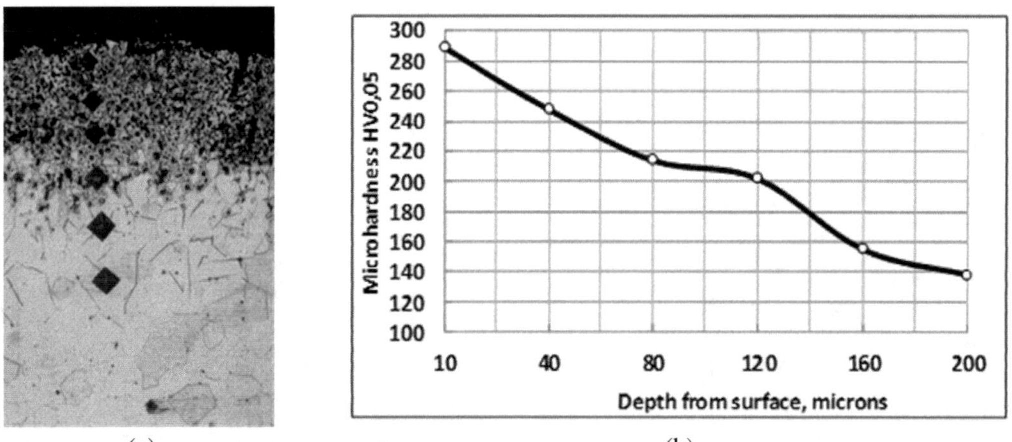

(a) (b)

Figure 7. Micrograph: x500 (a), and microhardness profile (b) of steel "B" after Cr+Ni diffusion saturation (1000ºC, 7h).

Diffusion metallization by chromium and nickel expands the passivating interval of low-carbon steel "B" to the level of this parameter for the alloyed steel "C" (table 2). A comparison of electrochemical parameters shows that corrosion resistance of the carbon steel after diffusion Cr+Ni saturation is higher than corrosion resistance of the alloyed steel as after nitriding and after combined saturation by zinc and nitrogen.

Table 2. Electrochemical parameters of investigated steels in 0.5M solution of H_2SO_4

No.	Treatment	E_p^s, mV	E_p^e, mV	E_p^e- E_p^s, mV
C-1	as-received	529	1635	1106
C-2	Nitriding: 540ºC/24 h, NH_3	432	486	54
C-9	Zn+N-saturation: 900ºC/4 h, NH_3	774	963	189
B-1	Cr+Ni-saturation: 900ºC/6 h, NH_4Cl	491	1571	1080

4. Conclusion

Forming the strengthened modified layers in low-carbon steels and low-alloyed steels is experimentally determined as the result of combined TCT processes: nitriding of steels with zinc coatings, joint diffusion saturation by zinc and nitrogen, and joint diffusion saturation by chromium and nickel.

An alloyed Cr+Ni layer in carbon steel shows better corrosion resistance in the 0.5M H_2SO_4 solution compared to the corrosion resistance of steel after classical nitriding and joint diffusion saturation by zinc and nitrogen. Electrochemical parameters of surface alloyed carbon steel are close to the parameters for low-alloyed steel.

The corrosion resistance of nitrided zinc coatings increases in a salt solution (NaCl), in salt fog, and slightly acidic environment due to the synergy of cathodic and waterproofing protection mechanisms. Full-scale tests of fixing parts in metal construction of the overpass have shown that nitrided bolts keep coatings integrity longer than the components with standard hot-plated coatings.

Acknowledgments
The material was prepared within the framework of scientific research under project no. FSFM-2020-0011 (2019-1342); experimental studies were carried out using the equipment of the MADI Centre of collective usage.

References

[1] Morozov Y D, Pemov I F, Matrosov M Y and Zin'ko B F 2020 Steels for bridge structures *Metallurgist* **9-10(63)** 933-950
[2] Petrova L G, Demin P E, Kosachev A V and Barabanov S I 2017 Application of diffusion techniques for formation of zinc coatings to improve corrosion resistance of structural steels *Polymer Science. Series D.* **2(10)** 179-184
[3] Biryukov I, Batmanova V, Galin R G et al 2019 The effect of the chemical composition of intermetallic phases on the corrosion of thermal diffusion zinc coatings *Surface and Coatings Technology* **372** 166-172
[4] Tseluikin V N and Koreshkova A A 2016 Corrosion resistance of composite coatings based on zinc *Chemical and Petroleum Engineering* **7-8(52)** 560-562
[5] Marchebois H, Touzain S, Joiret S et al 2002 Zinc-rich powder coatings corrosion in sea water: influence of conductive pigments *Progress in Organic Coatings* **4(45)** 415-421
[6] Belashova I S and Shashkov A O 2012 Kinetics of growth of the diffusion layer in nitriding by the thermogasocyclic method *Metal Sci. and Heat Treatment* **5-6(54)** 315-319
[7] Petrova L G, Demin P E, Timofeeva G Y et al 2019 Influence of nitriding on the wear resistance and adhesion of zinc coatings on steel *Russian Engineering Research* **8(39)** 696-698
[8] Aleksandrov V A 2010 Combined TCT process for surface hardening of steels *Metal Sci. and Heat Treatment* **5-6(52)** 214-218

Calculation of sealing pressures of shut-off valves

V K Pogodin[1], Yu I Belogolov[1], V E Gozbenko[1,2], S K Kargapoltsev[1], V A Olentsevich[1], A M Gladkih[3]

[1] Irkutsk State Transport University, 15, Chernyshevskogo str., Irkutsk, 664074, Russia
[2] Angarsk State Technical University, 60, Chaykovskogo str., Angarsk, 665835, Russia
[3] Irkutsk National Research Technical University, 83, Lermontova str., Irkutsk, 664074, Russia

E-mail: pogodin_vk@irgups.ru

Abstract. The proposed algorithm for calculating the sealing pressures of the shut-off valve allows one to take into account the structural features of its components and their elements that affect the amount of sealing pressures. These parts are divided into the mounting system and sealing system. Analytical dependencies for determining the forces acting in these systems allow us to determine the force values when tightening fasteners, during pressure buildup and release, and also when exposed to temperatures. When calculating the specific structure of the shut-off valve, it is shown that the forces in its sealing system and mounting system exceed the values determined by known methods of calculation, respectively, 1.4 and 1.2 times. The magnitude of these discrepancies for different structures of shut-off valves may differ from those indicated and depends on the number of components included in the stop valve station. In this regard, the proposed calculation method allows determining not only the forces in the shut-off valve, but also the optimal number of components and connections that will ensure safe operating conditions. It is shown that the method of calculating forces in the stop valve station used in practice is a special case in which the stiffness coefficient of the stop valve station $\alpha = 1$. The proposed calculation method makes it possible to more accurately determine the forces acting in the shut-off valve, the sealing conditions and the stress-strain state of the components. The introduction of the developed calculation method is connected with the development of a regulatory document on its basis to replace the existing methods of calculating the stop valve station in shut-off valves.

1. Introduction

The creation of optimal designs of shut-off valves involves the use of design methods that ensure their safe operation [1, 2]. As such methods, methods are used to calculate the strength of its components, as well as forces to seal the valve (seat – valve gate) and sealing gland [3-5]. The history of the design and study of shut-off valves showed that the determination of the sealing forces of its valve gate was mainly determined by calculation using simplified methods [6-8], developed on the basis of analytical or experimental studies of specific valve gate designs. In this case, the most relevant one was always the study of sealing conditions that determined the values of the forces excluding the emergence of leakage in the valve gate that is excess of standard values. Sealing conditions were mainly determined in valve gate models, since their determination in real shut-off valve designs presents a challenge associated with modeling their operating conditions [9-11].

The analysis of the calculation methods currently used in the design of shut-off valves showed that they do not take into account the conditions for the interaction of the components with each other, as well as the compliance of the components and their elements [12-13].

It can be assumed that for this reason, in the shut-off valves designed according to well-known methods, the values of forces required to seal the valve gate when applied under operating conditions often do not coincide with the calculated ones. To eliminate such discrepancies, the article proposes to consider a methodology for determining the sealing forces in a shut-off valve that allows one to take into account its design features and deformations of all parts and their elements.

To more fully take into account the design features and flexibility of the components and their shut-off valve elements (Figure 1), when determining the sealing forces, the introduction of a new concept "stop valve station (SVS)" was required. This, in addition to the valve gate components (seat-valve gate), includes the components and their elements that influence its sealing ability at different stages of the interaction of these components.

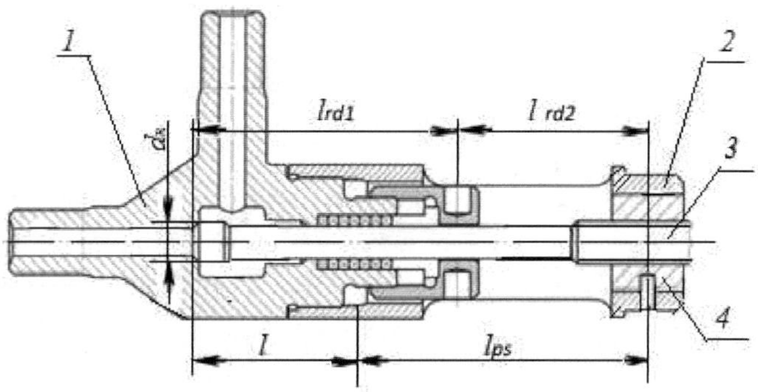

Figure 1. The shut-off valve SVS diagram for high operating parameters: 1 – valve housing; 2 – post; 3 – spindle; 4 – threaded sleeve.

From the diagram (Figure 1) it can be seen that the shut-off valve SVS can be represented as an option of a detachable joint (DJ) [9, 14] with a positive seal, the design of which is shown in Figure 2 *a*.

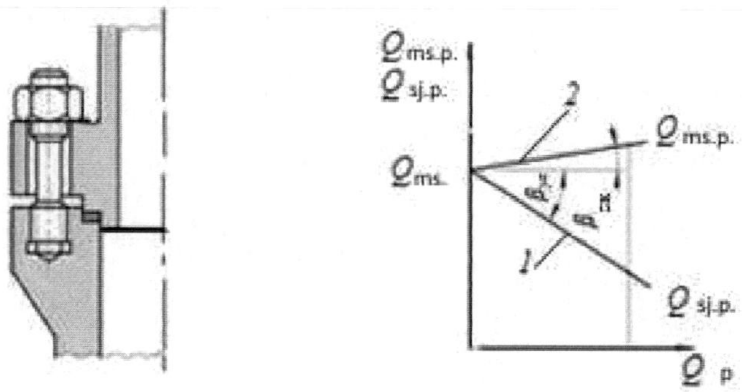

(a) (b)

Figure 2. A DJ with positive seal: a – design; b – computational force diagram: 1– change in the forces on the sealing ring; 2 – change in the forces on studs.

It was established in [9, 13, 14] that the determination of forces in the DJ designs without taking into account the deformations of components and their elements leads to significant errors that can be avoided with the conditions of their sealing system and mounting system deformation compatibility.

The SVS sealing system includes components and elements, the preliminary compression and bending deformations of which decrease with increasing pressure of the medium. This system should include components and their elements that are directly involved in ensuring the preliminary loading of the valve gate sealing joint (SJ). In the design of the shut-off valve (Figure 1), this system includes a spindle, seat, housing, and also a SJ in the contact zone of their sealing surfaces.

The SVS mounting system includes components and their elements that prevent the unloading of the SJ in the SVS when the valve is loaded with pressure of the medium. In the design of the shut-off valve, a diagram of which is shown in Fig. 1, this system includes: a post, a threaded sleeve, as well as threaded joints "spindle-threaded sleeve", "threaded sleeve-post", "post-housing".

2. Theoretical research

The relationship between the forces perceived by the mounting system Q_{ms} and the sealing system Q_{sj} in the works [3, 15] is presented in the form:

- in the tightened state

$$Q_{ms.t} = Q_p + Q_f \pm T ; \tag{1}$$

- under pressure

$$Q_{ms.p} = Q_{sj.p} + Q_f \pm T , \tag{2}$$

where $Q_p = \frac{\pi d_c^2}{4} p_w$ is the axial force from the action of working pressure p_w on the rod with the diameter of its contact with the housing seat d_c; $Q_f = \pi d_c^2 \cdot B \cdot q_1$ is the axial force required to seal the valve gate in the closed position of the shut-off valve with a contact diameter of d_c, a width of B of the contact “valve-seat” and contact pressure q_1; T is the axial force to overcome friction in the sealing gland during the rod movement taking into account the pressure of the medium p_m (+ when closing, – when opening the valve).

In DJ and SVS, during the transition from the tightened state to the working one under pressure, the sealing system compensates for the deformation of the mounting system due to its exposure to the medium, while maintaining the sufficient contact pressure in the SJ valve gate to ensure the specified valve tightness. That is, the condition for the compatibility of deformations must be observed [16]:

$\delta_{i\,ms} = \sum \delta_{i\,sj}$, where $\delta_{i\,ms}$, $\delta_{i\,sj}$ are the movement of the SVS components included in the mounting system and the sealing system, respectively.

Taking into account that $\delta_i = \lambda_i Q$, where λ_i is the unit axial compliance of the component, its element or contact zone of components, the relationship between the total deformations of the sealing system and the mounting system can be presented by expressions:

$\sum \delta_{i\,ms} = \sum \lambda_{i\,ms.p} \cdot Q_{ms.p} - \sum \lambda_{i\,ms.t} \cdot Q_{ms.t}$; $\sum \delta_{i\,sj.p} = \sum \lambda_{i\,sj.p} \cdot Q_{sj.p} - \sum \lambda_{i\,sj.t} \cdot Q_{sj.t}$.

For the shut-off valve SVS with the forced nature of ensuring tightness (Figure 1), providing for the deformations of its components and contact surfaces without taking into account the action of friction forces in the sealing gland, the relationship between the forces can be represented in a general form as dependencies:

$$Q_{sj.p} = \psi \cdot Q_{ms.t} - \alpha \cdot Q_d \tag{3}$$

$$Q_{ms.p} = \psi \cdot Q_{ms.t} + (1-\alpha) \cdot Q_d \tag{4}$$

where ψ is the transformation coefficient of force action:

$$\psi = \frac{\lambda_{ht} + \lambda_{sph} + \lambda_{sj.t} + \lambda_{pst} + \lambda_{slt} + \lambda_{ts.st} + \lambda_{ts.pt} + \lambda_{tp.ht}}{\lambda_{ht} + \lambda_{spp} + \lambda_{sj.p} + \lambda_{psp} + \lambda_{slp} + \lambda_{ts.sp} + \lambda_{slp.p} + \lambda_{tp.hp}} \tag{5}$$

α is the SVS stiffness coefficient:

$$\psi = \frac{\lambda_{psp} + \lambda_{slp} + \lambda_{ts.sp} + \lambda_{ts.pp} + \lambda_{tp.hp}}{\lambda_{spp} + \lambda_{hp} + \lambda_{sj.p} + \lambda_{psp} + \lambda_{slp} + \lambda_{ts.sp} + \lambda_{ts.pp} + \lambda_{tp.hp}} \quad (6)$$

λ_h, λ_{sp}, λ_{ps}, λ_{sl}, λ_{sj}, $\lambda_{ts.sl}$, $\lambda_{ts.ps}$, λ_{tpsh} are axial compliances, respectively, of the components of: housing, spindle, post, sleeve; of the sealing joint: "spindle–housing", as well as of threaded joints: "spindle–sleeve", "sleeve–post", "post–housing" during tightening (index "t") and pressure (index "p").

If the compliance values of components and joints are equal under the action of a tightening force and pressure of the medium $\psi = 1$, expressions (5) and (6) can be represented as:

$$Q_{sj.p} = Q_{ms.t} - \alpha \cdot Q_d \quad (7)$$

$$Q_{ms.p} = Q_{ms.t} - (1-\alpha) \cdot Q_d . \quad (8)$$

Graphically, equations (7), (8) are presented in Figure 2*b*, where line 1 reflects the change in force $Q_{sj.p}$, and line 2 reflects changes in force $Q_{ms.p}$ with an increase in the medium pressure in the SVS. Moreover, $\alpha = tg\ \beta_{sj}$; $tg\ (1-\alpha) = \beta_{ms}$.

With the calculation of the additional loading that acts on the component of shut-off valve with the presence of thermal effects, one can also use the regulations accepted with the determination of thermal load in the SJ [17]. In this case, the initial state after the valve gate tightening is accepted, in which the values of the linear expansion coefficients of the component materials α, temperature t and size l are respectively equal to: $\alpha_t = \alpha_0$; $t = 0$; $l_i = l_0$.

With these conditions, the additional thermal load, when the valve gate is closed and the components of the SG are heated or cooled for the design of the shut-off valve (Figure 1), can be determined by the formula:

$$Q_t = \frac{\sum_{i=1}^{n}(a_{ti}, t_i, l_i)}{(\lambda_{spt} + \lambda_{sj.t} + \lambda_{pst} + \lambda_{slt} + \lambda_{slp} + \lambda_{ts.st} + \lambda_{ts.pt} + \lambda_{tp.ht}} \quad (9)$$

where $\sum_{i=1}^{n}(a_{ti},\ t_i,\ l_i) = \alpha_{tps} \cdot t_{ps} \cdot l_{ps} + \alpha_{th} \cdot t_h \cdot (l_h - 0.5 d_h \cdot tg\gamma) - \alpha_{tsp1} \cdot t_{sp1} \cdot l_{sp1} - \alpha_{tsp2} \cdot t_{sp2} \cdot l_{sp2}$.

In this equation, the designated geometric parameters of the SVS components correspond to the designations indicated in Figure 1. The temperatures of the shut-off valve components with an internal heating source and the absence of thermal insulation can be taken equal to [14]: $t_{sp1} = t_{med}$; $t_{sp2} = 0.85\ t_{med}$; $t_{ps} = 0.8\ t_{med}$; $t_h = t_{med}$.

In the presence of thermal insulation, the temperatures in the parts are taken equal to t_{med}. The value of the load Q_t, if necessary, should be introduced into formulas (7), (8) with its own sign.

Using the obtained dependences, one can determine the specific contact forces q_{Np} per unit length of the seal perimeter, normal to the sealing surface at various medium pressure values. When a valve gate with conical sealing surfaces is used in the SJ, contact forces arise [13], the action of which is shown in Figure 3.

When the rod is displaced under the action of increasing pressure, the direction of the friction forces in the contact zone changes to the opposite compared to that which was during tightening and coincides with it with decreasing pressure [13].

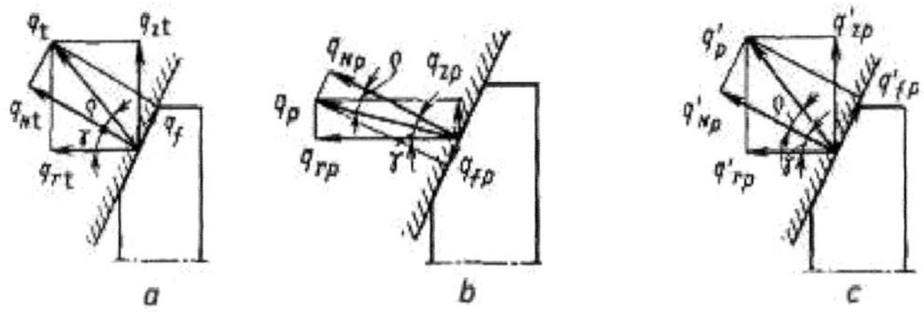

Figure 3. Diagrams of forces acting on conical sealing surfaces: a – during the tightening of fasteners; b – with increasing pressure of the medium; c – with a decrease in medium pressure.

The main relations between the components of the contact force q, acting per unit circumference of the contact D_c, for each of the diagrams shown in Figure 3:

$$q_{zt} = q_{Nt}\frac{\sin(\gamma+\rho)}{\cos\rho};\ q_{zp} = q_{Np}\frac{\sin(\gamma+\rho)}{\cos\rho};\ q'_{zp} = q'_{Np}\frac{\sin(\gamma+\rho)}{\cos\rho};$$

$$q_{rt} = q_{Nt}\frac{\cos(\gamma+\rho)}{\cos\rho};\ q_{rp} = q_{Np}\frac{\cos(\gamma+\rho)}{\cos\rho};\ q'_{rp} = q'_{Np}\frac{\cos(\gamma+\rho)}{\cos\rho}$$

$$q_{rt} = q_{rt}\mathrm{tg}\,(\gamma+\rho);\ q_{zp} = q_{rp}\mathrm{tg}\,(\gamma+\rho);\ q'_{zp} = q'_{rp}\,\mathrm{tg}\,(\gamma+\rho).$$

When using the flat sealing joint "rod–housing seat" in the design of the shut-off valve shown in Figure 1, contact pressures can be determined from the conditions:

$$\frac{\pi\,(d_{co}^2-d_{ci}^2)}{4}\cdot q_{1p} = \frac{\pi\,(d_{co}^2-d_{ci}^2)}{4}\cdot q_{1s} + \frac{\pi d_{mid}^2}{4}p_w - \frac{\pi d_{mid}^2}{4}p_{med},$$

where d_{co}, d_{ci}, d_{mid}, are, respectively, the outer, inner and middle contact diameters of the sealing surfaces of the spindle and housing seat; q_{1p}, q_{1s} are contact pressures on the sealing surface, respectively, under pressure and sufficient to ensure tightness under operating conditions; p_w, p_{med} are working pressure and current medium pressure under the valve spindle.

Then the contact pressure on the sealing surface under the action of pressure of the medium can be determined by the formula:

$$q_{1p} = q_{1s} + \frac{(d_{co}+d_{ci})}{4\,(d_{co}-d_{ci})}\cdot\,(p_w - p_{med}).$$

Let us analyze the influence of the deformations of components, sealing and threaded joints on the value of the axial sealing force that acts in the valve gate of the shut-off valve with the high parameters of its operation. For clarity, depending on (6), we replace the compliance by movements that occur in a particular shut-off valve (Figure 1), that is:

$$\alpha = \frac{\delta_{psp}+\delta_{slp}+\delta_{ts.sp}+\delta_{ts.pp}+\delta_{tp.hp}}{\delta_{spp}+\delta_{hp}+\delta_{sj.p}+\delta_{psp}+\delta_{slp}+\delta_{ts.sp}+\delta_{ts.pp}+\delta_{tp.hp}}. \quad (10)$$

Taking into account that the cross-sections of housing, sleeve, and post exceed the section of spindle by several times, let us take the values of the axial deformations δ_{hd}, δ_{psd}, δ_{sld}, of these components to be equal to zero. Then the expression (10) can be represented as:

$$\alpha = \frac{\delta_{ts.sp}+\delta_{ts.pp}+\delta_{tp.hp}}{\delta_{spp}+\delta_{sj.p}+\delta_{ts.sp}+\delta_{ts.pp}+\delta_{tp.hp}}. \quad (11)$$

The values of axial deformations in expression (11) are determined for a shut-off valve (Figure 1) with a flat SJ in its valve gate with the following parameters. The working pressure is p_w = 150 MPa; the average diameter of the contact of the spindle with the housing seat - d_{mid} = 12 mm; width of flat contact between spindle and housing seat - v =2 mm. The spindle diameter is d_w – 14 mm; spindle length - l = 120 mm; "spindle-sleeve" joint – M16; "sleeve-post" joint – M 32x2; "post-housing" joint – M64x6.

The effect of axial deformations of components and joints on the sealing ability of the shut-off valve (Figure 1) is analyzed taking into account the above parameters and contact pressures $_1$ = 250 MPa that are necessary in accordance with [3, 4] to ensure tightness in a flat "steel-to-steel" SJ. In this case, the force values in the stop valve station, determined in accordance with the procedure [3, 4] according to formula (1) without taking into account the friction forces T in it, have the following values: $Q_f = 1884$ kgf; $Q_d = 1695$ kgf; $Q_{sj.t}69 = 3579$ kgf.

The values of the forces show that in the shut-off valve with the values of Q_f, necessary to ensure the valve gate tightness, exceed the values of Q_d from the pressure of the working medium for this shut-off valve 1.1 times and is 0.526 $Q_{sj.t}$.

3. Conclusion

1. The calculation method of the forces acting in the shut-off valve is based on the consideration of the stop valve station in its design that includes a sealing system and a mounting system.

2. Consideration of the design features of the shut-off valve is based on the determination of the compliance of components, sealing and threaded joints and their elements included in the stop valve station.

3. It is shown that the method of calculating forces in the stop valve station used in practice is a special case in which the stiffness coefficient of the stop valve station $\alpha = 1$.

4. The proposed calculation method makes it possible to more accurately determine the forces acting in the shut-off valve, the sealing conditions and the stress – strain state of the components.

5. Using the proposed calculation method, it is possible to determine the optimal combination of components and joints that will ensure safe operating parameters of the shut-off valve.

6. The introduction of the developed calculation method is connected with the development of a regulatory document on its basis to replace the existing methods of calculating the stop valve station in shut-off valves.

References

[1] Shestov V V, Antipov V V and Ryabov D K 2017 Corrosion Resistance and Mechanical Properties of Layered Structural Material Based on Aluminum Alloy and Fiberglass Thin Sheets *Metallurgist* **60(11-12)** 1191-1196

[2] Ryabov D K, Antipov K V, Fesenko T V and Dynin N V 2017 Influence of cobalt alloying on structure and long-term strength of Al–Cu–Mg–Ag alloy system *Inorganic Mater.: Appl. Res.* **8(1)** 159-165

[3] Ryabov D K, Shestov V V, Buznik V M and Antipov V V 2016 Design concept of multifunctional layered composite materials for operation in arctic regions *Inorganic Mater.: Appl. Res.* **7(4)** 547-553

[4] Spiess H and Florian M 1989 *Optimal strategies. A new assignment model for transit networks. Transportation Research Part B: Methodological, 23 B (2)* pp 83-102

[5] Yolkin K S, Yolkin D K, Kolosov A D, Ivanov N A and Shtayger M G 2018 Technologies, which allow to reduce an impact of metal silicon production on the environment *IOP Conf. Ser.: Mater. Sci. Eng.* **411** 012028

[6] Kuzmin O V and Orkina K P 2006 Building error-correcting codes using a Pascal-type triangle *Bull. of the Buryat State University* **13** 32-39

[7] Solodusha S, Suslov K and Gerasimov D 2016 Applicability of Volterra integral polynomials in the control systems of electric power facilities *Proc. of 2016 Int. Conf. "Stability and Oscillations of Nonlinear Control Systems" (Pyatnitskiy's Conf.), STAB*

[8] Suslov K V, Solonina N N and Smirnov A S 2014 Distributed power quality monitoring *Proc. of Int. Conf. on Harmonics and Quality of Power, ICHQP*

[9] Buryanina N, Korolyuk Y, Koryakina M, Lesnykh E and Suslov K 2019 Algoritm of Current Protection Based on Three Instantaneous-Value Samples *Proc. of 2019 IEEE PES Innovative Smart Grid Technologies Europe, ISGT-Europe* 8905542

[10] Abrosimov N V, Ageev A I, Adushkin V V, Akimov V A et al 2015 *The security of Russia. Legal, socio-economic and scientific-technical aspects. Scientific foundations of technological safety* (Moscow)

[11] Kopylova T A and Mikhailov A Yu 2016 Evaluation of the functioning of intermodal nodes of

urban passenger transport *Transport systems of Siberia. The development of the transport system as a catalyst for the growth of the state economy. Int. sci. and pract. conf.* pp 528-532

[12] Lebedeva O A and Kripak M N 2016 Modeling freight transportation in the transport network *Bull. of Angarsk State Technical University* **10** 182-184

[13] Kolosov A D, Gozbenko V E, Shtayger M G, Kargapoltsev S K, Balanovskiy A E, Karlina A I, Sivtsov A V and Nebogin S A 2019 Comparative evaluation of austenite grain in high-strength rail steel during welding, thermal processing and plasma surface hardening *IOP Conf. Ser.: Mater. Sci. Eng.* **560** 012185

[14] Guseva E A, Kargapoltsev S K, Balanovskiy A E, Karlina A I, Shtayger M G, Gozbenko V E, Konstantinova M V and Sivtsov A V 2019 Comparative evaluation of corrosion resistance of wheel and rail steels in various media *IOP Conf. Ser.: Mater. Sci. Eng.* **560** 012181

[15] Karlina Yu I, Kargapoltsev S K, Gozbenko V E, Karlina A I and Leonovich D S 2020 Overview of electro physicochemical methods for deburring small-sized high-precision details of coaxial radio components *IOP Conf. Ser.: J. of Phys.* **1582** 012041

[16] Karlina Yu I, Kargapoltsev S K, Gozbenko V E and Karlina A I 2019 Removal of burrs from small-size high-precise parts for SHF electronics *IOP Conf. Ser.: Earth Env. Sci.* **378(1)** 012015

[17] Sazonov V N and Zagitov E D 2008 Russian Railways need modern effective solutions, not lightweight theories *World Railways* **4**

IOP Conf. Series: Materials Science and Engineering 1064 (2021) 012036 doi:10.1088/1757-899X/1064/1/012036

Method for determining the sealing forces in a shut-off valve

Yu I Belogolov[1], V K Pogodin[1], V E Gozbenko[1,2], S K Kargapoltsev[1], V A Olentsevich[1] and A M Gladkih[3]

[1] Irkutsk State Transport University, 15, Chernyshevskogo str., Irkutsk, 664074, Russia
[2] Angarsk State Technical University, 60, Chaykovskogo str., Angarsk, 665835, Russia
[3] Irkutsk National Research Technical University, 83, Lermontova str., Irkutsk, 664074, Russia

E-mail: belogolov_vui@irgups.ru

Abstract. The developed method for determining the sealing forces in a shut-off valve takes into account the design features of its components and their elements, which affect the amount of sealing pressure. The components and assemblies of the shut-off valve that provide the creation of sealing pressure are referred to the fastening system or the sealing system. Dependences for determining the forces acting in these systems are analytically revealed. The values of forces are determined when tightening fasteners, when increasing and releasing pressure, as well as when they are exposed to temperatures. The calculation of the specific structure of the shut-off valve shutdown showed that the forces in the sealing system of the check valve prototype are 1.4 times higher than the values obtained using known calculation methods, and the forces in the mounting system of this check valve are 1.2 times higher than the calculated values. The magnitude of the discrepancy between the calculated and actual values of the forces for different designs of shut-off valves may differ and depends on the number of components included in the station shut-off valve. The proposed method for determining the sealing forces in the shut-off valve makes it possible to determine not only the forces in the shut-off valve, but also the optimal number of components and connections that will ensure safe operating conditions. It is shown that the method of calculating the forces in the station shut-off valve, used in practice, is a special case in which the stiffness coefficient of the station shut-off valve is equal to unity. The proposed method for determining the sealing forces in the shut-off valve makes it possible to more accurately determine the forces acting on the shut-off valve, the sealing conditions and the stress-strain state of the components. To implement the developed methodology for determining the sealing forces in the shut-off valve of the stop station in the shut-off valves, it is necessary to develop regulatory documents formalizing the application of this method.

1. Introduction

Ensuring reliable operation of shut-off valves is an important aspect of design [1, 2]. For this, special design methods are used, such as methods for calculating the strength of components, calculating forces for valve sealing (seat valve plug) and packing gland [3, 4]. The practice of designing and researching stop valves has shown that the determination of the sealing forces of a valve gate is mainly determined by calculation using simplified methods [5-7], developed on the basis of analytical or experimental studies of a particular gate valve. However, the study of the sealing conditions that determine the values of the forces that excludes the occurrence of leakage in the valve gate, exceeding the standard values, remains an urgent task. In the existing calculation methods, the sealing conditions were mainly determined in the models of gate valves, since to determine the sealing conditions in real designs of stop valves, it is necessary to create a model taking into account their operating conditions [8-9].

Comparative analysis of the existing methods for calculating the forces acting in the shut-off valve showed that they do not take into account the conditions for the interaction of the components of the shut-off valve structure with each other, as well as the correspondence of the components and their elements [10, 11].

The exclusion of these factors from the calculation is the reason that in the calculations of the forces in the check valves, made in accordance with well-known methods, the value of the forces required to seal the check valve when used in service, often does not match the design. This article discusses a technique for determining the sealing forces in a shut-off valve, taking into account its design features and deformation of all parts and their elements. The use of this technique will eliminate the discrepancies between the calculated and real values of the forces and avoid design errors.

During the development of the methodology, the following well-known concepts were taken into account:

- a component – a product made of the material of the same brand without the use of assembly operations (welding, soldering, gluing, screwing);
- an element of the component – a part of the component intended to perform its definite (specific) characteristic function, for example, a sealing surface, a threaded surface, a supporting surface, etc.
- a station – a set of components interconnected by assembly operations in a structure designed to work together.

The introduction of the new concept "Station Shut-off Valve (SVS)" allows for more complete consideration of design features, component flexibility and disconnection of their valve elements (Figure 1), when determining the sealing forces. The station shut-off valve (SVS) includes not only the components of the gate valve (seat-gate valve), but also, in contrast to the existing design models, the components and their elements that affect its sealing ability at different stages of interaction of these components.

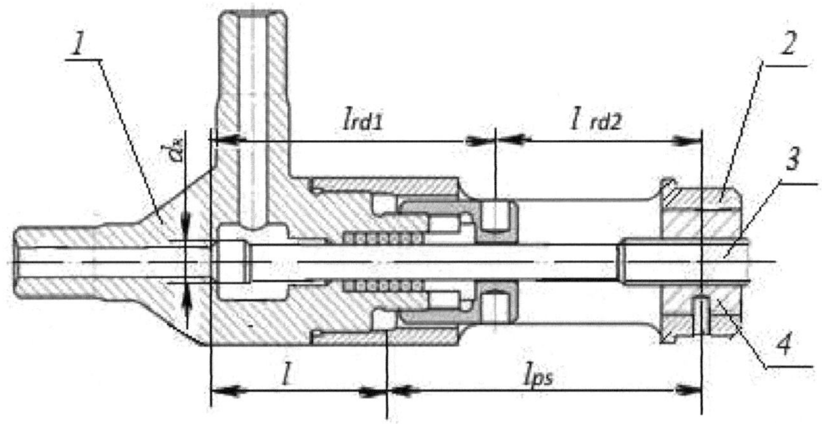

Figure 1. The shut-off valve SVS diagram for high operating parameters: 1 – valve housing; 2 – post; 3 – spindle; 4 – threaded sleeve.

2. Results and discussion

To determine the forces in the shut-off valve according to the proposed method, the values of α are determined at values of Q_f and $Q_{sj.t.}$ To do this, we define the movement values included in formula (11):

1. The axial movement of the spindle under pressure is determined by the formula:

$$\delta_{spp} = \frac{Q \cdot l}{E \cdot F}$$

where E is the modulus of elasticity of the spindle material; $F = \frac{\pi d_{sp}^2}{4}$ is the spindle sectional area.

In the considered design of the shut-off valve: for Q_f the value is $\delta_{sp} = 0.0737$ mm = 73.7 μm; for $Q_{sj.t}$ the value $\delta_{sp} = 0.117$ mm = 117 μm.

2. Full axial movement of the most loaded thread turn is defined as the sum of its individual components [12, 13]:

$$\delta_{th} = \delta_b + \delta_r + \delta_c + \delta_\mu ,$$

where δ_b is the axial movement from the bend of the turn; δ_r is the axial movement from radial movements of the turns of the threaded joint; δ_c is the deformation of the microroughness of the contacting surfaces with a roughness of Rz when exposed to contact pressures $p(H)$; δ_μ is the axial component from the lateral deformation of the spindle body due to a change in its length:

$$\delta_b = 2\frac{p(H)}{E} p \cdot (1 \pm 0.3 f_f),$$

$$\delta_r = p(H)\frac{t_2 \cdot d_2}{E \cdot P}\frac{d_0^2}{d_0^2 - d_2^2}(\mathrm{tg}\frac{\alpha}{2} - f_f)\mathrm{tg}\frac{\alpha}{2},$$

$$\delta_c = 0.371(1 + 0.0353 R_z) \cdot p(H)^{0.66} \cdot \cos\frac{\alpha}{2},$$

$$\delta_\mu = \mu\frac{2Q}{\pi \cdot E \cdot d_2}\mathrm{tg}\frac{\alpha}{2},$$

where P is the thread pitch; $f_f = 0.2$ is the coefficient of friction in a threaded pair (for steel threaded joints); t_2, d_2 are a height and an average diameter of the thread profile; d_o is the outer diameter of the female part of the threaded joint, α is the angle of the sectional profile of the metric thread ($\alpha = 60°$); R_z is the roughness of the contacting surfaces of the thread turns; μ is the Poisson's ratio; E is the elastic modulus of the material of threaded components.

In the above formulas, the relationship between the normal pressure $p(H)$ and the axial force adopted in the works [14, 15] is used:

$$p(H) = Q\frac{\nu}{f},$$

where $f = \pi d_2 \cdot t_2$ is the projection of the lateral surface of a thread turn of a height t_2 on a plane perpendicular to the axis of the threaded joint; Q is the axial load acting on a threaded joint;

$\nu = \sqrt{\frac{3.7(P/d)\cdot[1+1.2\cdot(P/d)]}{1.86+0.35(d/P)}}$ is the dimensionless coefficient.

When determining δ_b, the upper "+" sign should be used under load, the lower sign "–" when unloading a threaded joint.

In accordance with the studies [15, 16] for threaded joints of the shut-off valve at $p(H)$ = 250 MPa, the values δ_p can be taken: for thread M16x1.5 – $\delta_{ts.sp} = 24.5$ microns; for thread M36x2 – $\delta_{ts.pp} = 33$ μm; for thread M64x6 – $\delta_{tp.hp} = 63.1$ μm.

3. Microroughness deformation values in the connection of steel flat sealing surfaces with roughness $Rz = 9.6 \ldots 40$ μm when exposed to contact pressures q_1 in the range from 0 to 300 MPa were determined in accordance with [15, 17] using the formula:

$$\delta_\mu = 0.3711(1 + 0.0353 \cdot R_z) \cdot q_1^{0.66} .$$

When experimentally determining the compliance of components and joints, the researchers [15] recommend using the values obtained after 3-5 loading cycles in the calculations.

In the "rod– housing seat" sealing joint, the movements determined by this relationship with q_{1d} = q_{1f} = 250 MPa can have the following values: at Rz = 9.6 μm the value of $\delta_{sj.p}$= 19 μm; at Rz = 40 μm, the value $\delta_{sj.p}$= 34.237 μm.

After substituting in the expression (11) the displacement values $\delta_{ts.sp}$ = 24.5 μm, $\delta_{ts.pp}$ = 33 μm; $\delta_{tp.hf}$ = 63.1 μm, $\delta_{sj.p}$ = 19 μm и δ_{spp}, the value α= 0.565 was obtained. Then the forces acting in the sealing system and the mounting system, determined in accordance with dependences (7) and (8) at α= 0.565, $Q_{ms.t} = Q_{sj.t}$ = 3579 kgf, Q_d = 1695 kgf and p_w = 150 MPa, have the following values:

$$Q_{sj.p} = Q_{ms.t} - 0.565\ Q_d = 2651 \text{ kgf},$$

$$Q_{ms.p} = Q_{ms.t} - (1- 0.565)\ Q_d = 4316.32 \text{ kgf}.$$

The values of the forces acting in the sealing system and the mounting system of the shut-off valve differ from the values determined in accordance with the existing calculation procedure [3, 4], namely $Q_{sj.p}$ is 1.41 times higher than the values of Q_f, and $Q_{ms.p}$ is 1.2 times greater than $Q_{ms.t}$.

These differences can be explained by the fact that with the current design practice [3, 4], the value of the stiffness coefficient is taken equal to α = 1, at which $Q_{sj.p} = Q_{ms.t} - Q_d$ and $Q_{ms.p} = Q_{ms.t}$. This condition is acceptable only when the movements in the sealing system are completely absent ($\sum \delta_{i\ sj}$ = 0) or very small compared to movements in the mounting system.

Based on the analysis, when designing prospective models of shut-off valves and determining the forces in the memory, one should take into account the structural features of its components and their elements, as well as the mechanism of their interaction by determining the stiffness coefficients α and the transformation coefficients of force effects ψ.

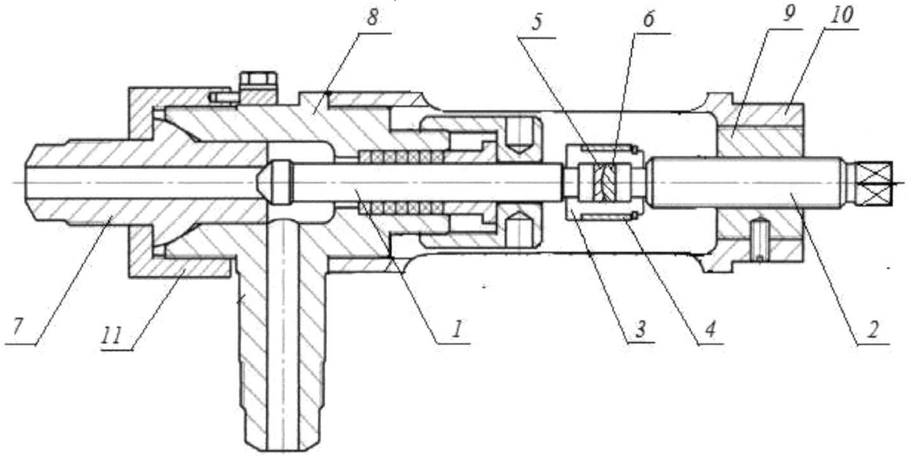

Figure 2. The diagram of a shut-off valve with a SVS, a replaceable nozzle and a split spindle: 1 – rod; 2 – spindle; 3 – half rings; 4 – outer ring; 5, 6 – supports; 7 – replaceable nozzle; 8 – housing; 9 – threaded sleeve; 10 – post; 11 – coupling nut.

In the design of the shut-off valve (Figure 2), when using SVS components and elements that differ from the design shown in Fig. 1, it is necessary to additionally introduce the ductility coefficients of the joints: "support-rod", "support-spindle"; threaded joints: "nut-housing" and joints: "replaceable nozzle-nut" when determining the stiffness coefficient α (10). Design features should also be taken into account when using replaceable seats and replaceable nozzle assemblies in the shut-off valve.

3. Conclusion

1. A new method is proposed for determining the sealing forces in a stop valve, based on consideration of the stop station valve in its design, which includes a sealing system and a fastening system.

2. The method takes into account the design features of the stop valve and includes determination of the conformity of components, sealing of threaded connections and their elements included in the stop valve of the station.

3. It is shown that the method of calculating the forces in the station shut-off valve used in practice has a limitation - the stiffness coefficient of the station shut-off valve is equal to one.

4. The proposed method for calculating the sealing forces in the shut-off valve makes it possible to eliminate the large error in calculating the forces acting in the shut-off valve, sealing conditions and the state of stress deformation of components.

5. The developed method for calculating the sealing forces in the shut-off valve makes it possible to determine the optimal combination of components and connections that will ensure safe operating parameters of the shut-off valve.

6. To implement the developed methodology for determining the sealing forces in the shut-off valve of the stop station in the shut-off valves, it is necessary to develop regulatory documents formalizing the application of this method.

References

[1] Suslov K V, Solonina N N and Smirnov A S 2014 Distributed power quality monitoring *Proc. of Int. Conf. on Harmonics and Quality of Power, ICHQP*

[2] Buryanina N, Korolyuk Y, Koryakina M, Lesnykh E and Suslov K 2019 Algoritm of Current Protection Based on Three Instantaneous-Value Samples *Proc. of 2019 IEEE PES Innovative Smart Grid Technologies Europe, ISGT-Europe* 8905542

[3] Suslov K, Solonina N and Gerasimov D 2018 Assessment of an impact of power supply participants on power quality *Proc. of Int. Conf. on Harmonics and Quality of Power, ICHQP* pp 1-5

[4] Vdovin K N, Feoktistov N A, Sinitskii E V, Gorlenko D A and Durov N A 2015 Production of high-manganese steel in arc furnaces. Part 1 *Steel in Translation* **45(10)** 729-732

[5] Shestov V V, Antipov V V and Ryabov D K 2017 Corrosion Resistance and Mechanical Properties of Layered Structural Material Based on Aluminum Alloy and Fiberglass Thin Sheets *Metallurgist* **60(11-12)** 1191-1196

[6] Ryabov D K, Antipov K V, Fesenko T V and Dynin N V 2017 Influence of cobalt alloying on structure and long-term strength of Al–Cu–Mg–Ag alloy system *Inorganic Mater.: Appl. Res.* **8(1)** 159-165

[7] Ryabov D K, Shestov V V, Buznik V M and Antipov V V 2016 Design concept of multifunctional layered composite materials for operation in arctic regions *Inorganic Mater.: Appl. Res.* **7(4)** 547-553

[8] Kuzmin O V and Orkina K P 2006 Building error-correcting codes using a Pascal-type triangle *Bull. of Buryat State University* **13** 32-39

[9] Kolosov A D, Gozbenko V E, Shtayger M G, Kargapoltsev S K, Balanovskiy A E, Karlina A I, Sivtsov A V and Nebogin S A 2019 Comparative evaluation of austenite grain in high-strength rail steel during welding, thermal processing and plasma surface hardening *IOP Conf. Ser.: Mater. Sci. Eng.* **560** 012185

[10] Guseva E A, Kargapoltsev S K, Balanovskiy A E, Karlina A I, Shtayger M G, Gozbenko V E, Konstantinova M V and Sivtsov A V 2019 Comparative evaluation of corrosion resistance of wheel and rail steels in various media *IOP Conf. Ser.: Mater. Sci. Eng.* **560** 012181

[11] Karlina Yu I, Kargapoltsev S K, Gozbenko V E, Karlina A I and Leonovich D S 2020 Overview of electro physicochemical methods for deburring small-sized high-precision details of coaxial radio components *IOP Conf. Ser.: J. of Phys.* **1582** 012041

[12] Karlina Yu I, Kargapoltsev S K, Gozbenko V E and Karlina A I 2019 Removal of burrs from small-size high-precise parts for SHF electronics *IOP Conf. Ser.: Earth Env. Sci.* **378(1)** 012015

[13] Nekrasov I V, Sheshukov O Y, Metelkin A A, Sivtsov A V, Tsymbalist M M and Egiazar'yan D K 2015 Ensuring Consistent Foamability in Electric-Furnace Slags *Metallurgist*

59(3-4) 300-304
[14] Kopylova T A and Mikhailov A Yu 2016 Evaluation of the functioning of intermodal nodes of urban passenger transport *Transport systems of Siberia. The development of the transport system as a catalyst for the growth of the state economy. Int. sci. and pract. conf.* pp 528-532
[15] Chriqui C and Robilland P 1975 Common bus lines *Transportation Sci.* **9** 115-121
[16] Lebedeva O A and Kripak M N 2016 Modeling freight transportation in the transport network *Bull. of Angarsk State Technical University* **10** 182-184
[17] Sazonov V N and Zagitov E D 2008 Russian Railways need modern effective solutions, not lightweight theories *World Railways* **4**

Methodology for the implementation of a technological solution, taking into account the optimization of production business processes based on simulation

I I Shanin

Voronezh State University of Forestry and Technologies named after G.F. Morozov, 8, Timiryazeva St., Voronezh, 394087, Russia

E-mail: kingoao@mail.ru

Abstract. The current production activity of modern commercial organizations is focused on the use of effective technological solutions. The main goal of any commercial organization that carries out production and sales activities is the release of a certain type of industrial product that meets certain consumer and quality characteristics. Any commercial activity is focused on improving the efficiency of technological processes using rational approaches to save production resources to further maximize profits. When building a technological production chain, it is necessary to analyze production business processes, taking into account both the individual needs of consumers in industrial products and taking into account the main goal of a commercial organization, taking into account the introduction of modern technological solutions. The study proposes to introduce an effective technological solution into the production process, based on the results of simulation (for example, furniture production). Therefore, to assess the performance of technological processes and the effectiveness of the use of machinery and equipment, it is necessary to apply an approach to the implementation of an effective technological solution. For this, it becomes necessary to develop an algorithm for the implementation of an effective technological solution, taking into account the optimization of production business processes based on simulation, based on increasing the efficiency of using machines and equipment and reducing downtime in the production process chain.

1. Introduction

It should be noted that the key task of any commercial organization that produces furniture products is to improve production business processes for further development of financial and production activities. It is necessary to improve production business processes on the basis of a reorientation of production towards the transition from serial technological processes to an optimized type of furniture production. The application of this approach will improve the efficiency of production business processes and the volume of products. Simulation methodology must be applied as a key component of a lean manufacturing approach. This approach involves the improvement of technological processes, in which a number of the most frequently used assortment units of parts and furniture components must be transferred to technological processes for the production of serial furniture products and their storage in a warehouse. Rarely used parts and components used for complex design features to order must be manufactured as part of an individual technological process based on modelling.

In modern scientific literary sources, there are no unified methodological foundations that would clearly distinguish between serial technological processes and technological production processes for individual orders. Production systems of mass, batch and unit types of production characterize only a part of production business processes, which do not take into account the indicators of volumetric and production criteria, as well as the specific features of the manufacture of furniture products by commercial organizations. In this direction, the most optimal one is a scientifically grounded approach to the application of a business model of an industrial-technological method, which requires optimization of serial technological processes and production processes for individual orders, in the aggregate of a single configuration of production business processes of a commercial organization. The proposed approach to the study is based on the construction of mathematical models using simulation modelling of technological processes of production and sales of products. A single industrial method involves the manufacture of furniture products on a large scale, by freeing up resources (material, raw materials, labour).

2. Materials and methods

The research methodology was based on developments on the research topic, tested and presented in modern scientific works of researchers in the area under study. We also took into account the empirical results of practical implementation at industrial enterprises and commercial organizations of methodological approaches to assessing the applied production business processes in the financial and production environment and the possibility of improving the ongoing technological processes [1, 2].

The study used methods of simulation, computer data analysis, mathematical modelling, systemic, economic and stochastic analysis, and a graphical method.

The efficiency of the flow of production business processes of a commercial organization, including furniture, is understood as a process in which technological processes proceed according to the planned indicators and the strategy of the commercial organization. The study, among other things, used an improved method for comparing the quality of technological processes based on the reporting documentation of a furniture commercial organization. The quality category of technological processes is of a comparative nature, therefore, it is expressed in quantitative terms and is determined by the formula (1) [3, 4]:

$$S = \sum_{i=1}^{n} T_i X_i, \qquad (1)$$

where T_i is an indicator of the quality of a separate technological process of the i-th group; X_i is the value of the coefficient of significance (weight) of the i-th group of quality indicators of technological processes.

3. Results and Discussion

The result of the study is an approach to improve the efficiency of production of products on the example of furniture enterprises, which combines all the necessary elements to ensure ongoing business processes in the production system of a commercial organization. This approach makes it possible to determine the reserves for increasing the efficiency of using all available resources in financial, economic, production and technological activities, as well as to determine possible ways aimed at achieving the maximum economic effect for a commercial organization and society as a whole.

Calculations of the passage of technological processes, carried out as part of a study of the financial and production activities of furniture commercial organizations in the Voronezh region of the Russian Federation, made it possible to determine the possibility of obtaining a potential economic effect in the process of implementing the proposed approach. As indicators-indicators of technological processes, indicators were used that determine the volume of production, indicators of the efficiency of using fixed assets, and also the ratio of the turnover of working capital was calculated according to a group criterion (Table 1).

The calculations shown in Table 1 were obtained based on the results of the study based on simulation modelling of technological production processes, as part of the analysis of the indicators provided in the reporting by the enterprise-object that produces and sells furniture products [5,6].

The results of calculations of simulation modelling (Table 1) made it possible to conclude that in a commercial organization for the production and sale of furniture products due to the improvement of technological processes, based on the transition to an optimized production process, it becomes possible to increase production by 25.9%.

Table 1. Comparative values of indicators of technological processes during the transition to an optimized production process of furniture products

Indicator name	Serial technological processes (first condition) - S1	Technological processes on orders (second condition) - S2	Optimized production process (third condition) – S_{opp}	Change of the third condition in relation to the first two, %
Time required for processing the technological process, h	27.4	32.8	22.7	-20.7 - first condition; -44.5 - second condition
The degree of utilization of production equipment (workshop for cutting materials and manufacturing of blanks, assembly workshop), %	0.58-0.69	0.47-0.51	0.73-0.78	25.9 - 1st condition; 55.3 - 2nd condition
The value of the average stagnation time of work pieces between production units, h	16.3	23.1	12.9	-20.9 - 1st condition; - 44.2 - 2nd condition
The value of the average stagnation time of work pieces, % of the time required for processing the technological process	59.5	70.4	56.8	–2.7 - 1st condition; -13.6 - 2nd condition
Average production volume per year, units.	196	112	334	138 - 1st condition; 222 - 2nd condition. On average by 25.9%

On the basis of the methodology of simulation modelling, an algorithm for introducing an approach to improving technological processes is proposed, based on increasing the efficiency of production business processes and the volume of products.

In the process of constructing the algorithm in the study, the following tasks were set: reducing labour costs in the interval of stagnation of blanks between production units; increasing the degree of utilization of production equipment (workshop for cutting materials and making blanks, assembly workshop) with the number of new orders; reducing the cost of products. To solve the set tasks, it is necessary to optimize all technological processes in general, namely, to change all time tasks and plans, using an imitation-mathematical approach to solving the tasks, aimed at increasing the possibility of fulfilling the number of orders [7,8].

Let us suppose that the vector in the data array $\vec{A}$ - number of orders received, then:

$$\vec{S} = [\vec{S_i}, t_i^1, t_i^2, T_i^3, T_i^{4}], \quad (2)$$

where $\vec{S_i}$ is a vector that determines a set of parameters of the i-th type of resources used in the technological process (raw materials, materials, components): t_i^1 - time required for loading and unloading operations of the i-th type of resources; t_i^2 - minimum time of the technological process of the i-th part; T_i^3 - value of the average execution time of the technological process for a specific nomenclature unit of the finished product; T_i^4 - value of the average execution time of the technological process for a specific nomenclature unit of blanks and furniture components.

Changing the necessary temporary tasks and plans, using a simulation-mathematical approach to solving the tasks, aimed at increasing the possibility of fulfilling the number of orders, is performed in several levels:

At the first level, orders are sorted and distributed among departments that carry out a control function in production business processes of various levels, with the determination of the lead time for a specific order in conditions of optimal planning of all necessary resources [9, 10].

At the second level, adjustments are made to planned indicators associated with the possibility of reducing the order execution in time, within the order period. It is worth considering the need to group orders to reduce the time of presence in production plans of homogeneous types of used resources by leading technological processes. The selection of the required elements of the data array $\vec{S}$ must be carried out on the basis of the following criteria:

1) Strategic lead time for one order:

$$t_i^1 \in \left[t_i^{left};\ t^{right}\right], \quad (3)$$

where t_i^{left} is the value of the initial (left) border of the interval for selecting the required elements of the data array for a specific order with the serial number I;

$$t_i^{\text{left}} = t_i^{\text{current}} + T_i^{mean} - T_i^{\text{safety stock}}, \quad (4)$$

$t_i^{current}$ – value of the current lead time for one order of the i-th product part; T_i^{mean} – value of the average lead time of one order of the i-th part of the product; $T_i^{safety\ stock}$ – value of additional time with a possible increase in the lead time of one order; t^{right} – value of the final (right) border of the selection interval for the data array elements for a specific order contained in the production plan;

$$t^{right} = t_0 + T_k; \quad (5)$$

t_0 – value of the initial lead time for one order; T_k – optimal planning value, minutes;

$$T_i^3 < t_i^{mean} - t_0 + T_i^{\text{safety stock}}, > T_i^4; \quad (6)$$

where $t_i^2 > t_0$.

2) Reducing the amount of resources in warehouses and production sites:

$$RWP(t) \Leftarrow \text{RWP}_i^{plan} \quad (7)$$

where $RWP(t)$ – amount of resources in warehouses and production sites at the current i-th part of the product; $PRWP_i^{plan}$ – amount of resources in warehouses and production sites according to the plan for the production of the i-th part of the product;

$$P_{making} \Leftarrow P_{making}^{plan} \quad (8)$$

P_{making} – actual value of all necessary resources required to complete the order in fact; P_{making}^{plan} – standard value of all the necessary resources required to fulfil the order according to the plan.

Figure 1 shows an algorithm for the transition from serial technological processes to an optimized type of furniture production, based on the implementation of an approach to improve technological processes, using the methodology of simulation.

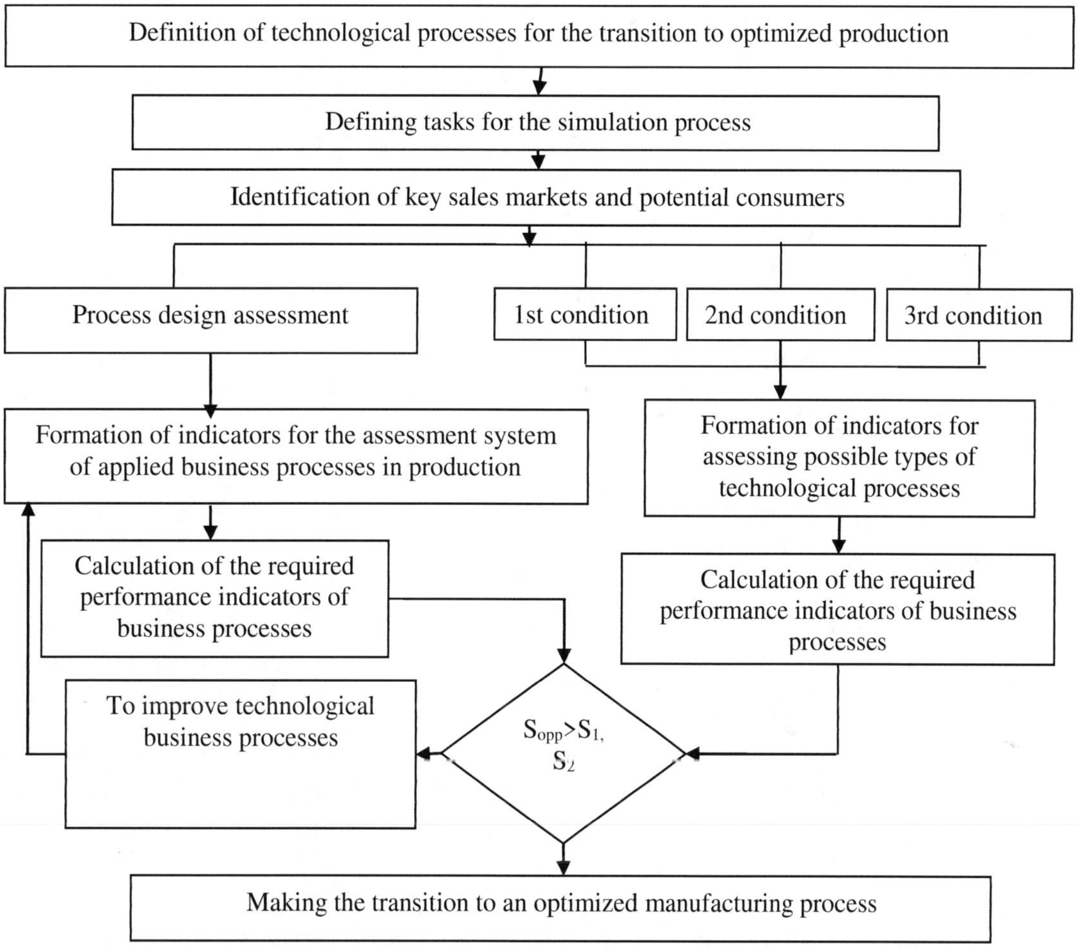

Figure 1. Algorithm for the implementation of an approach to improve technological processes based on simulation.

The proposed approach can be applied provided that all technological processes are carried out with the full availability of the necessary equipment, human resources, raw materials and materials, production areas have an appropriate capacity and the volume of orders does not exceed the production capacity. It should be noted that the time of order receipt and the order of the technological processes do not affect the processes in the algorithm.

When carrying out the optimal planning of all the necessary resources, as the basis for the optimality criterion, it is necessary to use the value of the minimum duration of the technological process of manufacturing a product, since on its basis most of the production indicators are calculated that affect the efficiency of production business processes and the volume of products, the speed of order fulfilment.

4. Conclusion

In the course of the implementation of the mathematical approach based on simulation modeling and the proposed algorithm, the search for optimality for finding the vector for solving the problem conditions can be represented as:

$$\left((S_1, S_2, S_3, \dots, S_i, \dots, S_n); (x_1, x_2, x_i, \dots, x_n); (t_1, t_2, xT_3, T_4)\right)$$

where x_i is the simulation indicator corresponding to the technological number of the production operation, determined in the i-th by the number in the technological process, while n is the total number of products sent to production. Where S_i is the parameters of a certain type of resources used in the technological process, $T_{3,4}$ is the value of the average execution time of the technological process for a specific nomenclature unit of the finished product and blanks and furniture components, respectively.

The result of the study at furniture enterprises was the identification of the main ineffective production business processes and the possibility of optimizing current technological processes in the context of increasing the efficiency and innovation of the production of manufactured furniture products. Based on the results of the application of the developed algorithm, it is possible to identify technological processes that require their optimization to improve production efficiency. The proposed approach allows you to optimize the use of material resources and increase the level of technological processes in modern commercial organizations engaged in the production and sale of industrial products, including furniture.

References

[1] Drapalyuk M V, Bartenev I M, Midges M A, Druchinin D Yu, Markov O B and Klubnichkin E E 2012 Mathematic model of process of giving and emission of soil by working bodies of the combined car for suppression of forest fires *Polythematic network electronic scientific magazine of the Kuban state agricultural university* **84** 232-246

[2] Morkovina S S, Rezanov V K, Panyavina E A and Sukhova V E 2018 Function value analysis in forestry practice *Innovation Management and Education Excellence through Vision 2020 Proc. of the 31st Int. Business Information Management Association Conf. (IBIMA)* vol 1 pp 4419-4425

[3] Drapalyuk M V, Bezrukova T L, Shanin I I and Bezrukov B A 2019 Methodology of probabilistic modelling of the current activity of industrial enterprises *J. of Phys.: Conf. Ser.* **1333** 072022

[4] de Melo F, Maslennikov V V, Popova E V, Bezrukova T L and Kyksova I V 2015 Quantitative analysis in economics based on wavelet transform: a new approach *Asian Social Science* **20** 66-73

[5] Yakovleva E A and Subkhonberdiev A Sh 2019 Implementation of "green" economy principles in the forest sector *IOP Conf. Ser.: Earth Env.* **392** 012016

[6] Bezrukova T L, Igolkin I S, Salikov Yu A, Smolyaninova I V and Akhmedov A E 2019 Innovation approach to diversification of activities of a modern university on the basis of remote education *The Int. J. of Educational Management* **33(3)** 486-493

[7] Morkovina S S, Drapalyuk M V, Sibiryatkina I V and Torzhkov I O 2017 Priorities of diversification in forest complex growth *Proc. of the 30th Int. Business Information Management Association Conf.* (ed Khalid S Soliman) vol 1 pp 2856-2862

[8] Shanin I I 2019 Modeling of technological processes at enterprises of timber processing industry *IOP Conf. Ser.: Mater. Sci. Eng.* **560** 012042

[9] Shanin I I and Boris O A 2018 Modeling operation of mechanism of holistic management of technological processes at enterprise *IOP Conf. Ser.: Mater. Sci. Eng.* **327** 022095

[10] Vasiltsova V M, Dyatlov S A, Vasiltsov V S, Bezrukova T L and Bezrukov B A 2015 Methodology of management innovation hypercompetition *Asian Social Science* **20** 165-169

Geotechnical monitoring of frozen soils: problems and possible solutions

A I Potapov, A I Shikhov and E N Dunaeva

St. Petersburg Mining University, 2, 21 line of Vasilyevsky Island, St. Petersburg, 199106, Russia

E-mail: shihov-gol@mail.ru

Abstract. A great part of the territory of the Russian Federation is located in the zone of permafrost soils. Safe operation management of buildings and structures, in permafrost conditions, requires geotechnical monitoring of subgrade soils. The increase in air temperature recorded during numerous monitoring leads to the degradation of permafrost soils. Therefore, the problem of conducting geotechnical monitoring becomes more urgent than ever. This paper analyzes the existing methods and systems for geotechnical monitoring of permafrost soils. The authors reveal the drawbacks of the applied methods and the reasons that do not allow assessing the state of frozen soils with sufficient accuracy. In order to solve these problems, it is proposed to use an active acoustic method of non-destructive testing and create a system that combines an engineering-geological method determining the deformation properties of frozen soils and active acoustic method.

1. Introduction

Mining and processing of minerals plays an important role in Russian economy. A significant part of the mineral deposits produced in Russia are located in the permafrost zone. The data of many-year monitoring of the temperature regime of the atmospheric air and soils in the areas of the permafrost zone showed a trend towards the increase in the temperature of both the atmospheric air and the temperature of frozen soils [1-3].

As a result of the degradation of frozen soils, the bearing capacity of soil subgrades decreases, which leads to disruptions in the normal operation of buildings and structures. The illustrative case of the impact of permafrost degradation on buildings and engineering structures is the Republic of Sakha. According to the data of the work [4], as a result of changes in the temperature regime of permafrost zones, more than 60% of the buildings in Yakutsk are in emergency condition.

In addition, the deformation of thawed soils leads to breaks of pipelines of urban communications, as a result of which the subgrades of buildings are undermined. In addition to the decrease in the bearing capacity of the subgrades, the increase in the layer of seasonal thawing of soils leads to the formation of heaving mounds of frozen soils and karsts. These two very factors, according to the authors of the work [5], led to the formation of a crater 33 m deep and 40 m in diameter, not far from Bovanenkovo oil and gas condenser deposit.

A non-collapse karst was formed under the heaving mound where gas accumulated. As a result, when the critical pressure was reached, a rupture of the mound and the formation of a funnel occurred.

In this situation, it is obvious that, in addition to the measures to maintain the temperature regime of frozen soils and eliminate the consequences of technogenic accidents, it is necessary to conduct geotechnical monitoring of buildings and structures constructed on frozen soils. According to SP 305.1325800.2017 “Buildings and structures. Rules for Conducting Geotechnical Monitoring During Construction” geotechnical monitoring is a mandatory procedure both during the construction of a structure and during its operation. The geotechnical monitoring project should include measures to control the temperature regime of soils, elevation points of geodetic marks and monitoring the level of groundwater.

In general monitoring of the temperature regime of frozen soils is carried out in thermometric wells equipped with thermometric streamers that perform periodic temperature measurements, and a logger in which the measurement data are stored. There is a significant number of autonomous systems for continuous monitoring of soil temperature, described in detail in the works [7-9].

The general problem of temperature monitoring of soils is that, despite the high accuracy of measurements and the possibility of continuous monitoring, according to the measurement data it is not always possible to predict and prevent emergency situation, since the freezing-thawing temperature of frozen soils is not constant and depends on various factors such as salinity, lithological and granulometric composition of frozen soils.

As in the case of temperature monitoring, there are a large number of methods and systems for geodetic monitoring [10-12]. It is necessary to note that geodetic monitoring is possible remotely, using aerial photography and GLONASS and GPS systems. The obvious disadvantage of geodetic monitoring is that the obtained data on the change in the altitude position of the geodetic point are a consequence of the changes in the deformation properties of frozen soils that have already occurred and it is impossible to directly assess these changes using geodetic monitoring.

From the above mentioned aspects, we can conclude that despite the presence of a large number of automated systems and means of geotechnical monitoring, at the moment, the problem of monitoring frozen soils has not been finally solved. This is evidenced by the increased frequency of emergencies [4].

According to the authors, this problem can be solved using the seismic-acoustic methods of non-destructive testing. This use of them allows controlling the change in the deformation properties of frozen soils directly in the area of influence of the structure on the soils.

2. Results

Seismoacoustic methods for the study of soils are based on the research of the propagation of elastic waves in a soil massif. Depending on the frequency of the elastic wave, seismoacoustic methods can be divided into seismic (frequency up to 1 kHz) and geoacoustic (frequency over 1 kHz) [13]. Seismoacoustic methods are widely used for the solution of various problems of engineering geology, exploration of mineral reserves, delineating ore bodies and mining tasks [14-16].

The method of acoustic crosswell sounding is used directly for the study of frozen soils. In particular, it is used to control the formation of ice walls, when using a special method of construction - freezing [17, 18].

The method of acoustic crosssounding is based on the use of active acoustic control method: elastic wave in the soil mass is excited by external sources. In this case, the informative parameters are the speed and time of passage of elastic wave through the array of frozen soils. It is possible both acoustic sounding (Figure 1), when the emitting and receiving transducers are located in different wells and the time of passage of elastic wave between them is recorded, as well as registration of a reflected elastic wave by a combined transducer (Figure 1).

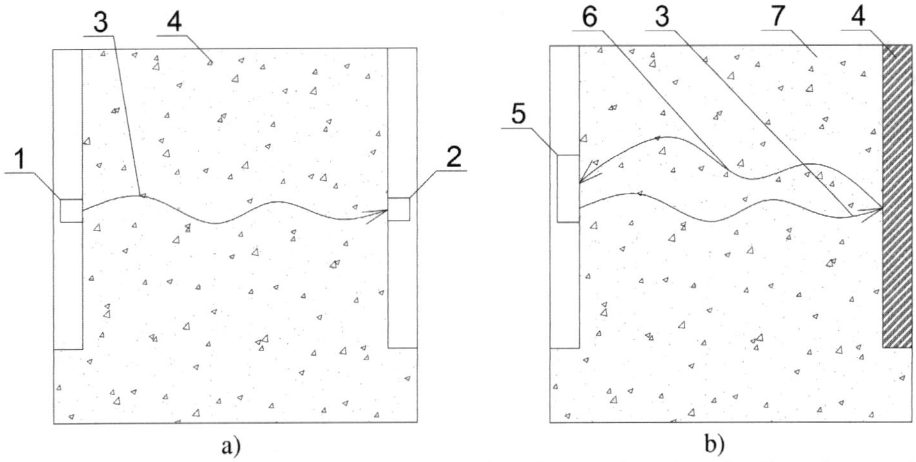

Figure 1. Acoustic sounding: a) – acoustic sounding; b) – registration of reflected wave. 1 – emitting transducer; 2 – receiving transducer; 3 – elastic wave, excited in the array; 4 – frozen soil; 5 – combined transducer; 6 – elastic wave reflected from the frozen soil mass; 7 – the array of thawed soil.

According to the results of measurements of the propagation rate of elastic wave, it is possible to determine the deformation properties of frozen soils:

$$E_D = V_P \cdot \rho \cdot (1+\mu) \cdot (1-2\mu)/(1-\mu);$$

$$\mu = (V_P^2 - 2V_S^2)/(2 \cdot (V_P^2 - V_S^2));$$

V_P - P-wave velocity; V_S - S-wave velocity; μ - Poisson's ratio; E_D - modulus of subgrade stiffness.

Unlike the passive method of acoustic control, considered in the work [19], the use of the active method of acoustic control will allow detecting not only changes in the deformation properties of soils during thawing, but also determining the thawing zone.

It is necessary to note that the deformation modulus calculated from the results of acoustic sounding is a dynamic modulus of soil deformation and it can significantly differ from the static modulus of soil deformation obtained from the results of engineering and geological surveys. The dynamic modulus of soil deformation is always greater than the static one.

During the design of the foundations of buildings and structures, static values of the deformation properties of soils are used. Therefore, to conduct geotechnical monitoring of changes in the deformation properties of frozen soils, it is necessary to establish a relationship between the static and dynamic values of deformation properties.

In publications [20, 21], the authors made the attempt to establish a universal relationship between the static and dynamic modulus of deformation for different groups of soils. In our opiniom, this approach has obvious disadvantages, which, can lead to serious errors when using the obtained universal dependencies for processing the results of geotechnical monitoring.

Firstly, it is the complexity in the preservation of frozen soil samples in undamaged condition for laboratory tests. Secondly, it is the so-called large-scale effect. Soils are a three-phase system, and in the case of frozen soils the present a four-phase system, which has a spatial heterogeneity of physicomechanical, thermophysical and physical properties. In the course of tests, the scale effect expresses this heterogeneity.

In other words, a soil sample selected at a construction site will not have the same properties as the array in general. Taking into account the scale effect, the obtained values of soil characteristics in the laboratory can be both underestimated and overestimated. For example, the increase in fracture or porosity in rock mass leads to the fact that laboratory tests overestimate the strength and deformation properties of soils in comparison with a laboratory sample. The inclusion of rocks in the soil massif, on the contrary, will lead to underestimated results in the course of laboratory studies.

In the case of frozen soils, the areas of saline soils will contain a greater amount of unfrozen water, which will affect the deformation and strength properties of frozen soils. The large-scale effect can manifest itself not only in laboratory studies. It is necessary to take into account in all cases when the volume of the studied soil is less than the soil mass located in the zone of influence of the structure. Therefore, we believe that, the establishment of any universal correlation dependence that gives reliable evaluation of the results of geotechnical monitoring of the deformation properties of frozen soils is impossible.

3. Conclusion

This review article presents the analysis of the applied methods of geotechnical monitoring. The analysis revealed obvious problems of the applied methods of geotechnical monitoring, which do not allow predicting and preventing emergencies. This is evidenced by the more frequent emergencies associated with the degradation of frozen soils. In our opinion, the use of active methods of acoustic control can solve most of the problems revealed during the analysis.

According to the results of acoustic sounding of frozen soils, it is possible to determine the deformation properties of soils, directly, in the zone of influence of the structure with regard to stress-strain condition. However, we should note that in order to conduct geotechnical monitoring of frozen soils, it is necessary to create a system that combines geotechnical methods for the determination of the deformation and acoustic properties of soils. It is mentioned in the article that the use of the obtained universal dependencies between static and dynamic values of deformation properties for processing the results of geotechnical monitoring is unacceptable. Therefore, it is necessary to establish a correlation between the static and dynamic values of the deformation properties of soils, directly under the conditions of a particular construction site.

References

[1] Varlamov S P, Skachkov Yu B and Skryabin P N 2017 Results of 35-year monitoring studies of the permafrost zone at the Chabyda station (Central Yakutia) *Science and education* **2** 65-73

[2] Ponomareva O E, Gravis A G, Blyakharchuk T A, Bochkarev Yu N, Ustinova E V, Berdnikov N M and Moskalenko N G 2016 Reaction of the insular permafrost zone of the northern taiga of Western Siberia to climate change *Mater. of the fifth conf. of geocryologists of Russia (Moscow State University named after M.V. Lomonosov)* (Moscow: University book) vol 2 pp 80-86

[3] Konischev V N 2009 Response of permafrost to climate warming *Moscow University Bull. Series 5. Geography* **4** 10-19

[4] Alekseeva O I, Balobaev V T, Grigoriev M N, Makarov V N, Zhang R V, Shats M M and Shepelev V V 2007 On the problems of construction in the permafrost zone (on the example of Yakutsk) *Cryosphere of the Earth* **XI(2)** 76-83

[5] Perlova E V, Miklyaeva E S, Tkacheva E V and Ukhova Yu A 2017 Yamal crater as an example of a rapidly developing cryogenic process in the conditions of climate warming in the Arctic *News of Gas Science* **3(31)** 292-297

[6] 2017 SP 305.1325800.2017 *Buildings and structures. Rules for conducting geotechnical monitoring during construction* (JSC "Research Center" Construction" - NIIOSP named after N.M. Gersevanov, CJSC "Triada-Holding")

[7] Bogdanov A I and Edigaryan A R 2006 Automated system for monitoring the state of columnar bases of bridges (ACS-SOU) operated according to principle I on high-temperature permafrost soils *Problems of the development of the railway network* **3** 140-145

[8] Amosova E V, Kropachev D Yu and Pazderin D S 2011 Temperature monitoring system for extended objects in permafrost *Equipment and technologies for the oil and gas complex* **2** 67-70

[9] Gromadsky A N, Arefiev S V, Volkov N G, Kamnev Y K and Sinitsky A I 2019 Remote

monitoring of the temperature regime of permafrost soils under the buildings of Salekhard *Sci. bull. of the Yamalo-Nenets Autonomous Okrug* **3** 17-21

[10] Sergina O V and Kuryachaya E A 2018 Geodetic monitoring of the sediment of foundations and foundations of buildings and structures of oil and gas condensate fields *Bull. of Omsk State Agrarian University* **2** 75-82

[11] Elizarova A S and Kurchatova A N 2020 Monitoring of aboveground pipelines using global navigation satellite systems *Bull. of Siberian State University of Geosystems and Technologies* **1** 28-42

[12] Ustavich G A, Avrunev E I, Salnikov V G and Popov V K 2018 Peculiarities of deformation monitoring of engineering structures in permafrost conditions *Bull. of Tomsk Polytechnic University. Engineering of georesources* **329(4)** 97-109

[13] Kuryanov Yu A, Kukharenko Yu A and Rock V E 2002 *Seismoacoustics of porous and fractured geological media. Volume 1: Theoretical models in seismoacoustics of porous-fractured elastic media* (Moscow: VNIIgeosystem)

[14] Shvakin M V, Mulev S N, Skakun A P and Rabota E N 2012 Experience in the application of geophysical methods for monitoring the stress-strain state of a rock mass to change design solutions in the elimination of flooding of an emergency section of mine No. 1-5 of the Barentsburg mine *Notes of the Mining Institute* **198** 137-143

[15] Babkin A I and Sanfirov I A 2011 Geophysical monitoring of the out-of-tubing space *Mining information and analytical bull.* **1** 213-218

[16] Shabalin V A, Zhuravlev A Yu and Bordyugov M D 2012 Determination of the depth and density of concrete pillars at bridge construction sites using the seismic acoustics method (as exemplified by the construction of a bridge across the Golden Horn Bay in Vladivostok) *Sci., technical and economic cooperation of the Asia-Pacific countries in the XXI century* vol 1 pp 207-211

[17] Sanfirov I A, Yaroslavtsev A G, Chugaev A V, Babkin A I and Baibakova T V 2020 Control of the formation of an ice wall of a mine shaft using a complex of surface and borehole seismic methods *Physical and technical problems of the development of mineral resources* **3** 34-46

[18] Yaroslavtsev A G 2019 Reflection of shaft ice formation in a seismic wavefield *Mountain echo* **3** 52-57

[19] Novikov E A, Shkuratnik V L and Zaitsev M G 2019 Manifestation of acoustic emission in frozen soils with simultaneous influence on them of variable mechanical and thermal effects *Notes of the Mining Institute* **238** 383-394

[20] Pioro E V 2014 *Deformation and acoustic properties of clay soils based on the results of laboratory geological engineering and ultrasonic research:* candidate dissertation (Moscow State University named after M. V. Lomonosov) pp 52-61

[21] Nikitin N V 1962 On the relationship between dynamic and static elastic moduli of rocks *Exploration and production geophysics* **45** 37-42

Method for determining the reduced parameters of the mixer gear drive

Y I Podgornyj[1], V Y Skeeba[1], T G Martynova[1], P Y Skeeba[1], D V Lobanov[2] and N V Martyushev[3]

[1] Novosibirsk State Technical University, 20, K. Marksa Ave., Novosibirsk, 630073, Russia

[2] Ulianov Chuvash State University, 15, Moskovskiy Ave., Cheboksary, 128010, Russia

[3] National Research Tomsk Polytechnic University, 30 Lenina Ave., Tomsk, 634050, Russia

E-mail: skeeba_vadim@mail.ru

Abstract. The article presents the method of determining the reduced moments of a mixer mechanism with an epicyclic gear and a rocker arm. The method consists of using a mathematical model of the mechanism; determining the change of process load on the device working shaft; determining the dependence of the moments change on inertia forces and their values and reducing them to the main shaft of the device. It also includes determining the dependence of the change in the total moments values on the forces of useful resistance and on the moments of inertia forces and reducing them to the main shaft of the device.

1. Introduction

Existing designs of continuous mixers operate mainly in a stationary mode, i.e. at a constant angular velocity of the work member.

When preparing bulk mixtures, the coincidence of the mixture movement speed and that of the work member may result in components deposits in the mixing zone, which leads to the final product quality deterioration. At the same time, the production time, as well as energy consumption, increases [1-3].

The production of liquid and low-viscosity mixtures in paddle mixers is also associated with the gradual alignment of the mixture movement speeds and those of the mixer's work member paddle. As a result, the mixture moves in layers, i.e. there is a laminar flow mode, in which the mixture layers do not mix with each other, which negatively affects the quality.

In both cases, a variable rotation speed of the mixer's working shafts solves these problems.

When a variable angular velocity is transmitted to the working body, the actual instantaneous velocity of the mixture (liquid or bulk) ω_m consists of the time-average value of the mixture velocity ω_{av} and the deviation of the actual velocity from the averaged $\Delta\omega$.

$$\omega_m = \omega_{av} + \Delta\omega, \tag{1}$$

In this case, the deviation of the actual speed from the average value is an increment of the work member rotation speed, which will be different in value and direction. This will lead to turbulent phenomena, i.e. the transition of elementary mixture masses from one layer to another, which contribute to improving the quality and intensity of the mixing process [2, 4–6].

The turbulent mode of particle motion allows achieving uniformity of the mixture. Thus, the use of continuous mixers with the work member variable angular rotation speed is the most efficient method to achieve a high-quality final product. The lever and gear drive mechanisms can transmit the specified type of the work member movement providing for the set objective.

Increasing the performance of mixers is possible both by changing the geometric parameters of the working chamber and by increasing the speed of the actuators within the requirements of the manufacturing process. Increasing the speed modes of the kneading device work members with a significant length (2 m or more) has a considerable impact not only on the kinematics, but also on the dynamics of the mechanism, which imposes higher requirements on individual elements and units of continuous mixers. Determining the dynamic characteristics of the synthesized mechanism [2] requires knowing the moment of useful resistances, the driving moment, and the inertia forces moment. During downtime of the mixing equipment, some mixtures change their chemical, biological, physical and mechanical properties; therefore, the load on the work member increases over time which leads to the electric motor overloads, as well as to the damage in the elements of the actuating and transmission mechanisms. Lowering the starting torque before restarting the mixer requires releasing the working chamber from the mixture (often done manually), which leads to additional loss of time and possibly raw materials. When a mixer comes into regular operation after a downtime, accelerating the unit and filling the working chamber with mixed raw materials takes certain time, which affects the performance and, consequently, the efficiency of the equipment. The research into determining the reduced parameters of inertia and useful resistances forces components will help establishing the energy costs for the useful operation of the mixer. The objective of the study is to increase the performance of food mixing equipment.

2. Theory and methods

The kinematic scheme of the unit is shown in Figure 1.

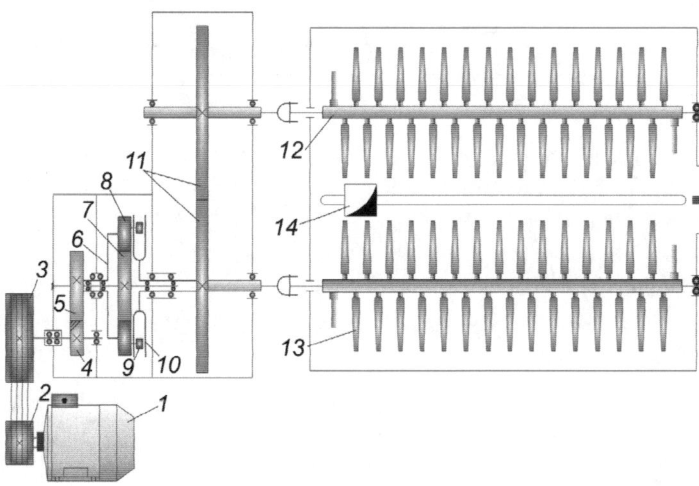

Figure 1. Kinematic diagram of a mixer driven by an epicyclic transmission: 1 - electric motor (P = 11 kW, n_m = 1460 min^{-1}); 2 - driving pulley (D = 140 mm); 3 - driven pulley (D = 560 mm); 4 - gear wheel (z = 19, m = 4); 5 - gear wheel (z = 71, m = 4); 6 - driver; 7 - gear wheel (z = 44, m = 4); 8 - planet gear (z = 22, m = 4); 9 - slot stone; 10 - rocker arm; 11 - gear drive (z = 88, m = 5); 12 - kneading shaft; 13 - blade; 14 – loading hopper; 15 - discharge hole.

The mechanism workflow: the V-belt transmission 2-3 transmits the movement from the electric motor shaft 1 to the shaft with gear 4. From it, via a gear cylindrical transmission 4-5, the movement is transmitted to the driver which moves the planet gears 8 along the fixed solar wheel 6. On the planet gears, there are fixed axes that carry slot stones 9, which engage with the slots of the rocker arm 10 and turn it at a certain angle. The shaft carrying blades 13 rotates together with rocker arm. The second shaft of the mixer is rotated by a gear drive 11.

For technical tasks, the graphic form is the most convenient one for representing the reduced moment of resistance forces. The work of the reduced resistance forces can be represented by the expression $A_{Mr} = \int_0^{\varphi} M_r \cdot d\varphi$, with A_{Mr} being the resistance forces work; M_r - resistance forces moment; φ – angle of the main shaft rotation.

The work of the reduced resistance forces is determined by the product of the area bounded by the curve $M_r = M_r(\varphi)$ and the axis φ by the angle φ, taking into account their scales.

In this view, the most acceptable form of research is to describe the behavior of the system through kinetic energy. In this case, the expression $T = T(J_{red})$ establishes the dependence of the kinetic energy on the reduced moment of inertia.

In the article, the reduced process load was determined at different angles of the blades placing [2]. It is the sum of the process loads on each blade placed in a separate row. The typical graphs of the reduced process load on each row of blades are shown in Figure 2.

The obtained data is the basis for plotting the change in the total loads from each row of blades reduced to the main shaft depending on the rotation angle of the working shaft. A typical graph is shown in Figure 3.

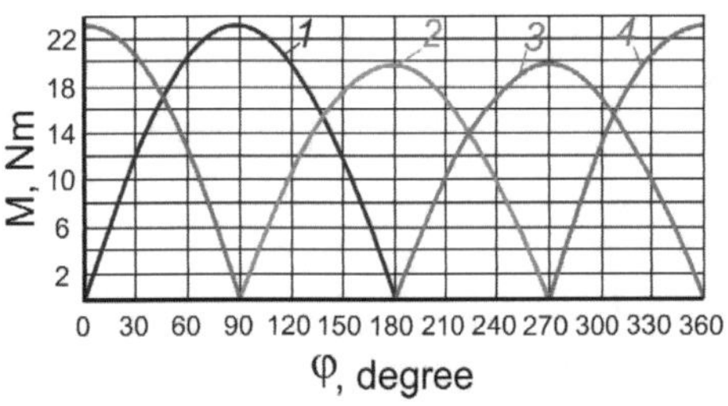

Figure 2. Graphs of changes in process loads reduced to the main shaft for each row of blades depending on the rotation angle of the working shaft at a speed of 120 min^{-1}: 1, 2, 3, 4 are the loads on the corresponding rows of blades

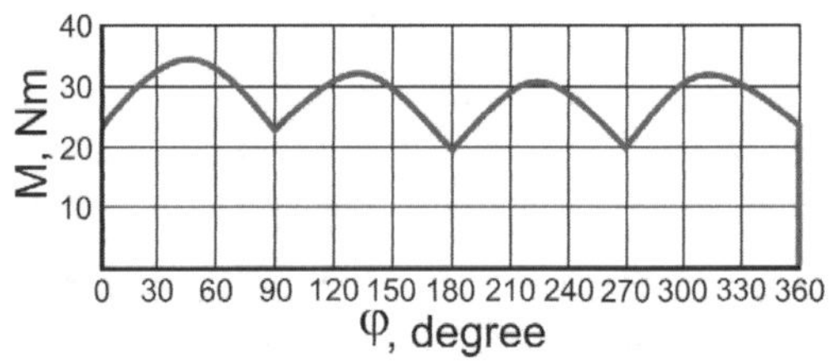

Figure 3. Graph of changes in the total load on the working shaft reduced to the main shaft depending on the rotation angle of the working shaft at a speed of 120 min^{-1}

The reduced moment of inertia was determined in accordance with the kinematic scheme shown in Figure 4 and the calculation formula:

$$M_i = J_{red} \frac{d\omega}{dt}, \quad (1)$$

where J_{red} is the reduced moment of inertia, kg·m^2; ω is the angular velocity of the main shaft, s^{-1}.

The angular velocity of the main shaft is a variable value and is substantiated in [2, 7–14].

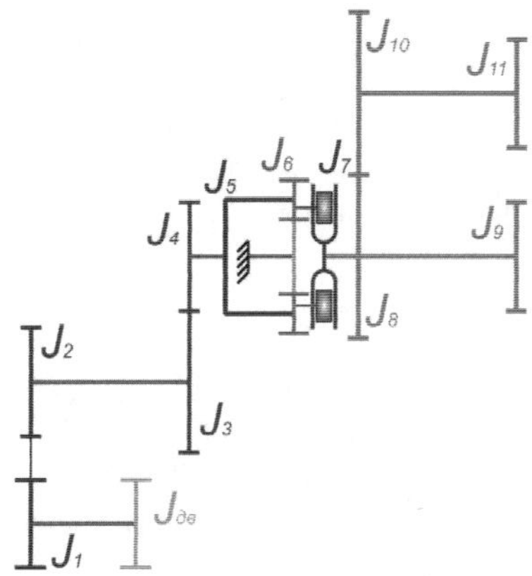

Figure 4. Design diagram of the mixer

To determine the reduced moment of inertia of the masses, we used the kinetic energy theorem [3, 7, 14]:

$$T_{red} = T_{mr} + T_1 + T_2 + T_3 + T_4 + T_5 + T_6 + T_7 + T_8 + T_9 + T_{10} + T_{11}, \qquad (2)$$

where T_{mr} – kinetic energy of the electric motor rotor; T_1 – kinetic energy of the drive pulley; T_2 – kinetic energy of the driven pulley; T_3 – kinetic energy of the gear shaft; T_4 – kinetic energy of the driven wheel; T_5 – kinetic energy of the driver; T_6 – kinetic energy of the planet gear and slot stone; T_7 – kinetic energy of the rocker arm; T_8 – kinetic energy of the drive wheel for the first working shaft; T_9 – kinetic energy for the first working shaft; T_{10} – kinetic energy of the drive wheel for the second working shaft; T_{11} – kinetic energy for the second working shaft; T_{12} – kinetic energy of the slot stone caused by its translational motion.

The reduced moment of mixture inertia for the entire device reduced to the main shaft was determined in accordance with the expression:

$$J_{red} = \frac{(J_{mr} + J_1)\cdot \omega_a^2}{\omega_a^2} + \frac{(J_2 + J_3)\cdot \omega_b^2}{\omega_a^2} + \frac{(J_4 + J_5)\cdot \omega_c^2}{\omega_a^2} + \frac{J_6 \cdot \omega_{\text{pl.}}^2 + m_6 \cdot V_{b\,\text{pl.}}^2}{\omega_a^2} +$$
$$+ \frac{(J_7 + J_8 + J_9 + J_{10} + J_{11})\cdot \omega_d^2}{\omega_a^2} + \frac{m_{\text{sl.st.}} \cdot V_{\text{sl.st.}}^2}{\omega_a^2}.$$

In expression (3), all notations correspond to the notation of the kinetic energies represented in formula (2), with the exception of the expression $m_{\text{sl.st.}} \cdot V_{\text{sl.st.}}^2 / \omega_a^2$, where $m_{\text{sl.st.}}$ is the mass of the slot stone; $V_{\text{sl.st.}}$ is the speed of the slot stone; ω_a is the angular velocity of the drive shaft.

3. Results and discussion

In accordance with the mentioned expressions and the design methodology for continuous mixers described in [3, 5, 7, 8, 10–14], the process load at the rotation angles of the working shaft equal to 45, 135, 225, 315 degrees was reduced to the main shaft of the mechanism presented in [2]. Taking into account the gear ratio, the process load is reduced to the main shaft. For one revolution of the work member, the motor shaft makes 16.069 revolutions. The results of the calculations are shown in Figure 5.

We obtained the inertial mass characteristics of the proposed mechanism's components to determine the reduced moments of the inertia forces (Table 1).

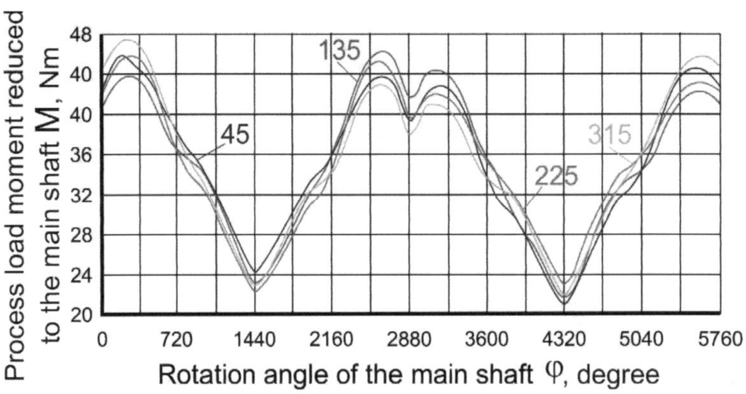

Figure 5. Process load reduced to the main shaft.

Table 1. Inertial mass characteristics of the working shaft drive components of the double agitator.

Name	Indication in Figure 4	Weight, m, kg	Moment of inertia, J, kg·m^2
The rotor of the motor	J_{mr}	83.500	0.183957
Drive pulley (D = 140 mm)	J_1	8.032	0.019629
Driven pulley (D = 560 mm)	J_2	37.333	2.122987
Shaft gear (m = 4, z = 19)	J_3	6.327	0.003069
Gear wheel (m = 4, z = 71)	J_4	27.587	0.329693
Driver	J_5	10.139	0.096606
Planet gear (m = 4, z = 22)	J_6	2.751	0.002739
Rocker arm	J_7	15.466	0.130923
Slot stone		0.454	0.000317
A gear wheel (m = 5, z = 88)	J_8, J_{10}	56.524	1.457055
Mixer's shaft	J_9, J_{11}	65.405	0.041337

The nature of the change in the mechanism's reduced moment of inertia is shown in Figure 6.

The angular velocity of the main shaft, taking into account the gear ratio, was determined by the formula $\omega_{red} = \omega_c \cdot i_{a-c}$.

Considering [2] and bearing in mind that $\varepsilon(\varphi) = \frac{d}{d\varphi}\omega(\varphi)$, we get the following dependence of angular acceleration which is graphically shown in Figure 6.

As a result, the reduced moment of forces calculated by the formula (1) has the following form (Figure 6).

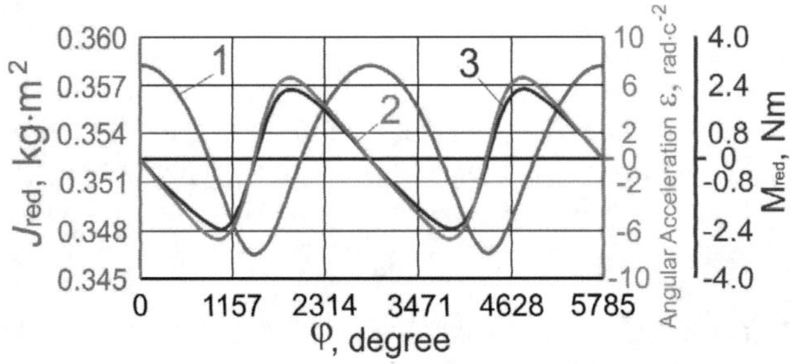

Figure 6. Graph of the change in: 1 - moment of inertia reduced to the main shaft; 2 - angular acceleration; 3 - reduced moment of inertia.

Summing up the graph of changes in the reduced moment of forces (Figure 6), and the graph of changes in the process load (Figure 5) yields a graph of changes in the total load reduced to the main shaft (Figure 7).

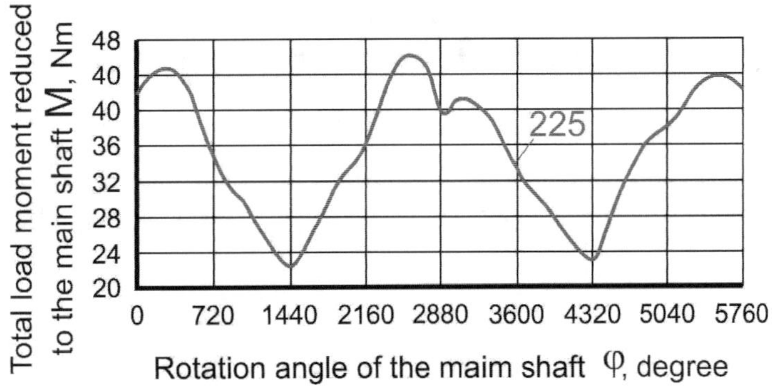

Figure 7. Graph of changes in the total load on the main shaft.

4. Conclusion

We performed the study of the reduced parameters of the epicyclic transmission drive with a rocker arm for the following engine parameters. The reference diameter of the larger wheel was d_l = 176 mm; the reference diameter of the small wheel was d_2 = 88 mm. The gear ratio of epicyclic transmission was i = 0.5; the driver length (center distance) was l = 132 mm; the eccentricity was e = 14 mm. The detailed substantiation of these parameters is presented in [2].

The paper presents a method for determining the reduced moments of the epicyclic transmission drive with a rocker arm, including:

- a mathematical model of the mechanism that allows calculating the reduced moment of inertia for the drive masses. The maximum values of the reduced moment of the masses inertia were 0.359 kg·m^2, the minimum – 0.357 kg·m^2;
- determining the nature of changes in the process load acting on the working shafts of the device; determining the values and nature of the reduced moments of these forces to the main shaft of the device. The maximum values of the total moment of the useful resistance forces were 44-47 N·m, the minimum – 22-24 N·m; the values depend on the blades placing angle on the working shafts;
- determining the dependence of the moments changes on the forces of inertia and their values and reducing them to the main shaft of the device. The maximum values of the inertia moment were 2.4 N·m, the minimum – 2.4 N·m;
- determining the dependence of changes in the total values of moments on the useful resistances forces as well as the dependence of moments on the inertia forces and reducing them to the main shaft of the device. The maximum values of the total moment of inertia forces were 45 N·m, the minimum – 22 N·m.

Acknowledgments
This work was financially supported within the framework of the NSTU Research and Development Thematic Plan (project No. TP-PTM-1_20 project).

References

[1] Mudrov A G 1999 *Development of spatial mixing devices of a new generation used in agriculture and industry* (Al'nopress Publ.) p 44
[2] Martynova T G et al 2020 *IOP Conf. Ser.: Mater. Sci. Eng.* **843(1)** 012006
[3] Hsieh J F 2014 *Mechanism and Machine Theory* **81** 155-165
[4] Podgornyj Y I et al 2017 *IOP Conf. Ser.: Earth Environ. Sci.* **87** 082039
[5] Antipov S T 2001 *Machines and equipment for food production* (High School Press.) p 703

[6] Martynova T G et al 2018 *IOP Conf. Ser.: Earth Environ. Sci.* **194(2)** 022022
[7] Zhoua C, Hua B, Chenb S and Mac L 2016 *Mechanism and Machine Theory* **104** 118-129
[8] Myszka D H 2012 *Machines & Mechanisms: Applied Kinematic Analysis* (Pearson Publ.) p 576
[9] Podgornyj Y I et al 2019 *Obrabotka metallov* **21(4)** 47-58
[10] Vulfson I 2015 *Dynamics of Cyclic Machines* (Springer Int. Publ.) p 390
[11] Podgornyj Y I, Skeeba V Y, Kirillov A V et al 2016 *Obrabotka metallov* **4** 24-33
[12] Podgornyj Y I et al 2019 *IOP Conf. Ser.: Earth Environ. Sci.* **378(1)** 012025
[13] Matyashin A V 1998 *Justification of the parameters and operating modes of the feed mixer of periodic action* (KGSKhA Publ.) p 24
[14] Mott R L 2013 *Machine Elements in Mechanical Design* (Pearson Publ.) p 816

Studying the dynamics of contact interactions during machining based on a system of nonlinear piecewise linear differential equations

D V Vasilkov[1], A S Alexandrov[1], V V Golikova[1], and T B Kochina[2]

[1] Baltic State Technical University VOENMEH, 1, 1st Krasnoarmejskaja street, St. Petersburg, 190005, Russia
[2] Nizhnevartovsk State University, 56, Lenina street, Nizhnevartovsk, Khanty-Mansiiskii avtonomous okrug - Iugra, 630073, Russia

E-mail: vasilkovdv@mail.ru

Abstract. When manufacturing parts for thermal control systems for operation in especially difficult conditions of the North and in spacecraft, titanium and complex alloyed alloys are used. They are subject to increased quality requirements. They are achieved through simulation modeling. The solution is based on a dynamic model in which the contact interaction between the tool flank and the workpiece is differentiated. The process of contact interaction is considered as two-phase in the form of a sequence of states of adhesion and sliding. In the space of state variables, a system of nonlinear differential equations of piecewise linear type is constructed based on a set of rheological models. Based on this model, simulation modeling was carried out in the form of a computational experiment. A periodic solution is obtained which is formed as a result of switching the sliding and adhesion phases. In the graphs presented in the paper, the transition from the adhesion phase to the sliding phase is accompanied by a surge in displacement. In this case, in the phase portrait, the phase trajectory reaches the limit cycle with a limited amplitude. At the moment of transition from the adhesion phase to the sliding phase, a characteristic deviation of the phase trajectory from the limit cycle with an increased amplitude is observed, followed by a return to the limit cycle. With a wide variation in cutting speed, the conditions for technical stability are determined. During the simulation, a dynamic manifestation of the formation of chip elements was found, which did not have a serious impact on the nature of dynamic processes. The results obtained make it possible to study the dynamics of contact interactions during cutting in the framework of nonlinear dynamic models. This makes it possible to assess the level of vibration amplitudes and determine the regulated parameters of processing accuracy and surface roughness of manufactured products.

1. Introduction

Devices and systems operating in the harsh conditions of the north, as well as in spacecraft serving the northern territories, operate using thermal control systems that ensure their normal operation [1]. In the manufacture of the drive part of these systems, the big problem is thin-walled parts such as bodies of revolution from complex alloyed and titanium alloys, such as shells, bushings, flanges, housings, etc. Some thin-walled elements have a thickness of 0.3 ... 0.9 mm [2].

In the manufacture and operational loads, the accuracy of the size, shape and relative position of the surfaces of these parts must be maintained in accordance with the requirements of the drawing and design documentation. Prediction of output characteristics during manufacturing is carried out by simulation based on nonlinear dynamic systems. The study of the dynamics of contact interactions during cutting has become widespread in domestic and foreign developments [3-11]. An analysis of the processes occurring during the contact interaction of the cutting tool and the workpiece made it possible to present this process as two phases: seizure and slide [12-14].

The molecular-mechanical processes that occur during the interaction of the cutting tool and the workpiece in the space of state variables can be described on the basis of a system of differential equations

$$\dot{u} = Du + S(u) \ , \tag{1}$$

where u is the vector of state variables; D is transformation matrix with constant elements; $S(u)$ is a vector function with piecewise linear components.

Model solutions as applied to the system of differential equations (1) are considered in detail in [12, 14, 15]. Its feature is that the vector function $S(u)$ is a piecewise linear function.

2. Materials and methods

Complex alloyed, heat-resistant and aluminum alloys are prone to the formation of adhesive bonds in a wide range of cutting speeds. In [5], it was confirmed that in the zone of contact interaction between the cutting tool and the workpiece, sequential formation (seizure) and destruction (sliding) of adhesive bonds continuously occur. Each of these states is described by a simple dynamic model. The transitions between the slip and seizure phases are determined by the kinematic and force characteristics embedded in the model and are monitored by the switching function $Sg1$, which can be written in the form of the relation [12]

$$Sg1 = \begin{cases} 1, -seizure\ phase\ ; \\ 0 - sliding\ phase\ . \end{cases} \tag{2}$$

The considered dynamic system (1) in the space of state variables is nonlinear piecewise linear system. Its solution is represented by expressions described in detail in [12-14]. Based on this dynamic system, it is possible to study the frequency composition and level of the amplitudes of oscillations with a wide variation of the model parameters.

3. Results

To assess the nature of the movements, we formulate the conditions of a computational experiment.

Workpiece parameters: diameter d = 90 mm, offset is 135 mm, material is XH65BMTJu (CrNi65BCuTeAl). Tool parameters: trailing angle $\alpha = 8°$; rake angle $\gamma = -8°$; the main entering angle $\varphi = 70°$; auxiliary entering angle $\varphi_1 = 20°$, material of the cutting part of the tool: at V = 30-70 m/min GC4235 “Sandvik” (Sweden); at V = 80-200 m/min M101S "Tungaloy" (Japan); section of the holder is 40x20; holder offset is 95 mm. Cutting modes: feed S = 0.19 mm/rev, cutting depth t = 1.0 mm; cutting speed V = 30 ... 220 m/min. Processing without cooling.

In all cases considered, the movement in the model begins with the seizure phase. This is evidenced by the graph of the switching function (time diagram 1 in Fig. 1a). The relative displacement along the normal to the forming point of the tool in the y direction (curve 2 in Fig. 1a) was obtained as a result of calculation according to model (1).

The periodic solution, which is formed as a result of switching the sliding and seizure phases, is clearly visible on it. In the graphs, the transition from the seizure phase to the sliding phase is accompanied by a surge in the displacement y (curve 2 in Fig. 1a). In this case, the phase trajectory exits the phase trajectory to the limit cycle with an amplitude of 2–5 μm (curve 3 in Fig. 1b). At the moment of transition from the seizure phase to the slip phase, a characteristic deviation of the phase trajectory from the limit cycle with an amplitude of 3-12 μm (curve 4 in Fig. 1b) is observed, followed by return to the limit cycle. This characteristic manifestation of a dynamic system is observed in the range of cutting speeds $V = 30 \ldots 50$ m/min (pos. 1-3, Table 1) and in the range of cutting speeds $V = 70 \ldots 200$ m / min (pos. 5-12, Table 1). In the indicated modes, the duration of the seizure phase was on average 10^{-4} s, and the sliding phases were an order of magnitude longer.

At a cutting speed of $V = 60$ m/min, there is a sharp increase in the level of vibration with a transition to the second limit cycle with an amplitude of 30 μm (pos. 5, Table 1). This mode is unacceptable in all cases and excluded from the list for possible implementation.

A comparison of the quantitative results of the calculation from the standpoint of technical stability is presented in Table 1. The range of acceptable modes is defined for tools made of hard alloys and mineral ceramic. Moreover, cermets are most effective at a temperature in the cutting zone, when there is a drop in the strength and yield strengths for a given processed material. This is evidenced by a decrease in the level of vibration amplitudes of less than 5 μm.

In the system of differential equations (1), the main mechanism of the adhesive-deformation interaction in the contact connection between the tool and the workpiece in the cutting zone, formed by a sequence of slip and seizure phases, is laid down. In this case, the phase switching function is formed synergistically by the system itself in accordance with expression (2).

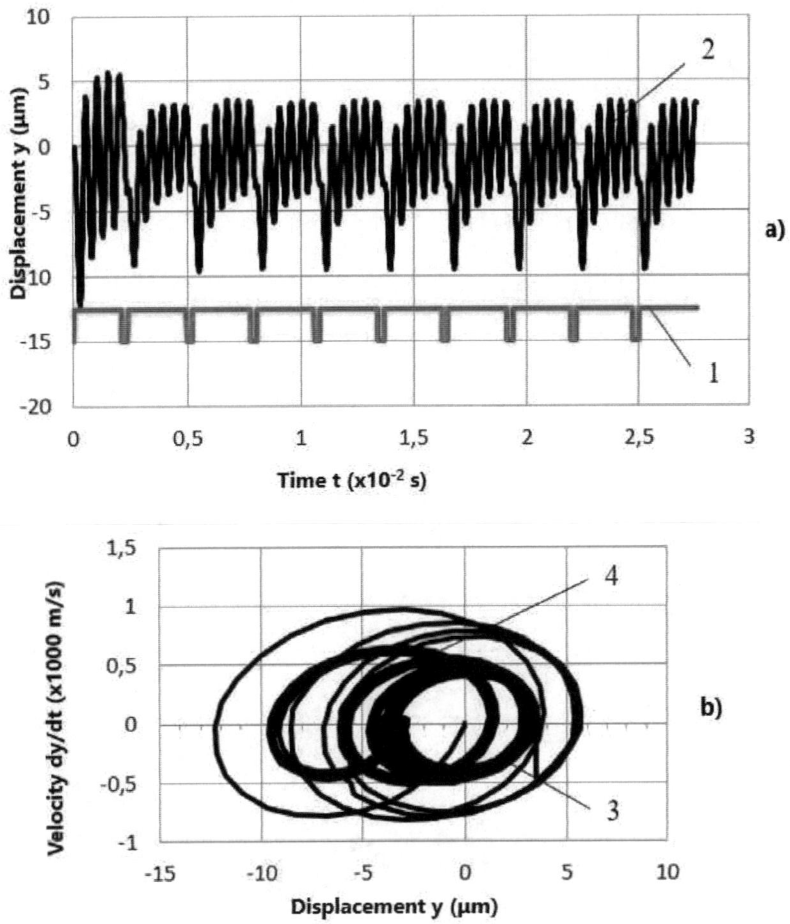

Figure 1. The calculated vibration displacements normal to the shaping point of the tool at $V = 40$ m/min: a - graph of displacements in the y direction, μm; time diagram of the function of switching dynamic modes; b - phase portrait of a dynamic system

In addition to this self-adjusting mechanism of a dynamic system, the mechanism for the formation of chip elements integrated in expressions (1) can be considered and dynamically modeled. In accordance with the accepted initial assumption, this mechanism is insignificant in terms of impact in comparison with the mechanism of adhesive-deformation interaction. It was assumed that the complication of the model by taking into account this mechanism will lead to unreasonable complication of the calculation results.

Omitting the mathematical calculations, we present the results of modeling a dynamic piecewise linear system, in which only switching, which reflects the formation of chip elements, is taken into account. Figure 2 presents the simulation result, taking into account only the formation of chip elements at $V = 200$ m/min. The graph is built at high magnification. In this case, the switching function displays the formation of chip elements. This result has an important theoretical meaning, since it allows identifying the frequency of formation of chip elements in the model solution and determine the conditions of elemental and adiabatic chip formation.

Table 1. Calculated data on the stability of a dynamic system

Pos. No.	Cutting speed [m/min]	Technical stability	max amplitude of vibrations [μm]	Tool material (+/-) Hard alloy	Mineral ceramics
1	30	SLC, a = 3 μm	12	+	-
2	40	SLC, a = 4 μm	9	+	-
3	50	SLC, a = 5 μm	10	+	-
4	60	NSLC, a = 30 μm	30	+	-
5	70	SLC, a = 6 μm	12	+	-
6	80	SLC, a = 4 μm	9	-	+
7	120	SLC, a = 3.5 μm	8	-	+
8	140	SLC, a = 3 μm	8	-	+
9	160	SLC, a = 3 μm	7	-	+
10	180	SLC, a = 2 μm	3	-	+
11	190	SLC, a = 2 μm	3	-	+
12	200	SLC, a = 2 μm	3	-	+

SLC – stable limit cycle; NSLC – non-stable limit cycle; a – limit cycle amplitude.

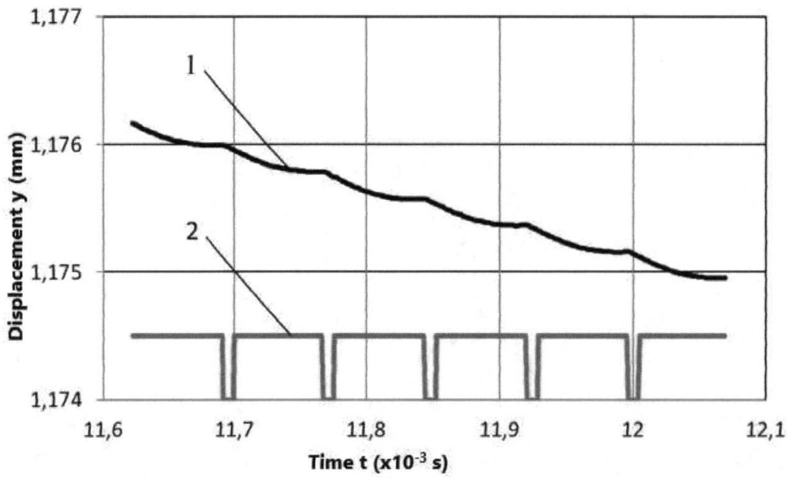

Figure 2. Calculated vibration displacements normal to the forming point of the instrument at high resolution (V = 200 m/min): 1 - graph of displacements in the y direction, μm; 2 is a time diagram of the dynamic mode switching function

The graph clearly shows small deviations (Fig. 2) associated with the formation of chip elements. They indicate the formation of chip elements, but do not have a serious impact on the nature of dynamic processes. This is also confirmed by the fact that the frequency of formation of chip elements is 11.4 kHz.

4. Conclusion

The results obtained enable the study of the dynamics of contact interactions during machining in the framework of nonlinear dynamic models. This allows assessing the level of oscillation amplitudes and determine the regulated parameters of processing accuracy and surface roughness of manufactured products.

Acknowledgments

The work was carried out at BSTU «VOENMEH» as part of the project «The creation of a high-tech import-substituting production of the elements of executive device systems with high life for transport and aerospace equipment, ensuring the development and use of the World Ocean, Arctic and Antarctic» with financial support from the Ministry of Science and Higher Education of the Russian Federation (agreement No. 075-11-2019-077 as of 13.12.2019) in accordance with the government decree of the Russian Federation No. 218 as of 09.04.2010.

References

[1] Bobkov A V and Tsvetkov E O 2011 Features of the balance of power losses in electric pump units of spacecraft thermal control systems *Bull. of the Samara Scientific Center of the Russian Academy of Sciences* **13(1(2))** 290-292
[2] Vasilkov D V, Kuznetsova Z A and Nikitin A V 2020 Technological quality support when turning thin-walled parts made of titanium alloys of electric pump units of the spacecraft temperature control system *Metalworking* **1(115)** 3-9
[3] Kudinov V A 1967 *Dynamics of machines* (Moscow: Mechanical Engineering) 359 p.
[4] Elyasberg M E 1993 *Self-oscillations of metal-cutting machines: Theory and practice* (St. Petersburg: OKBS) 180 p.
[5] Vasilkov D V, Weitz B L and Shevchenko V S 1997 *Dynamics of the technological system of mechanical processing* (Saint-Petersburg: Publishing Tool) 230 p.
[6] Maksarov V 2015 Improving the accuracy of manufacturing of hydraulic power cylinders using vibration-proof cutting tool *Agronomy Res.* **13(3)** 671-679
[7] Olt J, Liivapuu O, Maksarov V, Liyvapuu A and Tärgla T 2016 Mathematical modelling of cutting process system *Springer Proc. in Mathematics and Statistics* **178** 173-186
[8] Maksarov V V and Efimov A E 2018 Simulation modeling of dynamic characteristics of machining in NI LabView software environment to improve processing technique of a rod component *IOP Conf. Ser.: Earth Env. Sci.* **194(2)** 022021
[9] Olt J, Liyvapuu A, Madissoo M and Maksarov V 2016 Dynamic simulation of chip formation in the process of cutting *Int. J. of Materials and Product Technology* **53(1)** 1-14
[10] Maksarov V V, Khalimonenko A D and Matrenichev K G 2017 Stability analysis of multipoint tool equipped with metal cutting ceramics *IOP Conf. Ser.: Earth Env. Sci.* **7(8)** 082030
[11] Olt J and Maksarov V V 2015 Cutting process simulation on the basis of rheological properties of metals *Annals of DAAAM and Proc. of the Int. DAAAM Symp.*, pp 229-237
[12] Vasilkov D V and Kochina T B 2013 Modelling of contact interactions taking into account the rheology in the cutting zone at high-speed processing of products from heat resisting alloys *Metalworking* **4(76)** 2-10
[13] Vasilkov D V and Kochina T B Dynamic features of formation of conditions of contact interaction in a cutting zone on the basis of piecewise and linear approximation *Metalworking* **5-6(77-78)** 2-3
[14] Nikitin A V, Tarikov I Ya and Vasilkov D V 2019 Determination of technological residual

stresses in the surface layer of parts with thin-walled elements during turning *J. of Advanced Res. in Dynamical and Control Systems* **11(08)** 2926-2932
[15] Maksarov V, Efimov A and Golikov T 2020 Treatment of Titanium Alloys Based on Preliminary Plastic Impact *Key Engineering Materials Submitted* **836** 63-70

Modeling of contact interactions in lathe turning with accounts for rheology in the cutting zone

D V Vasilkov[1], V S Cherdakova[1] and M S Bundur[2]

[1] Baltic State Technical University VOENMEH, 1, 1st Krasnoarmejskaja street, St. Petersburg, 190005, Russia
[2] Peter the Great St. Petersburg Polytechnic University, 29, Polytekhnicheskaya street, St. Petersburg, 195251, Russia

E-mail: vasilkovdv@mail.ru

Abstract. This article discusses the dynamic modeling of contact interactions in the technological system of machining in the turning process based on piecewise linear approximation. A feature of the model is that it differentiates between the conditions of contact interaction between the front surface of the tool and the coming off chips, as well as between the back surface of the tool and the workpiece being processed. The consideration is based on the two-term characteristic of friction, realized in the form of a sequence of sliding and adhesion phases. In this case, the process of contact interaction in each phase is described by a set of rheological models. The phase switching conditions are determined by the rate of relative contact interactions, as well as the contact pressures between the front surface of the tool and the chips, between the back surface of the tool and the workpiece. Differentiated description of contact interactions from the front and back surfaces of the tool made it possible to build a tool for studying the processes of self-regulation of contact interactions in dynamic contours and the connectivity between them, which allows us to determine the conditions for the formation of self-oscillations that the dynamic system imposes on itself. The performed experimental studies have shown the correspondence between the nature of the movements during modeling and real cutting.

1. Introduction

During mechanical tooling, issues arise in ensuring product quality (dimensions, micro and macro of surfaces, residual post-processing stresses). They may be efficiently resolved by studying dynamic properties of the tooling processing system, in particular, the contact interaction in the chip formation zone [1-6].

Analysis of processes taking place during the interaction between the chip and the frontal surface of a tool and between the processed surface and the back surface of the tool allows representing this process as a two-stage one, with a binding stage and a gliding stage [7-9].

Molecular processes taking place in the chip formation zone are represented as rheological models that link together a blank subsystem and a cutting tool subsystem [8-10]. In the model representation, the processes in question are reflected as a viscoelastic Voigt medium, an ideal Hooke's medium, an ideal viscose Newton's medium and an ideal plastic Saint-Venant medium (a trigger-shaped key) (Figures 1, 2), adapted to the problem in question. The modeled conditions include the interaction of elastic dissipative and inertial characteristics of the processing system, and the viscoelastic and plastic characteristics in the chip formation zone.

2. Materials and methods

Complex, refractory and aluminum alloys tend to form adhesive links at a broad range of cutting speeds. In [5], it has been confirmed that in the contact interaction area between the cutting tool and the blank there is a continuous formation (binding) and destruction) gliding of adhesive links.

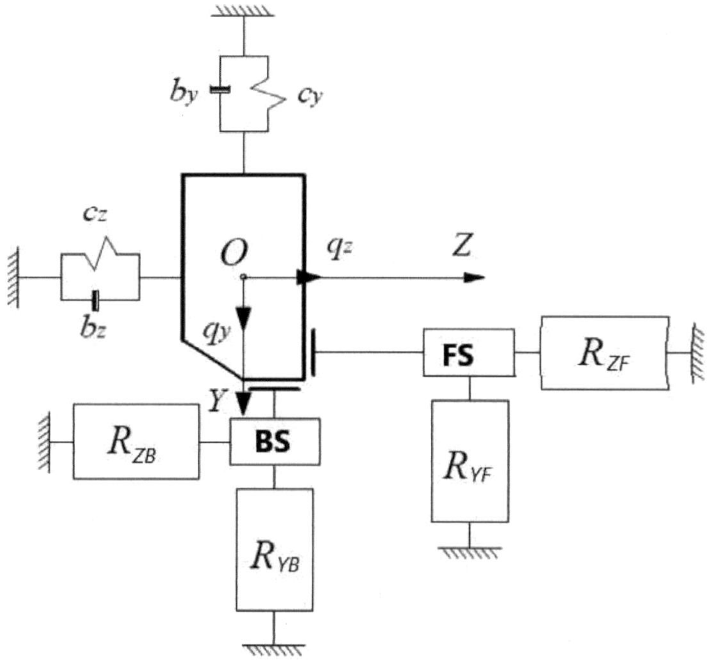

Figure 1. Rheological representations of force interactions in the cutting area

Conditions for transition between a gliding stage and a binding stage in interaction of the front and back surfaces of the tool with a chip and produced surface respectively are formed by the system itself, which is potentially auto-oscillating.

In Figure 1: R_{ZF}, R_{YF} is a group of rheological elements along the front surface in the direction of *OZ*, *OY* axis; R_{ZB}, R_{YB} is a group of rheological elements along the back surface in the direction of *OZ*, *OY* axes; z, y are generalized coordinated in the direction of *OZ*, *OY* axes, c_z, c_y are stiffness coefficients, b_z , b_y are dissipation factors.

In Figure 2: c_{zf1}, c_{zf2}, c_{yf1}, c_{yf2}, b_{zf1}, b_{zf2}, b_{yf1}, b_{yf2} are elasticity and dissipation factors of the rheological elements along the front surface of the tool in the direction of *OZ*, *OY* axes; c_{zb1}, c_{zb2}, c_{yb1}, c_{yb2}, b_{zb1}, b_{zb2}, b_{yb1}, b_{yb2} are elasticity and dissipation factors along the frontal surface of the tool in the direction of *OZ*, *OY* axes; Sv_{zb1} и Sv_{zb2} are Saint-Venant's tribological elements in the tangential outline along the back surface of the tool, Sv_{yb1} and Sv_{yb2} are Saint-Venant's tribological elements in the normal outline along the back surface of the tool; Sv_{zf1} and Sv_{zf2} are Saint-Venant's tribological elements in the tangential outline along the front surface of the tool, Sv_{yf1} и Sv_{yf2} are Saint-Venant's tribological elements in the normal outline along the frontal surface.

On the back surface of the tool, there is a viscoelastic interaction with the processed surface. Rheology of near-surface zone of the product material is modeled with the elements R_{YB} and R_{ZB} (Figure 1) [11]. From the frontal side of the tool, there is a viscoelastic interaction in the chip formation zone, which is modeled with the elements R_{YF} and R_{ZF} (Figure 1). The rheological models of the force interactions in the cutting zone and at the back surface of the tool are in detail shown in Figure 2.

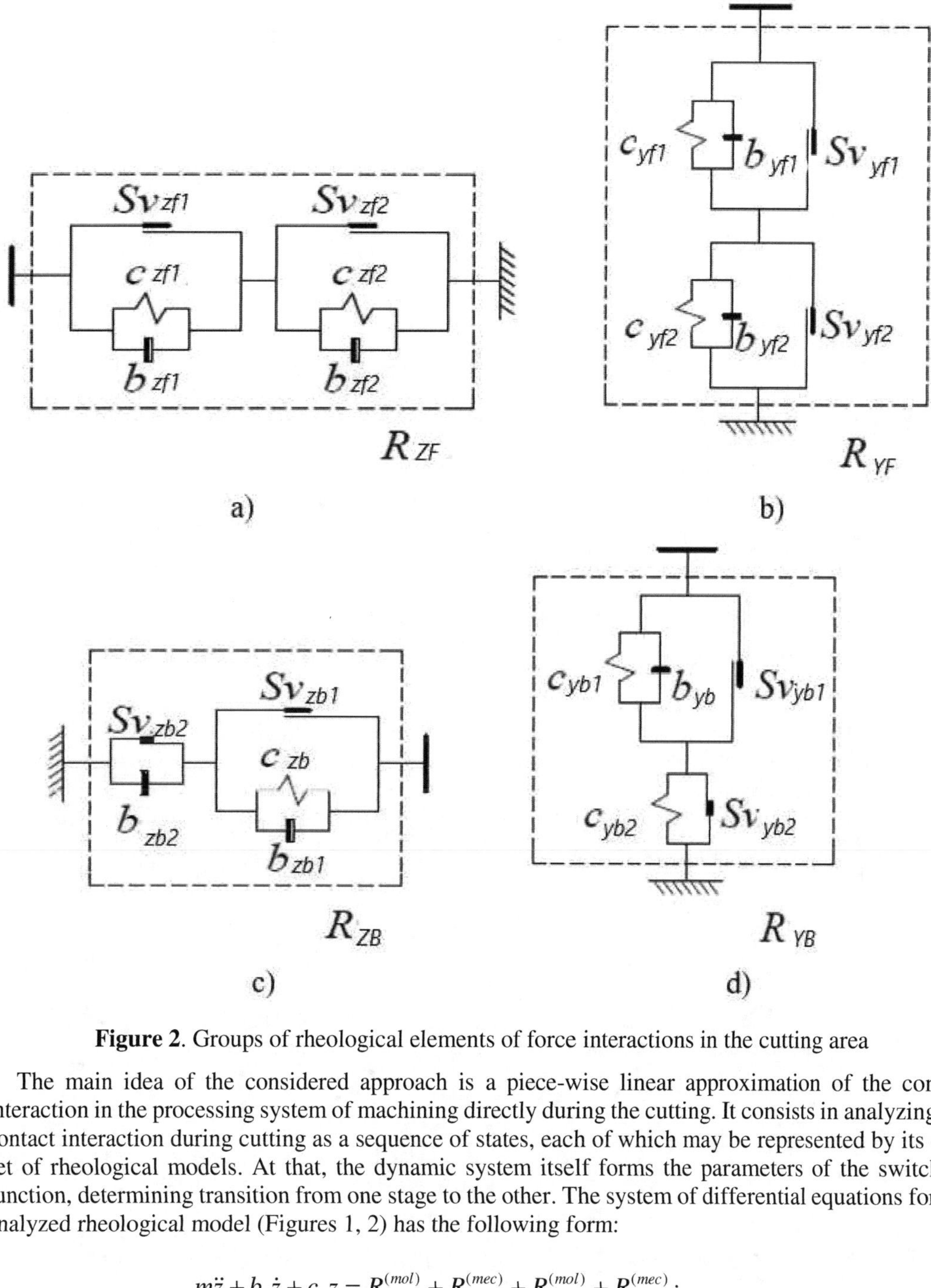

Figure 2. Groups of rheological elements of force interactions in the cutting area

The main idea of the considered approach is a piece-wise linear approximation of the contact interaction in the processing system of machining directly during the cutting. It consists in analyzing the contact interaction during cutting as a sequence of states, each of which may be represented by its own set of rheological models. At that, the dynamic system itself forms the parameters of the switching function, determining transition from one stage to the other. The system of differential equations for the analyzed rheological model (Figures 1, 2) has the following form:

$$m\ddot{z} + b_z\dot{z} + c_z z = R_{zf}^{(mol)} + R_{zf}^{(mec)} + R_{zb}^{(mol)} + R_{zb}^{(mec)}; \quad (1)$$

$$m\ddot{y} + b_y\dot{y} + c_y y = R_{yf}^{(mol)} + R_{yf}^{(mec)} + R_{yb}^{(mol)} + R_{yb}^{(mec)}; \quad (2)$$

$$T_{zf}^{(mol)}\dot{R}_{zf}^{(mol)}+R_{zf}^{(mol)}=-(b_{zf1}\dot{z}+c_{zf1}z)Sv_{zf1}; \quad (3)$$
$$T_{yf}^{(mol)}\dot{R}_{yf}^{(mol)}+R_{yf}^{(mol)}=-(b_{yf1}\dot{y}+c_{yf1}y)Sv_{yf1}; \quad (4)$$
$$T_{zf}^{(mec)}\dot{R}_{zf}^{(mec)}+R_{zf}^{(mec)}=-(b_{zf2}\dot{z}+c_{zf2}z)Sv_{zf2}; \quad (5)$$
$$T_{yf}^{(mec)}\dot{R}_{yf}^{(mec)}+R_{yf}^{(mec)}=-(b_{yf2}\dot{y}+c_{yf2}y)Sv_{yf2}; \quad (6)$$
$$T_{zb}^{(mol)}\dot{R}_{zb}^{(mol)}+R_{zb}^{(mol)}=-(b_{zb1}\dot{z}+c_{zb}z)Sv_{zb1}; \quad (7)$$
$$T_{yb}^{(mol)}\dot{R}_{yb}^{(mol)}+R_{yb}^{(mol)}=-(b_{yb}\dot{y}+c_{yb1}y)Sv_{yb1}; \quad (8)$$
$$T_{zb}^{(mec)}\dot{R}_{zb}^{(mec)}+R_{zb}^{(mec)}=-\mu c_{yb2}y\ Sv_{zb2}; \quad (9)$$
$$T_{yb}^{(mec)}\dot{R}_{yb}^{(mec)}+R_{yb}^{(mec)}=-b_{zb2}\dot{z}\ Sv_{yb2}, \quad (10)$$

where $T_{\text{zf}}^{(mol)}$, $T_{yf}^{(mol)}$, $T_{\text{zf}}^{(mec)}$, $T_{yf}^{(mec)}$, $T_{\text{zb}}^{(mol)}$, $T_{yb}^{(mol)}$, $T_{yb}^{(mec)}$, $T_{yb}^{(mec)}$ are time constants of the working processes in the direction of *OY, OZ* axes; $R_{zf}^{(mol)}$, $R_{zf}^{(mec)}$, $R_{zb}^{(mol)}$, $R_{zb}^{(mec)}$, $R_{yf}^{(mol)}$, $R_{yf}^{(mec)}$, $R_{yb}^{(mol)}$, $R_{yb}^{(mec)}$ is a projection of the cutting force in the form of dynamic characteristics from the sides of the front and back surfaces of the tool in the direction of the *OY, OZ* axes.

During the contact interaction, transition from the binding stage to the gliding stage and back is modeled by movable Saint-Venant's tribological elements with characteristics of Sv_{zf1}, Sv_{zf2}, Sv_{zb1}, Sv_{zb2} in the tangential outline, and Sv_{yf1}, Sv_{yf2}, Sv_{yb1}, Sv_{yb2} in the normal outline. The switching function for transition between the gliding stage and the binding stage may be written down in the following form

$$\begin{cases} when\ Sv_{zf1}=Sv_{yf1}=Sv_{zb1}=Sv_{yb1}=0,\ Sv_{zf2}=Sv_{yf2}=Sv_{zb2}=Sv_{yb2}=1\ -\text{binding}; \\ when\ Sv_{zf1}=Sv_{yf1}=Sv_{zb1}=Sv_{yb1}=1,\ Sv_{zf2}=Sv_{yf2}=Sv_{zb2}=Sv_{yb2}=0-gliding. \end{cases} \quad (11)$$

Characteristics of movable tribological elements $Sv_i=1$ (correspondingly for outlines z and y) point to the fact that this movable tribological element is closed and its rheological unit is a perfectly solid body. It means that only one rheological unit participates in the contact interaction. The switching function (11), simultaneously with the system of differential equations (1)-(10) performs sequential switching of the tribological elements between the binding and gliding stages during the monitoring of kinematic characteristics of relative oscillations in the dynamic system in question.

A truly adhesive process, where binding happens, takes place when the following conditions hold:

$$\delta_{cz}>[\delta_{cz}], \quad (12)$$

where δ_{cz} is a normal contact deformation; $[\delta_{cz}]$ is the value of the normal contact deformation after which an adhesive process starts, and

$$V_z<\frac{l_c}{T_{yf}^{(mol)}}=[V_z], \quad (13)$$

where V_z is a relative tangential speed in the contact coupling; $[V_z]$ is the value of the relative tangential speed after which adhesive processes cease; l_c is the chip formation route.

The system of equations (1)-(10) with a broad variation of parameters was used as a foundation for modeling the processing system of tooling by lathe turning: The processed material was 20Kh13 steel, the tool material was R6M6 rapid tool steel, the cutting speed was from 15 to 200 m/min, a width of cut was 0.5 – 5 mm, the plan approach angle was 60 degrees, blank diameter was 60 mm. The results of separate comparative calculations and experimental studies are shown in Figure 3.

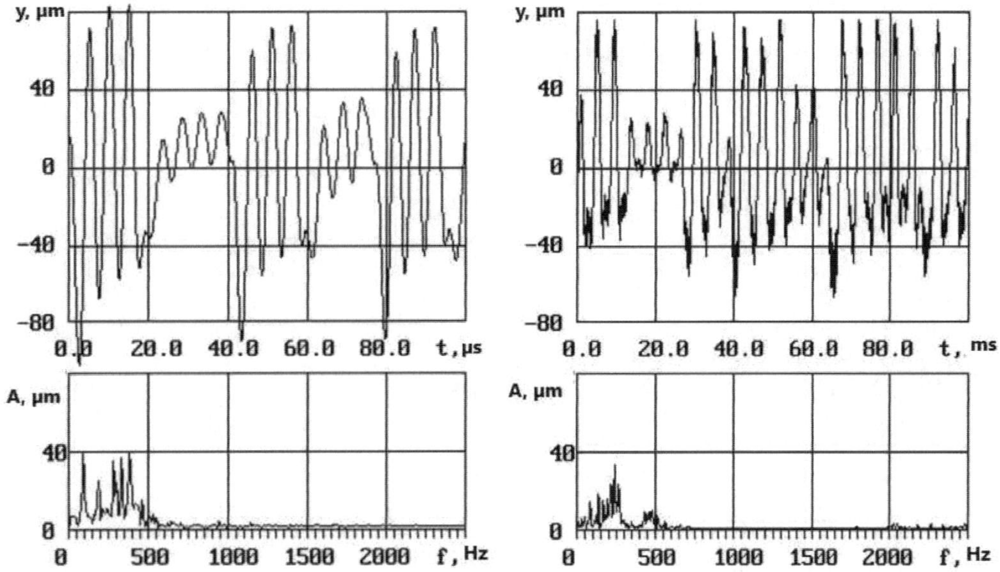

a) b)

Figure 3. Design (a) and experimental (b) oscillograph charts of vibrational motion during the lathe turning (*V*=30 m/min, the tool material – R6M5, the blank material – 20Kh13)

In both design and experimental plots of vibrational motions along the normal to the formative point of the blank at a cutting speed of 30 m/min, there are typical pulse motions that may be interpreted as sequential transitions between the gliding and binding stages. Divergence between the design and the experimental data does not exceed 122 percent, which is satisfactory.

3. Conclusion

In this modeling representation, the processes in the cutting zone are considered on the basis of a rheological approach, allowing reflecting the interaction of elastic-dissipative and inertial characteristics of the processing system in the cutting zone within the framework of the general dynamic model. The obtained differential equations significantly simplify the method for determining the number of transitions between the gliding and binding stages in the area where self-sustained oscillations exist. The method allows largely determining the conditions under which the self-sustained oscillations form.

4. Acknowledgments

The work was carried out in BSTU «VOENMEH» as part of the project «The creation of a high-tech import-substituting production of the elements of executive device systems with high life for transport and aerospace equipment, ensuring the development and use of the World Ocean, Arctic and Antarctic» with financial support from the Ministry of Science and Higher Education of the Russian Federation (agreement № 075-11-2019-077 of 13.12.2019) in accordance with the government decree of the Russian Federation of 09.04.2010 no. 218.

References

[1] Bobkov A V and Tsvetkov E O 2011 Features of the balance of power losses in electric pump units of spacecraft thermal control systems *Bull. of the Samara Scientific Center of the Russian Academy of Sciences* **13(1(2))** 290-292

[2] Vasilkov D V, Kuznetsova Z A and Nikitin A V 2020 Technological quality support when turning thin-walled parts made of titanium alloys of electric pump units of the spacecraft temperature control system *Metalworking* **1(115)** 3-9

[3] Kudinov V A 1967 *Dynamics of machines* (Moscow: Mechanical Engineering) 359 p.

[4] Elyasberg M E 1993 *Self-oscillations of metal-cutting machines: Theory and practice* (St. Petersburg: OKBS) 180 p.

[5] Vasilkov D V, Weitz B L and Shevchenko V S 1997 *Dynamics of the technological system of mechanical processing* (Saint-Petersburg: Publishing Tool) 230 p.

[6] Maksarov V 2015 Improving the accuracy of manufacturing of hydraulic power cylinders using vibration-proof cutting tool *Agronomy Res.* **13(3)** 671-679

[7] Olt J, Liivapuu O, Maksarov V, Liyvapuu A and Tärgla T 2016 Mathematical modelling of cutting process system *Springer Proc. in Mathematics and Statistics* **178** 173-186

[8] Maksarov V V and Efimov A E 2018 Simulation modeling of dynamic characteristics of machining in NI LabView software environment to improve processing technique of a rod component *IOP Conf. Ser.: Earth Env. Sci.* **194(2)** 022021

[9] Olt J, Liyvapuu A, Madissoo M and Maksarov V 2016 Dynamic simulation of chip formation in the process of cutting *Int. J. of Materials and Product Technology* **53(1)** 1-14

[10] Maksarov V V, Khalimonenko A D and Matrenichev K G 2017 Stability analysis of multipoint tool equipped with metal cutting ceramics *IOP Conf. Ser.: Earth Env. Sci.* **7(8)** 082030

[11] Olt J and Maksarov V V 2015 Cutting process simulation on the basis of rheological properties of metals *Annals of DAAAM and Proc. of the Int. DAAAM Symp.,* pp. 229-237

[12] Vasilkov D V and Kochina T B 2013 Modelling of contact interactions taking into account the rheology in the cutting zone at high-speed processing of products from heat resisting alloys *Metalworking* **4(76)** 2-10

[13] Vasilkov D V and Kochina T B 2013 Dynamic features of formation of conditions of contact interaction in a cutting zone on the basis of piecewise and linear approximation *Metalworking* **5-6(77-78)** 2-3

[14] Nikitin A V, Tarikov I Ya and Vasilkov D V 2019 Determination of technological residual stresses in the surface layer of parts with thin-walled elements during turning *J. of Advanced Res. in Dynamical and Control Systems* **11(08)** 2926-2932

Automated calculation method effect values in load securing elements fixed on a rolling stock

N V Vlasova[1], V A Olentsevich[2], V Yu Konyukhov[3], D A Lysenko[3]

[1] East Siberian Branch of the JSC "Russian Railways", 12/1, Vtoraya Zheleznodorozhnaya str., Irkutsk, 664005, Russia
[2] Irkutsk State Transport University, 15, Chernyshevskogo str., Irkutsk, 664074, Russia
[3] Irkutsk National Research Technical University, 83, Lermontova str., Irkutsk, 664074, Russia

E-mail: olencevich_va@irgups.ru

Abstract. The article presents a method for the automated calculation of the magnitude of the efforts in the means of securing cargo fixed to the rolling stock in accordance with the Technical Conditions for Stowing and Securing Cargo in Wagons and Containers - a regulatory document of JSC Russian Railways (hereinafter referred to as TU). The method provides for all possible factors of the influence of the forces acting on the transported cargo as a result of its movement from the point of loading to the point of unloading, including the necessary technological operations performed along the route, using the example of a cargo with a flat base using the capabilities of the computing environment. The obtained research results are supposed to be used by shippers in the development of the optimal method for placing and securing transported goods on the rolling stock. The automated calculation method will allow obtaining the predicted value of the cargo shift value both along and across the car from the total effect of longitudinal, transverse and vertical forces acting on the cargo along the way. These values make it possible to optimally determine the place of installation of thrust bars on the car floor, depending on the geometric parameters of the transported cargo and the chosen method of fastening it to the rolling stock.

1. Introduction

Today the safety of the transportation process is the most significant requirement for the organization of railway transportation. On the one hand, this type of transport is used to move a huge number of people; on the other hand, railways are a strategically important transport system for the country, on the smooth operation of which the economic condition of many, many industries, and the economy as a whole, depends.

Any "contingency" situation can provoke serious problems for entire sectors of the economy, not to mention individual enterprises or citizens. Therefore, ensuring the safety and continuity of the transportation process in all areas of activity is a multipurpose task of the railway transport system (RTS). Increasing requirements of ZDTS clients to the quality and the list of services provided determines the need to use advanced world technologies. In the short term, in order to reduce the distance between major regional centers, it is planned to develop high-speed traffic and increase the speed of passenger trains to 160-200 km / h. The main condition for the organization of traffic with the smooth operation of the railway system is to observe the safety of train passage. Unimpeded and safe passage of trains can be called quality execution of the train schedule.

The management of the Russian Railways has adopted and is successfully implementing a draft of a new concept for improving traffic safety in the field of transportation, based on the use of multifunctional integrated systems for regulating train traffic [1]. The transition to the use of technical safety equipment in the Russian railway transport system based on the latest information technologies - satellite navigation, digital radio channel, centralized route management and diagnostics of technical means - will allow the devices distributed along the tracks with their safety functions to be transferred to railway stations and train locomotives. This will significantly increase the level of safety of the transportation process and the customer focus of the industry. Since the introduction of modern security systems improves not only protection against hazardous situations, but also reduces system failures and failures, which makes it possible to increase the speed of trains, increase the throughput of railway facilities, and reduce the number of unscheduled repairs of technical equipment. That is, this time, the solution of security issues in the railway transport system was treated comprehensively.

2. Elimination of the negative impact of the human factor on traffic safety by automating production processes

The normal functioning of railway transport systems is possible only through coordinated and efficient work of groups of subsystems: infrastructure, commercial, freight, shunting and train, which are peculiar technological processes with specific features for each of them. Each of the subsystems has its own operating technology, its own technical equipment, fixed assets, specially trained technical personnel. Common things to all subsystems and the system as a whole are:

- subject of labor - transportation of goods (passengers);
- goal to generate income and profit;
- tasks ensuring the safety of functioning in order to reduce non-productive costs and the safety of fixed assets of the railway transport system and each of its subsystems.

All subsystems are interconnected by internal links and enter into technical, technological, financial and economic relations, without which the normal course of the transportation process is impossible. Clear and uninterrupted operation of the system, elimination of cases of risks, achievement of the goal and solution of the assigned tasks can be achieved only through close interaction of subsystems with each other, compliance with the requirements of regulatory documents, compliance with norms and standards of work.

In order to identify the causes most dependent on the occurrence of risk cases, it is advisable, first of all, to analyze the activities of the cargo subsystem. It is the control of the procedure for accepting cargo for transportation that is fundamental in organizing the transportation process, since the safety of the functioning of the railway transport system as a whole and its individual subsystems largely depends on the correct preparation of cargo and rolling stock in the transportation process [2, 3].

According to the reporting data of the Transportation Management Department of the Russian Railways Holding, the industry has been continuously increasing its traffic volumes in recent years. This has a positive effect on the financial and economic performance, but along with the growth in traffic volumes, the number of traffic safety risks associated with disorders of loading of goods loaded in wagons and containers. To identify the main factual causes of security violations in the railway transport system associated with violations of the loading disorder of goods loaded in wagons and containers, to improve the quality of the transportation process, providing a systematic approach to determine the root causes of this problem, based on data on the state of safety on Russian railways in 2010-2019 [1, 4], the Ishikawa diagram is constructed (figure 1).

As the analysis of the risks threatening traffic safety associated with violations of the disruption of loading of goods loaded in wagons and containers shows, the role of the human factor can be seen in the cause of almost each of them. Therefore, considering the priority tasks in the field of traffic safety, it is necessary to pay attention to the work of people.

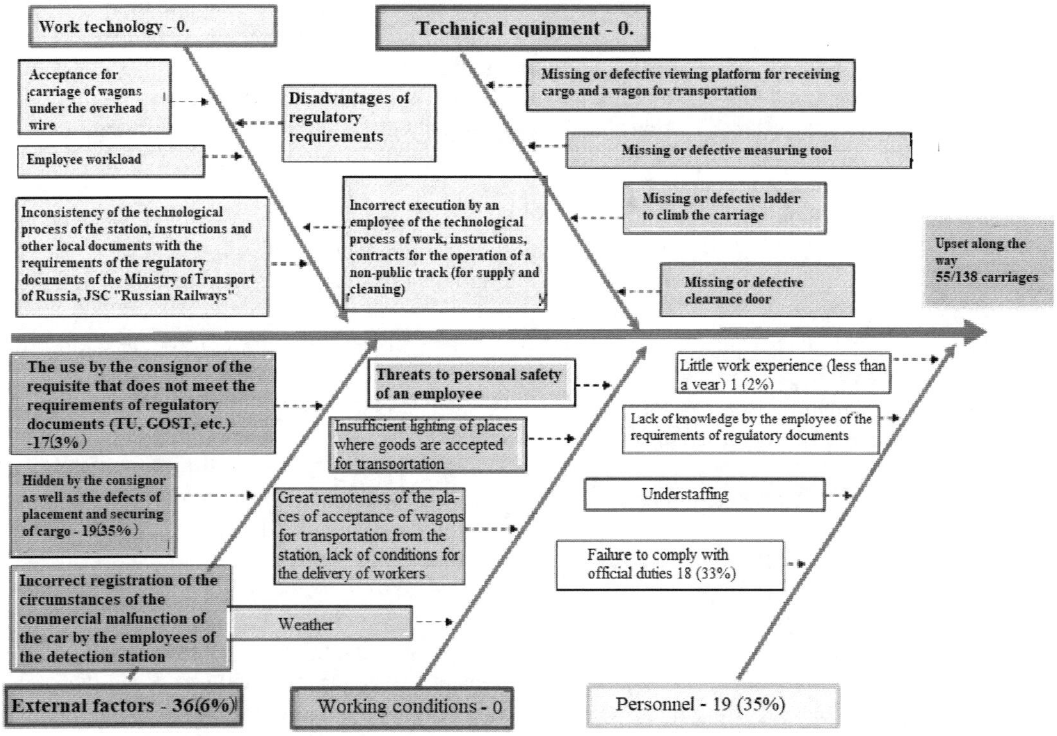

Figure 1. Diagram of causal relationships of violations of loading disorder of goods loaded in wagons and containers.

The new concept of improving traffic safety in the field of organizing transportation is focused on solving a number of tasks, the main of which is to eliminate the negative impact of the human factor on traffic safety by automating management processes and introducing additional control over the actions of operating personnel and users of railway services.

Coordinated development and improvement of existing technical means of security in certain areas of activity, taking into account the complex application of new systems and technologies in specific areas or sections of railways, the use of modern information technologies in the development of new technical means, creation of conditions for a comprehensive solution of transport security issues when accepting cargo to transportation are the main strategic tasks of the railway transport system.

3. The method of automated calculation of the magnitude of the forces in the means of securing the cargo fixed on the rolling stock

The article proposes a method for automated calculation of the magnitude of efforts in the elements of securing cargo fixed on the rolling stock in accordance with the Technical Conditions for Stowing and Securing Cargo in Wagons and Containers - a regulatory document of JSC Russian Railways (hereinafter - TU) [4, 5]. The method provides for all possible factors of influence of the forces acting on the cargo as a result of its movement from the point of loading to the point of unloading. This includes the necessary technological operations performed along the route, using the example of a cargo with a flat base using the capabilities of the computing environment.

The compiled system of equations [6-8], taking into account the nature of the change in the friction force, is easily realizable in a computing environment. The use of it makes it possible to compile calculation programs for the direct determination of the forces in the means of securing the transported cargo, the normal reaction of the connection and the coordinate of the point of its application, the values of the load displacement along the transverse the axis of the car, rotation around the vertical axis [9-11].

The approach proposed by the authors to the preparation of initial data is one of the crucial stages of computational experiments. This subsequently simplifies the compilation of a mathematical model and its implementation in an electronic environment, the use of which makes it possible to automate the calculation to determine the forces in the means of securing goods transported in wagons and containers. At the same time, in the mathematical model, the forces in the fastening means are taken into account only by positive signs, regardless of the direction of the projection of these forces relative to the adopted coordinate axes. All this is already taken into account automatically, and, observing the generally accepted rules of signs, external forces and moments acting on the system "transported cargo-securing means-rolling stock" with a flat base [12-14].

The approach developed using the basic provisions of general mechanics to describe the equilibrium conditions of the system "transported cargo-securing means-rolling stock" made it possible to compose a generalized mathematical model of this system using the example of transportation of cargo having a flat base, in the form of equations. This made it possible to use a computing environment to determine predictive values of the magnitude of the forces in the load securing braces at different values of the coefficient of friction, longitudinal, transverse and vertical accelerations, arising wind loads [2, 15-16].

The obtained research results are supposed to be used by shippers in the development of the optimal method for placing and securing transported goods on the rolling stock. The automated calculation method will allow obtaining the predicted value of the cargo shift value both along and across the car from the total effect of longitudinal, transverse and vertical forces acting on the cargo along the way [9, 17, 18]. The obtained values make it possible to optimally determine the place of installation of thrust bars on the car floor, depending on the geometric parameters of the transported cargo and the chosen method of fastening it to the rolling stock. This, in turn, will make it possible to analytically substantiate the installation locations of the fastening elements in the form of thrust bars with joint fastening cargo with guy wires, straps and stubborn bars. In addition, in an automated way, calculations can be performed to determine the number of fasteners to secure these thrust bars [9, 19].

Figure 2 shows, for example, the design model required to determine the required number of fasteners to secure the thrust bar on the car floor.

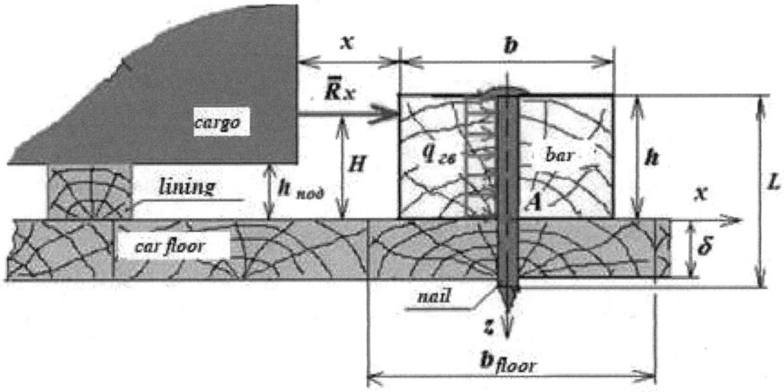

Figure 2. Scheme of fastening the bar to the car floor under the action of a longitudinal force.

Figure 2 has the following notation:

Rx - force acting from the side of the cargo on a thrust bar along the carriage, kN;

H - distance from the car floor to the point of application of the force Rx, m;

x or *y* - distance between the ends of the load and the thrust bar, m, with $0 \leq x$ or $y \leq 0.04$;

A - point of intersection of the plane of the car floor and the axis of the nail;

b and *h* - width and height of the bar, m;

δ - thickness of the car floor, m;

IOP Conf. Series: Materials Science and Engineering 1064 (2021) 012042 doi:10.1088/1757-899X/1064/1/012042

L - nail length, m,

b_{floor} - width of one carriage floor board, m.

For a specific load with the accepted number of securing means, the predicted value of the load shift can be determined using the capabilities of the presented computing environment. Computational experiments to determine the forces in the fasteners of the transported cargo, the values of the shifts of the cargo along and across the carriage are performed in the following sequence:

- input of initial data, which include: technical and operational characteristics of the rolling stock selected for transportation; geometric parameters of fastening means, cargo, carriage; coefficient of friction of the contacting surfaces of the cargo and the floor of the car; efforts of preliminary spins of fastening means;
- direct calculation of efforts in the fastening means, predicted values of possible shifts of the load along and across the car.

The source and other related data for the example copied from the layout document are shown in Figure 3.

Cargo parameters

$M := 60 \cdot 10^3$ - weight of cargo, kg $TOL := 10^{-5}$ $g := 9.81$

$h := 2.1$ - cargo height, m м

$b := 2$ - width of cargo, m $b1 := 1$ - lateral distance between loops for extreme stretch marks, m

$l := 9.1$ - length of cargo, m

$bп := 2.87$ - transverse distance between brackets, m

$f := 0.55$ - adhesion coefficient of friction (reinforced concrete on wood)

Parameters of flexible fasteners

$E := 1 \cdot 10^7$ - modulus of elasticity of flexible fastening elements, taking into account the wiring line, kN / m2

$d := 0.006$ - wire diameter of flexible fastening element, m

$n := 8$ - number of threads in flexible fastening elements, pcs

$A := n \cdot \pi \cdot \frac{d^2}{4}$ $A = 2.262 \times 10^{-4}$ - cross-sectional area of the wire of the flexible fastening element, m2

$S0 := 2.05 \cdot 9.81$ - initial tension of flexible fastening element, kN

Figure 3. Initial and other related data for computational experiments.

Below are the results of computational experiments to determine the forces acting on the side of the transported load on the thrust bar and the required number of fasteners of the thrust bar required to hold the load against shear when the longitudinal and transverse forces are applied to the system "bar - fastener - car floor".

4. Results of computational experiments

The calculation of the movement of the transported cargo, depending on the time it is on the route, is shown in Figure 4. Analysis of the dependences of the displacement of the load on time shows that the displacement of the load from the action of the portable inertial force of the load and the wind load increases with time.

IOP Conf. Series: Materials Science and Engineering 1064 (2021) 012042 doi:10.1088/1757-899X/1064/1/012042

$t := 0, 0.0005 .. 0.0065 \qquad x(t) := C1 \cdot (\cos(k0x \cdot t) - 1) + \frac{v1}{k0x} \cdot \sin(k0x \cdot t)$

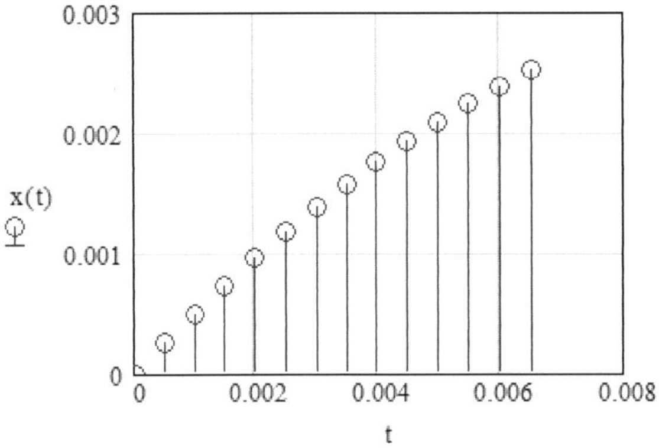

x(t) =	t =
0	0
$2.547 \cdot 10^{-4}$	$5 \cdot 10^{-4}$
$4.994 \cdot 10^{-4}$	$1 \cdot 10^{-3}$
$7.339 \cdot 10^{-4}$	$1.5 \cdot 10^{-3}$
$9.583 \cdot 10^{-4}$	$2 \cdot 10^{-3}$
$1.173 \cdot 10^{-3}$	$2.5 \cdot 10^{-3}$
$1.377 \cdot 10^{-3}$	$3 \cdot 10^{-3}$
$1.571 \cdot 10^{-3}$	$3.5 \cdot 10^{-3}$
$1.755 \cdot 10^{-3}$	$4 \cdot 10^{-3}$
$1.928 \cdot 10^{-3}$	$4.5 \cdot 10^{-3}$
$2.092 \cdot 10^{-3}$	$5 \cdot 10^{-3}$
$2.246 \cdot 10^{-3}$	$5.5 \cdot 10^{-3}$
$2.389 \cdot 10^{-3}$	$6 \cdot 10^{-3}$
$2.522 \cdot 10^{-3}$	$6.5 \cdot 10^{-3}$

Figure 4. Dependence of cargo movements on the time spent on the route.

The results of calculating the force acting from the load on the thrust bar are shown in Figure 5. The analysis of the obtained results shows that the force acting from the side of the transported cargo on the bar, depending on the movement of the cargo, decreases in a linearly decreasing manner, which is logical. Since this increases the restoring force of the load securing means and the friction force, the impact of the wind load is taken into account in the calculations.

$c0x = 2.609 \times 10^3 \quad f = 0.55 \quad Wx = 1.25 \quad \Phi ex = 470.88 \qquad \Phi ez = 388.476$

$R0x(x0) := -c0x \cdot x0 - f \cdot N0(x0) - Wx + \Phi ex$

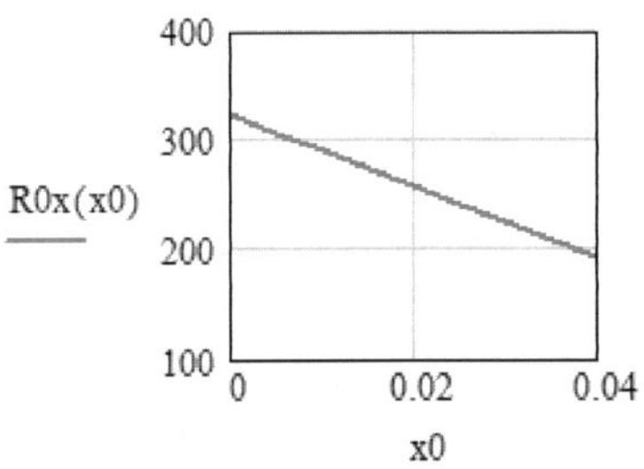

R0x(x0) =	x0 =
322.662	0
306.528	$5 \cdot 10^{-3}$
290.393	0.01
274.259	0.015
258.124	0.02
241.99	0.025
225.855	0.03
209.72	0.035
193.586	0.04

Figure 5. Graphical dependence of the force acting on the side of the transported cargo on the bar when moving the cargo.

The results of computational experiments to determine the number of fasteners required to secure the thrust bar, depending on its location in the car, are shown in Figure 6. The analysis of the given dependencies shows that the number of fasteners required securing the thrust bar, given the initial data, decreases in proportion to the increase in the distance to the transported cargo. In addition, the number of fasteners is significantly influenced by such physical and mechanical characteristics of wood as tensile strength and elastic modulus, which largely depends on the degree of its moisture content, i.e. on the climatic conditions in which the cargo is transported.

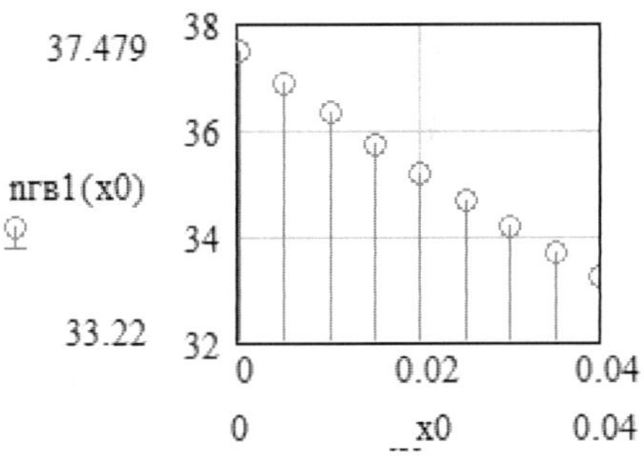

x0 =	пгв1(x0)
0	37.479
$5 \cdot 10^{-3}$	36.885
0.01	36.305
0.015	35.741
0.02	35.195
0.025	34.669
0.03	34.163
0.035	33.68
0.04	33.22

Figure 6. Dependence of the number of fasteners on the movement of the load.

Analysis of the graphical dependences of the force of impact on the bar shows that with the given initial data, the load hits the bar located at a given distance of 40 mm with a force of 193.57 kN (19.4 tf). When installing a thrust bar close to the end of the transported cargo, a force equal to 322.66 kN (32.3 tf) acts on it, which is 1.6 times more than the standard value established by the technical conditions. From the action of a force of such magnitude, a breakdown of the stubborn bar occurs, a violation of the entire securing system of the transported cargo, the probability of a risk event increases.

5. Conclusion

The continuous increase in the volume of work at railway stations makes it necessary to improve the technology and means of automating the processes of accepting cargo for transportation, as well as improving the quality of operational management of operational work in all technological links. The systems of automation of control processes developed to achieve the set goals should provide high reliability, accelerated response, locality, relative independence and "survivability" as part of a general automated control system for technological processes in railway transport systems.

The automated approach developed based on the use of the basic provisions of general mechanics made it possible to determine the force acting from the side of the load on a thrust bar located at a certain distance from the end of the load. This is the basis for calculations to determine the number of fasteners of this bar that retains the load from shear. The results of computational experiments showed the possibility of automatically calculating the required number of fasteners depending on the geometric parameters of the transported cargo, the size of the bar, fasteners and the car floor.

References

[1] 2018 *Development strategy of Federal Passenger Company JSC until 2030* Reporting materials in accordance with the terms of reference

[2] Gozbenko V E, Olentsevich V A and Belogolov Yu I 2018 Automation of individual operations of the transportation process in order to ensure sufficient conditions for the optimal functioning of digital transport and logistics *Modern technologies. System analysis. Modeling* **4(60)** 125-132

[3] Kargapoltsev S K, Gozbenko V E, Kuznetsov B O, Karlina Yu I and Karlina A I 2019 The effect of the periodic driving force on a system with two degrees of freedom *J. of Phys.: Conf. Ser.* **1333(5)** 052009

[4] Gozbenko V E, Kargapoltsev S K, Kuznetsov B O, Karlina A I and Karlina Yu I 2019 Determination of the principal coordinates in solving the problem of the vertical dynamics of the vehicle using the method of mathematical modeling *J. of Phys.: Conf. Ser.* **1333(5)** 052007

[5] Kuznetsov B O, Gozbenko V E, Kargapoltsev S K, Karlina Yu I and Karlina A I 2019 Dynamic vibration protection of the railway carriage *J. of Phys.: Conf. Ser.* **1333(5)** 052018

[6] Kargapoltsev S K, Gozbenko V E, Karlina Yu I and Karlina A I 2019 Introduction of additional

inertial couplings in mathematical models of problems of vibration protection and vibration isolation *IOP Conf. Ser.: Earth Env. Sci.* **378(1)** 012059
[7] Konyuhov V Yu, Gladkih A M, Galyautdinov I I and Shchadova E I 2020 Calculations of efficiency in implementing progressive mold forming methods *IOP Conf. Ser.: Mater. Sci. Eng.* **760(1)** 012027
[8] 2003 *Technical conditions for stowing and securing cargo in wagons and containers* (Moscow: Yurtrans) p 544
[9] Turanov Kh T, Pserovskaya E D and Koroleva D Yu 2002 Calculation of cargo fastening in case of collisions of wagons *Railway Transport* **3** 43-44
[10] Turanov Kh T, Vlasova N V and Korneev M V 2003 Modeling the force of impact of the load on the bar along the longitudinal axis of the car in the system "load - fastening - car" *Coll. of Works of Int. Sci.-Tech. Conf. Actual problems of reliability of technological, power and transport machines. Dependability-2003* (Moskow: Mechanical Engineering) vol 2 pp 274-280
[11] Bolotin V A and Doronin F A 1981 *Interaction of a group of goods and elements of their fastening with a car during shunting collisions* p 21
[12] Turanov Kh T and Vlasova N V 2003 Development of a rational method for placing fasteners of cargo transported on a wagon with the simultaneous action of longitudinal, transverse and vertical forces *Coll. of Works of Int. Sci.-Tech. Conf. Actual problems of reliability of technological, power and transport machines. Reliability-2003* (Moscow: Mechanical Engineering) vol II pp 268-274
[13] Gergert A P, Molostova I I and Ryabov D K 2013 Development of 1960 auminium alloy for gas centrifuges *Nonferrous Metals* **(9)** 33-36
[14] Vlasova N V 2004 Results of computational experiments on the calculation of thrust elements of cargo securing in the system "cargo - fastening element - car floor" *Mater. of the Int. Sci. Conf. Fundamental and Applied Research in Technology* (Rostov-on-Don: Rostov University) pp 65-73
[15] Suslov K, Solonina N and Gerasimov D 2018 Assessment of an impact of power supply participants on power quality *Proc. of Int. Conf. on Harmonics and Quality of Power* pp 1-5
[16] Ilyushin P and Suslov K 2019 Operation of automatic transfer switches in the networks with distributed generation *IEEE Milan PowerTech* 8810450
[17] Voropai N, Ukolova E, Gerasimov D, Suslov K, Lombardi P and Komarnicki P 2019 Simulation approach to integrated energy systems study based on energy hub concept *IEEE Milan PowerTech* 8810666
[18] Yolkin K S, Yolkin D K, Kolosov A D, Ivanov N A and Shtayger M G 2018 Technologies, which allow to reduce an impact of metal silicon production on the environment *IOP Conf. Ser.: Mater. Sci. Eng.* **411** 012028
[19] Korovyakovsky E K 2003 *Methodological foundations of placement and fastening of cargo with flat support on the rolling stock of railways,* author's abstract of dissertation (St. Petersburg: Emperor Alexander I St. Petersburg State Transport University) p 24

Influence of artificial filler made of calcium hydrosilicates of various basicity on the performance characteristics of non-autoclave hardened silicate materials

A A Volodchenko

Belgorod State Technological University named after V.G. Shukhov, 46, Kostyukova str., Belgorod, 308012, Russia

E-mail: alex-0904@mail.ru

Abstract. The work established the nature of the effect of an artificial filler made of calcium hydrosilicates of various basicity on the physical and mechanical properties of non-autoclave silicate composites based on unconventional aluminosilicate raw materials, represented by clay rocks of an unfinished stage of mineral formation. It is determined that the optimal content of an artificial filler of the composition $CaO:SiO_2$ is 1:1 in the raw mixture is in the range of 1–3 wt.%. The available interval makes it possible to vary the composition of the initial mixture and obtain products with the required properties, which is especially important when the material composition of the feedstock fluctuates, as well as in case of violation of the production technology. The rational content of the artificial filler of the composition $CaO:SiO_2$ = 2:1 is 1 wt. %. It is shown that the addition of an artificial filler promotes the formation of a cementitious substance from neoplasms of a higher degree of crystallization, which contributes to an increase in the strength and water resistance of the material. Mathematical models are proposed for the selection and optimization of the compositions of non-autoclave silicate materials based on unconventional aluminosilicate raw materials and modified with an artificial filler represented by calcium hydrosilicates of various basicity to obtain materials with a compressive strength of 18-23 MPa with an average density of finished products of 1750-2000 kg/m^3.

1. Introduction

Currently, an important scientific and technical task is the development and implementation of innovative approaches in the field of energy conservation and resource conservation, as well as the rational use of raw materials. It is necessary to introduce new production of building materials [1–4] with minimal impact on the environment, which will allow the sustainable development of the existing technological base and implement the policy of import substitution in order to increase the competitiveness of the economy of the regions and the country as a whole.

In the construction of buildings and structures, many wall materials are used, including silicate products for various functional purposes [5–9]. In the traditional technology for the production of such materials, the hardening process occurs due to the formation of a crystal structure in the CaO-SiO2-H2O system, which, depending on the conditions of hydrothermal treatment, is represented mainly by calcium hydrosilicates of various basicity [10–12]. There are various ways to modify the crystalline intergrowth of the cementitious substance, one of which is the use of certain components acting as crystallization centers, in particular, the use of natural and artificial calcium hydrosilicates. Research on improving the properties of building composites with synthetic hydrosilicates of calcium is mainly associated with cement systems and materials based on them [13–15].

To increase the strength characteristics of silicate products, the breakage of this material, as well as a certain kind of crystal seeds, can be used. Using breakage may not have a different effect, because it is a multicomponent system consisting of calcium hydrosilicates with unknown properties. Therefore, calcium hydrosilicates of a certain composition are best obtained synthetically.

Thus, in order to improve the operational characteristics of non-autoclave silicate materials based on unconventional aluminosilicate raw materials, represented by clay rocks of the unfinished stage of mineral formation, the issue of using artificial fillers from calcium hydrosilicates, including those of different basicity, in the technology of their production, is relevant and practically unexplored.

2. Materials and methods

The production of silicate materials of non-autoclave hardening uses unconventional clay rocks of an incomplete stage of mineral formation and characterized by a certain set of properties as an aluminosilicate raw material [16] and calcium oxide (construction quicklime) acts as a binder. In this work, aluminosilicate raw materials from the region of the Kursk Magnetic Anomaly (sandy-clay rock, according to the plasticity number related to sandy loam) were used as the main component.

Artificial calcium hydrosilicates, in these studies, were obtained by hydrothermal synthesis at elevated temperatures and pressures (P = 1 MPa, T = 175 °C). Calcium oxide and quartz sand of the Volsk deposit were used as the initial components. To increase the chemical activity of quartz sand, it was preliminarily crushed (S_{sp} = 300 m^2/kg).

The obtained hydrosilicates of calcium were introduced into the raw mixture by dry mixing with a binder component and clay rock in a laboratory mixer. After adding the required amount of water, followed by keeping the mixture for complete quenching of calcium oxide in the mixture, the samples were molded on a hydraulic press at a pressure of 20 MPa. The hardening process of the samples was carried out in a heat-and-moisture treatment chamber (HMC), at a water vapor temperature of 95 °C. The isothermal holding time of the samples was 9 hours.

The study of the phase composition, microstructure of the initial components and forming neoplasms was carried out by appropriate research methods (XRF, DTA, SEM).

3. Results

At present, both in national and international literature, there are many studies in the field of artificially obtaining compounds in the $CaO–SiO_2–H_2O$ system and their influence on the properties of composites [10–15]. The properties of such a system mainly depend on the molar ratio between $CaO:SiO_2$, the type and properties of SiO_2, as well as the parameters at which the synthesis of calcium hydrosilicates occurs. Of the many forms of artificial calcium hydrosilicates in the $CaO–SiO_2–H_2O$ system, low-base and dibasic hydrosilicates are the most stable.

In the course of research, two types of artificial filler (AF) represented by calcium hydrosilicones were obtained: AF-1 ($CaO:SiO_2$ = 1:1) and AF-2 ($CaO:SiO_2$ = 2:1). The results of X-ray phase and thermographic analysis (Figure 1) of the obtained artificial fillers (AF-1, AF-2) represented by the $CaO–SiO_2–H_2O$ system indicate the formation of calcium hydrosilicates of CSH(I) and C_2SH(A) types.

On the thermogram of AF-2, in comparison with AF-1, there is no exothermic effect in the range of 830–860 °C, which is characteristic of low-basic calcium hydrosilicates CSH(I). The presence of a single endo-effect at 460 °C is noted, which may indicate the presence of C_2SH(A) at C/S = 2, while C/S > 2 forms a double endo-effect at 430–480 °C.

In order to determine the influence of CaO content and an artificial filler represented by calcium hydrosilicates of various basicity on the performance characteristics of products, an experiment was carried out using the method of mathematical planning of the experiment. Two experiments were carried out using AF-1 and AF-2.

The data for the experiments are shown in Tables 1 and 2.

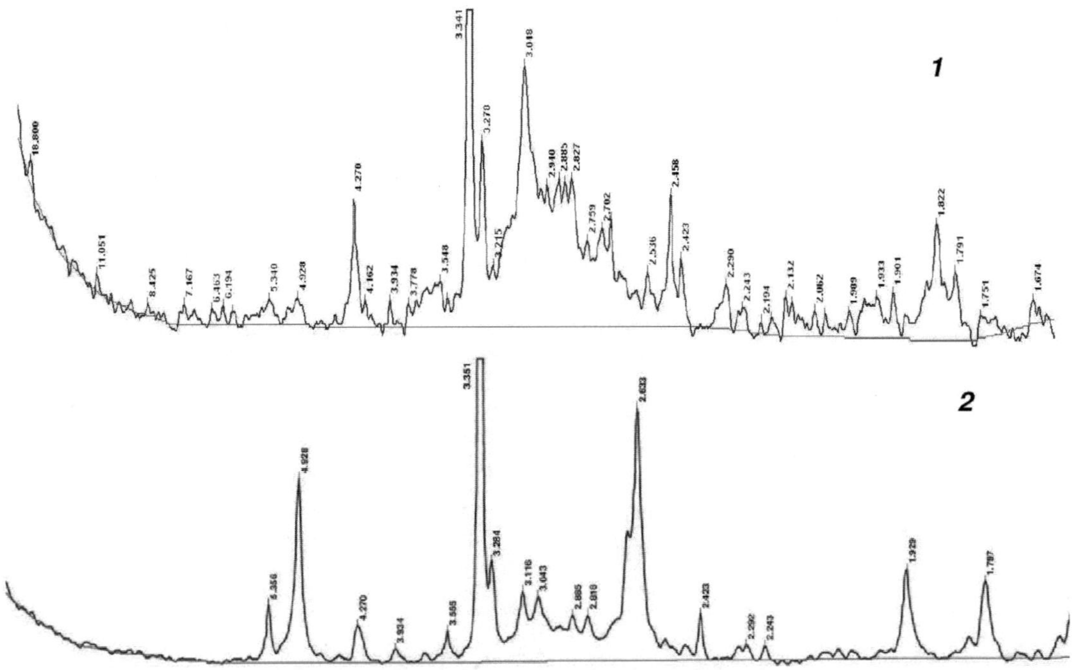

(a)

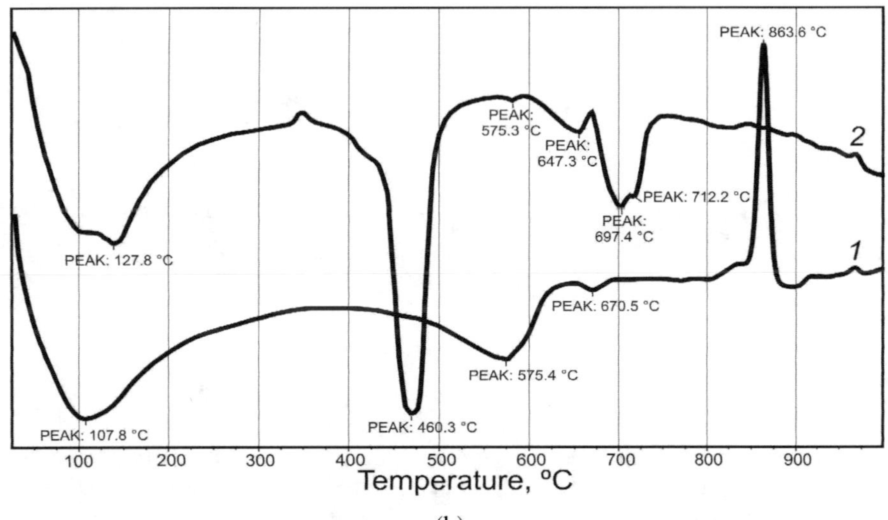

(b)

Figure 1. Radiographs (a) and thermograms (b) of the obtained artificial filler: 1) AF-1 ($CaO:SiO_2$ = 1:1); 2) AF-2 ($CaO:SiO_2$ = 2:1)

Table 1. Data for the experiment by the method of mathematical planning using AF-1

Factors		Variation levels			Variation interval
Natural form	Coded form	-1	0	+1	
CaO, wt.%	x_1	4	8	12	4
Artificial filler AF-1, wt.%	x_2	0.5	3	5.5	6

Table 2. Data for the experiment by the method of mathematical planning using AF-2

Factors		Variation levels			Variation interval
Natural form	Coded form	-1	0	+1	
CaO, wt.%	x_1	4	7	10	3
Artificial filler AF-2, wt.%	x_2	0.5	1.5	2.5	1

After analyzing and processing the initial data, mathematical regression equations were obtained that describe the changes in the compressive strength index (R), as well as the water resistance index (K) and average density (ρ).

The regression equations are (using AF-1):

$$R_{compr} = 22.41 - 1.75 \cdot x_2 - 2.42 \cdot x_1^2 - 1.5 \cdot x_2^2$$

$$\rho = 1879.43 - 42.51 \cdot x_1 - 45.84 \cdot x_2 - 7.644 \cdot x_1^2 - 7.644 \cdot x_2^2 + 3.75 \cdot x_1 \cdot x_2$$

$$K = 0.802 - 4.06 \cdot 10^{-2} \cdot x_1^2 - 5.06 \cdot 10^{-2} \cdot x_2^2$$

The regression equations are (using AF-2):

$$R = 23.23 + 0.283 \cdot x_1 - 0.517 \cdot x_2 - 2.33 \cdot x_1^2 - 0.532 \cdot x_2^2$$

$$\rho = 1944.72 - 49.34 \cdot x_1 - 11.339 \cdot x_1^2 - 5.6 \cdot x_2^2$$

$$K = 0.836 - 3.987 \cdot 10^{-2} \cdot x_1^2 - 1.487 \cdot 10^{-2} \cdot x_2^2$$

After processing the available regression equations, graphical dependences of the properties of non-autoclave silicate materials on the content of CaO and an artificial filler represented by calcium hydrosilicates of various basicity were built (Figures 2 and 3).

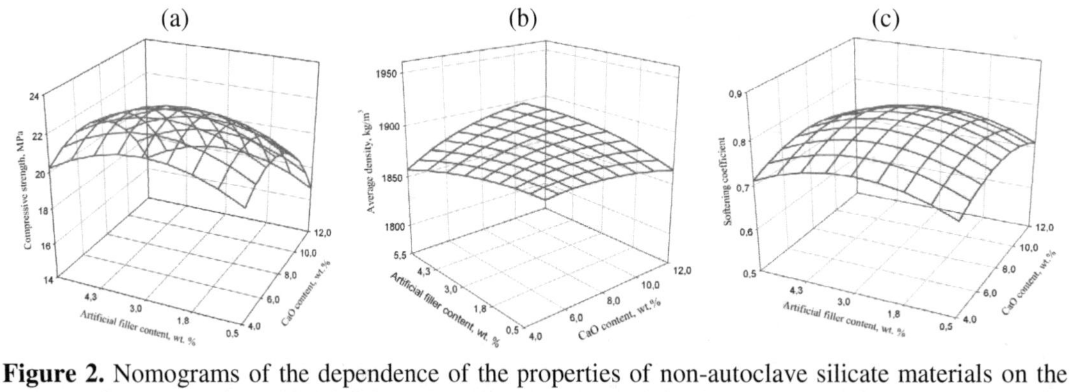

Figure 2. Nomograms of the dependence of the properties of non-autoclave silicate materials on the content of CaO and the artificial filler AF-1: (a) nomogram of the dependence of the ultimate strength in compression; (b) nomogram of the dependence of the average density indicator; (c) nomogram of the dependence of the softening coefficient

Based on the data obtained, it was found that at a CaO content of 4–12 wt.% it is possible to obtain durable non-autoclave silicate materials based on unconventional aluminosilicate raw materials and artificial filler (AF) represented by hydrosilicates of various basicity (Figures 2 and 3).

The rational content of AF-1, which ensures the achievement of maximum strength indicators (22 MPa), is 2 wt.% with a CaO content in the mixture in the amount of 8 wt.%. An increase in the content of AF-1 in the raw mixture to 5.5 wt.% decreases the ultimate strength in compression (Figure 2).

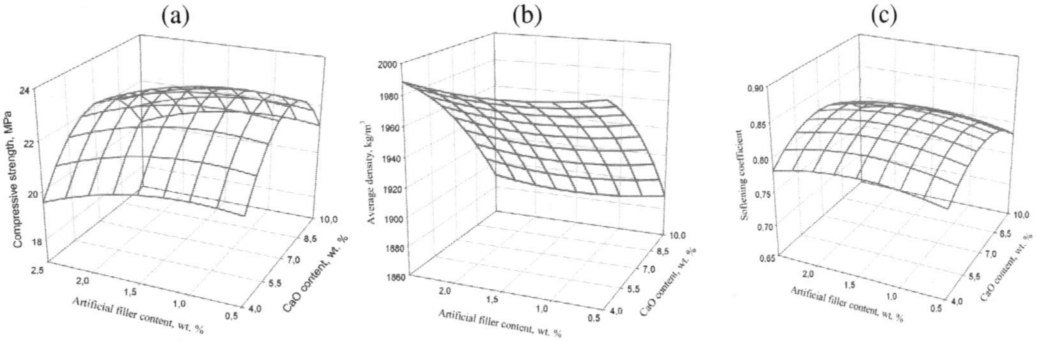

Figure 3. Nomograms of the dependence of the properties of non-autoclave silicate materials on the content of CaO and artificial filler AF-2: (a) nomogram of the dependence of the ultimate strength in compression; (b) nomogram of the dependence of the average density indicator; (c) nomogram of the dependence of the softening coefficient

The nature of the effect of the AF-2 artificial filler on the strength properties of the samples differs from the AF-1. In this case, the specimens reach their maximum strength (23 MPa) when the content of AF-2 in the mixture is 1 wt. % and that of CaO is 8 wt.% (Figure 3). A further increase in the content of AF-2 to 2.5 wt.% decreases the ultimate strength in compression down to 19.7 MPa.

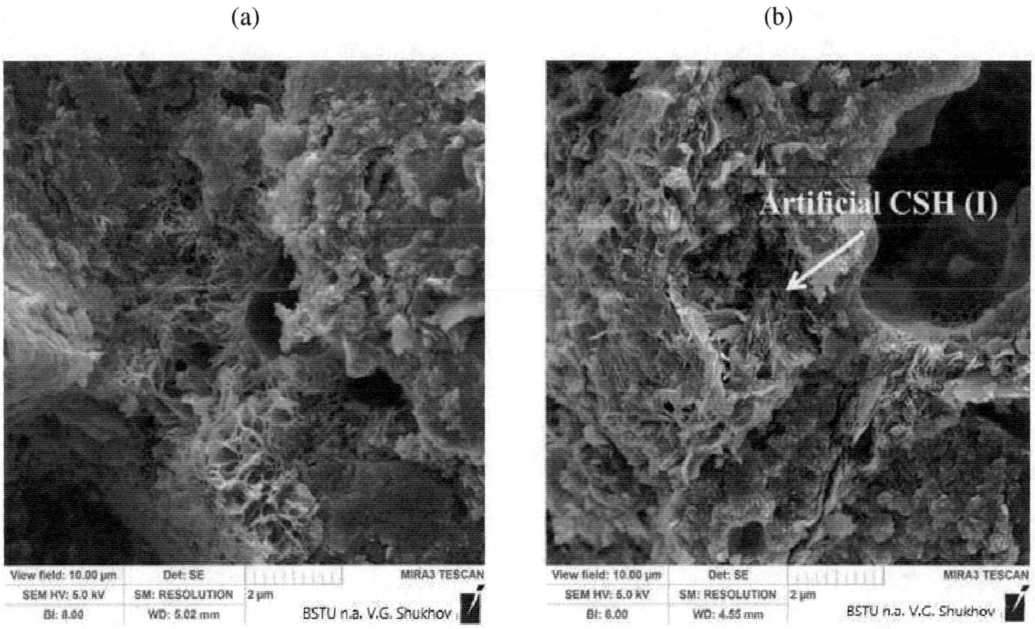

Figure 4. Effect of the artificial filler on the microstructure of the sample with 8 wt.% of CaO: (a) control composition; (b) content of AF-1 is 3 wt.%.

The least effect of the addition of artificial filler AF-1 and AF-2 has on samples with a mass content of CaO in an amount of 10%.

The indicator of the average density of the material with an increase in the mass content of the binder component (CaO) decreases, and the more, the more the artificial filler is added (Figures 2 and 3). The addition of an artificial filler increases the water resistance of the products. Samples with a CaO content of 8 wt.% and mass content of AF-1 of 2 wt.% and AF-2 of 1 wt.% have good water resistance (softening coefficient is not less than 0.8).

When examining the microstructure of the sample of the control composition, the presence of a crystalline weakly crystallized network of neoplasms was observed (Figure 4a).

In the samples with the addition of an artificial filler, a similar structure of neoplasms has a more crystallized appearance in the total mass of which crystals of an artificial filler are distributed (Figure 4b), which serve as a substrate that intensifies the process of synthesis of neoplasms, including at the early stages of hardening, which contributes to an increase in the operational properties of non-autoclave materials.

4. Conclusion

Thus, based on the data obtained, the nature of the influence of AF-1 and AF-2 on the performance indicators (strength, density, water resistance) of wall silicate materials of non-autoclave hardening based on aluminosilicate raw materials, characterized by the presence of certain clay rocks of an unfinished stage of mineral formation, was established. The optimal content of AF-1 ($CaO:SiO_2 = 1:1$) in the raw mixture is in the range of 1–3 wt.%. The available interval makes it possible to vary the composition of the initial mixture and obtain products with the required properties, which is especially important when the material composition of the feedstock fluctuates, as well as in case of violation of the production technology. The rational content of AF-2 (CaO:SiO2 = 2:1) is 1 wt.%, and with an increase in the percentage, the strength characteristics of the finished products decrease.

Mathematical models are proposed for the selection and optimization of the compositions of non-autoclave silicate materials based on unconventional aluminosilicate raw materials and modified with an artificial filler represented by calcium hydrosilicates of various basicity to obtain materials with a compressive strength of 18-23 MPa with an average density of finished products of 1750-2000 kg/m^3.

Acknowledgments
The work is realized in the framework of the grant Russian Science Foundation (project No. 19-79-00185).

References

[1] Lesovik V S and Fomina E V 2019 *Vestnik MGSU* **14(10)** 1241–57
[2] Fomina E V, Lesovik V S, Kozhukhova N I and Chulenyov A S 2019 *Mater. Sci. Forum* **974** 168-174
[3] Mandili B, Taqi M, Bouari A Ei and Errouaitic M 2019 *Construction and Building Materials* **228** 117097
[4] Nelyubova V V, Babayev V B, Alfimova N I, Usikov S A and Masanin O O 2019 *Construction Materials and Products* **2(2)** 4-9
[5] Fomina E V, Lesovik V S, Kozhukhova M I and Solovyova E B 2019 *Mater. Sci. Forum* **974** 224-230
[6] Volodchenko A N and Strokova V V 2017 *Bull. of BSTU named after V.G. Shukhov* **1** 138-43
[7] Huiwen Wan, Yong Hu, Gang Liu and Yuan Qu 2018 *Construction and Building Materials* **184** 20-26
[8] Yunliang Zhao, Yimin Zhang, Tiejun Chen, Yongliang Chen and Shenxu Bao 2012 *Construction and Building Materials* **28(1)** 450-455
[9] Hamdy El-Didamony, Ahmed A Amer, Mona S Mohammed and Mahmmoud Abd El-Hakim 2019 *J. of Building Engineering* **22** 528-538
[10] Liu B, Ray A, Thomas P S, Klimesch D and Guerbois J P 2007 *J. of Solid Waste Technology and Management* **33(2)** 61-66

[11] Nonat A 2010 *Cement, Wapno, Beton* **15(6)** 315-326
[12] Dachowskia R and Stępieńa A 2011 *Procedia Engineering* **21** 1173-78
[13] Kalla P, Misra A, Gupta R C, Csetenyi L, Gahlot V and Arora A 2013 *Construction and Building Materials* **40** 1142-50
[14] Karpikov E G, Lukuttsova N P, Soboleva G N, Golovin S N and Cherenkova Yu S 2019 *Construction Materials and Products* **2(6)** 20-28
[15] Yarusova S B, Gordienko P S, Kozin A V, Zhevtun I G and Perfilev A V 2018 *IOP Conf. Ser.: Mater. Sci. Eng.* **347** 012041
[16] Volodchenko A A 2018 *Bull. of BSTU named after V.G. Shukhov* **12** 12-20

Hydraulic control valve spool restoration by composite galvanic coating

E M Yudina, M R Kadyrov, A S Serguntsov and G E Samurganov

Kuban State Agrarian University named after I.T. Trubilin, 13, Kalinina str., Krasnodar, 350044, Russia

E-mail: elena_yudina1963@mail.ru

Abstract. Methods of restoration of hydraulic control valve spools by galvanic coatings are considered. Methods for preparing parts for applying electroplated composite coatings in order to increase the adhesion strength of electrolytic iron to alloy steel are proposed. The optimal modes of electrolysis for obtaining high-quality galvanic coatings are selected, the composition of the electrolyte is substantiated and the optimal content of the dispersed phase in it is selected. The design features of the coating bath for active mixing of the dispersed phase in the electrolyte are substantiated. The parameters of the technological process for spools restoration by applying composite galvanic iron-based coatings are proposed.

1. Introduction

In the technical equipment of the agro-industrial complex of the Russian Federation, tractors and agricultural machines that are obsolete are currently involved. Large agro-industrial enterprises with sufficient financial resources acquire imported equipment, while small farms are forced to use existing equipment. In this regard, the problem of restoring and increasing the durability of parts, components and assemblies of the agricultural machinery in use is very urgent. The maintenance and servicing base still has a wide network of repair enterprises, the rhythmic work of which is hampered by the routine maintenance of spare service parts and components, machine tool aging and technological equipment. At repair bases, it is necessary to organize the restoration and manufacture of repair units of a wider range of products.

The range of remanufactured parts also includes parts for precision pairs of hydraulic units. As structural materials of plungers and spools, the chromium, chromium-nickel, chromonickel-tungstic steels with beryllium content are used.

Among the various methods for restoring these parts, the technological processes that leave unchanged the microstructure of the parts and increase their strength characteristics, should be chosen. The main task of these processes is to protect the remanufactured parts from corrosion. These methods include the restoration of parts by electroplating. Most often, parts of precision pairs are restored by aging and chromium plating. When comparing these processes, it should be noted that the chromium plating process is inefficient, even when using high-speed electrolytes and intensive processes in chromium plating practice [1–4]. When applying coatings of standard chromium plating electrolytes, it is difficult to achieve a constant concentration of components during the coating process, which often leads to heterogeneity of the formed layers.

A universal approach to the selection of a method for restoring parts was given by A.N. Batishchev [5]. He proposed to use the energy criterion $\varphi_{эi}$ for this purpose. It should be noted that the assessment of the methods of restoring parts according to this criterion showed that the most rational of them is ironing ($\varphi_э$ = 0.36). This testifies to the high efficiency of this galvanic method of restoring parts with low wear. To increase the wear resistance of parts, it is recommended to use composite galvanic coatings (hereinafter – CGC) or iron-based alloys [6, 7].

2. Materials and methods

Since CGC are designed to increase the resource of agricultural machinery parts, they can be used not only to restore parts that work during abrasive wear, but also to be applied to new or restored parts in various ways. The hardenable parts must be preliminarily brought to the nominal dimensions and roughness. Thus, when applying CGC on the front side of the plowshare, it is recommended to sharpen its blade from the back side and put it in reverse. Cleaning parts from dirt, scale, rust and various technical media must be carried out by quartz treatment with metal brushes, followed by rinsing in hot water.

The pre-machining is necessary to restore the original shape of the part and remove the top oxidized, destroyed layer that lost its original structure. It has no peculiarities and is carried out according to the repairing drawing requirements in accordance with the parts production specifications. The hydraulic control valve spool should be grounded to remove any signs of wear before the ironing. The surface roughness should be no more than R_a = 0,32 mkm. After the grounding, the spools are washed from oil in a 10% solution of caustic soda at a temperature of 343 ... 353 K and in hot water. It is advisable to degrease parts electrochemically in a solution of composition (kg/m^3) of sodium hydroxide – 30 ... 40; soda ash – 30 ... 40; trisodium phosphate – 40 ... 50; liquid glass - 3 ... 5 at anode current density - D_c = 7 ... 8 A/m^2, temperature - T = 333 ... 343 K during 10 minutes. The special hanging tools are used, designed to process 60 spools simultaneously. After degreasing, the hanging tools are fed for rinsing with hot running water, which should ensure complete removal of the solution remains from the parts. Then the parts are washed with cold running water, dismantled from the hangers for degreasing, and installed in working devices. If a batch of degreased parts is gathered before the rest of the process, they should be stored after degreasing without rinsing. In this case, the washing is carried out before assembling the parts into the working device.

The hanging tool for coating can handle 12 spools simultaneously. Since it is necessary to apply coatings on finished parts, the surfaces that are not subject to restoration must be isolated from corrosive effects of electrolytes, for example, protective covers made of vinyl plastic or acid-resistant rubber. Since the spools are made of chromium steel, it is advisable to carry out their anode treatment in a 30% sulfuric acid solution at a current density of 60 ... 80 A/dm^2 for 30 ... 50 sec (a higher current density - D_c = 70 ... 80 A/dm^2 is used for etching spools) [7–9]. Along with 30% sulfuric acid for chromium steel, it is possible to recommend anode treatment in a saturated solution of aluminum sulfate ($Al_2(SO_4) \cdot 18H_2O$ – 350 kg/m^3), containing 20 kg/m^3 of sulfuric acid [10]. The etching in this solution is carried out at D_c = 60 ... 70 A/dm^2, T = 291 ... 295 K for 45 ... 60 sec.

In an electrolyte tank for anode treatment with a volume of 1 m^3, 5 hangers with spools are hung (when the tank is powered from a VAKR-6300/12 rectifier or other similar power). The parts are washed after anode treatment in warm (313 ... 323K) running water for 20 ... 30 sec. Practical experience shows the time of the rinsing operation in warm water is closely related to the "current interruption" technique. If the flushing time is increased, the interruption time should be reduced and conversely.

Composite coatings are applied with a chloride electrolyte of composition (kg/m^3) of: $FeCl_2 \cdot 4H_2O$ – 550 ... 600; white alumina of M14 type – 90 ... 100; pH = 0,7 ... 1,0 at a temperature of 313 K. Coating begins with holding the part in the electrolyte without current for 30 ... 45 sec, which is quite enough to equalize the temperature difference between the solution and the part, and activate the anodically processed surface. The deposition of CGC is carried out at an initial current density D_{cn} = 1 ... 1,5 A/dm^2 for 5 ... 7 minutes, then the current density is increased to 5 ... 7 A/dm^2 and the process is carried out for 5 ... 7 minutes, followed by a gradual increase in the current density, the working density (20 ... 25 A/dm^2). The electrolysis time at the working current density in applying coatings with a thickness of 0,40 ... 0,75 mm is 2,0 ... 3,5 h. The electrolyte acidity during the deposition of CGC is adjusted within pH = 0,7 ... 1,0 using diluted hydrochloric acid on a 1:5 ratio. To reduce the electrolyte oxidation rate during electrolysis breaks, it is acidified to pH = 0,3 ... 0,5, and the low-carbon steel plates are left in it.

If it is necessary to restore Fe (III), the electrolyte is processed under current with an increased of 3 ... 5 times of the cathode surface in comparison with the anode and current density of 4 ... 8 A/dm^2.

3. Results and discussion

The spools operating under conditions of hydroabrasive wear and made of chromium steel, heat-treated to a hardness of HRC = 56 ... 63, in accordance with the previously given data, should be restored with coatings containing dispersed phase (hereinafter – DP) in an amount of 15 ... 30% (vol.). Therefore, the concentration of the DP in the suspended electrolyte (hereinafter – SE) can be reduced to 60 ... 80 kg/m^2. The initial current density considering the material of the part should be reduced to 0,75 ... 1,0 A/dm^2.

The working volume of the tank for ironing (1m^3) allows its loading with hangers with a total area under the covering of 75 ... 80 dm^2, at a volumetric current density of 1,5 ... 2,0 A/dm^3. The main factor that ensures the obtaining of high-quality CGC with a uniform content of solid particles in the coating, regardless of the location of the part relative to the tank walls, is ES mixing. The studies on determining the impact of the stirring type on the uniformity of DP distribution in the suspended electrolyte and coating showed that mechanical stirring using one stirrer, regardless of its rotation speed, did not provide the required distribution of particles in the SE and CGC. At a stirrer rotation speed of the order of 3 s^{-1}, the upper electrolyte layer (10 ... 15 mm) had practically no solid particles in its composition.

Two stirrers with their synchronous rotation also did not allow achieving the desired quality of SE, while their counter-rotation at a rate of about 12,5 s^{-1}, ensured sufficient uniformity of DP distribution in the electrolyte, but solid particles were not evenly included in the coating. A further increase in the rotation speed reduced the number of inclusions in the coatings, probably due to the particles washing off from the cathode surface by the SE flow.

The installation of four mixers with their counter and synchronous rotation made it possible to achieve a more uniform distribution of particles in the SE and CGC. Thus, the counter-rotation of the mixers located diagonally of the tank, at a stirring speed of 12,5 s^{-1}, ensured the uniformity of the DP distribution in the SE at the level of 20% deviation from the average value. And at a stirrer rotation speed of 25 s^{-1}, the deviation was about 13%. At the same time, the particles were more intensively washed off by the flow, and their content in the coating was less than that required for a significant increase in the wear resistance of the CGC.

The best in terms of the uniformity of DP distribution in the electrolyte turned out to be jet mixing with the supply of a jet from the mixer along the reflector under the perforated bottom-partition of the tank. In stirring at a rate of 25 s^{-1}, the dispersed phase was diffused throughout the entire volume of the tank, and was rather evenly included in the iron electrolytic deposits. The deviation in the content of DP in the CGC along the height and length of the samples simulating the full height and width of the hanger was about 10 ... 15%. An increase in the stirring rate to 50 s^{-1} led to a decrease in the DP content in the coating due to the particles washout.

Jet mixing turned out to be less energy-intensive, does not require special gearboxes to reduce the engine speed, and made it possible to quickly transfer particles from the day of the electrolyzer to a suspended state after electrolysis breaks. Therefore, in industrial practice, the use of jet mixing is preferable. Thus, to obtain a uniform SE, it is possible to recommend jet mixing when the SE is supplied through a special compartment under the perforated bottom of the tank at a stirrer rotation speed of 25 s^{-1}.

After the coating has been applied, the parts are washed and neutralized in accordance with the usual requirements of ironing technology. For neutralization, a NaOH solution (60 ... 80 kg/m^3) is used at a temperature of 338 ... 348 K for 5 ... 10 minutes. Parts that have undergone neutralization are washed with hot water, dried and controlled by the quality of the applied coating. If necessary, it is passivated in a solution containing 15 ... 20 kg/m^3 of triethanolamine and 5 ... 7 kg/m^3 of sodium nitrite, and for long-term storage it is preserved with oils and special papers.

The composition of the ironing line is determined by the technological operations of the process, and includes a set of tanks for electrolytic degreasing, hot washing, cold washing, anode treatment, ironing, and neutralization. In addition to the tanks, the line is equipped with a monorail lifting device (Q = 0,25 t), a reserve tank, a drying box, a mounting table, a rack for equipment, a tank control post, a table for controlling coatings, a unit for filtration and pumping solutions.

It should be noted that for the spools finishing treatment, restored with CGC, it is possible to use the existing recommendations for the "pure iron" on centerless grinding machines of 3M184A, PC12, PC12S types. However, considering the increased hardness of CGC (due to the inclusion of a filler), instead of abrasive disk of 1 500x150x305 25A 40-P CM1 K6 type, the disks on a carborundum or cubonite base with a soft or medium soft bond, should be used.

The disk rotation speed should be at least 30 ... 35 m/s, and the supply of coolant-lubricant should be increased, in comparison with conventional grinding, 1,5 times. In this case, a flat and wide tip should be used, the width of which should be at least 1,25 times the width of the circle.

4. Conclusion

On the basis of the above recommendations, a technological process for the restoration of hydraulic control valve spools with iron-based composite coatings was developed for repair production.

To restore parts subject to hydroabrasive wear, it is possible to use galvanic coatings based on iron alloys with the inclusion of white alumina as a filler, as well as composite coatings with subsequent laser treatment, which have a number of advantages over, in particular, plasma spraying [11].

References

[1] Mironov V V 2015 *Restoration of plungers of in-line diesel fuel pumps by applying galvanic gas-phase chromium,* abstract of candidate dissertation (Ryazan, 2001) 12 p

[2] Yilbas B S, Patel F and Karatas C 2015 *Laser Surface Engineering* (Waugh J.L.G. Woodhead Publishing) pp 97-105

[3] Semenova E E 2004 *Technology of restoration and hardening of surfaces of spools of hydraulic boosters of steering by galvanic gas-phase chromium plating*, candidate dissertation (Penza, 2004) 170 p

[4] Ionov P A 1999 *The choice of optimal modes for the restoration of worn-out parts by electrospark surfacing: By the example of the valve of the hydraulic distributor R-75*, candidate dissertation (Saransk) 198 p

[5] Batishchev A N 1992 Justification of a rational way to restore parts *Mechanization and electrification of agriculture* **9(12)** 30-31

[6] Kisel Yu E 2001 *Increasing the durability of high-wear parts of agricultural machinery with composite electrochemical coatings based on iron alloys*, abstract of dissertation (Moscow) 18 p

[7] Kisel Yu E, Kisel P E, Guryanov G V and Yudina E M 2009 Scattering of micro-hardness of electroplated coatings *Proc. of the Kuban State Agrarian University* **19** 219-222

[8] Melkov M P 1971 *Solid rest of automotive parts* (Moscow: Transport) 224 p

[9] Guryanov G V and Kisel Y E 2006 *Antifriction and wear-resistant electrochemical coatings*, monograph (Bryansk)

[10] Shaidullin A M 1990 *Increasing the adhesion strength of electrolytic iron with alloy steel when restoring parts of agricultural machinery*, abstract of dissertation (Chisinau) 19 p

[11] Yudina E M, Guryanov G V, Kisel Yu E and Lysenko A N 2015 Laser hardening of composite electrochemical coatings *Rural mechanic* **2** 38-39

Mathematical modelling of operation of the hydraulic support system of the powered support sections with impulse-free continuous regulation of its resistance to the roof rock lowering

V V Gabov, N V Babyr and D A Zadkov

Saint-Petersburg Mining University, 2, 21st Line of Vasilevsky Island, St Petersburg, 199106, Russia

E-mail: dzadkov@yandex.ru

Abstract. The article describes the mathematical model of a hydraulic support system of the powered support sections with impulse-free continuous regulation of its resistance to the roof rock lowering. Safety valves of the hydraulic support perform only protective functions. Mathematical modeling was carried out using the Runge-Kutta method and the Wolfram Mathematica computer algebra system. The results of modeling the transient process of the hydraulic support system with a built-in block for regulating its resistance during impulse loading of the support section with a collapsed roof rock block at "Polosukhinskaya" mine are presented.

1. Introduction

The efficiency of coal mining in complex-mechanized working faces of mines depends on the operation of the powered support section during the long-term loading in a rock pressure control mode, that is, in a mode of adjustable resistance to the subsidence of roof rocks. This confirms the validity of regular research aimed to improve the operating modes of hydraulic props for powered support sections and devices for regulating their resistance [1-12].

Of interest are long-term operating modes of the hydraulic support system of powered support sections with continuous pulseless regulation of their resistance to the rock roof lowering by special devices - pulseless resistance control units [3, 4]. The choice of values of the parameters is associated with many factors, in particular, transient processes in the hydraulic system during cyclic switching of the operating modes of the objects under consideration, changes in external loads on the powered support and pressure of the working fluid in the pressure line of the powered support hydraulic system.

The most severe mode of operation of the hydraulic system corresponds roof rock block collapses, which last for a fraction of a second and occur rarely, although they can have severe consequences, especially when safety valves are blocked. Determination of the main regularities of transient processes in the hydraulic system with impulse-free continuous regulation of tits resistance to the roof rock lowering under prolonged loading is a difficult task that depends on time, design and operating parameters of the hydraulic prop, control unit and hydraulic system of the powered support sections. In view of the nonlinearity of the system, it is advisable to solve the problem by the method of numerical mathematical modeling [12-18].

The aim of the study is to develop a mathematical model of force interaction of the elements of the "roof - hydraulic rack - control unit - pressure line" system with continuous regulation of the resistance of hydraulic props to lowering the rocks of the immediate roof and imposition of a pulse surcharge simulating the collapse of the roof rock block.

2. Materials and methods

Mathematical modeling is carried out by the Runge-Kutta method [19], which makes it possible to reduce mining conditions and design parameters to a universal computational scheme, easily implemented by the Wolfram Mathematica computer algebra system.

Due to the fact that the system has a high degree of nonlinearity and a large number of logical conditions, mathematical models are developed for each element. In addition, the mathematical model is flexible, therefore, control or disturbing actions can be implemented. The system contains variable coefficients of differential equations that are associated with external loads that vary in time in a random manner. The method of mathematical modeling takes into account the required number of parameters with a high level of accuracy. The stages of mathematical model development are: development of a calculation scheme; formation of a flowchart based on the calculation scheme and making assumptions; mathematical description based on accepted conditions.

To draw up the design scheme, a hydraulic diagram of the hydraulic support of a support section with adjustable resistance and energy recovery into the pressure line of the powered mining complex support has been analyzed (Figure 1).

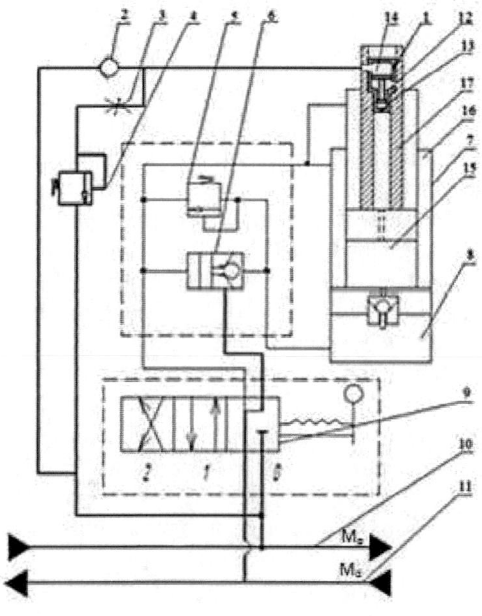

Figure 1. The hydraulic system of the hydraulic racks with a control unit

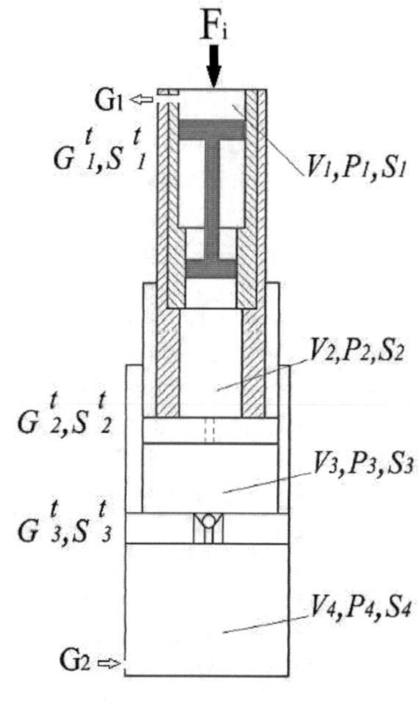

Figure 2. The calculation scheme of hydraulic props with a control unit

Torque converter 1 of the control unit is built into cavity 17 of the rod of the second stage of hydraulic prop 7. The inlet cavity of the torque converter is connected by hole 13 with piston cavity 15 of the second stage of the hydraulic prop, and outlet cavity 14 is connected to delivery line 10 of the powered support by a channel with throttle 3 and back-up valve 4.

The principle of operation of the control unit is that when the pressure in piston cavity 15 of the second stage of the hydraulic prop reaches the control level, the piston block of torque converter 1 begins to move upward (Figure 1) displacing the working fluid from cavity 14 through throttle 3 and pressure valve 4 into pressure head line 10 of the hydraulic support system. When moving the support section, the pressure in the piston cavities of its hydraulic props is less than the pressure in the pressure line, the liquid from which through check valve 2 enters cavity 14 of the torque converter, transferring it to a charged state.

On the basis of the hydraulic scheme, the calculation was made (Figure 2). V_1, P_1 and S_1 are volume, pressure and sectional area of the chamber of the large piston of the torque converter; V_2, P_2 and S_2 are volume, pressure and sectional area of the small piston chamber of the torque converter; V_3, P_3 and S_3 are volume of the piston cavity of the second stage of the hydraulic prop, pressure in the cavity and area of the working surface of its piston; V_4, P_4 and S_4 are volume of the piston cavity of the first stage of the hydraulic prop, pressure and area of the working surface of its piston.

On the basis of the calculation scheme, a block diagram of the system has been built (Figure 3). It is a set of blocks and their connections corresponding to the elements and connections of the real system.

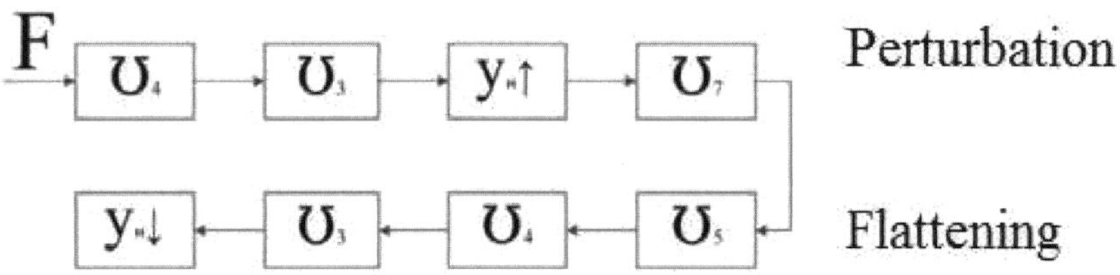

Figure 3. The block diagram of propagation of the shock wave during the transition period

Each structural element of the hydraulic props system contributes to the formation of the general process. This contribution can be represented by a differential equation describing the continuous process of pulseless regulation of resistance of hydraulic props to the subsidence of roof rocks in a complex mechanized working face with recovered rock pressure energy in the complex hydraulic system. The process of operation of the hydraulic system and its hydraulic props is a sequence of transient processes, caused by regular switching of the operating modes of the support sections, changes in the working face associated with the flank and frontal movement of the face, changes in rock pressure during primary and secondary roof sediments or block-type roof rocks collapse, etc. The most unfavorable case for hydraulic props of support sections is their impulse loading as a result of block collapses of main roof rocks.

When studying the transient process in the event of impulse surcharge of hydraulic props of a powered support section with collapsed rock blocks of the main roof in a complex mechanized working face, it is difficult to take into account all the determinants. Therefore, to simplify the problem being solved, factors whose influence is insignificant are not taken into account. The following assumptions are taken into consideration:

- there is no dry friction between the moving parts;
- the sliding friction force is proportional to the speed of movement, while the coefficient of proportionality remains constant;
- the values of coefficients of the flow rate of the working fluid through the valves are constant;
- the compressibility of liquid in the drain pipeline is neglected;
- the elasticity of system elements is constant.

3. Mathematical model

To establish the basic regularities of the transient process, the process of force interaction of the hydraulic props of the section with roof rocks in a complex-mechanized working face was modeled by the Runge-Kutta method.

The movement of the hydraulic rack rod under the action of the collapsed rocks is described by the differential equation of forces acting on it:

$$m_s \cdot \ddot{y}_l = -\mu \cdot \dot{y}_l + \mathrm{P}_3 \cdot S_3 - F_i - mg, \quad (1)$$

where F_i is the disturbing force of the impulse impact of the collapsed roof rock block on the hydraulic support, kN; m is the mass of immediate roof rocks, kg; g is the acceleration of gravity, $g = 9.81\, m/s$; m_s is the reduced mass of moving parts of the support section and hydraulic props, taking into account the mass of fluid in the hydraulic props, kg; y_l - movement of the hydraulic rack piston, mm; μ is the coefficient of fluid friction.

The results of mine research confirm that the change in the force acting on the support under the dynamic action of the roof occurs in a very short period (0.05-0.3 s). Considering the most severe operation of the support, we assume that the load on the support changes in the form of a rectangular force impulse.

According to Skochinsky [2, 20] and a number of foreign authors, the maximum load on the support should be calculated assuming that the stratification and possible destruction of the roof occur at a height equal to five thicknesses of the removed layer. Other researchers suggest taking into account the resistance of friction and thrust forces between the blocks. In this case, the adhesion coefficient can be taken equal to $K_f = 0.7$.

We assume that when the roof collapses, the block acting on the support rests on the collapsed rocks. To determine the maximum possible load on the support section at maximum values of the parameters of the main roof collapse, let us set the conditions of the Polosukhinskaya Mine JSC and use support M-138/2 of face 26-325 for the calculation.

Based on the passport data, we take the layer thickness m = 1.8 m, the main roof thickness H = 12 m, the secondary step of the collapse of the main roof L = 120 m, the specific gravity of sandstone γ = 2.65 t/m^3, the width of the upper support section is taken to be equal to the pitch of the sections along the lava b = 1.5 m.

We accept the load on the support section and its hydraulic supports without taking into account the friction forces:

$$F = \frac{1}{2} Lb\gamma H = 5400\ \kappa N \quad (2)$$

The effort per one hydraulic prop of the M-138/2 support section is:

$$F' = \frac{F}{2} = 2700\ \kappa N \quad (3)$$

The effort per one hydraulic prop, taking into account friction forces is:

$$F_f' = F'(1 - K_f) = 810\ \kappa N \quad (4)$$

The working resistance of one hydraulic support M-138/2 is equal to 1500 kN when the safety valve is adjusted to the response pressure P = 32.5 MPa.

Thus, the force acting on the hydraulic props of the M-138/2 support section during the dynamic movement of the roof varies from 1500 to 2700 kN.

When developing equations describing the transient process in the piston cavity of the hydraulic rack under the dynamic load, we consider the hydraulic rack with a control unit and a torque converter.

The fluid flow in the hydraulic prop is

$$G_l = S\dot{y}_l \quad (5)$$

The flow rate in the rack, taking into account the liquid compression and the cylinder wall deformation is

$$G = \beta V \frac{\mathrm{dp}}{\mathrm{dt}}, \quad (6)$$

where V is the volume of fluid in the hydraulic rack, mm^3; $\beta = \frac{d_l}{E_l \delta} + \frac{1}{E_f}$ is the coefficient of volumetric compression; d_l is the inner diameter of the hydraulic rack cylinder, mm; δ is wall thickness of the the hydraulic cylinder, mm; E_l is modulus of elasticity of the cylinder material; E_f is modulus of elasticity of the fluid.

The torque converter piston block movement is

$$m_b \cdot \ddot{y_b} = -\mu \frac{dy_b}{dt} - P_1 \cdot S_1 + P_2 \cdot S_2 \tag{7}$$

where m_b is mass of the torque converter piston block, taking into account the liquid mass, kg; y_b - displacement of the torque converter piston block, mm; μ is the coefficient of fluid friction.

The fluid flow in the chamber above the large converter piston is

$$\frac{V_1}{E_f} \cdot \frac{dp_1}{dt} = \dot{y_b} \cdot S_1 - G_1 \tag{8}$$

where $V_1 = S_1(y_b^{max} - y_b) + V_{o1}$ is the volume of the large piston chamber, mm^3; $G_1 = \varphi S_1 \sqrt{\frac{2}{\rho}(P_1 - P_p)}$ is the flow rate of the working fluid through the hydraulic line, mm^3/min; y_b^{max} is maximum displacement of the converter piston block, mm; P_p is pressure in the pressure line, MPa; V_{o1} is the volume of the harmful space (*or dead volume*) of the torque converter, mm^3.

The liquid flow rate in the chamber under the small piston of the torque converter is

$$\frac{V_2}{E_f} \cdot \frac{dp_2}{dt} = -G_1^t - \dot{y_b} \cdot S_2 \tag{9}$$

where $V_2 = S_2(y_b^{max} - y_b) + V_{o2}$ is the volume of the small piston chamber of the torque converter, mm^3; $G_1^t = \varphi \cdot S_1^t \cdot \sqrt{\frac{2}{\rho}(P_1 - P_2)}$ is the flow rate of the working fluid through the throttle, mm^3/min; V_{o2} is the volume of the inner cavity of the second stage rod, mm^3.

The liquid flow rate in the piston cavity of the second stage of the hydraulic rack is

$$\frac{V_3}{E_f} \cdot \frac{dP_3}{dt} = G_1^t + G_2^t, \tag{10}$$

where $V_3 = S_3(y_l^{max} - y_l) + V_{o3}$ is the volume in the second stage piston cavity, mm^3; $G_2^t = \varphi \cdot S_3^t \cdot \sqrt{\frac{2}{\rho}(P_3 - P_2)}$ is the flow rate of the working fluid through the throttle, mm^3/min; V_{o3} is the volume of the harmful space (*or dead volume*) of the second stage piston cavity, mm^3.

The liquid flow rate in the piston cavity of the first stage of the hydraulic rack is

$$\frac{V_4}{E_f} \cdot \frac{dp_4}{dt} = G_2 - G_3^t \tag{11}$$

where $V_4 = S_4 \cdot (y_l^{max} - y_l) + V_{o4}$ is the volume in the first stage piston cavity, mm^3; $G_3^t = \varphi \cdot S_3^t \cdot \sqrt{\frac{2}{\rho}(P_4 - P_3)}$ is the flow rate of the working fluid through the throttle, mm^3/min; $G_2 = \varphi \cdot S_2 \cdot \sqrt{\frac{2}{\rho}(P_p - P_4)}$ is the flow rate of the working fluid in the pressure line, mm^3/min; V_{o4} is the volume of the harmful space (*or dead volume*) of the first stage piston cavity, mm^3.

Changes in the volume of the hose of the hydraulic line are

$$G_1 - G_2 = \beta \cdot V_p \frac{dp_p}{dt} \tag{12}$$

where β is the expansion coefficient of the sleeve; V_p is the volume of the pressure line, mm^3.

The logical conditions are

$$G_3^t = \begin{cases} G_3^t, \text{ if } P_4 > P_3 \\ 0, \text{ if } P_4 - P_3 \leq 0 \end{cases} \tag{13}$$

When combining equations 7-13, we have a system of equations describing the transient process in the hydraulic system of the rack with a torque converter as a result of the impulse action of disturbing force F_i. In this system, an appropriate set of differential, auxiliary equations and logical conditions can be used.

The system of basic differential equations is as follows:

$$\begin{cases} m_s \cdot \ddot{y}_l = -\mu \cdot \dot{y}_l + \mathrm{P}_3 \cdot S_3 - F_i - mg; \\ m_b \cdot \ddot{y}_b = -\mu \frac{dy_b}{dt} - \mathrm{P}_1 \cdot S_1 + \mathrm{P}_2 \cdot S_1; \\ \frac{S_1(y_b^{max} - y_b) + V_{01}}{\mathrm{E}_f} \cdot \frac{dp_1}{dt} = \dot{y}_b \cdot S_1 - \varphi \cdot S_1 \cdot \sqrt{\frac{2}{\rho}(\mathrm{P}_1 - \mathrm{P}_p)}; \\ \frac{S_2(y_b^{max} - y_b) + V_{02}}{\mathrm{E}_f} \cdot \frac{dp_2}{dt} = -\varphi \cdot S_1^t \cdot \sqrt{\frac{2}{\rho}(\mathrm{P}_1 - \mathrm{P}_2)} - \dot{y}_b \cdot S_2; \\ \frac{S_3(y_l^{max} - y_l) + V_{03}}{\mathrm{E}_f} \cdot \frac{dp_3}{dt} = \varphi \cdot S_1^t \cdot \sqrt{\frac{2}{\rho}(\mathrm{P}_1 - \mathrm{P}_2)} + \varphi \cdot S_2^t \cdot \sqrt{\frac{2}{\rho}(\mathrm{P}_3 - \mathrm{P}_2)}; \\ \frac{S_4(y_l^{max} - y_l) + V_{04}}{\mathrm{E}_f} \cdot \frac{dp_4}{dt} = \varphi \cdot S_2 \cdot \sqrt{\frac{2}{\rho}(\mathrm{P}_p - \mathrm{P}_4)} - \varphi \cdot S_3^t \cdot \sqrt{\frac{2}{\rho}(\mathrm{P}_4 - \mathrm{P}_3)}; \end{cases}$$

$$\varphi \cdot S_1 \cdot \sqrt{\frac{2}{\rho}(\mathrm{P}_1 - \mathrm{P}_\mathrm{p})} - \varphi \cdot S_2 \cdot \sqrt{\frac{2}{\rho}(\mathrm{P}_p - \mathrm{P}_4)} = \beta \cdot V_\mathrm{p} \cdot \frac{dp_\mathrm{p}}{dt}.$$

Table 1. Initial data for modeling

Parameter	Description	Quantity	Unit measure
Liquid in the pressure line coal face system:			
- pressure	P_{nom}	16-32	MPa
Disruption valve:			
- fluid flow	G_{max}	8	l/min
- pressure	P_{pk}	30-32	MPa
Choke:			
- pressure	P_{nom}	32	MPa
- fluid flow	G_{nom}	25	l/min
- diameter	$Ø_{pr}$	0.5-1.5	mm
Check valve:			
- pressure	P_{nom}	32	MPa
- fluid flow	G_{nom}	63	l/min
- diameter	$Ø_{pr}$	3	mm
Torque convertor:			
- small piston diameter	D_{mp}	20	mm
- large piston diameter	D_{bp}	45	mm
- length of stroke	Lx.п	800	mm
Hydraulic line:			
-diameter	$Ø_r$	7	mm
-length	L_r	3000	mm
Actuation fluid:			
- viscous friction coefficient	m_y	0.1	d/q
- flow coefficient in throttle	γ	0.8	d/q
- liquid density	ρ_o	1000	kg/m^3

Due to the nonlinearity of the system of equations (Figure 2) and the need for cyclical triggering of the logical conditions describing the movement of sliding elements of the hydraulic rack, the classical four-stage Runge-Kutta method is used. The task was solved by the Wolfram Mathematica computer algebra system.

When solving the system of equations 7-13, force F_i can be specified: with a smooth change in the load and with an instantaneous impulse loading of the system.

For the calculation, the operating characteristics of real serial equipment used as a prototype were taken as the initial data (Table 1).

4. Results and Discussion

Figures 4-6 show the results of computer modeling. Pressure changes in the piston cavity are insignificant. The duration of the first pulse was no more than 0.01 seconds and did not exceed 2.5 MPa, which reflects an increase in damping.

The bearing capacity of the hydraulic props of the powered support section with continuous impulseless control units under various dynamic loads (from A(t) = 1600 kN to A(t) = 3200 kN) can be estimated from the pressure change in the rod cavity of the second stage of the hydraulic prop (Figures 4-6) under loading.

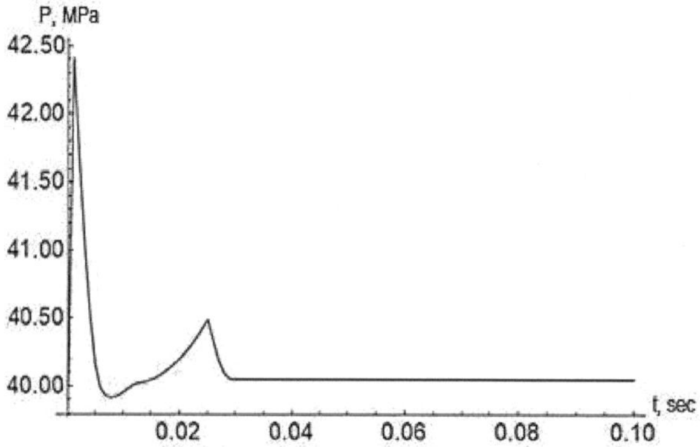

Figure 4. The transient process during impulse loading of the hydraulic rack force A(t) = 1600 kN

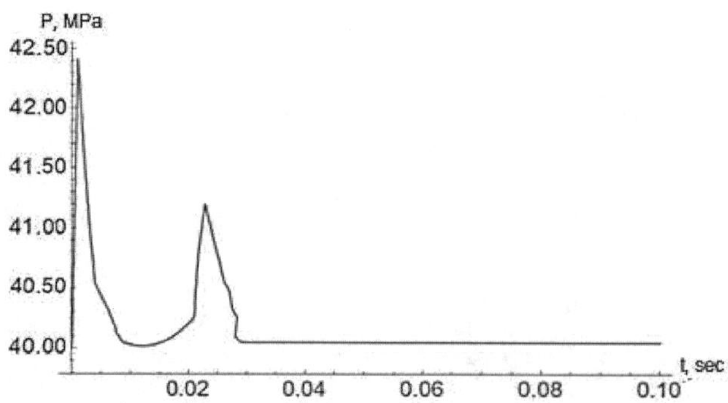

Figure 5. The transient process during impulse loading of the hydraulic rack force A(t) = 2500 kN

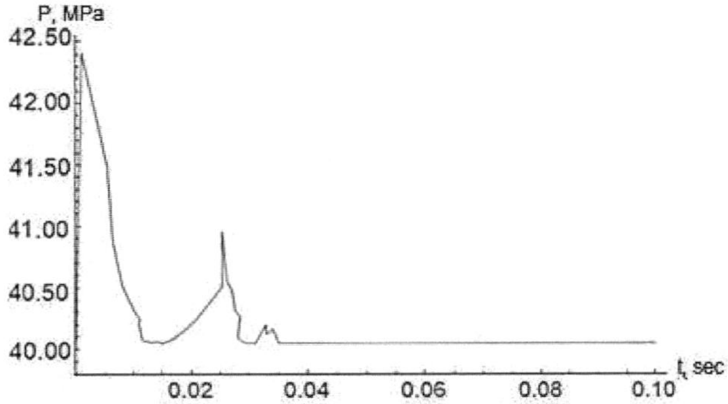

Figure 6. The transient process during impulse loading of the hydraulic rack force A(t) = 3200 kN

The modelling results (Figures 4-6) confirm adequacy of the mathematical model of transient processes when moving the elements of the gyro-resistant section of the powered support when changing the acting forces. This provides significant damping of power impulses and pressures in the elements of the hydraulic rack and its hydraulic system. The comparison of theoretical results and experimental data on tests of the MKYu 2Sh-13/27 hydraulic support with a control unit at Krasny Oktyabr plant [20] confirmed adequacy of the mathematical model with an increase in the external load on the hydraulic support until the safety valve was triggered (point C on the pressure line, Figure 7). There was a steady displacement of pistons of the torque converter of the control unit with displacements of the working fluid into the discharge line of the pumping station.

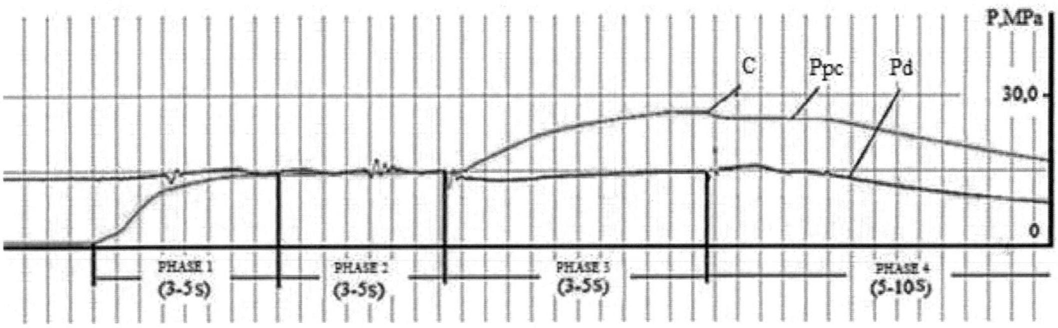

Figure 7. The setting mode of the experiment at "Red October" plant: P_{pc} - pressure in the piston cavity of the hydraulic prop; P_d - pressure in the hydraulic system drain line.

At the same time, no pronounced increase in the pressure amplitude or oscillatory phenomena in the hydraulic system were observed. This confirms adequacy of the mathematical model of the transient processes of movement of elements of the hydraulic support section of the powered support to the real processes when the external forces change.

From the analysis of Figures 4-6, it can be concluded that the nature of the process coincides with the experimental one (Figure 7) [20], and errors of the calculated values do not exceed 10%.

5. Conclusion

The mathematical model of the transient process in the hydraulic system of the rack equipped with a control unit for its resistance to roof rock lowering under the prolonged and impulse loading makes it possible to assess the influence of changes in the structure and values of the parameters of the control unit on the efficiency of hydraulic systems and possibilities of choosing their rational values at the stages of design development, which is confirmed by a number of established factors:

- displacement of the working fluid from the piston cavity of the second stage of the hydraulic prop into the pressure line takes place without oscillatory processes, i.e. the control unit is a damping element;
- with an impulse load surge on the hydraulic rack, the duration of the transient process is no more than 0.03 seconds, and the pressure drop is no more than 2.5 MPa, which corresponds to the permissible values.

References

[1] Jin Z 2008 Mechanical model study on roof control for fully-mechanized coal face with large mining height *Chinese J. of Rock Mechanics and Engineering* **27** 193-198

[2] Babyr N V, Korolev A I and Neupokoeva T V 2018 Enhancement of powered cleaning equipment with the view of mining and geological conditions *IOP Conf. Ser.: Earth Env. Sci.* **194** 032004. doi:10.1088/1755-1315/194/3/032004

[3] Stebnev A V and Buevich V V 2017 Improvement of Performance Indicators of Hydraulic Drive of Props of Powered Support Units of Heading Complexes *Zapiski Gornogo instituta* **227** 576-581. DOI: 10.25515/PMI.2017.5.576

[4] Gabov V V, Zadkov D A and Stebnev A V 2016 Evaluation of structure and variables within performance rating of hydraulically powered roof support legs with smooth roof control *Eurasian Mining* **2** 37-40. DOI: 10.17580/em.2016.02.09

[5] Zeng X T, Meng G Y and Zhou J H 2018 Analysis on the pose and dynamic response of hydraulic support under dual impact loads *Int. J. of Simulation Modelling* **17(1)** 69-80. DOI: 10.2507/IJSIMM17(1)412

[6] Byakov M A, Buyalich G D, Buyalich K G and Uvakin S V 2020 Radial strains in two-stage hydraulic extension legs *MIAB. Mining Inf. Anal. Bull.* **1** 133-140. DOI: 10.25018/0236-1493-2020-1-0-133-140.

[7] Pavlenko M V, Khaidina M P, Kuziev D A, Pihtorinskiy D and Muratov A Z 2019 Impacts of the combine harvester in the production of coal to increase methane recovery array in the workspace lava *Ugol'* **4(1)** 8-11. DOI: 10.18796/0041-5790-2019-4-8-11

[8] Zubov V P 2017 Status and directions of improvement of development systems of coal seams on perspective Kuzbass coal mines *Zapiski Gornogo instituta* **225** 292-297. DOI: 10.18454/pmi.2017.3.292

[9] Buyalich G D, Buyalich K G and Umrikhina V Yu 2016 Study of Falling Roof Vibrations in a Production Face at Roof Support Resistance in the Form of Concentrated Force *IOP Conf. Ser.: Mater. Sci. Eng.* **142** 012120. DOI: 10.1088/1757-899X/142/1/012120

[10] Kazanin O I 2015 About the design of underground mining of series of flat gas-bearing coal seams *Zapiski Gornogo instituta* **215** 38-45

[11] Kazanin O I and Sidorenko A A 2017 Interaction between gas dynamic and geomechanical processes in coal mines *ARPN J. of Engineering and Appl. Sci.* **12(5)** 1458-1462

[12] Buyalich G D, Byakov M A, Buyalich K G and Uvakin S V 2018 Model to analyze hydraulic legs with two-stage extension in mines *Mining Informational and Analytical Bull.* **S65** 21-28. DOI: 10.25018/0236-1493-2018-12-65-21-28

[13] Zuev B Y, Zubov V P and Smychnik A D 2019 Determination of static and dynamic stresses in physical models of layered and block rock masses *Gornyi Zhurnal* **7** 61-66. DOI: 10.17580/gzh.2019.07.02

[14] Gospodarikov A P 2016 Nonlinear math model development and numerical model of strain deformed rock mass conditions prognosis *Zapiski Gornogo instituta* **219** 382-386. DOI 10.18454/PMI.2016.3.382

[15] Gospodarikov A P, Vykhodtsev Y N and Zatsepin M A 2017 Mathematical Modeling of Seismic Explosion Waves Impact on Rock Mass with a Working *Zapiski Gornogo institute* **226** 405-411. DOI: 10.25515/PMI.2017.4.405

[16] Gabov V V and Zadkov D A 2018 Mathematical model of simple spalling formation during coal cutting with extracting machine *J. of Phys.: Conf. Ser.* **1015** 52007. DOI:10.1088/1742-6596/1015/5/052007
[17] Khazanovich G S, Voronova E Yu and Otrokov A V 2017 Simulation of the performance formation process of the loader with wedge-like working elements as a part of the blast and bulk tunneling complex *Procedia Engineering* **206** 457-464
[18] Ershov D Y, Zlotnikov E G and Timofeev D Y 2019 Analysis of proper fluctuations of technological systems *IOP Conf. Ser.: Mater. Sci. Eng.* **560** 012015. DOI:10.1088/1757-899X/560/1/012015
[19] Ascher U M and Petzold L R 1998 *Computer Methods for Ordinary Differential Equations and Differential-Algebraic Equations* (Philadelphia: SIAM Publications)
[20] Stebnev A V, Muhortikov S G, Gabov V V and Babyr N V 2018 Test of the unit of non-impact regulation of resistance of hydraulic stands of mechanical roof support to lowering of breeds of hanging wall *Mining Informational and Analytical Bull.* **S48** 416-425. DOI: 10.25018/0236-1493-2018-11-48-416-425

IOP Conf. Series: Materials Science and Engineering 1064 (2021) 012046 doi:10.1088/1757-899X/1064/1/012046

Standardization and control of thread parameters of a roller-screw gear

V S Zavyalov[1,2], M E Fedosovsky[1,2], N K Maltseva[1], V V Degtyarev[3], I P Boldyrev[2], A G Sharshov[2]

[1]ITMO University, 49A, Kronverkskiy avenue, Saint Petersburg, 197101, Russia
[2]AO Diakont, 2, Uchitelskaya street, Saint Petersburg, 195274, Russia
[3]Peter the Great St. Petersburg Polytechnic University, 29, Politekhnicheskaya street, Saint Petersburg, 195251, Russia

E-mail: zavyalov_v@diakont.com

Abstract. Among the mechanisms that transform rotational motion into translational motion, helical gears of rolling, which provide higher efficiency and speed, are currently dominant. These include ball-screw gears and roller-screw gears. A comparative analysis of these showed that the latter has a number of limitations compared to the former. Serial production requires the development of automated technologies for machining, assembly and testing, the use of high-precision metrological equipment to control functional surfaces with minimal time costs in digital automated systems. Modern measuring instruments, such as contour meters allow for element-by-element control of the parameters of threaded surfaces. In the manufacture of high-tech products such as an electromechanical drive, the use of element-wise control is the only way to ensure product quality. However, not enough standardization of the parameters of the thread of the roller-screw gears requires the development of a set of requirements that provide the specified operational characteristics of the product. The work is devoted to the standardization of the parameters of the special roller-screw gears thread used in the electromechanical drive manufactured by AO Diakont. This article discusses issues related to the standardization of the accuracy parameters of the profiles of high-precision roller-screw gears, metrological support tools, a set of techniques and software based on a systematic approach to inputting, outputting and transmitting information about measurement results and evaluating their validity. The application of the proposed control methodology allowed improving the quality of products and reducing its cost.

1. Introduction

The development of computer technologies used in manufacturing management systems leads to the emergence of new requirements for the functioning of these systems. One of the promising areas is the use of electromechanical drives (EMD) as part of control systems for synchronous motors. This type of drive has found application in the automotive industry, aerospace, steam turbine control systems and many other industries [1-2]. The use of modern technologies of mechanical and heat treatment of parts of a roller-screw transmission can improve the operational characteristics of the product [3-5]. The control of the special internal thread of the roller-screw drive is a complex technological and metrological task. At the moment, there is no generally available international or national standard for threading a roller-screw drive. The closest standards are GOST 11708-82 [6] and ISO 3408-1: 2006 [7]; however, they do not cover all the parameters necessary to ensure the operation of the gear. In [8], an

IOP Conf. Series: Materials Science and Engineering 1064 (2021) 012046 doi:10.1088/1757-899X/1064/1/012046

analysis was made of the methods for measuring the parameters of the internal thread; however, none of them provides control of the full range of requirements for the thread of a roller-screw gear.

2. Terms and definitions

Based on the functional purpose for describing the profile of threaded surfaces, it is proposed to use the following definitions.

Contour of the threaded surface of a roller-screw gear is the identified profile of the threaded surface of the roller-screw gear in axial section, obtained as a result of one measuring pass of the contour meter described by an array of points with coordinates xi, yi.

Turn is a section of a thread contour related to one full revolution of the points of the helical line of the thread relative to the axis.

Thread generator (O1, O2) is a segment of the contour of the threaded surface connecting the top and bottom of the thread.

Top of the thread (A1, A2) is a segment of the contour of the threaded surface connecting the generators of adjacent turns along the top of its protrusion.

Bottom of the thread (V) is a segment of the contour of the threaded surface connecting the generatrixes belonging to one thread along the bottom of its groove.

Inscribed circle is a circle of a given diameter dc, tangent to the generators of one thread at two points with centers described by the coordinates Xi, Yi.

Screw pitch (SP) is the distance between the centers of adjacent inscribed circles of one contour.

Average diameter of the thread (D2) is the distance between the centers of the nearest inscribed circles for opposite thread contours.

Scan length is the stroke length of the contour meter.

Controlled length (Lu) is the length of the identified thread profile after exclusion of incomplete turns.

Roller length (LR) is a segment of geometric length in contact with the mating surface.

Actual interval of variation of the average diameter over the entire controlled length (OD2) is the difference between the largest and smallest values of the average diameter of the thread over the entire controlled length.

Actual interval of variation of the average diameter on the length of the roller (OD2R) is the difference between the largest and smallest values of the average diameter of the thread along the length of the roller.

Actual translation is the actual Y coordinate of the center of the inscribed circle.

Actual mean translation is a straight line plotted through the values of the actual displacement by the method of least squares.

Actual value of the error of the average translation within the measurement length (Ea) is the difference between the actual average translation on the measurement length and the nominal movement for a given length.

Actual interval of the error of the average translation within the measurement length (Vua) is the value of the interval of the accumulated error of steps within the measurement length.

Tolerance of a given thread profile is the distance along the normal to the nominal thread profile from the nominal thread profile to the controlled point of the identified thread profile.

3. Determining identified profile

In view of the foregoing, Figure 1 shows a diagram for determining the identified thread contour using a Mitutoyo SV-C4500 contour meter:

- Point 0 – the starting point of the control program;
- Point 1 – start of measurement of the contour *C1*;
- Point 2 – end of the measurement of contour *C1*, change of direction of the measuring force
- Point 3 – start of measurement of the contour *C2*;
- Point 4 – end of contour measurement *C2*;

– Point 0 – change of direction of the measuring force, completion of the control program.

The figure shows the following coordinate axes used in the measurement:

– *Z1* – axis of the column of the contour meter
– *Z2* – axis of the probe
– *X* – longitudinal axis of the contour meter
– *x, y* – part coordinate system

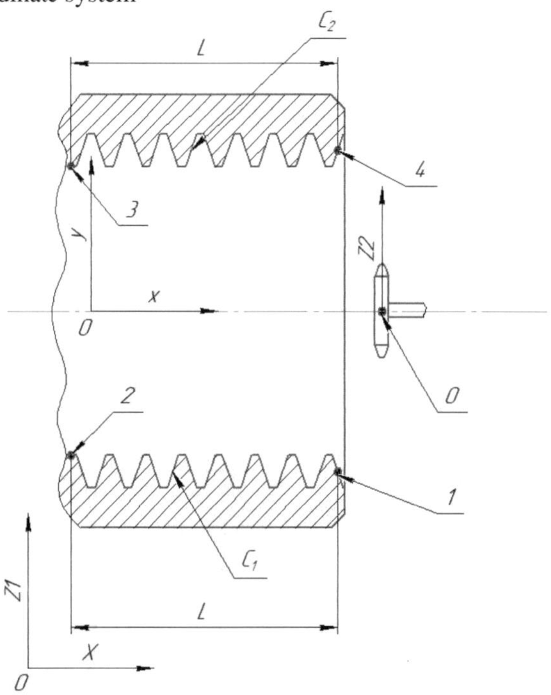

Figure 1. Scheme of thread profile scanning

The part is mounted on the contour meter, ensuring that the axis of symmetry of the part is located on the coordinate plane *XZ1* of the contour meter. The measuring probe is brought to point 0 in the position *Z2* = 0, the probe is lowered until the measuring tip contacts the surface of the part at point 1, after which the circuit C_1 is scanned. At the endpoint 2 of circuit C_1, the direction of the measuring force is changed, the probe touches the surface of the part at point 3, the measurement of the circuit C_2 begins, the measurement of the circuit ends at point 4, after which the probe is taken to the starting point and the direction of the measuring force is changed. Scan results are saved on the server as a text file with .dat extension.

The file with the coordinates of the profile points contains the following information: the units of measurement used (usually mm) are recorded on the first line of the file, the number of points making up one thread contour from the start to the end point of the scan is indicated on the second line, and the coordinates xi, yi of every point are written in the following lines. If there are several loops in the same file, after the last point of the first loop, information on the remaining loops is similarly recorded. Each measured circuit contains up to 10,000 points.

After reading points from the file into the data.point table, each point is assigned the name of the contour to which it refers. Each contour is divided into intervals with a length equal to the thread pitch. Each turn is divided into five segments: A1, O1, V, O2, A2. The scheme for dividing a thread into segments is shown in Figure 2. Section V is defined as points having a y coordinate less than yV specified in the nominal thread data.

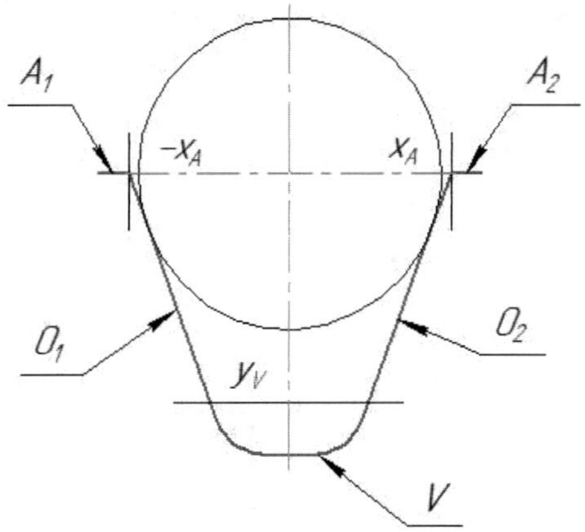

Figure 2. Dividing thread turn profile into segments

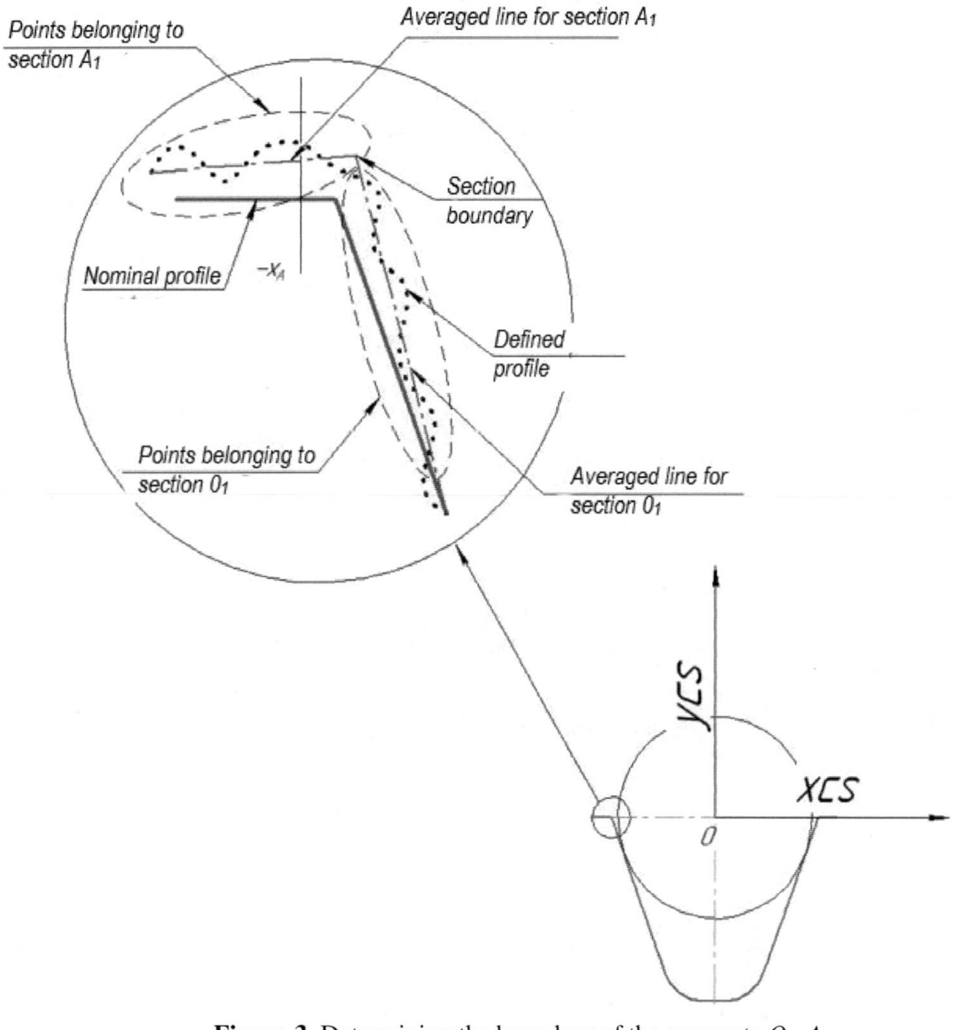

Figure 3. Determining the boundary of the segments O_1, A_1

IOP Conf. Series: Materials Science and Engineering 1064 (2021) 012046 doi:10.1088/1757-899X/1064/1/012046

The definition of the segments A_1, O_1 and A_2, O_2 is shown in Figure 3. The points that are guaranteed to belong to these segments are preliminarily selected. Points located to the left of the $-x_A$ coordinate of the part specified in the nominal data are accepted as segment A_1, those located to the right, are accepted as segment O_1. After that, for each section, the equation of the line is calculated using the least squares method; the coordinates of the intersection of these lines are found. This point is taken as the border of two segments, the points located to the left of the border belong to the segment A_1, those to the right, belong to the segment O_1. Plots A_2, O_2 are determined symmetrically.

4. Calculating the centers of the inscribed circles

The coordinates of the center of the inscribed circle X_0, Y_0 of a given radius are calculated for several iterations for each thread turn. For the first one, the average coordinates of the profile points are taken as the center coordinate.

$$X_0 = \frac{\sum_{i=m}^{n} x_i}{n-m}; \; Y_0 = \frac{\sum_{i=m}^{n} y_i}{n-m},$$

where m is the minimum point number related to this turn, n is the maximum point number related to this turn.

After that, the gap e_1, e_2 between the inscribed circle and the revealed thread profile is calculated by the following equation:

$$e_1; \, e_2 = \sqrt{\min((X_0 - x_i)^2 + (Y_0 - y_i)^2)} - d_c$$
$$e = \min(e_1, e_2)$$

After calculating the size of the gap, the coordinates of the center are recalculated by the following equation:

$$X_0 = X_0 + e \cdot \sin\alpha$$
$$Y_0 = Y_0 - e \cdot \sin\alpha \, ,$$

where α is the half of the thread profile angle in axial section.

When the value of e □ 10^{-10} is reached, the circle is considered inscribed, and the values X_0, Y_0 are entered in the data.center table as X_i, Y_i, where i is the current thread turn number.

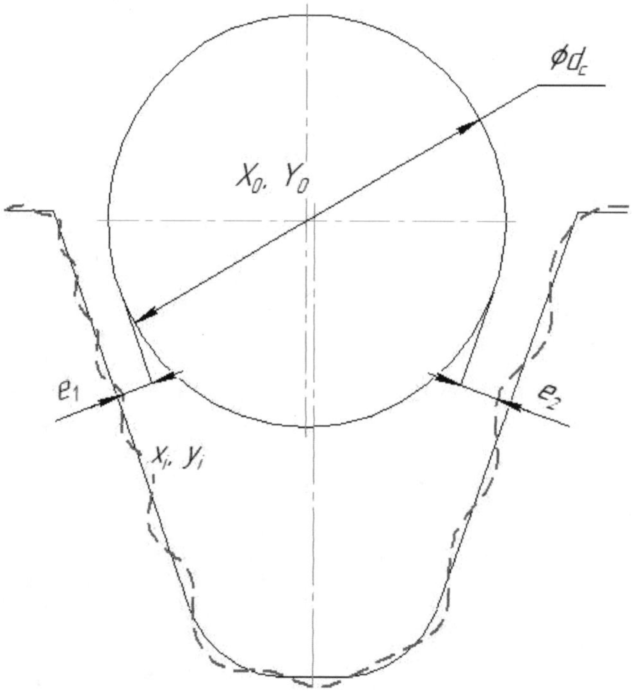

Figure 4. Calculation of the center of the inscribed circle

5. Assessment of the geometric parameters of the thread

5.1. Thread pitch estimation

For each thread turn within the contour, the actual values of the thread pitch are calculated according to the following equation:

$$SP_i = X_{i+1} - X_i,$$

where SP_i is the actual value of the i-th thread pitch on the lower/upper detected contour, X_i, $X_{(i+1)}$ are the coordinates along the abscissa of the centers of the inscribed circles that are calculated separately for the contour C_1 and C_2.

For each turn, the value of the accumulated error of the thread pitch is calculated using the following equation:

$$CSP_i = CSP_{i-1} + (SP_i - SP_{nom}),$$

where CSP_i, CSP_{i-1} is the accumulated error on the i-th and the preceding thread pitches, respectively, on the lower/upper detected contour within the measurement length, SP_{nom} is the nominal value of a thread pitch according to the design documentation.

Having determined the pitch values, we can calculate the accumulated pitch error. To do this, according to the obtained values of SP_i by the least squares method, the dependence $CSP(X)$ of the trend line for the accumulated error is determined:

$$Ea = k \cdot X,$$

where Ea are the errors of average displacement within the measurement length, k is the tangent of the line slope of the actual average translation.

For the resulting equation, the maximum positive r_1 and negative r_2 error are calculated:

$$r_1 = |\min(CSP_i - k \cdot x_i)|$$
$$r_2 = |max(CSP_i - k \cdot x_i)|.$$

Then the value of the error interval of the average translation within the measurement length (V_{ua}) is calculated by the following equation:

$$V_{ua} = r_1 + r_2$$

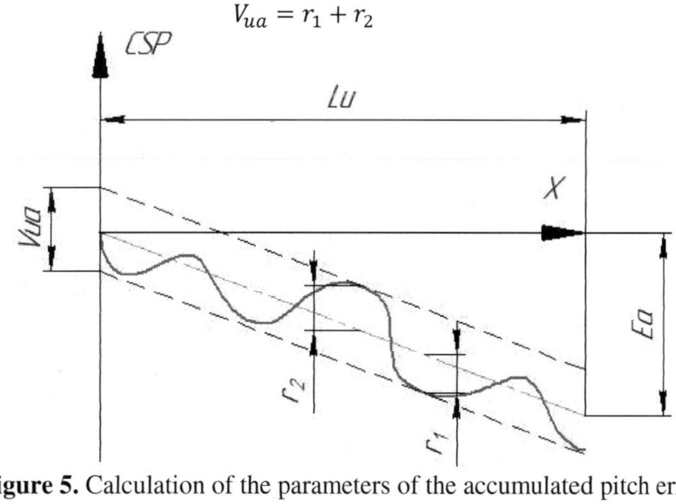

Figure 5. Calculation of the parameters of the accumulated pitch error

5.2. Estimating average thread diameter

The calculation of the average diameter of the thread mimics a three-wire measurement scheme used to control the external thread. Figure 6 illustrates the calculation of the average diameter. The calculation of the average diameter for one inscribed circle of the lower contour involves three inscribed circles for the upper contour, this calculation scheme is adopted based on the characteristics of a multistart thread. For each thread, we calculate the value of the average diameter:

$$D2_i = \frac{Y_{i-1}^{c2} + Y_i^{c2} + Y_{i+1}^{c2}}{3} - Y_i^{c1},$$

where Y_i^{c1} are the coordinates of the centers of the inscribed circles for the contour C_1, Y_i^{c2} are the coordinates of the centers of the inscribed circles for the contour C_2.

To exclude the influence of fluctuations in the Y coordinate of adjacent inscribed circles on the value of D2 and to increase the stability of the result, the filtering procedure for the obtained values is performed according to the following equations:

$$D2_1^n = \frac{D2_1^{n-1} + D2_2^{n-1}}{2}$$

$$D2_i^n = \frac{0.5 \cdot D2_{i-1}^{n-1} + D2_i^{n-1} + 0.5 \cdot D2_{i+1}^{n-1}}{2}$$

$$D2_m^n = \frac{D2_{m-1}^{n-1} + D2_m^{n-1}}{2},$$

where i from 1 to m is the thread turn number, n from 1 to 5 is the number of filtration cycle.

For further convenience, the value of the average diameter of the thread is taken as $D2_i = D2_i^5$. The value of the interval of fluctuation of the average diameter over the measurement length Lu is calculated using the equation:

$$OD2 = \max(D2) - \min(D2)$$

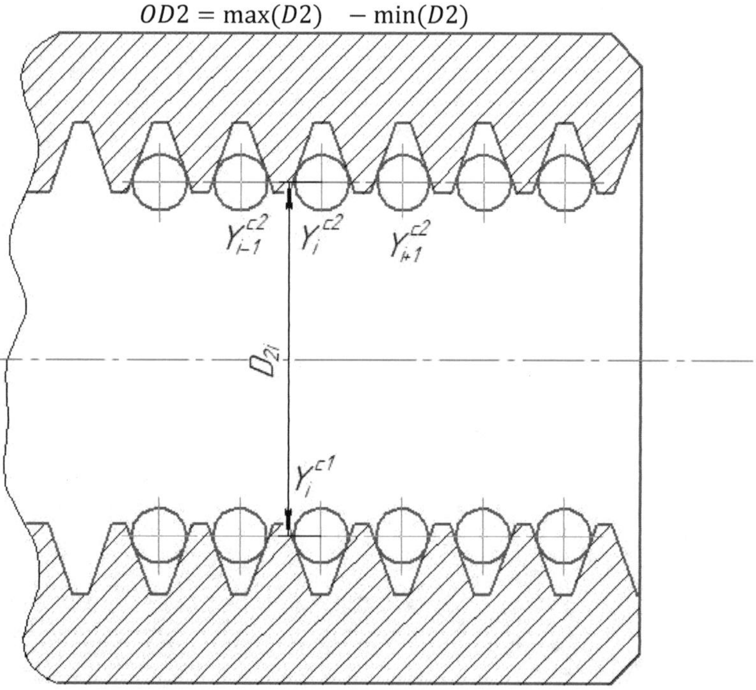

Figure 6. Calculation scheme for estimating the average diameter

For the roller length (L_R) value specified in the design documentation, the number of full turns n that fit this length is calculated:

$$n = \frac{L_R}{SP},$$

then for each turn, starting with n, the actual value of the interval of change in the average diameter on the length of the roller is calculated according to the formula:

$$OD2_{Ri} = \max_{i-n \to i}(D2) - \min_{(i-n) \to i}(D2)$$

where i is the turn number.

5.3. Calculating the actual tolerance a given profile shape

Based on the functional purpose of the part, the tolerance of the shape of a given profile has different values on different segments of the thread contour. It is normalized in the form of three circuits: nominal, external maximum permissible circuit and internal permissible circuit. The assessment of the actual value of the profile tolerance is carried out by calculating the actual value of the distance from the corresponding point of the nominal profile to the point of the detected profile and comparing this point with two valid contours. For each turn, the coordinate system is transferred to the center point of the inscribed circle, the coordinates of the profile points in the new system xcs_i, ycs_i.

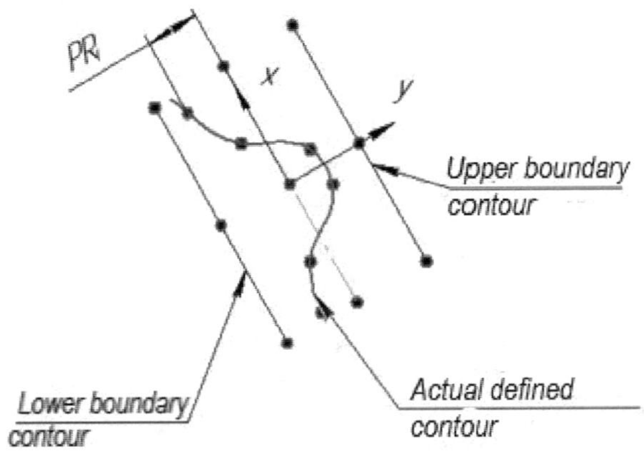

Figure 7. Assessing the tolerance of a given profile shape

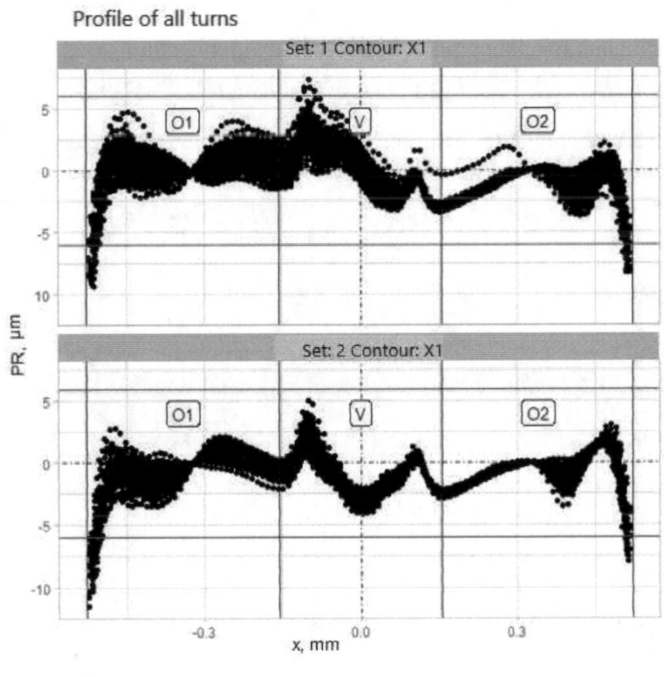

Figure 8. Result of the tolerance assessment of a given profile

For each point of the measured contour, we find the two closest points of each contour of the nominal profile. At the points of the nominal contour, we flatten the coordinate system (Figure 7). The *y* coordinate of the obtained point shows the actual value of the tolerance of the shape of the given profile PR_i, and the *y* coordinate of the upper and lower contours show the permissible deviation for this point. The obtained value and the decision on the suitability or non-suitability of the point are written in separate columns of the data.point table. For each segment of the contour, we determine the percentage of points within the tolerance and the maximum and minimum deviations. The control results are displayed in the form of a profile scan (Figure 8) where the coordinate of the x point relative to the center of the inscribed circle (in mm) is plotted along the abscissa axis, and the actual tolerance value of the given *PR* profile in micrometers is plotted along the ordinate axis. The flatten coordinate system is built separately for each thread start and thread contour.

6. Conclusion

The developed technique for standardizing the accuracy parameters of the roller-screw gear made it possible to provide an integrated approach to the design of structures based on their operational characteristics of bearing capacity, the number of operation cycles under various conditions in the CAD structure.

The productivity and reliability of the processes of metrological evaluation of the parameters of the roller-screw gear are increased, so the operational time of control operations is reduced by 45 percent. Currently, roller-screw gears are promising for converting rotational motion into translational motion, and surpass similar solutions in most operational parameters.

Their operational characteristics: stiffness, bearing capacity, kinematic accuracy, positioning accuracy, are most affected by the error in the manufacture of the internal thread profile in terms of pitch, the errors in the interval of change in the actual dimensions of the average diameter and tolerance values for a given profile. Therefore, high-quality standardization of thread profile parameters and full objective metrological control of the thread became relevant and highly responsible.

The article is devoted to the method of standardizing the parameters of threaded profiles implemented in the production environment for the purpose of their application in the conditions of CAD of roller-screw gears and metrological control of the parameters when measuring profiles in the axial plane on a Mitutoyo SV-C4500 contour meter

To process the measurement results obtained in the process of metrological control, mathematical and software have been developed that are used to analyze and evaluate the quality of mass-produced products.

The developed software is universal for threads with a symmetrical profile of the side surface formed by both straight and arc segments, and therefore can be used in metrological evaluations of screws and rollers of roller-screw gears.

References

[1] Yuryeva R A et al 2015 Design Features of an Electromechanical Drive *Science and Business: Development Paths* **10** 21-23

[2] Giangrande P, Galassini A, Papadopoulos S, Al-Timimy A, Calzo G L, Degano M, Gerada C et al 2019 Considerations on the development of an electric drive for a secondary flight control electromechanical actuator *IEEE Transactions on Industry Applications* **55(4)** 3544-3554

[3] Ivanenko A O, Tulkova I A and Uvarov M M 2018 Technological features of nitrogening thread surfaces of critical parts of electromechanical drive *J. of Instrument Engineering* **61(4)** 360-367

[4] Wang Yulin et al 2020 Modeling and optimization of dynamic performances of large-scale lead screws whirl milling with multi-point variable constraints *J. of Materials Processing Technology* **276** 116392

[5] Ivanenko A O, Degtyarev V V et al 2017 Experience in the use of diamond-like coatings to increase the life of a roller-screw gear box *Science and Business: Development Paths* **10** 5-14

[6] GOST 11708-82 State standard of the USSR. Basic standards of interchangeability. Thread.

Terms and definitions
[7] ISO 3408-1: 2006 Ball screws – Part 1: Vocabulary and designation
[8] Zavyalov V S and Maltseva N K 2019 Measurement of the parameters of the internal thread of a nut of a roller-screw transmission of an electromechanical drive *Bull. of Higher Educational Institutions. Instrument Making* **62(8)** 749-757

IOP Conf. Series: Materials Science and Engineering 1064 (2021) 012047 doi:10.1088/1757-899X/1064/1/012047

Recognition of structural elements of mechanical engineering as a solution to data formalization problem

I V Fokin, N V Podrez, A N Smirnov and A S Govorkov

Irkutsk National Research Technical University, 83, Lermontova str., Irkutsk, 664074, Russia

E-mail: govorkov_as@istu.edu

Abstract. The paper discusses a typical example of recognizing a sheet metal part containing 2 structural elements. According to the developed recognition algorithm, a software module was implemented for automated recognition of structural elements in a 3-dimensional model of parts with the * .stp extension, which allowed working not only with models made in the Siemens NX system, but also, in general, with almost any 3D models exported in this format. The result of recognition can be used to develop systems that design technological processes for the production of mechanical engineering products, including at aircraft production enterprises.

1. Introduction

The classification of parts produced at factories includes thousands of different types, shapes and sizes. Every year their number only grows, under the influence of the development of technology for the production of new details. In this regard, we can state that there is a need to systematize all the variety of details. The process of systematization becomes possible through the implementation of the classification of existing and new elements of the model. Classification is important in the management of the quality and variety of parts, as it promotes the systematic study of parts, rational organization of production, effective quality control work and ultimately leads to the decrease in the cost of the final product.

2. Results

The algorithm for the qualitative assessment of products can be conditionally divided into 2 stages:

1. the analysis of the geometric characteristics of structural elements included in the structure of the product;
2. the analysis of the capabilities of an enterprise having a production basis for the manufacture of individual structural elements and the product as a whole.

The idea of product analysis is based on the assumption that any product (diaphragms, ribs, angle bars, etc.) can be formalized as a set of structurally connected in a certain way and having its own set of characteristics by analogy with an assembly unit, which consists of individual products connected in a certain way.

Thus the sheet part shown in Figure 1 can be described by a set of structural elements included in it: wall + flanging, flanging.

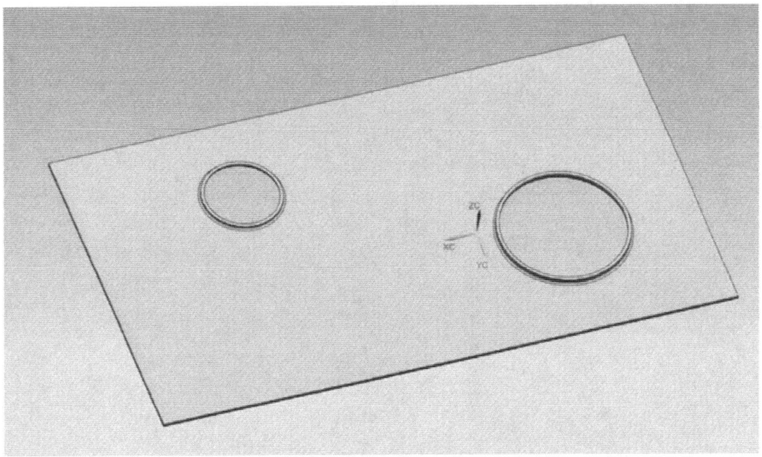

Figure 1. Wall with 2 flanges

In order to ensure the possibility of an unambiguous definition of a product through its structural elements, it is necessary to develop a classifier of possible structural elements.

Table 1. Typical classifier of structural elements

Code of structural elements	Name of structural elements of a part
01	Flanging on a flat surface
02	Flanging on a curved surface
03	Blind flanging on a flat surface
04	Blind flanging on a curved surface

In this classifier, only one structural element (SE) is presented. However, the list of SEs that should be included in the classifier is individual parameters for each production, taking into account their production capabilities.

Such a representation of a part through a set of structural elements is enough for human understanding, but not enough for automating recognition by a program method, since the very concept of a "structural element" is empirical.

The electronic model of the product is presented in the form of a hierarchical structure in Figure 2.

In this case, the structural elements are uniquely defined through the graph set of "edges-surfaces", where the edges of the 3d-model act as the vertices of the graph and the surfaces of the 3d-model act as the edges of the graph.

The flanging is a structural element that increases the rigidity of the wall of the part, in addition, it can reduce the weight of a part. In the aircraft industry, the following types of flanges are distinguished: type 1, type 2 and type 3.

The process of formation during flanging is performed by thinning the material. When making flanges of type 1 and type 2, there is uniaxial tension in the circumferential direction, the degree of deformation is proportional to the increase in the diameter of the pre-formed hole. The examples of flanges are shown in Figure 3.

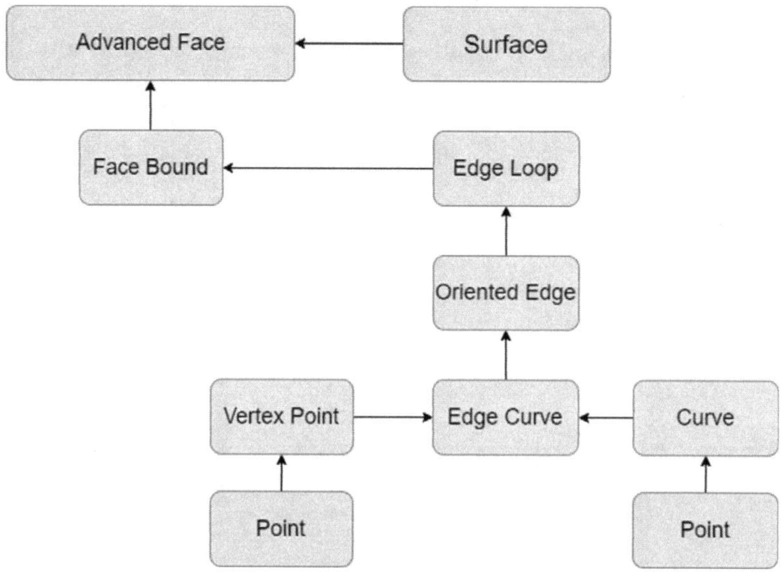

Figure 2. Electronic representation of a 3D product model

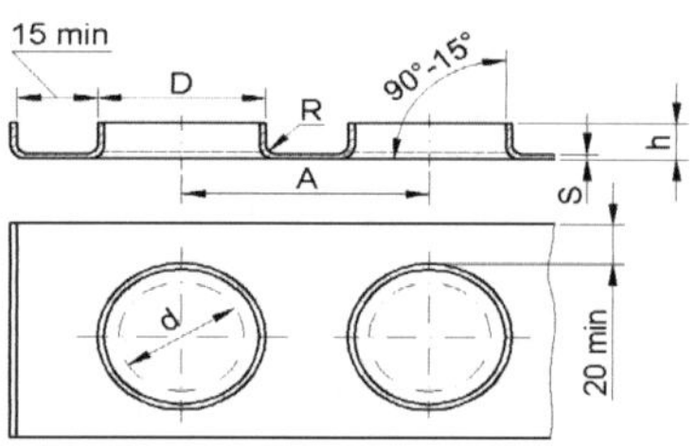

a) normal flanging, type 1

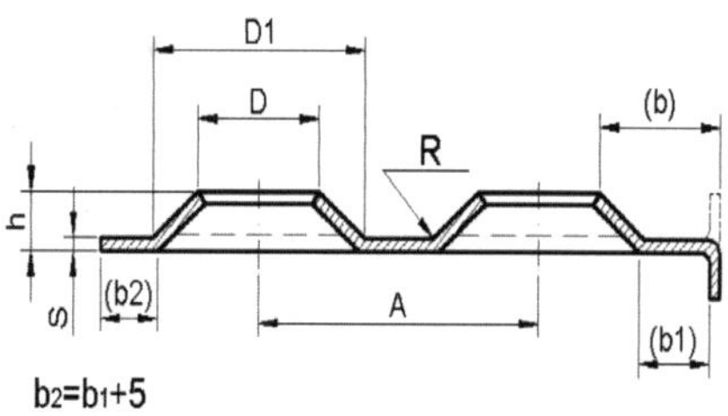

b) flanging at an angle of 60°, type 2

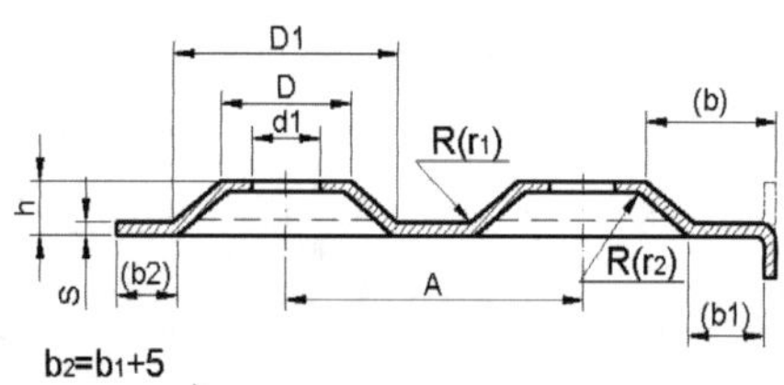

c) disc flanging, type 3

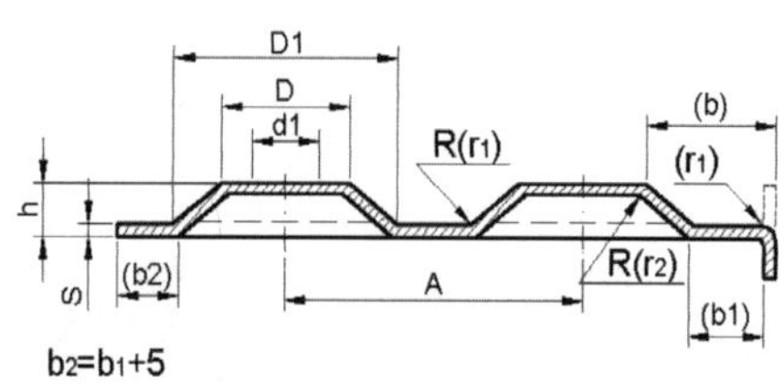

d) blind flanging, type 4

Figure 3. Type of flanging

During the production of disc flanging of type 3, the hole is punched after the flanging cavity has been formed. When a blind flanging is formed, uniaxial stretching occurs in the meridian direction. The process of formation is carried out by thinning the workpiece material, including the areas surrounding the flanging. This aspect must be taken into account when placing adjacent structural elements when assigning dimension A (Figure 3), in order to prevent excessive thinning of the wall.

It is determined more unambiguously despite the geometrically more complex shape than, for example, near the wall from a technical point of view. The structure of flanging of type 1 consists of a flat surface bounded by two circles, which, together with two more circles, form cylindrical surfaces of the beads flowing into toroidal surfaces. Toroidal surfaces have a common center, therefore, for automatic recognition, it is sufficient to find two toroidal surfaces with a common center in the part structure. Next we consider the classification by the type of flanginf. The graphical structure of flanging of type 1 is shown in Figure 4.

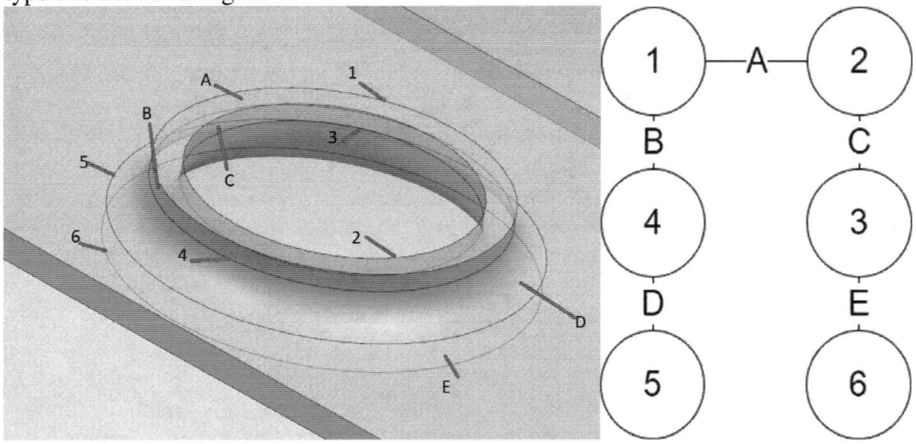

Figure 4. Structural element of flanging of type 1

In fact, the above mentioned aspects are relevant for all structural elements with a similar production method, such as a rift.

According to the developed theory, a program was implemented that allows recognizing flanging of type 1, chamfers, fillets and holes in parts with the extension .stp, which allows working not only with Siemens NX models, but also, in general, with almost any 3D models exported into this format. Figures 5, 6 show the operation of the program.

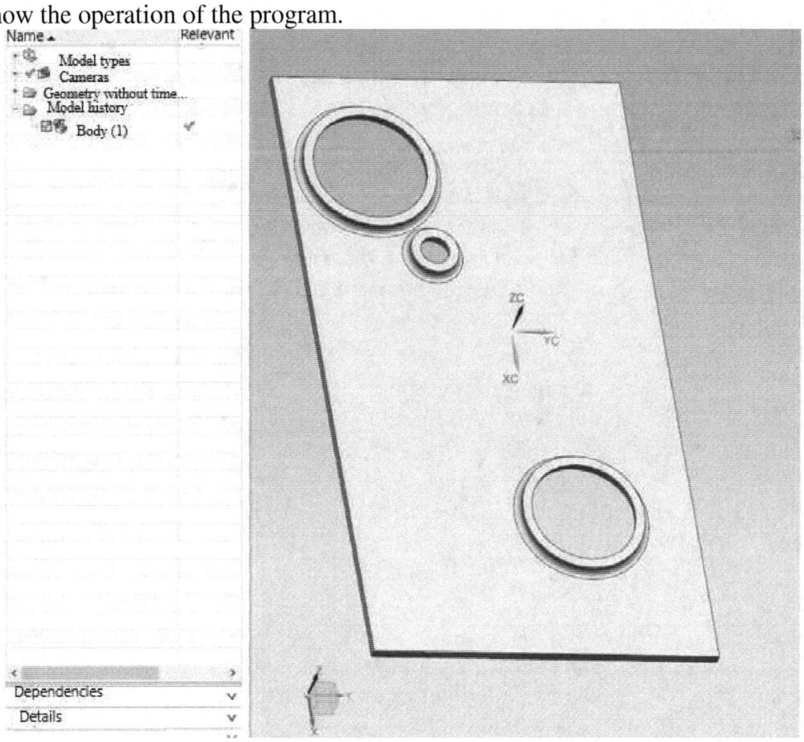

Figure 5. Studied model

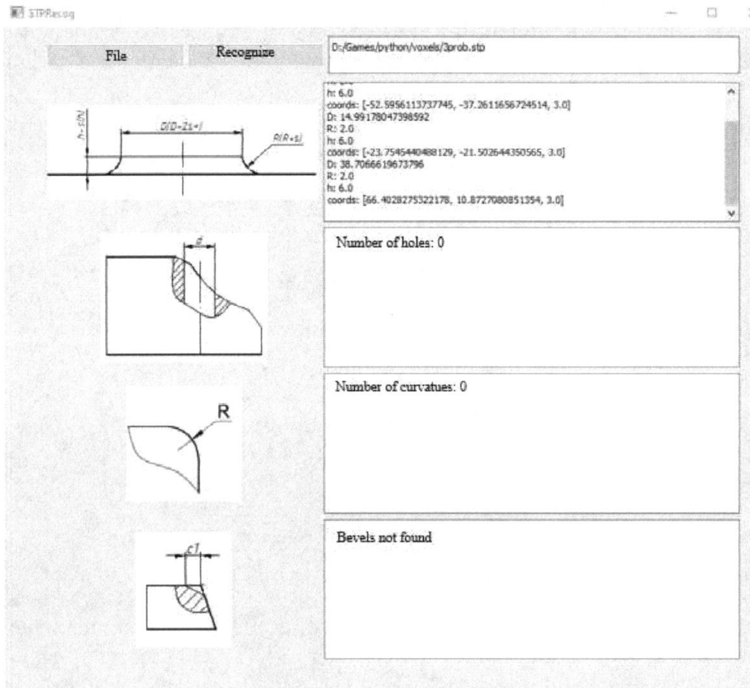

Figure 6. Program interface

As we can see from Figure 6, the program interface includes a file selection button, a button for starting the recognition process, graphic diagrams indicating the type of the recognized element and the parameters to be determined, the blocks where information about the recognized elements is given and a block indicating the path to the current file.

3. Conclusion

The given example shows correct recognition of three different flanging on a part that does not have a construction tree. This approach can be implemented not only on parts made of semi-finished sheet products, but also on any other SE used in other industries. According to the results of the study, a Certificate for computer program No. 2019665155 dated 20.11.2019 was issued.

References

[1] Appius - PDM: Project management system and electronic archive (for the 1C: Enterprise 8.0 platform) (Moscow: Appius) Retrieved from: www.appius.ru
[2] Boureau Y-L, Bach F, LeCun Y and Ponce J 2010 Learning mid-level features for recognition *IEEE Conference on Computer Vision and Pattern Recognition* pp 2559-2566
[3] MacKay D J C 1992 A practical Bayesian framework for backpropagation networks *Neural Computation* **4(3)** 448-472
[4] Gonzalez R C and Woods R E 1992 *Digital Image Processing* (Boston, MA: Addison-Wesley Publishing Company) 528 p
[5] Hagan M T and Menhaj M 1994 Training feedforward networks with the Marquardt algorithm *IEEE Transactions on Neural Networks* **5(6)** 989-993
[6] Marquardt D 1963 An Algorithm for Least-Squares Estimation of Nonlinear Parameters *SIAM J. Appl. Math.* **11(2)** 431-441
[7] Mashor M Y and Sulaiman S N 2001 Recognition of Noisy Numerals using Neural Network *IJCIM* **9(3)** 158-164

[8] Nguyen D and Widrow B 1990 Improving the learning speed of 2-layer neural networks by choosing initial values of the adaptive weights *Int. Joint Conf. on Neural Networks* **3** 21-26

[9] Poland J 2001 *On the Robustness of update strategies for the Bayesian hyperparameter alpha.* Retrieved from: http://www-alg.ist.hokudai.ac.jp/~jan/alpha.pdf

[10] Wilamowski B M, Chen Y and Malinowski A 1999 Efficient algorithm for training neural networks with one hidden layer *Int. Joint Conf. on Neural Networks (IJCNN'99)* **3** 1725-1728

[11] Galkin V 2008 FeatureCAM - an effective solution for multitasking CNC machines *CAD and graphics* **6** 85-87

[12] Izotov P Yu, Sukhanov S V and Golovashkin D L 2010 Implementation technology of a neural network algorithm in the CUDA environment using the example of handwritten digit recognition *Computer Optics* **34(2)** 243–252

[13] Kulikov A A, Nebyshinets Yu V, Sidorova A V and Balanovsky A E 2020 3D printing technology for metal products: from a computer-aided design system to a real part *Irkutsk State Technical University Bull.* **24(4)** 728–739. DOI: https://doi.org/10.21285/1814-3520-2020-4-728-739.

[14] Larchenko A G 2019 Assessment of the quality of products made of polymer materials for mechanical engineering purposes *Irkutsk State Technical University Bull.* **23(3)** 463–471. DOI: 10.21285 / 1814-3520-2019-3-463-471.

[15] Pavlov Yu A 2008 *Software control of technological equipment in a flexible automated stone-processing production:* Textbook in 3 books (Moscow: Publishing house of MGGU)

[16] Pavlov Yu A and Tkach V R 2006 *Organization of stone-processing production using information technology* (Moscow: Catalog) 358 p

[17] 2008 Russian complex of programs T-Flex CAD / CAM / CAE / PDM (Moscow: Top Systems) Retrieved from: www.topsystems.ru.

IOP Conf. Series: Materials Science and Engineering 1064 (2021) 012048 doi:10.1088/1757-899X/1064/1/012048

Formation of "undercuts" on titanium case parts

M V Lavrentyeva and V V Mironenko

Department of Aircraft Construction and Maintenance, Institute of Aircraft Construction, Mechanical Engineering and Transport, Irkutsk National Research Technical University, 83, Lermontova str., Irkutsk, 664074, Russia

E-mail: mira.amazon@gmail.com

Abstract. The paper presents a variant of the formation of a part from a hard-to-form case. The calculation of the required area of the movable clamp depending on the stresses arising during the formation of the "undercut" is performed. The results of real and virtual experiments are presented. The conclusions about the applicability of the technology and the advantages that the use of a movable clamp gives are made.

1. Introduction

In sheet forging, there is an acute issue of the formation of sheet parts of their hard-to-form cases. These cases have physical and mechanical characteristics that allow them to work in very difficult conditions (such as high temperatures and loads). The problem is that these cases are either poorly processed in the cold state (realizable deformation in the cold state is from 5 to 7%), or require large formation actions to obtain the desired form. The most common solution is the formation of parts from heated cases. However, the introduction of heating during formation complicates the process and makes it necessary to ensure that a case does not absorb harmful impurities from the air during heating. In particular, titanium cases should not be heated in an uncontrolled environment, since there is a high risk of an alpha case and cracking of the part during operation.

The aircraft construction industry is an industry that widely uses sheet metal forging. It is believed that the parts for aircraft are the most complex in geometry and have the most stringent requirements. The industry also often uses sheet metal parts made of hard-to-form cases. In particular, they often try to use titanium case parts in loaded parts because they combine the best strength-to-weight ratio. However, most of the sheet metal parts are connected with other parts by means of such a complex element for formation as "undercut".

2. Results

According to the articles [2–4], the main problem in the formation "undercut" is the appearance of two defects such as "underforging" and "fold". This paper will show the use of technology with a movable clamp for the formation of a part from a hard-to-deform case OT-4 (Figure 1).

Figure 1 shows the characteristics of the "undercut" which allows determining the place of this part on the nomogram of the normative document regulating the formation of "undercuts".

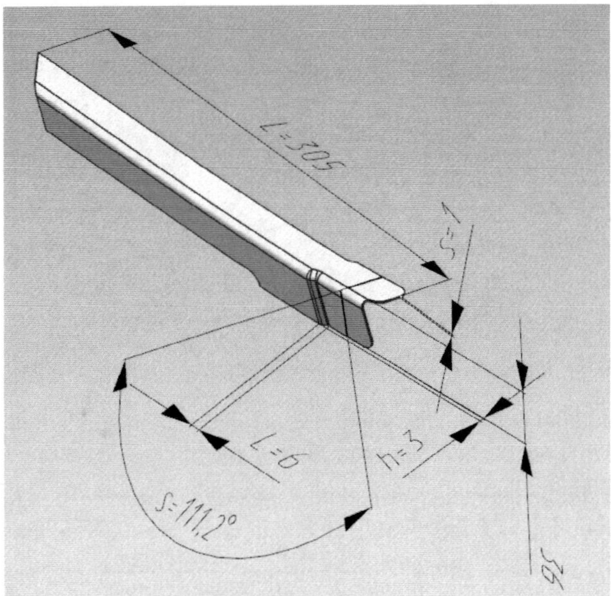

Figure 1. Geometric model of a part with geometric characteristics

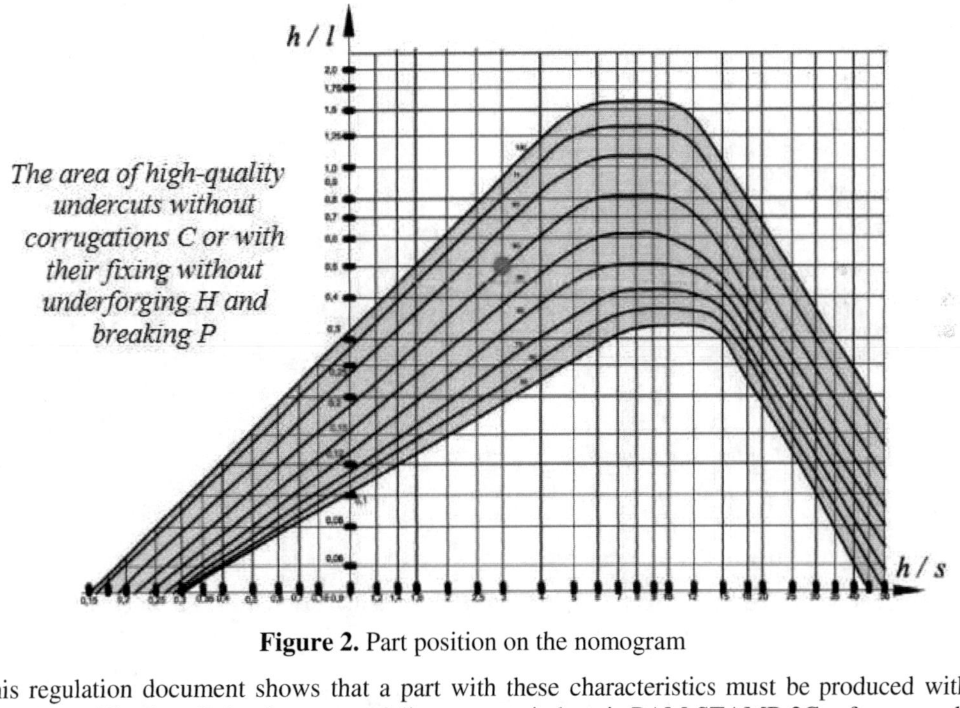

Figure 2. Part position on the nomogram

This regulation document shows that a part with these characteristics must be produced without defects. For verification, finite element modeling was carried out in PAM-STAMP-2G software package of French company ESI-Group. The process of elasto-formation in the cold state at a pressure of 100 MPa was simulated. As a result of modeling, a defect of the "underforging" type was found (Figure 3).

This simulation shows that this part, due to the hard-to-deform case has a defect appeared in the upper part of the nomogram. It is recommended to apply to this part the technology of the formation of a movable clamp described in the article [3].

IOP Conf. Series: Materials Science and Engineering 1064 (2021) 012048 doi:10.1088/1757-899X/1064/1/012048

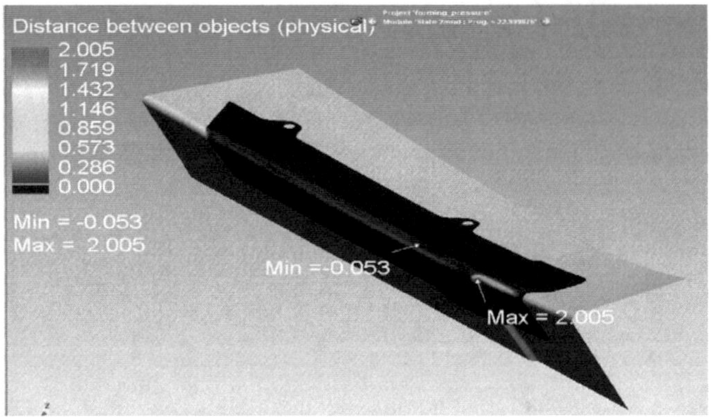

Figure 3. Distribution of "underforging" defect over the part (maximum value 2.005 mm.)

In order to calculate the characteristics of the movable clamp, first it is necessary to determine the total deformation according to the method from the paper [3] (Formula (1), Figure 4):

$$\varepsilon_{tot} = \frac{S3-S4}{S3} \cdot 100\% = \frac{1710{,}334\ mm^2 - 1511{,}767\ mm^2}{1710{,}334\ mm^2} \cdot 100\% = 11{,}61\% \quad (1)$$

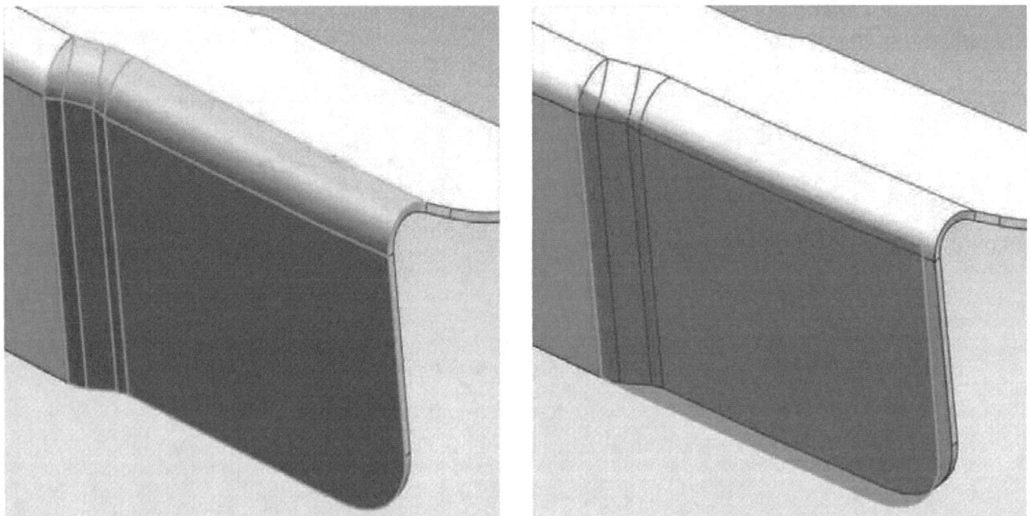

Surface S3=1710,3337 mm²   Surface S4=1511,7675 mm²

Figure 4. Surface required for the calculation of total deformation

In order to find the required pressure, it is necessary to calculate the stresses in the plastic part of the flow curve [3] (for OT-4 K = 845.103 MPa, n = 0.07543; ε_0 = 0.00000001 and $\varepsilon_p = \varepsilon_{tot}$ / 100%) (Formula (2)):

$$\sigma_{mov} = 845{,}103\ MPa(0{,}00000001 + 0{,}318)^{0{,}07543} = 718{,}405\ MPa \quad (2)$$

As it can be seen from the stresses during undercut formation, this part can be produced only with the use of additional accessories such as a movable clamp []. Using the stresses in the sub-section, we calculate the required formation force on the basis of fact that all the effort is directed to the "undercut" area and at different moments of formation it is different and increases from the top to the full area (formula (3).

$$F_{rec} = \sigma_{mov} \cdot (S_{1_{mov}} + S_{2_{mov}} + S_{3_{mov}} + S_{4_{mov}} + S_{5_{mov}} + S_{6_{mov}}) =$$
$$845{,}103\text{MPa} \cdot \begin{pmatrix} 18{,}147\text{mm.}^2 + 87{,}356\text{ mm.}^2 + 132{,}193\text{ mm.}^2 + 143{,}472\text{ mm.}^2 \\ +183{,}387\text{ mm.}^2 + 211{,}52\text{ mm.}^2 \end{pmatrix} =$$
$$845{,}103\text{MPa} \cdot 956{,}315 mm.^2 = 687021{,}769\text{H} \quad (3)$$

This value allows determining the required surface for the distribution of pressure on the movable clamp in order to create this force (Formula (4). Forming pressure is 20.5 MPa.

$$S_{rec} = \frac{F_{rec}}{q_{rec}} = \frac{687021{,}769\text{H}}{20{,}5\text{ MPa}} = 33633\ mm^2 \quad (4)$$

The following curve is set as the profile for the reference curve of the movable clamp (Figure 5)

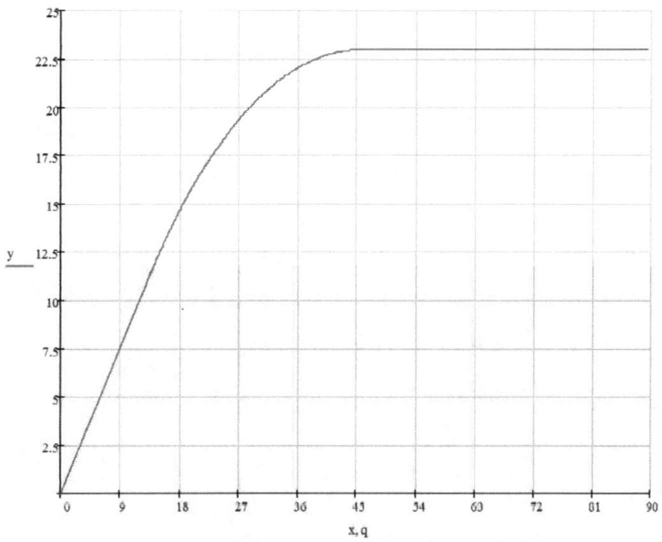

Figure 5. Reference curve of the movable clamp

Using the method of least squares, we interpolate the profile of the curve by a polynomial of the 4th order and compose the equation of the surface for the movable clamp (formula (5):

$$z(x,y) = -0{,}076615 + 0.812644 \cdot x + 0.013748 \cdot x^2 - 0{,}00104381 \cdot x^3 + 0{,}0000188464 \cdot x^4 \quad (5)$$

As a result, according to these data, a set of equipment for the formation of the studied part with a movable clamp was designed (Figure 6).

Using CAD system and the built-in system of integral calculus of the surface, the surface of the movable clamp was calculated (it was 33363.3009 mm^2, which minimally diverged from the calculated one) (Figure 7).

Using this geometry of the tool, finite element modeling was carried out using a movable clamp with a pressure of 20 MPa. As a result, "underforging" was 0.111 mm. in fact, it was absent (Figure 8).

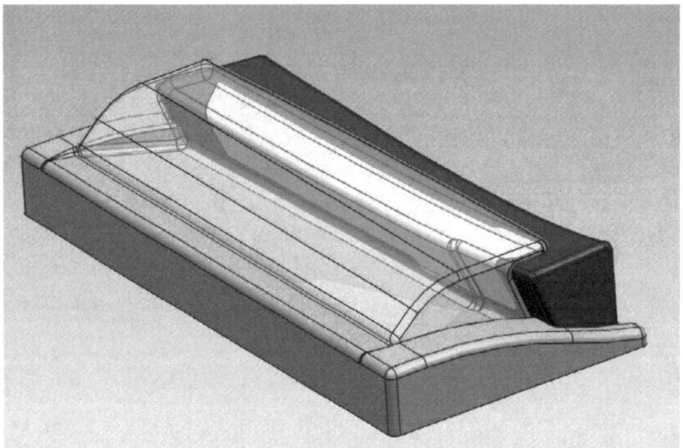

Figure 6. Formation tool kit

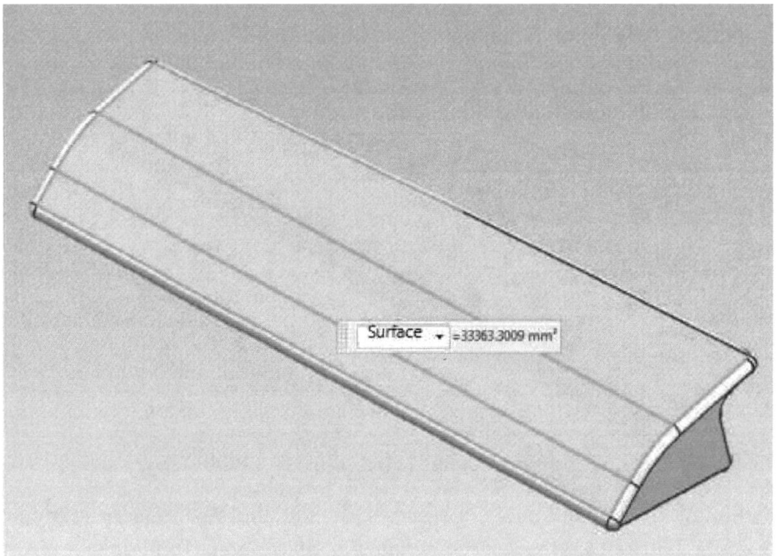

Figure 7. Contact surface of the movable clamp

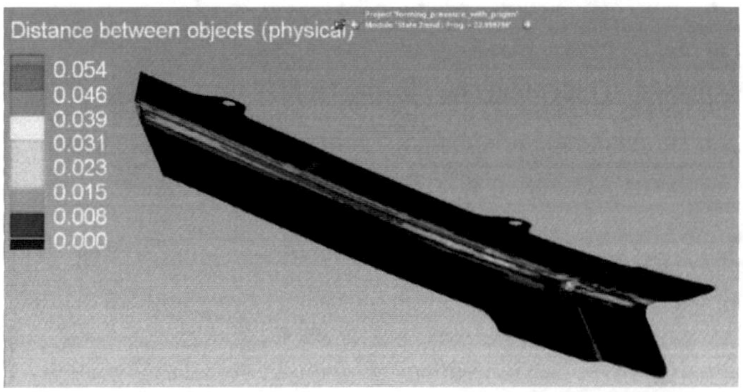

Figure 8. Distribution of the "underforging" defect over the part (maximum value was 0.054 mm.)

According to the analytical calculation and created geometric models, a set of equipment was produced (Figure 9).

Figure 9. Tool kit for the formation of a titanium part

As a result of the formation using QFC 1.2x3-1000 press at a pressure of 20.5 MPa. the part was molded using a movable clamp (Figure 10).

In order to confirm the efficiency of the method and to prove the impossibility of the production of this part without a movable clamp, a real experiment was carried out on a QFC 1.2x3-1000 production unit at a pressure of 100 MPa (Figure 11).

Visually we can estimate that the undercut zone is not formed even at maximum pressure (Figure 10), while with the proposed technology the pressure is 5 times less and the part is produced (Figure 11).

Figure 10. Titanium formed part with a movable clamp

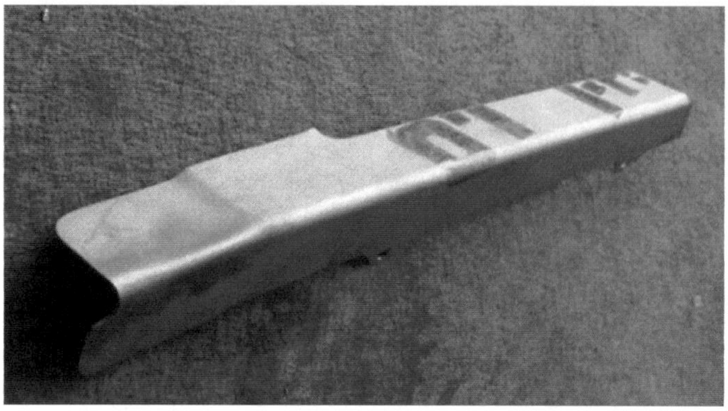

Figure 11. Formed titanium part without a movable clamp

3. Conclusion

As a result, it is shown that the proposed technology can form undercuts on parts made of hard-to-deform cases without temperature exposure.

References

[1] Govorkov A S, Lavrenteva M V and Fokin I V 2018 Mathematical modeling of making mechanical engineering products based on an information model *MATEC Web of Conferences* **224** 02022. DOI: https://doi.org/10.1051/matecconf/201822402022

[2] Chumachenko E, Aksenov S and Logashina I 2012 Optimization of superplastic forming technology *Metal 2012 – Conf. Proc., 21st int. conf. on metallurgy and materials* pp 295-301

[3] Sevastyanov G, Chernomas V, Marin S and Sevastyanov A 2015 Numerical simulation features of continuous casting process form ad31 (a (greek passage)31) alloy using finite-difference and finite-element models *Non-Ferrous Metals* **2015(2)** 25-29

[4] Govorkov A S, Zhilyaev A S and Fokin I V 2018 Method of transition from 3d model to its ontological representation in aircraft design process *J. of Phys.: Conf. Ser.* **1015(3)** 032046

[5] Lavrentyieva M and Govorkov A 2017 Using a discrete product model to determine the design element junctures *MATEC Web of Conferences* **129** 03002. DOI: 10.1051/matecconf/201712903003

[6] Lavrentieva M V, Chimitov P E and Govorkov A S 2019 Development of a computer aided design system for assembly equipment of MC-21 aircraft *IOP Conf. Ser.: Mater. Sci. Eng.* **632** 012099. DOI:10.1088/1757-899X/632/1/012099

[7] Fedorchuk M and Pak I 2005 Rigidity and polynomial invariants of convex polytopes *Duke Math. J.* **129(2)** 371-404. DOI:10.1215/S0012-7094-05-12926-X.

[8] Pokrasin M, Semashko N, Krupskii R and Kupov A 2004 Effect of electric-pulse treatment on the dislocation structure of ot4 titanium alloy *Russian metallurgy (Metally)* **2004(6)** 595-600

[9] Mironenko V V and Larionova Y N 2019 Calculation of process parameters of pieces containing irregular lateral cuttings, which forming leads to such defect as "corrugation forming" *IOP Conf. Ser.: Mater. Sci. Eng.* **632** 012102

[10] Mironenko V V and Larionova Y N 2019 Mathematical model to calculate key parameters of forming irregular lateral cuttings on sheet articles to eliminate such defects as "under-forging" *IOP Conf. Ser.: Mater. Sci. Eng.* **632** 012101

[11] Mokritskii B, Vereshchagin V, Mokritskaya E, Pyachin S, Belykh S and Vereshchagin A 2016 Composite hard-alloy end mills *Russian Engineering Research* **36(12)** 1030-1032

[12] Pal-Val P, Natsik V, Pal-Val L, Loginov Y, Demakov S and Illarionov A 2014 Unusual young's modulus behavior in ultrafine-grained and microcrystalline copper wires caused by texture changes during processing and annealing *Mater. Sci. Eng. A* **618** 9-15

[13] PAM-STAMP 2012 User's Guide, ESI Group, 2012 – 960
[14] Fabík R, Kliber J, Kubina T, Mamuzic I and Aksenov S 2012 Mathematical modelling of flat and long hot rolling based on finite element methods (FEM) *Metalurgija (Sisak, Yugoslavia)* **51(3)** 341-344
[15] Khusainov R, Sabirov A and Mubarakshin I 2017 Study of deformations field in the working zone of vertical milling machine *Procedia Engineering* **206** 1069-1074
[16] Belykh S and Perevalov A 2013 Modelling of bending rolls of extruded nonsymmetric in the msc marc *European Researcher* **5-1(48)** 1140-1146
[17] Bourkine S, Korshunov E, Loginov Y, Shakhapazov E, Nassibov A and Babailov N 1999 Projection of steel wire producing technology beyond the continuous-casting of an ingot using a direct combination of casting and metal forming *J. of Materials Processing Technology* **86(1-3)** 278-290
[18] Sokolnikov R A, Bozheeva T V and Govorkov A S 2020 Development of methodology for formalized selection of technological operations when designing technological process manufacturing of machinery *J. of Phys.: Conf. Ser.* **1582(1)**
[19] Roshchupkin V, Pokrasin M, Chernov A, Sobol N, Semashko N, Krupskii R and Kupov A 2005 Deformation of VT20 and ot4 titanium alloys subjected to a high-density pulse current during static loading *Russian metallurgy (Metally)* **2005(4)** 350-354
[20] Roshchupkin V, Semashko N, Krupskii R, Kupov A and Shport V 2003 Temperature and strain changes in vt20 titanium alloy under electric-pulse effect *High Temperature* **41(5)** 633-638
[21] Roshchupkin V, Krupskii R, Levchuk T and Semashko N 2002 Methodical aspects of exciting acoustic emission of magnetostriction *Russian metallurgy (Metally)* **4** 361-362
[22] Mironenko V and Shmakov A 2018 Movable-clamp undercutting of sheet metal parts *Advances in Engineering Research* **158** 271-275
[23] Erisov Y, Grechnikov F and Surudin S 2016 Yield function of the orthotropic material considering the crystallographic texture *Structural Engineering and Mechanics* **58(4)** 677-687
[24] Erisov Y, Surudin S, Shlyapugin A and Grechnikov F 2016 The end-to-end computer simulation of casting and subsequent metal forming *Key Engineering Materials* **685** 167-171
[25] Isachenkov E I 1967 *Rubber and liquid forging* (Moscow: Mashinostroenie) 367 p

Pressure calculation for preliminary formation of undercuts prone to defect of fold type

M V Lavrentyeva and V V Mironenko

Department of Aircraft Construction and Maintenance, Institute of Aircraft Construction, Mechanical Engineering and Transport, Irkutsk National Research Technical University, 83, Lermontova str., Irkutsk, 664074, Russia

E-mail: mira.amazon@gmail.com

Abstract. The paper presents the calculation of the pressure at the first transition during the formation of the "undercuts" on which corrugation takes place. The formulas in general form for the stresses at the first transition and the required pressure depending on the length of the sweep are derived. The results with the finite element analysis of formation and field experiments are compared. The conclusions about the verification of the results are drawn.

1. Introduction

Fold is one of the most unfavorable defects that may appear during formation in sheet metal stampings.

This is explained by the fact that if this defect is on the detail, then it can not be eliminated even by the increase in the laboriousness of producing the detail (for example, a defect of undercuts type can often be eliminated by manual fixing). In this regard, the researchers pay special attention to the details on which this defect may arise. A defect of fold type is preceded by such a process as corrugation. A fold occurs if a detail is uncontrollably molded during orbital formation.

The most difficult thing is to control this process if the corrugation occurs on details with such an element as undercuts. However, there is a method to eliminate this defect and prevent the formation of defects on such details [3]. The movable clamp allows molding details of this kind and eliminating defects. However, if at the first transition we do not select the correct pressure at which preliminary formation will be performed, the height of the corrugation will be small and it will be too rigid and even the use of a movable clamp will not ensure high-quality production of the detail.

2. Results

In this article the calculation of the preliminary pressure for formation will be shown through the example of one detail. The detail for consideration is shown in Figure 1.

In order to carry out this calculation, we will express the length of the sweep, through the value of which it will have radial compression and will decrease by the length ΔL in the zone where the corrugation will be formed (Figure 2).

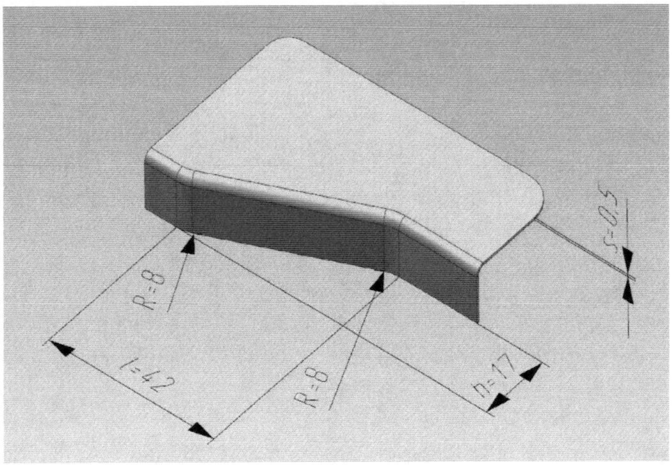

Figure 1. Studied detail

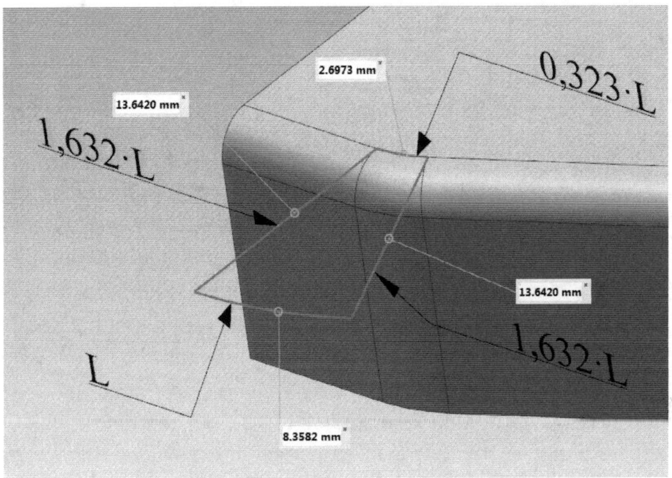

Figure 2. Variable L on sweep

Let us assume that all sides decrease during fixing keeping the shown proportion at the initial moment of formation. Then, according to Brahmagupta formula [3], the area will be equal (Formula (1):

$$S_{init.} = \frac{1}{4} \cdot \sqrt{-\begin{vmatrix} 0{,}322 \cdot L & 1{,}632 \cdot L & L & -1{,}632 \cdot L \\ 1{,}632 \cdot L & 0{,}322 \cdot L & -1{,}632 \cdot L & L \\ L & -1{,}632 \cdot L & 0{,}322 \cdot L & 1{,}632 \cdot L \\ -1{,}632 \cdot L & L & 1{,}632 \cdot L & 0{,}322 \cdot L \end{vmatrix}} = 0{,}25 \cdot \sqrt{17{,}8 \cdot L^4} \quad (1)$$

In general, this formula will have the following form (Formula (2):

$$S_{init.} = 0{,}25 \cdot \sqrt{L^4 \cdot (4 \cdot t^2 \cdot y^2 - 1 \cdot t^4 + 2 \cdot t^2 + 8 \cdot t \cdot y^2 + 4 \cdot y^2 - 1)} \quad (2)$$

where t - coefficient of dependence on L of the larger side of the sweep (in our case, 1.632); y - coefficient of dependence on L of the smaller side of the sweep (in our case 0.322).

Let us express the stress through the function "Krupkowsky law" [3]. In this case, we express plastic deformations as the ratio of the areas of the scan area in the corrugation region to the current corrugation area. As a result, we get (Formula (3):

$$\varepsilon_p = \frac{S5 - S_{init.}}{S5} = \frac{S5 - 0{,}25 \cdot \sqrt{L^4 \cdot \begin{pmatrix} 4 \cdot t^2 \cdot y^2 - 1 \cdot t^4 + 2 \cdot t^2 + \\ 8 \cdot t \cdot y^2 + 4 \cdot y^2 - 1 \end{pmatrix}}}{S5} = \frac{73{,}75\ mm^2 - 0{,}25 \cdot \sqrt{17{,}8 \cdot L^4}}{73{,}75\ mm^2} \quad (3)$$

where K - mathematical constant of a given material; n - work hardening coefficient; ε_0 – deformation of the start of plastic deformations; S5 - scan area at the initial moment before formation;

For D16AM K= 324, 17 MPa, n= 0,2183; ε_0=0,0003;

Then for the stress we have (Formula (4):

$$\sigma = K(\varepsilon_0 + \varepsilon_p)^n = K\left(\varepsilon_0 + \left(\frac{S5 - 0{,}25 \cdot \sqrt{L^4 \cdot \begin{pmatrix} 4 \cdot t^2 \cdot y^2 - 1 \cdot t^4 + 2 \cdot t^2 + \\ 8 \cdot t \cdot y^2 + 4 \cdot y^2 - 1 \end{pmatrix}}}{S5}\right)\right)^n$$

$$= K\left(\frac{S5 + S5 \cdot \varepsilon_0 - 0{,}25 \cdot \sqrt{L^4 \cdot \begin{pmatrix} 4 \cdot t^2 \cdot y^2 - 1 \cdot t^4 + 2 \cdot t^2 + \\ 8 \cdot t \cdot y^2 + 4 \cdot y^2 - 1 \end{pmatrix}}}{S5}\right)^n$$

$$= 324{,}17\ \mathrm{MPa}\left(\frac{73{,}75\ mm^2 + 73{,}75\ mm^2 \cdot 0{,}0003 - 0{,}25 \cdot \sqrt{17{,}8 \cdot L^4}}{73{,}75\ mm^2}\right)^{0{,}2183} \quad (4)$$

Since we are considering the initial moment of formation, the value of L changes slightly. From this it follows that the change in this function will occur in a small range. Consider the range 7.5 to L = 8.3582 (Figure 3).

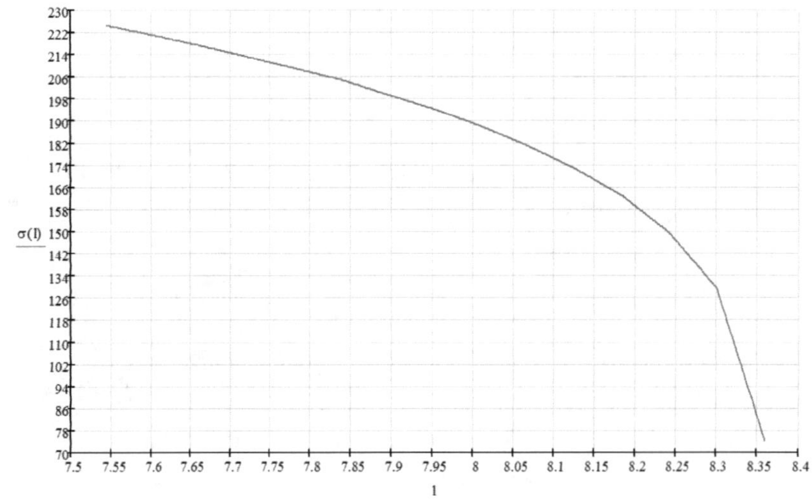

Figure 3. Stress dependence diagram from L

With the value of pressure, we can calculate the value of the required pressure at the first transition based on the fact that the formation force is distributed over the entire area of the detail (Figure 4, Formula (5) (in our case, 3116.0741mm^2).

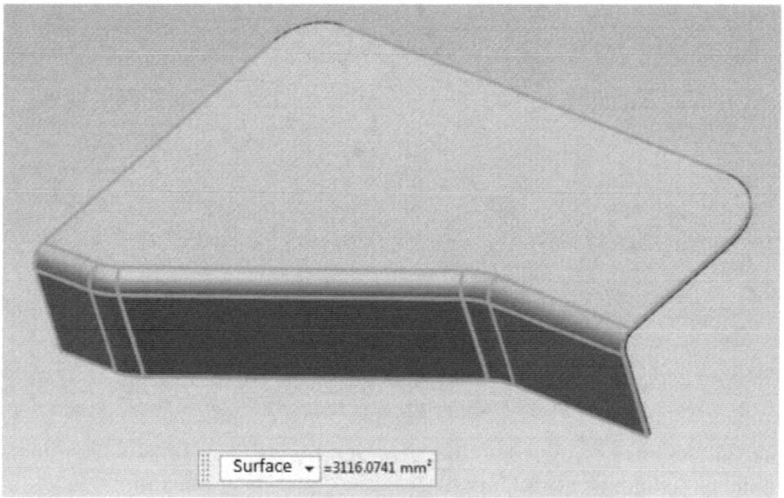

Figure 4. Contact area of detail

$$q_{rec} = \frac{\sigma \cdot S_{init.}}{S_d}$$

$$= \frac{K\left(\varepsilon_0 + \left(\frac{S5 - 0{,}25 \cdot \sqrt{L^4 \cdot \left(\begin{matrix}4 \cdot t^2 \cdot y^2 - 1 \cdot t^4 + 2 \cdot t^2 + \\ 8 \cdot t \cdot y^2 + 4 \cdot y^2 - 1\end{matrix}\right)}}{S5}\right)\right)^n \cdot 0{,}25 \cdot \sqrt{L^4 \cdot \left(\begin{matrix}4 \cdot t^2 \cdot y^2 - 1 \cdot t^4 + 2 \cdot t^2 + \\ 8 \cdot t \cdot y^2 + 4 \cdot y^2 - 1\end{matrix}\right)}}{S_d}$$

$$= \frac{0{,}25^{n+1} \cdot K \cdot \sqrt{L^4 \cdot \left(\begin{matrix}4 \cdot t^2 \cdot y^2 - 1 \cdot t^4 + 2 \cdot t^2 + \\ 8 \cdot t \cdot y^2 + 4 \cdot y^2 - 1\end{matrix}\right)} \cdot \left(\frac{4 \cdot S5 + 4 \cdot S5 \cdot \varepsilon_0 - \sqrt{L^4 \cdot \left(\begin{matrix}4 \cdot t^2 \cdot y^2 - 1 \cdot t^4 + 2 \cdot t^2 + \\ 8 \cdot t \cdot y^2 + 4 \cdot y^2 - 1\end{matrix}\right)}}{S5}\right)}{S_d}$$

$$= \frac{0{,}25^{n+1} \cdot K \cdot \sqrt{17{,}8 \cdot L^4} \cdot \left(\frac{4 \cdot S5 + 4 \cdot S5 \cdot \varepsilon_0 - \sqrt{17{,}8 \cdot L^4}}{S5}\right)}{S_d}$$

$$= \frac{0{,}25^{0{,}2183+1} \cdot 324{,}17\ MPa \cdot \sqrt{17{,}8 \cdot L^4} \cdot \left(\frac{4 \cdot 73{,}75\ mm^2 + 4 \cdot 73{,}75\ mm^2 \cdot 0{,}0003 - \sqrt{17{,}8 \cdot L^4}}{73{,}75\ mm^2}\right)}{3116{,}0741 mm^2} \quad (5)$$

As a result we get the following diagram (Figure 5).

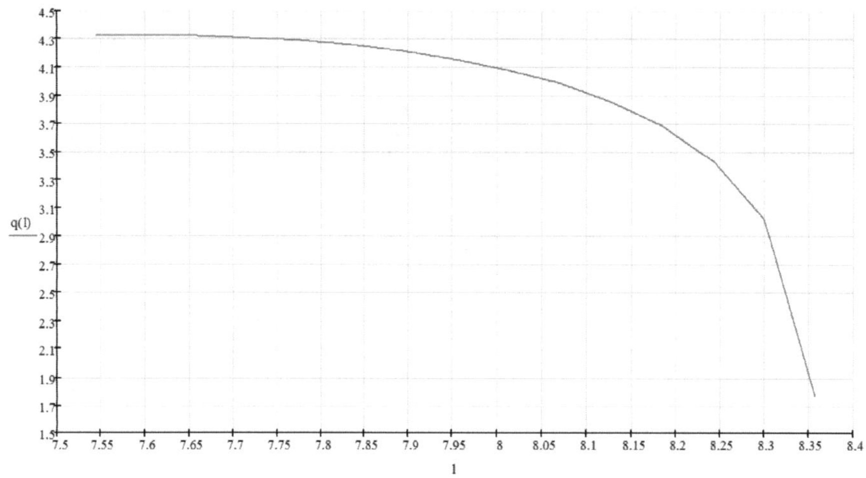

Figure 5. Graph of the required pressure from L for the studied detail

For example, in this case for L = 8 = the required pressure is 3.627 MPa. The stresses at this pressure in the detail according to the formula (4) are equal to σ = 159.357 MPa. As a result, the function for the calculation of the required pressure on the preliminary formation in general form is presented in Formula (5):

$$q_{rec}(L) = \frac{0{,}25^{n+1} \cdot K \cdot \sqrt{17{,}8 \cdot L^4} \cdot \left(\frac{4 \cdot S5 + 4 \cdot S5 \cdot \varepsilon_0 - \sqrt{17{,}8 \cdot L^4}}{S5} \right)}{S_d} \quad (6)$$

In order to verify the results, finite element modeling the formation of this detail in two transitions was carried out:

- 1st transition: pre-formation with the pressure that was calculated
- 2nd transition: final molding with a movable clamp;

As a result we have:

– equivalent stresses calculated by the formula of the energy theory of strength [3].

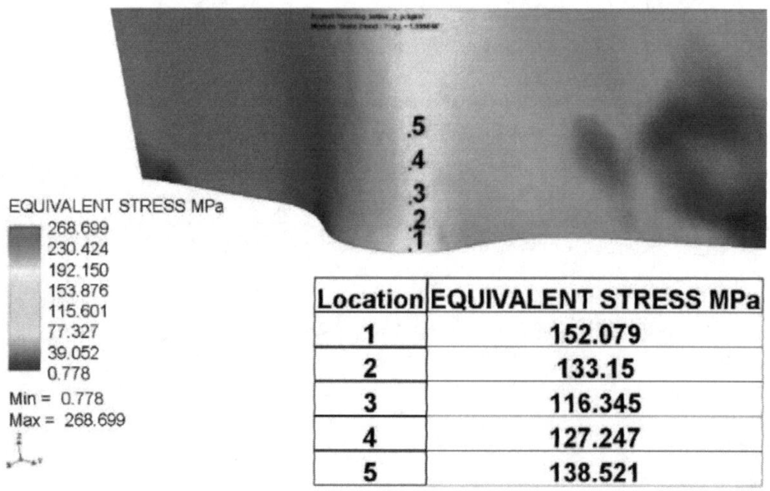

Location	EQUIVALENT STRESS MPa
1	152.079
2	133.15
3	116.345
4	127.247
5	138.521

Figure 6. Stress gradient for 1 transition

The difference between $\sigma \quad = 159{,}357$ MPa (Formula (4) calculated analytically and $\sigma_{\text{eq 1 trans}} = 152{,}079$ (Figure 7) is 4,6% thus the results can be considered identical.

- The length of the sweep during corrugation (L = 8.064 mm.) which is less than the calculated one (8.2 mm.) by 1.7% thus the results can be considered identical (Figure).

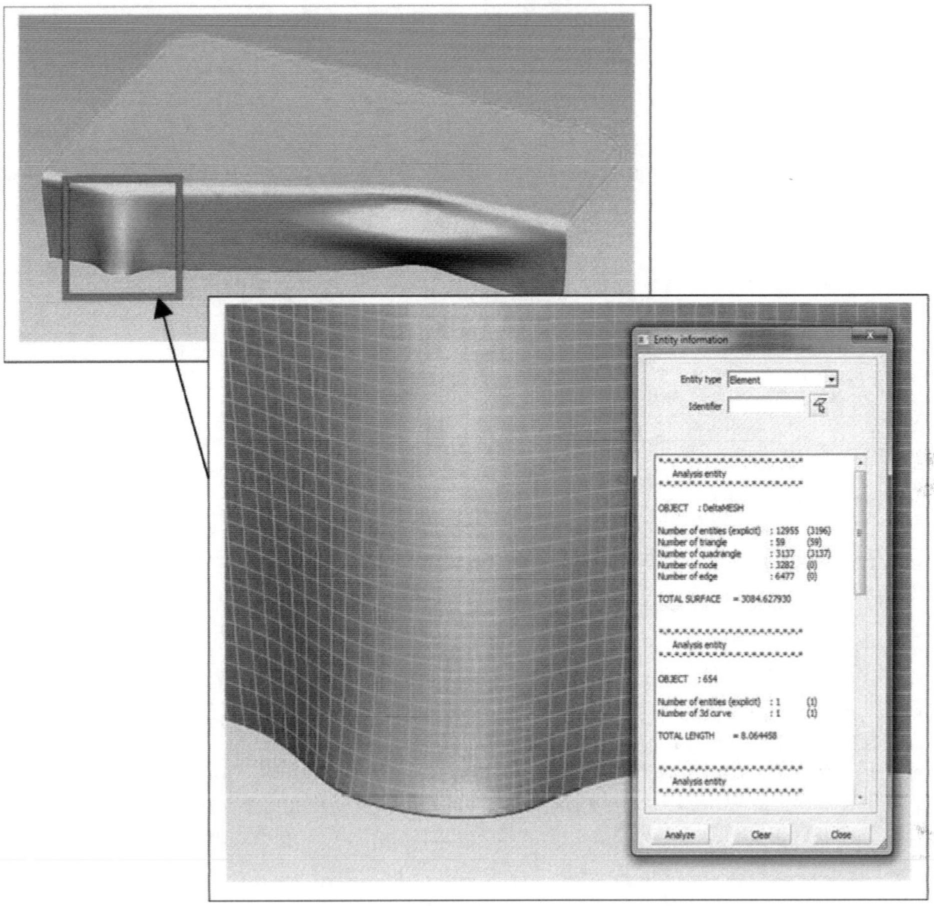

Figure 7. Sweep length at pressure 3,627 MPa

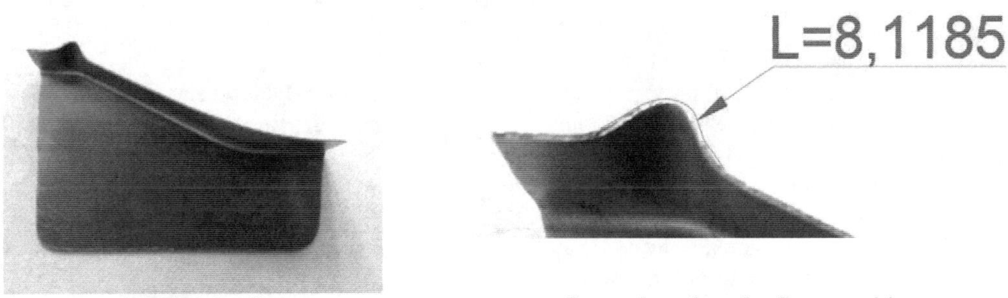

Molded detail Sweep length at the first transition

Figure 8. Detail in real experiment and the length of the sweep at the first transition

A real experiment was carried out, which also proved convergence. The sweep length was L = 8.2 mm. calculated analytically according to the proposed method and the length of the sweep on the real detail L_{real} = 8.1185 mm. (Figure 8) differed by 0.994% so the results can be considered identical.

After the second transition with a movable clamp, a detail is produced without defects [2] (Figure 9).

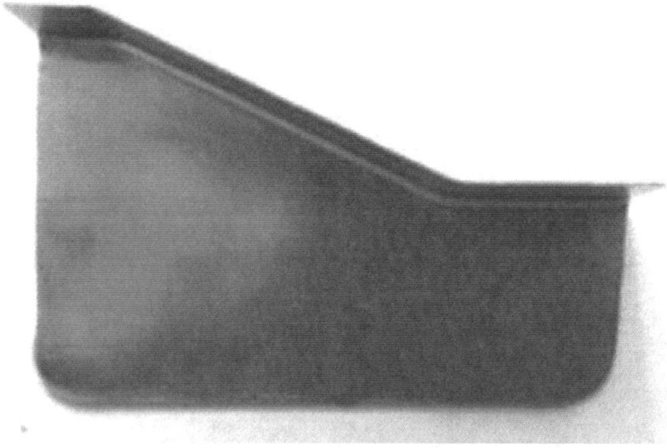

Figure 9. Molded detail after the second transition

3. Conclusion

As a result, the authors proposed the method for the calculation of the pressure at the first transition during the formation of undercuts on which corrugation can take place. The results of real and virtual experiments prove the adequacy of the proposed formulas.

References

[1] Govorkov A S, Lavrenteva M V and Fokin I V 2018 Mathematical modeling of making mechanical engineering products based on an information model *MATEC Web of Conferences* **224** 02022. DOI: https://doi.org/10.1051/matecconf/201822402022

[2] Chumachenko E, Aksenov S and Logashina I 2012 Optimization of superplastic forming technology *Metal 2012 – Conf. Proc., 21st int. conf. on metallurgy and materials* pp 295-301

[3] Sevastyanov G, Chernomas V, Marin S and Sevastyanov A 2015 Numerical simulation features of continuous casting process form ad31 (a (greek passage)31) alloy using finite-difference and finite-element models *Non-Ferrous Metals* **2015(2)** 25-29

[4] Govorkov A S, Zhilyaev A S and Fokin I V 2018 Method of transition from 3d model to its ontological representation in aircraft design process *J. of Phys.: Conf. Ser.* **1015(3)** 032046

[5] Lavrentyieva M and Govorkov A 2017 Using a discrete product model to determine the design element junctures *MATEC Web of Conferences* **129** 03002. DOI: 10.1051/matecconf/201712903003

[6] Lavrentieva M V, Chimitov P E and Govorkov A S 2019 Development of a computer aided design system for assembly equipment of MC-21 aircraft *IOP Conf. Ser.: Mater. Sci. Eng.* **632** 012099. DOI:10.1088/1757-899X/632/1/012099

[7] Fedorchuk M and Pak I 2005 Rigidity and polynomial invariants of convex polytopes *Duke Math. J.* **129(2)** 371-404. DOI:10.1215/S0012-7094-05-12926-X.

[8] Pokrasin M, Semashko N, Krupskii R and Kupov A 2004 Effect of electric-pulse treatment on the dislocation structure of ot4 titanium alloy *Russian metallurgy (Metally)* **2004(6)** 595-600

[9] Mironenko V V and Larionova Y N 2019 Calculation of process parameters of pieces containing irregular lateral cuttings, which forming leads to such defect as "corrugation forming" *IOP Conf. Ser.: Mater. Sci. Eng.* **632** 012102

[10] Mironenko V V and Larionova Y N 2019 Mathematical model to calculate key parameters of forming irregular lateral cuttings on sheet articles to eliminate such defects as "under-forging" *IOP Conf. Ser.: Mater. Sci. Eng.* **632** 012101
[11] Mokritskii B, Vereshchagin V, Mokritskaya E, Pyachin S, Belykh S and Vereshchagin A 2016 Composite hard-alloy end mills *Russian Engineering Research* **36(12)** 1030-1032
[12] Pal-Val P, Natsik V, Pal-Val L, Loginov Y, Demakov S and Illarionov A 2014 Unusual young's modulus behavior in ultrafine-grained and microcrystalline copper wires caused by texture changes during processing and annealing *Mater. Sci. Eng. A* **618** 9-15
[13] PAM-STAMP 2012 User's Guide, ESI Group, 2012 – 960
[14] Fabík R, Kliber J, Kubina T, Mamuzic I and Aksenov S 2012 Mathematical modelling of flat and long hot rolling based on finite element methods (FEM) *Metalurgija (Sisak, Yugoslavia)* **51(3)** 341-344
[15] Khusainov R, Sabirov A and Mubarakshin I 2017 Study of deformations field in the working zone of vertical milling machine *Procedia Engineering* **206** 1069-1074
[16] Belykh S and Perevalov A 2013 Modelling of bending rolls of extruded nonsymmetric in the msc marc *European Researcher* **5-1(48)** 1140-1146
[17] Bourkine S, Korshunov E, Loginov Y, Shakhapazov E, Nassibov A and Babailov N 1999 Projection of steel wire producing technology beyond the continuous-casting of an ingot using a direct combination of casting and metal forming *J. of Materials Processing Technology* **86(1-3)** 278-290
[18] Sokolnikov R A, Bozheeva T V and Govorkov A S 2020 Development of methodology for formalized selection of technological operations when designing technological process manufacturing of machinery *J. of Phys.: Conf. Ser.* **1582(1)**
[19] Roshchupkin V, Pokrasin M, Chernov A, Sobol N, Semashko N, Krupskii R and Kupov A 2005 Deformation of VT20 and ot4 titanium alloys subjected to a high-density pulse current during static loading *Russian metallurgy (Metally)* **2005(4)** 350-354
[20] Roshchupkin V, Semashko N, Krupskii R, Kupov A and Shport V 2003 Temperature and strain changes in vt20 titanium alloy under electric-pulse effect *High Temperature* **41(5)** 633-638
[21] Roshchupkin V, Krupskii R, Levchuk T and Semashko N 2002 Methodical aspects of exciting acoustic emission of magnetostriction *Russian metallurgy (Metally)* **4** 361-362
[22] Mironenko V and Shmakov A 2018 Movable-clamp undercutting of sheet metal parts *Advances in Engineering Research* **158** 271-275
[23] Erisov Y, Grechnikov F and Surudin S 2016 Yield function of the orthotropic material considering the crystallographic texture *Structural Engineering and Mechanics* **58(4)** 677-687
[24] Erisov Y, Surudin S, Shlyapugin A and Grechnikov F 2016 The end-to-end computer simulation of casting and subsequent metal forming *Key Engineering Materials* **685** 167-171
[25] Isachenkov E I 1967 *Rubber and liquid forging* (Moscow: Mashinostroenie) 367 p

On the issue of compensation of spring fittings for parts made by longitudinal stretching

V V Mironenko and A S Govorkov

Institute of Aircraft Engineering and Transport, Department of Aircraft Engineering and Operation of Aviation Equipment, Irkutsk National Research Technical University, 83 Lermontova str., Irkutsk, 664074, Russia

E-mail: mironenko_vv@istu.edu, govorkov_as@istu.edu

Abstract. The paper considers the problem related to the compensation of springing in the fitting of parts produced by longitudinal stretching. The imitating modeling of shaping by longitudinal stretching using elasto-plastic model of material taking into account anisotropy was carried out on the example of one element. The adequacy of modeling was checked according to the following criteria: deformation ratio; thinning; Keller-Godwin's chart and required shaping force. Conclusions were drawn on the adequacy of modeling. Implicit calculation of springing gradients in an element was carried out and thus the iterative calculation of compensation of springing in the fitting. The negative effect of springing compensation at such scheme of shaping is shown. Change of the springing size in the course of iterative calculation of springing compensation in the fitting is also shown.

1. Introduction

One of the most difficult tasks in the production of aircraft sheet parts is the manufacture of aircraft outline parts. The specifics of such manufacturing is that the deformation of the formation in the conventional manufacturing process of such parts is sufficiently small and there is a high probability that the springs will be very large after the formation. This, in turn, leads to an increase in manual refinement so much that in fact it turns out that the part is made manually. In order to avoid this, a forming process was proposed – stretching. By adding tension at the end of the process, additional deformations are applied to an element that prevent its large elastic response. However, in most cases it is not possible to completely avoid springing, and compensating for the springing in the tooling remains an urgent task.

2. Problems arising in the compensation of springing of sheet parts produced by longitudinal stretching

Stretching of sheet metal parts may be of two types – transverse and longitudinal. The typical representatives of parts for transverse and longitudinal stretching are shown in Figure. 1.

a – part made by transverse stretching

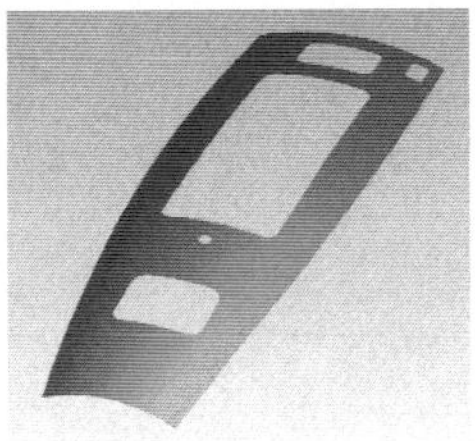

b – part made by longitudinal stretching

Figure 1. Parts made with stretching

It is known that the loading scheme in parts made by transverse and longitudinal stretching is different [10]. Hence, the nature of springing will be different. Based on this, it can be concluded that the compensation algorithms that exist at the moment cannot effectively compensate for the tooling for longitudinal stretching. Let us prove this with an example of one element.

3. An element made by longitudinal stretching and compensation of its springing in tooling

Let us consider an element made by longitudinal stretching from aluminum alloy 1163AM (Figure. 2).

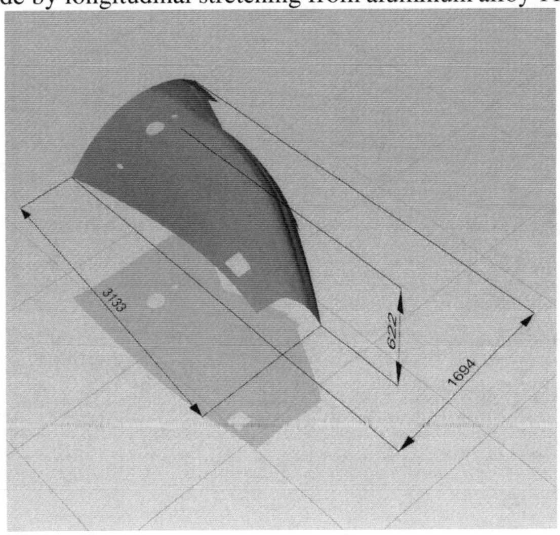

Figure 2. Considered element

Let us simulate the forming process. The software product PAM-STAMP 2G by the French company ESI Group will be used as a program for finite element analysis. The material model in the software product was set by the following parameters:

- Young modulus – 72.05 GPa;
- Poisson coefficient – 0.323;
- density – 2.6 kg/mm3;
- anisotropy coefficients r0°, r45°, r90° – 0,534, 0,834, 0,59;
- plastic part of the flow curve is set by the Krupkowsky law function, which looks as follows (Formula 1):

$$\sigma = K(\varepsilon_0 + \varepsilon_p)^n \tag{1}$$

Function constants for the material: K=0.65204 GPa, n=0.2949, $\varepsilon_{ps0} = 0.0045$.

A fitting model was built for element forming (Figure. 3).

IOP Conf. Series: Materials Science and Engineering 1064 (2021) 012050 doi:10.1088/1757-899X/1064/1/012050

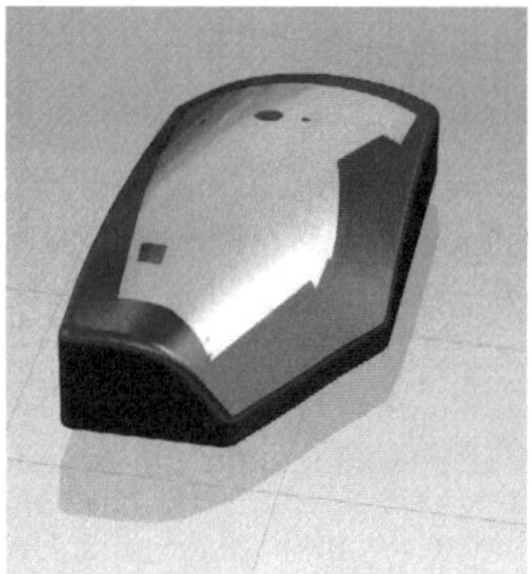

Figure 3. Fitting model for longitudinal stretching

As in real life, the formation process was divided into stages:
• first stage – lengthwise curvature (most often more than the necessary shape of an element);
• second stage – fitting from the bottom of an element upwards and its shaping by "unbending".
The simulation results are shown in Figure 4.

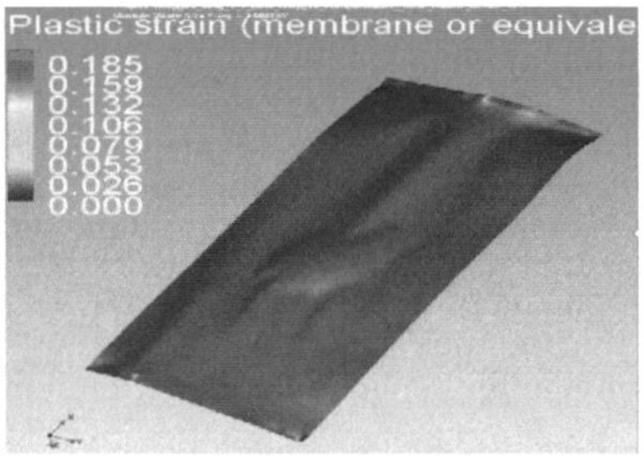

a – lengthwise curvature

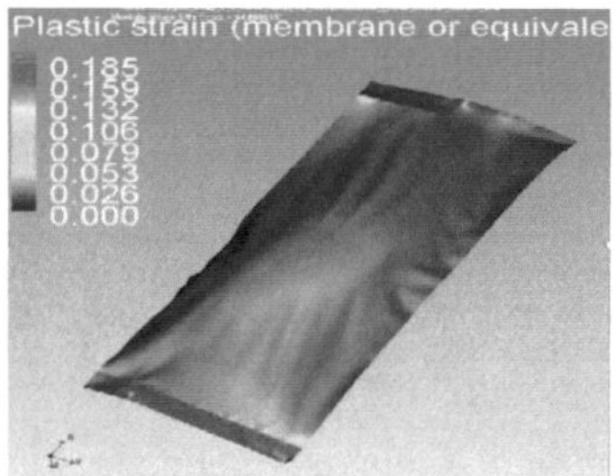

b – fitting from the bottom of an element upwards and its shaping by "unbending", stage 1

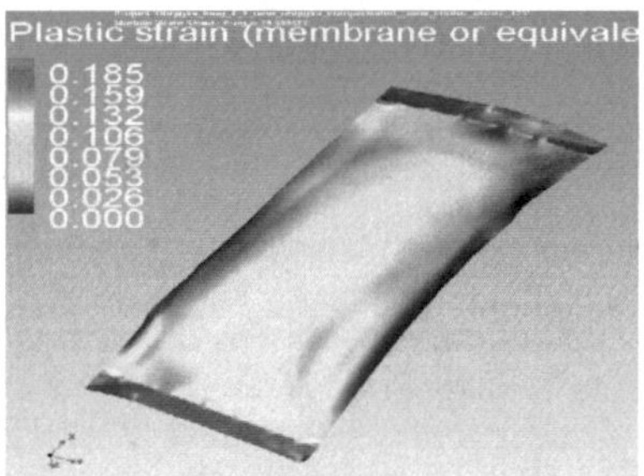

c – fitting from the bottom of an element upwards and its shaping by "unbending", stage 2

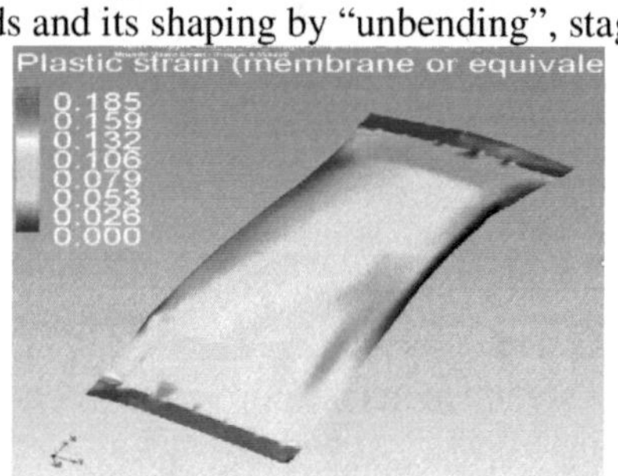

d – fitting from the bottom of an element upwards and its shaping by "unbending", stage 3

Figure 4. Distribution of plastic deformations during shaping

To verify the adequacy of simulation, deformation and thinning were checked in the element area on a workpiece. As a result, it can be seen that deformation and thinning do not exceed 20% (Figure. 5) [12]. The deformation capacity of the material on a workpiece in the contour of the element was also checked according to the Keller-Goodwin limit deformation diagram [9] in the material model. Hence, no dangerous areas of deformation were found. Based on the above results it is possible to conclude that the element can be made with longitudinal stretching according to the above flow chart.

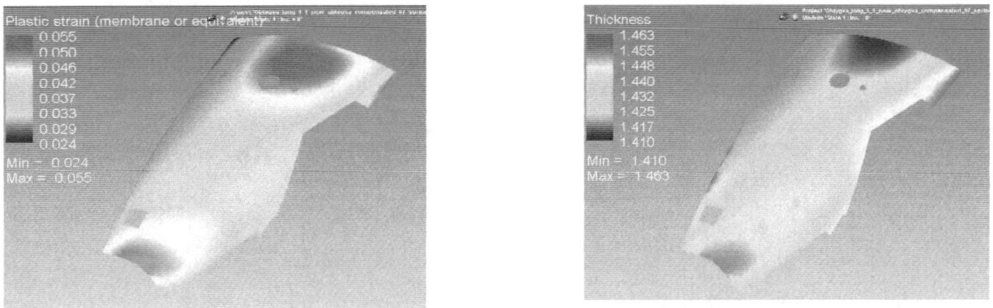

a – distribution of plastic deformations in the element contour on a workpiece, maximum deformation – up to 5.5%, < 20%

b – thickness distribution in the element contour on a workpiece, minimum thickness (blue) – 1.410 mm, thinning – up to 6%, less than 20%

Figure 5. Results in the element contour on a workpiece

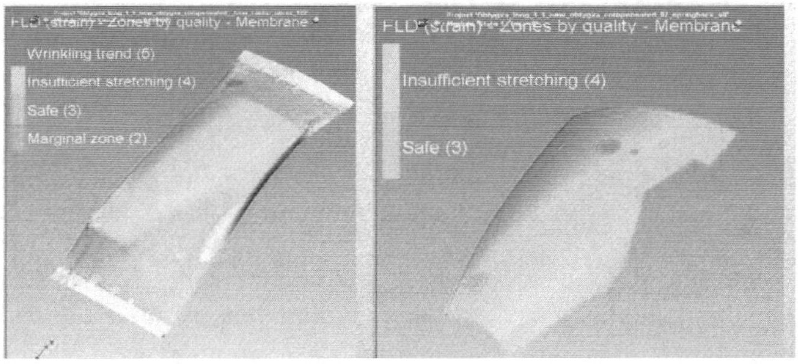

Figure 6. Results of the Keller-Goodwin diagram on a workpiece and in element contour on a workpiece

To check convergence, the tensile force was calculated elementwise according to formula 2:

$$F = \sigma \cdot s \cdot a \tag{2}$$

where σ – equivalent tension in an element, MPa; s – thickness in the element, mm; a – length of the edge of the element, mm.

The total force at the ends of the workpiece is calculated by elementwise addition. Thus, the following results were obtained (Figure. 7).

The total forming force is equal to 575 kN + 547 kN = 1122 kN = 112.2 tons. Besides, the dimensions of the workpiece for stretching are 5700 mm x 1800 mm. The PO-3M press [13] has a force of 2700 kN = 270 t and the largest dimensions of the workpiece: 8000 mm x 1800 mm x 10 mm, i.e. it is possible to manufacture this element using this press. All of the above indicates the adequacy of simulation.

A springing gradient was also calculated for the area of the workpiece in contact with fitting (Figure. 8 a). The thesis from the section of this paper was confirmed (section "Problems of springing compensation of sheet parts formed with longitudinal stretching) (Figure. 8 b). With such a sequential loading scheme, as with longitudinal stretching, the element will spring inward due to the fact that in the second forming step it is formed to reduce curvature. As a consequence, the elastic components of deformation will tend to return it to its original state i.e. bending.

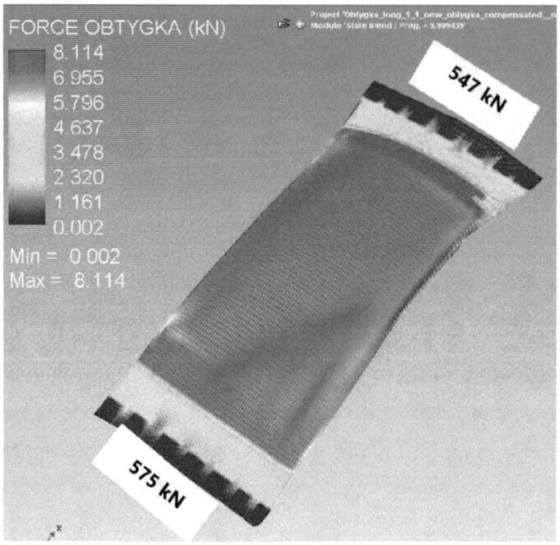

Figure 7. Results of shaping force

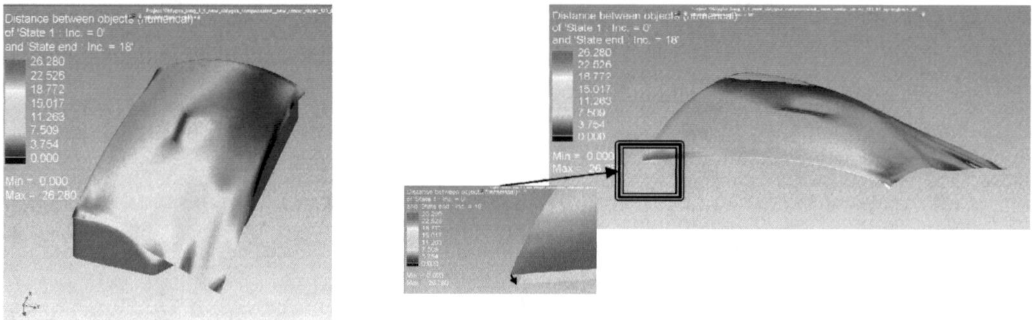

a – spring distribution for the area of the workpiece in contact with fitting, the maximum distance of the workpiece from the fitting surface is 26.2 mm.

b – area of the workpiece in contact with fitting before and after the springing effect, white surface – workpiece after elastic response (springing caused the edges of the workpiece to deflect inward).

Figure 8. Distribution of longitudinal stretching springing

The springing compensation was carried out, which showed that this calculation does not make it possible to fight springing because it compensates for the springing to the side opposite the springing, i.e. reduces the curvature of the element (Figure 9).

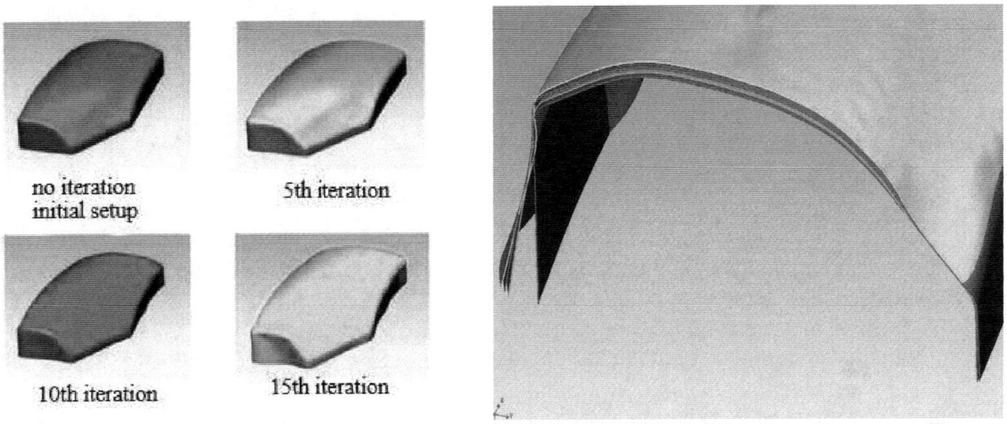

a – color distribution showing fitting at different iterations of springing compensation

b – configuration of fitting options according to springing compensation iterations in one section

Figure 9. Results of fitting compensation for longitudinal stretching

Reduced curvature results in reduced deformation during workpiece deformation. As a result, we have springing increase, which makes the springing compensation inefficient (Figure. 10).

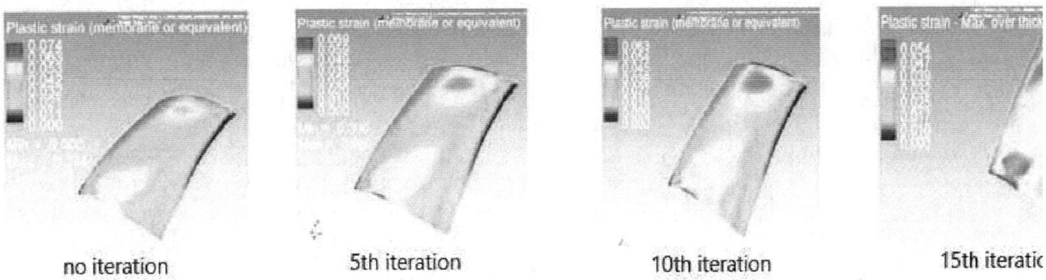

a – distribution of plastic deformation for the area of the workpiece in contact with fitting according to iterations of spring compensation (plastic deformation decreases)

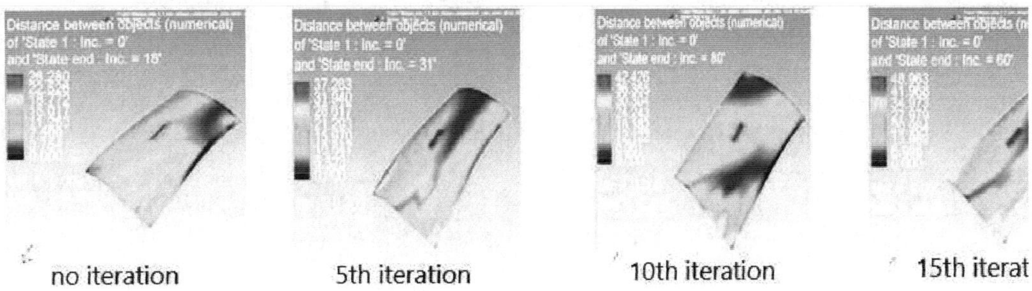

b – distribution of springing for the area of the workpiece in contact with fitting according to iterations of springing compensation (springing increases)

Figure 10. Results of longitudinal stretching compensation

4. Conclusion

Thus, on the example of one element, a finite element analysis of forming the element by longitudinal stretching was carried out. The simulation results of the shaping process were estimated by the degree of deformation, thinning, Keller-Goodwin diagram and the required shaping force. The results showed that the element is formed without defects and can be made on the PO-3M press. The springing distribution gradient in the element was also calculated, which showed that the element springs inward when springing. In the future, the results of the iterative calculation of springing compensation in the fitting showed that spring compensation with such a loading scheme is not effective.

References

[1] Govorkov A S, Zhilyaev A S and Fokin I V 2018 A method of transition from 3d model to its ontological representation in aircraft design process *J. of Phys.: Conf. Ser.* 032046

[2] Govorkov A S, Lavrenteva M V and Fokin I V 2018 Mathematical modeling of making mechanical engineering products based on an information model *MATEC Web of Conferences* **224** 02022. DOI: https://doi.org/10.1051/matecconf/201822402022 .

[3] Lavrentyieva M and Govorkov A 2017 Using a discrete product model to determine the design element junctures *MATEC Web of Conferences* **129** 03002. DOI: 10.1051/matecconf/201712903003

[4] Fedorchuk M and Pak I 2005 Rigidity and polynomial invariants of convex polytopes *Duke Math. J.* **129(2)** 371-404. DOI:10.1215/S0012-7094-05-12926-X.

[5] Pokrasin M, Semashko N, Krupskii R and Kupov A 2004 Effect of electric-pulse treatment on the dislocation structure of ot4 titanium alloy *Russian metallurgy (Metally)* **6** 595-600

[6] Khusainov R, Sabirov A and Mubarakshin I 2017 Study of deformations field in the working zone of vertical milling machine *Procedia Engineering* 1069-1074

[7] Sokolnikov R A, Bozheeva T V and Govorkov A S 2020 Development of methodology for formalized selection of technological operations when designing technological process manufacturing of machinery *J. of Phys.: Conf. Ser.* **1582(1)**

[8] Erisov Y, Surudin S, Shlyapugin A and Grechnikov F 2016 The end-to-end computer simulation of casting and subsequent metal forming *Key Engineering Materials* **685** 167-171

[9] Averkiev A Yu 1985 *Methods for evaluating the stampability of sheet metal* (Moscow: Machine building) 176 p

[10] Gorbunov M N 1981 *Technology of procurement and stamping operations in the production of aircraft,* textbook for universities, 2nd ed., revised and added (Moscow: Engineering) 224 p

[11] Isachenkov E I 1967 *Stamping with rubber and liquid* (Moscow: Engineering) 367 p

[12] 1973 PI 1.4.1977-2006 *Parts of procurement and stamping production. General technical requirements* (Moscow: Standardinform) 29 p

[13] Industrial Equipment Directory. Retrieved from: http://www.8e.ru/print/10023.php.

[14] Kulikov A A, Nebyshinets Yu V, Sidorova A V and Balanovsky A E 2020 3D printing of metal products: from an automated design system to a real part *Bull. of Irkutsk State Technical University* **24(4)** 728-739. DOI: https://doi.org/10.21285/1814-3520- 2020-4-728-739.

[15] Larchenko A G 2019 Evaluation of the quality of products from polymer materials of machine-building purpose *Bull. of Irkutsk State Technical University* **23(3)** 463–471. DOI: 10.21285/1814-3520-2019-3-463-471.

Identification of problem parts made by longitudinal stretching when compensating for springing

V V Mironenko and A S Govorkov

Institute of Aircraft Engineering and Transport, Department of Aircraft Engineering and Operation of Aviation Equipment, Irkutsk National Research Technical University, 83 Lermontova str., Irkutsk, 664074, Russia

E-mail: mironenko_vv@istu.edu, govorkov_as@istu.edu

Abstract. The paper considers problem related to the compensation of springing of parts produced by longitudinal stretching. It proposes the method of eliminating this effect by replacing the shaping method with transverse stretching. A finite element simulation of transverse stretching for the selected element produced by longitudinal stretching was carried out. The analysis of results revealed elimination of negative phenomenon of springing. After changing the scheme, the nature of the workpiece movement during springing became classic. The paper describes the procedure for the geometry of the element to determine problems in the compensation of fitting related to springing. The method takes into account the geometric parameters of the curvature radii of the element in different directions and the relations between them. The criteria for identifying problem parts are presented.

1. Introduction

One of the most difficult tasks in the production of aircraft sheet parts is the manufacture of aircraft outline parts. The specifics of making these parts includes the application of a double effect. As a result, at the last stage, when the element is unbent by moving the tooling upwards, a stressed deformed state occurs that makes the workpiece spring inwardly. Hence, the compensation of springing becomes ineffective because it reduces the curvature radius of the element and as a consequence increases springing. This problem is a very serious obstacle to the application of this shaping technology.

2. Considered element

Let us consider an element made by longitudinal stretching from aluminum alloy 1163AM, which forming causes problems with the above springing compensation (Figure. 1).

Let us simulate the forming process. The software product PAM-STAMP 2G by the French company ESI Group will be used as a program for finite element analysis. The material model in the software product was set by the following parameters:

- Young modulus – 72.05 GPa;
- Poisson coefficient – 0.323;
- density – 2.6 kg/mm3;
- anisotropy coefficients r0°, r45°, r90° – 0,534, 0,834, 0,59;
- plastic part of the flow curve is set by the Krupkowsky law function, which looks as follows (Formula 1):

$$\sigma = K(\varepsilon_0 + \varepsilon_p)^n$$

Function constants for the material: K= 0.65204 GPa, n= 0.2949, ε_{ps0} = 0.0045

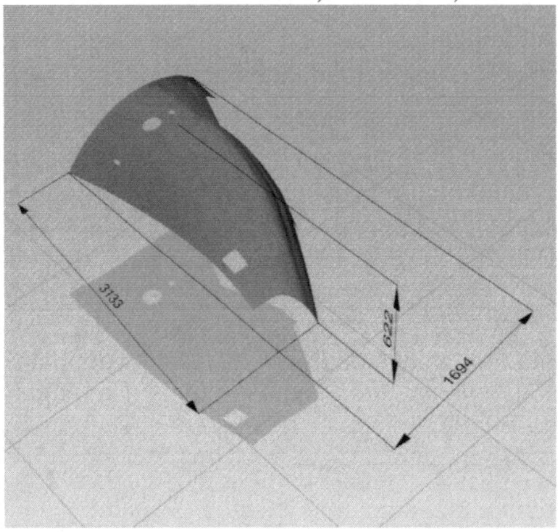

Figure 1. Considered element

3. Replacement of the forming method with transverse stretching and compensation of its springing in fitting

In order to eliminate the negative effect of elastic response and make springing compensation effective, it is necessary to change the loading pattern during shaping. It is proposed to change the type of stretching from longitudinal to transverse because there is no shape change in two directions. With the same parameters of the material model, a simulation was conducted for transverse stretching, according to which it can be concluded that an element can be made this way (Figure. 2).

A springing gradient was calculated for the area of a workpiece in contact with fitting and in this case the springing is aimed at reducing curvature (Figure. 3, Figure. 4). As a result, the compensation becomes effective since the curvature of the element increases, and hence, the deformation should increase or at least not change. In our case, the deformation remained constant, therefore, the springing is constant and will be equal to 0 iterations (Figure. 5 a). The shape will change and become similar to the element shape. This value should decrease with proper compensation progress (Figure. 5 b).

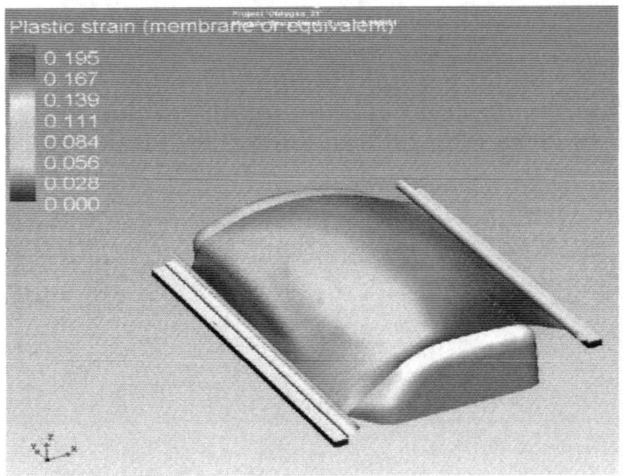

Maximum deformation – 19.5%, less than 20%

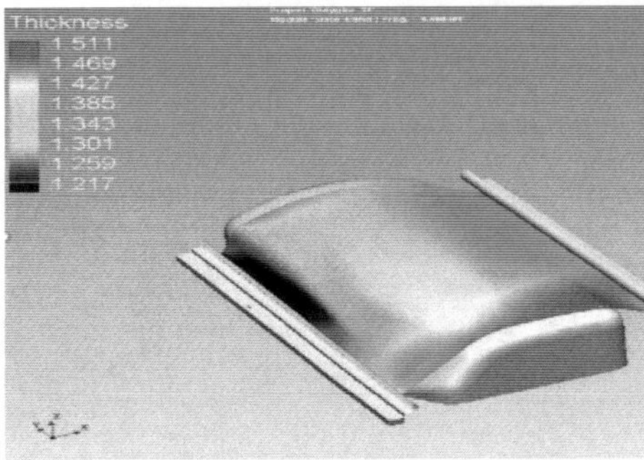

Minimum thickness (blue) – 1.217 mm, thinning 18%, less than 20%

a – distribution of deformation and thickness on a workpiece

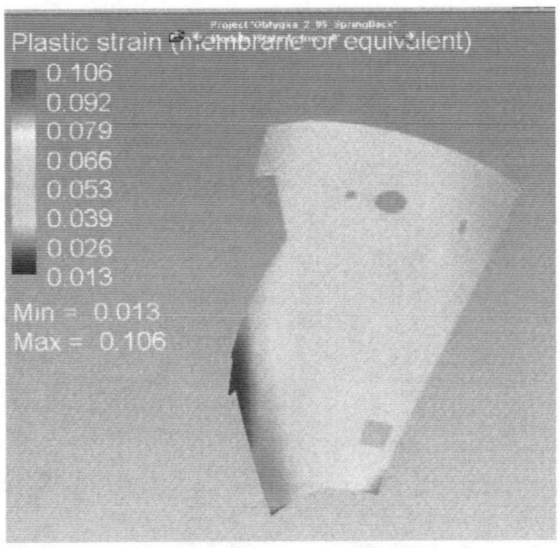

Maximum deformation – 10.6%, less than 20%

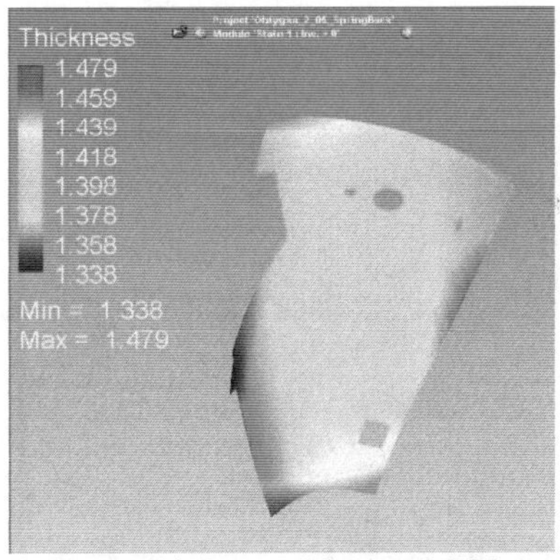

Minimum thickness (blue) – 1.338 mm, thinning 10.8%, less than 20%

b – distribution of deformation and thickness in the element contour on a workpiece

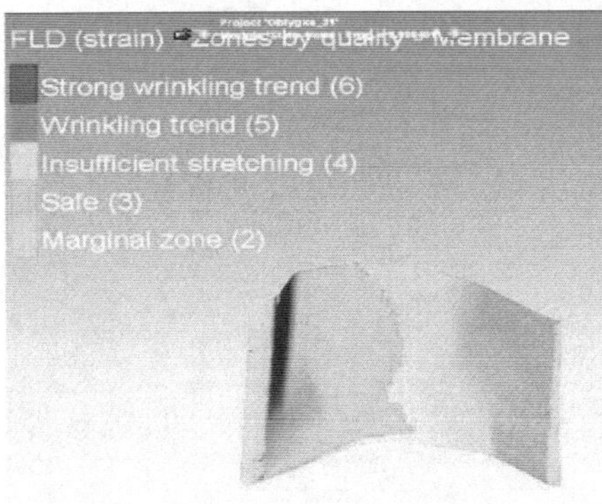

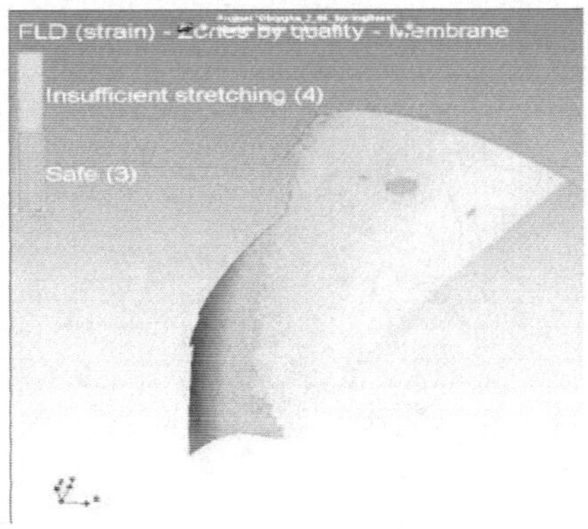

there are no blue zones on the workpiece and parts indicating the presence of dangerous deformation zones in which cracks can occur

c – Keller-Goodwin limit deformations diagram in the element contour of a workpiece and on the workpiece as a whole

Figure 2. Results of transverse stretching simulation

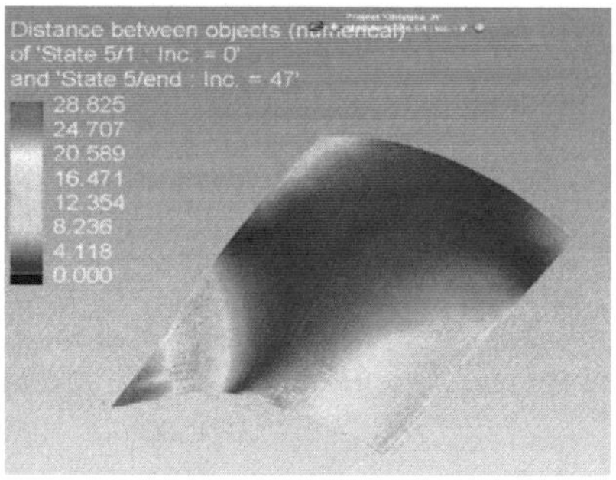

a – spring distribution for the area of the workpiece in contact with fitting, the maximum distance of the workpiece from the fitting surface is 28.8 mm

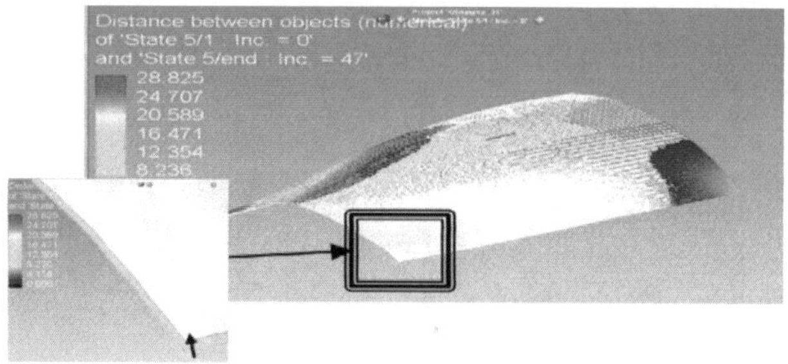

b – area of the workpiece in contact with fitting before and after the springing effect, white surface – workpiece after elastic response (springing caused the edges of the workpiece to deflect outward)

Figure 3. Distribution of transverse stretching springing

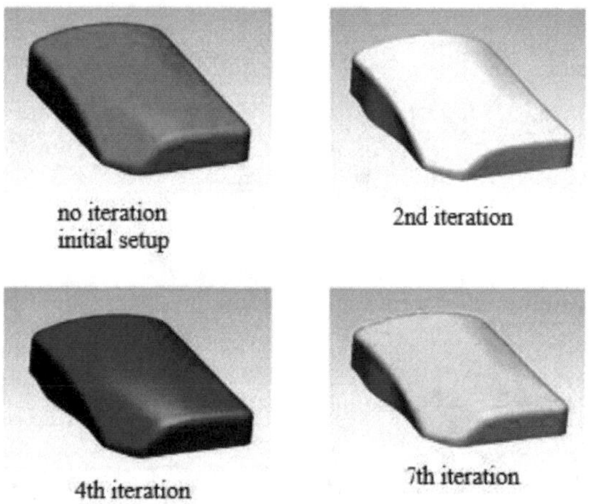

a – color distribution showing fitting at different iterations of springing compensation

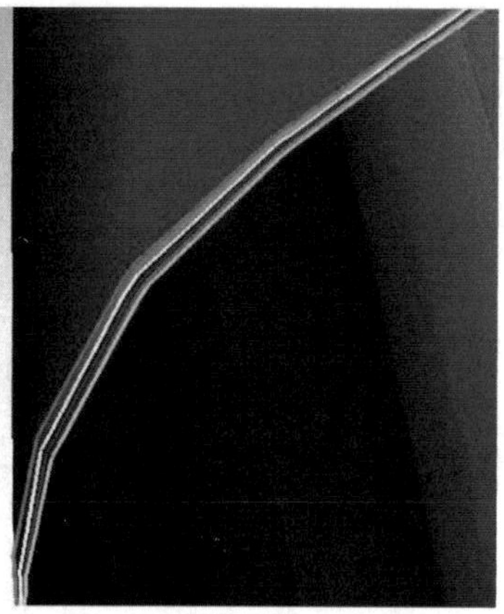

b – configuration of fitting options according to springing compensation iterations in one section

Figure 4. Results of fitting compensation for transverse stretching

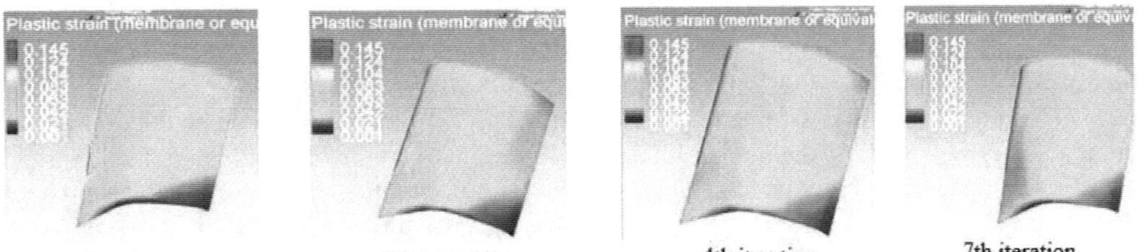

no iteration 2nd iteration 4th iteration 7th iteration

a – distribution of plastic deformation for the area of the workpiece in contact with fitting according to iterations of spring compensation (plastic deformation decreases)

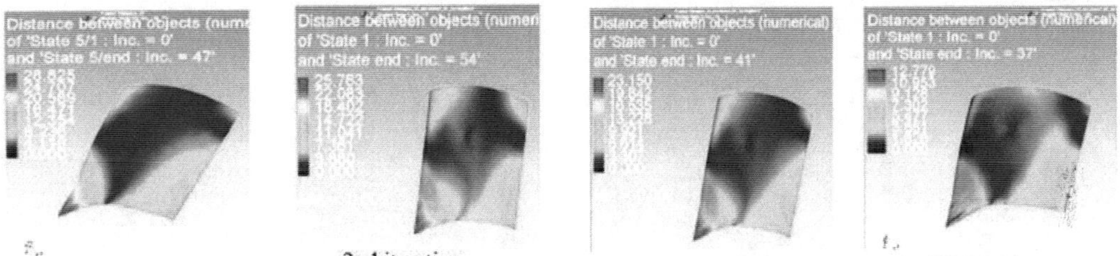

no iteration 2nd iteration 4th iteration 7th iteration

b distribution of springing for the area of the workpiece in contact with fitting according to iterations of springing compensation (the value decreases, i.e. the shape becomes similar to the shape of the element)

Figure 5. Results of transverse stretching compensation

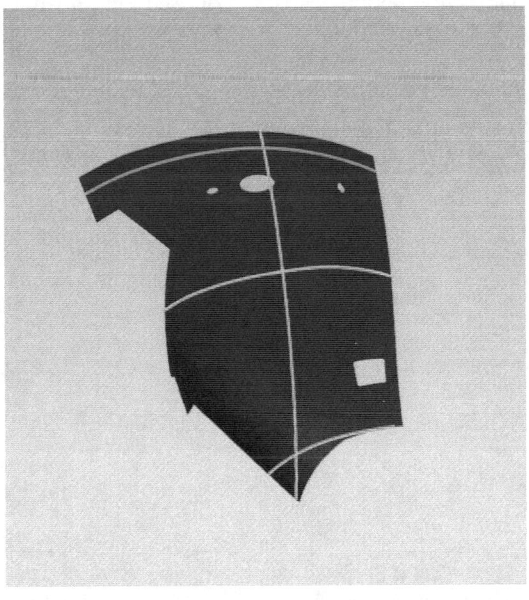

Figure 6. Sections to define the curvature radius (red line indicates section for analysis)

All of the above indicates the correct course of the compensation process and the elimination of the negative springing effect.

4. Identification of elements with problem shape made by stretching for springing compensation

However, it remains a problem to identify the element with a problem shape. Using the example of the element to simulate the shaping process, let us analyze its shape for curvature. To do this, let us split it into sections in places of pronounced change of the curvature radius. As a result, we have 4 sections (three in one direction and one in the other) (Figure. 15). Using the example of one section, let us calculate the curvature radius of (Figure. 6).

To calculate, let us move the section to the origin of the coordinate system and plot the function.

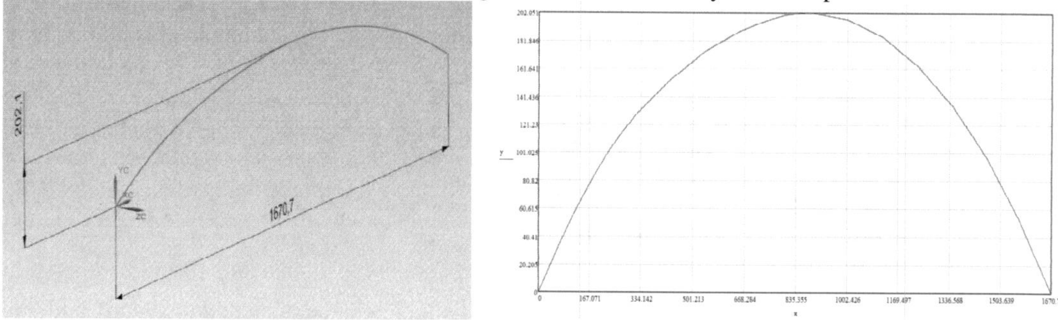

a – section in CAD system b – section in the form of a graph

Figure 7. Section for analysis

Since a curve is similar in appearance to a parabola, it makes sense to find its function. To do this, let us use the least square method in setting the approximation to a power function $a \cdot x^2 + b \cdot x + c$ [13]. As a result, we have the following function describing our section (Formula 1):

$$y(x) = -0.00028297731220943 \cdot x^2 + 0.47393259942476196 \cdot x + 4.589376870719401 \quad (1)$$

The graphic view of the two graphs shows that the match is not ideal, therefore, it is necessary to evaluate the correlation (Figure. 8). To do this, let us calculate the Pearson coefficient [11]. The Pearson coefficient eventually turned out to be 0.763. According to the Chaddock table [10] we estimate the correlation force and we see that it is high, i.e. the error between functions is not significant.

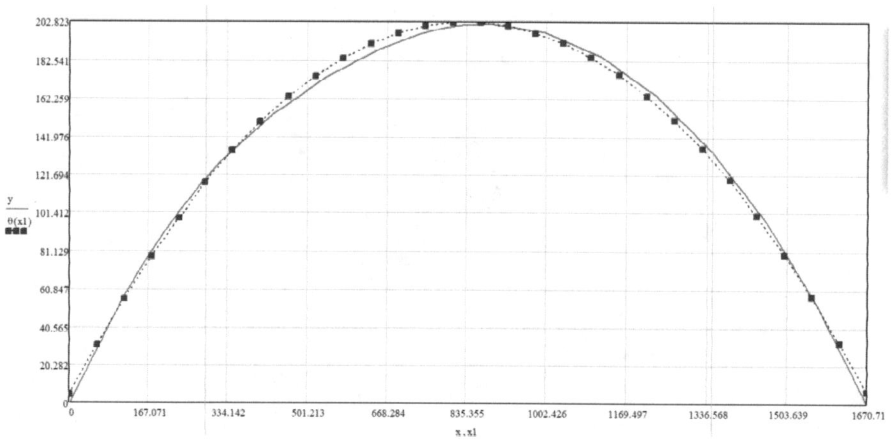

Figure 8. Function graphs

Now we need to calculate the curvature parameter of the function using the formula 2 [10]:

$$K = \frac{\ddot{y}}{(1 + (\dot{y})^2)^{3/2}} \qquad (2)$$

The first derivative has the form – 0.00028297731220943 · x + 0.47393259942476196, and the second is equal to the scalar value – 0.00056595462441886. The function for curvature calculation takes the form blow (Formula 3):

$$K = \frac{-0.00056595462441886}{(1 + (-0{,}00028297731220943 \cdot x + 0{,}47393259942476196)^2)^{3/2}} \qquad (3)$$

Having calculated its values in the range from 0 to 1670.71 (i.e. in the dimensions along the x-axis), we take the minimum value modulo. As a result, we get 0.0004176. The curvature radius is equal to R = 1/K [17]. Hence, we have the curvature radius equal to 2394.636. For verification in the Siemens NX 12 CAD system, let us build a curvature radius envelope. Since the envelope is not constant (it has a pronounced "hole", i.e. there it is necessary to make a denser sample), we take the average value of the curvature radius according to the values indicated below (Figure. 9). As a result, we have the curvature radius of 2340.2 and the difference between the calculated value (2394.636) according to the proposed method is not more than 3%.

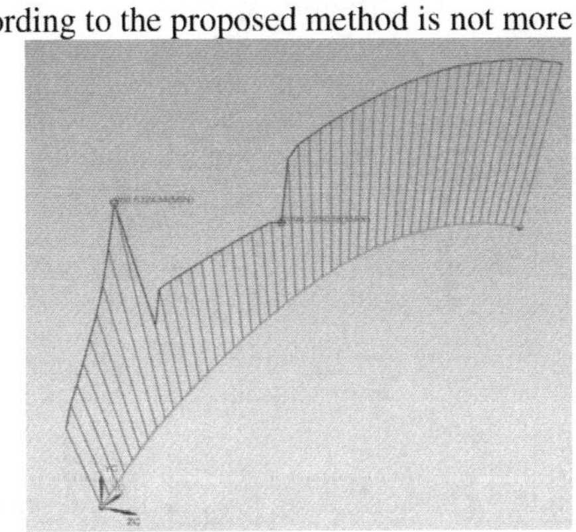

Maximum value – 2996,2250
Minimum value – 960.632

a – radius of curvature diagram constructed in Siemens NX 12 with maximum and minimum value

Intermediate values from maximum to minimum epure 2718.753; 2552,354; 2475,909

b – radius of curvature diagram constructed in Siemens NX 12 with intermediate values in the range of interest

Figure 9. Radius of curvature for section in Siemens NX 12

For the remaining sections, according to the proposed method, the curvature radii were found (Figure. 10).

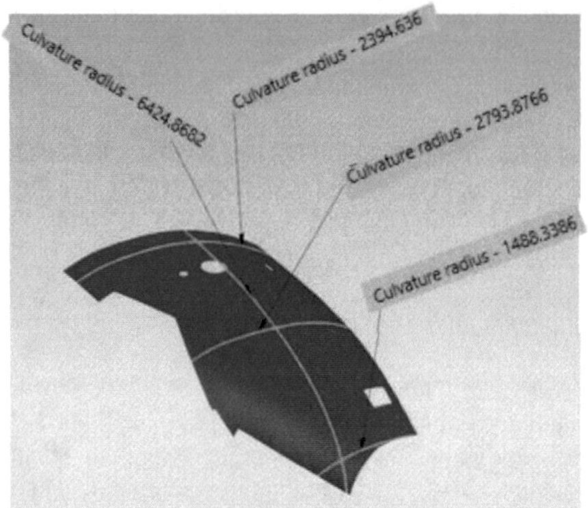

Figure 10. Section curvature radii according to the proposed methodology

Now it is necessary to relate curvature in one direction and another alternately. In our case, there are more sections in one direction, therefore, it is previously necessary to take the arithmetic mean of these curvature radii and then deal with this value. As a result, we get the value of two coefficients, the difference between which in percentage will show the curvature value. As a result, we get (Formula 4):

$$\Xi_1 = \frac{6424{,}8682}{\left(\frac{2394{,}636 + 2793{,}8766 + 1488{,}3386}{3}\right)} = 2{,}887;$$

$$\Xi_2 = \frac{\left(\frac{2394{,}636 + 2793{,}8766 + 1488{,}3386}{3}\right)}{6424{,}8682} = 0{,}346 \quad (4)$$

Now let us calculate the percentage difference between these values taking the maximum value as the basis [16]. As a result, we obtain the value of the curvature proportionality in the element Ψ = 88%. Having analyzed other elements manufactured by longitudinal stretching (with which there are no similar problems with springing compensation), the value of curvature proportionality is more than 90% (Figure 11).

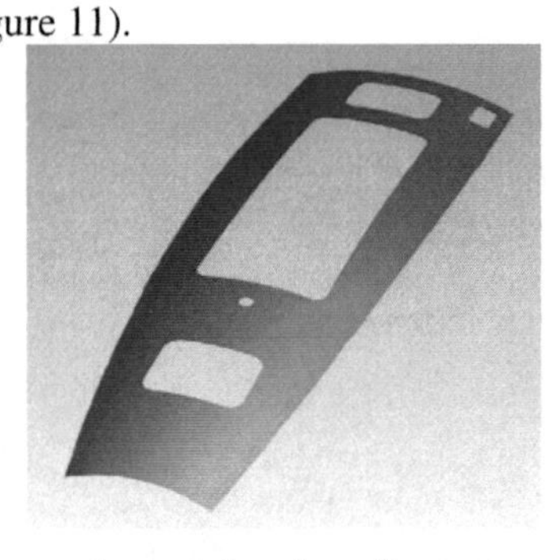

a – element 1, value of curvature proportionality 95%

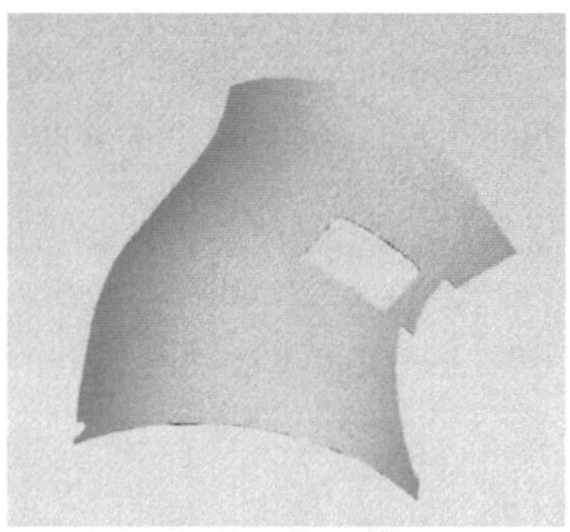

b – element 2, value of curvature proportionality 99%

Figure 11. Values of curvature proportionality

If all three values (88%, 95%, 99%) are analyzed (from the point of view that the first value is critical and it is not recommended to go below it) and a mathematical expectation is found, it will be 94%. Variance and standard deviation are 20.667 and 4.546%, respectively. Based on this, it can be assumed that the threshold value after which the proportionality value indicates compensation problems will be equal to the mathematical expectation minus the mean square deviation, i.e. 94% – 4.546% = 89.454% [11].

As a result, in the general form of the formula for the coefficients of the curvature ratio and the value of the curvature proportionality will take the following form (Formula 5):

$$\Xi_1 = \frac{\left(\sum_n^0 R1\right)/_n}{\left(\sum_i^0 R2\right)/_i};$$

$$\Xi_2 = \frac{\left(\sum_n^0 R2\right)/_n}{\left(\sum_i^0 R1\right)/_i};$$

$$\Psi = \frac{\max(\Xi_1, \Xi_2) - \min(\Xi_1, \Xi_2)}{\max(\Xi_1, \Xi_2)} \cdot 100\% \qquad (5)$$

where R1, R2 – curvature radii in different directions of the element of n- and i-section (respectively); i, n – number of sections in different directions of the element.

5. Conclusion

As a result, the paper describes the problem in springing compensation in the fitting of parts manufactured by longitudinal stretching. It was proposed to solve this problem by replacing the shaping scheme with transverse stretching. The study suggests the method of assessing the element on problems with springing compensation. A threshold value of the suggested criterion is shown, which helps to determine whether the element will have problems with springing compensation.

References

[1] Govorkov A S, Zhilyaev A S and Fokin I V 2018 A method of transition from 3d model to its ontological representation in aircraft design process *J. of Phys.: Conf. Ser.* 032046

[2] Govorkov A S, Lavrenteva M V and Fokin I V 2018 Mathematical modeling of making mechanical engineering products based on an information model *MATEC Web of Conferences* **224** 02022. DOI: https://doi.org/10.1051/matecconf/201822402022 .

[3] Lavrentyieva M and Govorkov A 2017 Using a discrete product model to determine the design element junctures *MATEC Web of Conferences* **129** 03002. DOI: 10.1051/matecconf/201712903003

[4] Fedorchuk M and Pak I 2005 Rigidity and polynomial invariants of convex polytopes *Duke Math. J.* **129(2)** 371-404. DOI:10.1215/S0012-7094-05-12926-X.

[5] Pokrasin M, Semashko N, Krupskii R and Kupov A 2004 Effect of electric-pulse treatment on the dislocation structure of ot4 titanium alloy *Russian metallurgy (Metally)* **6** 595-600

[6] Khusainov R, Sabirov A and Mubarakshin I 2017 Study of deformations field in the working zone of vertical milling machine *Procedia Engineering* 1069-1074

[7] Sokolnikov R A, Bozheeva T V and Govorkov A S 2020 Development of methodology for formalized selection of technological operations when designing technological process manufacturing of machinery *J. of Phys.: Conf. Ser.* **1582(1)**

[8] Erisov Y, Surudin S, Shlyapugin A and Grechnikov F 2016 The end-to-end computer simulation of casting and subsequent metal forming *Key Engineering Materials* **685** 167-171

[9] Averkiev A Yu 1985 *Methods for evaluating the stampability of sheet metal* (Moscow: Machine building) 176 p

[10] Ventzel E S 1963 *Probability Theory* (Moscow: Fizmatgiz) 366 p

[11] Gmurman V E 2004 *Probability theory and mathematical statistics,* textbook for universities, 10th edition (Moscow: Higher school) 479 p
[12] Gorbunov M N 1981 *Technology of procurement and stamping operations in the production of aircraft,* textbook for universities, 2nd ed., revised and added (Moscow: Engineering) 224 p
[13] Guter R S and Ovchinskiy B V 1970 *Elements of numerical analysis and mathematical processing of experimental results* (Moscow: Science) 432 p
[14] Isachenkov E I 1967 *Stamping with rubber and liquid* (Moscow: Engineering) 367 p
[15] 1973 PI 1.4.1977-2006 *Parts of procurement and stamping production. General technical requirements* (Moscow: Standardinform) 29 p
[16] Prokhorov A M 1978 Harmonic proportion, in: *Great Soviet Encyclopedia*, 3rd ed (Moscow: Soviet Encyclopedia) 766 p
[17] Rashevsky P K 1950 *Course of differential geometry,* 3rd edition (Moscow: GITTL) 428 p
[18] Industrial Equipment Directory. Retrieved from: http://www.8e.ru/print/10023.php.
[19] Kulikov A A, Nebyshinets Yu V, Sidorova A V and Balanovsky A E 2020 3D printing of metal products: from an automated design system to a real part *Bull. of Irkutsk State Technical University* **24(4)** 728-739. DOI: https://doi.org/10.21285/1814-3520- 2020-4-728-739.
[20] Larchenko A G 2019 Evaluation of the quality of products from polymer materials of machine-building purpose *Bull. of Irkutsk State Technical University* **23(3)** 463–471. DOI: 10.21285/1814-3520-2019-3-463-471.

Assessment of strength of elements produced using 3D printing technology

A Yu Perelygina, V B Raspopina, A V Rudykh and P A Grigorov

Irkutsk National Research Technical University, 83, Lermontova str., Irkutsk, 664074, Russia

E-mail: perelygina@ex.istu.edu, raspopina_vb@ex.istu.edu

Abstract. In some cases, 3D-printed products must be able to resist the stresses caused by their functionality. The design of bionic artificial hand is considered as a construction made using 3D printing technology, the strength of which must be assessed. The artificial limb is made using extrusion plastic printing. In order to assess the strength of the artificial limb, the experimental study of the samples made using 3D printing technology was carried out. The influence of three parameters on the strength characteristics of the samples was studied: the percentage of filling, the filling pattern and the direction of the material layering in the sample.

1. Introduction

The powerful possibilities of additive technologies make them relevant in the modern world of digital production [6, 12, 13, 22]. Nowadays their functionality has increased significantly from prototyping to structures that accept workloads [12, 13, 15, 19, 21, 25]. Moreover, it covers such high-tech spheres of production as aircraft construction, shipbuilding, mechanical engineering, instrument production and the field of medical engineering.

In particular, the technology of three-dimensional printing or 3D-printing, called FDM - printing (Fused Deposition Modeling) is of great interest. It is based on extrusion printing, the essence of the method of which is that under the influence of an extruder, the source material is heated to a predetermined temperature and squeezed out through a nozzle, forming the required product layer by layer in accordance with the electronic model. This technology is interesting for its simplicity in operation, the ability to create parts of almost any geometry complexity and availability for a wide range of users.

The spectrum ranges from industrial to household users. This is due to the offer on the market of 3D printers of any price category with different technical capabilities that determine different functionality. In this regard, this direction of 3D printing is rapidly developing and already covers various products, the reliability criteria of which include the criteria of strength and rigidity. This means that the issues of the assessment of the strength of such products are becoming relevant [13–15, 19, 20, 26].

In particular, in the field of medicine, additive technologies have become extremely popular in many areas [20–24] for example, in the production of implants, prosthetics, dentistry, etc. Here the prototype model should take into account its physical and mechanical characteristics along with the geometric parameters of the prototype. Therefore, the important aspect of 3D production is to ensure the strength and stiffness corresponding to the prototype.

2. Problem statement

The students of Irkutsk National Research University in the student innovation bureau “Aviator” are developing a bionic artificial hand for disabled people who have no wrist joint. Along with a wide functionality of the limb, one of the main advantages of the development is affordable price. It was decided to make the details of the limb using additive technology - plastic extrusion printing. For the production of the parts of the limb, inexpensive materials available to a wide range of consumers are proposed: ABS plastic, PetG plastic.

The performance characteristics of bionic artificial hand necessitate a responsible attitude towards its strength reliability. The design of the limb (Fig. 1) is a complex structure, the elements of which are simultaneously in different stress-strain states. Moreover, for an obvious reason, these states change during the operation of the limb. According to the classical theory of strength, the assessment of the parameters of the stress-strain state of a body requires knowledge of the mechanical characteristics of the material from which it is made [4]. At the same time, the theory is based on a number of hypotheses, principles and assumptions. Their performance has been proven experimentally. The important component of the theory is hypotheses about the properties of the material, in particular, about the homogeneity and continuity of the material within the volume of the studied body [3]. At the same time, they do not forget about the natural discreteness of the material. The schematization of the internal forces of a deformed body is based on this model of material properties.

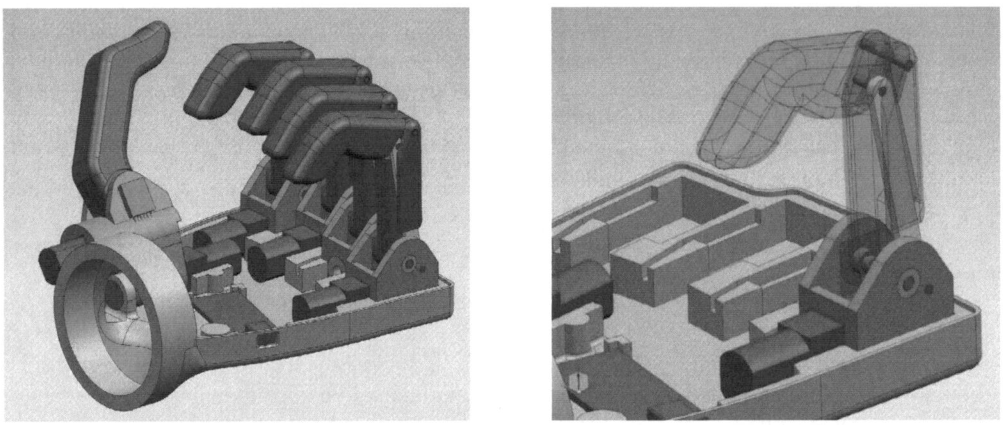

Figure 1. Design of bionic artificial hand

The combination of this theory, the information on the mechanical characteristics of materials and modern technologies for the automation of engineering calculations of structures allows carrying out the calculations of a structure of almost any complexity with a high degree of reliability of the results. However, the additive technology used for the production of the main parts of the structure of bionic artificial hand does not allow assessing the parameters of the stress-strain state of the structure, based on the classical theory of strength. Considering how the content of the volume occupied by the body is formed [1, 17], the application of hypotheses about the homogeneity and continuity of the material is queried. In this case, it is necessary to take into account the structure that is formed from the filament of the corresponding material [7]. At the same time, the solution to this problem is carried out not at the micro, but at the macro level. Since the assessment of the structural components is correlated with the categories comparable to the diameter of the extruder bore.

In order to assess the parameters of strength and stiffness of a product, it is necessary to know the mechanical characteristics of the filament materials, that is, the monofilament of the raw material. As a rule, the producer declares them. However, this information is insufficient to assess the strength and stiffness of the product itself. Practice shows that these characteristics of a filament and a product do not coincide. In particular, the strength of the product is lower than the strength of the monofilament of the raw material [4, 7]. Moreover, the value of this deviation depends on the factors both reasoned by the printing mode and the structure that fills the volume of the product [5, 8–10, 14, 18].

There are enough thermoplastics available for non-industrial 3D printing. Moreover, practice shows that the same thermoplastic from different producers has different characteristics. The characteristics may differ even when they are produced by one and the same producer and from one and the same thermoplastic but differently colored. Since the colorant of the material can affect the characteristics of the plastic.

The information on the research of the parameters of the stress-strain state of 3D-printed products is not uniform and, accordingly, not systematized. As a rule, these studies are carried out on the basis of field tests. These are tests of the samples or finished products with the violation of integrity and fixation of the value of the corresponding parameter of the state of strength or stiffness at this stage and the use of the terminology relevant within the framework of the hypothesis of the homogeneity and continuity of the material [8, 10, 11, 18].

The specificity of the design of the artificial hand is such that full-scale experiments that provide information on the strength and stiffness of the structure under all possible loading methods are impossible. Apart from other reasons they are impossible only because the number of experimental samples due to the number of loading cases is large.

However, the guarantee of product reliability is essential. As a result, a deep analysis of the structure is required, containing a reliable calculation of strength and stiffness. Therefore, a correct description of the material behavior of the loaded structure is required. In this case, the experimental studies of the behavior of loaded samples are needed.

There is information on the strength characteristics of molded plastic samples, while the design of bionic artificial hand requires the characteristics of samples layer-by-layer printed on a 3D printer. At this stage of the creation of bionic artificial hand, the information on the behavior of materials under static loads is relevant. In order to obtain this kind of information, the samples were tested for axial tension.

3. Input data for the experiment

The sizes of the sampless for testing for axial tension were taken in accordance with GOST 11262-2017 [16]. Figure 2 shows a standard sample of A type.

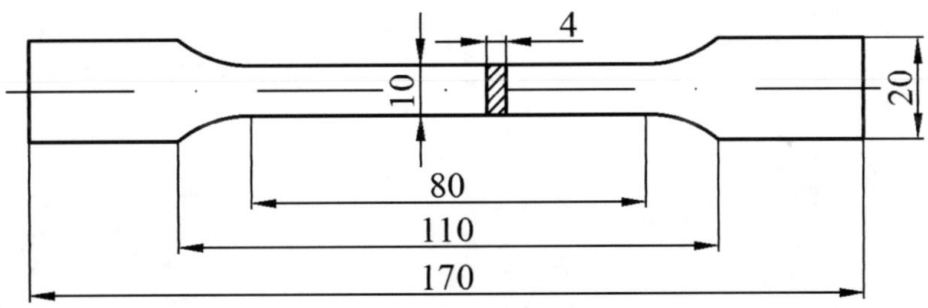

Figure 2. Sample of A type

The samples were made from PetG thermoplastic, produced by ABS Maker. The samples were printed on Anycubic Chiron printer with a print speed of 50 mm / sec, a working table temperature of 75 ° C and a nozzle temperature of 220 ° C. We tested the samples printed with two filling patterns: lines inclined at an angle to the longitudinal axis of the sample (Fig. 3, a) and triangles (Fig. 3, b). In the first case, the direction of the printing filaments periodically changed from layer to layer. Accordingly, the layers were attached to each other point wise, namely, in the limited areas of contact of the printing filaments. The location of these zones can be schematically characterized as the position of intersection points of the directions of printing filaments. In the second case, the structure of each layer can be characterized as a uniform grid with triangular cells. Moreover, the position of each triangle does not change from layer to layer. Therefore, the layers are not attached to each other point wise, but along the entire path of the printing filament of a layer.

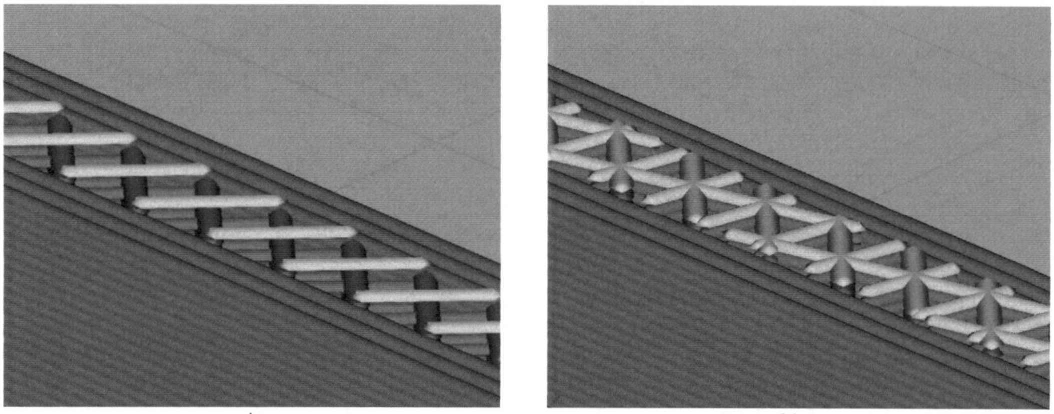

a) b)

Figure 3. Filling patterns

The above mentioned analysis of the structure of the arrangement of the filaments when filling the sample volume allows assuming that the behavior of externally identical samples may differ under the action of the same loading scheme with different filling patterns.

Moreover, we can assume that it will differ for externally identical samples with the same filling pattern, but with a different direction of layer formation. In one case, the layers are formed in the direction of the minimum cross-sectional size of the sample, in the other case, in the direction of the maximum cross-sectional size of the sample (Fig. 4). Here, there is different amount of material in the cross sections of the samples. Within the framework of the theory of axial tension-compression, this factor determines different parameters of strength for externally identical samples. Therefore, the tests were carried out with two types of samples, printed flat (Fig. 4, a) and standing on the edge (Fig. 4, b).

10 samples were printed for each position, filling density and filling pattern. The tests were carried out on INSTRON 5982 tension testing machine for static testing with a loading rate of 1 mm / min.

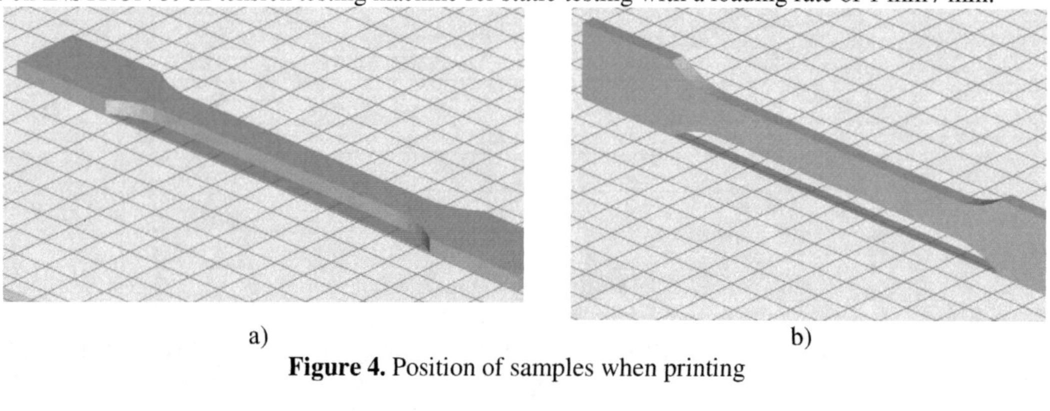

a) b)

Figure 4. Position of samples when printing

4. Research results

Table 1 and Figure 5 shows the results of tension tests of the samples with the considered filling patterns (Fig. 3), the samples were printed with 60% filling density. The following designations are adopted in the tables of the article:

N_{max} - maximum effort, N;

N_{max}^{aver} - average maximum effort, N;

Δl - absolute elongation at maximum load, mm;

IOP Conf. Series: Materials Science and Engineering 1064 (2021) 012052 doi:10.1088/1757-899X/1064/1/012052

Δl^{aver}- - average value of absolute elongation at maximum load, mm.

The abscissa shows the absolute elongation values at the moment of loading the samples. We mentioned above that 10 samples were printed for each filling pattern, but at the time of testing, some samples were destroyed outside the working area and were selected.

Table 1. Characteristics of samples with different filling patterns

Filling pattern	No. sample	N_{max}	N_{max}^{aver}	Δl	Δl^{aver}
Triangular	2	1212	1196±(60)	3,2	3,1±(0,2)
	4	1214		3,2	
	6	1086		2,5	
	8	1237		3,3	
	9	1230		3,2	
Linear	1	1285	1378±(69)	3	3,4±(0,2)
	2	1383		3,4	
	3	1369		3,5	
	5	1474		3,8	

According to Table 1 and Figure 5, the value of the average maximum tension force endured by the samples with a linear filling pattern exceeded by 13% the value of the average maximum tension force for samples with triangular elements. As for the stiffness characteristics, the average value of absolute elongation at maximum load for the samples with a linear pattern was 9% higher than for the samples with triangular filling. Perhaps the differences in strength and stiffness characteristics were associated both with the nature of the attachment of layers to one another during printing and with their structure, as it was mentioned earlier.

In further experiments, all the samples were printed with a triangular filling pattern as the sample with the most stable mechanical properties during testing. Let us consider the results of the experiments with different position of the samples during printing (flat and standing on the edge (Fig. 4)).

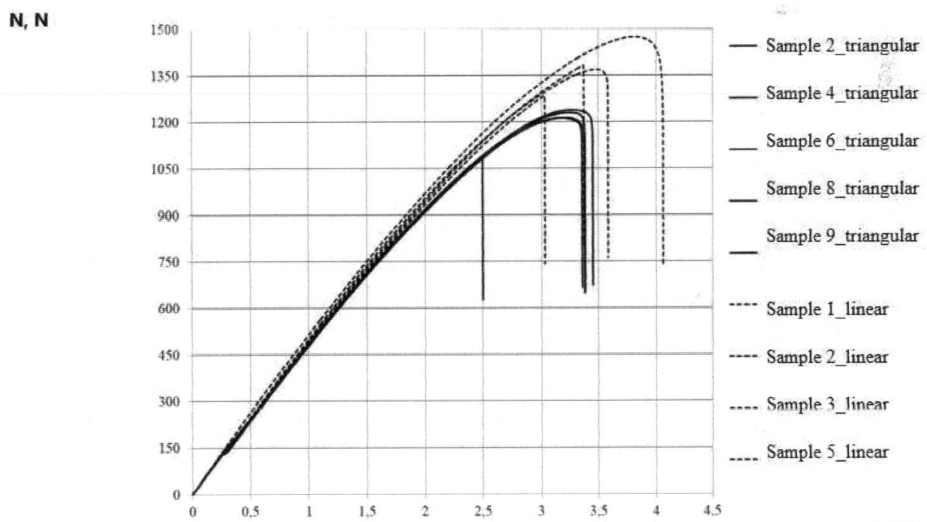

Figure 5. Tension diagrams of the samples with different filling patterns

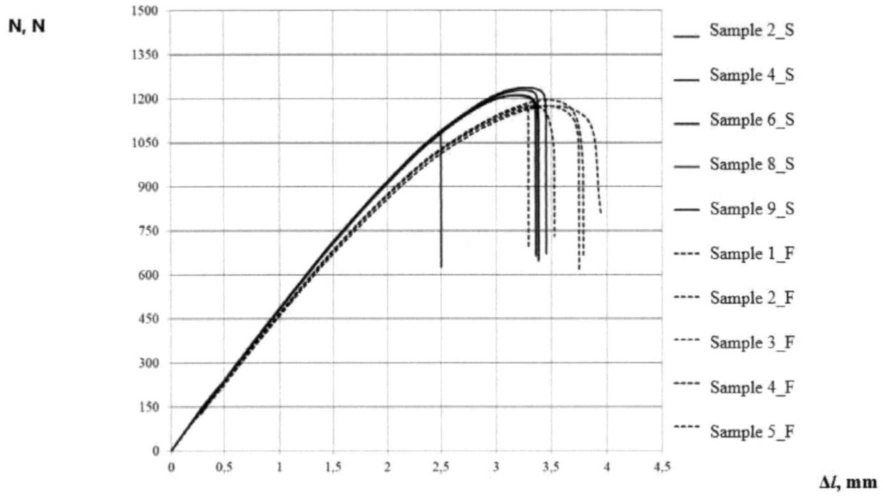

Figure 6. Tension diagrams of the samples with different position of samples when printing

According to Table 2 and Figure 6, the position of the samples when printing does not affect the characteristics of the samples, since the deviation between the values of tension forces when printing flat and standing is 1%. However, the stiffness characteristic when printing samples flat is 9% higher when printing standing on the edge.

In order to assess the influence of the filling density of the samples, we also conducted tests, the results of which are shown in Table 3 and Figure 7.

Table 2. Characteristics of the samples with different position of samples when printing

Position of samples	No. sample	N_{max}	N_{max}^{aver}	Δl	Δl^{aver}
	2	1212		3,2	
	4	1214		3,2	
Standing on the edge	6	1086	1196±(60)	2,5	3,1±(0,2)
	8	1239		3,3	
	9	1230		3,2	
	1	1177		3,5	
	2	1197		3,5	
Flat	3	1177	1181±(59)	3,3	3,4±(0,2)
	4	1176		3,5	
	5	1177		3,3	

Table 3. Characteristics of the samples depending on the % of filling

Filling density of the samples	No. sample	N_{max}	N_{max}^{aver}	Δl	Δl^{aver}
	2	1212		3,2	
	4	1214		3,2	
60%	6	1086	1196±(60)	2,5	3,1±(0,2)
	8	1237		3,3	
	9	1230		3,2	
	1	1391		3,1	
	3	1299		2,9	
100%	6	1350	1378±(69)	3,1	3±(0,2)
	8	1427		3	
	9	1424		3,1	

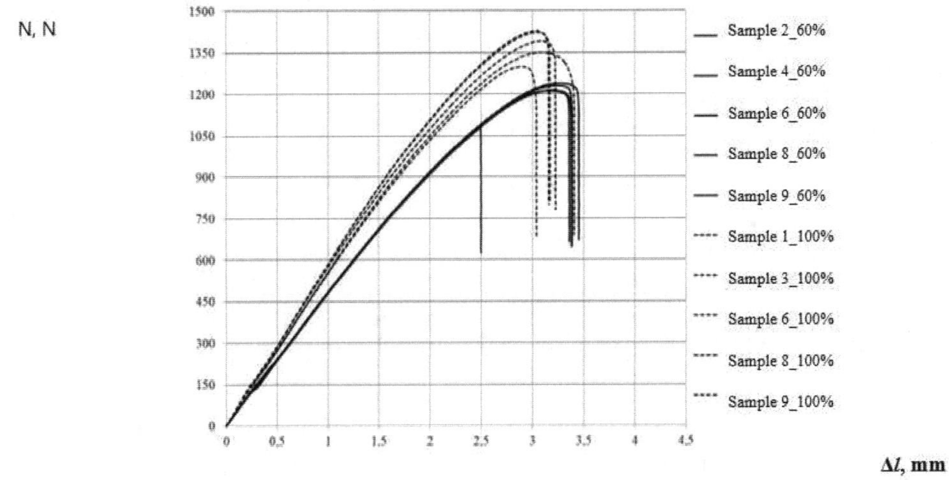

Figure 7. Tension diagrams of the samples at different filling densities

Thus, according to Table 3 and Figure 7, the average maximum tension force maintained by the samples with a filling density of 100% exceeds by 13% the average tension force for the samples with a filling density of 60%. At the same time, the mean value of absolute elongation at maximum load differs by only 3%, which is included in the confidence interval.

5. Conclusion

In this article, the authors studied three parameters of influence on the strength and stiffness characteristics of the samples: sample filling pattern, percentage of sample filling, direction of layering in the sample (the position of the samples when printing). Moreover, each influence parameter had two values, which were shown in Table 4. It summarized the results of all experiments, presented through the average values of the corresponding parameters of strength and stiffness.

Table 4. Experiment results

	Value of influence parameters			
Filling pattern	Inclined lines	Triangles	Triangles	Triangles
Filling percentage	60%	60%	60%	100%
Layering direction in the sample	Edge	Edge	Flat	Edge
Average maximum effort, N	1378±(69)	1196±(60)	1181±(59)	1378±(69)
Average absolute elongation at maximum load, mm	3,4±(0,2)	3,1±(0,2)	3,4±(0,2)	3,0±(0,2)

The experimental results allow assuming that with two equal parameters of influence (filling pattern and direction of layering), the strength characteristics of the samples depend on the density of their filling (60% and 100%), namely, with the increase in density, the maximum load endured by the samples increases. In addition, it affects strength characteristics and filling pattern (printing with lines or triangles) when the other two influence parameters are the same. However, the location of the samples during printing, that is, the direction of layering (standing on the edge or flat), with the other two equal parameters, does not affect the value of the average maximum tension force.

Regarding the stiffness characteristics of the samples, the experiments showed a significant effect of the filling patterns and the position of the samples when printed on the average absolute elongation.

The filling density with two equal parameters does not affect the specified characteristic (Table 4). Moreover, during the tests it was found that the samples printed with a linear pattern and a filling density of 60% and the samples printed with a triangular pattern and a filling density of 100% have equal values of the average tension force endured by the samples. A similar situation is observed for the characteristic of stiffness (Table 4). That is, it can be assumed that in order to achieve the required strength and stiffness characteristics several influence parameters should be combined in a certain way provided that the filament is saved.

It is necessary to note that the volume of experiments carried out during the preparation of this article does not allow making unambiguous conclusions about the strength and stiffness of the samples. As a result, at the moment it does not allow assessing the strength of a bionic artificial hand (Fig. 1). However, the results of the experiments make it possible to see some trends in the development of this topic and require further research with a large number of samples.

References

[1] Balashov A V, Cherdantsev A O, Novikovsky E A, Ananyin S V and Beloplotov S V 2016 Study of the strength of products obtained by 3D printing *Polzunovsky Bull.* **2** 61-64

[2] Kocheshkov I V, Kamonichkina N V and Kamonichkin D T 2019 Prospects for the use of 3D printing with polymeric materials in the manufacture and repair of engineering products *Innovative Renovation Technologies in Mechanical Engineering: Coll. of I67 Proc. of the Int. Sci. and Tech. Conf. dedicated to the 150th anniversary of the Faculty of Mechanical Engineering Technologies and the Department of Materials Processing Technologies of M. N.E. Bauman* (4-5 February 2019, Moscow) (Moscow: IIU MGOU) pp 95-99

[3] Rabotnov Yu N 1988 *Mechanics of a deformable solid body*, training manual for universities, 2nd ed (Moscow: Science) 712 p

[4] Dragunov Yu G, Zubchenko A S, Kashirsky Yu V et al 2014 *Grade of steels and alloys,* 4th ed. (Moscow) 1216 p

[5] Petrov V M, Bezpalchuk S N and Yakovlev S P 2017 About the influence of structure on the strength of plastic products obtained by 3D printing *Bull. of the State University and River Fleet named after Admiral S. O. Makarov* **9(4)** 765-776. DOI: 1021821 / 2309-5180-2017-9-4-765-776.

[6] Dizon J R C, Espera A H, Chen Q and Advincula R C 2017 Mechanical characterization of 3D-printed polymers *Additive Manufacturing* **20** 44-67. DOI: 10.1016 /j.addma.2017.12.002.

[7] Balachov A V and Makarova M I 2019 Study of the structure and properties of products obtained by 3D printing *Engineering J. of Don* **1**

[8] Wu W, Geng P, Li G, Zhao D, Zhang H and Zhao J 2015 Influence of Layer Thickness and Raster Angle on the Mechanical Properties of 3D-Printed PEEK and a Comparative Mechanical Study between PEEK and ABS *Materials* **8** 5834-5846. DOI: 10.3390 / ma809527.

[9] Kovan V, Tezel T, Camurlu H E and Topal E S 2018 Effect of printing parameters on mechanical properties of 3D printed PLA / carbon fiber compos *Materials Science. Non-Equilibrium Phase Transformations* **4** 126-128.

[10] Cantrell J, Rohde S, Damiani D et al 2017 Experimental Characterization of the Mechanical Properties of 3D-Printed ABS and Polycarbonate Parts *Advancement of Optical Methods in Experimental Mechanics* **3** 89-105. DOI: 10.1007 / 978-3-319-41600-7_11.

[11] Divyathej M V, Varun M and Rajeev P 2016 Analysis of mechanical behavior of 3D printed ABS parts by experiments *Int. J. of Scientific & Engineering Research* **7(3)** 116-124

[12] Nikitina L L and Gavrilova O E 2015 Prospects for the use of modern technologies of 3D printing in the production of light industry products from polymer materials *Bull. of the Technological University* **18(7)** 224-226

[13] Kocheshkov I V, Kamonichkina N V and Kamonichkin D T 2019 Prospects for the use of modern 3D printing technologies with polymeric materials in the manufacture and repair of

machine-building products *Proc. of the Int. Sci. and Tech. Conf. "Innovative Renovation Technologies in Mechanical Engineering"* pp 95-99

[14] Coogan T J and Kazmer D O 2020 Prediction of interlayer strength in material extrusion additive manufacturing *Additive Manufacturing* **35** 1-16. DOI: 10.1016 /j.addma.2020.101368.

[15] Balashov A V, Zabortseva M N, Elyubaeva A S and Akhmetkali B K 2018 Improving the accuracy of gear wheels obtained by 3D printing *Proc. of the IX Int. Sci. and Pract. Conf. "Innovations in Mechanical Engineering"* pp 326-333

[16] 2018 GOST 11262-2017 (ISO 527-2: 2012) *Plastics. Tensile test method. Interstate standard* (Moscow: Standartinform) 44 p

[17] 2018 GOST R 57558-2017 (ISO / ASTM 52900: 2015) *Russian Federation. Additive technological processes. Basic principles. Part 1. Terms and definitions. National standard* (Moscow: Standartinform) 16 p

[18] Kartashova E D and Muizemnek A Yu 2018 Identification of the parameters of the ABS-plastic model obtained by layer-by-layer fusion *Models, systems, networks in economics, technology, nature and society* **1(25)** 130-140

[19] Munasinghe N, Woods M, Miles L and Paul G 2019 3-D Printed Strain Sensor for Structural Health Monitoring *IEEE Int. Conf. on Cybernetics and Intelligent Systems (CIS) and the IEEE Int. Conf. on Robotics, Automation and Mechatronics (RAM), Bangkok.* DOI: 10.1109 / CIS-RAM47153.2019.9095826

[20] Popescu D, Baciu F, Vlăsceanu D, Cotruţ C and Marinescu R 2020 Effects of multiple sterilizations and natural aging on the mechanical behavior of 3D-printed ABS *Mechanics of Materials* **148** 1-12. doi.org/10.1016/j.mechmat.2020.103423

[21] Javaid M and Haleem A 2018 Additive manufacturing applications in orthopedics: A review *J. of Clinical Orthopedics and Trauma* **9** 202-206

[22] Reddy R D P 2020 Additive manufacturing in drug delivery applications: A review *Int. J. of Pharmaceutics* **589** 1-34

[23] da Silva L A, Medola F O, Rodrigues A C T and Sandnes F E 2019 Interdisciplinary-Based Development of User-Friendly Customized 3D Printed Upper Limb Prosthesis, in: Ahram T, Falcao C (eds) *Advances in Usability, User Experience and Assistive Technology. AHFE 2018. Advances in Intelligent Systems and Computing*, vol 794 (Springer, Cham) DOI: https:// doi.org/10.1007/978-3-319-94947-5_88.

[24] da Silva B B, Porsani R N, Hellmeister L A V, Medola F O and Paschoarelli L C 2020 Design and Development of a Myoelectric Upper Limb Prosthesis with 3D Printing: A Low-Cost Alternative, in: Di Nicolantonio M, Rossi E, Alexander T (eds) *Advances in Additive Manufacturing, Modeling Systems and 3D Prototyping. AHFE 2019. Advances in Intelligent Systems and Computing,* vol 975 (Springer, Cham) DOI: https://doi.org/10.1007/978-3-030-20216-3_30.

[25] Kulikov A A, Nebyshinets Yu V, Sidorova A V and Balanovsky A E 2020 3D printing technology for metal products: from a computer-aided design system to a real part *Irkutsk State Technical University Bull.* **24(4)** 728–739. DOI: https://doi.org/10.21285/1814-3520-2020-4-728-739.

[26] Larchenko A G 2019 Assessment of the quality of products made of polymer materials for machine-building purposes *Irkutsk State Technical University Bull.* **23(3)** 463–471. DOI: 10.21285 / 1814-3520-2019-3-463-471.

[27] Govorkov A S, Lavrenteva M V and Fokin I V 2018 Mathematical modeling of making mechanical engineering products based on an information model *MATEC Web of Conferences* **224** 02022. DOI: https://doi.org/10.1051/matecconf/201822402022 .

[28] Lavrentyieva M and Govorkov A 2017 Using a discrete product model to determine the design element junctures *MATEC Web of Conferences* **129** 03002. DOI: 10.1051/matecconf/201712903003

Optimization of the gas collection and treatment system of the Beregovoye gas condensate field

A A Karimova, E M Almukhametova, and F F Fakhrutdinov

Ufa State Petroleum Technological University, Branch of the University in the City of Oktyabrsky, 54a, Devonskaya str., Oktyabrsky, Republic of Bashkortostan, 452607, Russian Federation

E-mail: karimova.nastya2011@mail.ru, info@of.ugntu.ru

Abstract. At present, the experience of exploitation of various fields shows that a lot of associated water and gas is extracted during the oil production process. A liquid is extracted from wells, which is a complex mixture of various hydrocarbons in various aggregate states, various impurities and water. The water extracted from the reservoir negatively affects the equipment, causing its abrasion. At the same time, the period of major repairs is being shortened. The object of the current research is a gas condensate field. The purpose of this work is to increase the efficiency of the gas dehydration process by improving mass-transfer, as well as to analyze the gas collection and treatment system. At the same time, the analysis of the arrangement of technological schemes of equipment for gas treatment and absorption is carried out.

1. Introduction

The gas condensate complex of the Beregovoye field is a centralized scheme for collecting gas from gas condensate wells at the primary gas treatment facilities (UCPG-V).

To date, 3 clusters of gas condensate wells with the number of wells in the cluster from 2 to 7 have been equipped and are in operation at all three gas collection points. On average, 3-4 wells are placed on the cluster. The minimum wellhead pressure was 10.3 MPa for the area of the UCPG-V in 2006. The existing scheme of gas collection is mainly radial, a single pipeline is laid from each well cluster to the inlet manifold of the collection point [11].

By the VNTP 01-81 ("Standards for Technological Design of Gas Production Facilities and Underground Gas Storage Stations"), the binding of production wells of the cluster must include:

- the ability to work both on the oil well tubing (NCT), and on the annular space;
- automatic disconnecting of wells in the case of a pipeline-plume rupture or increasing the pressure above the working pressure with the installation of a cut-off valve;
- using thermowells to measure the gas temperature at the well mouth, before and after the pressure regulator;
- installation of safety valves to protect the gas plume due to the excess pressure;
- the possibility of conducting works related to the well killing operations, the hydraulic fracturing of the formation, the salt-acid treatment, etc., as well as to the well research;
- gas discharging to the flare when the wells are blown-off and when the pressure safety valves release [10].

2. Materials and methods

The field gas collection system is designed to collect and transport the extracted gas from the well clusters to the complex gas treatment facility (UCPG-V). The majority of gas condensate fields in the Yamal-Nenets Autonomous District has adopted a linear gas collection scheme, which provides higher reliability of the system and effective control to operate on the clusters and wells in comparison with the collector collection scheme. However, due to the long distances from the producing well clusters to the gas treatment facilities and, consequently, the high metal content of the gas collection networks, the parameters of plumes for collector-linear and linear systems to collect well products were calculated to provide comparative economic analysis [5].

The selection of the plume diameters is based on the condition of the permissible gas velocity in the pipe is equal to 25 m/s and the pressure at the inlet to the UCPG is not lower than the gas pressure in the existing and suitable plumes.

The calculation of the required diameters of the plumes was carried out with a limit of the maximum diameter of no more than Df = 300 mm. If it is necessary to lay a pipeline with a larger diameter, a decision was made to build a parallel pipeline and the year of its implementing was determined.

The operation mode of most of the heat-insulated plumes of above-ground laying from clusters and wells in the winter season is hydrated, so the supply of methanol to the clusters of wells from the UCPG is provided.

Strength calculations of gas collection networks were performed following the requirements of SP 34-116-97 using a computer program "Excel" for a gas pressure of 16 MPa. Gas collection networks are classified as category II [7, 12].

The analysis of the hydraulic and thermal calculations of the gas inter-field transport system and methanol consumption in dynamics from the beginning of the well clusters placing in operation shows that the specific amount of metal of gas collection systems accepted for calculation is approximately the same, but the total length of gas collection networks increases by more than 30% under the linear scheme.

To connect the newly introduced well clusters, it is necessary to provide for the expansion of the inlet manifold at the UCPG-V by constructing a separate block with connecting it to the common collector. At other gas collection points, there are reserves for connecting additional plumes [5].

3. Results and Discussion

For evaluating the effectiveness of the field technology, a new and very interesting universal indicator is additionally proposed – the distribution coefficients of components and narrow fractions of extracted fluids by the products of its field preparation – drained gas and unstable condensate (the percentage of a selection of components and fractions of extracted fluid into condensate or gas). This indicator is easily calculated based on the technological model of the UCPG, adapted on the results of its survey using this methodology. The distribution coefficients (or selections) are represented as the ratio, in absolute percentages, of the amount of a component (or narrow fraction) in one of the flows (gas or condensate) to the amount of the same component (or fraction) in the extracted fluid. The information on the distribution coefficients of hydrocarbons obtained from the results of the integrated surveys of the UKPG-V is presented in table 2.

These distribution coefficients should not be confused with the absolute residual content of liquid hydrocarbons (C5+) in the commercial gas. The distribution coefficients are relative balance values and mean the mass fraction (in percentage terms) of the corresponding component of the extracted fluid, which leaves the field as a part of the commercial gas and unstable condensate. The remaining part of the component (trace amount) can be withdrawn from the field as part of a water-methanol solution [9].

The distribution coefficients characterize the clarity of separation of components and fractions of the extracted fluid along the gas and condensate flows at each specific field. These coefficients depend on the applied field technology, the composition of the extracted fluid, and thermodynamic parameters in technological devices [7, 12].

The commercial products of the UCPG-V are the following:

- natural gas corresponding to OST 51.40-93, supplied to the inter-field collector with a temperature of minus 2-5°C and pressure up to 6.0 MPa;
- unstable gas condensate corresponding to TU 05751745-02-88, accumulated in the PVC-1000.

The gas treatment process utilizes:

- methanol as a hydrate formation inhibitor;
- cold condensate of the 1st separation stage as a hydrocarbon absorbent.

The characteristics of the reservoir gas-condensate mixture are shown in table 1.

Table 1. The characteristics of the reservoir gas condensate mixture

Components	Molecular weight, kg/mol	% molar
N2	28	0.68
CO2	44	0.3
CH4	16	89.64
C2N6	30	4.54
C3H8	44	1.99
i-C4H10	58	0.48
n-C4H10	58	0.47
C5+	115	1.9

The raw feedstock for the production of commercial gas and condensate at the UCPG-V is the formation gas of the Valanginian deposits.

The specific yield of unstable gas condensate when using the low-temperature absorption process, a pressure of 6.0 MPa and a temperature of minus 25°C ~ 120 g/m^3.

The disadvantage of the technological scheme is that the input and output collectors of the gas air-cooling unit are made of non-cold-resistant steel. To prevent the destruction of collectors at temperatures below minus 20°C, the electric heating of collectors is provided, carried out by a heating cable of the KMJ type, which provides the above-zero temperature of the walls of collectors and pipes at ambient temperatures up to minus 54°C.

A mixture of gases with a pressure of 5.5-6.4 MPa is sent through the common input collector into the absorbers A-2/1-3.

The remainder of the gas flow after C-3/1-3, which is not used in ejectors, is throttled, passing the control valves KRD-2 and also enters the absorbers A-2/1-3.

The General Designer of Geotransgaz Company – Institute "YUZHNIIGIPROGAZ" allowed to conduct the process of low-temperature absorption at a temperature not lower than minus 30°C in the absorbers for extracting heavy hydrocarbons A-2/1-3.

The schematic diagram of the technological line of the field is shown in Figure 1.

According to the information provided by VNIIGAZ LLC, the performance indicators of the main UCPG-1.V devices are characterized as follows. Primary separators C-1 operate at a pressure of 8.7-8.5 MPa in the range of gas loads from 240 ths m^3/h (5.8 mln m^3/day) to 330 ths m^3/h (7.9 mln m^3/day) and all of them, without exception, are characterized by low efficiency: liquid entrainment with gas reaches 10 g/m^3, which significantly exceeds the passport value (0.02 g/m^3). The expertise of the device under variable load conditions allowed us to assume that there was a mechanical violation of the integrity of the separation devices in the units. It is necessary to revise and replace the structural elements.

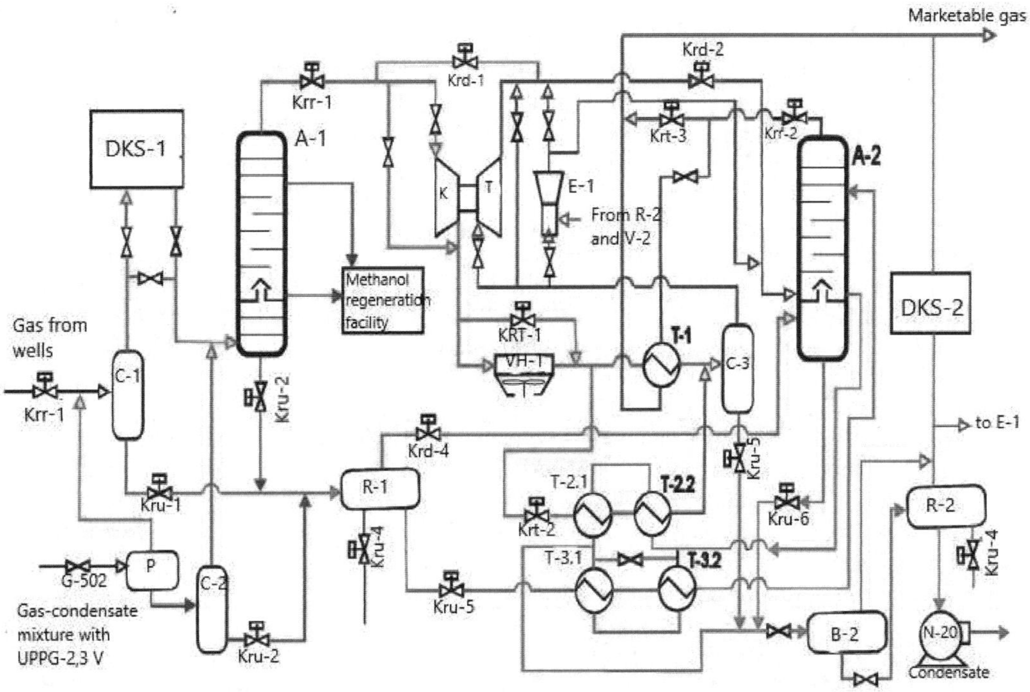

Figure 1. The schematic diagram of the technological line of the UCPG-V of the Beregovoye field

Table 2. The distribution of components of the total extracted fluid (%) of the Beregovoye field by the products of field preparation

The distribution of components	CH4	C2N6	C3N8	1C4N10	PC4N10	1C5N12	pC5H12	C5+	C6+	Total
into commercial gas	99.15	89.73	63.97	36.51	27.34	13.24	10.39	3.01	1.46	87.43
into an unstable condensate	0.85	10.27	36.03	63.49	72.66	86.76	89.61	96.99	98.54	12.57

4. Conclusion

Based on the mentioned above characteristics of the technology and equipment to improve the efficiency of operation of UCPG-V, in our opinion, it is necessary to carry out the following tasks: to organize the work of the primary separators C-1 and separators DKC-1V to increase the efficiency of the separation to the desired value (the liquid entrainment should not be more than 20 mg/m^3); to complete reconstruction of the binding site of the final condensate degassing (separators R-2) and the heat exchanger T-2 to reduce the methanol loss with the product of unstable condensate by reducing the temperature of the final degassing of unstable condensate to minus 15-20°C; to work on improving the phase separation of secondary recovery methods – the condensate in separator R-2, which, consequently, will reduce the direct methanol losses.

It is also necessary to reconstruct the internal devices of the A-2 absorber to achieve maximum gas flow efficiency and create conditions for stopping the first stage for reconstruction.

It should be noted that it is necessary to conduct calculation studies of the hydrate formation prevention system and to determine the optimal degree of methanol blow-off in the A-1 absorber, to develop and implement recommendations for achieving the optimal degree of blow-off and regulating the concentration of methanol on the lower chimney tray A-1.

References

[1] Almukhametova E M 2019 Changes in Formation Pressure and Technological Efficiency of Unsteady Flooding in Double Permeability Reservoirs Having Different Productivity Rates *IOP Conf. Ser.: Earth Environ. Sci.* **378** 012036

[2] Batalov S A, Andreev V E, Lobankov V M and Mukhametshin V Sh 2019 Numerical Simulation of the Oil Reservoir with Regulated Disturbances. Oil Recovery Stability Simulation *J. of Phys.: Conf. Ser. (ITBI 2019 – Int. Conf. "Information Technologies in Business and Industry")* **1333(3)** 032007. DOI:10.1088/1742-6596/1333/3/032007

[3] Uleberg K and Hoier L 2002 Miscible Gas Injection in Fractured Reservoir *SPE/DOE IOR Symp.* (13-17 April 2002, Tulsa, Oklahoma) (Oklahoma: Society of Petroleum Engineers) p 15

[4] Ogdena J, Myers Jaffe A, Scheitrum D, McDonald Z and Millera M 2018 Natural Gas as a Bridge to Hydrogen Transportation Fuel: Insights from the Literature *Energy Policy* **115** 317–29

[5] Dai J 2017 Ch. 7. Coal-Derived Gas Fields and the Gas Sources of Offshore China Giant Coal Derived *Gas Fields and their Gas Sources in China* **48** 447–549

[6] Rogachev M K, Mukhametshin V V and Kuleshova L S 2019 Improving the Efficiency of Using Resource Base of Liquid Hydrocarbons in Jurassic Deposits of Western Siberia *J. of Mining Institute* **240** 711-715. DOI: 10.31897/PMI.2019.6.711

[7] Guzhov A I 1973 *Joint Collection and Transportation of Oil and Gas* (Moscow: Nedra) p 63

[8] Sudip M and Sumit G 2017 *Oil and Gas Exploration: Methods and Application* (Special Publications by American Geophysical Union) p 10

[9] Saha P, Abolarin J and Parsa A 2014 NUGGETS Gas Field: Pushing the Operational Barriers Publisher *Society of Petroleum Engineers: SPE Production & Operation* p 10

[10] Katz D L 1959 *Handbook of Natural Gas Engineering* (New York: McGraw-Hill Higher Education) p 49

[11] Sloan E D Jr and Koh C A 2008 *Clathrate Hydrates of Natural Gases,* 3rd ed (Boca Raton, Florida: Chemical Industries, CRC Press) 119 p 758

[12] 2009 *Development Project of the Beregovoe Gas Condensate Field* (Tyumen: ROSNEDRA) p 2

Theoretical and practical aspects of absorbing layers insulation in the conditions of hydraulically perfect modes of grouting solutions injection

A P Averyanov[1], V E Andreev[2,3], A P Chizhov[2,3], V Sh Mukhametshin[4], and A V Chibisov[2,3]

[1]Mechanical Engineering Research Institute of the Russian Academy of Sciences, 4, Bardina St., Moscow, 119334, Russian Federation
[2]SASI "Institute for Strategic Research of the Republic of Bashkortostan", 129/3, Oktyabrsky Ave., Ufa, Republic of Bashkortostan, 450075, Russian Federation
[3]Ufa State Petroleum Technological University, 1, Kosmonavtov St., Ufa, Republic of Bashkortostan, 450062, Russian Federation
[4]Ufa State Petroleum Technological University, Branch of the University in the City of Oktyabrsky, 54a, Devonskaya St., Oktyabrsky, Republic of Bashkortostan, 452607, Russian Federation

E-mail: averyanov@rambler.ru, vsh@of.ugntu.ru

Abstract. The article presents the results of theoretical and experimental studies of insulation of absorbing intervals during penetration of rocks. The proposed systematic approach makes it possible to effectively apply insulation methods in various mining conditions of well construction. The methods for controlling the absorption of drilling fluids in the conditions of abnormally low layer pressures are combined in a systematic technology.

1. Introduction
It is known that the main indicators of performance of field development management are largely determined by the efficiency of well drilling [1-5].

The main qualitative and technical-economic indicators of drilling and completion of deep wells largely depend on the effectiveness of the methods used to prevent complications (absorption, gas and oil ingress, hydraulic fracturing of rocks, caving formations, interlayer flows, etc.). However, the greatest time spent in the overall balance of drilling wells in the oil and gas industry (22-23 %) and material and financial resources (4-8%) are associated with the fight against the absorption of drilling and grouting solutions [6-8].

To solve this complex technical problem, various methods and technologies are used: regulation of the parameters of drilling and grouting solutions, the supply of drilling pumps, the concentration of clogging fillers, injection of highly structured clay solutions of pastes and paste plugs, viscoelastic systems, injection of cement, gel-cement and polymer-cement mixtures, etc. But all these methods, with an undeniable difference, have the same significant disadvantages – low efficiency (performance indicator – 20-50 %) and low quality of insulation works [6, 7]. The current situation is explained by the methodological underdevelopment of technologies and the lack of systematic approaches and solutions in the theory and practice of absorption prevention. First of all, this is due to the lack of techniques for controlling and regulating the hydro-mechanical processes of interaction between grouting solutions and absorbing rocks when implementing various mechanisms for their insulation [6, 7].

2. Methods and materials

In practice, specialists choose technological parameters of the modes of injection of grouting mixtures into the channels of absorbing layers, without taking into account the mechanisms for reducing the intensity of absorbing rocks, the type, grouting and technical properties and volumes of waterproofing solutions and hardening mixtures, fillers and chemicals, which, as a rule, are not always adequate to the geological and physical conditions of insulation works [6, 7]. As a result, there is a natural decrease in the quality and technical and economic indicators of insulation works and a reduction in the scope of their effective application.

Waterproofing of absorbing layers is a non-stationary process of hydro-mechanical interaction of injected visco-plastic liquids with permeable rocks at the radius of penetration into the filtration channels under the influence of external factors in time; fluid flow and injection pressure. Despite the wide application of technology in commercial practice, the mechanism of the hydro-mechanical process of insulation of highly permeable absorbing rocks is still insufficiently and fully covered in professional literature [6, 8]. The present article is intended to fill this gap and show how systematic approaches can be used to solve the set tasks in drilling technology.

From the methodological viewpoint, implementation of the systematic approach in the methods of insulation of absorbing intervals includes several interrelated stages of work, which can be represented as a diagram. Figure 1 shows the diagram for the system organization and performance of insulation works under hydraulically perfect modes of injection of grouting mixtures into the bottom-hole zone of absorbing layers. However, the present article outlines the second stage and focuses on the mechanisms of insulation of absorbing intervals with grouting solutions.

Stage I	*Stage II*	*Stage III*	*Stage IV*	*Stage V*
Estimation of the hydrodynamic state of the well and geological-physical parameters of the absorbing layer ($H_к$, h_{in}, P_{pl}, h_{st}, h_d, T_{pl}, K)	***Substantiation of*** the mechanism of absorption insulation or their combination (coagulant structure formation, dehydration and crystallization hardening, effect of wedging pressure)	***Choice*** of the type and properties of grouting mixtures, ***calculation*** of the required volume	***Calculation*** of the parameters of the insulation mode of the absorbing layer, ***choice*** of the technical scheme of works (Q, P, T open barrel, filling pipes, disconnecting devices)	***Control*** and ***regulation*** of the mode of grouting of the absorbing layer (K_{dyn}, δ, Q, P, T)

Figure 1. The block diagram of realization of the system organization and execution of the insulation works: N_k – depth of the layer roof; h_{in} – thickness of permeable rocks; P_{pl} – layer pressure; h_{st} – depth of static level of liquid in the well; h_d – depth of the dynamic level; T_{pl} – layer temperature; Q – flow rate; K – coefficient of acceleration of the permeable layer; P – injection pressure; T – time of grouting mixture injection; K_{dyn} – coefficient of dynamic acceleration of absorption; δ – coefficient of hydraulic perfection of the mode of injecting grouting mixture

3. Results and discussion

Restoration of impermeability under abnormal low pressure is carried out by injecting grouting solutions into the bottom-hole zone of the absorbing layers. This process can be implemented in different injection modes: lateral, transient and bottom [6, 7]. Each of the modes differs in the mechanism of influence on permeable rocks, the parameters of technological process and the scope of effective application.

The mode of lateral injection of grouting mixtures into the absorption zone is achieved by pressing the layer roof by regulating the initial structural and mechanical properties of grouting mixtures and their injection modes from the borehole to the insulated interval (pump supply and injection pressure)

(Figures 2, a; 3). Under these hydraulic conditions, the movement of grouting solutions injected into the bottom-hole zone occurs simultaneously along the entire thickness of the absorbing rocks. This injection mode is considered to be hydraulically perfect and allows implementing a number of important technological effects (internal factors) that provide a non-linear increase in the quality and technical and economic indicators of insulation operations [8, 10]. These factors include:

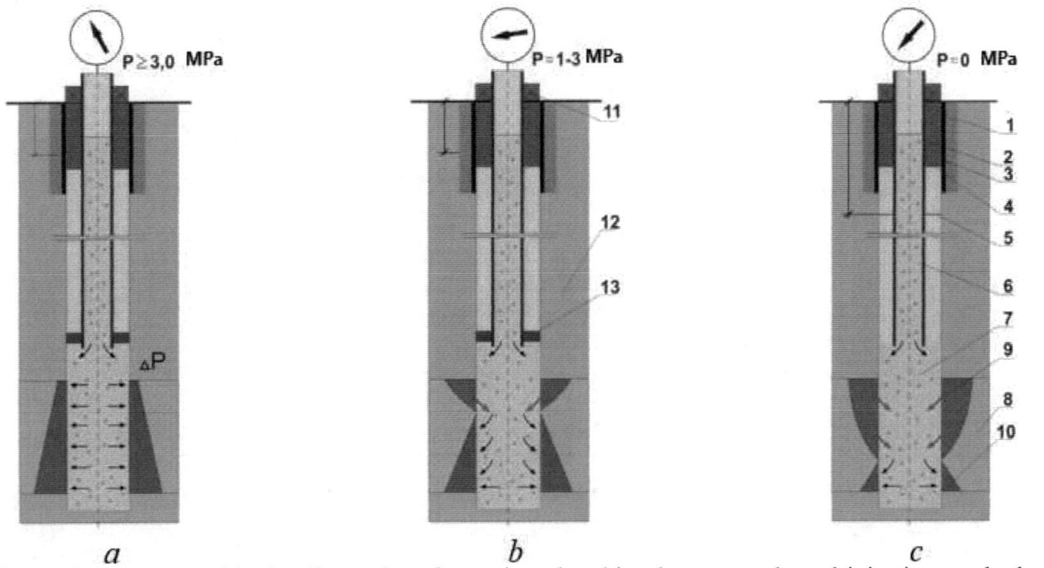

Figure 2. Diagrams of hydraulic modes of grouting absorbing layers: a – lateral injection mode, b – transient injection mode; c – bottom injection mode. Numbers indicate: 1 – technical column; 2 – drill pipe column; 3 – well; 4 – dynamic liquid level; 5 – static liquid level; 6 – drilling solution; 7 – displacement liquid; 8 – grouting solution; 9 – low pressure zone; 10 – absorbing layer; 11 – preventer; 12 – rock mass; 13 – packer

The first effect (internal factor) consists in "piston" displacement and replacement of layer fluid at the radius of penetration of grouting solution simultaneously throughout the thickness of the interval of absorption, which eliminates adverse change in original properties of the grouting solution and the occurrence of intra-layer flows during and after the insulating operation.

The second internal factor is the intensification of the dehydration process of hardening grouting mixtures and the growth of their structural-mechanical and grouting-technical properties (Figures 4, 5, Table 1). The mechanism of dehydration of hardening mixtures accelerates the process of structure formation of grouting solutions, in which for 2-3 minutes the initial plastic viscosity of the cement solution increases by 4–6 times (Figure 3), and the maximum shear stress – by 6.0–7.5 times (Figure 4). It should be mentioned that according to laboratory research, the start time of setting of viscous-plastic liquid at water-cement ratio (W/C) of 0.35–0.40 reduces by 1.5–1.9 times, and the end time — by 1.4–1.5 times.

IOP Conf. Series: Materials Science and Engineering 1064 (2021) 012054 doi:10.1088/1757-899X/1064/1/012054

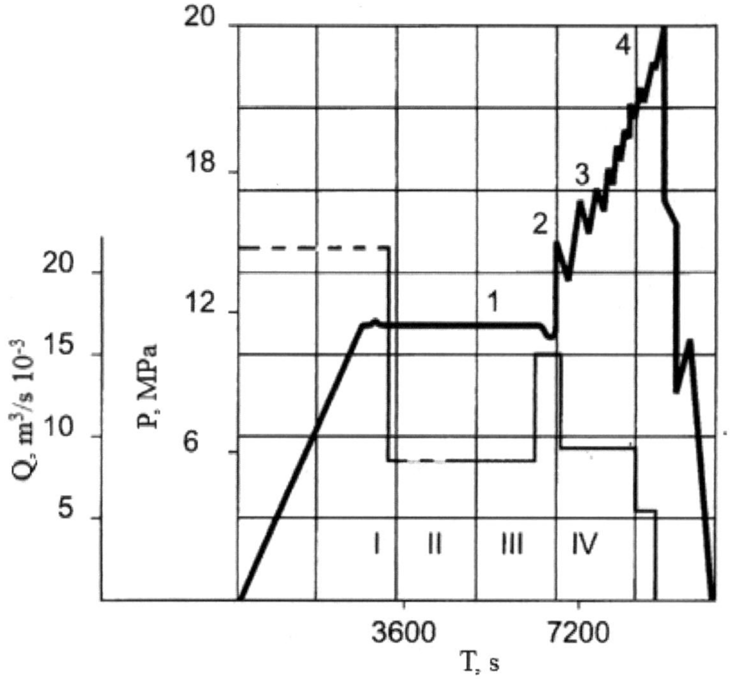

Figure 3. Insulation of absorption with intensity up to 70 m³/h in the mode of lateral injection of grouting mixtures: 1–2 – clay paste; 2–4 – cement solution η

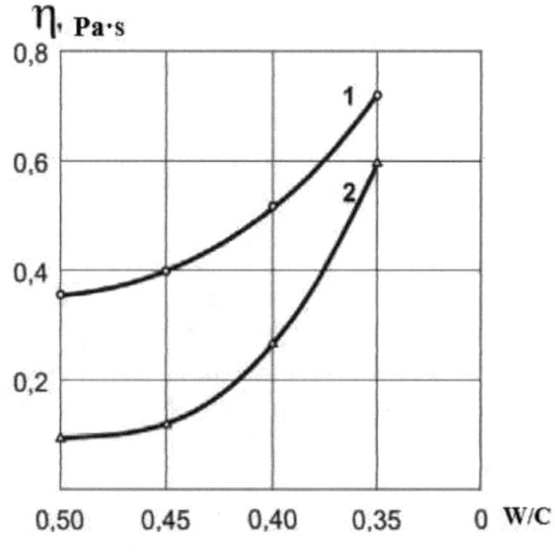

Figure 4. Dependence of plastic viscosity on water-cement ratio:
1– in 20 minutes after closing;
2 – in 4 minutes after closing

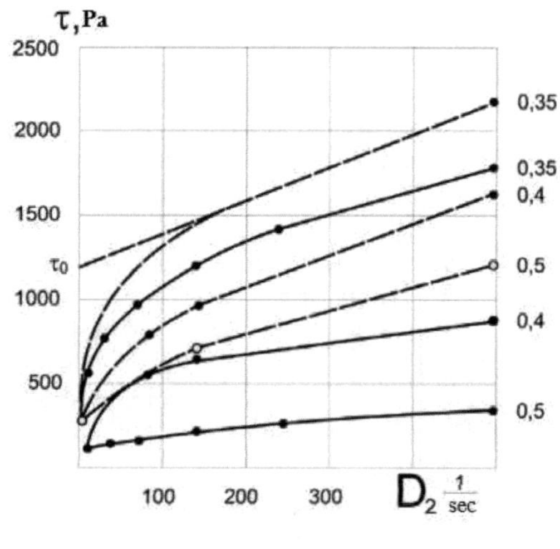

Figure 5. Dependence of the shear stress on the velocity gradient for different water-cement ratios:
——— – in 4 minutes after closing;
- - - - - - – in 20 minutes after closing

Table 1. Dependence of grouting and technical properties of cement mortar-stone on the water-cement ratio

Water-cement ratio, (W/C)	Spreadability, cm	Density, kg/m^3	Plastic viscosity, Pa*s	Shear stress, Pa	Setting time, hour-min		Ultimate strength, MPa	
					start	end	at bending	at compression
0.50	22	1830	0.10	300	6-19	2-06	2.22	4.5
0.45	20	1880	0.12	–	5-05	2-05	2.63	7.5
0.40	15	1950	0.26	800	4-12	1-45	3.50	11.0
0.35	12	2030	0.60	1750	3-20	1-33	4.60	18.3

Table 2. Indicators of improved methods of insulation of absorbing layers

Insulation technology	Analyzed volume		Efficiency coefficient	Time spent per absorption, h	Expenditure of grouting materials and reagents per absorption			
	wells	insulation operations			cement, t	clay solution, m^3	fillers, t	calcium chloride, t
Serial	74	156	0.474	28.53	24.4	11.6	6.5	0.3
Improved	87	104	0.836	15.42	16.4	10.3	1.0	0.1
Performance indicators			+1.76	-1.85	-1.49	-1.13	-5.9	-3.0

The third factor is continuous monitoring and operational regulation of the insulation process when injecting solidifying solutions into the absorption channels.

The practical implementation of the above-mentioned intra-system effects in the lateral injection scheme allows insulating absorption with intensity from 40 to 80 m^3/h by performing a single operation (in most cases) and reducing material and financial costs for preventing absorption by 1.5–1.85 times [6, 8]. The comparative technological efficiency of traditional and advanced absorption insulation methods is presented in Table 2.

Transient mode of injection (Figures 2, *b*; 6) is characterized by two consecutive processes. At the beginning of the flow of the grouting solution into the bottom-hole zone of the layer, the bottom flow mode is formed at pressures in the roof of the layer close to the values of the layer or less (item 2 in Figure 6). At the second stage, in the process of insulation of some filter channels (cracks) and occurrence of repression in the roof of absorbing layers, bottom flow mode switches to a lateral mode of grouting mixture injection, where the indicator of hydraulic perfection of the mode δ is determined as:

$$\delta = \Delta P_{rf}/\Delta P_{sl},$$

where ΔP_{rf} – pressure drop on the roof of the absorbing layer during injection of the grouting solution, MPa; ΔP_{sl} – pressure drop on the sole of the absorbing layer during injection of the grouting solution, MPa.

A characteristic feature of the technological process in this scheme is represented with the periodic intake of absorbing rocks from the roof in the insulation interval of layer fluids (water, gas). This leads to dilution of grouting solutions and violation of their initial functional properties. The indicator of hydraulic perfection of injection modes for grouting mixtures is 0.4–0.6 (Figure 7). The area of effective application of the transient mode of the viscous-plastic liquid injection is associated with catastrophic absorption intensity of more than 80-100 m^3/h.

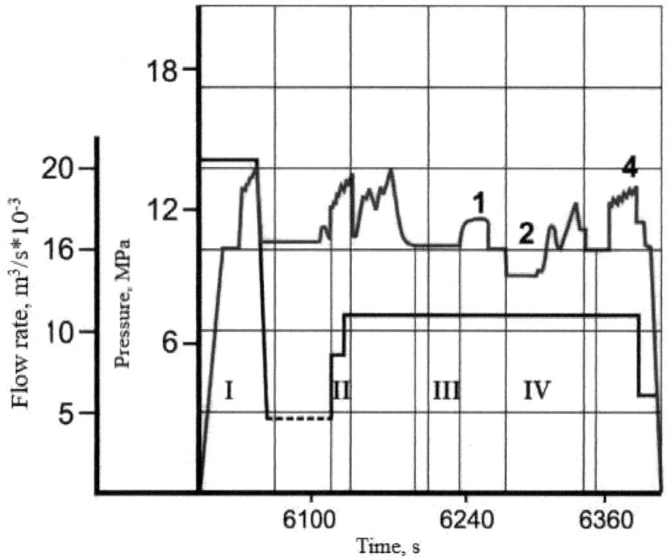

Figure 6. Insulation of absorption with intensity up to 80 m^3/h or more in the mode of lateral injection of grouting mixtures: 1–2 – clay paste; 2–4 – cement solution

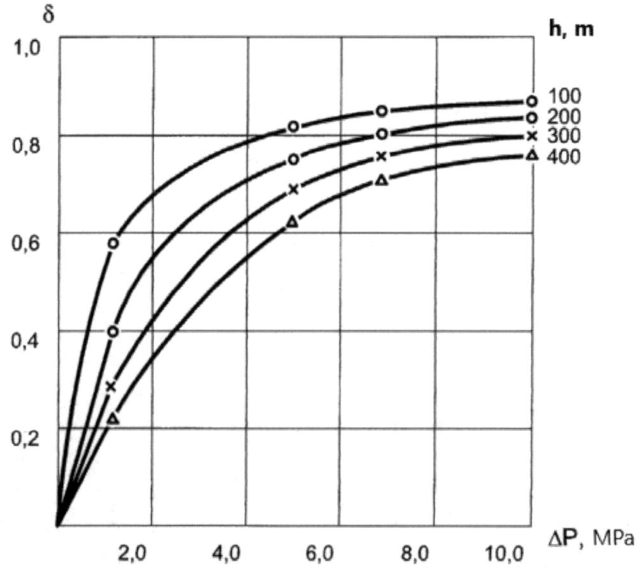

Figure 7. Dependence of the coefficient of hydraulic perfection of the solution injection mode on the thickness of the absorbing rocks (h) and the pressure drop (ΔP): $\delta_1 = 0.1–0.4$ – bottom injection mode; $\delta_2 = 0.4–0.6$ – transient injection mode

The mode of bottom injection of the viscous-plastic liquid into the absorbing layer occurs in cases when the applied technical means (pumps, packers, etc.) and technological solutions (parameters of injection, types and properties of grouting mixtures, clogging fillers, the methods for regulating the setting and hardening of grouting mixtures, etc.) do not lead to reprisals on the roof of absorbing rocks (Figure 8). In such hydraulic conditions of insulation works, the movement of grouting solutions in the absorption intervals is accompanied by a constant influx of layer fluids from the roof part (intra-layer flow) and all the negative consequences that follow from this process. The indicator of hydraulic perfection of the insulation mode does not exceed 0.4.

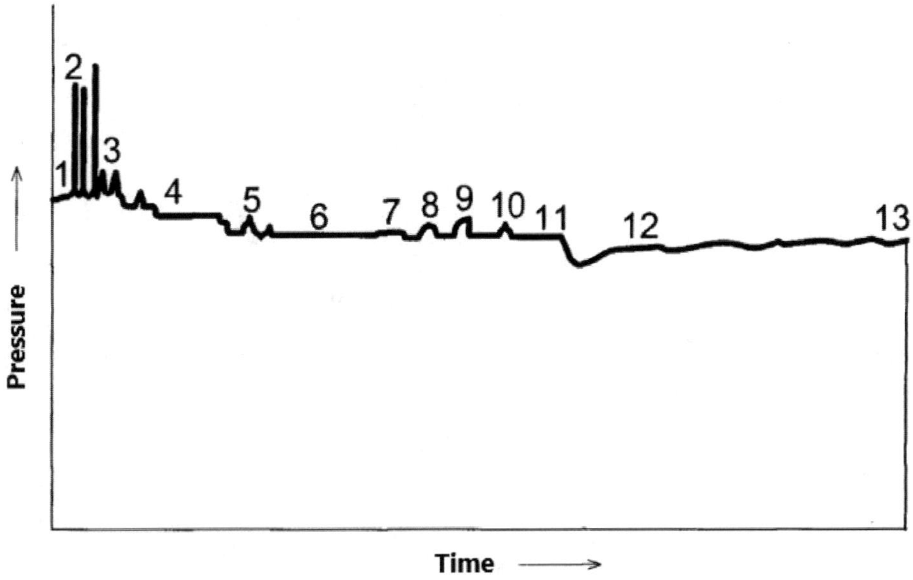

Figure 8. Insulation of absorption with intensity of 100 m^3/h or more in the bottom mode of injection of grouting mixtures: 1–9 – preparation of insulation works; 10 – clay paste; 11–12 – cement solution; 12, 13 – movement of mixtures during the setting period for cement solidification

4. Conclusion

It should be pointed out in the conclusion that the mass application of advanced technology for insulation of absorbing layers under hydraulically perfect injection modes of grouting solutions in the fields of Bashkortostan, Tatarstan, Eastern Siberia, Orenburg region with the use of non-deficit materials (clay powder, grouting cement, polymers and chemicals) has led to a non-linear increase in the quality and efficiency of insulation works (1.5-2.0 times higher than the indicators of traditional methods) and to the 30-50% reduction in the cost and time spent for solving the problem of absorptions with intensity ranging from 40 m^3/h to 150 m^3/h.

References

[1] Yakupov R F, Mukhametshin V Sh, Khakimzyanov I N and Trofimov V E 2019 Optimization of Reserve Production From Water Oil Xones of D3ps Horizon of Shkapovsky Oil Field Using Horizontal Wells *Georesursy* **21(3)** 55–61. DOI: 10.18599/grs.2019.3.55-61

[2] Stenkin A V, Kotenev Yu A, Mukhametshin V Sh and Sultanov Sh Kh 2019 Use of Low-Mineralized Water for Displacing Oil from Clay Productive Field Formations *IOP Conf. Ser.: Mater. Sci. and Eng.* **560(1)** 012202. DOI: 10.1088/1757-899X/560/1/012202

[3] Valeev A S, Kotenev Yu A, Mukhametshin V Sh and Sultanov Sh Kh 2019 Substantiation of the Recovery of Residual Oil from Low-Productive and Heterogeneous Formations in Western Siberia by Improving the Waterflood System Using Gas and Water-Gas Impacts *IOP Conf. Ser.: Mater. Sci. and Eng.* **560(1)** 012204. DOI: 10.1088/1757-899X/560/1/012204

[4] Mukhametshin V V and Kuleshova L S 2019 Justification of Low-Productive Oil Deposits Flooding Systems in the Conditions of Limited Information Amount *SOCAR Proc.* **2** 16–22. DOI: 10.5510/OGP20190200384

IOP Conf. Series: Materials Science and Engineering 1064 (2021) 012054 doi:10.1088/1757-899X/1064/1/012054

[5] Akhmetov R T, Kuleshova L S and Mukhametshin V V 2019 Application of the Brooks-Corey Model in the Conditions of Lower Cretaceous Deposits in Terrigenous Reservoirs of Western Siberia *IOP Conf. Ser.: Mater. Sci. and Eng.* **560(1)** 012004. DOI: 10.1088/1757-899X/560/1/012004

[6] Polyakov V N, Postnikov S A, Chizhov A P and Blinov S A 2013 Technology of absorbing layers isolation in case of hydraulically perfect modes of cement mixtures injection *Construction of Oil and Gas Wells on Land and Sea* **10** 14–18

[7] Polyakov V N, Sayapin E D and Valyamov R G 1977 Technological influence of flow rate and pressure of injection of oil-well mixture on efficiency of operations on isolation of absorbing layers *Drilling* **9** 20–22

[8] Polyakov V N, Vyakhirev V I and Ippolitov V V 2003 *System solutions of technological problems of well construction* (Moscow: Nedra–Biznestsentr) p 240

[9] Polyakov V N, Andreev V E, Chizhov A P, Averyanov A P and Postnikov S A 2017 System improvement and development of the technology applied when drilling and operation of oil and gas wells *Construction of Oil and Gas Wells on Land and Sea* **1** 4–6

[10] Bone K, Bradley B and Kale I 2019 Next-Generation Hybrid Drill Bit Produces Exceptional Drilling Dynamics and Time-Saving Improvements *Offshore Technology Conf.* (Houston, Texas, 6-9 May) 18 p. DOI: 10.4043/29440-MS

[11] Dolezal T, Felderhoff F and Holliday A 2011 Expansion of Field Testing and Application of New Hybrid Drill Bit *SPE Annual Tech. Conf. and Exhibition (Denver, Colorado, 30 October-02 November)* 12 p. DOI: 10.2118/146737-MS

[12] Dupriest F E and Koederitz W L 2005 Maximizing Drill Rates with Real-Time Surveillance of Mechanical Specific Energy *SPE/IADC Drilling Conf.* (Amsterdam, The Netherlands, 23-25 February) 10 p. DOI: 10.2118/92194-MS

[13] Dykstra M W, Armenta M A, Fitzerald A M, Adesokan O, Schornick T L, Ashabikash R Ch and Allain M D 2018 Converting Power to Performance: Gulf of Mexico Examples of an Optimization Workflow for Bit Selection, Drilling System Design and Operation *Offshore Tech. Conf.* (Houston, Texas, 30 April–3 May) 29 p. DOI: 10.4043/29065-MS

Development regulation of oil and gas reservoirs based on effective geological and geophysical information

D A Batalov[1], V E Andreev[2,3], V V Mukhametshin[2] and L S Kuleshova[4]

[1] CDO LLC "LUKOIL–Western Siberia", 20, Baltic St., Kogalym, Tyumen region, Khanty-Mansi Autonomous Area – Yugra, 628486, Russian Federation
[2] Ufa State Petroleum Technological University, 1, Kosmonavtov st., Ufa, Republic of Bashkortostan, 450062, Russian Federation
[3] Institute for Strategic Research of the Republic of Bashkortostan, 129/3, Oktyabrsky Ave., Ufa, Republic of Bashkortostan, 450075, Russian Federation
[4] Ufa State Petroleum Technological University, Branch of the University in the City of Oktyabrsky, 54a, Devonskaya St., Oktyabrsky, Republic of Bashkortostan, 452607, Russian Federation

E-mail: intnm@yandex.ru, vv@of.ugntu.ru

Abstract. The paper studies the dynamics of fracture reservoir management during the solution of the problems of the increase in the efficiency of development of oil deposits with hard-to-recover reserves. The directions of the creation of models of multichannel telemetry systems are studied from the point of view of the proposed concepts, as well as the conditions under which the study of well conditions is carried out. The results show that these telemetry systems can function in a well when the sensor elements are integrated, which are based on a single-core armored cable.

1. Introduction

At present, the increase in oil production in the classical definition of continuous oil displacement [1] is carried out by the improvement of secondary and tertiary methods of the influence on productive formations [2–6]. In this case, it is important to improve the flowmetric control methods and equipment for geophysical well logging (GWL) [7]. The increase in the accuracy of the obtained geological and geophysical material is provided due to the reliability of the metrological analysis of the used well logging equipment [8].

The use of digital telemetry systems in gas [9] and oil [10] industries greatly facilitates the implementation of processes in the surgace operating conditions. The example of high efficiency is digital telemetry systems in the implementation of the processes of seismic exploration, drilling, preparation and transportation of oil. However, digital telemetry systems have significant drawbacks for in-depth studies of production wells during continuous development of oil-and-gas pool (OGP). First of all, these drawbacks are the large dimensions of well logging equipement in terms of diameters, low thermo- and radioactive resistance, etc. Such telesystems partially satisfy the requirements of classical technologies of oil recovery [1], characterized by low values of the oil recovery factor ($ORF_{max} \rightarrow 0.43$ only for terrigenous layers).

The solution of the problem of further increase of oil recovery factor leads to the need for extreme regulation of production facilities in reservoir sections when creating a metatechnology for the development of hard-to-recover hydrocarbon reserves (HRHR) [8, 9]. In this case, the initial requirements for the used multichannel telemetry systems (MTS) become more stringent. Therefore, the main purpose of these studies is the creation of well logging equipment with increased telemetry capabilities of geological and geophysical information.

2. Materials and methods

In order to determine the initial parameters of MTS, we consider the necessary conditions for the implementation of MTS. These conditions are defined in the specifics of HRHR metatechnology [8, 9]. The essence of the implementation of new oil recovery technology lies in the procedures for the production of residual oil throughout the entire 3D volume of the production facility.

In the prehistory of the development of the production block of an object, many technological lines of functions of the reservoir parameters *R* (porosity, permeability, oil saturation, etc.) and the properties of fluids *N* (viscosity, density, gas saturation and others) are formed according to the data of core studies in the processes of drilling and well arrangement F_{arran}^{well} (*R*, *N*) in conjunction with geological and technological research. At the same stage, according to directional survey data, the length of the trajectory l_{traj} between the deep zones of well perforations is calculated.

After the primary hydraulic fracturing of the formation (for example, $P_{hydr} \simeq 100$ MPa), an interwell flow tube is formed, that is, a formation channel with a primary trajectory KVP_1. In the experimental-industrial pilot production of wells, their integral and differential profiles of injectivity and inflows are determined [11, 12] on the basis of standard geophysical (gamma and neutron), as well as hydrodynamic (flow metering, barometric, moisture metering and other) methods. According to the Dupuis dependence, the minimum $R_{ex.min}$, maximum $R_{ex.max}$ and the nominal value of the WAS_1 pressure during operation $R_{ex\text{-}n}$ are determined.

First, the Dupuis-Forchheimer dependence is used for the initial fluid filtration rates, and taking into account the recorded transit time of tracer marks, the length of KVP_1 trajectory can be adjusted with high accuracy in $l_{ran} > l_{traj}$. This is physically implemented in discontinuous flow control of the working agent (water) and portions of tracing marks.

Thus, the combined cycles of preliminary research and operation $\tau^{pelim}_{in.cycle\ i}$ are carried out under the nominal pressure Reks.n in KVP.

In meta-technology, it is possible to combine the operating modes of KVP with the hydraulic pulse modes of preliminary, refined and redefined studies. The revealed spatio-temporal coordinates of unproductive areas Δl_{ua} are subjected to the isolation mode (plugging) in the evolutionary development of KVP.

Further, it is planned to perform hydraulic pulse modes of preliminary studies of the coordinates of the sections Δl_{ua} in the ith cycles $\tau^{pelim}_{in.cycle\ i}$ during the delivery of marks $\tau^{pelim}_{in.cycle\ i} = \tau^{pelim}_{in.cycle\ i\text{-}1}$ through a quantum of time τ_{qu} of filtration process under the pressure of operation $R_{ex.max}$ at equidistant points along the entire length of KVP_1 through the quantum of length $\Delta l_{qu} << \Delta l_{ua}$.

After the completion of the pulse of the duration of the delivery of marks in each cycle, the pauses are fixed in the implementation of the nominal operation of KVP under the $R_{eks.n.}$ pressure. The results of such studies predetermine the approximate number of plugging points $\xi = \overline{1, o}$ as a function of the static characteristics of their parameters depending on the errors in the control of hydrodynamic quantities with well logging devices [7], provide the identification of the coordinates of plugging points along the KVP_1.

Similarly to preliminary studies at the late stage of oil recovery, the modes of refined and redefined studies are carried out in relations of the delivery pressures of the marks $R_{ex.max} < P^{def}_{init} < P^{redef}_{init} < P_{\tau min} < P_{hydr}$, qhere P_{hydr} - hydraulic fracturing pressure; $P_{\tau min}$ - minimum value of plugging pressure in KVP_1. In this case, the number of plugging points $\xi = \overline{1, d}$ is specified as a function of dynamic characteristics, taking into account the parameters of transient processes in the KVP_i, where i = is the number of possible trajectories in the cube of the operational object.

As the first plugging point Δl_{ua}, the d-point in KVP_1 is selected which is close to the production well zone. Delivery of plugging material into it is carried out during the time τ^1_{dpmd} under plugging pressure $P^1_{tmin}<P^{predef1}_{init.\,d}<P^1_{t\,d}= P^1_{t\,max} \simeq P_{hydr}$. This leads to the first compounding (enveloping) connection in KVP_1 and the origin of the KVP_2 trajectory. Similarly, the modes of additional determination of studies and plugging of the remaining d-j points in KVP_1 are carried out, which leads to the formation of the 2nd trajectory of KVP_2, etc. until the trajectory ψ is developed throughout the entire volume of the operational object.

3. Development of a conceptual scheme of MTS at the macrolevel of of oil and gas pool

For the practical realizability of MTS under the influence of a complex of non-informative influencing factors, it is necessary to introduce at least one KP_0 with an exemplary measure. In accordance with the developed algorithm for the two-channel transformation of the resistances of the sensors in the telemetry system, a schematic diagram of the conceptual model of an analog MTS with a small-sized and heat-resistant transmitting part was developed (Figure 1).

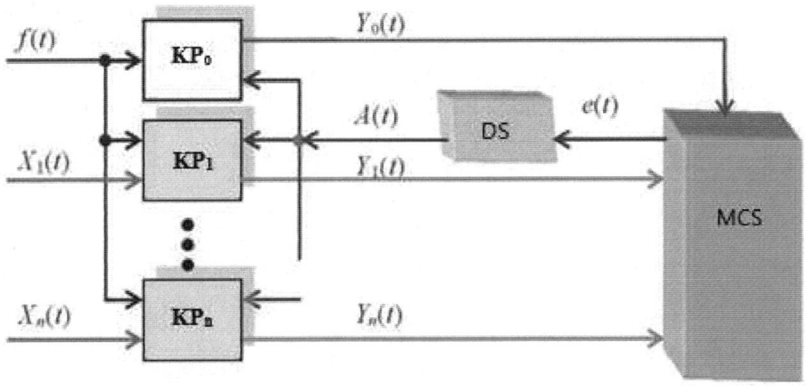

Figure 1. Diagram of a conceptual model of an analog MTS: DS - disturbance source; MCS - measurement and calculation system.

In this diagram, the organization of the multichannel structure of the telemetry system is carried out in the time division mode with the inclusion of additional passive elements in the measuring circuit, when the resistances of the sensors are subjected to a pulse effect *A (t)* in the form of current or voltage pulses. The induced action through the communication line (CL) in the form of a functional KP is fed to the resistance converter of the sensor in KP_i or an exemplary resistor in KP_0. The converted values $Y_0(t)$ and $Y_i(t)$ are the basis for the development of the invariant transformation scale during the elimination of interference signals and influencing factors in the form of variable values of the resistances of a two-wire CL.

The implementation of the invariant algorithm for the transformation of the resistances of the sensors becomes possible when deriving a system of independent transformation equations. The analysis of this conclusion can be performed using the transformation equation for an exemplary resistor, following the example of a two-channel MWD system

$$U_0 = I(R_0 + R_{si} + R_{core} + R_{armour}) + E_p, \quad (1)$$

where U_0 – the amplitude of the voltages of unipolar strobe pulses at the MTS ground station, obtained as a result of the conversion of the standard resistor; $R_{si} \approx R_{s0}$ – resistive resistances of semiconductor junctions, equally matched elements of RRS switch and R_0; R_{core} and R_{armour} – resistance of the core and armor of a cable, respectively; E_p – polarization potentials applied to the cable armor.

In another time interval, when connected to the measuring circuit of RRS with variable R_{xi} values, the following relation is fulfilled

$$U_{xi} = I(R_{xi} + R_{si} + R_{core} + R_{armour}) + E_{п}, \quad (2)$$

where U_{xi} – the amplitude of the voltages of unipolar strobing pulses at MTS ground station, obtained as a result of the RRS conversion.

Subtracting equation (1) from equation (2), we obtain the equation of the transformation scale for any RRS in the form

$$\Delta U_{xi} = I(R_{xi} - R_0) = I\Delta R_{xi}, \quad (3)$$

The resulting equation (3) shows the invariant ratio of the converted values of RRS with respect to the complex of interference induced in a two-wire CL, as well as the changes in the resistance of CL under the influence of the station lift collector brush and temperatures. Thus, the above mentioned diagram is the basis for the creation of a complex device in the transmission part of MTS.

4. Development of a small-sized MTS diagram with RRS

The analysis of the compound equations in the developed model indicates the possibility of simultaneous measurements of such physical quantities as temperature, pressure, velocity and composition of impurities of multiphase flows in the well intervals of perforations. Figure 2 shows the MTS circuit [12] with direct connection of RRS group and the exemplary resistance R_0. For the given diagram, the following designations are used: PSU - power supply unit; SG - signal generator; AS - amplitude selector; C - counter; DE - delay element; PG - pulse generator; ISSU - initial state setting unit; Distrib - impulse distributor; CU - computing unit; Reg - registrar. A group of reed (magnetically controlled) contacts of relays installed opposite a rotating shaft with a permanent magnet is used as a switch of the transmitting part.

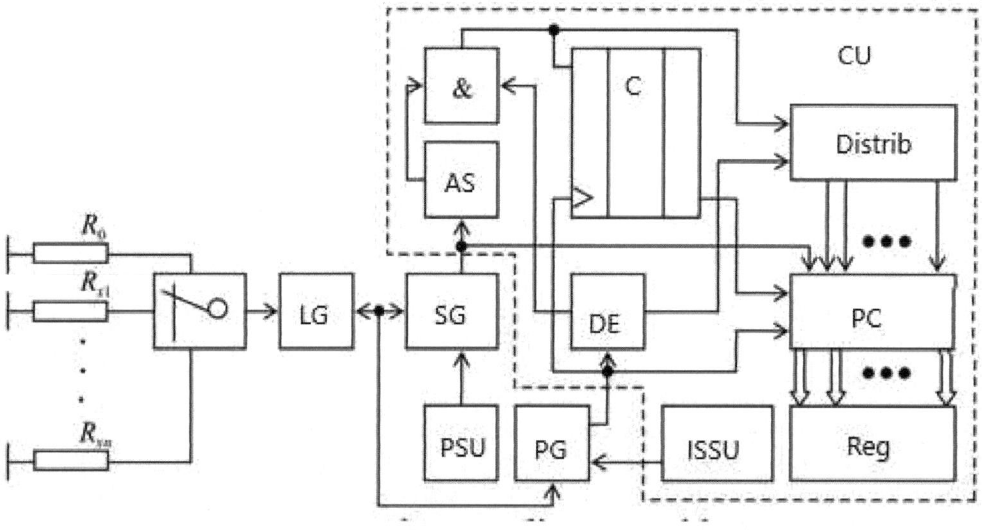

Figure 2. Block MTS diagram with RRS

In addition, a clock mechanism or a pulse distributor can also be used as a switch. Figure 3 shows timing diagrams explaining the principle of operation of MTIS with RRS.

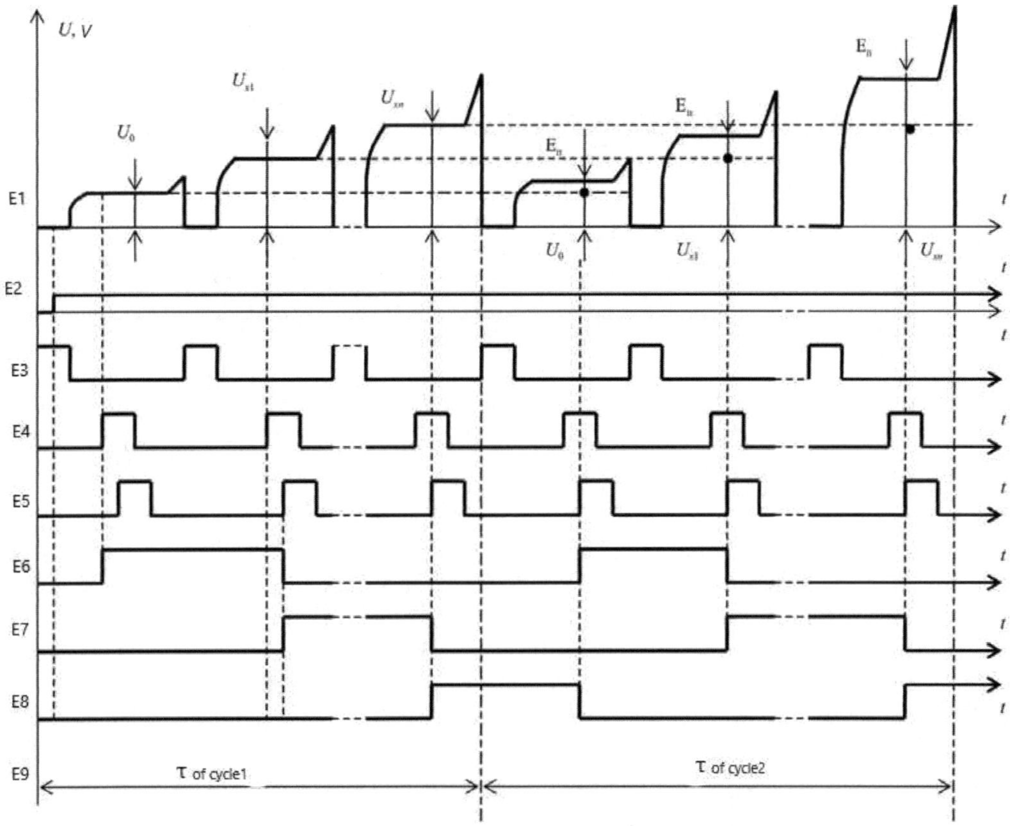

Figure 3. Timing diagrams of MTS with RPSD

The principle of the system is as follows. When the power supply unit of the transmitting part is turned on, as well as the power supply unit of the equipment, a short pulse is generated at the output of ISSU (diagram E2 in Figure 3), which is then allocated by PG to perform the initial reset. In the case when the unit connects an exemplary resistor R_0 to SG to LG, the value of the exemplary voltage U_0 appears at its input, which is formed by AS into a single enabling signal for passing through the I element of the delayed pulse from the first output of DE. Therefore, the device uses the initial gating at the second input of the element I with the initial pulse from ISSU unit (diagram E3 in Figure 3) through PG and DE at each initial switching on of the device equipment.

The single pulse generated by the element I in each measurement cycle sets the counter and the distributor of impulses to the initial zero state, when a single signal is generated from the first bit output of distributor (2 °) during the entire first sub-cycle of measurement (diagram E6 in Figure 3).

If one of the resistors Rxi of RRS is connected to SG in the transmitting part through LG, the I element does not generate enabling pulse and the beginning of the next measurement cycle will not begin until a standard resistor is connected to the switching unit. At the moment of switching the first RRS by the switching unit through LG to FG (diagram E1 in Figure 3), the last one selects a pulse (diagram E3 in Figure 3), which first transfers Counter pulses to a new position, enters the CU input, and then (after a time τ_{31}, diagram E4 in Figure 3) enters the first input of element I.

After some time (τ_{32}) from the second output E3, the same impulse transfers the distributor to a new state, when its single signal from the first bit output moves to its second bit output (E5 and E7 in Figure 3). In this case, a voltage is generated from SG output that is different from the standard one $U_{xi} \neq U_{stand}$). Therefore, the prohibiting signal will be constantly removed from AS output by a zero level. At the moment of pulses delay (diagrams E4 and E5 in Figure 3), the processing and transformation of information is carried out in CU in accordance with the specified algorithm of the invariant transformation. The corrected signals are then recorded in the recorder.

After some time, from the second output of E3, the pulse in the third sub-cycle of measurement transfers the distributor to a new state, when a single signal appears at its third bit output (diagram E8 in Figure 3). Similarly the reception and transmission of measuring signals are carried out in the last sub-cycle, when a single signal (diagram E9 in Figure 3) is generated from the fourth bit output of the distributor to the input of CU for processing and converting information signals in the last KP, formed by the last sensitive element of sensory elements in the transmitting part of MTS. In this case, rewriting of information signals of each MTS channel from CU output memory block to the channel inputs Reg is possible for the periods of each measurement sub-cycle.

The second measurement cycle is carried out similarly to the first, that is, with the connection of a standard resistor, etc. The organization of the corresponding number of channels of KP in MTS is ensured by double synchronization of the operation of the controlled switch in the transmitting part. Such synchronization is performed by means of the address connection of a standard resistor and RRS group. On the other hand, such synchronization is also provided due to address reproduction in the transmitting part of the values of the standard resistor and RRS in the form of digital sweeps in CU at different measurement sub-cycles.

The MTS measurement sub-cycles are clocked by sync pulses generated in the transmitting part due to sequential commutation of the standard resistor R_0 and RRS. The sequence of generated service sync pulses is used in the process of analog-to-digital conversion and the execution of invariant algorithms of computational procedures.

As sensory elements in the described structure of well logging device of MTS, tungsten thermoresistances and strain gauges on a substrate of the "Sapphire" type are widely used. For additional types of measurements, the integration of RRS group with the groups of inductive reactance sensors (IRS) is provided.

5. Methods for combining various sensory elements in MTS

In analogy with the circuit implementation of MTS with RRS according to Figure 2, its asynchronous structure was developed with the possibility of the integration of various types of IRS [13], for example, with an additional implementation of an induction resistivity meter (Figure 4).

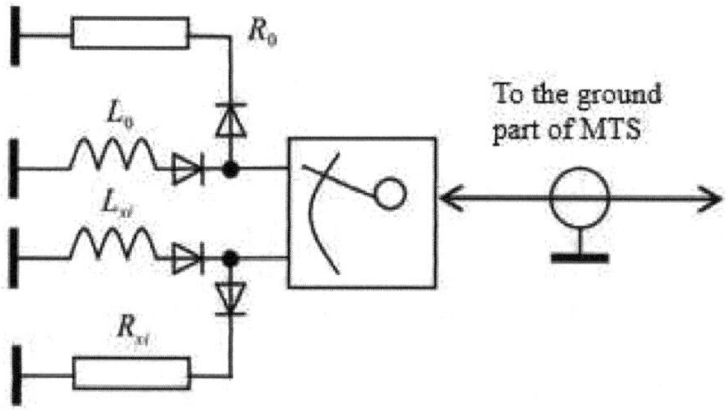

Figure 4. Block diagram of the deep part of MTS with IRS and RRS

Similarly, the organization of MTS is ensured by two-way synchronization of the deep and ground parts in the transformation mode at negative levels of IRS transformations and at positive levels of RRS transformations.

In the above mentioned diagram, the synchronization of the operation of the ground and deep parts of the SP with IRS is carried out also due to the address connection of the standard inductance L_0 and resistance L_{xi} under the influence of frequency gating in different ranges of the IRS operation (Ad L_0 || ... | | Ad L_{xi}). In the ground part, analog-to-digital conversion and separation of SP are carried out according to the results of the analysis of digital sweeps (Ad Z_0 || ... || Ad Z_{xi}). In this case, the results of the primary transformations of the exemplary inductance and IRS in the deep part are then corrected in MTS taking into account the amplitude-frequency characteristic (AFC).

In analogy with expressions (1) - (3), the transformation scale equation is obtained in frequency form as

$$\Delta X_{Li} = \Delta U(F)/I \quad (4)$$

where $\Delta X_{Li} = 2\pi F_i$ – inductive reactance of the sensor at different frequencies of the gated signals in the frequency-pulse form.

6. Discussion

Various combinations of models of linear and non-linear parts of the expression of Dupuis-Forchheimer, Boussinesque, Poiseuille and Poisson [8, 9] were used as the basis for the formalization of fluid filtration processes in discontinuous-coordinate reservoir control, both before the water cut front of oil displacement and after it. Therefore, for GIS-control of reservoir parameters in remote zones, the most appropriate are the complexes of flowmetric methods.

The considered structures of telemetry equipment create the possibility of its most complete implementation in a multichannel mode of operation, which is especially important for the real-time mode of metatechnology. In this case, the basis of the applied sensory elements are temperature-sensitive resistances and strain resistances of sensors of the "Sapphire" type, as well as a turbine flow sensor. The use of IRS creates the prerequisites for the most accurate control of tracing marks in various modes of reservoir exploration. For the regime of reservoir operation using new technology, traditional physicochemical, hydrodynamic and thermal methods of impact on productive formations are used similarly to classical ones [2–6].

7. Conclusion

The use of only sensory elements in a complex SP leads to a number of significant advantages in the implementation of MTS: 1) the small size of SP in its diameter at the level of 10–22 mm; 2) thermal stability (up to 300°C and above); 3) radioactive stability (typical for Mangyshlak deposits) during the use of only one passive transformation chains; 4) pressure resistance (up to 80 MPa and higher) due to the implementation of small dimensions, which also leads to the conditions of durability of well logging equipment (from 12-14 hours per day to several months). A distinctive feature of the created MTS is the joint integration of not only RRSs, but also various IRS groups.

References

[1] Barenblatt G I, Entov V M and Ruzhik V M 1990 *Theory of fluid flows Through natural Rocks* (Dordrecht, Boston, London: Kluwer academic publishers) 395 p

[2] Yakupov R F, Mukhametshin V Sh, Khakimzyanov I N and Trofimov V E 2019 Optimization of reserve production from water oil zones of D3ps horizon of Shkapovsky oil field by means of horizontal wells *Georesursy* **21(3)** 55-61. DOI: 10.18599/grs.2019.3.55-61

[3] Stenkin A V, Kotenev Yu A, Mukhametshin V Sh and Sultanov Sh Kh 2019 Use of low-mineralized water for displacing oil from clay productive field formations *IOP Conf. Ser.: Mater. Sci. Eng.* **560(1)** 012202. DOI: 10.1088/1757-899X/560/1/012202

[4] Valeev A S, Kotenev Yu A, Mukhametshin V Sh and Sultanov Sh Kh 2019 Substantiation of the recovery of residual oil from low-productive and heterogeneous formations in Western Siberia by improving the waterflood system using gas and water-gas impacts *IOP Conf. Ser.: Mater. Sci. Eng.* **560(1)** 012204. DOI: 10.1088/1757-899X/560/1/012204

[5] Mukhametshin V V and Kuleshova L S 2019 Justification of Low-Productive Oil Deposits Flooding Systems in the Conditions of Limited Information Amount *SOCAR Proceedings* **2** 16–22. DOI: 10.5510/OGP20190200384

[6] Akhmetov R T, Kuleshova L S and Mukhametshin V V 2019 Application of the Brooks-Corey model in the conditions of lower cretaceous deposits in terrigenous reservoirs of Western

Siberia *IOP Conf. Ser.: Mater. Sci. Eng.* **560(1)** 012004. DOI: 10.1088/1757-899X/560/1/012004

[7] Lobankov V M 2009 *Fundamentals of metrology of geophysical measurements. Graduation, calibration and verification of borehole geophysical equipment* (Ufa: Novyy stil') 152 p

[8] Stolyarov V E, Eremin N A, Eremin Al N and Basnieva I K 2018 Digital gas wells: state and prospects *Oilfield Engineering* **7** 48-55. DOI: 10.30713/0207-2351-2018-7-48-55

[9] Nazarov A V, Kozyrev G I, Shitov I V et al 2007 *Modern telemetry in theory and practice* (St Petersburg: Nauka i tekhnika) 672 p

[10] Batalov S A 2016 Modeling of the Initial Parameters in the Adjustment of an Oil Recovery Process Control System. Part 2. Determination of Limits on the Vectors of the State Variables and Disturbances *Chemical and Petroleum Engineering* **52(7)** 452-459. DOI: 10.1007/s10556-016-0213-6

[11] Batalov S A, Andreev V E, Dubinskiy G S, Khuzin R R and Khuzin N I, RF Patent No. 2628343 CPC E 21 B 43/20 *Method for producing formations with hard to recover hydrocarbon reserves* (16 August 2017)

[12] Batalov S A and Kolovertnov Yu D 1989 A. s. 1487088 SSSR CPC G 08 C 15/08 Multichannel telemetry device *Discoveries. Inventions* **22**

[13] Batalov S A and Kolovertnov Yu D 1991 A. s. 1622841 SSSR CPC G 01 R 27/02 Device for separate measurement of parameters of complex quantities *Discoveries. Inventions* **3**

Scientific and methodological basic principles for determining design of clay acid treatments applied to wells

D I Akhmetshina[1], V V Mukhametshin[2], L S Kuleshova[1] and D A Batalov[3]

[1]Ufa State Petroleum Technological University, Branch of the University in the City of Oktyabrsky, 54a, Devonskaya St., Oktyabrsky, Republic of Bashkortostan, 452607, Russian Federation
[2] Ufa State Petroleum Technological University, 1, Kosmonavtov st., Ufa, Republic of Bashkortostan, 450062, Russian Federation
[3] CDO LLC "LUKOIL–Western Siberia", 20, Baltic St., Kogalym, Tyumen region, Khanty-Mansi Autonomous Area – Yugra, 628486, Russian Federation

E-mail: of.ugntu204@mail.ru, vv@of.ugntu.ru

Abstract. A new methodological approach is presented in determining design of current clay acid treatments of wells. The proposed scientific and methodological approach allows differentiating for different groups of objects in the Langepass region to determine certain technological parameters of impact on the bottom hole zone of formations, taking into account the features of geological and physical properties of formations and fluids saturating them. In special cases, it is suggested to use acid retarders for reaction with rock. The authors proposed algorithm for determining wells where the maximum effect can be obtained at a specific time.

1. Introduction

Acid treatment technologies have a long history of application as well stimulation technologies. The development of new approaches to well stimulation and development of new technologies and compositions can transfer them into the category of geological and technical measures (GTM) that will solve problems of developing hard-to-recover oil and gas reserves (TRIZ) associated with complex pore-fractured reservoirs. Salt-acid treatments (SAT) and their modifications (various compositions and technological features) are widely used for production fields [1]. The number of SAT is so large that total additional oil production from the deposit due to SAT and their modifications can be compared with methods of enhanced oil recovery (MUN). Clay acid treatments are also widely used and, if applied appropriately, can take a more important position among well stimulation methods.

Recently, there has been a trend to increase the share of TRIZ in the overall world balance, including in Russia. Significant volumes of TRIZ are confined to low-permeable mudd-off terrigenous reservoirs, which have their own features of geological structure. During well operation, the reservoir's filtration and capacity properties (FEC) deteriorate in the bottom hole formation zone (BZF) in comparison with the remote zone. Because of deterioration of FES in BZF in primary and secondary opening of the reservoir, and after that, during well operation and due to objective physical phenomena [2, 3], here the main consumptions of produced energy takes place, it is used for the movement of fluids in the formation.

Proper well operation implies keeping FEC of BZF close to natural ones; so works are also performed to improve filtration and reservoir properties of the bottom hole zone of the formation. Thus, the need to restore and improve FEC arises from the task of maintaining and intensifying hydrocarbon production.

2. Materials and methods

A large number of factors that reduce permeability require an individual approach to the choice of technology and active agent to influence BZF. In order to correctly plan the impact on productive formations and predict its (impact) effectiveness, it is necessary to understand the processes occurring in BZF, and determine possible responses to the impact. For this purpose, there is a methodology for predicting and designing impact, the mandatory element of which is modeling [4, 5].

To justify the choice of oil production intensification method, the authors analysed geological and physical characteristics and current state of operational facilities development of the Langepass group of fields (West Siberian NGP). All objects of the named group are characterized by high compartmentalization of productive layers with high clay content (up to 14 %) and low carbonate content (less than 1%). There are differences in facial conditions and, as a result, differences in mineralogical composition and reservoir properties.

The effectiveness of clay acid treatments (CAT) for terrigenous formations is conventional. In accordance with methodological approach of metotechnology of physical and chemical impact on reservoirs [6], we will conduct impact modeling to predict the efficiency of CAT for conditions of the Langepass group of fields.

To build a mathematical model of distribution of clay acid solution in a porous medium, it is necessary to take into account, first of all, mineral composition of rocks. Presence of minerals and compounds of elements that can form an insoluble sediment during the reaction with clay acid is of great importance. Sandy rock treated with clay acid must have a very low content of carbons and ferruginous compounds. While the high content of clays is not an obstacle to CAT success. Therefore, mineralogical composition of terrigenous oil formation UV1/1 of the Las-Eganskoye field has been studied. The average composition of rocks of this formation is shown in table 1.

Table 1. Mineralogical composition of rock formation UV 1/1, Las-Eganskoye field

Minerals	Percentage, %
Quartz	43
Feldspar	34
Micas	6
Clays	11
Carbons	6

3. Results and discussion

Table 1 shows that the formation under consideration contains a large amount (up to 34 %) of feldspar, clay (up to 11 %) and carbons (6 %). In order to increase the flow rates of wells that open UV1/1 formation, it is possible to use treatment of wells with clay acid solution (or mud acid solution – (HF + HCl)).

The theory of calculating the main parameters of clay acid treatment (depth of acid solution penetration and increase in productivity of the well after treatment) is given in the paper [7]. The proposed theory considers one-dimensional axisymmetric filtration of acid solution in a porous medium involving chemical reactions. It reviews acid solution injection into a layered formation with hydrodynamically isolated interlayers. The values of desired parameters of treatment are found from the condition of mass balance of reacting components on frontal advance chemical reactions, provided that the process is in equilibrium. The model takes into account: mineral composition of rocks forming the formation, composition of acid solution, stoichiometric coefficients of main reactions of rock dissolution by acids, structure of the formation (permeability, porosity, thickness of interlayers composing the formation). The obtained value of depth of acid solution penetration has been used to assess relative productivity of the well after treatment.

The authors considered the process of equilibrium dissolution of a porous medium containing soluble minerals with acid solution (a mixture of hydrochloric and hydrofluoric acids). The model ignores the diffusion transfer of components in comparison with convective, gravitational pressure drop at the bottom of the well in comparison with hydrodynamic one.

The system of equations describing the filtration process of a clay acid solution and reaction products obtained as a result of dissolution consists of continuity equations for each component of solution and the equation of motion:

$$\frac{\partial m\cdot\rho_i^0\cdot c_{HF}}{\partial t}+\frac{1}{r}\frac{\partial}{\partial r}r\cdot m\cdot\rho_i^0\cdot c_{HF}\cdot\upsilon=-K_f\cdot J_f-K_{cl1}\cdot J_{cl1};$$

$$\frac{\partial m\cdot\rho_i^0\cdot c_{HCL}}{\partial t}+\frac{1}{r}\frac{\partial}{\partial r}r\cdot m\cdot\rho_i^0\cdot c_{HCL}\cdot\upsilon=-K_c\cdot J_c-K_{cl2}\cdot J_{cl2};$$

$$\frac{\partial m\cdot\rho_i^0\cdot c_1}{\partial t}+\frac{1}{r}\frac{\partial}{\partial r}r\cdot m\cdot\rho_i^0\cdot c_1\cdot\upsilon=-K_{1f}\cdot J_f-K_{11cl}\cdot J_{cl1};$$

$$\frac{\partial m\cdot\rho_i^0\cdot c_2}{\partial t}+\frac{1}{r}\frac{\partial}{\partial r}r\cdot m\cdot\rho_i^0\cdot c_2\cdot\upsilon=-K_{2c}\cdot J_c-\rho\cdot K_{22cl}\cdot J_{cl2}; \quad (1)$$

$$c_w=1-(c_1+c_2+c_{HCL}+c_{HF});$$

$$J_f=-\rho_f^0\frac{\partial\alpha_f}{\partial t};\quad J_{cl1}=-\rho_{cl}^0\frac{\partial\alpha_{cl}}{\partial t};\quad J_c=-\rho_c^0\frac{\partial\alpha_c}{\partial t};$$

$$J_{cl2}=-\rho_{cl}^0\frac{\partial\alpha_{sol.cl}}{\partial t};\quad r_w\cdot\upsilon\cdot m=\frac{Q}{2\pi\cdot h}$$

where m is the porosity of the rock; $\rho_{i,R,f,cl,c}^0$ – density of water, rock, rock minerals, respectively; $c_{HF,HCL,1,2,w}$ – concentration of hydrofluoric, hydrochloric acids, salts obtained as a result of dissolution reaction, water in solution; – flow rate of the liquid; $K_{f,cl1,c.cl2,1f,11cl,2c,22cl}$ – stoichiometric coefficients of the dissolution reaction; $J_{f,cl1,c,cl2}$ – dissolution reaction rate of feldspars, clays with hydrofluoric acid and carbonates and parts of clays with hydrochloric acid; $\alpha_{f,cl,c,sol.cl}$ – volume content of feldspars, clays, carbonates and parts of clays soluble in hydrochloric acid ($\alpha_I = c_i\cdot\rho_R/\rho_i$); r_w – radius of well; Q – flow rate; h - thickness of formation.

It is assumed that the composition of feldspar includes K-feldspar and Na-feldspar. The clays imply kaolinite, smectite, illite, montmorillonite, bentonite. Carbons are represented by calcite and magnesite-dolomite. Stoichiometric coefficients were taken as average for each group of minerals.

Initial and boundary conditions for different regions of flow as follows: in the initial moment of time in the formation there is water, concentration of minerals in the rock and porosity is equal to original values; zero lower bound on concentration of acids in solution is equal to original values, the formation contains quartz and carbons, porosity is equal to $m_1 = 1 - \mu_R - \mu_K$, for the first boundary in the solution there are salts obtained in the reaction, and the changed amount of hydrochloric acid, in the formation there is quartz, feldspars and part of clay, $m_2 =$

$1 - \alpha_R - \alpha_f - (\alpha_{cl} - \alpha_{sol.cl})$, in the second boundary in the solution there are salts obtained by the reactions of dissolution, in the formation – all of the original minerals, the porosity is equal to the original value.

A hyperbolic system of equations admits availability of discontinuous solutions. The values of unknown parameters at the break obtained from mass balance conditions on the frontal advance of chemical reactions provided that the process is in equilibrium taking into account the initial and boundary conditions are expressed as:

$$r_1^2 = \frac{V_n \cdot J(\pi \cdot h)\rho_i^n \cdot c_{HF}}{m_i \cdot \rho_i^n \cdot c_{HF} + K_f \cdot \rho_f^n \cdot \alpha_f + K_{cl1} \cdot \rho_{cl}^n(\alpha_{cl} - \alpha_{sol.cl})};$$

$$\begin{aligned} V_1 = {} & (V_0 / (\pi \cdot h)\rho_i^0(1 - c_{HF}) - r_1^2(m_1 \cdot \rho_i^0(1 - c_{HF}) - K_c \cdot \rho_c^0 \cdot \alpha_c + K_{wc} \cdot \rho_c^0 \cdot \alpha_c - m_2 \cdot \rho_i^0 - \\ & - K_{wf} \cdot \rho_f^0 \cdot \alpha_f - K_{wcl1} \cdot \rho_{cl}^0(\alpha_{cl} - \alpha_{sol.cl}) - K_{1f} \cdot \rho_f^0 \cdot \alpha_f - K_{11cl} \cdot \rho_{cl}^0(\alpha_{cl} - \alpha_{sol.cl}))\frac{\pi \cdot h}{\rho_i^n}; \end{aligned} \quad (2)$$

$$c_{HCL} = \frac{r_1^2(m_1 \cdot \rho_1^0 \cdot c_{HCL} - K_c \cdot \rho_c^0 \cdot \alpha_c) - V_0 / (\pi \cdot h)\rho_1^0 \cdot c_{HCL}}{r_1^2 \cdot m_2 \cdot \rho_1^0 - V_1 / (\pi \cdot h)\rho_1^0};$$

$$r_2^2 = \frac{V_1 / (\pi \cdot h)\rho_1^0 \cdot c_{HCL}}{m_2 \cdot \rho_1^0 \cdot c_{HCL} + K_c \cdot \rho_c^o \cdot \alpha_c + K_{cl2} \cdot \rho_2^0 \cdot \alpha_{sol.cl}},$$

where $r_{1.2}$ is depth of penetration of acid solution (hydrofluoric and hydrochloric, respectively); $V_{0.1}$ – volume of injection initial and after the reaction; C_{HCL} – decrease in the concentration of hydrochloric acid in solution due to water released as a result of the reaction of dissolution of rock minerals with hydrofluoric acid.

Change in the permeability of treated formation zones was calculated according to The Kozeny–Carman law:

$$k(r)/k_0(r) = (m/m_0)^n, \quad (3)$$

where $k(r)$, $k_0(r)$ is formation permeability after and before acid treatment; n is degree indicator (n = 10).

Relative productivity of interlayers in the layered formation was calculated according to the formula:

$$Qi / \Delta P = \frac{2\pi \cdot h_i}{\mu_i} \frac{1}{\int_{r_w}^{R} \frac{dr}{r \cdot k_i(r)}} \quad (4)$$

where $Q/\Delta P$ is productivity of interlayer; k is permeability of interlayer after clay acid treatment; i- corresponds to the number of interlayers in the layered formation.

To assess the impact efficiency, a model well was considered with a geological and statistical section based on the data from table 2, shown in table 3, and rock composition, shown in table 1.

To calculate the efficiency of acid treatment of selected wells, they used the following composition of main reagent of working solution: 1% HF + 7% HCl. The concentration of acids in the solution was selected based on the requirements for minimizing corrosion of equipment and structure of wells being treated. The volume of acid solution injection was selected depending on productive capacity of the formation: at the rate of 20 m^3 of solution per meter of capacity [8, 9].

The results of forecasting the treatment efficiency for a model well are summarized as follows. The total volume of recommended injection of clay acid solution is 225 m^3 and 144.5 m^3 for facilities. Depending on conductivity of interlayers, this volume was distributed over the cross-section of formations according to the number of interlayers as follows: 205.14; 19.9 and 80.7; 45.5 and 18.26

m^3. The estimated depth of dissipation of contamination through the interlayers for wells under consideration, taking into account the drop in their injection capacity during operation, was: 0.66; 0.339; 0.66; 0.387 and 0.261 m. Radii of acid solution penetration through the interlayers in the analyzed wells, calculated according to the proposed method, are as follows: 0.41; 0.231; 0.582; 0.355 and 0.23 m. Changes in fluid flow after acid treatment of wells were calculated for each interlayer using the generalized Dupuis formula, taking into account zonal heterogeneity of the bottom hole zone of wells (zones of acid leaching and contamination of rock). The distribution of injection capacity of model wells across the interlayers before and after acid treatment is shown in figure 1.

Table 2. Formation models of typical objects used for mathematical modeling

Formation, field	Interlayer	Thickness, m	Absolute permeability μm^2	Porosity, %
YuV 1/1, Las-Yeganskoye	1	3	0.014	16
	2	1.5	0.005	16
AV1/3, Yuzhno-Pokachevskoe	1	0.54	0.099	19
	2	0.95	0.045	19
	3	1.4	0.020	19

Table 3. Data on structure and injection capacity of wells, selected for clay acid treatment

Formati on	Permeability of the interlayer, mD	Interlayer thickness, m	Maximum well flow rate for the entire operation time Q_{max}, t / day	Current well flow rate Q, t / day
UV 1/1	0.014	3.0	40	8
	0.005	1.5		
AV1 / 3	0.099	0.54	120	80
	0.045	0.95		
	0.02	1.4		

The purpose of acid treatment of productive formation, as noted above, is to clean the colmataged filtration channels, create new channels and expand existing ones by dissolving the rock matrix.

The process of creating new channels in reservoirs during acid treatment can be regulated by increasing the injection rate and reducing the diffusion rate (time required for acid to reach the rock surface). In domestic and foreign practice, various methods are used to reduce diffusion rate. The basis for such methods are compositions containing special chemicals that cause the diffusion process to slow down acid frontal advance, these reagents are called "retarders". Simultaneously with slowing down process when using such compositions, also a deviation takes place. The "delayed" composition retains its ability to dissolve rock and can be pushed further into the formation, so there is deepening and deviation of acid frontal advance. The effect of such deviations increases the coefficient of formation coverage with treatment and increases acid treatment efficiency.

The acid application with a acid moderator (ZSK-1M) reduces the rate of dissolution of rock in the water saturated part of reservoir formations, and simultaneously increases the rate and efficiency of dissolution of hydrocarbon-saturated interlayers. The active reagent ZSK-1M makes it possible to hydrophobize the surface of the pore-crack area of the productive formation. The developed technology using a complex selective action reagent (CSAR) based on an acid moderator (ZSK-1M) has the properties described above of deflecting acid frontal advance and slowing down the rate of its reaction with rock. Successful tests of the technology were carried out [10-12], they showed the prospects of its application in various geological and physical conditions.

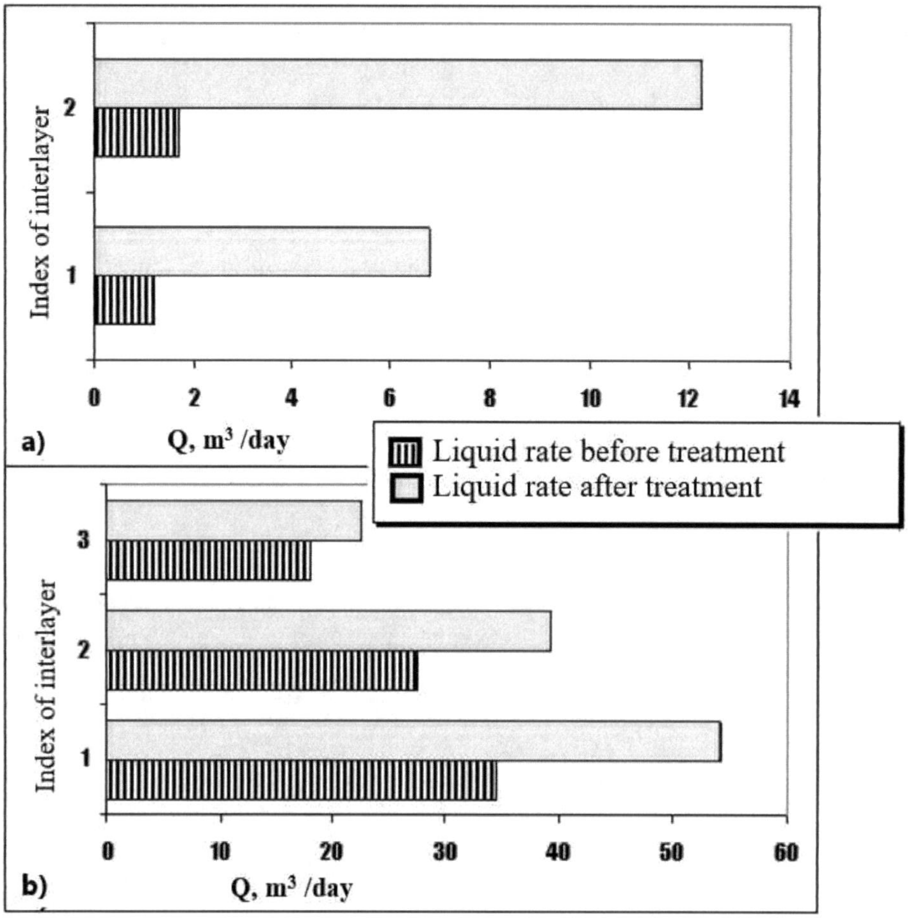

Figure 1. Diagrams of distribution of injection capacity on interlayers before and after acid treatment of model wells of formations: a) group I, b) group II

The technology makes it possible to intensify the following categories of wells:
– wells that exploit terrigenous and carbonate formations with poor reservoir properties;
– wells characterized by a weak reaction to repeated acid treatments.

Features of "CSAR" impact are as follows:
– reduces the rate and degree of dissolution in watered part of reservoir (about 3 times);
– increases the rate and degree of dissolution of oil-saturated reservoir (1.5 times);
– the composition has a low viscosity (up to 2 MPa·s);
– does not form stable emulsions;
– the composition has a reduced corrosion activity (0.293 g / m^2·hour).

The treated facilities are:
- wells that exploit terrigenous formations with low reservoir properties;
- wells characterized by a weak reaction to repeated acid treatments;
- reagent "CSAR " – for wells with a water content of no more than 40%;
- solution "ZSK-1" in a solvent for wells with a water content of more than 40 %.

To increase the treatment efficiency, "CSAR " reagent can be used for static or hydrodynamic baths in the well in combination with wave action or implosion action. It is possible to use "CSAR" technology with an additional flow-deflecting or water-insulating gel screen.

4. Conclusion

1) For a specific well-formation system, impact modeling and design should be performed.

2) Effective use of CAT in terrigenous formations is possible in various geological and technological conditions, if the criteria for impact applicability are met and correct design for the impact is performed.

3) To increase the efficiency of the process of secondary acid treatments, it is possible to use composite acid solutions of selective action.

References

[1] Mishhenko I T 2003 *Downhole oil production* (Moscow: Neft' i Gaz publishing house; National University of Oil and Gas “Gubkin University”) 816 p

[2] Dubinskij G S, Andreev V E, Mijassarov A Sh, Huzin R R and Huzin N I 2013 Algorithm of geological and technological justification of application of targeted technologies for development of hard-to-recover oil reserves *Oil and gas technologies and new materials. Problems and solutions: collection of proceedings* (Ufa: Monografija) **2(7)** 253-262

[3] Mukhametshin V V and Kuleshova L S 2019 Justification of Low-Productive Oil Deposits Flooding Systems in the Conditions of Limited Information Amount *SOCAR Proceedings* **2** 16–22. DOI: 10.5510/OGP20190200384

[4] Andreev V E, Dubinskij G S, Fedorov K M and Andreev A V 2014 Modeling of acidic effects on carbonate reservoirs using composite compositions that regulate the return and pick-up profiles of wells *Oil and gas technologies and new materials. Problems and solutions: collection of proceedings* (Ufa: Monografija) **3(8)** 216-223

[5] Akhmetov R T, Kuleshova L S and Mukhametshin V V 2019 Application of the Brooks-Corey model in the conditions of lower cretaceous deposits in terrigenous reservoirs of Western Siberia *IOP Conf. Ser.: Mater. Sci. Eng.* **560(1)** 012004. DOI: 10.1088/1757-899X/560/1/012004

[6] Andreev V E and Dubinskij G S 2012 Metro technological approach as a scientific methodology for solving problems in oil and gas production *Oil and gas technologies and new materials. Problems and solutions: collection of proceedings* (Ufa: Monografija) **1(6)** 36-44

[7] Kadochnikova L M 1999 *Non-stationary filtration in the bottom-hole zone of wells in the presence of inhomogeneities in the formation,* author's dissertation abstract (Tyumen) 18 p

[8] Yakupov R F, Mukhametshin V Sh, Khakimzyanov I N and Trofimov V E 2019 Optimization of reserve production from water oil zones of D3ps horizon of Shkapovsky oil field by means of horizontal wells *Georesursy* **21(3)** 55-61. DOI: 10.18599/grs.2019.3.55-61

[9] Ovsjukov A V, Blinov S A, Chuprov N M, Levkin V A and Garifullin M G 1998 Study of reservoir, physico-chemical and petrophysical properties of reservoirs in the region in their interaction with chemical reagents, in: *Increasing the level of oil production at the Noyabrskneftegaz OJSC deposits* (Moscow VNIIOÈNG Publ.) pp 333-338

[10] Stenkin A V, Kotenev Yu A, Mukhametshin V Sh and Sultanov Sh Kh 2019 Use of low-mineralized water for displacing oil from clay productive field formations *IOP Conf. Ser.: Mater. Sci. Eng.* **560(1)** 012202. DOI: 10.1088/1757-899X/560/1/012202

[11] Andreev V E and Dubinskij G S 2014 Experimental Field Works in Regulating Oil Recovery Profile in Wells of the Kozhasay Field Using Component Composition *Bulletin of the Academy of Sciences of the Republic of Bashkortostan* **19(2)** 79-84

[12] Valeev A S, Kotenev Yu A, Mukhametshin V Sh and Sultanov Sh Kh 2019 Substantiation of the recovery of residual oil from low-productive and heterogeneous formations in Western Siberia by improving the waterflood system using gas and water-gas impacts *IOP Conf. Ser.: Mater. Sci. Eng.* **560(1)** 012204. DOI: 10.1088/1757-899X/560/1/012204

IOP Conf. Series: Materials Science and Engineering 1064 (2021) 012057 doi:10.1088/1757-899X/1064/1/012057

Modeling of clay acid effects on the bottomhole zone of wells

A R Safiullina[1], V V Mukhametshin[3] L S Kuleshova[1] and D A Batalov[2]

[1]Ufa State Petroleum Technological University, Branch of the University in the City of Oktyabrsky, 54a, Devonskaya St., Oktyabrsky, Republic of Bashkortostan, 452607, Russian Federation
[2]TPE “Povhneftegaz” LLC “LUKOIL–Western Siberia”, 40, Shirokaya St., Kogalym, Tyumen region, Khanty-Mansi Autonomous Area – Yugra, 628482, Russian Federation
[3]Ufa State Petroleum Technological University, 1, Kosmonavtov st., Ufa, Republic of Bashkortostan, 450062, Russian Federation

E-mail: safiullina.a.r@yandex.ru, vv@of.ugntu.ru

Abstract. Typical wells have been identified in the conditions of the Langepass group deposits, according to which modeling of the clay-acid process on the bottom-hole formation zone has been performed. There was a significant increase in the efficiency of operations based on the use of developed models, which allows choosing the most optimal technologies and exposure parameters depending on the properties of the productive formations.

1. Introduction

The last decade has shown an increase in the share of hard-to-recover oil reserves in Russia and around the world [1–5]. Significant volumes of hard-to-recover oil reserves are concentrated in low-permeability clayed terrigenous reservoirs of complex geological structure. In the process of development, there is a deterioration in the reservoir properties of the reservoir in the bottomhole formation zone. The Bottom-hole formation zone is distinguished as part of a formation whose properties may differ from those of the rest of the formation. Due to the deterioration of the reservoir properties of the bottomhole zone during the opening of the formation in the future, as well as due to objective physical phenomena [6, 7], the main share of the energy spent on the movement of oil and gas in the formation is lost.

The main factors that worsen reservoir properties are as follows: swelling of clay cement of reservoir rocks, introducing the smallest mechanical particles into the bottomhole zone during the operation of the well and its repair, rock compaction, and adsorption layers of various compositions on the surface of pores and cracks. The reservoir properties of the bottom-hole formation zone also deteriorate due to the precipitation of salts and a variety of reaction products after chemical reagents are injected. Another reason for deterioration is the collector's water saturation and phase permeability to oil decrease when mineralized water changes to foreign or fresh water. The radius of the bottom-hole zone is from several tens of centimeters to several meters.

An important task in the operation of wells is to maintain reservoir properties of the bottomhole zone in a state as close to the initial one as possible or to improve its characteristics. As a result, there is a need to restore and improve reservoir properties in the bottom-hole formation zone to intensify oil production.

The natural diversity of the mineral composition of the reservoirs, the difference in the structure of the pore space and the variety of factors influencing the reduction in permeability require an individual approach to the selection of an active agent for treating the bottom-hole formation zone [8, 9].

2. Materials and methods

For the basic justification of the choice of the method of intensification of oil production in this paper, we carried out a comparative analysis of the geological and physical characteristics and the current state of development of production facilities of the Langepass group of fields confined to the West Siberian oil and gas field. All of the studied objects are characterized by increased clay content (up to 14%), low carbonate content (less than 1%), and high fragmentation of productive formations. The differences lie in facies conditions and, as a consequence, the difference in mineralogical composition and reservoir properties.

To predict and analyze the effectiveness of the application of enhanced oil recovery methods, simplified models of the studied formations have been developed. We performed the grouping of objects and the selection of polygon objects by known methods. Based on the analysis of averaged geological and statistical sections of productive thicknesses processed by mathematical methods, the main layers and their parameters were identified. The values of these parameters are such that the average values of the permeability of the reservoir model, porosity and productive thickness correspond to the real values of the characteristics of the reservoirs. The properties of the selected interlayers are constant along strike. Parameters of models of polygon objects are summarized in table 1.

Table 1. Models of reservoirs of polygon objects used for mathematical modeling

Plast, field	Interlayer Number	Thickness, m	Absolute Permeability, mD	Porosity,%
ЮВ1/1, Las Yegansk	1)	3	14	16
	2)	1.5	5	16
BV8, Uryevskoe	1)	4.5	302	21
	2)	2.5	120	21
	3)	1.2	50	21
BV6, Stream	1)	4.26	463	20
	2)	5.3	180	20
AB1/3, South Pokachaevskoe	1)	0.54	99	19
	2)	0.95	45	19
	3)	1.4	20	19

3. Results and Discussion

The first object is characterized by low sandiness so that the ratio of the vertical and horizontal components of permeability is 100. The interlayers are assumed to be hydrodynamically coupled, since only in this case the use of cyclic influence has a positive effect. The distribution of oil and water saturation in the reservoir is determined by solving the two-dimensional problem of oil displacement by water to the water cut of the resulting product equal to the average current water cut (72%).

In other accepted models, the interlayers are considered to be hydrodynamically disconnected. The distribution of oil-water saturation is also given by solving the displacement problem to the values of the average current water cut of the product: 88.3%, 97.3% and 55.6% for the second, third and fourth landfill facilities.

The method of stimulation of injection and production wells with clay acid - a composition of hydrochloric and hydrofluoric acids with additives stabilizing against insoluble salts (organic acids) is widely used to treat the bottom-hole zone of wells. The solution acts on rock minerals (carbonates and clays), process fluids and deposits falling into the bottomhole formation zone. Dissolution of deposits leads to an increase in porosity and permeability of the bottomhole zone. A limitation is the high carbonation of the formation, in which case leaching takes place in the form of "wormholes".

Surfactant-acid exposure (PCV) is also common. It has a complex effect: it removes hydrocarbon deposits, acts on the skeleton of the formation (mainly clays and carbonates), process fluids and deposits falling into the bottomhole zone of the formation. There are no application restrictions, with the exception of high water cut in production wells.

Oil-bearing reservoirs of the YV 1/1 layer of the Las Yeganskoye field were considered. This layer contains a large amount (up to 34%) of field spars, clay (up to 11%) and carbonates (6%). In order to increase the flow rate of wells that open up the YV 1/1layer, you can use well treatment with a clay acid solution.

The main chemical reactions of the interaction of acids with rock minerals are described by the following schemes:

$$\underset{\text{quartz}}{SiO_2}+6HF \rightarrow 2H^+ + SiF^-_6 + 2H_2O;$$

$$\underset{\text{K-feldspar}}{KAlSi_3O_8}+20HF \rightarrow K^+ + 4H^+ + AlF^+_2 + 3SiF^{-2}_6 + 8H_2O;$$

$$\underset{\text{Na-feldspar}}{NaAlSi_3O_8}+20HF \rightarrow Na^+ + 4H^+ + AlF^+_2 + 3SiF^{-2}_6 + 8H_2O;$$

$$\underset{\text{kaolinite}}{Al_2Si_2O_5(OH)_4}+16HF \rightarrow 2H^+ + 2AlF^+_2 + 2SiF^{-2}_6 + 9H_2O;$$

$$\underset{\text{illite}}{KAl_3Si_3O_{10}(OH)_2}+24HF \rightarrow K^+ + 2H^+ + 3AlF^+_2 + 3SiF^{-2}_6 + 12H_2O;$$

$$\underset{\text{carbonate}}{CaCO_3}+H^+ \rightarrow Ca^{+2} + HCO^-_3.$$

The article presents the theory of calculating the main parameters of clay-acid treatment (the depth of penetration of the acid solution and the increase in well productivity after treatment) [10]. It studies one-dimensional axisymmetric filtration of an acid solution in a porous medium in the presence of chemical reactions. The injection of an acid solution into a layered reservoir with hydrodynamically isolated interlayers is also under study. The values of the desired processing parameters are found from the condition of the mass balance of reacting components on the fronts of chemical reactions, subject to the equilibrium of the process. The model takes into account: the mineral composition of the rocks forming the formation, the composition of the acid solution, stoichiometric coefficients of the main reactions of dissolution of the rock with acids, the structure of the formation (permeability, porosity, thickness of the layers that make up the formation). The obtained value of the penetration depth of the acid solution was used to assess the relative productivity of the well after treatment.

To assess the effectiveness of the impact, a model well with a geological and statistical section, based on the data in tables 1 and 2, and the composition of the rocks were considered.

Table 2. Data on the structure and injectivity of wells selected for clay processing

Plast	Permeability interlayers, mD	Interlayer power, m	The maximum well production rate for the entire operation period Q_{max}, t / day	Well production rate Q,t / day
ЮB1/1	0.014 0.005	3 1.5	40	8
AB1/3	0.099 0.045 0.02	0.54 0.95 1.4	120	80

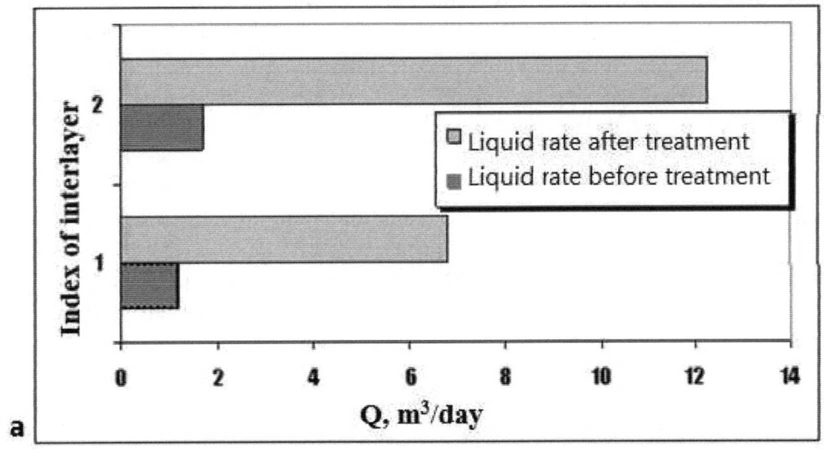

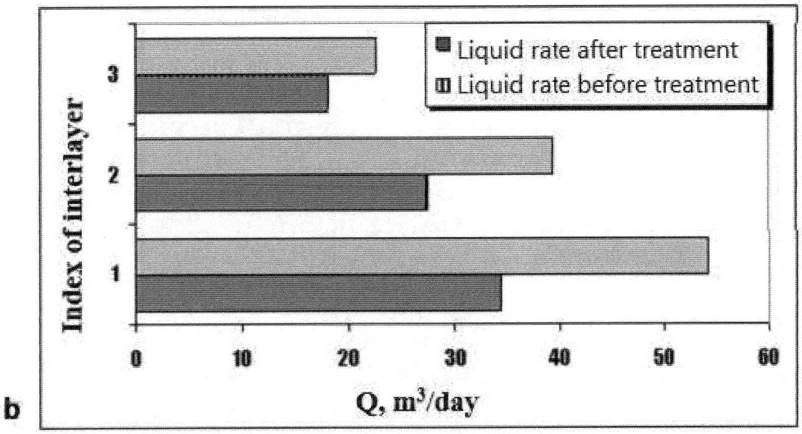

Figure 1. Charts of the distribution of injectivity in the interlayers before and after the acid treatment of model wells of the formations: a) group I; b) II group

The main criteria for selecting wells for treatment are the following: a drop in fluid consumption during operation by more than a factor of two, low flow rates for liquid at this time (for injection wells below 200 m^3 / day, for producers below 40 m^3 / day), low reservoir heterogeneity in terms of power (dissection coefficient of the order of 3). According to the data on the decrease in well flow rates, the size of the bottomhole contamination zone was estimated.

To calculate the efficiency of acid treatments of selected wells, the following composition of the main reagent of the working solution was adopted: 1% HF + 7% HCl. The concentration of acids in the solution was selected based on the requirements to minimize corrosion of the treated wells. The volume of injection of the acid solution was selected depending on the productive capacity of the formation: from the calculation of 20 m^3 of solution per meter of power.

The forecast of the efficiency of processing a well model is as follows. The total volume of the recommended injection of the acid solution is 225 m^3 and 144.5 m^3 for the facilities. Depending on the conductivity of the layers, the indicated volumes are distributed over the section of the layers according to the numbers of the layers as follows: 205.14, 19.9 and 80.7, 45.5 and 18.26 m3. The estimated depths of penetration of contamination through the interlayers, for the considered wells, taking into account the drop in their injectivity during operation, are as follows: 0.66, 0.339, 0.66,

0.387 m and 0.261 m. 0.41, 0.231, 0.582, 0.355 and 0.23 m. Changes in fluid flow after acid treatment of the bottom-hole formation zone were calculated for each layer using the generalized Dupuis formula taking into account the zonal inhomogeneity of the bottom-hole zone of the wells (leach zones Bani acid rock and dirt). The distribution of injectivity of model wells across the interlayers before and after acid treatment is shown in the figure.

4. Conclusion

Thus, we simulated the process of acid exposure on the bottom-hole formation zone in order to intensify the influx of oil and gas and made a forecast of efficiency. Such modeling should be carried out when planning the impact on the bottom-hole formation zone and predicting the technological effect, which is an indispensable element of the recovery technology and increasing well productivity and oil recovery.

References

[1] Fjelde I 2018 Sulfate in Rock Samples from Carbonate Reservoirs *Int. Symp. of the Society of Core Analysts, SCA2008-19 (Abu Dhabi, UAE, 29 October-2 November 2008)* p 12

[2] Akhmetov R T, Mukhametshin V V and Kuleshova L S 2019 Simulation of the absolute permeability based on the capillary pressure curves using the dumbbell model *J. of Phys.: Conf. Ser.* **1333(3)** 032001. DOI: 10.1088/1742-6596/1333/3/032001.

[3] Yakupov R F, Mukhametshin V Sh, Khakimzyanov I N and Trofimov V E 2019 Optimization of reserve production from water oil zones of D3ps horizon of Shkapovsky oil field by means of horizontal wells *Georesursy* **21(3)** 55-61. DOI: 10.18599/grs.2019.3.55-61.

[4] Shuchart C E and Buster D C 1995 Determination of the Chemistry of HF Acidizing with the Use of F NMR Spectroscopy *SPE Int. Symp. on Oilfield Chemistry (San Antonio, Texas, 14-17 February 1995)* p 12. DOI: 10.2118/28975-MS.

[5] Rogachev M K, Mukhametshin V V and Kuleshova L S 2019 Improving the efficiency of using resource base of liquid hydrocarbons in Jurassic deposits of Western Siberia *J. of Mining Institute* **240** 711-715. DOI: 10.31897/PMI.2019.6.711.

[6] Stenkin A V, Kotenev Yu A, Mukhametshin V Sh and Sultanov Sh Kh 2019 Use of low-mineralized water for displacing oil from clay productive field formations *IOP Conf. Ser.: Mater. Sci. Eng.* **560(1)** 012202. DOI: 10.1088/1757-899X/560/1/012202.

[7] Gomari K A R, Karoussi O and Hamouda A A 2006 Mechanistic Study of Interaction between Water and Carbonate Rocks for Enhancing Oil Recovery *SPE Europec/EAGE Annual Conf. and Exhibition (Vienna, Austria, 12-15 June 2006)* p 8. DOI: 10.2118/99628-MS.

[8] Valeev A S, Kotenev Yu A, Mukhametshin V Sh and Sultanov Sh Kh 2019 Substantiation of the recovery of residual oil from low-productive and heterogeneous formations in Western Siberia by improving the waterflood system using gas and water-gas impacts *IOP Conf. Ser.: Mater. Sci. Eng.* **560(1)** 012204. DOI: 10.1088/1757-899X/560/1/012204.

[9] Yakupov R F, Mukhametshin V Sh and Tyncherov K T 2018 Filtration model of oil coning in a bottom water-drive reservoir *Periodico Tche Quimica* **15(30)** 725-733

[10] Webb K J, Black C J J and Tjetland G 2005 A Laboratory Study Investigating Methods for Improving Oil Recovery in Carbonates *Int. Petroleum Technology Conf. (Doha, Qatar, 21-23 November 2005)* p 7. DOI: 10.2523/IPTC-10506-MS.

Geological and technological justification of the parameters of acid-clay treatment of wells

A Kh Gabzalilova[1], D A Batalov[2], V Sh Mukhametshin[3] and V E Andreev[3,4]

[1] Ufa State Petroleum Technological University, Branch of the University in the City of Oktyabrsky, 54a, Devonskaya St., Oktyabrsky, Republic of Bashkortostan, 452607, Russian Federation

[2] TPE "Povhneftegaz" LLC "LUKOIL–Western Siberia", 40, Shirokaya St., Kogalym, Tyumen region, Khanty-Mansi Autonomous Area – Yugra, 628482, Russian Federation

[3] Ufa State Petroleum Technological University, 1, Kosmonavtov st., Ufa, Republic of Bashkortostan, 450062, Russian Federation

[4] SASI "Institute of strategic research of the Republic of Bashkortostan", 129/3, October avenue, Ufa, Republic of Bashkortostan, 450075, Russian Federation

E-mail: alfiragabzalilova@mail.ru, vsh@of.ugntu.ru

Abstract. The article shows that modelling and design definition of the impact requires the differentiation for a specific well-formation system. The authors found that the effective use of acid-clay exposure in terrigenous strata is possible in various geological and technological conditions considering the criteria for the impact applicability and the implementation of a specific impact design. They proposed to use selective acidic solutions of selective action to increase the efficiency of secondary acid treatments.

1. Introduction

Acid treatment technologies have a significant history of application as stimulating wells. The development of new approaches to wells stimulation and the development of new technologies and compositions can transfer them to the category of geological and technical measures (GTM) that will solve the problems of developing hard-to-recover reserves (HRR) of oil and gas associated with complex pore-fractured reservoirs. Oil and gas fields widely use hydrochloric acid treatments (HAT) and their modifications: various compositions and technological features [1–4]. So, the amount of HATs is so large that the total additional oil production from the reservoir due to hydrochloric acid treatment and their modifications is such that it is possible to compare the method with methods for enhanced oil recovery (EOR) of formations [5–7]. Acid-clay treatments are also widely used, and their corresponding application may occupy a more important position among well stimulation methods.

Recently, there has been an upward trend in the share of HRR in the global balance sheet, including in Russia. Significant volumes of HRR are associated with low permeability clayed terrigenous reservoirs with their own geological features. During well operation, the permeability and porosity properties (PPP) of the reservoir deteriorate in the bottom hole formation zone (BHFZ) in comparison with the remote zone. Since deterioration of the reservoir properties of BHFZ occurs during the primary and secondary drilling, as well as further during the well operation and due to objective physical phenomena. Here, there are the main costs of reservoir energy used to move fluids in the formation.

The degradation of PPP results from the swelling of clay cement of reservoir rocks, the precipitation of salts and various reaction products (after injection of chemicals, etc.), an increase in the water saturation of the reservoir and a decrease in the phase permeability for oil, the penetration of the smallest mechanical particles into the BHFZ as from a well, and from the remote zone of the formation, there is the compaction of rocks, the adsorption layers of various origin on the surface of the pore-fissure space. Most experts estimate the radius of the bottom hole zone from several tenths to several meters.

Proper operation of the wells implies keeping the PPP of BHFZ close to the natural state, as well as the work on improving the PPP of BHFZ. Thus, the need to restore and improve PPP arises from the task of maintaining and intensifying hydrocarbon production.

2. Materials and methods

Numerous influence factors that reduce permeability require an individual approach to the choice of technology and an active agent for influencing the BHFZ. To correctly plan the impact on productive formations and to predict its effectiveness, it is necessary to understand the processes occurring in the BHFZ and to determine possible responses to the impact. For this, there is a methodology for forecasting and designing impacts with modelling as a mandatory element [8–11].

To justify the choice of the stimulating oil production method, we analyzed the geological and physical characteristics and the current state of developing production facilities of the Langepass group of fields (West Siberian oil and gas production). All objects of this group have a high number of permeable intervals of formations with increased clay content (up to 14%), low carbonate content (less than 1%). There are differences in facies conditions and, as a consequence, differences in mineralogical composition and reservoir properties.

The effectiveness of acid-clay treatments (ACTs) for terrigenous strata is well known. We conducted modelling of the impact to forecast the effectiveness of ACTs for the conditions of the Langepass group of fields by the methodological approach of the methotechnology of physicochemical effects on the formations.

3. Results and Discussion

It should consider ACTs used to increase the injectivity of injection wells and the productivity of producing wells as a method aimed at cleaning the pore space from technogenic products brought into the BHZ during the opening or operation of wells. Thus, the success of the ACTs is largely dependent on the composition of cement, binding grains of the rocks of the reservoir and the composition of the clogging substances.

It is necessary to consider, first, the mineral composition of the rocks to build a mathematical model of the distribution of the acid-clay solution in a porous medium. The presence of minerals and compounds of elements which in the process of reaction with acid-clay can form an insoluble precipitate has a great influence. Sandy rock treated with acid-clay must have a very low content of carbonates, glandular compounds. While the high content of clay is not an obstacle to the success of ACTs. Therefore, we studied the mineralogical composition of the oil-bearing terrigenous stratum JV 1/1 of the Las Yeganskoye field. Table 1 shows the average composition of the rocks forming this formation.

Table 1 demonstrates that the formation contains a large amount of feldspar (up to 34%), clay (up to 11%) and carbonates (6%). It is possible to use well treatment with the acid-clay solution (HF + HCl) to increase the production rates of wells drilling in the JV 1/1 formation.

Table 1. The mineralogical composition of the formation rock JV1/1 of the Las Yeganskoe field

Minerals	Percentage, %
Quartz	43
Feldspars	34
Micas	6
Clays	11
Carbonates	6

The effectiveness of ACTs depends on the presence of minerals in the rock cement that are capable to dissolute. The main chemical reactions of the interaction of acids with rock minerals have the following formulas:

$SiO_2+6HF \rightarrow 2H^++SiF^-_6+2H_2O$;
Quartz

$KAlSi_3O_8+20HF \rightarrow K^++4H^++AlF^+_2+3SiF^{-2}_6+8H_2O$;
K-Feldspar

$NaAlSi_3O_8+20HF \rightarrow Na^++4H^++AlF^+_2+3SiF^{-2}_6+8H_2O$;
Na-Feldspar

$Al_2Si_2O_5(OH)_4+16HF \rightarrow 2H^++2AlF^+_2+2SiF^{-2}_6+9H_2O$;
Kaolinite

$KAl_3Si_3O_{10}(OH)_2+24HF \rightarrow K^++2H^++3AlF^+_2+3SiF^{-2}_6+12H_2O$;
Illite

$CaCO_3+H^+ \rightarrow Ca^{+2}+HCO^-_3$.
Carbonate

The problems of processing secondary sedimentation reactions in the formation during acid treatments have a partial solution by introducing special additives into the working solution, as well as using various technological methods.

During processing, the feldspars and clays dissolve with hydrofluoric acid and the carbonates and some clays dissolve with hydrochloric acid. In these two stages of the dissolution of the reservoir rock, the feldspars and clays are completely dissolved by hydrofluoric acid and their concentration decreases, then the carbonates and some clays dissolve in another area and the hydrochloric acid concentration decreases. We conventionally divide the model layer into three flow areas of the injected solution: the 1st area corresponds to the flow of a solution containing hydrofluoric and hydrochloric acids (feldspars and clay dissolve); 2nd area corresponds to the flow of an acid solution containing hydrochloric acid and salts obtained as a result of a dissolution reaction (carbonates and part of clays dissolve); 3rd area corresponds to the flow of a solution of reaction products along with formation with initial filtration parameters.

We considered the process of equilibrium dissolution of a porous medium containing soluble minerals with an acid solution (a mixture of hydrochloric and hydrofluoric acids). The model neglects the diffusion transfer of components compared to the convective, gravitational pressure drop across the bottom of the well compared to the hydrodynamic one.

The system of equations describing the process of filtering a clay-acid solution and reaction products obtained as a result of dissolution consists of continuity equations for each component of the solution and the equation of motion:

$$\frac{\partial m \cdot \rho_i^0 \cdot c_{\mathrm{HF}}}{\partial t} + \frac{1}{r}\frac{\partial}{\partial r} r \cdot m \cdot \rho_i^0 \cdot c_{\mathrm{HF}} \cdot v = -K_{sh} \cdot J_{sh} - K_{g1} \cdot J_{g1};$$
$$\frac{\partial m \cdot \rho_i^0 \cdot c_{\mathrm{HCL}}}{\partial t} + \frac{1}{r}\frac{\partial}{\partial r} r \cdot m \cdot \rho_i^0 \cdot c_{\mathrm{HCL}} \cdot v = -K_k \cdot J_k - K_{g2} \cdot J_{g2};$$
$$\frac{\partial m \cdot \rho_i^0 \cdot c_1}{\partial t} + \frac{1}{r}\frac{\partial}{\partial r} r \cdot m \cdot \rho_i^0 \cdot c_1 \cdot v = -K_{1sh} \cdot J_{sh} - K_{11g} \cdot J_{g1}; \quad (1)$$

$$\frac{\partial m \cdot \rho_i^0 \cdot c_2}{\partial t} + \frac{1}{r}\frac{\partial}{\partial r} r \cdot m \cdot \rho_i^0 \cdot c_2 \cdot v = -K_{2k} \cdot J_k - \rho \cdot K_{22g} \cdot J_{g2};$$
$$c_v = 1 - (c_1 + c_2 + c_{\mathrm{HCL}} + c_{\mathrm{HF}});$$
$$J_{sh} = -\rho_{sh}^0 \frac{\partial \alpha_{sh}}{\partial t};$$
$$J_{g1} = -\rho_g^0 \frac{\partial \alpha_g}{\partial t};$$
$$J_k = -\rho_k^0 \frac{\partial \alpha_k}{\partial t};$$
$$J_{g2} = -\rho_g^0 \frac{\partial \alpha_{\mathrm{sol.g}}}{\partial t};$$
$$r_c \cdot v \cdot m = \frac{Q}{2\pi \cdot h},$$

where m is the porosity of the rock, $\rho_{i,\,R,\,sh,\,g,\,k}^0$ are the densities of water, rocks, minerals of the rock, respectively, *with* $_{HF,\,HCL,\,1,2,v}$ are the concentrations of hydrogen fluoride, hydrochloric acid, and salts obtained as a result of the reaction dissolution of water in the solution, v is the fluid flow rate, $K_{sh,\,g1,\,k.g2,1sh,\,11g,\,2k,\,22g}$ are the stoichiometric dissolution reaction coefficients, $J_{sh,\,g1,\,k,\,g2}$ are the reaction rates of the dissolution of feldspars, hydrogen fluoride clay acid and carbonates and parts of clays with hydrochloric acid, $\alpha_{sh,\,g,\,k,\,sol.\,g}$ are volume content of feldspars, clay, carbonates and parts of clays that are threaded in hydrochloric acid ($\alpha_I = c_i \cdot \rho_R / \rho_i$), r_c is the well radius, Q is the fluid flow rate, and h is the formation thickness.

It is assumed that feldspars include K-feldspar and Na-feldspar. The clay incluses kaolinite, smectite, illite, montmorillonite, bentonite. Carbonates include calcite and magnesite-dolomite. We took the stoichiometric coefficients as the average in each group of minerals.

For different flow areas, there are the following initial and boundary conditions: at the initial time, there is water in the formation, the concentration of minerals in the rock and porosity are equal to the initial values; at the zero boundary, the concentration of acids in the solution is equal to the initial, the formation contains quartz and carbonates, the porosity is $m_1 = 1 - \alpha_R - \alpha_K$, at the first boundary there are salts from the reaction and a changed amount of hydrochloric acid in the formation, quartz, feldspars and a part of clays, $m_2 = 1 - \alpha_R - \alpha_{sh} - (\alpha_g - \alpha_{sol.\,g})$, at the second boundary in the solution there are salts obtained as a result of dissolution reactions, in the formation there are all the original minerals, porosity is equal to the initial value.

The hyperbolic system of equations allows discontinuous solutions. The values of the unknown parameters at the break obtained from the conditions of mass balance at the fronts of chemical reactions under the condition of equilibrium of the process, considering the initial and boundary conditions, have the following form:

$$r_1^2 = \frac{V_n \cdot J(\pi \cdot \square)\rho_i^n \cdot c_{HF}}{m_i \cdot \rho_i^n \cdot c_{HF} + K_{sh} \cdot \rho_{sh}^n \cdot \alpha_{sh} + K_{g1} \cdot \rho_g^n(\alpha_g - \alpha_{sol.g})};$$
$$V_1 = (V_0/(\pi \cdot h)\rho_i^0(1 - c_{\mathrm{HF}}) - r_1^2(m_1 \cdot \rho_i^0(1 - c_{\mathrm{HF}}) - K_k \cdot \rho_k^0 \cdot \alpha_k + K_{\mathrm{vk}} \cdot \rho_k^0 \cdot \alpha_k - m_2 \cdot \rho_i^0 -$$
$$-K_{\mathrm{vsh}} \cdot \rho_{sh}^0 \cdot \alpha_{sh} - K_{\mathrm{vg1}} \cdot \rho_g^0(\alpha_g - \alpha_{\mathrm{sol.g}}) - K_{1sh} \cdot \rho_{sh}^0 \cdot \alpha_{sh} - K_{11g} \cdot \rho_g^0(\alpha_g - \alpha_{\mathrm{sol.g}}))\frac{\pi \cdot h}{\rho_i^n}; \quad (2)$$
$$c_{HCL} = \frac{r_1^2(m_1 \cdot \rho_1^0 \cdot c_{HCL} - K_k \cdot \rho_k^0 \cdot \alpha_k) - V_0/(\pi \cdot \square)\rho_1^0 \cdot c_{HCL}}{r_1^2 \cdot m_2 \cdot \rho_1^0 - V_1/(\pi \cdot \square)\rho_1^0};$$
$$r_2^2 = \frac{V_1/(\pi \cdot \square)\rho_1^0 \cdot c_{HCL}}{m_2 \cdot \rho_1^0 \cdot c_{HCL} + K_k \cdot \rho_k^o \cdot \alpha_k + K_{g2} \cdot \rho_2^0 \cdot \alpha_{sol.g}},$$

where $r_{1,2}$ is the penetration depth of the acid solution (hydrofluoric and hydrochloric, respectively), $V_{0,1}$ is the initial injection volume after the reaction, C_{HCL} is the decrease in the concentration of hydrochloric acid in the solution due to the water released as a result of the dissolution of rock minerals with hydrofluoric acid.

We calculated the change in the permeability of the treated zones of the formation according to the

Cozen-Karman law:

$$k(r)/k_0(r) = (m/m_0)^n, \quad (3)$$

where $k(r)$, $k_0(r)$ is the permeability of the formation after and before acid treatment, n is an exponent ($n = 10$).

We calculated the relative productivity for the interlayers in the stratified formation according to the formula:

$$Q_i/\Delta P = \frac{2\pi \cdot h_i}{\mu_i} \frac{1}{\int_{r_{cr}}^{R} \frac{dr}{r \cdot k_i(r)}}, \quad (4)$$

where $Q/\Delta P$ is the productivity of the layer, k is the permeability of the layer after acid-clay treatment, i is the number of the layer in the stratified formation.

To assess the effectiveness of the impact, we considered a model well with a geological and statistical section built on the data from Table 2, shown in Table 3 and the composition of the rocks from Table 1.

Table 2. Models of formations of typical objects used for mathematical modelling

Formation, field	Interlayer	Thickness, m	Absolute permeability, mm^2	Porosity, %
JV 1/1, Las Yeganskoye	1	3.0	0.014	16
	2	1.5	0.005	16
AV 1/3, Yuzhno-Pokachaevskoye	1	0.54	0.099	19
	2	0.95	0.045	19
	3	1.4	0.020	19

Table 3. Data on the structure and injectivity of wells selected for acid-clay treatment

Formation	Permeability interlayers, MD	Layer thickness, m	The maximum production rate of the well for the entire period of operation Q_{max}, t / day	Well production rate Q, t / day
JV 1/1	0.014	3.0	40	8
	0.005	1.5		
AV 1/3	0.099	0.54	120	80
	0.045	0.95		
	0.02	1.4		

The main criteria for selecting wells for treatment are the following factors: a drop in fluid consumption during operation by more than twice, low injectivity/flow rates for the fluid at a given time (for injection wells below 200 m³/day, for producers below 40 m³/day), low formation heterogeneity of thickness (number of permeable intervals of at least 2–3). The data on the decrease in the flow rate of the wells allowed to estimate the size of the bottom-hole damage zone.

To calculate the effectiveness of acid treatments of selected wells, we used the following composition of the main reagent of the working solution: 1% HF + 7% HCl. We selected the concentration of acids in the solution based on the requirements to minimize corrosion of equipment and the design of the wells being treated. We chose the volume of injection of the acid solution depending on the productive formation thickness based on 20 m^3 of solution per meter of thickness.

The results of forecasting the effectiveness of processing modelling well include the following: The total volume of the recommended injection of the acid-clay solution is 225 m^3 and 144.5 m^3 according to the facilities. Depending on the conductivity of the interlayers, this volume is distributed across the section of the layers according to the number of interlayers as follows: 205.14; 19.9 and 80.7; 45.5 and 18.26 m^3. The estimated depths of the spread of pollution through the interlayers for the considered wells, considering the drop in their injectivity during operation, amounted to 0.66; 0.339; 0.66; 0.387

and 0.261 m. The penetration radius of the acid solution through the interlayers in the analyzed wells, calculated by the proposed method, was 0.41; 0.231; 0.582; 0.355 and 0.23 m. We calculated changes in fluid flow after acid treatment of wells for each layer according to the generalized Dupuis formula considering the zonal heterogeneity of the bottom-hole zone of the wells (zones of rock leaching with acid and contamination). The figure represents the injectivity distribution of model wells in the interlayers before and after acid treatment

The purpose of acid treatment of the production formation, as noted above, is to clean up the cremated filtration channels, create new channels and expand the existing ones by dissolving the rock matrix.

Fulfilment of the purpose leads to an increase in the effective diameter of the well and the bottom-hole formation zone. ACTs with proper planning and implementation, for the most part, contribute to increasing well productivity by reducing the skin factor in the BHFZ to negative values. Using the methotechnological approach, the correct planning of acid treatment requires modelling processes in the BHFZ and processing design.

When processing of BHFZ with acid without any modifiers, its maximum effect on the rock occurs in the wellbore zone of the formation. Acid loses most of its activity at the entrance to the formation, and the remote zones of the formation are practically not processed. This mechanism of acid exposure leads to the very intensive formation of the maximum number of new dissolution channels and expansion of old channels in the near-wellbore zone, and even at a small distance from the wellbore in the formation, the formation of dissolution channels and expansion of old ones practically does not occur. Laboratory experiments and field practice confirm that, during the process of being injected into the formation, the acid moves mainly along the same channels and cracks and leaves a significant part of the productive formation without impact. More permeable intervals are processed, and low permeability ones remain without improvement. Therefore, the efficiency of acid treatment drops rapidly during repeated treatments in one well.

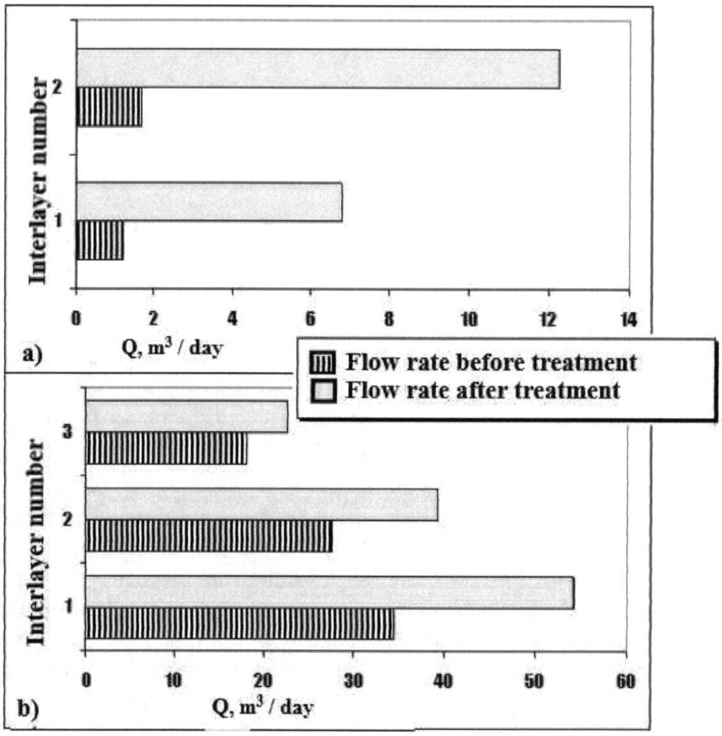

Figure 1. Charts of the distribution of injectivity in the interlayers before and after acid treatment of model wells of the formations: a) group I, b) group II

The increase in the injection rate and a decrease in the diffusion rate, the time required for the acid to reach the rock surface, make it possible to control the process of creating new channels in the collectors during acid treatment. Russian and foreign practice use various methods aimed at reducing the diffusion rate. The basis of such methods includes compositions containing special chemicals, that cause a slowdown of the diffusion process when the acid front moves, these reagents are called decelerators. The deviation occurs simultaneously with the deceleration process when using such formulations. The decelerated composition retains its ability to dissolve the rock and can be pressed further into the formation, therefore, the active acid front deepens and deflects. This deviation effect will increase the coverage factor of the formation and increase the efficiency of acid exposure.

The acid with a decelerator of the reaction (ZSK-1M) reduces the rate of rock dissolution in the water-saturated part of reservoirs, and simultaneously increase the speed and efficiency of dissolution of hydrocarbon-saturated interlayers. The active reagent ZSK-1M allows hydrophobization of the surface of the pore-fissure space of the reservoir. The developed technology using an integrated selective reagent (CRID) based on an acid decelerator (ZSK-1M) has the above-described properties of deflecting the acid front and decelerating down the rate of its reaction with the rock. Researchers successfully tested the technology, which showed the promise of its use in various geological and physical conditions.

The technology allows to intensify the work of the following categories of wells:

- wells operating terrigenous and carbonate formations with poor reservoir properties;
- wells characterized by a weak reaction to repeated acid treatments.

The effects of CRID have the following features:

- the speed and degree of dissolution in the waterflooded part of the formation reduces by about 3 times;
- increases the speed and degree of dissolution of the oil-saturated reservoir by 1.5 times;
- the composition has a low viscosity, up to 2 MPa · s;
- does not form persistent emulsions;
- the composition has a reduced corrosion activity [0.293 g / (m^2 · h)];

The objects of influence include:

- wells operating terrigenous formations with low reservoir properties;
- wells characterized by a weak reaction to repeated acid treatments;
- CRID used for wells with water cutting in production of not more than 40%;
- ZSK-1 solution in a solvent for wells with water cutting in production of more than 40%.

The use of CRID reagent for a static or hydrodynamic bath in a well with a wave action or an implosive effect will increase the processing efficiency. It is possible to use the CRID technology with an additional flow-deflecting or waterproofing gel screen.

4. Conclusion

1. The specific well-formation system requires the individual modelling and design of the impact.
2. The effective ACTs use in terrigenous formations is possible in various geological and technological conditions, considering the criteria for the applicability of the impact and the implementation of the correct design of the impact.
3. It is possible to use composite acid solutions of selective action to increase the efficiency of the process of secondary acid treatments.

References

[1] Hartman R L, Lecerf B, Frenier W W, Ziauddin M E and Fogler H S 2006 Acid-Sensitive Aluminosilicates: Dissolution Kinetics and Fluid Selection for Matrix-Stimulation Treatments *SPE Production & Operations* **21(2)** 11. DOI: 10.2118/82267-PA

[2] Rogachev M K and Mukhametshin V V 2018 Control and regulation of the hydrochloric acid treatment of the bottomhole zone based on field-geological data *J. of Mining Institute* **231** 275-280. DOI: 10.25515/PMI.2018.3.275

[3] Yakupov R F, Gimazov A A, Mukhametshin V Sh and Makaev R I 2018 Analytical method for estimating efficiency of oil recovery technology in case of bottom water-drive reservoir, verified on the hydrodynamic model *Oil Industry* **6** 66-69. DOI: 10.24887/0028-2448-2018-6-66-69

[4] Stenkin A V, Kotenev Yu A, Mukhametshin V Sh and Sultanov Sh Kh 2019 Use of low-mineralized water for displacing oil from clay productive field formations *IOP Conf. Ser.: Mater. Sci. Eng.* **560(1)** 012202. DOI: 10.1088/1757-899X/560/1/012202

[5] Kuleshova L S, Kadyrov R R, Mukhametshin V V and Safiullina A R 2019 Design changes of injection and supply wellhead fittings operating in winter conditions *IOP Conf. Ser.: Mater. Sci. Eng.* **560(1)** 012072. DOI: 10.1088/1757-899X/560/1/012072

[6] Gdanski R D 1999 Kinetics of the Secondary Reaction of HF on Alumino-Silicates *SPE Production & Facilities* **14(4)** 9. DOI: 10.2118/59094-PA

[7] Rogachev M K, Mukhametshin V V and Kuleshova L S 2019 Improving the efficiency of using resource base of liquid hydrocarbons in Jurassic deposits of Western Siberia *J. of Mining Institute* **240** 711-715. DOI: 10.31897/PMI.2019.6.711

[8] Valeev A S, Kotenev Yu A, Mukhametshin V Sh, and Sultanov Sh Kh 2019 Substantiation of the recovery of residual oil from low-productive and heterogeneous formations in Western Siberia by improving the waterflood system using gas and water-gas impacts *IOP Conf. Ser.: Mater. Sci. Eng.* **560(1)** 012204. DOI: 10.1088/1757-899X/560/1/012204

[9] Mukhametshin V V and Andreev V E 2018 Increasing the efficiency of assessing the performance of techniques aimed at expanding the use of resource potential of oilfields with hard-to-recover reserves *Bull. of the Tomsk Polytechnic University. Geo Assets Engineering* **329(8)** 30–36

[10] Mukhametshin V V 2017 Eliminating uncertainties in solving bottom hole zone stimulation tasks *Bull. of the Tomsk Polytechnic University. Geo Assets Engineering* **328(7)** 40–50

[11] Lestz R S, Clarke J N, Plattner D and Byrd A C 2002 Perforating for Stimulation: An Engineered Solution *SPE Drilling & Completion* **17(1)** 8. DOI: 10.2118/76812-PA

Oil and gas well drilling technology based on the system approaches and the results of application

R R Gazizov[1], A P Chizhov[1,2], V E Andreev[1,2], A V Chibisov[1,2], V V Mukhametshin[2] and L S Kuleshova[3]

[1] SASI "Institute for Strategic Research of the Republic of Bashkortostan", 129/3, Oktyabrsky Ave., Ufa, Republic of Bashkortostan, 450075, Russian Federation
[2] Ufa State Petroleum Technological University, 1, Kosmonavtov St., Ufa, Republic of Bashkortostan, 450062, Russian Federation
[3] Ufa State Petroleum Technological University, Branch of the University in the City of Oktyabrsky, 54a, Devonskaya St., Oktyabrsky, Republic of Bashkortostan, 452607, Russian Federation

E-mail: gazizoff.renat@mail.ru, vv@of.ugntu.ru

Abstract. The article analyzes the reason and factors of declining in key indicators of quality and efficiency of well construction and operation, which have led to a long-term stagnation in drilling technology. To handle the urgent issues, a systematic approach to implement the complex of hydro-mechanical hardening of an open-hole well by the synchronous and coordinated action of hydraulic monitoring strings of drilling fluids is proposed. The indicators of quality and efficiency of the complex compared to the traditional ones, which exceed the former by two or more times, are provided.

1. Introduction

The steady tendency for the increasing complexity of the geological and technical conditions needed to construct and operate the oil and gas wells of both natural and man-made origins is increasingly negatively affecting their quality and efficiency [1, 2]. The current situation is due to two circumstances.

The first one is a logical decrease in the efficiency of traditional drilling technologies and well completion in abnormal and changing geological, technical and thermodynamic conditions. Based on the principles of maintaining hydraulic equilibrium in the "well – rock mass" system as a factor in preventing gas and oil occurrences, emissions and stabilizing the production conditions of technological operations, these approaches cause the non-stationary technical state of the borehole and the uncertainty of the hydraulic behaviour of wells, characteristic features of those are close to chaotic (for the processes of hydro-mechanical and physical-chemical interaction of the rock mass, process fluids and reservoir fluids).

The second circumstance is the lack of effective methods to control and manage technological processes of drilling and operation of wells (it refers to the regulation of mechanisms of hydro-mechanical impact on permeable rocks of the near-well and remote zones: absorbing rocks, gas-oil-water-saturated layers and unstable rocks).

The negative impacts of these circumstances are logical decreases in key indicators of drilling and operation of wells: the deterioration of filtration characteristics of the productive formations, the violation of the sealing capacity of the support, the consolidation of fluid-saturated layers of the effective thickness into a single filter, the differentiation of the current reservoir pressure by the section and area of the developed oil and gas deposits. In the final analysis, the result is a decrease in the integral indicator of the efficiency of oil and gas deposits development – the oil recovery coefficient (ORC), as well as an increase in one of the key negative technological indicators of field development – the water-cut of the extracted products [3-7].

2. Methods and materials

Long-standing monitoring has shown that the use of traditional and newly developed technologies does not lead to an effective solution to key technological problems. It was the reason for the long stagnation in the field of construction and well operation. However, among the majority of specialists, there is a strong opinion that the existing technological problems in drilling (the prevention of complications: absorption, gas and oil occurrences, rock instability, preservation of natural reservoir properties of the effective formations, increasing the sealing capacity of the borehole annulus of the support, etc.) can be successfully solved by the changing of the rheological properties and parameters of process fluids (drilling, grouting and special solutions), the optimizing of drilling modes, support setting and completion of wells, etc.

However, the results of field research and experience show that this opinion does not have sufficient scientific justification, is not confirmed by practice, and is very controversial in many circumstances [1]. This is primarily due to the unsteadiness of hydraulic drilling conditions, which are not subject to control and regulation of the processes of hydro-mechanical and physical-chemical interaction of drilling fluids and the opening drilling of a rock mass by an irreversible violation of the initial properties of drilling and grouting solutions in the well.

Field studies of the recent years show that the cause of most complications in the well is an active hydraulic connection of drilling fluids and fluid-saturated layers in the array of rocks opened by drilling with different filtration and strength characteristics. The final result of these non-stationary, uncontrolled thermodynamic processes is a violation of technology, a decrease in the quality and efficiency of drilling operations, and well operation.

3. Results and discussion

To handle these complex field issues (the stabilization of the hydraulic state and behaviour of wells, the preservation of the natural reservoir properties of productive layers and their isolation from the fluid-filled reservoirs of the productive thickness, the increase of sealing capacity of the borehole annulus) the complex of technologies has been designed and developed (table 1) [1], which in turn is closely connected with the key problems of the increasing efficiency of development of oil and gas deposits, the maintenance of high rates of hydrocarbon production, the prolongation of waterless operation of wells, the increase of ORC [2, 9-12].

The methodological basis of the proposed technology is the simultaneous strengthening of the borehole with the drilling process by water jets directed at the walls of well [8]. The near-wellbore forming waterproofing screen occurs when the synchronous and coordinated occurrence of hydrodynamic processes and transforming the kinetic energy of the water jet in the quantum of the pressure pulse at which the force of the dynamic impact of the jet is 0.15–0.40 t, the contact time with the rock is 0.012–0.017 seconds, the temperature of the processes is 500-800 °C (approximately [8]). Formed in these thermodynamic conditions, the near-wellbore waterproofing screen is characterized by the parameters: the thickness of the adhesive layer on the borehole wall is 3-5 mm; the depth of the zone of clogging in the matrix permeable rocks – 10-30 mm; hydromechanical strength of the borehole under the action of repression – 0.018–0.025 MPa/m (gradient of rock pressure); the sealing capacity of the borehole during the depression – 0.010–0.015 MPa/m [1, 8].

Table 1. The complex of advanced technology for drilling and completion of wells

Methods of the technological complex	Purpose of complex elements	Final results of the complex implementation
1. The technology of water jetting of the borehole in the drilling process and selective effects	Restoration of waterproofing of the exposed absorbing and oil-and-gas-saturated reservoirs from the wellbore	Operational evaluation of technical condition indicators of the wellbore
2. Method of the rapid assessment of the technical condition of the borehole (the sealing capacity and the strength of the walls)	Monitoring the sealing capacity and strength of the borehole, assessment of filtration characteristics of fluid-saturated reservoirs	Prevention of complications and optimization of hydraulic drilling and completion processes
3. Method of isolation of absorbing layers with an intensity of more than 50 m^3/hour.	Restoring the sealing capacity and strength of the borehole in the conditions of abnormally low formation pressure (ALFP)	Complete isolation of high-intensity absorbing layers (50 m^3/h or more)
4. The technique of selection and implementation of the mechanism for reducing reservoir permeability, calculation of optimal parameters of the isolation process	Operational control and regulation of permeable rock isolation mechanisms and strengthening of unstable rocks	Non-linear improvement of quality, technical, economic, environmental indicators and safety of drilling operations

Features of the proposed technology are the following:

- creating a fluid-proof clogging screen when the sections with the absorbing and oil-water-gas comprising layers and the rocks which are prone to hydraulic fracturing are opened;
- preventing scree and failure of unstable rocks due to the formation of a highly-structured clogging layer;
- requiring minimal energy consumption to pump the solution with simultaneous creating a highly-structured wall layer and a layer in the clogging zone (figure) due to its unique rheological properties;
- applying is recommended with clay-based drilling fluids, including cement.

The analysis of cavernogramms and the comparison of actual cavernosity ratio during drilling wells with the use of hydro-mechanical hardening technology allowed ensuring significant stability of the borehole within the range of the verey horizon, which is prone to scree and failure.

The transition to the system level of development of traditional technologies is confirmed by the results of non-linear improvement of technological, technical, economical, environmental indicators, as well as the safety of drilling operations.

The integral indicators for reducing the time of the well construction were 20-25 %, increasing the current flow rate by 1.6-3.0 times and decreasing the water-cut of extracted products by 2.6 times. A more detailed assessment of the technological indicators of opening the productive layer using hydro-mechanical strengthening of the borehole of well is presented in Table 2, in the period from 2002 to 2004 in JSC "Tatneft" [8].

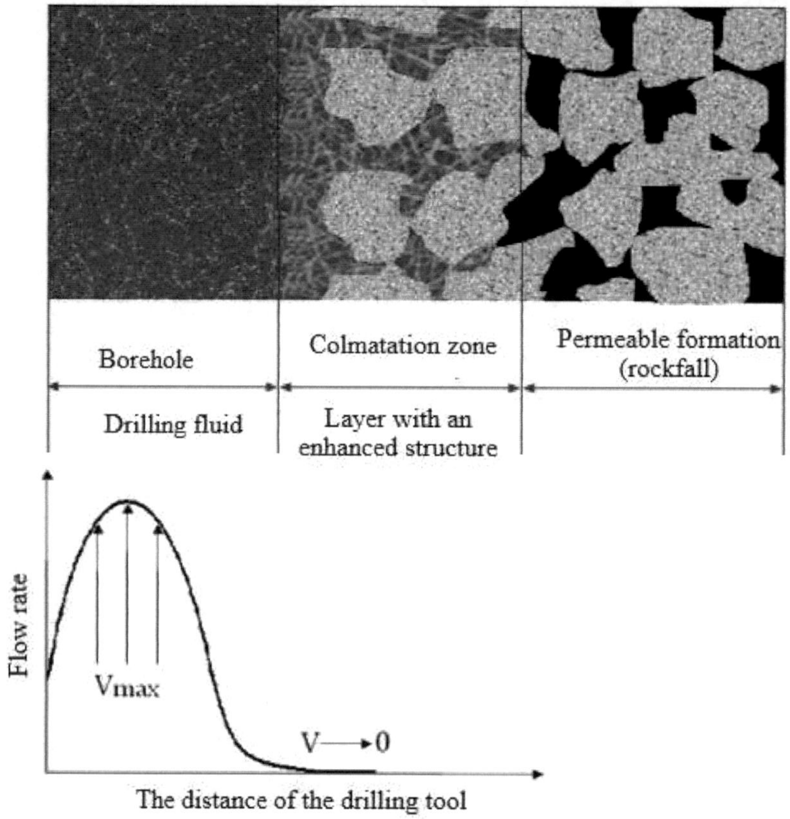

Figure 1. Distribution of the drilling fluid flow rate in the well and the near-wellbore zone of drilled reservoir rocks [8]

Table 2. The assessment of qualitative and technological indicators of the well completion through a system technology complex

Implementation volume, wellbore	The thickness of the layer (perforated), m	Average operating indicators			The dynamic pressure gradient between w/s and o/s layers, MPa/m	Specific-productivity index, $m^3/(day \times MPa \times m)$
		oil flow rate, t/day	water-cut,%	depression, MPa		
Total: 56 wells	1.0–2.0	8.49	37.7	3.7	5.4	1.79
Total: 70 wells	2.1–4.0	9.6	30	4.3	5.6	1.13
Total: 41 wells	more than 4.0	15.3	35.4	2.7	4.2	0.68
Average values for test wells (the sample of wells drilled utilizing the complex technology)		11.1	34.4	3.6	5.07	1.20
Indicators of base wells		4.3	88.0	5.7	3.2	0.33
The ratio of indicators of test and base wells		+2.6	–2.6	–1.6	+1.8	+3.6
Comment: w/s, o/s – water-saturated and oil-saturated layers, respectively.						

4. Conclusion

The comparative analysis of technical and economical indicators of the drilled interval in the well with the same interval previously drilled nearby wells on the area demonstrated the high efficiency of the applied technology. In particular, the opening of the verey horizon at the well without the hydro-mechanical strengthening of the borehole, which further led to the destabilization of the borehole, specifically required the installation of a hardening cement bridging in this interval. When drilling through this technology, these problems were avoided and the installation of a cement bridging in this interval was excluded, respectively, which significantly reduced the time for construction of the well.

The analytical assessment presented in the article allows concluding that the transition to the system development and the improvement of technologies of drilling and operating oil and gas wells is one of perspective scientific and technical directions of non-linear improving the quality, efficiency, environmental protection and production safety in the drilling and operation of wells and increasing current oil production rate and reduce water-cut of the extracted product.

When using the technology, the following technical and economical indicators were achieved: the technical drilling speed was increased by 8.65%; the commercial drilling speed was increased by 6.5%; the cavernosity ratio was reduced by an average of 6.2%. The application of technology made it possible to exclude the installation of a cement bridging in the interval of the verey horizon.

The conclusions about the effectiveness of the proposed technology are stated to be preliminary and require the developed technology to be undergone further field tests. The application of technology has allowed to carry out high-quality wiring of the borehole under the production string and to increase the commercial speed of construction.

References

[1] Polyakov V N, Kuznetsov Yu S, Sagidullin L A, Shulgina N Yu, Dubrovsky V S, Khusainov V M, Khaminov N I, Akhmetzyanov R G, Vildanov A A and Starov O E 2005 The Solving of Problems of Tailing-in and Wells Operation in Abnormal Thermodynamic Conditions *Oil Industry* **5** 104–110

[2] Polyakov V N, Mnatsakanov V A, Fokin V V and Averyanov A P 2009 Analytical Assessment of the Current State of Well Construction Technologies and Prospects for their Improvement *Construction of Oil and Gas Wells on Land and Sea* **3** 10–13

[3] Yakupov R F, Mukhametshin V Sh, Khakimzyanov I N and Trofimov V E 2019 Optimization of Reserve Production From Water Oil Xones of D3ps Horizon of Shkapovsky Oil Field Using Horizontal Wells *Georesursy* **21(3)** 55–61. DOI: 10.18599/grs.2019.3.55-61

[4] Stenkin A V, Kotenev Yu A, Mukhametshin V Sh and Sultanov Sh Kh 2019 Use of Low-Mineralized Water for Displacing Oil from Clay Productive Field Formations *IOP Conf. Ser.: Mater. Sci. Eng.* **560(1)** 012202. DOI: 10.1088/1757-899X/560/1/012202

[5] Valeev A S, Kotenev Yu A, Mukhametshin V Sh and Sultanov Sh Kh 2019 Substantiation of the Recovery of Residual Oil from Low-Productive and Heterogeneous Formations in Western Siberia by Improving the Waterflood System Using Gas and Water-Gas Impacts *IOP Conf. Ser.: Mater. Sci. Eng.* **560(1)** 012204. DOI: 10.1088/1757-899X/560/1/012204

[6] Mukhametshin V V and Kuleshova L S 2019 Justification of Low-Productive Oil Deposits Flooding Systems in the Conditions of Limited Information Amount *SOCAR Proceedings* **2** 16–22. DOI: 10.5510/OGP20190200384

[7] Akhmetov R T, Kuleshova L S and Mukhametshin V V 2019 Application of the Brooks-Corey Model in the Conditions of Lower Cretaceous Deposits in Terrigenous Reservoirs of Western Siberia *IOP Conf. Ser.: Mater. Sci. Eng.* **560(1)** 012004. DOI: 10.1088/1757-899X/560/1/012004

[8] Ishbaev G G, Dil`miev M R, Mileyko A A, Mamaeva O G, Garayev A V, Gabdrafikov R V, Polyakov V N and Chizhov A P 2017 Development and Experience of the Cement Gel Mud Gel-Drill Application on Tatyshlinskaya Field of the Republic of Bashkortostan *Drilling and Oil* **4** 23–26

[9] Romero S, Monroy R R, Johnson C, Cardenas F and Abraham G A T 2006 Preventing Lost Circulation by Use of Lightweight Slurries With Reticular Systems: Depleted Reservoirs in Southern Mexico *SPE Drilling & Completions J.* **21(3)** 185–192
[10] Roth J, Reeves C J, Johnson C R, DeBruijn G G, Bellabarba M, Le Roy-Delage S and Bulte-Loyer H 2008 Innovative Hydraulic Isolation Material Preserves Well Integrity *IADC/SPE Drilling Conf. (Orlando, Florida, March 4-6, 2008)* 14 p. DOI: 10.2118/112715-MS
[11] Rodgers J D 2006 *Report on Assessment of Technologies for Environmental Friendly Drilling Project: Land-Based Operations* (Houston Advanced Research Draft 3 Version 2 March 23 2006) 147 p
[12] Brown T J, Idoine N E, Raycraft E R, Shaw R A, Hobbs S F, Everett P, Deady E A and Bide T 2018 *World Mineral Production 2012-2016* (Keyworth: Nottingham British Geological Survey) 87 p

Process modeling of gas displacement of oil on the example of oil fields in Bashkortostan

E R Efimov[1], A P Chizhov[1,2], V Sh Mukhametshin[3], A V Chibisov[1,2], and V E Andreev[1,2]

[1] Institute of strategic research of the Republic of Bashkortostan, 129/3, October avenue, Ufa, Republic of Bashkortostan, 450075, Russian Federation
[2] Ufa State Petroleum Technological University, 1, Kosmonavtov st., Ufa, Republic of Bashkortostan, 450062, Russian Federation
[3] Ufa State Petroleum Technological University, Branch of the University in the City of Oktyabrsky, 54a, Devonskaya St., Oktyabrsky, Republic of Bashkortostan, 452607, Russian Federation

E-mail: intnm@ya.ru, vsh@of.ugntu.ru

Abstract. The article presents the results of the process modeling of gas displacement of oil. The proposed model takes into account the features of the state and changes in the physical and chemical state of formation systems on exposure to formation fluids with various gases-carbon dioxide and hydrocarbons. Numerical experiments made it possible to predict changes in technological parameters of impact objects and determine the incremental value in the oil recovery factor for Devonian fields of several oil-fields in Bashkortostan.

1. Introduction

It is known that when oil-pool developing under natural conditions, the oil recovery factor (ORF) is on average 10-15%. When using secondary and tertiary methods, the values of this indicator increase to 60-70 %. In some cases, this increase is absent, and if the wrong method of increasing oil recovery (EOR) is chosen, the ORF may be lower than when oil-pool developing under natural conditions, other things being equal [1-5].

In this regard, the justification and choice of a particular recovery method should be justified and should take into account the features of the geological structure of fields.

Methods based on the oil displacement from a porous medium by gas are considered to be one of the most promising EOR.

The oil displacement processes from a porous medium by gas, solvent and gas banks, water-gas mixture, carbon dioxide (CO_2) banks and water are accompanied by complex phenomena. Individual phases move at different speeds. The formation fluid composition changes in the collaborative filtering area. The mixture components are redistributed between the phases. This leads to composition changes, hence to the physical and chemical properties of the phases and under certain conditions to phase transitions (appearance, development or disappearance of individual phases). This process is described by the seepage theory of multicomponent mixtures [6].

To calculate the three-phase filtration process, the asymmetric Masket-Meres model is often used, in which the oil phase is represented by two components: gas (the volatile oil part) and oil (the nonvolatile part) [7, 8]. This model is not suitable in our case. It is based on the assumption that the saturation pressure and the volume factors are constant and phase viscosities and solubility of the gas in oil depending on the pressure.

When solving development problems using methods of gas injection on the bank, the use of the Masket-Meres model leads to significant errors [9, 10].

Compositional models describe the process most fully [11], but they are usually used for solving one-dimensional problems.

2. Materials and methods

To describe special cases of the studying process are executed symmetrical models of three-phase filtration of ternary mixtures, which are a generalization of the Masket-Meres model for the case with the time value of saturation pressure [11]. In our case, models are not suitable for calculations for several reasons: they have a limited dimension; they do not take into account the irregularity of distributions of geological and physical parameters; they also neglect the gravity segregation of phases. And oil is represented by only two pseudo-components: high volatile (gas) and nonvolatile (commercial oil).

In this paper, we propose a general three-dimensional model of three-phase filtration of quaternary mixtures, which, in turn, is a special case of the compositional model. The model allows you to take into account complex geological and physical, physical and chemical parameters and the technology of formation stimulation:

- heterogeneity in terms of permeability, porosity, the saturation thickness with formation fluids;
- pressure-temperature reservoir conditions, occurrence depth and total thickness of the field;
- presence fully or partially isolated interlayers in the formation;
- area of marginal and bottom water;
- presence of a gas cap;
- elastic properties of formation fluids and gas and porous medium;
- three-dimensional three-phase filtration of a multicomponent mixture with phase transitions and changes in the physical and chemical properties of these phases;
- non-stationary injection of water or gas, or a water-gas mixture with an arbitrary gas proportion in the mixture;
- non-stationary withdrawal of fluid and gas;
- different technologies for injection banks of gas, solvents, CO_2, and water;
- continuous, sequential, alternating, and joint.

The hydrodynamic filtration model of multicomponent mixtures can be divided into two parts. The first is based on the laws of material balance and fractional flows, the second - the laws of the phase behavior of the mixture. The conservation equations and phase equilibria are supplemented with initial and limiting conditions that reflect the initial state of the formation system and the technology of influencing it during development.

Let's make the following assumptions. Let filtration take place under isothermal conditions, and phase equilibria in each selected element (cell) of the filtering area occur instantly. We believe that the movement speed of individual phases obeys the generalized Darcy law, and the number of jointly filtered phases can be from one to three. We neglect the diffusion and capillary forces.

We consider a three-dimensional flow in a heterogeneous stratum. Let's choose a rectangular coordinate system. Let x, y be horizontal, and z be a vertical coordinate that coincides with the vector direction of the gravity acceleration g. The heterogeneity of the formation thickness is taken into account by specifying the distribution of absolute permeability $k = k(x, y, z)$, porosity $m = m(x, y, z)$, no-flow boundaries within the filtration area, initial pressure distributions, saturation and phase compositions.

A complex mixture of hydrocarbon and non-hydrocarbon components (formation oil) is represented as a mixture of three pseudo – components: I – high volatile pseudo-component (methane, nitrogen) k = 2-volatile NGLs, condensate, CO_2), k = 3-low-volatile (actually commercial oil is higher boiling hydrocarbons). We believe that the second pseudo – component can be contained in oil and gas, the first – in all three, and the third (water) - only in the water phase. Taking into account the accepted assumptions and designations, based on the general seepage theory of multicomponent mixtures, the process under consideration can be represented as the following simultaneous equations:

$$\nabla[\sum_{\alpha=1}^{3} K\, K_\alpha \mu_\alpha^{-1} c_{k\alpha}\rho_\alpha(\nabla P + \rho_\alpha g \nabla z)] + \sum_{j=1}^{1}\sum_{\alpha=1}^{3} q_k \rho_\alpha c_{k\alpha}\delta(\vec{r} - \overrightarrow{r_{cj}}) = \frac{\partial}{\partial t} mUc_k, \quad (1)$$

$$k = 1,2,3, \quad x \in [0, L], \qquad y \in [0, B], \qquad z \in [0, H]$$

$$\sum_{k=1}^{1} \mathrm{Ck} = 1, \quad (2)$$

where k_a – relative permeability of the phase α; μ_a – phase viscosity α; c_{ak} – weight content of the k component in the phase α; P – pressure; p_a – phase density α; q_a – source density (flow) of the phase α; σ – Dirac delta function; r_{cj}-coordinates of the j source (flow) (j-sources number); C_k - weight component of the k component in the mixture; L, B, H-characteristic length, width and thickness of the formation element; k – component index, k = 1, 2, 3 – pseudo – components of the oil and gas part of the mixture, K =4 – water; α – phase index, α =1-3 – gas, oil, and water, respectively.

The equation coefficients (1)k_a, μ_a, p_a, c_{ak} are pressure functions, compositions, and phase saturation, which are uniquely determined from the solution of the equations of phase equilibria, as will be shown below, by pressure setting and the ratio of the mixture. Therefore, the relations are valid for the values:

$$k_\alpha = k_\alpha(P, C_1, C_2, C_3);$$
$$\mu_\alpha = \mu_\alpha(P, C_1, C_2, C_3);$$
$$\rho_\alpha = \rho_\alpha(P, C_1, C_2, C_3);$$
$$c_{k\alpha} = c_{k\alpha}(P, C_1, C_2, C_3); \quad (3)$$

$$U = \sum_{\alpha=1}^{3} \rho_\alpha S_\alpha = U(P, C_1, C_2, C_3)\ , \quad \alpha = 1,2,3; \quad k = \overline{1,4},$$

where S_a is the saturation of the porous medium with the phase α, and

$$\sum_{\alpha=1}^{3} S\alpha = 1. \quad (4)$$

Thus, the equation system (1), (2) taking into account the relations (3) is formally complete concerning the independent variables P and C_k (k = 1.4).

The formality is that direct obtaining of partial characteristics (3) is possible only in some exceptional cases (for example, when filtering a single phase fluid, by piston swabbing one fluid by another).

In general, to construct dependencies (3), it is necessary to have an equation system that relates the pressure and ratio of the mixture to the composition and volume ratios of the phases. This property has an equation system describing phase equilibria.

The thermodynamical equilibrium condition of a multiphase mixture at plane interfaces and in the absence of external influence is the minimum of the Gibbs free energy. This means a mechanical (pressure balance in the phases), thermal (temperature balance) and chemical (chemical potentials balance or volatility of each component in the phases) equilibrium [6]. Under the above assumptions, the first two conditions are met, and the third can be written in the equivalent form:

$$x_{k1} = {}_{k12}\, x_{k12}, \quad k = 1.2,$$
$$x_{12} = æ_{123}\, x_{13}, \quad (5)$$

where x_{ka} is the mole fraction of the k component in the α phase; $æ_{ka\beta}$ is the distribution coefficient (equilibrium constant) of the k component between the α and β phases.

The use of relations (5) is convenient if the values of $æ_{ka\beta}$ can be determined experimentally or computationally (for example, using methods based on the principle of corresponding states [6] or state equations [6]. Note that calculating the characteristics of phase equilibria based on the ratios (5) reduces the volume of calculations by about two orders of magnitude in comparison with the state equations. System (5) should be supplemented with the following restrictive relations:

$$\sum_{k=1}^{3} x_{k2} = 1\,, x_{13} + x_{33} = 1, \quad (6)$$

$$\sum_{\alpha=1}^{3} L_\alpha = 1, \quad Z_k = \sum_{\alpha=1}^{3} x_{k\alpha} L_\alpha\,, \ k = 1,2,3, \quad (7)$$

where L_α is the mole fraction of the phase and in the mixture; Z_k Z_k is the mole fraction of component k in the mixture.

Mole and weight fractions are related by the following ratios:

$$Z_k = \frac{C_k}{M_k}\left(\sum_{l=1}^{4}\frac{C_l}{M_l}\right), \tag{8}$$

$$c_{k\alpha} = x_{k\alpha}M_k/(\textstyle\sum_{l=1}^{4} x_{l\alpha}M_{l\alpha}), \tag{9}$$

where M_k M_k is the molecular weight of the component k.

Thus, the mole fractions of the components in the mixture are calculated first from the ratio (8) based on the pressure setpoints and the mixture weight composition. Then, using a given binary phase diagram using the ratios (5)–(7), the mole fractions and phase compositions are calculated. Finally, the component weight fractions in the phases are calculated from the ratios (9).

Phase densities and viscosities are calculated either using known dependencies [6] or directly from experimental data. In the latter case, the experimental data can be approximated as follows.

The phase density:

$$\rho_1 = \rho_{10}\frac{P}{P_{st}}\frac{T_{st}}{T}\frac{1}{ZZ(P, C_{11})};$$

$$\rho_2 = \begin{cases} \rho_{20}3^{-B_2(P-P_{cm})}, & for\ P \le P^*_{sat} \\ \rho_2(P^*_{sat}) + \beta_2(P - P^*_{sat}), & for\ P > P^*_{sat} \end{cases};$$

$$\rho_3 = \rho_{30}[1 + \beta_3(P - P_{st})], \tag{10}$$

where $\rho_{\alpha 0}$– phase density α under standard conditions; P_{st}, T_{st}-standard pressure and temperature, P_{st} = 0.1013 MPa; T_{st} = 288.06 K.; T_{fm}-formation temperature, K; ZZ ZZ - real gas factor; B_2, β_2, β_3 B_2, β_2, β_3 - experimentally determined coefficients, MPa^{-1}; P_{sat} – current saturation pressure, a function of the mixture composition (determined by a binary phase diagram).

The phase viscosity:

$$\mu_1 = \mu_{10}(P - P_{st}^{A1});$$

$$\mu_2 = \begin{cases} \mu_{20}e^{-A_2(P-P_{cm})}, & for\ P \le P^*_{sat} \\ \mu_2(P^*_{sat}) + A_3(P - P^*_{sat}), & for\ P > P^*_{sat} \end{cases};$$

$$\mu_3 = \mu_{30}, \tag{11}$$

where $\mu_{\alpha 0}$ is the phase viscosity α under standard conditions; A_1, A_2 ,A_3 are experimentally determined coefficients, MPa^{-1}.

The mass density is related to the mole (p^*_α) ratio

$$\rho^*_\alpha = \rho_\alpha/M_\alpha, \tag{12}$$

where M_α is the molecular weight of the phase α, which can be calculated using the well-known k-rule:

$$M_\alpha = \textstyle\sum_{k=1}^{3} x_{k\alpha}M_k, \tag{13}$$

where M_k is the molecular weight of the component *k*.

The volume ratio (saturation) of phases are calculated from the ratio:

$$S_\alpha = \frac{L_\alpha}{\rho^*_\alpha}\Big/\sum_{\beta=1}^{3}\frac{L_\beta}{\rho^*_\beta}, \alpha = 1, 2, 3. \tag{14}$$

To problem solution of fluids and gas filtration on a limited formation volume, you must set the initial and boundary conditions.

As initial conditions, it is advisable to set the pressure distribution, compositions, and phase saturation:

$$P(x, y, z, o) = P^O(x, y, o) + \int_O^H \left(\sum_{\alpha=1}^{3} \rho_\alpha S_\alpha\right) g dz;$$

$$c_{k\alpha}(x, y, z, o) = c^o_{k\alpha}(x, y, z, o);$$

$$S_\alpha(x, y, z, o) = S^o_\alpha(x, y, z, o);$$

$$x \in [o, L],\ y \in [o, B], \quad z \in [o, H], \tag{15}$$

where P° is the initial pressure on the reference surface.

Boundary conditions are determined, as a rule, by the impermeability conditions of the roof and sole of the oil formation. The lateral surfaces of the formation can be isolated (discharge, wedging) or can communicate with a marginal water basin. The property of conditional isolation of the lateral surfaces has a so-called symmetrical section of the formation, which is characterized by zero pressure gradient at the border with neighboring elements (there are no flows between the elements). The impermeability condition of surfaces is a limiting condition of the 2nd kind, it can be written as follows:

$$q_\alpha^* = \frac{K\, K_\alpha}{\mu_\alpha} \frac{\partial P}{\partial n}\bigg|_G = 0, \qquad \alpha = 1, 2, 3, \tag{16}$$

whereq_α^* is the phase flow α through a certain surface G.

At the contact of an oil formation with a strong aquifer, the condition described by the limiting condition of the 3rd kind is fulfilled:

$$-a \frac{\partial P}{\partial n}\bigg|_{G-0} = b(< P > -P)\bigg|_{G-0}, \tag{17}$$

where a is the mobility ratio; b is the transmission coefficient; $<P>$ is the medium pressure in the strong aquifer.

Boundary conditions on the well can be of the 1st kind (set pressure) and the 2nd kind (set flow rate).

In the 1st case, using the well-known Dupuis formula, we can write:

$$q^* = \frac{2\pi HK\, K_\alpha \chi}{\mu_\alpha} \frac{p\,(r_c, t) - p(r_c, t)}{\ln(r_c/r_n)}, \quad \alpha = 1, 2, 3, \tag{18}$$

where χ is the participation share of the well in the selected formation element; r_c is the radius of the downhole area (cell) [11]; r_c , r_n is the well radius (absolute and reduced, respectively).

In the second case the phase flow rates can be approximated by the formula:

$$q_\alpha^* = q^* \frac{K_\alpha}{\mu_\alpha} \Big/ \left(\sum_{\beta=1}^{3} \frac{K_\beta}{\mu_\beta} \right), \qquad \alpha = 1, 2, 3, \tag{19}$$

where q* is the total production rate (injection capacity) of the well.

The source density (flow) appears in an equation (1) is equal to the ratio of the flow rate (injection capacity) to the volume of the selected near-well area:

$$q_\alpha = \frac{q_\alpha^*}{\pi (r_c^2 - r_c^2) H}. \tag{20}$$

The nonlinearity of 2nd-order partial differential equations describing the filtration of multicomponent mixtures, the three-dimensional problem formulation, and complex boundary conditions determine the possibility of problem solution only by numerical methods. One of the most effective of them is, in this case, the finite-difference method [9].

3. Results and Discussion

To solve this problem, a finite-difference scheme is constructed, which differs from the known ones [6, 9] in the following features. A completely conservative finite-difference scheme that approximates the filtration equations belongs to the tin "implicit pressure – explicit saturation". It is calculated based on the effective Newton iteration method not only for pressure but also for the composition of the mixture. Also, the minimum derivative principle is used for approximating the mobility ratios, a ternary phase diagram for calculating the partial characteristics of phase equilibria, and a block for automatic step selection of the time integration. All this significantly improves the efficiency of the numerical model.

To predict the technological efficiency of oil formation development with terrigenous, carbonate (reef) reservoirs, the influence methods of hydrocarbon and carbon dioxide on the displacement of hard-to-recover and remaining oil reserves are considered.

Theoretical researches to determine the optimal parameters of the hydrocarbon gas exposure technology was performed on two formation models D_I and $D_{II}$of the Tuymazinskij fields-with and without taking into account vertical flows of formation fluids. The research results after the corresponding generalization were as follows: the total flow gas 27-51% of reservoir void space, the specific technological efficiency of 0.31 tons of oil per 1 thousand m^3 of gas, the increase in ORF – 5.8–6.1 points.

Studies on the displacement of remaining oil of formations of the Vvedenovsky field have established that the maximum value of the displacement coefficient will reach 0.92 UF when: pressure build-up to the initial values of 13 MPa by pumping hydrocarbon gas, which will significantly restore the original physical and chemical properties of the oil; pumping the bank of the NGLs with a volume of 5.1 % of the pore volume; the displacement rate should not exceed the critical speed of phase separation.

Numerical modeling of the oil displacement process has shown that when dry gas is injected into the formation, the agent's influence covers almost the entire object. The central and middle parts of the object are produced until the residual oil saturation, and the peripheral parts are produced only partially. Calculations have shown that the technological efficiency will reach 2.1 t / t; the increase in ORF will be 9.5–10.1 points.

Determination of the optimal parameters of the CO_2+ water injection technology and calculation of the development parameters on the terrigenous reservoirs of the formations D_{KH}, D_I, D_{II}, D_{IV} of the Sergeyevsky field under normal (basic) flooding and using the method is performed using the proposed numerical model. As a result of calculations, it is established that the ratio of bank volumes of CO_2 and water on average should be 1:1.23; the total flow of CO_2 is 14% of the reservoir void space. The application of the proposed method will increase the ORF by 8.3–9.7 points, the specific efficiency will be 0.51 t/t.

4. Conclusion

Thus, the mathematical model of gas injection on hard-to-recover and remaining oil reserves was adapted taking into account the geological and field conditions for applying the impact, which allowed reducing the amount of calculations.

The results of mathematical modeling of the oil displacement process from the porous medium of reservoir rocks of the Republic of Bashkortostan fields due to the use of gas injection are presented. The incremental value in recoverable reserves for Devonian production facilities is in the range of 5-10 %.

References

[1] Mukhametshin V V 2018 Bottomhole formation zone treatment process modelling with the use geological and geophysical information *IOP Conf. Ser.: Earth Env. Sci.* **194(2)** 022024. DOI: 10.1088/1755-1315/194/2/022024

[2] Akhmetov R T and Mukhametshin V V 2018 Estimation of displacement coefficient with due account for hydrophobization of reservoir using geophysical data of wells *IOP Conf. Ser.: Earth Env. Sci.* **194(6)** 062001. DOI: 10.1088/1755-1315/194/6/062001

[3] Batalov D A, Mukhametshin V V, Dubinskiy G S and Andreev V E 2018 Laboratory grounding of waterproofing sealant based on acrylic polymers *IOP Conf. Ser.: Earth Env. Sci.* **194(4)** 042003. DOI: 10.1088/1755-1315/194/4/042003

[4] Malyarenko A M, Bogdan V A, Kotenev Yu A, Mukhametshin V Sh and Umetbaev V G 2019 Wettability and formation conditions of reservoirs *IOP Conf. Ser.: Earth Env. Sci.* **378(1)** 012040. DOI: 10.1088/1755-1315/378/1/012040

[5] Yakupov R F, Mukhametshin V Sh and Tyncherov K T 2018 Filtration model of oil coning in a bottom water-drive reservoir *Periodico Tche Quimica* **15(30)** 725-733

[6] Rosenberg M D, Kundin S A and Kurbanov A K 1969 *Filtration of carbonated liquid and other multicomponent mixtures in oil reservoirs* (Moscow: Nedra) 453 p

[7] Ptashko O A, Andreev V E, Chibisov A V and Chizhov A P 2014 Development of a complex surfactant for oil production processes, in: *Oil and Gas technologies and new materials.*

Problems and solutions: collection of proceedings (Ufa) pp 235-238

[8] Katz P M 1982 On a difference method for solving multiphase masket-Meres filtration equations *Proceedings of VNIPIneft* **79** 42-51

[9] Nguyen Q P, Rossen WR, Zitha P L J and Currie P K 2005 Determination of Gas Trapping With Foam Using X-Ray Computed Tomography and Effluent Analysis *SPE Journal* **14(02)** 222–236. DOI: 10.2118/94764-PA

[10] Ojukwu C, Onyekonwu M O, Ogolo N A and Ubani C 2013 Alkaline Surfactant Polymer (Local) Enhanced Oil Recovery: An Experimental Approach *SPE Nigeria Annual Int. Conf. and Exhibition* DOI: 10.2118/167529-MS

[11] Levi B I and Shakirov H G 1985 Numerical simulation of filtration of multicomponent mixtures *Mechanics of liquid and gas* **4** 101-110

IOP Conf. Series: Materials Science and Engineering 1064 (2021) 012061 doi:10.1088/1757-899X/1064/1/012061

Geological, technological and technical justification for choosing a design solution for drilling wells under different geological conditions

V G Mukhametshin[1], G S Dubinskiy[2,3], V E Andreev[2,3], V V Mukhametshin[3] and L S Kuleshova[4]

[1]SC «NizhnevartovskNIPIneft», 5, Lenina St., Nizhnevartovsk, Nizhnevartovsk region, Khanty-Mansi Autonomous Area – Yugra, 628616, Russia
[2] SASI "Institute for Strategic Research of the Republic of Bashkortostan", 129/3, Oktyabrsky Ave., Ufa, Republic of Bashkortostan, 450075, Russian Federation
[3]Ufa State Petroleum Technological University, 1, Kosmonavtov St., Ufa, Republic of Bashkortostan, 450062, Russian Federation
[4]Ufa State Petroleum Technological University, Branch of the University in the City of Oktyabrsky, 54a, Devonskaya St., Oktyabrsky, Republic of Bashkortostan, 452607, Russian Federation

E-mail: MuhametshinVG@nvnipi.ru, vv@of.ugntu.ru

Abstract. The oil companies focus on the prevention of leakage of the producing well to ensure the long-term well viability. The article considers the rationale for preventive measures to reduce the risks to the integrity of the strings and ensure the operation of the wells for the duration of their use. The authors propose tools for analysing the condition of wells as well as technical and technological solutions for extending the operation of the well without damage.

1. Introduction

The well fund that is operable for a long period and preferably throughout all field development process will provide the efficient development of oil and gas fields. It requires preservation of the porosity and permeability properties of the bottom-hole formation zone in a state as close as possible to natural parameters [1-7].

The well design must meet environmental protection conditions and exclude possible formation water pollution and pore fluid flows not only during drilling and operation but also after well completion and abandonment [8-10]. It is, therefore, necessary to ensure conditions for the high-quality and efficient separation of the strata. This is one of the main factors.

Thus, geological factors should fist determine the principles of designing well structures. Geological factors such as estimated and actual lithology, stratigraphy and section tectonics, rock thicknesses with different strengths, permeability, porosity, presence of fluid-saturated formations in the section and formation pressures have the greatest influence on well characteristics. When designing wells, the geological structure of the section of the rock is a constant factor.

During the development of hydrocarbon deposits, the initial formation characteristics change because the recovery rate of fluid, the methods of intensifying production and maintaining the formation pressures, the use of different influences on the productive horizons for fuller extraction of oil and gas from the subsoil, the duration of exploitation modifies the formation pressure and temperature. Designing well should consider all these factors due to the mentioned processes [8-10].

To ensure the best drilling conditions, the most efficient wiring technology and prevent possible complications, it is necessary to consider the following factors [9-13]:

- description of the rocks exposed by the well in terms of possible landslides, screes, cavity formation;
- rock hardness;
- rock permeability and formation (pore) pressure;
- zones of possible gas, oil and water show and absorption of the washing fluid and the conditions for the occurrence of these complications;
- temperature of rocks along the bore;
- angles of dips and the frequency of their alternation in hardness.

The detailed examination of the first four factors can justify the determination of the length (depth of descent) of the casing strings, i.e. to make a correct design decision.

2. Purpose

The study aims to show the justification of design solutions when choosing the wellbore trajectory and its structure based on the geological features of the field and technological processes for drilling and operating wells.

3. Methods and materials

Since it is necessary to study the influence of geological, technological and technical factors on the choice of the trajectory and design of the well and the justification for the choice, the justification of such recommendations requires methods of geological field analysis and modelling of trip processes during drilling and well operation. Analysis and modelling used information and materials from two oil fields located in Western Siberia and one in Bashkortostan.

4. Results and discussion

The study examined and analysed three objects with different geological and physical conditions: the Kipskoe field located on the territory of Bashkortostan, the Kusvoe field and the Samskoe field in the West Siberian region.

Design lithological-stratigraphic sections of deposits are based on data from prospecting and exploration. The section of the Kusvoe deposit contains permafrost rocks in the range of 180-400 m. Researches examined depths of up to 1833 m: quaternary sediments are soft, and hard rocks are deeper. The researchers examined the section of the Kipskoe deposit to a depth of 2920 m. The main rocks of the Quaternary are sand, sandy loam, clay, loam grey, dark grey soils. Deeper, there are mainly clay rocks, siltstones, less often limestone, sandy rocks (regionally oil-bearing). The lower part of the J3vs formation is represented by grey to black mudstones with rare interbeds of sandstones and siltstones.

The profile of well drilling should address the requirements for drilling cluster wells, the strength characteristics of the rocks that make up the geological section of the field, the methods and technical means used in the well operation. The well profile of the Kusvoe field was calculated by the Temporary technological regulation for the control and design of profiles at the fields of LLC LUKOIL-Western Siberia.

A well with a six-interval profile (Figure 1) has following parts: 1) the vertical section; 2) the first section of the set of zenith angle; 3) the stabilization section of the zenith angle; 4) the second section of the set of zenith angle; 5) the stabilization section of the zenith angle; 6) the section of low-intensity reduction of the zenith angle.

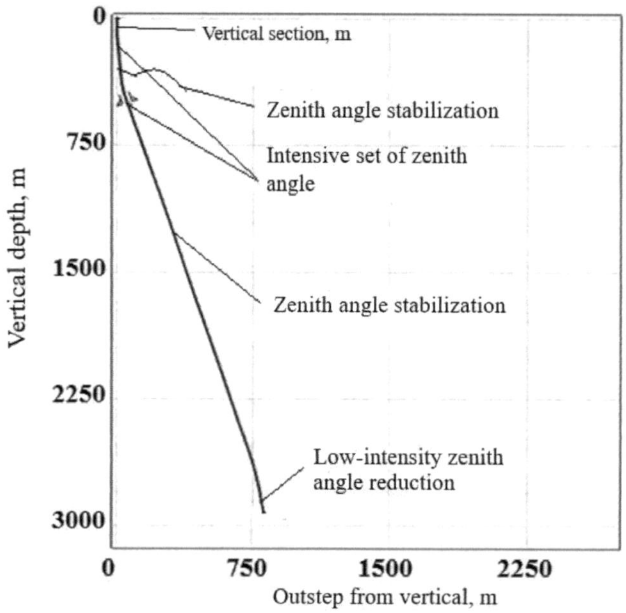

Figure 1. Design profile of the directional well bore of the Kusvoe field

When drilling the Kusvoe field, all work related to the set and adjustment of the zenith and azimuth angles of the directional bore should not extend beyond the interval of the planned depth of descent of the downhole pumping equipment (DPE). The necessity allows performing these tasks below the descent interval of the DPE.

Because in the section of the designed well of the Kusvoe field, there are no horizons with abnormally high reservoir pressures and zones with significant complications, then the geological section represents the area of compatible drilling conditions. The presence of permafrost, water-saturated formations in the context of the designed wells determines the choice of structures considering these features.

The well design solution of the Kusvoe field resulted from the planned tasks and the requirements of the "Safety Regulations in the Oil and Gas Industry".

The design of the production well should provide:

- maximum possible use of the formation energy of the productive horizons during operation due to the selection of the optimal bottom-hole designs and production string diameter;
- use of effective equipment, optimal methods and modes of operation, maintaining formation pressure, heat exposure and other methods of increasing oil recovery;
- conditions for safe work without accidents and complications at all stages of well construction and operation;
- obtaining the necessary geological information on the opened section;
- conditions for protecting the subsoil and the environment, primarily due to the strength and durability of the lining of the well, the tightness of the production strings and annular spaces, as well as the isolation of fluid-containing horizons from each other, from permeable formation and the daylight surface;
- maximum standardization by casings and wellbore sizes.

The geological features of the section of the Kusvoe field as well as the planned research complex, include the following well design features:

The direction with a diameter of 324 mm descends to a depth of 50 m to ensure reliable overlapping of the bedding interval of unstable deposits prone to screes and landslides and prevent erosion of the wellhead. The cement lifting to a depth of 50 m to the wellhead.

A well conductor with a diameter of 245 mm descends to a depth of 500 (680) m vertically or 505 (694) m along the wellbore to ensure reliable overlapping of the bed interval of permafrost rocks and unstable rocks prone to collapse of the upper part of the Lyulinvor formation. It requiers two portions of cementation.

The production string with an external diameter of 146 mm descends to a depth of 2915 m vertically (3036 m along the wellbore). It serves to attach the well walls, separate the production horizons and test them. The height of the cement slurry above the roof of the productive horizon is at least 150 m The distance from the wellhead to the level of cement slurry rise behind the production string for injection wells is 260 m.

Drilling through a layer of permafrost, loose sandstones and unstable clay sediments requires solving the following main tasks: strengthening of well walls, reducing the thawing effect of the drilling fluid, increasing the drilling mud capacity at a relatively low ascending flow velocity. The solution of these problems needs the clay drilling fluids with a high content of active clay phase, highly effective polymers - structural agents, and the use of flocculating chemicals maintaining a low temperature. For drilling under the conductor, the project envisages drilling solution prepared from bentonite clay powder treated with chemical reagents. To open the productive formation, it is necessary to use a solution with a minimal negative effect. Therefore, in the conditions of the Kusvoe field, the clay solution for the initial opening of the oil formation is the suspension of high-active bentonite clay with the addition of chalk treated with soda ash, lubricants and surfactants to reduce the surface tension of the filtrate and hydrophobize the surface of the pore channels of the reservoir formation.

The geological features of the Samskoe field are similar to those of the Kusvoe field, and therefore the wellbore profile has similar sections of the wellbore (Figure 2). Drilling features are also similar except for the presence of separate absorption zones.

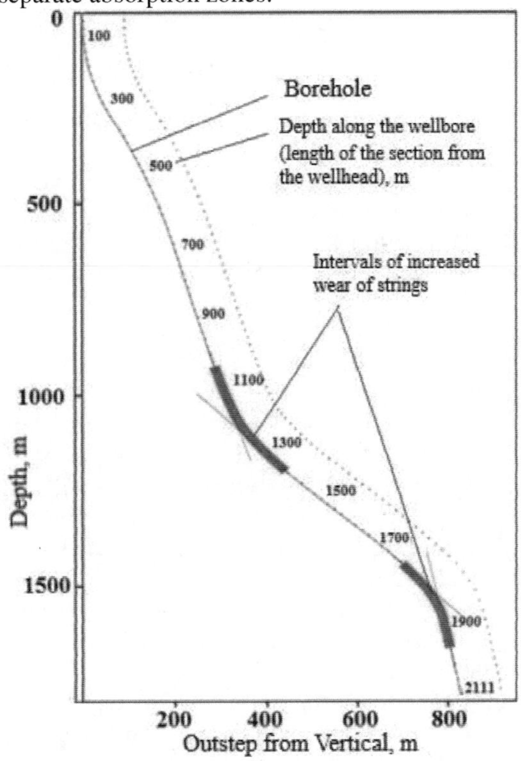

Figure 2. Profile of the directional wellbore of the Samskoe field

The geological features of the Kipskoe field have led to the selection of a three-interval profile (Figure 3): 1) vertical section, 2) set of zenith angle, 3) holed section. The sections of the profile are in a vertical plane passing through the design bottom hole and the wellhead. This orientation creates an optimal profile for the economical and rational operation of wells with sucker rod pumping units.

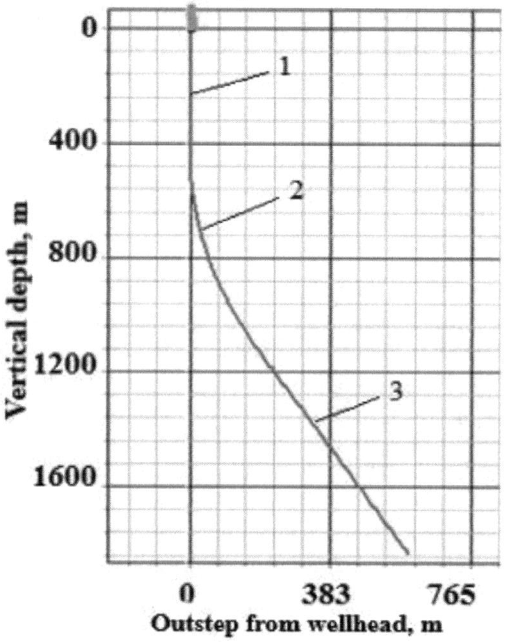

Figure 3. Scheme of the three-interval profile of the Kipskoe field

The geological features of the Kipskoe field section, the formation pressure and hydraulic fracturing determined the following well design:

- the direction (50 m) isolates the absorbing horizon and the freshwater complex, fixes the unstable upper rocks and prevents the sticking of the drill string;

- a conductor (200 m) overlaps the freshwater complex in the interval of 50-100 m, isolates the horizons, fixes the areas of scree and landslides and prevents the sticking of the drill string in the interval of 50-190 m;

- intermediate string (to a depth of 400 m) overlaps the saline deposits of the Kungur formation in the interval of 260–290 m;

- production string (to a depth of 1826 m) for fixing zones of scree and landslides in the intervals of 1222–1258 m, 1696–1702 m, overlapping absorbing horizons in the intervals of 467–487 m, 1025–1110 m, 1390–1670 m.

The project of the Kipskoe field provides the production strings with a diameter of 168 mm. Solid and stable limestones composing the production facility, the absence of high-pressure horizons, bottom water and gas caps lead to a lack of fastening and it remains open. This design of the well structure increases the well perfection coefficient.

Even under different geological conditions, the paths of the wellbores have bending zones (changes in azimuth and zenith angles). Of course, the bends of the wellbore predetermine the appearance of leakages in the strings due to the increased load during tripping and lowering operations (TLO) during drilling, especially when drilling below the string shoe, and well repair.

To increase the efficiency of well drilling and improve design methods for designing profiles of directional and horizontal wells, it is necessary to model the wells and operations in them [14]. For the correct design of the trajectory of the barrel and the string, preservation of its stability and reliability during the entire period of operation, it is necessary to evaluate the possible state of the pipe string during the TLO.

To simulate the stress-strain state of the production string (PS), we used the software in the field of drilling applications of the Landmark Halliburton company: COMPASS, WELLPLAN. The determination of the distribution of lateral forces along the wellbore used the module "Analysis of tension and moments" which allows predicting and analysing the moments and axial loads that occur in the drill string, casing string or in the shanks during the TLO.

The authors used the fixed-ended column model that considers the position of the drill pipes relative to the axis of the well, determined the contact points of the drill pipes with the casing string which permits calculating lateral forces with the position of the contact points (couplings, locks). For our case, the calculated contact points of the drill pipe with the casing will be close to the behaviour of the tubing during tripping operations of the current workover and capital workover operations.

In a virtual well with a minimum deviation from the vertical of 150 m (radius of curvature), we modelled the stress-strain state of drill pipes for a well with the design shown in Figure 4.

An analysis of the results of modelling a virtual well in the conditions of the Samskoe field showed that at depths of 1000–1180 m in the virtual well there will be contact between the drill pipes and the production string, that is, damage in the form of the wiping troughs.

When modelling the stress-strain state at a real well No. 8082 of the Samskoe field, we found that the main loads on the string are on the upper part of the well. In all cases, we also distinguish the lower zone as the second risk zone in the section of the string 1400–1600 m. The position of the string during TLO indicates the possibility of the tool touching the walls of the PS and consequently causing friction and wear.

Geophysical studies in well 8082 revealed two leakage intervals in 1460–1485 m and 1597–1606 m

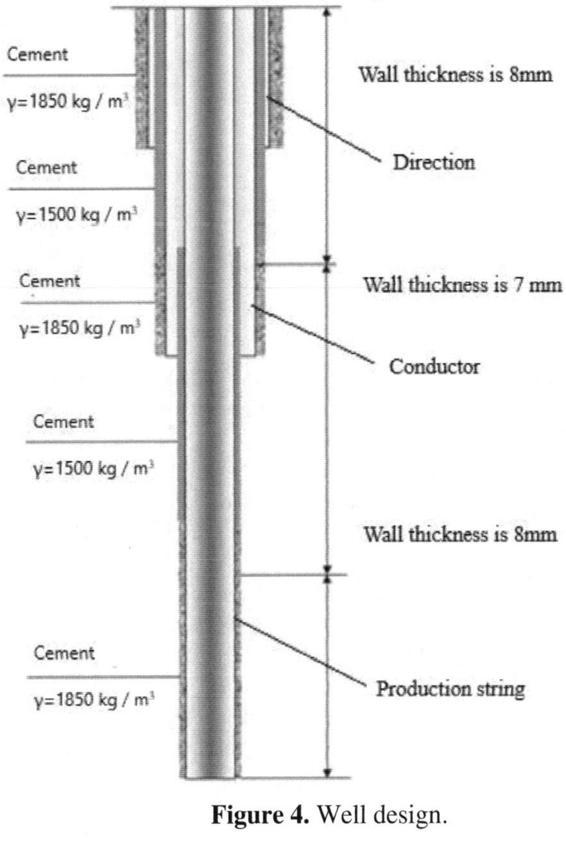

Figure 4. Well design.

We performed the same analysis of loads and deformations along the wellbore for other wells. For example, in well 16547 we identified:

- the main loads on PS arise in the upper part of the bore, all loads decrease with depth but at the same time remain significant to a depth of 1600 m;
- the position of the string when the tool lowers and raises indicates the possibility of the tool touching the PS walls in the depth interval of 1560–1680 m;
- according to the geophysical logging results in well 16547, we identified one leakage interval of 1557–1582 m.

Based on the analysis of 180 wells of the Samskoe field, it was revealed that inferiority in PS leakage in 49% of cases occurs due to the influence of the zenith angle of the wellbore (areas with pronounced touches of wellbores and straight paths in the interval of PS leakage), in 6% of the wells, the azimuthal angle, for other reasons - 45%.

It is obvious that at the points of bending of the barrel (changing the angles of the trajectory) there is an increased risk of leakage of the string which requires preventive measures to increase the efficiency of the wells.

The profiles of directional and horizontal wells in these fields must satisfy the following criteria: ensuring normal patency of the various configurations of the drilling tool and casing strings; the possibility of launching devices during the completion of wells and their operation; to ensure that the well and sidetracks get into the given circle and the tolerance corridor (vertically and horizontally); designing the trajectory and construction of the well should minimize the possibility of leakage of PS during further operation.

5. Conclusion

The article reveals the possibility of assessing the impact of a decent tool (drilling pipes, etc.) by touching-friction on the wells PS.

Modelling the stress-strain state of PS using the Landib Halliburton package allows predicting the risks that relate the tool operating in the well. Prediction of stress-strain state of PS allows designing a more stable and reliable design of wells.

Based on research and modelling for the Samskoe field, we propose to prevent the leakage of the PS of the new well stock from drilling wells with gently sloping trajectories, excluding the intersection of straight trajectories and production strings below 1300 m (below the dynamic level intervals of operating submersible electric centrifugal plants and/or increase the strength of the problem sections of the string.

References

[1] Mukhametshin V V and Kuleshova L S 2019 Justification of Low-Productive Oil Deposits Flooding Systems in the Conditions of Limited Information Amount *SOCAR Proceedings* **2** 16–22. DOI: 10.5510/OGP20190200384

[2] Akhmetov R T, Kuleshova L S and Mukhametshin V V 2019 Application of the Brooks-Corey Model in the Conditions of Lower Cretaceous Deposits in Terrigenous Reservoirs of Western Siberia *IOP Conf. Ser.: Mater. Sci. Eng.* **560(1)** 012004. DOI: 10.1088/1757-899X/560/1/012004

[3] Kuleshova L S, Kadyrov R R, Mukhametshin V V and Safiullina A R 2019 Design changes of injection and supply wellhead fittings operating in winter conditions *IOP Conf. Ser.: Mater. Sci. Eng.* **560(1)** 012072. DOI: 10.1088/1757-899X/560/1/012072

[4] Kuleshova L S, Kadyrov R R, Mukhametshin V V and Akhmetov R T 2019 Auxiliary equipment for downhole fittings of injection wells and water supply lines used to improve their performance in winter *IOP Conf. Ser.: Mater. Sci. Eng.* **560(1)** 012071. DOI: 10.1088/1757-899X/560/1/012071

[5] Yakupov R F, Mukhametshin V Sh, Khakimzyanov I N and Trofimov V E 2019 Optimization of Reserve Production From Water Oil Xones of D3ps Horizon of Shkapovsky Oil Field Using Horizontal Wells *Georesursy* **21(3)** 55–61. DOI: 10.18599/grs.2019.3.55-61

[6] Stenkin A V, Kotenev Yu A, Mukhametshin V Sh and Sultanov Sh Kh 2019 Use of Low-Mineralized Water for Displacing Oil from Clay Productive Field Formations *IOP Conf. Ser.: Mater. Sci. Eng.* **560(1)** 012202. DOI: 10.1088/1757-899X/560/1/012202

[7] Valeev A S, Kotenev Yu A, Mukhametshin V Sh and Sultanov Sh Kh 2019 Substantiation of the Recovery of Residual Oil from Low-Productive and Heterogeneous Formations in Western Siberia by Improving the Waterflood System Using Gas and Water-Gas Impacts *IOP Conf. Ser.: Mater. Sci. Eng.* **560(1)** 012204. DOI: 10.1088/1757-899X/560/1/012204

[8] Bulatov A I and Savenok O V 2010 *Completion of oil and gas wells. Theory and practice* (Krasnodar: Prosveshchenie-Yug) p 542

[9] Kalinin A G, Ganjumyan I R and Messer A G 2005 *Reference book of the engineer-technologist on drilling of deep wells* (Moscow: Nedra) p 808

[10] Milshtein V M 2003 *Cementation of drilling wells* (Krasnodar: Prosveshchenie-Yug) p 375

[11] Hossain M E and Al-Mejed A A 2018 *Fundamentals of Sustainable Drilling Engineering* (New York: John Wiley & Sons Limited (USD)) p 785

[12] Jimenez W C, Pang X, Benge M, Darbe R and Meadows D L 2017 Enhanced Mechanical Integrity Characterization of Oilwell Annular Sealants Under In-Situ Downhole Conditions: A Path to Optimized Wellbore Economics *Society of Petroleum Engineers* DOI: 10.2118/185341-MS 16 p

[13] Kendall Mcclendon 2014 *Drilling Technology* (Gardners Books) p 203

[14] Dvoynikov M V 2018 Designing of Well Trajectory for Efficient Drilling FOR by Rotary Controlled Systems *J. of Mining Institute* **231** 254-262. DOI: 10.25515/PMI.2018.3.254

IOP Conf. Series: Materials Science and Engineering 1064 (2021) 012062 doi:10.1088/1757-899X/1064/1/012062

Modeling the production of hard-to-recover oil reserves using thermal methods

A A Gizzatullina, T N Minnivaleev and I Sh Mingulov

Ufa State Petroleum Technological University, Branch of the University in the City of Oktyabrsky, 54a, Devonskaya St., Oktyabrsky, Republic of Bashkortostan, 452607, Russian Federation

E-mail: alina.gizzatullina87@mail.ru, info@of.ugntu.ru

Abstract. This article presents the theoretical and numerical implementation of the possibility of heating a reservoir with hard-to-recover oil. In a radial one-dimensional formulation, the article proposes a solution to the problem of hard-to-recover oil filtration with an increased viscosity coefficient. The procedure is performed in a reservoir using the thermal method of exposure in a single horizontal channel. Hot steam or water flows are used as a heat carrier. Only one channel is used as a heated and production well, which operates alternately.

1. Introduction

A specific feature of the current development stage of the Russian oil industry is that oil reserves are being replaced by hard-to-recover ones [1-3]. The lack of scientific and methodological foundations for the development of deposits with hard-to-recover reserves makes it urgent to solve the problems of modeling various processes in order to increase the efficiency of oil reserves recovery [6–8].

Among the reservoirs with hard-to-recover reserves, a large group of fields contains high-viscosity oils. In this regard, the problem of modeling oil production processes using thermal methods is an urgent issue.

2. Materials and methods

Let us consider the problem where a single horizontal channel operates in the process of thermal impact on a formation rock with hard-to-recover oil (Figure 1). The system consisting of the equations of continuity and heat flow for an oil-saturated reservoir ($r_c < r < \infty$) is as follows:

$$\frac{\partial m\rho_l}{\partial t} + \frac{1}{r}\frac{\partial}{\partial r}(rm\rho_l \upsilon) = 0,$$
$$\rho c\frac{\partial T}{\partial t} + \rho_l m c_l \upsilon \frac{\partial T}{\partial r} = \frac{\lambda}{r}\frac{\partial}{\partial r}\left(r\frac{\partial T}{\partial r}\right), \qquad (1)$$

where r is the radial coordinate, which is measured from the center of the well, λ is the thermal conductivity, υ is the oil velocity, ρc is the specific heat capacity of the rock, T is the temperature of the rock, r_c is the radius of the channel, m is the porosity, ρ_l and c_l are the density and heat capacity of oil.

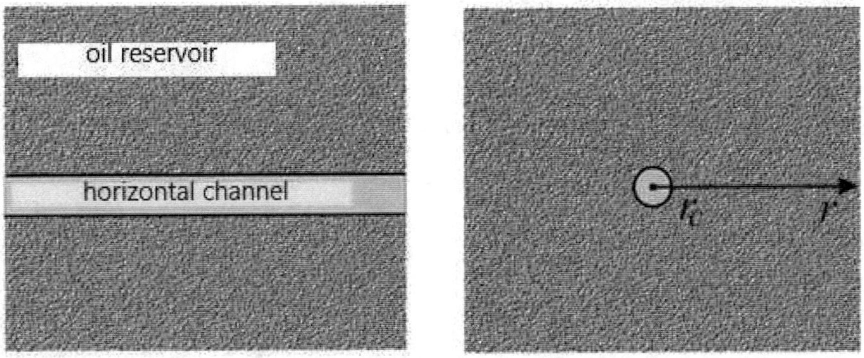

Figure 1. Scheme of using a horizontal channel in an oil-saturated reservoir as a heating and production well, operating alternately

3. Results and Discussion

Darcy's law helps to determine the rate of oil filtration in a porous medium in the following form:

$$m\upsilon = -\frac{k}{\mu(T)}\frac{\partial p}{\partial r}. \tag{2}$$

Here p is the pressure, k is the permeability of the porous medium.

We write the equation of state for oil in the linear approximation [8,12]

$$\rho_l = \rho_{l0}\left(1-\alpha^{(T)}(T-T_0)+\alpha^{(p)}(p-p_0)\right). \tag{3}$$

The subscripts 0, $\alpha^{(T)}$ and $\alpha^{(p)}$ correspond to the initial values of oil parameters - the coefficients of compressibility and thermal expansion of oil.

In this problem statement, the viscosity-temperature dependence of $\mu(T)$ for heavy or bituminous oil is as follows:

$$\mu(T) = \mu_0 e^{-\gamma(T-T_0)}, \tag{4}$$

where μ_0 is the oil viscosity parameter at the initial temperature of the medium T_0, γ is the temperature coefficient.

From expressions (1), considering (2) and (3), we obtain new expressions for temperature and pressure:

$$\frac{\partial p}{\partial t} = \frac{\alpha^{(T)}}{\alpha^{(p)}}\frac{\partial T}{\partial t} + \frac{k}{m\alpha^{(p)}r}\frac{\partial}{\partial r}\left(\frac{r}{\mu(T)}\frac{\partial p}{\partial r}\right),$$
$$\frac{\partial T}{\partial t} = \frac{k\rho_l c_l}{\mu(T)\rho c}\frac{\partial p}{\partial r}\frac{\partial T}{\partial r} + \frac{\nu}{r}\frac{\partial}{\partial r}\left(r\frac{\partial T}{\partial r}\right). \tag{5}$$

Here $\nu = \lambda/(\rho c)$ is the thermal diffusivity of the formation. The resulting system of expressions (5) makes it possible to simulate changes in temperature and pressure in an oil-saturated reservoir, thus making it possible to describe the process of oil recovery with an increased viscosity index in detail.

Development of a field with hard-to-recover oil involves continuous thermal impact on the formation and horizontal drilling [15]. The production of traditional easily recoverable oil has large-scale costs for the construction of sites where the field is being developed [10, 11]. Heavy oil makes everything different: the initial investment may not be impressive, but the costs during the operation of the well will be substantial. Monitoring heat consumption and oil production is an integral part of the development of a field with high-viscosity oil under thermal influence.

We formulate the equation for the total heat flux from the well to the reservoir:

$$q^{(T)} = 2\pi r_c \lambda \left(\frac{\partial T}{\partial r} \right)_{r_c} . \tag{6}$$

The total spent heat will be determined from the expression:

$$Q = \int_0^t q^{(T)} dt . \tag{7}$$

In this work, we used the formula described in [13, 14]:

$$\mu(T) = \mu_o e^{-\gamma(T - T_o)} , \tag{8}$$

where γ is the temperature coefficient, μ_0 is the oil viscosity at a temperature of T_0.

For the mass flow rate of oil from the well, considering Darcy's law, we write the equation

$$q^{(\upsilon)} = -2\pi r_c \rho_l \frac{k}{\mu(T)} \left(\frac{\partial p}{\partial r} \right)_{r_c} . \tag{9}$$

The total oil produced can be calculated from the expression

$$M = \int_0^t q^{(\upsilon)} dt . \tag{10}$$

Let us consider the 1st stage of thermal impact on the oil reservoir. At this stage of development, we can study the heating of oil-saturated rock ($t < t_1$), while the channel used is closed and oil withdrawal is not carried out. At the channel boundary ($r = r_c$), the pressure gradient is equal to zero $(\partial p / \partial r = 0)$ and a constant temperature is maintained at T_e far from the well($r \to \infty$). The pressure and temperature of the formation are equal to the initial values p_0, T_0. This condition means that the thickness of the oil-bearing layer is much greater than the area where filtration and temperature drops are observed for the time interval of interest to us. At the initial stage of development ($t = 0$) the pressure and temperature are p_0, T_0 respectively.

The accepted system of expressions (5) under the given boundary conditions was solved using an explicit scheme using the finite difference method. For the numerical implementation of the numerical solution, a uniform grid was constructed in the coordinate $r_i = r_c + (i-1) \cdot h$ (i=1, 2...N) and in time $t = n \cdot \tau$. After replacing the differential operators with their finite-difference analogs, we obtain a system of linear algebraic equations (5):

$$p_i^{j+1} = p_i^j + \frac{\alpha^{(T)}}{\alpha^{(p)}} \left(T_i^{j+1} - T_i^j \right) + \frac{k \cdot \tau}{m \alpha^{(p)} r_i h^2} \left(\frac{r_{i+1}}{\mu\left(T_{i+1}^j\right)} \left(p_{i+1}^j - p_i^j \right) - \frac{r_i}{\mu\left(T_i^j\right)} \left(p_i^j - p_{i-1}^j \right) \right),$$

$$T_i^{j+1} = T_i^j + \frac{k \rho_l c_l \tau}{\mu\left(T_i^j\right) \rho c} \cdot \frac{p_{i+1}^j - p_i^j}{h} \cdot \frac{T_{i+1}^j - T_i^j}{h} + \frac{\nu \tau}{r_i h^2} \cdot \left(r_{i+1} \left(T_{i+1}^j - T_i^j \right) - r_i \left(T_i^j - T_{i-1}^j \right) \right),$$

$$\mu\left(T_i^j\right) = \mu_0 e^{-\gamma\left(T_i^j - T_o\right)} \tag{11}$$

The boundary conditions will be:

$$p_1^j = p_2^j ; \; p_N^j = p_0 \; T_1^j = T_c ; T_N^j = T_0 . \tag{12}$$

The boundary condition at $r = \infty$ is shifted to the boundary of the grid $r = r_{(\infty)}$.

The Courant condition determined the stability conditions for the difference scheme. For this, the following values were calculated, the smallest of which is taken as the step of integration over the time coordinate, i.e. $\tau = \min(\tau_1, \tau_2, \tau_3)$:

$$\tau_1 = \frac{h^2}{2\nu}; \tau_2 = \frac{\mu h^2 \rho c}{2k\rho_l c_l}; \tau_3 = \frac{m\alpha^{(p)}\mu h^2}{2k} \tag{13}$$

The study performed numerical calculations for the following parameter values: $p_0 = 0.5\,MPa$, $T_c = 160^{\circ}\mathrm{C}$, $r_c = 0.05\ m$, $\lambda = 1.28\,W/(\mathrm{m}\cdot\mathrm{K})$, $\rho_l = 980\,kg/m^3$, $c_l = 1400\ J/(\mathrm{K}\cdot kg)$, $\rho c = 2.5\cdot 10^6\ J/(\mathrm{K}\cdot m^3)$, $\nu = 10^{-6}\,m^2/s$, $\alpha^{(T)} = 10^{-5}\,\mathrm{K}^{-1}$, $\alpha^{(p)} = 5\cdot 10^{-9}\ Pa^{-1}$, $k = 10^{-12}\,m^2$, $m = 0.3$.

Figure 2 is a graphical illustration of the viscosity-temperature dependence for oil products at an initial value of the viscosity index $\mu_0 = 10\ Pa\cdot s$ ($\gamma = 0.047\mathrm{K}^{-1}$). It follows from the graph that a high rate of decrease in viscosity is observed at the initial stage of temperature increase. Furthermore, the viscosity slowly decreases along with further heat exposure.

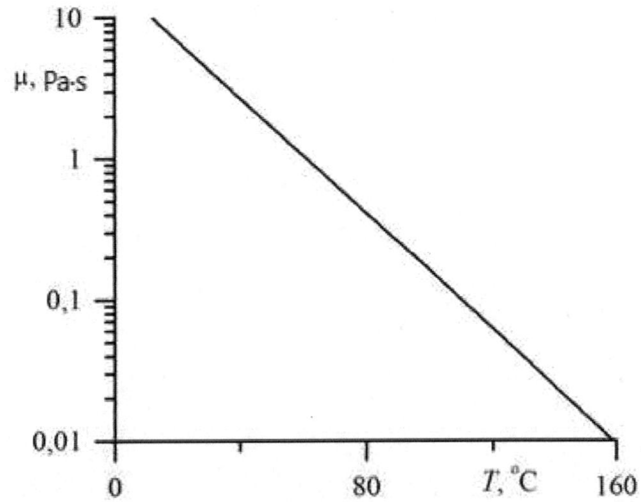

Figure 2. The dependence of the viscosity of oil μ on temperature T

It should be noted that the quantitative dependences of viscosity on temperature are very different for different oils. However, qualitatively they are approximately the same. Assuming that the viscosity of liquefied oil in the temperature range under consideration is 10 times higher than the viscosity of water, the expression γ is obtained for the coefficient $\gamma = 0{,}047\mathrm{K}^{-1}$. The maximum rate of decrease in the viscosity parameter is observed at the initial stage of temperature increase; the intensity of viscosity decrease slows down with further heating.

Figure 3 shows a graphical illustration of the field of pressure dynamics (a), temperature (b) and viscosity (c) in a reservoir with high-viscosity oil at different periods of heating time t_l = 10, 20, 30 days.

At this stage of the thermal effect on high-viscosity oil, a certain area of several meters is located near the heating boundary, which leads to an increase in pressure in the reservoir, the maximum value of which is reached at the boundary. The pressure increase due to thermal expansion of oil will be insignificant (less than 0.1 MPa). As the temperature rises in the reservoir, the viscosity of the produced product decreases in the boundary region of the well. The thickness of the oil-saturated layer determines the time of thermal impact on the reservoir t_1.

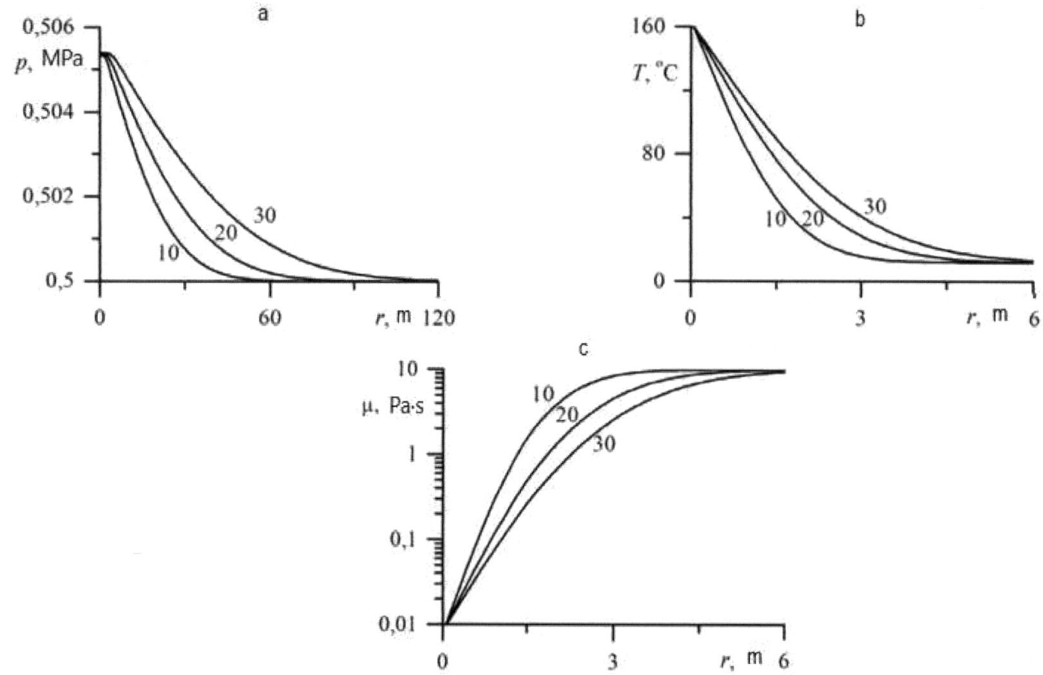

Figure 3. Distribution of pressure (a), temperature (b) and viscosity (c) along the coordinate r at different times of heating t = 10, 20 and 30 days

It is proposed to consider the holding stage, which can be proposed to increase oil recovery after the heating stage ($t_1 < t < t_2$). At this stage of reservoir development, we perform shutdown until a uniform temperature distribution in the reservoir is achieved. At this stage of the thermal effect, we reduce the risk of destruction of the bottomhole region of the well during its development in the "overheated" state. Then the absence of heat flux $\partial T/\partial r = 0$ at $r = r_c$ will be the boundary condition for the second equation from (5). The distributions of pressure $p = p(r)$ and temperature $T = T(r)$ obtained from the previous solution are the initial conditions at $t = t_1$.

In this case, we solve the standard system of linear algebraic equations (11) with the following boundary conditions:

$$p_1^j = p_2^j;\ p_N^j = p_0;T_1^j = T_2^j;T_N^j = T_0 \quad (14).$$

It should be noted here that the values τ_2 and τ_3 in formulas (13) are variables. Therefore, we use an uneven time grid in the numerical implementation of the problem for this stage, where the value τ is recalculated at each time step. This approach allows you to increase the speed of the task counting.

Figure 4 shows an illustration of the evolution of the pressure field (a), temperature (b) and viscosity (c) after stopping heating of the oil-saturated reservoir at different time intervals t_2 = 10, 20 and 30 days.

At the stage of holding during the considered time, there is an alignment of the heated area in the formation and an increase in the heated area with a reduced viscosity parameter. Effective use of the thermal method of influencing the oil reservoir and increasing oil recovery is achieved by increasing the heated zone with a reduced viscosity parameter.

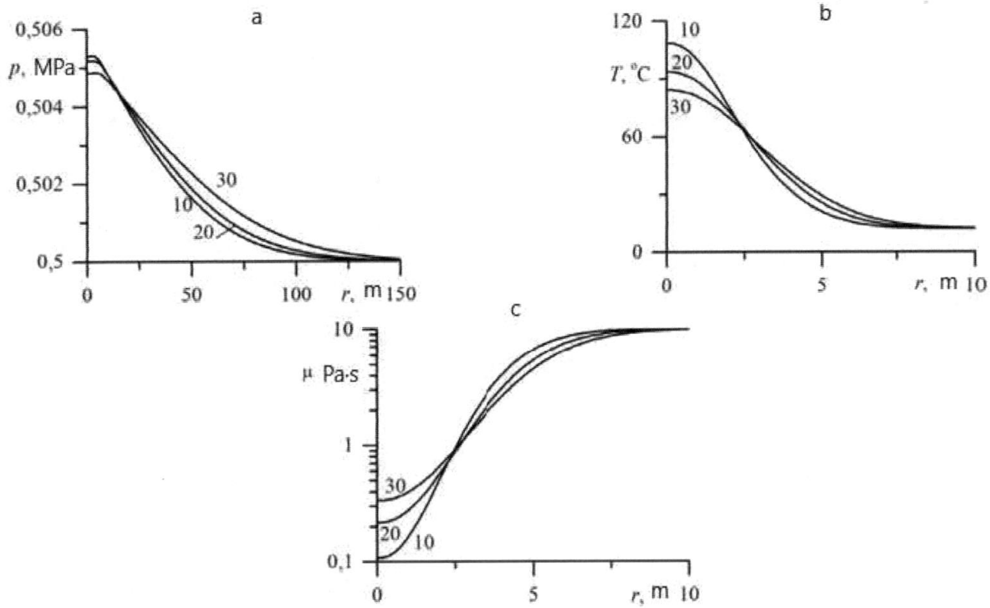

Figure 4. Distribution of pressure (a), temperature (b) and viscosity (c) along the coordinate r at different moments of the well holding timet = 10, 20 and 30 days

Let us consider the stage of channel functioning ($t > t_2$), where near the boundary $r = r_c$ there is a heated region with a reduced viscosity parameter, i.e. at the initial moment of time ($t = t_2$) the distributions of pressure *p=p(r)* and temperature $T = T(r)$ are taken from the previously obtained solutions for the heating and holding stage. At this stage of development, we will solve the system of equations for pressure and temperature (5) with the following boundary conditions: at the boundary $r = r_c$ the heat flux is absent $\partial T/\partial r = 0$ and constant pressure p_c ($p_c < p_0$), is maintained; at the section ($r \to \infty$) remote from the well, the initial parameters are temperature and pressure T_0, p_0.

After passing to finite-difference analogs, the system (11) is solved with the boundary conditions:

$$p_1^j = p_c^j;\, p_N^j = p_0;\, T_1^j = T_2^j;\, T_N^j = T_0. \qquad (15)$$

It also uses an uneven temporal grid.

Figure: 5 illustrates the results of calculations for the pressure field (a), temperature (b) and viscosity (c) at different filtration times of 2, 5, and 7 days. The well pressure takes the value $p_c = 0.3$ MPa, the heating time of the oil reservoir is $t_1 = 30$ days, and the holding time is $t_2 = 30$ days. The dashed line corresponds to the case of the formation heating during the time $t_1 = 30$ days

The propagation of the filtration wave front is much deeper than the temperature differences. Later on, a noticeable decrease in temperature happens in the developed part of the formation near the studied channel and the viscosity value increases. As a result, there is a decrease the rate of oil filtration in the porous medium.

Without the stage of keeping the oil reservoir in a closed state, the rate of temperature decrease increases near the well. If there is a holding stage, then the temperature spreads deep into the formation; an intensive selection of accumulated heat from the oil-bearing formation happens during filtration. The stage of formation holding significantly reduces the threat of destruction of the well bottomhole zone due to severe overheating and can contribute to an increase in the filtration period of products with a reduced viscosity index.

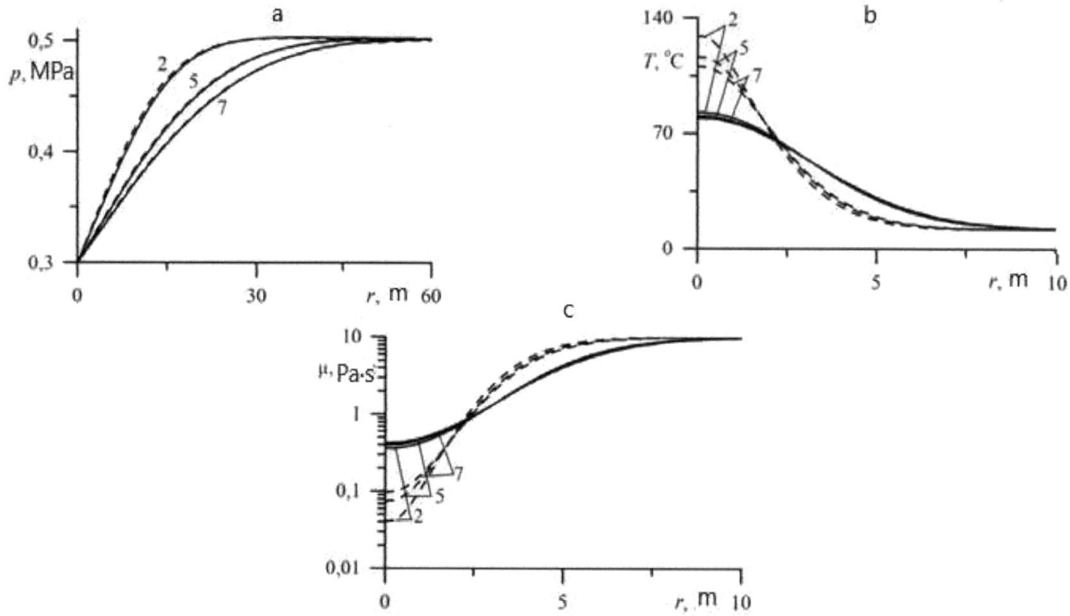

Figure 5. Distribution of pressure (a), temperature (b) and viscosity (c) along the coordinate at different times of filtration of 2, 5, and 7 days. The dashed line corresponds to the case without reservoir holding.

4. Conclusion

The article discusses a numerical study of the filtration process of high-viscosity oil through one horizontal well, which is used first as a heating channel, and then as a production well. A system of differential equations for pressure and temperature is obtained for given boundary conditions. The system of equations is solved numerically by the finite difference method according to an explicit scheme. The task is solved in three stages. At the first stage, the oil reservoir is heated, which makes it possible to control the heating of the reservoir. The heating time depends on the thickness of an underlying rock. At the second stage, holding takes place in closed conditions to enhance oil recovery until the most uniform temperature distribution in the oil reservoir is achieved. At the third stage of the well operation, heated oil with a reduced viscosity index is filtered through the horizontal channel under study. The proposed method improves the efficiency of the development of an oil reservoir with hard-to-recover oil due to the possibility of uniform heating of the studied field area.

References

[1] Mukhametshin V V and Kuleshova L S 2020 On uncertainty level reduction in managing waterflooding of the deposits with hard to extract reserves *Bull. of Tomsk Polytechnic University. Geo Assets Engineering* **331(5)** 140–146. DOI 10.18799/24131830/2020/5/2644

[2] Yakupov R F, Mukhametshin V Sh and Tyncherov K T 2018 Filtration model of oil coning in a bottom water-drive reservoir *Periodico Tche Quimica* **15(30)** 725-733

[3] Mukhametshin V V and Andreev V E 2018 Increasing the efficiency of assessing the performance of techniques aimed at expanding the use of resource potential of oilfields with hard-to-recover reserves *Bull. of Tomsk Polytechnic University. Geo Assets Engineering* **329(8)** 30–36

[4] Miller G A 2007 Some Perspectives on Various Methods of Oil Shale Extraction *27th Oil Shale Symposium Colorado School of Mines Golden* (Piceance Basin, Colorado) 14 p

[5] Andreev A V, Mukhametshin V Sh and Kotenev Yu A 2016 Deposit Productivity Forecast in

Carbonate Reservoirs with Hard to Recover Reserves *SOCAR Proc.* **3** 40-45. DOI: 10.5510/OGP20160300287

[6] Mukhametshin V V 2018 Rationale for trends in increasing oil reserves depletion in Western Siberia cretaceous deposits based on targets identification *Bull. of Tomsk Polytechnic University. Geo Assets Engineering* **329(5)** 117–124

[7] Tyncherov K T, Mukhametshin V Sh, Paderin M G, Selivanova M V, Shokurov I V and Almukhametova E M 2018 Thermoacoustic inductor for heavy oil extraction *IOP Conf. Ser.: Mater. Sci. Eng.* **327(4)** 042111. DOI:10.1088/1757-899X/327/4/042111

[8] Economides J M and Nolte K I 2000 *Reservoir stimulation* (West Sussex, England: John Wiley and Sons) 856 p

[9] Burger J, Sourieau P and Combarnous M 1988 *Recuperation assistee du petrole les methods thermiques* (Moscow: Nedra) 422 p

[10] Mirzadzhanzade A K, Kovalev A G and Zaytsev Y V 1972 *Features of exploitation of abnormal oil fields* (Moscow: Nedra) 200 p

[11] Antoniadi D G 1995 *Scientific bases of development of oil fields by thermal methods* (Moscow: Nedra) 314 p

[12] Butler R M 2010 *Horizontal wells for oil, gas and bitumen production* (Moscow: Institute of computer research, regular and chaotic dynamics) 536 p

[13] Shagapov V S, Tazetdinova Y A and Gizzatullina A A 2015 Development of high-viscosity oil deposits by heat methods *J. of Engineering Physics and Thermophysics* **91(5)** 1242–1249

[14] Minnivaleev T N 2015 About the dynamics of drilling unit on drilling indicators *Proc. of the 42nd Int. sci. and tech. conf. of young scientists, postgraduates and students: in 2 volumes,* vol 1, pp 87-91

Two-stage stimulation of the oil formation

N A Vorsina and G R Izmaylova

Ufa State Petroleum Technological University, Branch of the University in the City of Oktyabrsky, 54a, Devonskaya St., Oktyabrsky, Republic of Bashkortostan, 452607, Russia

E-mail: tatacina@mail.ru, info@of.ugntu.ru

Abstract. The article considers a two-stage exposure to the formation. At the 1st stage, the combined exposure of high-frequency (HF) electromagnetic (EM) and sonic fields on the oil formation are performed without simultaneous formation fluid withdrawal. At the 2nd stage, the HF EM field is affected on the formation and at the same time, a low-viscosity solvent is injected into it. At the 1st stage, the bottom-hole zone is strongly warmed up due to the total thermal effect of the fields and an increase in the effective thermal conductivity of the medium due to acoustic treatment on the formation. At the 2nd stage, the solvent injected into a formation withdraws heat from the bottom-hole formation zone (BHFZ) further into the field. The dependence of the oil displacement front by the solvent on the dispersion parameter of the porous medium is considered.

1. Introduction

A lot of research has been devoted to the heating method of productive oil-filled formations with an HF EM field. In the beginning, early works were justifying the application possibility of HF EM fields for heating oil formations and extracting high-viscosity oils and bitumens [1]. Experimental researches were conducted to determine the dielectric properties of productive formations and materials of oil technology in variable HF EM fields [2].

Many works have been published that suggest various methods and technologies for developing hydrocarbon accumulations by EM fields. Some works suggest a combined effect on the hydrocarbon formation. Thus, in [3], two stages of stimulation of the bituminous formation are envisaged. First, HF EM waves with a frequency of up to 20 MHz are emitted, then they are warmed up by high-frequency currents. In [2], simultaneous injection of the oxidizer into the formation and the effect of the HF EM field on it is proposed. Thus, an in-situ combustion source is created. Also of interest is the combined effect of HF EM radiation on the oil formation with simultaneous injection of a miscible agent into the formation. Experimental and computational work has shown that this combined effect in the case of high-viscosity oils gives the best results compared to other methods [4, 5]. In works [6–8] is proposed simultaneous exposure of the oil formation to EM and acoustic fields, since the acoustic field increases the effective thermal conductivity of the formation rocks and due to this, more uniform heating of the formation is realized.

2. Problem statement

In [2], it was noted that for heating the bottom-hole zone of wells using high-frequency electromagnetic oscillations, the most effective is a linear radiator placed on the well bottom-hole and providing radial electromagnetic waves propagation deep into the formation. As a transmitting source, you can use the extension of oil-well tubing below the boring casing. Heating will be most effective if all the energy of high-frequency electromagnetic oscillations supplied to the transmitting source is transferred to the surrounding environment.

As a sound-emitting device, a garland of piezoceramic transcribers is used, the elements of which are evenly distributed throughout the entire formation thickness.

Due to the transmission losses of EM energy from the ground generator to the EM wave emitter in the formation, the radiant power is determined by the formula:

$$N_{0e} = N_g \exp[-2(\alpha_1 + \alpha_2)H] \tag{1}$$

Here N_{e0}, N_g is the radiant power of the EM wave and the generator power of the EM wave located on the earth's surface; α_1, $\alpha_2$is the attenuation coefficient of EM waves in the annular tubes of the well - oil-well tubing (OWT) and boring casing; H is the occurrence depth of the heated formation.

The energy power W lost in the well is determined by the formula:

$$W = 0,225\frac{(N_g - N_{e0})\alpha_1}{\alpha_1 + \alpha_2} \tag{2}$$

The absorption coefficients of EM waves in the boring casing and oil-well tubing (OWT) can be determined from the formulas [3]:

$$\alpha_1 = \frac{R_s}{2Z \cdot \ln(R_3 / R_2)} \cdot \frac{1}{R_2}, R_s^2 = \frac{\pi f \mu_{am}}{\sigma_m}, \ \alpha_2 = \frac{R_s}{2Z \cdot \ln(R_3 / R_2)} \cdot \frac{1}{R_3}, \ Z^2 = \frac{\mu_0}{\varepsilon_0}, \tag{3}$$

where R_s is the active surface resistance of the pipes, Z is the air wave resistance located in the casing-tubing annulus of the well; R_2 is the outside radius of the tubing joint; R_3 is the pipes inside radius of the boring casing; $\mu_{a.m}$ is absolute permeability of the well pipes; σ_m is the specific electrical conductivity of the well pipes.

When calculating the heating of the solvent moving in the oil-well tubing (OWT) to the bottom hole of the well. It is assumed that the energy part released in the oil-well tubing (OWT) due to non-zero effective resistance of the pipe material is used to heat the rocks surrounding the well. Part of it is spent on solvent heating. In this case, the temperature value to which the solvent is heated is determined by the formula:

$$T_b = T_s + \frac{W}{c_s \rho_s g_s} \tag{4}$$

where c_s, ρ_s, g_s is specific heat capacity, density, and flow rate of the solvent injected into the formation; T_s is the initial temperature of the injected solvent. In this case, the flow rate of the g_s solvent is determined by the filtration rate at the well bottom:

$$g_s = 2\pi r_0 hv \tag{5}$$

Here h is the formation thickness; r_0 is the well radius; v is the injection capacity of the solvent at the bottom.

At the first stage of formation stimulation, only the thermal conductivity equation is solved without the convective term, since the formation fluid withdrawal is not selected simultaneously with the HF impact. Therefore, the equation solved at the 1st stage has the form:

$$\frac{\partial T}{\partial t} = \frac{1}{C_p r}\frac{\partial}{\partial r}\left(r\lambda_a \frac{\partial T}{\partial r}\right) + \frac{q_{ae}}{C_p}, \tag{6}$$

where q_{ae} is the total heat sources created by EM and acoustic influences on the formation; λ_a is the thermal conductivity factor, C_p is the specific heat per unit volume of the formation rocks.

At the 2nd stage of the formation stimulation, an equation system consisting of the equations of piezo-conductivity, thermal conductivity and diffusion is solved [2]:

$$\frac{\partial P}{\partial t} = k_1 \frac{1}{r}\frac{\partial}{\partial r}\left(\frac{r}{\mu_{mx}}\frac{\partial P}{\partial r}\right); \quad k_1 = \frac{k}{m\beta_{mx} + \beta_0} \tag{7}$$

$$\frac{\partial T}{\partial t} = \frac{1}{C_p}\frac{1}{r}\frac{\partial}{\partial r}\left(\lambda_0 r \frac{\partial T}{\partial r}\right) - \frac{v_{mx}\rho_{mx}c_{mx}}{C_p}\frac{\partial T}{\partial r} + \frac{q_e}{C_p}; \tag{8}$$

$$m\frac{\partial C_j}{\partial t} = \frac{1}{r}\frac{\partial}{\partial r}\left(Dr\frac{\partial C_j}{\partial r}\right) - v_{mx}\frac{\partial C_j}{\partial r} \tag{9}$$

$$D = \delta(D_0 + \lambda v) \tag{10}$$

Here j=1, 2 – component indices for the solvent and oil; c_{mx}, ρ_{mx} – heat capacity, the density of the oil mixture and solvent; P – pressure; C_j– component concentration, $C_1+C_2=1$; D –convective diffusion coefficient; m, k – porosity and permeability of the formation; λ – dispersion parameter of the porous medium; β_{mx}, β_0 – compression coefficients of the mixture and the rock matrix; $D_o << \lambda v$ – molecular diffusion coefficient; δ – an empirical coefficient, the value of which depends on the presence of an EM field, in the field absence δ=1. In the diffusion equation, the solution is sought concerning the solvent concentration C_1.

The filtration rate of the solvent mixture and oil in the formation is determined from Darcy's equation:

$$v_{mx} = -\frac{k}{\mu_{mx}}\frac{\partial P}{\partial r} \tag{11}$$

where μ_{mx} is the dynamic viscosity of the oil mixture and solvent, determined from the well-known Kendall expression:

$$\ln \mu_{mx} = C_1 \ln \mu_1 + C_2 \ln \mu_2 \tag{12}$$

The solvent viscosity and oil depend on the thermal field and are determined from the formulas:

$$\mu_1 = \mu_{10}e^{-\gamma_1\Delta T}, \; \mu_2 = \mu_{20}e^{-\gamma_2\Delta T} \tag{13}$$

where μ_{10}, μ_{20} are the initial solvent viscosities and oil (at temperature $T=T_o$); γ_1, γ_2 is the temperature coefficients of the solvent and oil; $\Delta T=T-T_0$.

The initial and boundary conditions for the equation system (6) – (13) have the form:
for the thermal conductivity equation:

$$T(r,0) = T_0, \; T(r_0,t) = T_b, \; \frac{\partial T(r_k,t)}{\partial r} = 0 \tag{14}$$

for the diffusion equation concerning the solvent concentration:

$$C_1(r,0) = 0, \; C_1(r_0,t) = 1, \; C_1(r_k,t) = 0 \tag{15}$$

for the piezo-conductivity equation:

$$P(r,0) = P_0, \; P(r_0,t) = P_b, \; P(r_k,t) = P_0 \tag{16}$$

3. Results and Discussion

The problem (9) – (13) with initial and boundary conditions (14) – (16) is solved by the numerical method of finite differences using an implicit scheme. Heat sources that occur in the formation when HF EM and acoustic waves propagate in it were calculated using the formulas:

$$q_{ae} = \frac{\alpha_a N_{a0}}{\pi r h} \exp\left(-2\alpha_a (r - r_0)\right) + \frac{\alpha_e N_{e0}}{\pi r h} \exp\left(-2\alpha_e (r - r_0)\right) \quad (17)$$

where $N_{a0}$is the radiant power of the acoustic waves.

Consider how the solvent concentration distribution in a porous medium is affected by different values of the dispersion parameter of the porous medium λ.

The following values were used in the calculations: T_0=12 ^{0}C; f_e=13.56 MHz;ε'=7.5; $tg\delta$=0.05; α_e =0.0194 m^{-1}; λ_0=1.28 W/(mK); A=0.000352 m/c; f_a=22 kHz; α_a=0.4758 m^{-1}; C_p=1378 kJ/(m^3·K); h=8 m; r_0=0.05 m; P_0=10 MPa; T_s=15 K; P_b=15 MPa; m=0.3; k=5 * 10-$^{-13}$m^2; Z=376.8 Ohm; H=700 m;β_{mx} =10^{-9} PA^{-1}; β_0=10^{-10} PA^{-1}; R_2=0.03015 m; R_3=0.05015 m; $\mu_{a.m}$=3.418 * 10^{-6} Hn/m; σ_m=3.4·10^6 Ohm^{-1} * m^{-1}; c_s=2057 j/(kg * K); ρ_s=769 kg/m^3; ρ_{mx}=918 kg/m^3; c_{mx}=2024 j/(kg * K); λ_0=1.28 W/(m * k); μ_{01}=1.73 * 10^{-3} PA * s; μ_{02}=0.2 PA * s; γ_1=0.0128 K^{-1}; γ_2=0.042 K^{-1}. The value of λ=0.1 m is the dispersion parameter of a porous medium. And its value was not neglected the molecular diffusion coefficient due to its proximity to zero. The value D_0=10^{-8}m^2/s was used. In this connection, comparative calculations of the solvent concentration distribution at λ=0.01 m and D_0=0 were performed. It turned out that with the HF EM field powerN_{e0}=40 kW and N_{a0}=10 kW – the acoustic field power, the temperature field, and the pressure field do not depend on the parameter value of the dispersion. The solvent concentration distribution is shown in Figure 1.

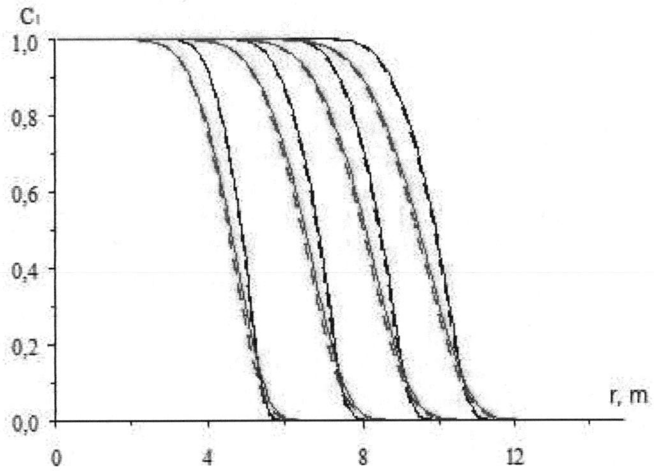

Figure 1. The solvent concentration distribution: N_{e0}=40 kW; N_{a0}=10 kW (full curves) and N_{a0}=0 (dotted line); λ =0.01 m, D_o =0 (black curves); λ =0.1 m, D_o =10^{-8}m^2/s (red curves). 1 – t=25 days; 2 – t=30 days; 3 – t=35 days; 4-t=40 days.

At λ=0.01 m and forgetfulness the molecular diffusion coefficient due to its smallness, the curves $C_1(r)$ in the case of valuesN_{a0}=0 and N_{a0}=10 kW merged, and at λ=0.1 and D_o=10^{-8} m^2/s, the curves $C_1(r)$ in the cases of values N_{a0}=0 and N_{a0}=10 kW differ slightly. At N_{a0}=10 kW, the solvent filtration is more intensive and it penetrates further into the depth of the formation. At λ=0.01 m and neglecting the molecular diffusion coefficient, the $C_1(r)$ curves are steeper and the solvent penetrates deeper into the formation at a lower depth. Approximately the same differences in the solvent concentration distribution are observed at another initial value of the oil viscosity – μ_{02}=0.8 PA * s.

4. Conclusion

The paper solves the problem of two-stage stimulation of the formation with high-viscosity oil. At the 1st stage, HF EM and formation acoustic heating are performed. At the 2nd stage, the formation continues to be heated by the HF EM field and at the same time, a low-viscosity solvent is injected. The solvent Injection into the formation leads to a more severe transfer of heat from the bottom to the formation, namely, an increase in the formation heating radius by 1.5-2 times, where the oil viscosity is reduced by 1.5-1.7 times. At the same time, the temperature distribution becomes more uniform and the temperature at the bottom of the well decreases, which in practice will prevent negative consequences associated with overheating of the well bore.

The dispersion parameter changing of a porous medium practically does not affect the temperature and pressure distribution process, which is due to the predominance of molecular transfer heat and pressure in this case, but there are noticeable differences in the distribution of the solvent concentration, which indicates the convective transfer of the substance (solvent and oil) in the formation.

References

[1] De Gauque P and Grudzinski R 1987 Propagation of Electromagnetic Waves Along a Drillstring of Finite Conductivity *SPE* **12943** 127-134

[2] Izmailova G R 2018 Research on the high-frequency electromagnetic acoustic heating of an oil reservoir with further pumping of a solvent *High temperature* **56(6)** 910-914

[3] Chakma A and Jha K N 1992 Heavy-Oil Recovery From Thin Pay Zones by Electromagnetic Heating *SPE* **24817** 525–534

[4] Izmailova G R, Kovaleva L A and Nasyrov N M 2016 Acoustic wave energy absorption and distributed heat sources upon an acoustic impact on media *High temperature* **54(1)** 56-61

[5] Vladimirov I V, Almuhametova E M and Veliyev E M 2016 Conditions of the effective use of hot water flooding for development of heavy oil deposits with non-uniform permeability *Oil Industry* **9** 62-65

[6] Davletbaev A, Kireev V, Kovaleva L, Zainullin A and Minnigalimov R 2016 Cold heavy oil production and production by radio-frequency electromagnetic radiation: Comparative numerical study *AIP Conf. Proc.: Int. Conf. of Computational Methods in Sciences and Engineering ICCMSE* **1790** 150021

[7] Batalov S A, Andreev V E, Lobankov V M and Mukhametshin V Sh 2019 Numerical simulation of oil formation with regulated disturbances. Oil recovery quality simulation *J. of Phys.: Conf. Ser.* **1333(3)** 032006. DOI: 10.1088/1742-6596/1333/3/032006

[8] Rogachev M K, Mukhametshin V V and Kuleshova L S 2019 Improving the Efficiency of Using Resource Base of Liquid Hydrocarbons in Jurassic Deposits of Western Siberia *J. of Mining Institute* **240** 711-715. DOI: 10.31897/PMI.2019.6.711

Simulation of the method of excitation of pulsations in a moving fluid flow

M L Galimullin, M Ya Khabibullin and R I Suleimanov

Ufa State Petroleum Technological University, Branch of the University in the City of Oktyabrsky, 54a, Devonskaya St., Oktyabrsky, Republic of Bashkortostan, 452607, Russian Federation

E-mail: minivaris02@mail.ru, info@of.ugntu.ru

Abstract. The article deals with an important issue of efficiency of using energy supplied into each of the elements of the Venturi-type pipe, i.e. the value of energy efficiency coefficient. The issue coverage requires conducting comparative studies of the specific surface area of bubbles in a gas-liquid system and the mass transfer coefficients in a pipe with a periodically changing cross-section and in a cylindrical pipe with other conditions being equal. The proposed method of excitation of pulsations in a moving fluid flow can be used for the processes of splitting droplets and bubbles in a continuous liquid phase, since this creates favorable conditions – significant amplitudes of pressure, speed and acceleration at a frequency of about dozens of Hertz.

1. Introduction

The article deals with an important issue of efficiency of using energy supplied into each of the elements of the Venturi-type pipe, i.e. the value of energy efficiency coefficient. The issue coverage requires conducting comparative studies of the specific surface area of bubbles in a gas-liquid system and the mass transfer coefficients in a pipe with a periodically changing cross-section and in a cylindrical pipe with other conditions being equal.

2. Results and Discussion

Several modes of two-phase flow in a horizontal channel that are characteristic of hydraulic resistance and heat exchange can be conventionally distinguished. This is a stratified flow that includes layered and wave flows, projectile and wave flows with jumpers, as well as flow modes that do not depend on the orientation of the channel – bubble and annular flows, including wavy-annular and dispersed-annular flows [1]. In an annular channel, flow modes are qualitatively consistent with the specified flow modes in the pipe. However, the channel's outer diameter is one of the determinant parameters for detecting the stratification conditions of a two-phase flow in an annular channel. Size of the annular gap affects the flow structure in a wavy-annular flow. Thus, reduction of the annular gap leads to the contact of the vertices of individual waves on the surface of liquid films at both walls of the channel with the formation of screw-like mutually overlapping liquid bridges. The distance, or pitch, between these bridges is about four outer diameters of the annular channel. It should be noted that the film of the channel's inner wall is thiner than that of the outer wall, and accordingly, the drying of the former is more speedy in a dispersed-annular flow. The analysis of literature and generalization of the results obtained on vapor-liquid flows of helium and nitrogen in the annular and round channels, as well as investigation of the known maps of flow modes for two-phase flows of water and freon, have allowed proposing calculated relations for determining the conditions of transition to the flow modes characteristic of resistance and heat exchange [2-4]:

The cyclic frequency of pulsations of pressure, speed, and acceleration in this case is determined by the formula

$$\omega = 2\pi / T. \quad (1)$$

The pressure distribution can be approximated with sufficient accuracy for engineering calculations using a one-dimensional model.

According to [1], the local resistance coefficient for the confuser is written as

$$\zeta_m = (-0.0125n_0^4 + 0.0224\, n_0^3 - 0.00723\, n_0^2 + 0.00444\, n_0 - 0.00745)\,(\,\alpha_p^3 - 2\pi\,\alpha_p^2 - -10\,\alpha_p), \quad (2)$$

where $n_0 = (d/D)^2$ – degree of narrowing of the flow; α_p – angle α expressed in radians.

The coefficient of friction loss in the confuser is determined by the formula

$$\zeta_{f.c} = \lambda_1\,(1 - n_0)^2 / 8tg(\alpha/2), \quad (3)$$

where λ_1 – coefficient of hydraulic friction, which is calculated using the Altschul formula in a section with a diameter D:

$$\lambda_1 = 0.11[(68/Re_D) + (\Delta/D)]^{0,25}; \quad (4)$$

$Re_D = w_2 D\rho/\mu$ – Reynolds number, which is calculated from the cross section with a diameter D; ρ – density of the liquid, kg/m^3; μ – coefficient of dynamic viscosity, PA*s; Δ – absolute roughness of the inner surface of the pipe, m.

The pressure loss in the confuser can be represented as [2]

$$\Delta p_k = (\zeta_m + \zeta_{f.c})(\,\rho\, w_1^2/2). \quad (5)$$

The impulse completeness coefficient for the operating diffuser is represented in [3] by the formula

$$\varphi = 3.2[tg(\beta/2)]^{1,25}, \quad (6)$$

The coefficient of resistance conditioned by expansion of the flow is represented as follows:

$$\zeta_{exp} = \varphi(1 - n_0)^2. \quad (7)$$

According to [4], the coefficient of friction loss in the diffuser is determined by the formula

$$\zeta_{f.d} = \lambda_{av}(1 - n_0)^2/8tg(\beta/2), \quad (8)$$

where λ_{av} – coefficient of hydraulic friction calculated as the arithmetic mean: $\lambda_{av} = (\lambda_1 + \lambda_2)/2$, and the coefficient of hydraulic friction λ_2 is calculated using the Altschul formula in a section with a diameter d:

$$\lambda_2 = 0.11[(68/Re_D) + (\Delta/D)]^{0,25}, \quad (9)$$

where $Re_d = w_1 d\rho/\mu$ – Reynolds number calculated from the cross section with a diameter d.

The pressure loss in the diffuser is represented in [5] as

$$\Delta p_d = (\zeta_{exp} + \zeta_{f.d})(\rho w_1^2/2). \quad (10)$$

Compression of the flow in the Venturi pipe necks can be ignored, which is confirmed by our finite element calculations.

Then the pressure loss in the narrow neck is determined by the formula

$$\Delta p_d = \lambda_2\,(L_2 / d)(\rho w_1^2/2), \quad (11)$$

and in the wide neck – by the formula

$$\Delta p_D = \lambda_1\,(L_4 / d)(\rho w_2^2/2), \quad (12)$$

Total irreversible pressure loss in a single Venturi-type element of length L is determined by the sum of losses

$$\Delta p = \Delta p_\kappa + \Delta p_d + \Delta p_d + \Delta p_D. \quad (13)$$

Assuming the flow being smoothly changing everywhere, it is not difficult to obtain formulas for calculating the pressures in sections 1-4 from the Bernoulli equation (Figure 1):

$$p_1 = P0 + (\rho w_2^2/2) - (\rho w_1^2/2) - \Delta p_\kappa; \quad (14)$$

$$p_2 = p_1 - \Delta p_d; \quad (15)$$

$$p_3 = p_2 + (\rho w_1^2/2) - (\rho w_2^2/2) - \Delta p_d; \quad (16)$$

$$p_4 = p_3 - \Delta p_D. \quad (17)$$

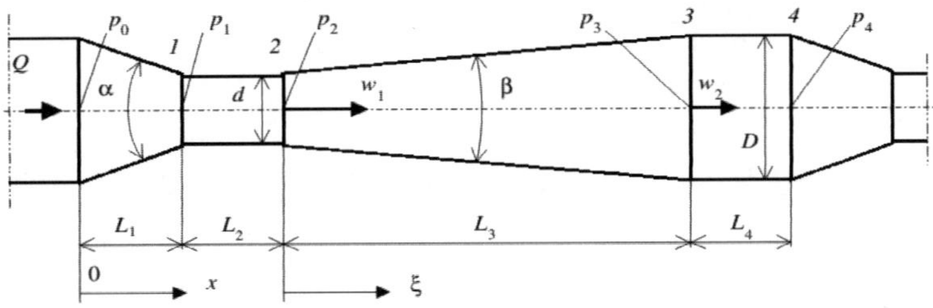

Figure 1. Diagram of the confuser-diffuser element of a pipe with a periodic profile

Graphs of the distribution of speeds, accelerations, and pressures along the length of a Venturi-type element, given to the x coordinate, are shown in Figure 2.

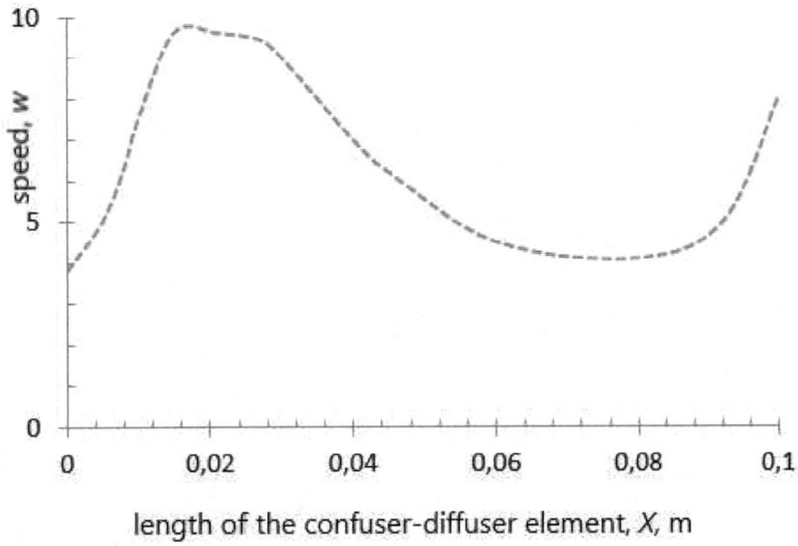

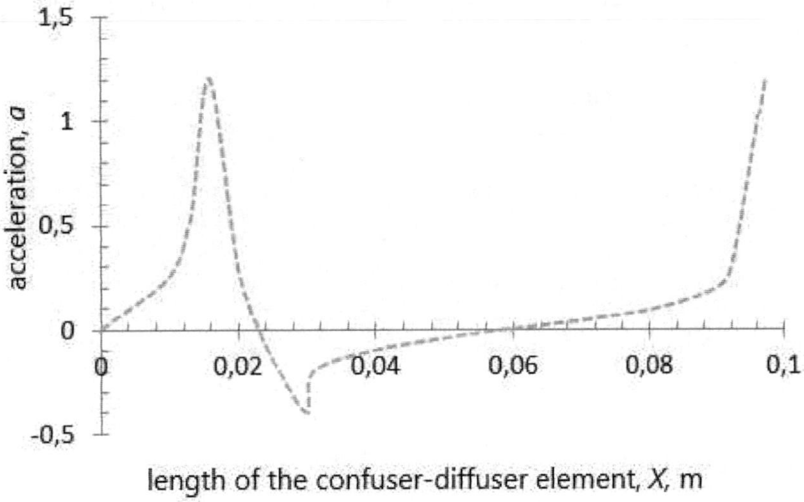

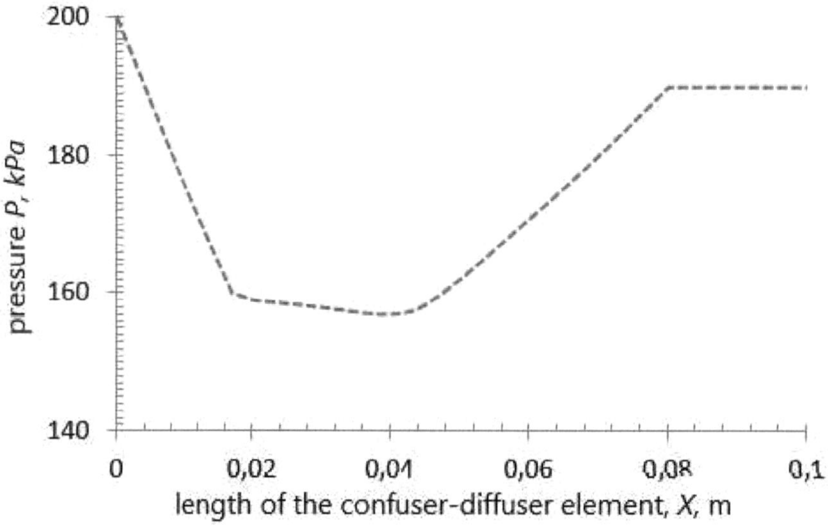

Figure 2. Theoretical distribution of speeds *w*, accelerations *a* and pressures *p* along the length *x* of the confuser-diffuser element (calculated parameters: d = 10 mm; D = 20 mm; L_1 = 15 mm; $L_2 = L_4$ = 10 mm; L_3 = 50 mm; α = 36°; β = 11.5°; Q = $7.8 \cdot 10^{-4}$ m^3/s)

Special attention should be paid to the fact that acceleration and deceleration maxima are located near the narrowest part of the pipe – the neck with a diameter d. Particles of the deformable dispersed phase, approaching the neck, start to accelerate, especially near the axis of the pipe, stretching in the axial direction. At the same time, they undergo a significant drop in hydrostatic pressure. After passing the neck, the liquid begins to brake sharply, while the pressure is restored almost to the initial value. The wall layers of the liquid are greatly affected by huge shear stresses due to large transverse velocity gradients at the neck walls. Combination of these factors can serve as a powerful means of deforming and splitting dispersed inclusions, as well as contribute to wave formation on their surface and improvement of internal mixing and, in general, accelerate mass transfer processes.

In addition, a drop in the pressure in the neck to the value of saturated vapors can lead to cavitation and associated effects [6, 7]. As for solid particles that have a density different from a liquid, their constant acceleration and deceleration should also enhance mass transfer. When processing capillary-porous particles, the acceleration of mass transfer inside the particle can be facilitated by pressure pulsation [8, 9], and outside – by periodic pulsation renewal of liquid near the surface of the particle [10].

Figure 3 shows the pressure distribution in a pipe with a periodically changing cross-section consisting of 10 consecutive elements of the Venturi-type pipe, which has been calculated using the formulas (14)–(17), taking into account the ratios (9)–(12).

The calculated parameters are shown in the caption to Figure 2. The results obtained have been tested experimentally on a glass model of such a pipe with a periodically changing cross-section. The flow rate of the liquid in the pipe varied in the range (1.16–2.98) × 10^{-4} m^3/s, the flow mode being turbulent (Re_d > 9500).

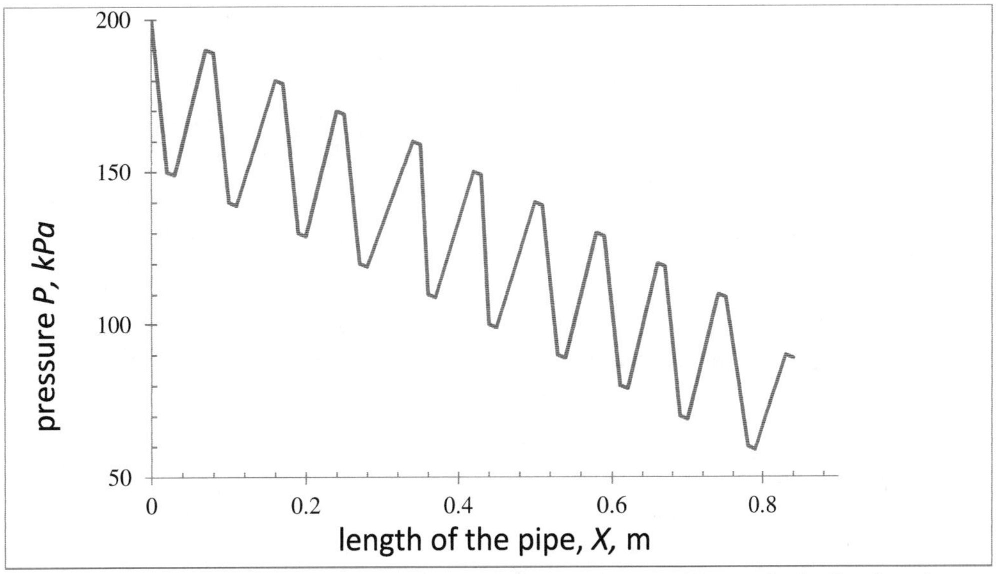

Figure 3. Theoretical distribution of pressure p along the pipe x with a periodically changing cross section consisting of 10 confuser-diffuser elements

The results of measurements and calculations have been generalized by a dependency of the form

$$\Delta p_0 = CQ^2, \quad (18)$$

where $\Delta p_0 = 10\Delta p$ – pressure loss in the pipe consisting of 10 consecutive elements of the Venturi type; C – pressure loss coefficient.

3. Experiment

According to experimental data, the coefficient C makes $2.61 \cdot 10^{11}$ kg·m^{-7}, while the calculation using formulas (9)–(13) at $Q = 2.8 \cdot 10^{-4}$ m^3/s gives a theoretical value of the coefficient C equal to $1.69 \cdot 10^{11}$ kg·m^{-7}, which is 1.54 times less than the experimental value. This discrepancy may be explained by the error in manufacturing of a complex-shaped glass tube, deviations of its linear and angular dimensions from the design (when the diameter of the neck deviates from the nominal by only 8%, the resistance of the pipe increases by 1.52 times); the error in measuring pressure and flow, as well as the fact that the coefficient of completeness of the impulse of the diffuser is calculated on the basis of the rectangular speed profile in a narrow section.

At the same time, the nature of the experimental dependence (19) coincides with the nature of the theoretical one, which allows recommending the proposed calculation method for engineering calculations of a pipe with a periodically changing cross-section.

It should be noted that for the cylindrical pipe with the diameter of 10 mm and the length of 10L = 0.85 m at $Q = 2.8 \cdot 10^{-4}$ m^3/s, the theoretical value of the coefficient C has made $1.76 \cdot 10^{11}$ kg·m^{-7}, which is 4.1% higher than the theoretical value of the coefficient for a pipe with a periodically changing cross-section. This is due to the fact that, firstly, its profile provides a continuous flow of liquid, and, secondly, in some areas, the flow rate of the liquid is lower. Thus, despite the complex shape of the pipe with a periodically changing cross-section, the overall energy loss in it is somewhat lower.

The theoretical frequency of pulsations according to formula (9) at $Q = 2.8 \cdot 10^{-4}$ m^3/s for the pipe of the considered shape is 130 rad/s. Unfortunately, the experimental verification of this ratio is difficult due to the need to move the pressure sensor inside the pipe along with the flow.

4. Conclusion

The proposed method of excitation of pulsations in a moving fluid flow can be used for the processes of splitting droplets and bubbles in a continuous liquid phase, since this creates favorable conditions – significant amplitudes of pressure, speed and acceleration at a frequency of about dozens of Hertz.

Combination of these factors can serve as a powerful means of deforming and splitting dispersed inclusions, as well as contribute to wave formation on their surface and improvement of internal mixing and, in general, accelerate mass transfer processes. These factors can also help speed up the mass transfer process when processing solid impervious and capillary-porous particles.

References

[1] Khabibullin M Ya and Suleimanov R I 2018 Selection of optimal design of a universal device for nonstationary pulse pumping of liquid in a reservoir pressure maintenance system *Chemical and Petroleum Engineering* **54(3-4)** 225-232. DOI: 10.1007/s10556-018-0467-2

[2] Haoran Zh, Yongtu L and Xingyuan Zh 2017 Sensitivity analysis and optimal operation control for large-scale waterflooding pipeline network of oilfield *J. of petroleum science and engineering* **154** 38-48

[3] Khabibullin M Ya, Suleimanov R I, Sidorkin D I and Arslanov I G 2017 Parameters of damping of vibrations of tubing string in the operation of bottomhole pulse devices *Chemical and Petroleum Engineering* **53(5-6)** 378-384. DOI: 10.1007/s10556-017- 0350-6

[4] Nyemb N V, Jacques H and Elkana P 2017 Numerical Simulation of Heat Distribution with Temperature-Dependent Thermal Conductivity in a Two-Dimensional Liquid Flow *Int. J. of nonlinear sciences and numerical simulation* **18(6)** 507-513

[5] Kuleshova L S, Kadyrov R R, Mukhametshin V V and Safiullina A R 2019 Design changes of injection and supply wellhead fittings operating in winter conditions *IOP Conf. Ser.: Mater. Sci. Eng.* **560(1)** 012072. DOI: 10.1088/1757-899X/560/1/012072

[6] Welsh E 1987 *Borehole Coupling in Porous Media,* PhD Thesis *(Colorado School of Mines, Golden, Colo)* 63 p

[7] Khabibullin M Ya 2018 Impulse non-stationary flooding of hydrocarbon deposits *Advances in Engineering Research (AER)* **157** 266-271. DOI: 10.2991/aime-18.2018.51

[8] Sun W and Mun-Hong H 2017 Forecasting and uncertainty quantification for naturally fractured reservoirs using a new data-space inversion procedure *European Assoc Geoscientists & Engineers Computational geosciences: 15th Conf. on the Mathematics of Oil Recovery (ECMOR) (Amsterdam, Netherlands)* **21(5-6)** 1443-1458

[9] Kuleshova L S, Kadyrov R R, Mukhametshin V V and Akhmetov R T 2019 Auxiliary Equipment for Downhole Fittings of Injection Wells and Water Supply Lines Used to Improve Their Performance in Winter *IOP Conf. Ser.: Mater. Sci. Eng.* **560(1)** 012071. DOI: 10.1088/1757-899X/560/1/012071

[10] Stenkin A V, Kotenev Yu A, Mukhametshin V Sh and Sultanov Sh Kh 2019 Use of low-mineralized water for displacing oil from clay productive field formations *IOP Conf. Ser.: Mater. Sci. Eng.* **560(1)** 012202. DOI: 10.1088/1757-899X/560/1/012202

Modelling viscous fluid filtration to select optimal bottom-hole filter designs

N N Soloviev[1], and M Ya Khabibullin[2]

[1] LLC Gazprom VNIIGAZ, Staraya Basmannaya St., 20, b. 8, Moscow, Russian Federation
[2] Ufa State Petroleum Technological University, Branch of the University in the City of Oktyabrsky, 54a, Devonskaya St., Oktyabrsky, Republic of Bashkortostan, 452607, Russian Federation

E-mail: N_Soloviev@vniigaz.gazprom.ru, info@of.ugntu.ru

Abstract. Determining the resistance value of filters with a filtering surface of mesh and block type is difficult mathematically. This is because many parameters affecting the shell filtration capacity. As for the mesh filters, depending on the type of weaving, the different thickness of weft and warp, the different size in the horizontal and vertical planes, the frame structure which is integral with the mesh have critical influence. As for block constructions, it is the size and uniformity of the composition of the particles, the thickness of the filter shell, the geometry of the filtration channels and other features that play the role. It is, therefore, more convenient for such filters to assess their permeability. The most promising filters are the frames with horizontal slots which have considerably less resistance and interference of the holes. In addition, the features of the frame-rod filter device make it easy to realize a greater inlet area than in other designs, with the same minimum size of the slots. Recommendations for choosing the inlet areas and the size of the filter slots should consider factors related to well production rate, the formation fluid viscosity, particle size distribution of the formations, granulometric composition of the formation, consolidation factor of the bottom-hole zone, clodding and bridging of the deposits.

1. Introduction

There is a shortcoming in quarrying wells equipped with filters, both in the degree and the nature of the formation exposing which depends on the filter construction installed in the formation [1-3]. According to [4], in the filtration calculation, the total filtration resistance is divided into private, corresponding to different filtration zones. If the well bottom construction does not affect the nature of the flow in the zone sufficiently far from the well, additional hydraulic losses appear in the filter zone and on the filter itself and must be seen as dimensionless parameters affecting the total resistance to the flow of produced fluid.

According to the common flow schematization, additional resistance shall be presented according to [4] as:

$$\zeta = \zeta_1 + \zeta_2, \tag{1}$$

where ζ_1 and ζ_2 are the resistance due to the degree and nature of the formation exposing.

Assessment of the well imperfection influence according to the formation exposing degree with the inflow to it depending on the ratio l/m (where l is the filter length and t is the total power of the productive formation) and graphs for determining the resistance ζ_1 are given in [5].

The resistance $\zeta_2$is a complex characteristic depending on the quality of well completion, the hydraulic features of the filter caused by construction as well as the filter shell clogging and the breakdown of the linear filtration mode. Therefore, the value $\zeta_2$can be determined in this way:

$$\zeta_2 = \zeta_{2f} + \zeta_{2d} + \zeta_{2dl}\,, \qquad (2)$$

where ζ_{2f},ζ_{2d} ,ζ_{2dl} are the resistances caused by the filter, deposit character, and deviation from the linear filtration mode.

The process of filtering a viscous fluid through a model of a porous medium with relatively small pore sizes obeys the Darcy linear law [5] and then the value of the generalized resistance can be determined by the formula:

$$\zeta_2 = \zeta_{2f} + \zeta_{2d}. \qquad (3)$$

From here it is clear that when choosing optimal constructions of filters and their elements, their hydraulic parameters play a decisive role. They are determined by the resistance of the filter itself, and the placement of filters in weak and soft formations or gravel package well leads to significant contact pressure losses in the filter-rock zone and the sand-holding ability of the filter shell.

2. Results and Discussion

The hydraulic parameters of the rock-filter system (under other conditions) are a consequence of the geometric parameters of the filtering shell forming the mutual position of the slots and influencing the resistance to movement of the produced fluid due to the breakdown of flow linearity, its diffraction, etc.

Increasing the inlet area of the filter shell with the same size of slots affects on reducing contact losses in the filter-rock interface by increasing porosity in the near filter zone. Also, the form and mutual position of the elements of the filtering shell influencing the creation of stable bridges from large fractions of sand or gravel packing ensures the retention of particles that make up the formation matrix with the simultaneous passage of the colmatant.

Various filter constructions are currently in use. There are filters with round holes, slotted, located in the vertical or horizontal plane, with different lengths, which can reach the size of the formation thickness, as well as with a filtering surface of meshes and block-type filters. For filters with round holes V.I. Shchurov [6] studied similar solutions of M. Mason and, as a result, constructed refined graphs of the dependence (Figure 1) ζ_{2f} on the parameters:

$$\alpha = d_0/D \; and \; \beta = nD\,, \qquad (4)$$

where d_0 is the diameter of the filter openings; *p* is the number of filter openings per unit length; D is the diameter of the filter.

So α and β parameters are indicators of relative inlet area, the graph shows that increasing the inlet area leads to decreasing the filter resistanceζ_{2f}.

M.R. Harris` theoretical research [7] shows that the total efficiency of the formation of round holes group in terms of flow coefficient depends on their mutual position. The graph (Figure 2) shows the results of some conclusions.

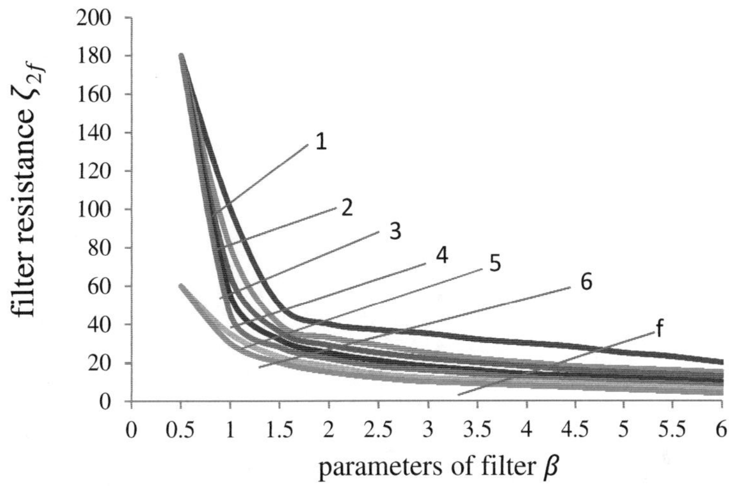

Figure 1. The dependence of ζ_{2f} on parameters of β= 0.01nD and α for filters with round holes: 1 - α = 0.03; 2 -α = 0.04; 3 - α = 0.05; 4 -α = 0.06; 5 -α = 0.07; 6 - α = 0.08; 7 -α = 0.09

Graphic dependencies are built considering the following conditions:
- supply contour radius is 200 m;
- outer radius of the cement column 76 mm;
- perforated channel length is 30.5 cm;
- perforated channel diameter is 12.7 mm;
- perforator bullet penetration behind a cement column is 300 mm.

The results show that the holes located evenly around the circumference in the horizontal plane lead to the lowest pressure drop at a given flow rate, and holes located vertically along one generating line lead to the largest pressure drop.

The resistance values ζ_{2f} for filters with a length that is equal to the formation thickness located vertically along the generating line of the pipe, have very similar results in various methods of determination. V. T. Cordwell and S.K. Dodson [8] proposed the expression:

$$\zeta_{2f} = \frac{2}{n} \cdot ln\frac{2}{\pi\eta}, \text{ with } =0,3, \quad (5)$$

where n is the number of slots; η is inlet area.

V.P. Pilatovsky [9] obtained the expression of dimensionless hydraulic resistance in the form:

$$\zeta_{2f} = \frac{4}{n} \cdot ln\, sin\frac{n^1\alpha}{2}, \quad (6)$$

where n^1 is the number of impermeable partitions; α is the central angle determining the size of the slot.

Graphic dependences (see Figure 1) performed according to the results of calculating formula (6) allow finding the value $\zeta_{2f}$of rod frames and filter structures similar to them.

The work [10] confirms that the graphical dependencies (Figure 1) allow determining the resistance value for filters with a limited length of vertical slots. These dependencies are valid for filters with holes made in the form of a circle. For rectangular holes, it is possible to interpolate to round holes considering the equivalent total cross-sectional area.

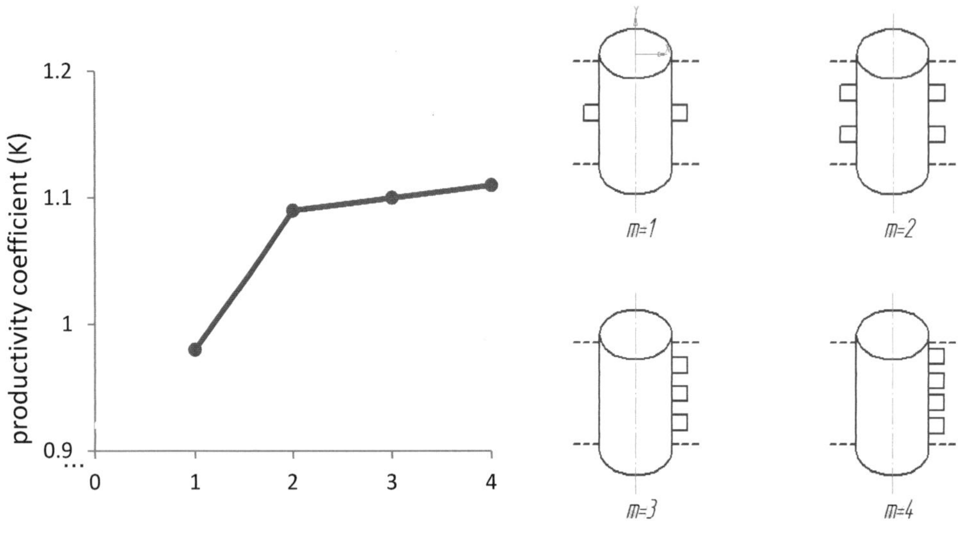

Figure 2. The dependence of the productivity coefficient (K) on the perforation pattern and the number of perforations (m) in the plan

3. Experiment

Recommended radii of reduction depend on the ratio of the length of the slot l to its width b. With the ratio $l/b \leq 3$, the reduction is preferable to produce on the area, and the radius of the reduced hole is expressed by the following dependence:

$$r_{red} = \sqrt{\frac{b \cdot l}{\pi}}, \tag{7}$$

and with $l/b > 3$ (long and narrow slots):

$$r'_{red} = \frac{b \cdot l}{2\pi}, \tag{8}$$

i.e the radius is determined along the perimeter.

Graphical dependences (Figure 3) show that the resistance of filters with vertical slots decreases significantly both with an increase in the inlet area and with a decrease in the width of the slot with the same inlet area of the filter surface.

For filters with slotted holes located in the horizontal plane, A.L. Hein gave a similar solution, with relatively large r/d values:

$$\zeta_{2f} = \frac{d^3}{\pi^2 \, r_c b^2} \left[6{,}835 \cdot \frac{b}{d}\left(1 - \frac{b}{d}\right) - 2L_i \cdot \left(\pi \cdot \frac{b}{d}\right) + \chi\left(\frac{b}{d}; \frac{2r_c}{d}\right)\right], \tag{9}$$

where d is the distance between the centres of adjacent slots; b is the height of the slot; r_c is the current radius; $\chi\left(\frac{b}{d}; \frac{2r_c}{d}\right)$is a tabulated function whose values are given in [4].

The work [10] gives solutions according to which the resistance value of filters with horizontal slots is approximately 20% less than that of filters with cylindrical openings shown in the graphical dependencies (Figure 1).

Determining $\zeta_{2f}$of filters with a filtering surface of mesh and block type is difficult mathematically. This is because many parameters affecting the shell filtration capacity. As for the mesh filters, depending on the type of weaving, the different thickness of weft and warp, the different size in the horizontal and vertical planes, the frame structure which is integral with the mesh have critical influence. As for block constructions, it is the size and uniformity of the composition of the particles, the thickness of the filter shell, the geometry of the filtration channels and other features that play the role. Therefore, for such filters it is more convenient to evaluate their permeability according to the Darcy formula [5].

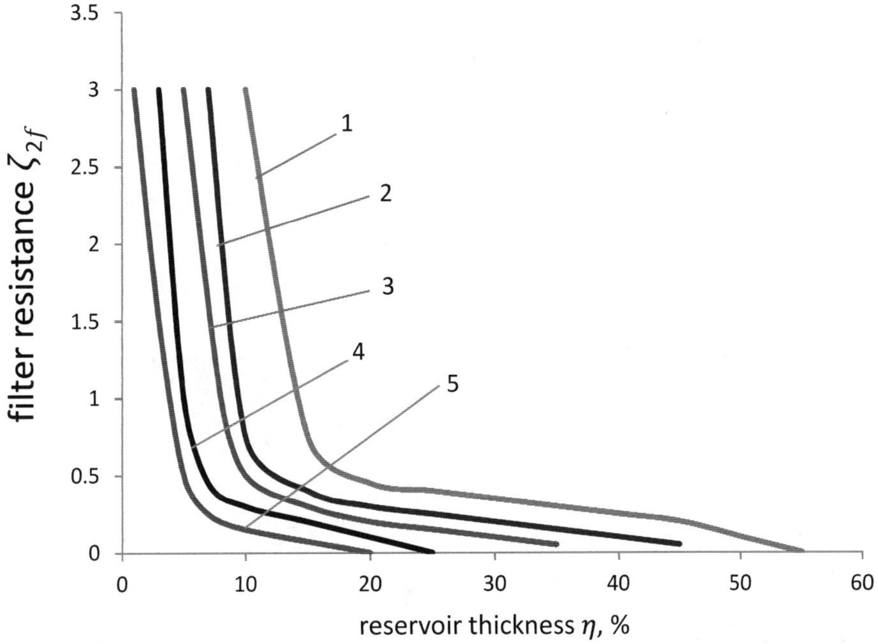

Figure 3. Graphic dependences ζ_{2f} of on inlet area η for the entire reservoir thickness for filters with slots with a diameter of 168 mm: slot width: 1 - 10 mm; 2 - 5 mm; 3 - 3 mm; 4 - 1 mm; 5 - 0.5 mm

For radial filtration through an annular sample:

$$K = \frac{\mu_g \cdot Q_g \cdot ln\frac{r_{out}}{r_{in}}}{20 \cdot \pi \cdot h(P_{out} - P_{in})}, \tag{10}$$

where Q_g is the flow rate, cm^3 / s; μ_g is the dynamic viscosity, MPa · s; P_{out}, P_{in} is the pressure difference between the outer and inner surfaces, MPa; r_{out}, r_{in} - outer and inner radii in cm; h is the height of the test sample in cm.

The formula shows that the permeability functionally depends on the differential pressure, which in turn is due to the filter resistanceζ_{2f}.

The work [6] studied the hydraulics of filters of various constructions. Figure 4 presents the results of the filter permeability assessment based on their inlet area.

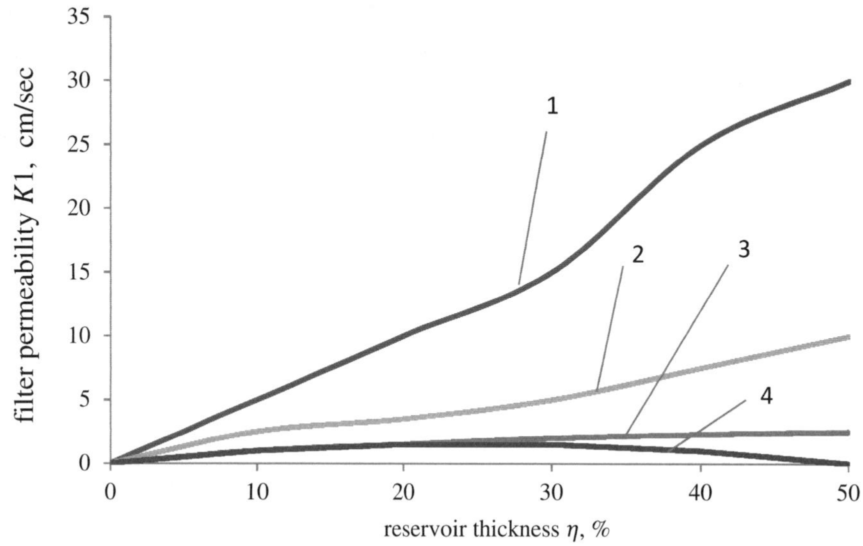

Figure 4. Graphic dependencies of filter permeability on inlet area according to Klotz: 1 - wire-frame; 2 - with bridge-shaped holes; 3 - slotted holes with open perforation; 4 - mesh filters with galloon weaving nets

4. Conclusion

As seen, the most perspective filters are filters - frames with horizontal slots which have significantly less resistance and the influence of an interference of the holes. In addition, the features of the frame-rod filter device make it easy to realize a greater inlet area than in other designs, with the same minimum size of the slots.

Recommendations for choosing the inlet areas and the size of the filter slots should consider factors related to well production rate, the formation fluid viscosity, particle size distribution of the formations, granulometric composition of the formation, consolidation factor of the bottom-hole zone, clodding and bridging of the deposits. Only the result of experimental studies in the laboratory and natural models will determine these parameters, as well as losses at the filter-rock contact.

References

[1] Meirmanov A and Nekrasova I 2013 Mathematical models of a hydraulic shock *J. of mathematical analysis and applications* **408(1)** 76-90

[2] Haoran Zh, Yongtu L and Xingyuan Zh 2017 Sensitivity analysis and optimal operation control for large-scale waterflooding pipeline network of oilfield *J. of petroleum science and engineering* **154** 38-48

[3] Kuleshova L S, Kadyrov R R, Mukhametshin V V and Safiullina A R 2019 Design changes of injection and supply wellhead fittings operating in winter conditions *IOP Conf. Ser.: Mater. Sci. Eng.* **560(1)** 012072. DOI: 10.1088/1757-899X/560/1/012072

[4] Kuleshova L S, Kadyrov R R, Mukhametshin V V and Akhmetov R T 2019 Auxiliary Equipment for Downhole Fittings of Injection Wells and Water Supply Lines Used to Improve Their Performance in Winter *IOP Conf. Ser.: Mater. Sci. Eng.* **560(1)** 012071. DOI: 10.1088/1757-899X/560/1/012071

[5] Heuzeroth F, Fritzsche J and Werzner E 2015 Viscous force - An important parameter for the modeling of deep bed filtration in liquid media *Powder technology* **283** 190-198

[6] Stenkin A V, Kotenev Yu A, Mukhametshin V Sh and Sultanov Sh Kh 2019 Use of low-mineralized water for displacing oil from clay productive field formations *IOP Conf. Ser.: Mater. Sci. Eng.* **560(1)** 012202. DOI: 10.1088/1757-899X/560/1/012202

[7] Kuleshova L S, Kadyrov R R, Mukhametshin V V and Akhmetov R T 2019 Auxiliary Equipment for Downhole Fittings of Injection Wells and Water Supply Lines Used to Improve Their Performance in Winter *IOP Conf. Ser.: Mater. Sci. Eng.* **560(1)** 012071. DOI: 10.1088/1757-899X/560/1/012071

[8] Khabibullin M Ya, Suleimanov R I, Sidorkin D I and Arslanov I G 2017 Parameters of damping of vibrations of tubing string in the operation of bottomhole pulse devices *Chemical and Petroleum Engineering* **53(5-6)** 378-384. DOI: 10.1007/s10556-017- 0350-6.

[9] Sun W and Mun-Hong H 2017 Forecasting and uncertainty quantification for naturally fractured reservoirs using a new data-space inversion procedure *European Assoc Geoscientists & Engineers Computational geosciences: 15th Conf. on the Mathematics of Oil Recovery (ECMOR) (Amsterdam, Netherlands)* **21(5-6)** 1443-1458

[10] Khabibullin M Ya and Suleimanov R I 2018 Selection of optimal design of a universal device for nonstationary pulse pumping of liquid in a reservoir pressure maintenance system *Chemical and Petroleum Engineering* **54(3-4)** 225-232. DOI: 10.1007/s10556-018-0467-2

IOP Conf. Series: Materials Science and Engineering 1064 (2021) 012066 doi:10.1088/1757-899X/1064/1/012066

Improving the efficiency of horizontal wells at multilayer facilities

R R Khuzin[1], V Sh Mukhametshin[2], D A Salikhov[1], V E Andreev[3,4], D V Pepelyaev[5] and Yu N Stefanovich[5]

[1] Karbon-Oil LLC, 25, Surgutskaya St., Almetyevsk, Almetyevskiy District, Republic of Tatarstan, 423450, Russian Federation
[2] Ufa State Petroleum Technological University, Branch of the University in the City of Oktyabrsky, 54a, Devonskaya St., Oktyabrsky, Republic of Bashkortostan, 452607, Russian Federation
[3] Institute for Strategic Research of the Republic of Bashkortostan, 129/3, Oktyabrsky Ave., Ufa, Republic of Bashkortostan, 450075, Russian Federation
[4] Ufa State Petroleum Technological University, 1, Kosmonavtov St., Ufa, Republic of Bashkortostan, 450062, Russian Federation
[5] Perm construction and technological Bureau of technical design and production organization LLC, 21, Akademika Koroleva St., Perm, 614013, Russian Federation

E-mail: karbon@tatais.ru, vsh@of.ugntu.ru

Abstract. The paper presents the results of testing of new production equipment of wells, as well as the operating modes of horizontal wells in the conditions of development of multilayer facilities. The paper also presents the results of drilling small-diameter wells, the methods of oil production intensification, and the principles of operation of pumping equipment with an electric valve with account for optimal draw-downs on reservoirs. The suggested approach to the operation of horizontal, sub-horizontal and directional wells of small diameter allows increasing the technical and economic indicators of the development of complex multilayer facilities. It is proposed to use the developed technology in the operation of two or more reservoirs in conjunction with the implementation of dual injection into the exploited reservoirs through injection wells in the center of the element of development.

1. Introduction
The intense development of small oil fields with hard-to-recover reserves, characterized by low profitability brings up the issue of searching for new technologies of drilling, well operation and facilities development with the aim of reducing the cost of production [1-11].

2. Materials and methods
Small oil company Karbon-Oil LLC has been developing six small fields in the Republic of Tatarstan since 2006. Deposits of high-viscosity oil are confined to the inner side zone of the Aksubaevo-Melekess draw-down. In terms of the geological structure, the deposit is classified as a complex with a high degree of heterogeneity. The main object of operation is represented with carbonate reservoirs of the Bashkir tier of medium carbon, which are characterized by low filtration and capacity properties, weak hydrodynamic connection with the aquatic part of the reservoir, as well as inter-well zones.

Today, there are no low-cost technologies that ensure high efficiency of development of such fields, so the rate of oil reserves extraction is low there.

In 2007-2008, the first horizontal wells for the Bashkir tier with open shafts were drilled at the Nekrasov field. Since 2014, the company has adopted a strategy of drilling fields with a system of horizontal wells and multi-hole horizontal wells. Construction, development and operation of wells are carried out only with the use of Russian equipment, including extractable deflector wedges. All shafts of a multi-hole horizontal well are located separately from each other. Since 2014, two-hole horizontal wells have been equipped with dual operation of two-lift structures with separate shafts and separate operation of each shaft. To reduce the cost of horizontal well construction, drilling was carried out with 155.6 mm bits, and 114 mm-diameter production columns were used. The construction of horizontal wells has allowed accumulating experience in using optimal well designs: in the filter section 200-300 m long, magnesium acid-soluble plugs (REA filters) are used, as the least expensive option in comparison with cumulative (on a hard cable) and hydro-mechanical perforation.

Various designs of horizontal well stems were used: open non-cased shafts; cased cemented shafts; cased non-cemented shafts, separated into intervals by column-type swollen packers. Complicating and increasing the cost of the well design does not always lead to an increase in well flow rates. Cased cemented shafts separated into a few intervals by REA filters (sections of horizontal well) are considered to be the most optimal option for well completion. Development of horizontal wells was also carried out in different variants: from simple openings of REI with acid bath installations followed by swab development to REI opening in dynamic modes with subsequent point-to-point bottom hole treatment with resetting the packer layouts opposite the sections of horizontal wells. The volume of acidic compounds used for development from drilling increased from 10 to 50-60 m^3. All this made it possible to increase start-up flow rates and reduce the rate of decline in flow rates after wells were put into operation. As of 1 December 2019, 37% of the production well stock is made up of horizontal wells, including 3% of multi-hole horizontal wells, with the total revenue from horizontal wells being 55% of the company's total. The transition to horizontal drilling allowed eliminating the drop in the average flow rate of wells, which was observed from 2006 to 2014, and even increasing by 10-15% over the last 2 years.

Further potential for increasing the average flow rate of the well stock and improving the performance of development projects consists in the introduction of returnable objects and the organization of separate operation of each object by one well in the new operating mode. The stock of wells drilled since 2014 is mainly represented with small-diameter wells. There are restrictions on the implementation of deep-pumping equipment layouts for separate operation.

3. Results and Discussion

In this regard, a new approach is proposed in the dual operation of beam pumping units or vertical sludge pumps with a packer and an electric valve MIXER for ORD produced by Tekhproekt company. Cons: the cost of a set of electric valves. Pros: the ability to run into the operating column with a diameter of 114 mm or more. Flexible adjustment of selections according to controlled parameters on-site (bottom-hole pressures, selection/accumulation time, pump drive operating frequencies). Additionally, the built-in manometer-thermometer allows taking the pressure build-up curve of each object and setting the optimal draw-down and selection time. The control station allows calculating the total flow rate of the well from the measuring device at the mouth of the time spent for each object of operation. This type of configuration of MIXER for ORD equipment allows operating a well, or rather two objects in the well, taking into account different properties in the mode of dual production.

In 2019, a successful output test was carried out on the variant of the downhole pumping equipment layout for the operation of a horizontal well with two objects in the dual production mode. The horizontal well drilled in 2016, exploiting the Bashkir tier with a product viscosity of 250 MPa·s was selected for the test. During the well overhaul, the upper return object, the verey horizon, was perforated and selectively developed. In accordance with the Program for pilot testing of production equipment MIXER for ORD produced by Tekhproekt LLC in Karbon-Oil LLC, pilot tests of dual production equipment MIXER for ORD had been conducted over the period from 29 May 2019 to 25 October 2019 in a well with the small diameter of 114x7.4 mm.

The tests allowed for:

– evaluating the possibility of using the equipment to implement the technology of dual operation of two objects in the conditions of oil fields of Karbon-Oil LLC in small-diameter wells with a large horizontal routing;

– checking the performance of the equipment in downhole conditions of oil fields of Karbon-Oil LLC;

– assessing the performance quality of the equipment;

– updating the criteria for equipment use;

– making decisions on the feasibility of putting into operation the equipment at the facilities of Karbon-Oil LLC and giving recommendations on its further use and acquisition.

Brief description of operation of the equipment: the electric valve MIXER for ORD provides operation of two objects separated by a packer, in the mode of dual (separate-alternate) production. The equipment is mounted between the packer and the pumping unit (Figure 1).

The operating mechanism of the equipment moves the spool of the electric valve by a signal coming from the wellhead control station via a geocable, providing a message to the pump reception of either the upper object, or the lower object, or 2 objects simultaneously. Switching between objects is carried out either by the parameter of the liquid extraction time (timer mode) or by the value of the minimum bottom-hole pressure for each of the objects (pressure mode). The pressure and temperature of the liquid at the pump intake are registered by a depth gauge-thermometer. After the objects are examined, the control station automatically supports the specified operating modes of the electric valve and pump, including control of the frequency of the pump drive depending on the object currently in operation.

To reduce losses, the optimal mode of operation of each object is adjusted according to the recorded pressure recovery curves from each object of operation. The mode of short-term selection and accumulation allowed increasing the pump filling coefficient. For carbonates with their double matrix and crack porosity, there is need for accumulation time to ensure flow from the matrix to the crack during operation. When switching selections to the accumulation mode, the reservoir inertia remains in operation with liquid accumulation in the production column and the bottom-hole part of the reservoir for subsequent intensive selection. Moreover, this mode in the well allowed not only preserving the flow rate of the lower, previously operated object, but also increasing its flow rate by 28%.

Test results include evaluation of performance criteria in accordance with the output test program, namely:

1. *Implementation of dual extraction of two objects with differentiated draw-downs for each, created by a single pump.*

Liquid was collected from two objects at the well. The liquid was lifted by a single 175-TNM pump. The mode of dual operation of the upper and lower objects in automatic modes "by timer" and "by pressure" is implemented. The frequency of operation of the pump drive changes in accordance with the set points on-site.

2. *Possibility of separation of the objects.*

The separation of objects is confirmed by changing the depth gauge values when switching between objects (Figure 2).

3. Possibility of researching one of the objects when the second object is stopped.

The equipment allows sampling of the fluid produced from each object under the following condition: one of the reservoirs disabled, the well bred to steady state selection, the final selection of not less than 3 volumes of the inner space of the tubing to the liquid outlet of the studied reservoir.

4. Possibility of controling bottom-hole pressure on objects in the dual production mode.

Pressure monitoring of each of the objects is performed by a single depth gauge in the process of liquid extraction from this object. It is possible to build pressure recovery curves for each of the objects when the well operation is stopped. To study the pressure recovery curve of one of the objects without stopping the second, it is necessary to introduce an additional sensor into the downhole pumping equipment, but this condition can only be implemented in wells with an operating column of 146 mm in diameter.

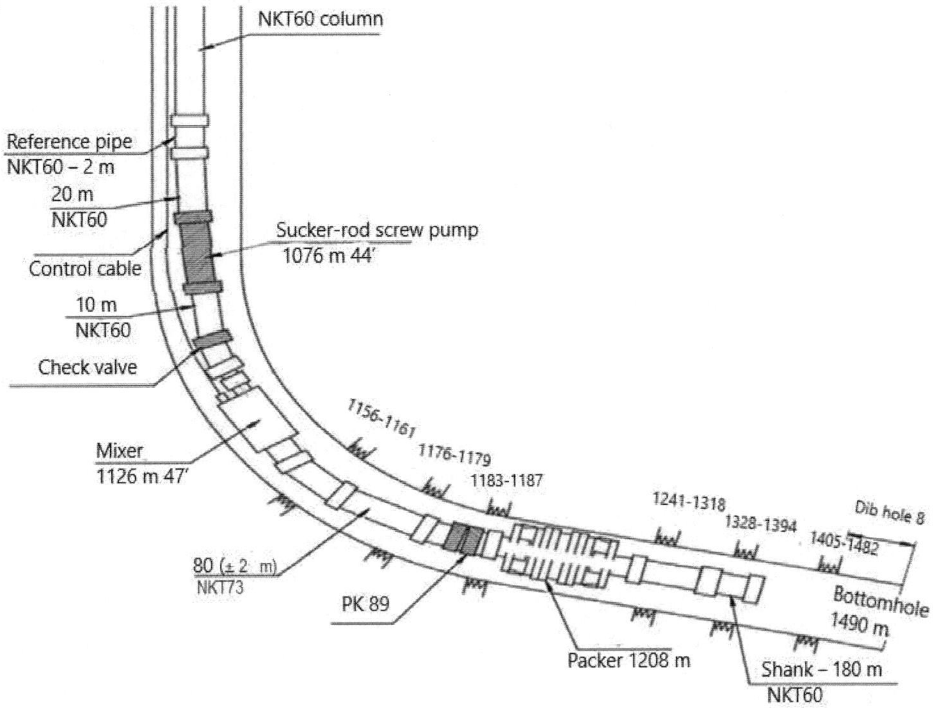

Figure 1. Diagram of setting of the electric valve Mixer as part of the downhole pumping equipment

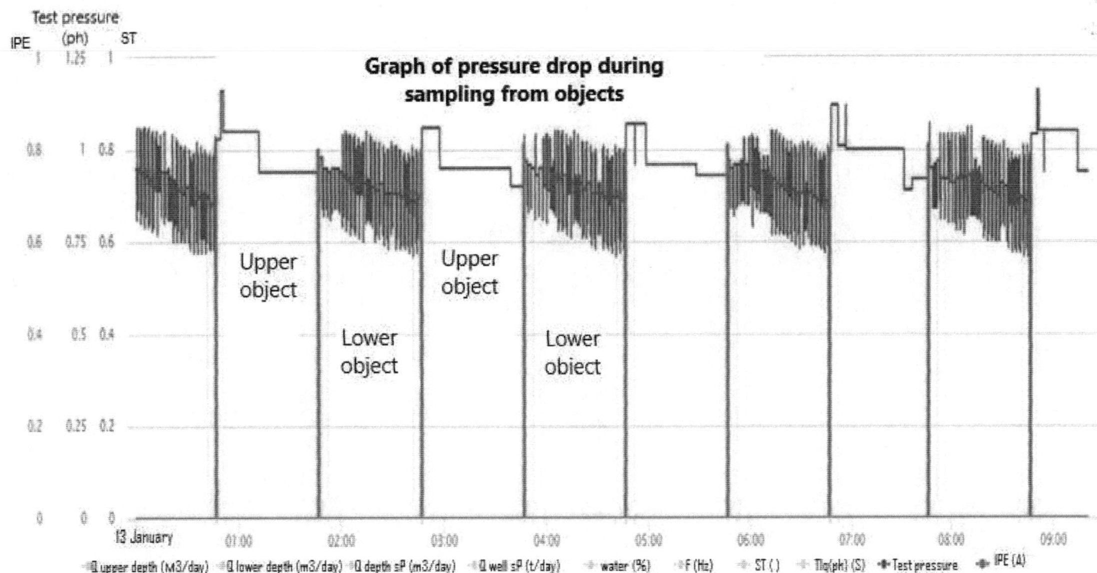

Figure 2. Graph of pressure changes during the operation of objects

5. Possibility of automatical switching between objects in the dual production mode by pressure parameter or by timer.

The mode of dual operation of the upper and lower objects in automatic modes "by timer" and "by pressure" is implemented. Switching is automatically controlled by timer according to the modes set at the control station. The mode is not quite suitable for objects with low properties in terms of pressure, since a few days of operation past and frequent automatic switching between objects begins. During the period of operation of one object, the second one didn't have enough time to accumulate the pressure required to start the cycle of operation, whereby periods of the selection cycle for each of the objects have been rapidly declining, and attempts to reduce the working frequency of the pump drive each of the objects have led to the decrease of the total flow rate of the objects.

6. Lack of refusals during the test period.

No equipment failures were detected during the test period. It should be noted that operation of the well where the equipment was tested, was complicated by the high viscosity of the extracted fluid.

7. Possibility of measuring the flow rate of the objects

By-object account of fluid withdrawal was carried out in real-time according to the counter BSCI (the RING). Accounting was carried out in the mode of dual operation without stopping the operation of the pumping unit.

8. Possibility of remote monitoring and change in the parameters of the solenoid valve and pump installation.

Remote monitoring and change in the parameters of the solenoid valve and pump installation were carried out via the control panel (Figure 3).

Using a special application that repeats all the functions of the control station on the well, operation of the equipment was adjusted (Figure 4).

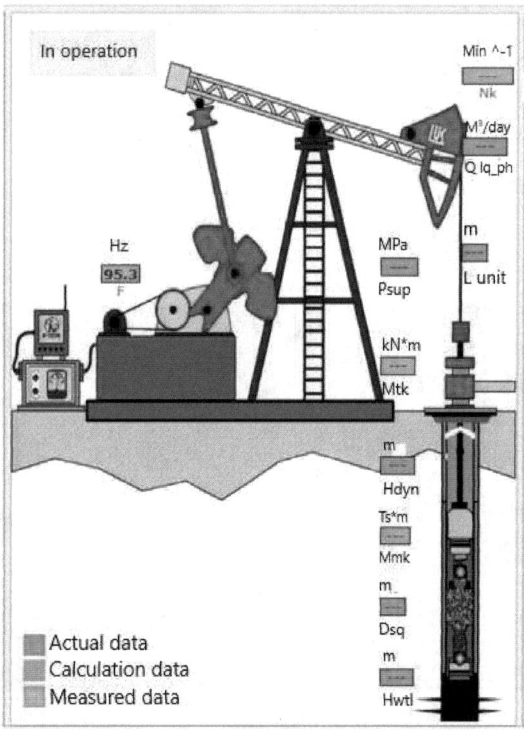

Parameter	Value	Unit of measure
Pressure Photon	0.74	MPa
Flow rate well up	1.22	t/day
Flow rate well down	0.73	t/day
Flow rate well sP	0	t/day
State of control system	Lower position	ST
Temperature Photon	25.61	C
MIXER motor current	0	A
Frequency	95.3	Hz

Figure 3. Interface of the well monitoring system and data archive

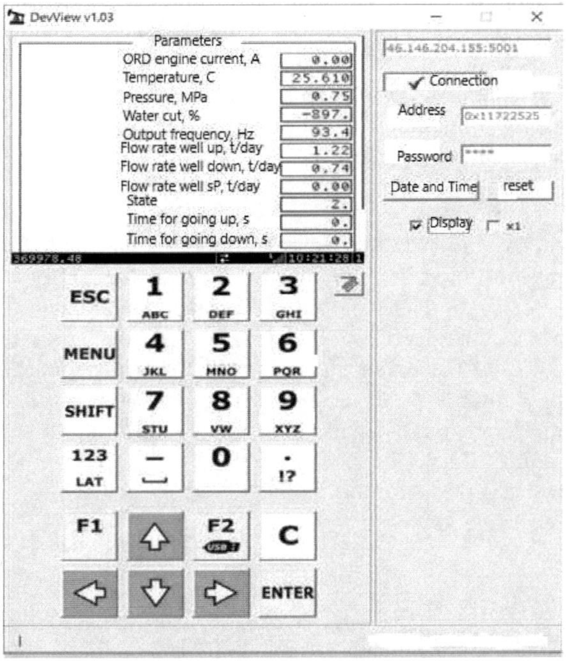

Figure 4. Remote control application for electric valve operation

The mode of dual production as a method of operation of a submerged downhole electric valve is a conceptually new mode of operation of downhole equipment, which was not previously described in the scientific literature, since it has not been used in practice until now due to the lack of technologies and appropriate equipment. The first attempts at such equipment operation were made in 2016 at two oil wells in the Perm region with a production column of 146 mm in diameter by the producer of the equipment MIXER for ORD, Tekhproekt LLC. At that time, the main operational characteristics of layouts and criteria for selecting wells for the introduction of innovative equipment were not formulated due to the imperfection of software, the failure to achieve the required operating time of equipment components, as well as minimal experience in selecting operating modes.

At the beginning of 2018, pilot tests of the MIXER for ORD equipment were successfully completed, key operating modes of the equipment were tested, including automatically supported modes, the criteria for selecting wells and downhole pumping equipment layouts were clarified.

As early as by the end of 2018, with the joint participation of Tekhproekt LLC and Karbon-Oil LLC, a set of MIXER for ORD equipment for small-diameter wells (114ˣ7.4 mm) was developed and put into the well. It should be noted that the pumping well under study was the first platform for successful testing of the method of well operation with one operational facility having a large horizontal distance, and the high viscosity of the produced fluid, where the lack of necessary technology determined the lack of opportunity for providing the necessary drainage remote from the pump portion of the horizontal section, the so-called well toe. The well was divided into two horizontal sections by a packer, which were dually connected to the pump by the MIXER for ORD valve, and the draw-downs were created differentially for each zone. Thus, an additional draw-down in the remote area of the horizontal well provided an additional influx. In practice, the result exceeded expectations and the increase in oil production made 48%.

The dual production mode was also appropriate at the next small-diameter well, where the previously undrained upper object was attached to the main production facility. When switching to the mode, it was noted that the new object with high reservoir pressure, and, consequently, when using a common lift, the lower object is silenced by the liquid of the upper object and does not operate. Flow and filling of reservoir pressure of the lower object take place. After switching to the dual mode, the pressure of the upper attached object remained stable above the lower one for some time, after which it decreased, and the pump operation mode was changed, and the lower object continued to work in the originally selected mode. Changes in the operation of the bundle: objects of operation – well and pump – are analyzed remotely from the archives, and the adjustment of the operation mode is made from a PC and takes no more than 3-5 minutes.

4. Conclusion

Thus, this approach to the operation of horizontal, sub-horizontal and directional wells of small diameter allows increasing the flow rate, reducing the payback period of wells and upgrading the technical and economic indicators of the development of complex multilayer facilities.

Application of the mode of dual operation of two reservoirs together with the implementation of the mode of dual injection into the exploited reservoirs through an injection well is the further perspective direction of development of this technology. Equipment in the design for SMD, necessary for the intelligent implementation of the mode of dual injection, is currently being developed by Tekhproekt LLC. Karbon-Oil LLC selected an experimental two-object site as a development element with an injection well in the center and surrounding production facilities. Implementation of this technology with a comprehensive approach to production and injection at this pilot site is planned between 2020 and 2021.

References

[1] Miller G A 2007 Some Perspectives on Various Methods of Oil Shale Extraction, Piceance Basin, Colorado *27th Oil Shale Symp. Colorado School of Mines Golden (Colorado, 15-19 October 2007)* p 14
[2] Andreev A V, Mukhametshin V Sh and Kotenev Yu A 2016 Deposit Productivity Forecast in Carbonate Reservoirs with Hard to Recover Reserves *SOCAR Proceedings* **3** 40-45. DOI: 10.5510/OGP20160300287
[3] Mukhametshin V V, Andreev V E, Dubinsky G S, Sultanov Sh Kh and Akhmetov R T 2016 The Usage of Principles of System Geological-Technological Forecasting in the Justification of the Recovery Methods *SOCAR Proceedings* **3** 46–51. DOI: 10.5510/OGP20160300288
[4] Downey M W, Garvin J, Lagomarsino R C and Nicklin D F 2011 Quick Look Determination of Oil-in-Place in Oil Shale Resource Plays *AAPG Annual Convention and Exhibition (Houston, Texas, USA, 10-13 April 2011)* p 21
[5] Yakupov R F, Mukhametshin V Sh, Khakimzyanov I N and Trofimov V E 2019 Optimization of reserve production from water oil zones of D3ps horizon of Shkapovsky oil field by means of horizontal wells *Georesursy* **21(3)** 55-61. DOI: 10.18599/grs.2019.3.55-61
[6] Mukhametshin V V and Kuleshova L S 2019 Justification of Low-Productive Oil Deposits Flooding Systems in the Conditions of Limited Information Amount *SOCAR Proceedings* **2** 16–22. DOI: 10.5510/OGP20190200384
[7] Lestz R S, Clarke J N, Plattner D and Byrd A C 2002 Perforating for Stimulation: An Engineered Solution *SPE Drilling & Completion* **17(1)** 8. DOI: 10.2118/76812-PA
[8] Rogachev M K, Mukhametshin V V and Kuleshova L S 2019 Improving the efficiency of using resource base of liquid hydrocarbons in Jurassic deposits of Western Siberia *J. of Mining Institute* **240** 711-715. DOI: 10.31897/PMI.2019.6.711
[9] Akhmetov R T, Kuleshova L S and Mukhametshin V V 2019 Application of the Brooks-Corey model in the conditions of lower cretaceous deposits in terrigenous reservoirs of Western Siberia *IOP Conf. Ser.: Mater. Sci. Eng.* **560(1)** 012004. DOI: 10.1088/1757-899X/560/1/012004

[10] Stenkin A V, Kotenev Yu A, Mukhametshin V Sh and Sultanov Sh Kh 2019 Use of low-mineralized water for displacing oil from clay productive field formations *IOP Conf. Ser.: Mater. Sci. Eng.* **560(1)** 012202. DOI: 10.1088/1757-899X/560/1/012202

[11] Kuleshova L S, Kadyrov R R, Mukhametshin V V and Akhmetov R T 2019 Auxiliary equipment for downhole fittings of injection wells and water supply lines used to improve their performance in winter *IOP Conf. Ser.: Mater. Sci. Eng.* **560(1)** 012071. DOI: 10.1088/1757-899X/560/1/012071

IOP Conf. Series: Materials Science and Engineering 1064 (2021) 012067 doi:10.1088/1757-899X/1064/1/012067

The choice of technologies for restoring the properties of the bottomhole zone of wells in deposits with low-viscosity oils

R F Yakupov, Sh G Mingulov and I Sh Mingulov

Ufa State Petroleum Technological University, Branch of the University in the City of Oktyabrsky, 54a, Devonskaya St., Oktyabrsky, Republic of Bashkortostan, 452607, Russian Federation

E-mail: rustem_geolog@mail.ru, info@of.ugntu.ru

Abstract. Based on the analysis of many years of experience in the operation of the Tuymazinskaya group of oil fields, it has been established that one of the most effective methods for restoring the injectivity of injection wells is chemical treatment of the bottomhole zone of wells with various acid compositions in conjunction with well reperforation and the use of coiled tubing technologies.

1. Introduction

The bottomhole zone (BHZ) condition is constantly changing during the entire period of well operation, from drilling to abandonment. As a rule, the permeability of the bottomhole formation zone is significantly lower than the permeability of the remote formation zone. Already at the stage of drilling, during the initial opening of the productive formation, irreversible processes occur in the near-wellbore zone, significantly changing the structure of rocks and their permeability.

The main factors affecting the pollution of productive formations are repression, the duration of its action, the composition and properties of the drilling fluid. Deterioration of reservoir properties occurs as a result of the penetration of various solutions, solid particles into the BHZ, as well as their interaction with formation fluids and rock-forming minerals of the formation.

Formation water is a concentrated solution of inorganic salts (55–270 g / l) with a very wide range of ions (Na^+, K^+, $NH4^+$, Ca^{2+}, Mg^{2+}, Cl^-, Br^-, HCO3, CO^{2-}, $SO4^{2-}$, etc.). Thus, the BHZ is clogged with swelling clay particles and the products of interaction of the drilling fluid filtrate with the formation fluids, and the formation rock is blocked by an oil-water emulsion, while the pore space is clogged with the solid phase of the drilling fluid.

Open reservoirs, underground mineralized waters of deep horizons are used as sources of water supply for the maintain formation pressure system, as well as industrial wastewater: a mixture of bottom water and fresh water from ground sources. The intensity of BHZ contamination significantly depends on the physicochemical properties and composition of the formation fluid, and fluid for reservoir pressure maintenance.

Water injected into the reservoir causes contamination of the bottomhole zone of injection wells. All this impairs the wells' connectivity with the formation and makes it difficult for oil to flow from the formation to the well and water to enter the formation from injection wells. When injecting wastewater from oil fields containing different amounts of oil and oil products into the reservoir, both the perforations and the porous medium of the bottomhole zone of the wells are contaminated, i.e., the oil saturation BHZ of the injection well increases, which must be reduced to improve the phase permeabilities for water.

The main pollutants of BHZ are salt deposits, asphalt-resin-paraffin deposits (ARPD), corrosion products, clays, silica sand and other minerals. The formation of pollution has a number of negative consequences: equipment failure; decrease in the productivity of wells and other technological facilities; increase in the number and duration of expensive and time-consuming repairs; leading to a significant shortage of oil, an increase in energy consumption; general deterioration of technical and economic indicators. Often, pollution in the BHZ is a conglomerate of various agents that form a complex system. Consequently, before carrying out geological and technical measures, it is necessary to analyze the causes of contamination blocking the BHZ of injection wells and draw up a program of geological and technical measures (GTM) to restore injectivity.

2. Materials and methods

The main tool for stabilizing production levels and improving the efficiency of oil field development is the GTM. The objectives of their implementation are maintaining the planned volumes of oil production, intensifying oil production, and increasing oil recovery [1–10].

With an increase in the well stock, the role of GTM increases. This is especially true for the treatment of the bottomhole zone, optimization of the technological modes of well operation, regulation of the water injection pressure in the formations and packs, separate and focal waterflooding.

The tasks to be solved within the framework of planning measures are determining the purpose of carrying out, the choice of technology for stimulating the well, the selection of the required resources. Operational control and monitoring are carried out at all stages of planning.

The sequence of stages of the GTM planning process can be distinguished [11-14]:

- identification of the foci of injection wells with the presence of potential in the production fund associated with an increase in the energy of the reservoir, as well as the study of materials for hydrodynamic and geophysical studies of wells;
- production forecasting after technological optimization and well stimulation;
- formation of a sequence of activities. A number of measures is carried out to increase or level the injectivity profile of the problem zone of formation development, as well as technological and economic efficiency.

To assess the technical condition and operation of the BHZ injection well, geophysical surveys of the wells are regularly carried out. This type of research allows you to evaluate:

- tightness of the production string, tubing string and packer fit;
- circulation behind the string of liquid during water injection;
- construction of the injectivity profile of the bottomhole formation zone;
- total acceleration, bottomhole pressure, plotting the pressure drop curve at the bottom of the well after shutdown.

Based on the analysis of the interpretation of well logging data, conclusions are drawn about the current state of the injection well. In the event of a leak in the production casing or underground equipment, measures are taken to restore the tightness. If fluid circulation behind the string is detected up or down the perforation interval, remedial work is carried out to eliminate this circulation behind the string by squeezing the slurry and re-perforating the bottomhole perforation.

Of special importance are the data obtained in the course of constructing the injectivity profile, which makes it possible to identify the intervals of perforation and interlayers of the production object with the lowest and highest permeability. As a result of the uneven operation of the perforated interval of the formation, flushed channels and cracks are formed, which lead to an increase in the water cut of the production wells. If necessary, works to level the acceleration profile by mechanical and chemical impact methods on the bottomhole zone of the well are carried out.

The mechanical method means cumulative perforation of the bottomhole zone of the injection well with a density of 10-15 holes per 1 m, which allows not only to increase the filtration characteristics of the formation, but also to include previously unused areas of the formation.

3. Results and Discussion

Nowadays, in the conditions of the Tuimazinskaya group of fields, in order to restore and increase the injectivity of injection wells, physicochemical methods are used to influence the BHZ:

- acid treatment by the well workover team;
- acid treatment using coiled tubing equipment;
- repeated perforating and blasting operations in combination with acid treatment.

When acid treatment is carried out by the workover team, the underground equipment is being extracted, gauging and pressure testing of the production string is being performed, the bottom of the well is being flushed, clay acid solution is being injected (24% HF). These actions are done in case of impact on terrigenous reservoirs of the Bobrikov-Radaevsky and Pashiysky horizons or hydrochloric acid (12% HCl) for carbonate reservoirs of the Tournaisian stage of the Kizelovsky horizon.

The advantages of this type of treatment are the ability to assess the technical condition of the well and revision of underground equipment, the disadvantages are the need to prepare the working site, discharge and kill the well, as well as the repair time, which is about 160 hours, depending on the depth of the production facility.

When conducting GTM with the use of coiled tubing equipment, continuous coiled tubing is lowered into the well through the tubing string, through which acid solutions are pumped into the BHZ of the injection well.

The advantage of this type of treatment is that there is no need to discharge and kill the well, as well as a significantly shorter time for processing - no more than 72 hours. The disadvantage is the inability to assess the technical condition of the well and revise the underground equipment.

When conducting GTM with a workover crew, it is possible to reperforate the production casing in the formation interval with subsequent acidizing, which allows increasing the permeability more significantly due to the formation of additional perforations, and it is also possible to include previously unused formation intervals.

This method is characterized by the same disadvantages as during acid treatment by the workover crew, and an increase in the duration of the repair, which will be about 200 hours, depending on the depth of the perforation interval. The advantage is a more efficient increase in the porosity and permeability of BHZ, the possibility of increasing the perforated thickness of the formation.

The study analyzes works on restoration and increase of injectivity of injection wells of Tuimazinskaya group of fields. As the initial information, it analyzed the effectiveness of work on 182 wells of the Tuymazinskaya group of fields. Reservoirs are divided into 3 generalized groups of wells with similar geological and physical characteristics:

- terrigenous reservoir of the Sbob + rad formation;
- carbonate reservoir of STkiz, Dfams formation;
- terrigenous reservoir of Dpash formation.

To assess the effectiveness of the method, an increase in the injectivity of the well after the work was taken in relation to the injectivity before the work, which makes it possible to exclude the influence of various parameters of the formation and the technology of work.

As can be seen from the diagram in the figure, the most effective is the production casing reperforation method in the formation interval followed by acid treatment, which is explained by the combination of various stimulation methods.

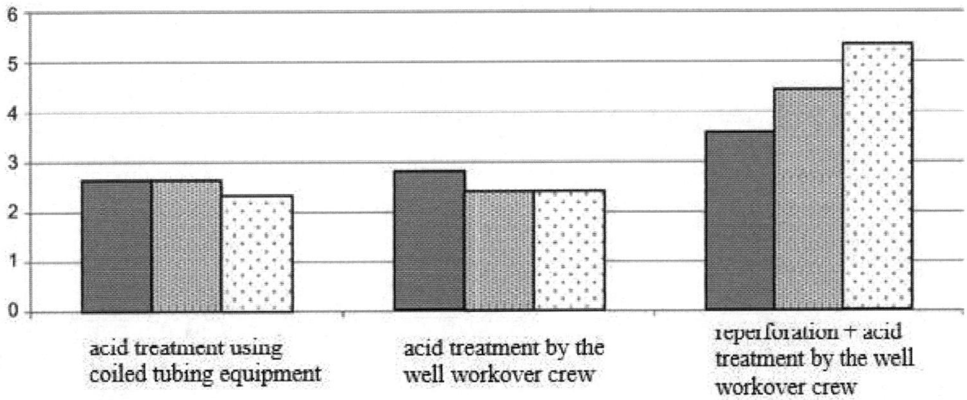

Figure 1. The effectiveness of methods for restoring the injectivity of injection wells: ■--an increase in acceleration, after GTM / up to GTM, fraction of units. Sbob + glad;▦– increase in injectivity, after GTM / before GTM, share of units STkiz, Dfams; ⊡– increase in injectivity, after GTM / before GTM, unit share of Dpash

After the measures are taken, the actual effect is calculated, schedules are built and analyzed, the main reasons for not achieving the planned indicators and measures for their compensation are determined. The following table presents the results of such an analysis.

Table 1. Comparative analysis of the used methods

GTM type	Increase in injectivity in layers, after GTM / before GTM, unit fractions		
	Sbob + glad	STkiz, Dfams	Dpash
Acid treatment using coiled tubing equipment	2.62	2.64	2.31
Acid treatment by the workover team	2.81	2.39	2.42
Reperforation + acidizing by workover team	3.57	4.42	5.34

4. Conclusion

Many years of experience in the operation of the Tuimazinskaya group of oil fields has established that one of the most effective methods for restoring the injectivity of injection wells includes chemical treatments of BHZ with various acid compositions in conjunction with well reperforation and the use of coiled tubing technologies.

References

[1] Muslimov R K 2005 *Modern methods of oil recovery increasing: design, optimization and performance evaluation* (Kazan: FEN) 688 p
[2] Yakupov R F, Mukhametshin V Sh, Khakimzyanov I N and Trofimov V E 2019 Optimization of reserve production from water oil zones of D3ps horizon of Shkapovsky oil field by means of horizontal wells *Georesursy* **21(3)** 55-61. DOI: 10.18599/grs.2019.3.55-61
[3] Mukhametshin V V and Kuleshova L S 2020 On uncertainty level reduction in managing waterflooding of the deposits with hard to extract reserves *Bull. of Tomsk Polytechnic University. Geo Assets Engineering* **331(5)** 140–146. DOI 10.18799/24131830/2020/5/2644
[4] Economides J M and Nolte K I 2000 *Reservoir stimulation* (West Sussex, England: John Wiley and Sons) 856 p
[5] Stenkin A V, Kotenev Yu A, Mukhametshin V Sh and Sultanov Sh Kh 2019 Use of low-mineralized water for displacing oil from clay productive field formations *IOP Conf. Ser.: Mater. Sci. Eng.* **560(1)** 012202. DOI: 10.1088/1757-899X/560/1/012202

[6] Alvarado V, Reich E M, Yunfeng Y and Potsch K 2006 Integration of a Risk Management Tool and an Analytical Simulator for Assisted Decision-Making in IOR *SPE Europec/EAGE Annual Conf. and Exhibition (Vienna, Austria, 12-15 June 2006)* p 6. DOI: 10.2118/100217-MS

[7] Rogachev M K and Mukhametshin V V 2018 Control and regulation of the hydrochloric acid treatment of the bottomhole zone based on field-geological data *J. of Mining Institute* **231** 275-280. DOI: 10.25515/PMI.2018.3.275

[8] Valeev A S, Kotenev Yu A, Mukhametshin V Sh and Sultanov Sh Kh 2019 Substantiation of the recovery of residual oil from low-productive and heterogeneous formations in Western Siberia by improving the waterflood system using gas and water-gas impacts *IOP Conf. Ser.: Mater. Sci. Eng.* **560(1)** 012204. DOI: 10.1088/1757-899X/560/1/012204

[9] Kuleshova L S, Kadyrov R R, Mukhametshin V V and Akhmetov R T 2019 Auxiliary equipment for downhole fittings of injection wells and water supply lines used to improve their performance in winter *IOP Conf. Ser.: Mater. Sci. Eng.* **560(1)** 012071. DOI: 10.1088/1757-899X/560/1/012071

[10] Bou-Mikael S, Asmadi F, Marwoto D and Cease C 2000 Minas Surfactant Field Trial Tests Two Newly Designed Surfactants with High EOR Potential *SPE Asia Pacific Oil and Gas Conf. and Exhibition (Brisbane, Australia, 16-18 October 2000)* p 12. DOI: 10.2118/64288-MS

[11] Tyncherov K T, Mukhametshin V Sh, Paderin M G, Selivanova M V, Shokurov I V and Almukhametova E M 2018 Thermoacoustic inductor for heavy oil extraction *IOP Conf. Ser.: Mater. Sci. Eng.* **327(4)** 042111. DOI:10.1088/1757-899X/327/4/042111

[12] Rogachev M K, Mukhametshin V V and Kuleshova L S 2019 Improving the efficiency of using resource base of liquid hydrocarbons in Jurassic deposits of Western Siberia *J. of Mining Institute* **240** 711-715. DOI: 10.31897/PMI.2019.6.711

[13] Yakupov R F, Mukhametshin V Sh and Tyncherov K T 2018 Filtration model of oil coning in a bottom water-drive reservoir *Periodico Tche Quimica* **15(30)** 725-733

[14] Briones M, Zambrano J A and Zerpa C 2002 Study of Gas-Condensate Well Productivity in Santa Barbara Field, Venezuela, by Well Test Analysis *SPE Annual Technical Conf. and Exhibition (San Antonio, Texas, 29 September-2 October 2002)* p 9. DOI: 10.2118/77538-MS

Geological and technological substantiation of waterflooding systems in deposits with hard-to-recover reserves

V Sh Mukhametshin, K T Tyncherov and N Rakhimov

Ufa State Petroleum Technological University, 54a Devonskaya St., Oktyabrsky, Republic of Bashkortostan, 452607, Russian Federation

E-mail: vsh@of.ugntu.ru, academic-mvd@mail.ru

Abstract. The study of the process of developing oil reserves in one of the fields with hard-to-recover reserves of the Tournaisian stage of the Volga-Ural oil and gas basin. A significant increase in the oil recovery factor was established when using in-circuit waterflooding as compared to the development of deposits in natural mode. The influence on the degree of reserves recovery is shown by the density of the grid of wells, geological heterogeneity, the sweep efficiency by injection by thickness, variations in the injectivity profile in injection wells. Models are proposed to assess the parameters of the efficiency of waterflooding systems.

1. Introduction

A significant number of studies have been devoted to the issue of increasing the efficiency of the development of oil deposits using waterflooding [1–6]. Most of them concern highly productive facilities. However, significant oil reserves are concentrated in low-producing reservoirs, where water injection into the reservoir does not always give positive results, requires studying the waterflooding process and creating a scientific basis for making management decisions [7–14].

For oil deposits in the carbonate reservoirs of the Volga-Ural oil and gas province, there are usually a deterioration of reservoir properties, an increase in oil viscosity near the oil-water contact, an oxidized oil layer, sometimes - almost complete isolation of the reservoir from the reservoir water drive system. Due to the tendency for significant deterioration of reservoir properties of the water-oil zone, the systems of boundary waterflooding have not found wide application due to their low efficiency.

As the experience of reservoir development using contour waterflooding systems shows, the effectiveness of this measure is quite high, especially in the conditions of relatively highly productive deposits of the Kuibyshev, Orenburg and Perm regions. In the conditions of low-productivity objects, the efficiency is much lower, which is due to increased geological heterogeneity, low reservoir properties, high oil viscosity, lenticular distribution of reservoir rocks. Moreover, the influence of various geological parameters on the waterflooding efficiency is rather ambiguous in the conditions of various groups of objects, which imposes increased requirements on the choice of injection wells, their interposition relative to the producing wells, to the justification of injection pressure, etc. For example, in the conditions of the Tournaisian (Znamenskoye, Meleuzovskoye, Mancharovskoye, Sergeevskoye fields), Kashiro-Podolsky (Arlanskoye), Bashkir-Vereysky (Igrovskoye, Yugomashevskoye, Chetyrmanskoye) deposits of the platform part of Bashkiria in areas with insignificant effective oil-saturated strata (at pressures at the bottom of injection wells less than 0.7 of the rock pressure),

IOP Conf. Series: Materials Science and Engineering 1064 (2021) 012068 doi:10.1088/1757-899X/1064/1/012068

production wells are often not affected by injection, even when they are at a distance of 400–500 m from the injection well. Apparently, under these conditions, the distances between the production and injection wells should be less, as indicated by the experience of development of the Khorskhed field, where the effective oil-saturated thickness of 1.7 m did not become an obstacle to successful water injection into the reservoir.

At the same time, the experimental water injection into contour wells in the conditions of the thick Sakmaro-Artinsky limestones of the Ishimbay group of fields showed that the waterflooding efficiency was very low. This is due to the fact that, with significant layer-by-layer heterogeneity of reef massifs, water quickly breaks through several high-permeability interlayers into adjacent production wells and floods them. At the same time, a significant part of the section is not mined.

Zonal heterogeneity in reservoir properties also has a significant impact on the efficiency of contour waterflooding. The study of carbonate strata in drilled fields shows that secondary processes that improve the reservoir properties of a carbonate stratum proceeded unevenly not only along the section, but also in the area of individual structural and lithological rock differences. These processes are most developed at the highs and steep sides of structures and along the directions of tectonic fracturing.

Most of the low-producing oil deposits in the carbonate reservoirs of the Tournaisian stage of the platform part of Bashkiria are developed practically in natural modes. The final oil recovery factor (predicted) will not exceed 0.15–0.20.

At the same time, as the experience of developing the Tournaisian oil reservoir of the Znamenskoye field shows, the use of in-loop waterflooding with a ratio of injection and production wells of about 0.5 and lower injection pressures (on average - 4.2 MPa) will increase the oil recovery factor by almost 2 times.

2. Materials and methods

In this regard, the oil recovery process was studied in seven sections of this field, separated from each other by cutting rows of injection wells. Within each section, there was a focal flooding carried out; the density of the grid of wells varies from 13.8 to 30.2 ha/well (on average it is 22 ha/well) the ratio of injection (N_i) wells to production (N_p) – from 0.35 to 0.67 (on average – 0.5).

The field is characterized by the following geological and physical parameters: net pay thickness - 4.7 m; permeability coefficients - 0.019 μm2, porosity - 0.12 fractional units, oil saturation - 0.8 fractional units; formation oil viscosity - 12.5 mPa · s; gas content of reservoir oil - 29.3 m3 / t; saturation pressure - 6.1 MPa; initial reservoir pressure - 14.9 MPa; occurrence depth - 1600 m.

For each of the sections, calculated parameters reflected the geological heterogeneity, the final oil recovery factors under the condition of development in natural mode (η) and using the existing in-circuit waterflooding (η_{wat}). Furthermore, the increase in the oil recovery factor was determined due to the use of waterflooding ($\Delta\eta$).

Prediction of the value of h was carried out using a model obtained earlier from similar deposits in carbonate reservoirs, the development of which was carried out in a natural mode:

$$\eta = 102\,(K_{rr}^{0.5} \cdot K_{comp}^{0.5} \cdot H_t)/(\mu_o^{0.5} \cdot S_g) - 108 \cdot H_t^{0.5}/(\mu_o^{0.5} \cdot S_g^{0.5}) + 38.8 \cdot H_t^{0.5}/(\mu_o^{0.5} \cdot S_g^{0.5}) - 31/t_{melt}, \quad (1)$$

where K_{rr} – the proportion of reservoir rocks in the total thickness of the formation; K_{comp} – coefficient of compartmentalization; H_t – average thickness of oil-saturated layers, m; μ_o – viscosity of reservoir oil, mPa · s; S_g – density of the grid of production and injection wells, ha / well; t_{melt} - initial formation temperature, K.

The η_{wat} value was predicted using the method of Kopytov A.V. on displacement characteristics. The gain was defined as the difference between η_{wat} and η. The values of the calculated parameters are given in the table 1.

Table 1. Values of technological parameters for plots of the Znamenskoye field

Parameter	Lot number						
	1	2	3	4	5	6	7
S_g, ha/well	30.2	14.8	14.0	36.6	13.8	20.8	22.2
N_i/N_p	0.35	0.67	0.50	0.35	0.49	0.40	0.48
η_{wat}, %	12.9	40.0	49.9	16.4	39.2	25.7	29.4
η, %	8.1	16.6	22.7	9.4	20.0	13.8	12.5
$\Delta\eta$, %	4.8	23.4	27.2	7.0	19.2	11.9	16.9

3. Results and discussion

It is clear that in areas with a denser grid of wells and more favorable geological characteristics $h_{зав}$ can reach 50%, although in general for the field with N_i/N_p=0.5 and S_g=22 ha / well, the oil recovery factor will be 30.2%. At the same time, the average value will be only 14.7%.

The largest increase in oil recovery is observed in areas with a denser well pattern and better reservoir characteristics.

When constructing models and identifying the most informative geological and technological parameters (29 parameters were considered), which have a prevailing effect on the oil recovery process, the study uses stepwise regression analysis. Models of the following form were obtained:

$$\eta_{wat} = 102 - 0.55W_{ot} - 1.12S_g + 7.13\ K_{sw}/W_{ip}, \quad (2)$$

$$\Delta\eta = 60.0 - 0.36W_{ot} - 0.68S_g + 2.75K_{sw}/W_{ip}, \quad (3)$$

where W_{ot} – variation in the thickness of oil-saturated layers,%; K_{sw}/W_{ip} is the ratio of the average value of the reservoir sweep efficiency by thickness to the variation of injectivity profiles (determined from the data of flow measurements in injection wells).

The values of the multiple correlation coefficients (R) are 0.95 and 0.94, respectively.

After excluding parameters reflecting geological heterogeneity, the following models were obtained:

$$\eta_{wat} = 26.7 + 8.6K_{prod} - 1.22\ S_g + 10.6\ K_{sw}/W_{ip}, \quad (4)$$

$$\Delta\eta = 26.6 - 5.68\ K_d + 5.84\ K_{prod} - 0.64\ S_g + 3.38\ K_{sw}/W_{ip}, \quad (5)$$

Where K_{prod} – is the productivity factor, t / day · MPa; K_d – coefficient of dissection.

The R values are 0.87 and 0.89, respectively.

Models (2) and (3) can be used to predict the oil recovery factor and its growth in deposits similar to the investigated and drilled with a sufficiently dense grid of wells (so that W_{ot} can be calculated). At the same time, in the absence of flow measurement studies of injection wells, the values of K_{sw}/W_{ip} can be estimated using the obtained empirical formula:

$$K_{sw}/W_{ip} = 2.63 + 0.049H_e + 0.73\sigma_{ot} + 0.31K_d, \quad (6)$$

Where H_e is the average value of the effective oil-saturated thickness, m; σ_{ot} – root-mean-square deviation of oil-saturated layers thickness, m.

Models (4) and (5) can be used at the stage of field withdrawal from exploration, and the ratio K_{sw}/W_{ip} can be calculated by the formula:

$$K_{sw}/W_{ip} = 1.81 + 0.22m - 0.47K_d, \quad (7)$$

where m is the average value of the porosity coefficient according to the geophysical survey data.
The R values of models (6) and (7) are 0.77 and 0.62, respectively.

4. Conclusion

The research makes the following main conclusions:

- the use of an intensive waterflooding system (close to areal) in the conditions of a low-productive Tournaisian reservoir of the Znamenskoye field will make it possible to increase the oil recovery factor by almost 2 times as compared to natural development;
- the density of the grid of production and injection wells, geological heterogeneity, the coefficient of coverage of the reservoir by injection in thickness and variation of the injectivity profile have a

significant impact on the oil recovery factor and its growth during the development of deposits using in-loop waterflooding;

- for the conditions of low-productivity deposits of the Tournaisian stage of the platform part of Bashkiria, the article proposes the models that allow predicting the effectiveness of the use of in-circuit waterflooding, as well as evaluating the coefficients of reservoir injection in thickness and variation of the injectivity profile using indirect geological and field data.

References

[1] Miller G A 2007 Some Perspectives on Various Methods of Oil Shale Extraction, Piceance Basin, Colorado *27th Oil Shale Symp. Colorado School of Mines Golden* (Colorado, October 2007) p 14

[2] Shaidullina R M, Amirov A F, Mukhametshin V Sh and Tyncherov K T 2017 Designing Economic Socialization System in the Educational Process of Technological University *European J. of Contemporary Education* **6(1)** 149–158. DOI: 10.13187/ejced.2017.1.149

[3] Rogachev M K and Mukhametshin V V 2018 Control and regulation of the hydrochloric acid treatment of the bottomhole zone based on field-geological data *J. of Mining Institute* **231** 275-280. DOI: 10.25515/PMI.2018.3.275.

[4] Hutton A C 1987 Petrographic classification of oil shales *Int. J. of Coal Geology* **8(3)** 203-231. DOI: 10.1016/0166-5162(87)90032-2

[5] Kuleshova L S, Mukhametshin V V and Safiullina A R 2019 Applying information technologies in identifying the features of deposit identification under conditions of different oil-and gas provinces *J. of Phys.: Conf. Ser.* **1333(7)** 072012. DOI: 10.1088/1742-6596/1333/7/072012

[6] Tyncherov K T, Mukhametshin V Sh, Paderin M G, Selivanova M V, Shokurov I V and Almukhametova E M 2018 Thermoacoustic inductor for heavy oil extraction *IOP Conf. Ser.: Mater. Sci. Eng.* **327(4)** 042111. DOI:10.1088/1757-899X/327/4/042111

[7] Kuleshova L S, Kadyrov R R, Mukhametshin V V and Akhmetov R T 2019 Auxiliary equipment for downhole fittings of injection wells and water supply lines used to improve their performance in winter *IOP Conf. Ser.: Mater. Sci. Eng.* **560(1)** 012071. DOI: 10.1088/1757-899X/560/1/012071

[8] Downey M W, Garvin J, Lagomarsino R C and Nicklin D F 2011 Quick Look Determination of Oil-in-Place in Oil Shale Resource Plays *AAPG Annual Convention and Exhibition* (Houston, Texas, USA, April 2011) p 21

[9] Akhmetov R T, Mukhametshin V V and Kuleshova L S 2019 Simulation of the absolute permeability based on the capillary pressure curves using the dumbbell model *J. of Phys.: Conf. Ser.* **1333(3)** 032001. DOI: 10.1088/1742-6596/1333/3/032001

[10] Kuleshova L S and Mukhametshin V V 2019 Elimination of uncertainties in predicting well interaction using indirect geological field information *IOP Conf. Ser.: Earth Env. Sci.* **378(1)** 012115. DOI: 10.1088/1755-1315/378/1/012115

[11] Mukhametshin V V and Kuleshova L S 2019 Prediction of production well flow rates using survey data *IOP Conf. Ser.: Earth Env. Sci.* **378(1)** 012116. DOI: 10.1088/1755-1315/378/1/012116

[12] Ghauri W K, Osborne A E and Magnuson W L 1974 Changing concerpts in carbonate waterflooding. West Texas Denver Unit prodect *J. of Petroleum Technology* **3** 595-606

[13] Mukhametshin V V 2018 Bottomhole formation zone treatment process modelling with the use geological and geophysical information *IOP Conf. Ser.: Earth Env. Sci.* **194(2)** 022024. DOI: 10.1088/1755-1315/194/2/022024

[14] Akhmetov R T and Mukhametshin V V 2018 Estimation of displacement coefficient with due account for hydrophobization of reservoir using geophysical data of wells *IOP Conf. Ser.: Earth Env. Sci.* **194(6)** 062001. DOI: 10.1088/1755-1315/194/6/062001

IOP Conf. Series: Materials Science and Engineering 1064 (2021) 012069 doi:10.1088/1757-899X/1064/1/012069

Using the method of canonical discriminant functions for a qualitative assessment of the response degree of producing wells to water injection during the development of carbonate deposits

V V Mukhametshin[1] and L S Kuleshova[2]

[1] Ufa State Petroleum Technological University, 1, Kosmonavtov St., Ufa, Republic of Bashkortostan, 450062, Russian Federation
[2] Ufa State Petroleum Technological University, Branch of the University in Oktyabrsky, 54a Devonskaya St., Oktyabrsky, Republic of Bashkortostan, 452607, Russian Federation

E-mail: vv@of.ugntu.ru, markl212@mail.ru

Abstract. The article proposes an algorithm for qualitative assessment of the response degree of producing wells to water injection into wells for the conditions of oil deposits of the Tournaisian stage of the Volga-Ural oil and gas province, based on the use of the method of canonical discriminant functions (CDF). The obtained CDF equations and distributions make it possible to quickly solve the issues of assessing success of a particular waterflooding system, select wells for transferring them to injection, evaluate the density of the grid of production and injection wells, pressure and volumes of water injection at the stage of drawing up the first design documents.

1. Introduction

An increase in the degree of development of hard-to-recover oil reserves is largely determined by the efficiency of the applied development system [1–8]. At the same time, it is important to determine not only during the development process, but also after the field leaves exploration, the degree of response of producing wells to water injection into injection wells [9–11].

The research presents [12] a method for determining the interaction of wells based on the use of the values of the total diagnostic coefficients (TDC) and the values of cross-correlation functions (R) for deposits in the carbonate reservoirs of the Volga -Ural oil and gas province.

One of the weak points of the results obtained in [12] is the presence of zones of uncertainty, which do not allow to give an unambiguous answer to the question of whether the producing well will react to water injection under certain conditions or not. The percentage of wells according to the calculation options that fell into the zones of uncertainty varies from 21 to 51%, averaging 34.3%, and the zones of uncertainty occupy from 8 to 68% of the total interval of changes in the values of the total diagnostic coefficients, averaging 53.4%, which is quite essential. In order to eliminate these shortcomings, namely, to increase the resolution of the method, simplify the calculations, reduce their labor intensity, as well as to control the results obtained and more clearly separate the wells, the article used the method of canonical discriminant functions (CDF).

2. Materials and methods

Calculations using this method were carried out according to six options:

1, 2 - using all parameters and parameters that are significant by the Kullback criterion, respectively;

3, 4 - using all parameters and parameters that are significant according to the Kullback criterion, respectively, and the ones that are characterizing geological and physical properties of formations in wells, physicochemical properties of formation fluids, initial conditions of occurrence and intensity of the waterflooding system;

5, 6 - using all parameters and parameters that are significant according to the Kullback criterion, respectively, characterizing technological features of the operation of wells and deposits, as well as intensity of the waterflooding system.

Variants with the use of parameters that characterize only the intensity of the waterflooding system were not considered, since their separate use without parameters reflecting features of the geological structure of objects and technological features of the operation of wells and deposits is not justified due to low resolution.

At the first stage, two groups of wells were considered:

1 - wells that reacted to water injection ($R \geq 0.5$);

2 – wells that did not respond to water injection ($R < 0.5$).

3. Results and Discussion

The distribution of wells shown in the figure in the axes of the first two canonical discriminant functions by calculation options showed a clearer division of wells into groups compared to the results that used values of the total diagnostic coefficients [12]. A comparison of the resolution capabilities of the TDC and CDF methods is presented in Table 1. Parameter Z_1 represents the percentage of wells that fell into the zone of uncertainty to the total number of wells that participated in the analysis in the TDC method, and the percentage of wells from the total number that fell into the zone of uncertainty and when this is not correctly assigned to a particular group in the CDF method. Parameter Z_2 is the ratio of the length of the interval of change in TDC values, to which the zone of uncertainty is confined, to the total interval of change in TDC values. In the CDF Z_2 method, the ratio of the area of the zone of uncertainty to the area within which all wells are concentrated.

Table 1. Relative values of intervals and areas of uncertainty zones, the number of wells in them by calculation options using TDC and CDF methods

Parameter	Method	Parameter values by variant					
		1	2	3	4	5	6
Z_1,%	TDC	21	23	28	30	44	40
	CDF	4	4	4	4	7	10
Z_2,%	TDC	8	57	60	59	61	46
	CDF	8	10	9	11	25	30

Table 1 shows that when using the method of canonical discriminant functions, the number of wells in the zone of uncertainty compared with the method of total diagnostic coefficients, according to the options presented, decreases by 4-7.5 times, and on average - 5.6 times, and the zone of uncertainty decreases by 3.1 times.

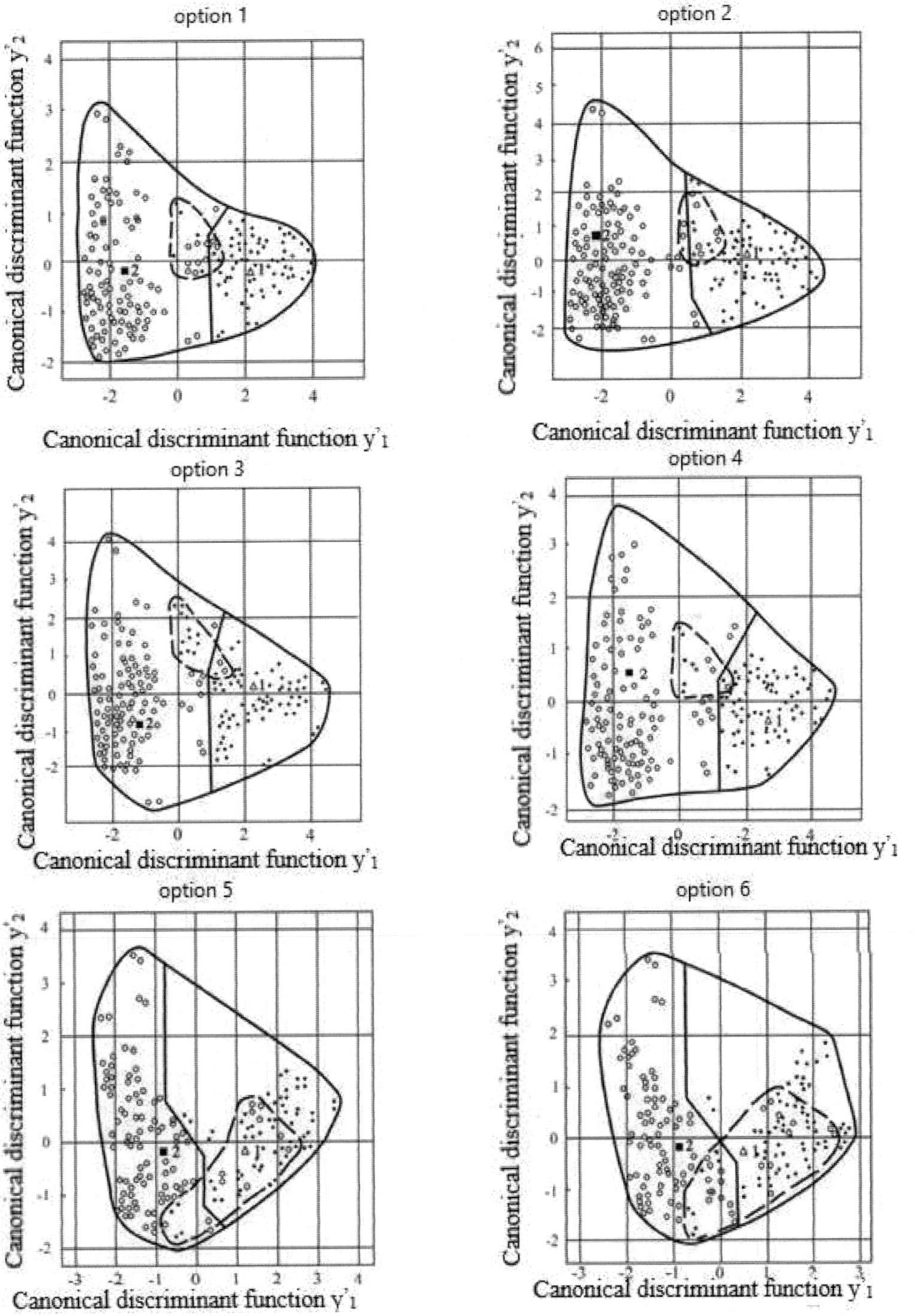

Δ1, ■2 – centroid and number of a group of wells that reacted (R ≥ 0,5) and did not respond (R< 0,5) to water injection, respectively; ●, ○ – wells of groups 1 and 2, respectively; ~ – boundary dividing groups of wells; O, ( ) – zone of concentration of all wells and zone of uncertainty, respectively

Figure 1. Distribution of wells in the axes of canonical discriminant functions

However, some of the conclusions obtained using the CDF method support the conclusions obtained using the TDC method. Table 1 shows that there is no need to use all the parameters under consideration when diagnosing. It is enough to restrict ourselves to only the parameters that are significant according to the Kullback criterion, since there is no significant loss of information in options 2, 4, 6 compared to options 1, 3, 5. At the same time, the parameters characterizing the intensity of the waterflooding system and the parameters reflecting features of the geological structure of the deposits have a prevailing influence on the separation of wells by interaction. The parameters characterizing technological features of wells and reservoirs have a lesser impact on the separation of wells. The resulting equations of the canonical discriminant functions have the following form:

- according to option 1:

$$
\begin{aligned}
y_1 = \; & -0{,}31\mu_0 + 0{,}08\rho_N + 0{,}30K_{mix} + 0{,}80K_{prod} - 0{,}02F + 2{,}67H_E^N + 1{,}30H_P^N + 0{,}55n^N - \\
& -0{,}54M_G^N + 0{,}42H_E^D + 0{,}87H_P^D - 0{,}13n^D + 0{,}05M_G^D - 3{,}07K_P^N - 1{,}46H_{tot}^N + 0{,}16K_P^D + \\
& +0{,}09H_{tot}^D - 0{,}48P_{inj}/P_{rock} - 0{,}16Q_N^D + 0{,}33Q_{inc}^D + 0{,}31Q_V^D - 0{,}30f_1 - 0{,}002Q_{inj} - \\
& -0{,}57t - 0{,}26Q_{N\max}^D,
\end{aligned} \tag{1}
$$

$$
\begin{aligned}
y_2 = \; & -1{,}18\mu_0 - 0{,}78\rho_N + 0{,}26K_{mix} - 0{,}26K_{prod} + 0{,}41F + 0{,}22H_E^N - 2{,}70H_P^N - 2{,}29n^N + \\
& +0{,}42M_G^N - 0{,}49H_E^D - 0{,}57H_P^D - 0{,}77n^D + 0{,}30M_G^D + 2{,}13K_P^N + 2{,}37H_{tot}^N + 0{,}98K_P^D \\
& +1{,}21H_{tot}^D + 1{,}31P_{inj}/P_{rock} + 0{,}46Q_N^D - 0{,}27Q_{inc}^D + 0{,}18Q_V^D - 0{,}06f_1 + 0{,}11Q_{inj} + 1{,}28t - \\
& -0{,}79Q_{N\max}^D,
\end{aligned} \tag{2}
$$

$$
\begin{aligned}
y_1' = \; & -9{,}23 - 0{,}14\mu_0 + 0{,}02\rho_N + 0{,}05K_{mix} + 1{,}12K_{prod} + 0{,}0001F + 0{,}84H_E^N + 1{,}17H_P^N + \\
& +0{,}37n^N - 0{,}41M_G^N + 0{,}13H_E^D + 0{,}84H_P^D - 0{,}08n^D + 0{,}03M_G^D - 21{,}1K_P^N - 0{,}28H_{tot}^N + \\
& +1{,}02K_P^D + 0{,}02H_{tot}^D - 5{,}54P_{inj}/P_{rock} - 0{,}001Q_N^D + 0{,}0001Q_{inc}^D + 0{,}003Q_V^D - 1{,}33f_1 + \\
& +0{,}0001Q_{inj} - 0{,}006t - 0{,}001Q_{N\max}^D,
\end{aligned} \tag{3}
$$

$$
\begin{aligned}
y_2' = \; & 174 - 0{,}54\mu_0 - 0{,}21\rho_N + 0{,}04K_{mix} - 0{,}37K_{prod} + 0{,}002F + 0{,}07H_E^N - 2{,}42H_P^N - \\
& -1{,}53n^N + 0{,}32M_G^N - 0{,}15H_E^D - 0{,}55H_P^D - 0{,}47n^D + 0{,}21M_G^D + 14{,}6K_P^N + 0{,}45H_{tot}^N + \\
& +6{,}43K_P^D + 0{,}22H_{tot}^D + 15{,}1P_{inj}/P_{rock} + 0{,}004Q_N^D + 0{,}0001Q_{inc}^D + 0{,}002Q_V^D - 0{,}25f_1 + \\
& +0{,}0001Q_{inj} + 0{,}01t - 0{,}004Q_{N\max}^D;
\end{aligned} \tag{4}
$$

- according to option 2:

$$
\begin{aligned}
y_1 = \; & 2{,}81H_E^N + 0{,}66H_P^N - 0{,}03F - 0{,}42M_G^N + 0{,}33H_E^D + 0{,}84H_P^D - 2{,}57K_P^N - 1{,}15H_{tot}^N + \\
& +0{,}25K_P^D + 1{,}48t_{melt} + 0{,}73K_{prod} - 0{,}27P_{inj}/P_{rock} - 0{,}04Q_{inj} - 0{,}72H_{bed} + 0{,}04M_G^D - \\
& -0{,}16Q_{N\max}^D - 0{,}08t,
\end{aligned} \tag{5}
$$

$$
\begin{aligned}
y_2 = \; & -2{,}57H_E^N - 0{,}23H_P^N + 0{,}36F + 1{,}03M_G^N - 0{,}19H_E^D + 0{,}10H_P^D + 1{,}98K_P^N + 2{,}27H_{tot}^N + \\
& +0{,}32K_P^D + 5{,}99t_{melt} - 0{,}38K_{prod} + 1{,}08P_{inj}/P_{rock} - 0{,}30Q_{inj} - 5{,}31H_{bed} - 0{,}20M_G^D - \\
& -0{,}22Q_{N\max}^D + 0{,}35t,
\end{aligned} \tag{6}
$$

$$y_1' = -164+0,89H_E^N+0,60H_P^N+0,0001F-0,32M_G^N+0,10H_E^D+0,81H_P^D-17,6K_P^N- -0,22H_{tot}^N+1,63K_P^D+0,60t_{melt}+1,03K_{prod}-3,08P_{inj}/P_{rock}+0,0001Q_{inj}-0,007H_{bed}+ +0,03M_G^D-0,001Q_{N\max}^D-0,001t, \quad (7)$$

$$y_2' = -670-0,81H_E^N-0,21H_P^N+0,002F+0,78M_G^N-0,06H_E^D+0,09H_P^D+13,6K_P^N+ +0,43H_{tot}^N+2,08K_P^D+2,45t_{melt}-0,53K_{prod}+12,5P_{inj}/P_{rock}+0,0001Q_{inj}-0,05H_{bed}- -0,14M_G^D-0,001Q_{N\max}^D+0,004t; \quad (8)$$

- according to option 3:

$$y_1 = -0,51\mu_0+0,11\rho_N+0,21K_{mix}+0,79K_{prod}-0,12F+2,25H_E^N+0,94H_P^N+0,42n^N- -0,44M_G^N+0,64H_E^D+0,64H_P^D-0,15n^D-0,13M_G^D-2,18K_P^N-1,17H_{tot}^N-0,03K_P^D+ +0,05H_{tot}^D-0,48P_{inj}/P_{rock}-0,004Q_{inj}, \quad (9)$$

$$y_2 = -0,67\mu_0-0,98\rho_N+0,28K_{mix}-0,29K_{prod}+0,33F+0,62H_E^N-1,89H_P^N-2,03n^N+ +0,31M_G^N-0,64H_E^D-0,10H_P^D-0,36n^D+0,12M_G^D+0,65K_P^N+2,00H_{tot}^N+0,95K_P^D+ +0,95H_{tot}^D+1,31P_{inj}/P_{rock}-0,03Q_{inj}, \quad (10)$$

$$y_1' = -17,8-0,23\mu_0+0,03\rho_N+0,04K_{mix}+1,10K_{prod}-0,001F+0,71H_E^N+0,84H_P^N+ +0,28n^N-0,33M_G^N+0,20H_E^D+0,62H_P^D-0,09n^D-0,009M_G^D-15,0K_P^N-0,22H_{tot}^N -0,17K_P^D+0,009H_{tot}^D-5,53P_{inj}/P_{rock}+0,001Q_{inj}, \quad (11)$$

$$y_2' = 221-0,31\mu_0-0,26\rho_N+0,05K_{mix}-0,40K_{prod}+0,002F+0,20H_E^N-1,70H_P^N-1,36n^N +0,23M_G^N-0,20H_E^D-0,10H_P^D-0,22n^D+0,08M_G^D+4,46K_P^N+0,38H_{tot}^N+6,21K_P^D+ +0,17H_{tot}^D+15,2P_{inj}/P_{rock}+0,0001Q_{inj}; \quad (12)$$

- according to option 4:

$$y_1 = 2,66H_E^N+0,58H_P^N-0,09F-0,40M_G^N+0,41H_E^D+0,74H_P^D-2,26K_P^N-1,08H_{tot}^N+ +0,11K_P^D+1,40t_{melt}+0,74K_{prod}-0,30P_{inj}/P_{rock}-0,01Q_{inj}-0,67H_{bed}+0,004M_G^D, \quad (13)$$

$$y_2 = -2,12H_E^N-0,03H_P^N+0,32F+0,91M_G^N-0,16H_E^D+0,09H_P^D+1,21K_P^N-2,03H_{tot}^N+ +0,33K_p^D+6,01t_{melt}-0,37K_{prod}+1,09P_{inj}/P_{rock}-0,27Q_{inj}-5,52H_{bed}-0,24M_G^D, \quad (14)$$

$$y_1' = -156+0,84H_E^N+0,52H_P^N+0,0001F-0,30M_G^N+0,13H_E^D+0,71H_P^D-15,5K_P^N- -0,21H_{tot}^N+0,73K_P^D+0,57t_{melt}+1,04K_{prod}-3,47P_{inj}/P_{rock}+0,0001Q_{inj}-0,007H_{bed}+ +0,002M_G^D, \quad (15)$$

$$y_2' = -666-0,67H_E^N-0,03H_P^N+0,002F+0,69M_G^N-0,05H_E^D+0,09H_P^D+8,32K_P^N+ +0,39H_{tot}^N+2,14K_P^D+2,46t_{melt}-0,52K_{prod}+12,6P_{inj}/P_{rock}+0,0001Q_{inj}-0,06H_{bed}- -0,17M_G^D; \quad (16)$$

- according to option 5:

$$y_1 = -0,17F+0,93P_{inj}/P_{rock}+0,21Q_N^D+0,12Q_{inc}^D+0,21Q_V^D-0,31f_1+0,15Q_{inj}+0,18t- -0,40Q_{N\max}^D, \quad (17)$$

$$y_2 = 1,04F + 0,33P_{inj} / P_{rock} - 0,01Q_N^D + 0,40Q_{inc}^D + 0,40Q_V^D - 0,38f_1 + 0,07Q_{inj} - 0,11t - -0,52Q_{N\max}^D, \quad (18)$$

$$y_1' = -4,77 - 0,001F + 10,8P_{inj} / P_{rock} + 0,002Q_N^D + 0,0001Q_{inc}^D + 0,002Q_V^D - 1,37f_1 + +0,0001Q_{inj} + 0,002t - 0,002Q_{N\max}^D, \quad (19)$$

$$y_2' = -3,79 + 0,005F + 3,79P_{inj} / P_{rock} + 0,0001Q_N^D + 0,001Q_{inc}^D + 0,004Q_V^D - 1,67f_1 + +0,0001Q_{inj} - 0,001t - 0,003Q_{N\max}^D; \quad (20)$$

- according to option 6:

$$y_1 = -0,18F + 0,95P_{inj} / P_{rock} + 0,06Q_{inj} + 0,08t - 0,05Q_{inc}^D, \quad (21)$$

$$y_2 = 1,02F + 0,34P_{inj} / P_{rock} + 0,02Q_{inj} + 0,14t - 0,25Q_{inc}^D, \quad (22)$$

$$y_1' = -4,97 - 0,001F + 11,0P_{inj} / P_{rock} + 0,0001Q_{inj} + 0,001t - 0,00001Q_{inc}^D, \quad (23)$$

$$y_2' = -4,27 + 0,005F + 3,95P_{inj} / P_{rock} + 0,0001Q_{inj} + 0,002t - 0,00001Q_{inc}^D. \quad (24)$$

The interpretation of the parameters of the equations is presented in [12].

The equations (1), (2), (5), (6), (9), (10), (13), (14), (17), (18), (21), (22) normalize the coefficients of canonical discriminant functions. The values of functions in centroids of the selected groups are presented in Table 2.

Table 2. Values of discriminant functions in group centroids

Option	Function	Well group number 1	2
1	y_1'	2.281	–1.579
	y_2'	0.025	–0.114
2	y_1'	2.181	–1.514
	y_2'	0.011	–0.073
3	y_1'	2.191	–1.518
	y_2'	0.021	–0.102
4	y_1'	2.155	–1.496
	y_2'	0.011	–0.072
5	y_1'	1.450	–1.023
	y_2'	–0.005	–0.012
6	y_1'	1.409	–0.993
	y_2'	–0.005	–0.011

The algorithm for determining the belonging of wells to one of the two groups will be as follows:

- according to one of the options (depending on the availability of the initial geological and field information), the research determines the values of the canonical discriminant functions;
- belonging of the wells to any group in the axes of canonical discriminant functions in the figure is determined.

The values of indicators characterizing the intensity of the waterflooding system are set based on the wishes of the user.

If the well has fallen into the zone of uncertainty, by changing the adjustable parameters (F, P_{inj} / P_{rock}, Q_{inj}), it is necessary to transfer it to the zone where the wells are affected by injection.

If the well in the axes of the canonical discriminant functions is outside the zone of wells concentration shown in the figure, it is important to determine which group of wells it is closer to. The search for such a group should be done as follows:

- values of the canonical discriminant functions of the well should be determined, as well as its location on the plane in the axes $y_1' - y_2'$;

- find the centroid of the group closest to the given object in the Euclidean space of two canonical discriminant functions based on the expression

$$d_j = \sqrt{\sum_{i=1}^{m} (y_i' - y_i^{j'})^2}, \qquad (25)$$

where d_j is the Euclidean distance between the object (well) and the j-th cetroid; y_i' – value of the i-th non-normalized canonical discriminant function of the well; $y_i^{j'}$ – the value of the i-th unnormalized canonical discriminant function of the j-th centroid (see table 2); m=2 is the number of unnormalized canonical discriminant functions.

4. Conclusion

1. For the conditions of oil deposits of the Tournaisian stage of the Volga-Ural oil and gas province, based on the use of the method of canonical discriminant functions (CDF), an algorithm is proposed for the qualitative assessment of the degree of response of producing wells to water injection into injection wells.

2. The obtained CDF and distribution equations make it possible to quickly solve the issues of assessing the success of a particular waterflooding system, select wells for transferring them to injection, evaluate the density of the grid of production and injection wells, pressure and volumes of water injection at the stage of drawing up the first design documents.

References

[1] Lungwitz B, Fredd C, Brady M, Miller M, Ali S and Hughes K 2004 Diversion and Cleanup Studies of Viscoelastic Surfactant-Based Self-Diverting Acid *SPE Int. Symp. and Exhibition on Formation Damage Control (Lafayette, 18-20 February 2004)* p 10. DOI: 10.2118/86504-MS

[2] Rogachev M K and Mukhametshin V V 2018 Control and regulation of the hydrochloric acid treatment of the bottomhole zone based on field-geological data *J. of Mining Institute* **231** 275-280. DOI: 10.25515/PMI.2018.3.275

[3] Stenkin A V, Kotenev Yu A, Mukhametshin V Sh and Sultanov Sh Kh 2019 Use of low-mineralized water for displacing oil from clay productive field formations *IOP Conf. Ser.: Mater. Sci. Eng.* **560(1)** 012202. DOI: 10.1088/1757-899X/560/1/012202

[4] Nasr-El-Din H A, Van Domelen M S, Sierra L and Welton T D 2007 Optimization of Surfactant-based Fluids for Acid Diversion *European Formation Damage Conf. (Scheveningen, The Netherlands, 30 May-1 June 2007)* p 11. DOI: 10.2118/107687-MS

[5] Valeev A S, Kotenev Yu A, Mukhametshin V Sh and Sultanov Sh Kh 2019 Substantiation of the recovery of residual oil from low-productive and heterogeneous formations in Western Siberia by improving the waterflood system using gas and water-gas impacts *IOP Conf. Ser.: Mater. Sci. Eng.* **560(1)** 012204. DOI: 10.1088/1757-899X/560/1/012204

[6] Yakupov R F, Mukhametshin V Sh and Tyncherov K T 2018 Filtration model of oil coning in a bottom water-drive reservoir *Periodico Tche Quimica* **15(30)** 725-733

[7] Tyncherov K T, Mukhametshin V Sh, Paderin M G, Selivanova M V, Shokurov I V and Almukhametova E M 2018 Thermoacoustic inductor for heavy oil extraction *IOP Conf. Ser.: Mater. Sci. Eng.* **327(4)** 042111. DOI:10.1088/1757-899X/327/4/042111

[8] Brian C and Metcalf S 2009 Intensification of oil production at the San Andreas site thanks to the use of weak acid *Oil and Gas technologies* **12** 21-24
[9] Zeigman Yu V and Mukhametshin V V 2009 Flooding systems correspondence to formation geological structure features *Oilfield Engineering* **5** 10–12
[10] Zeigman Yu V, Mukhametshin V Sh, Khafizov A R and Kharina S B 2016 Prospects of Application of Multi-Functional Well Killing Fluids in Carbonate Reservoirs *SOCAR Proceedings* **3** 33–39. DOI: 10.5510/OGP20160300286
[11] Shaidullina R M, Amirov A F, Mukhametshin V Sh and Tyncherov K T 2017 Designing Economic Socialization System in the Educational Process of Technological University *European J. of Contemporary Education* **6(1)** 149–158. DOI: 10.13187/ejced.2017.1.149
[12] Shuchart C E and Buster D C 1995 Determination of the Chemistry of HF Acidizing with the Use of F NMR Spectroscopy *SPE Int. Symp. on Oilfield Chemistry (San Antonio, Texas, 14-17 February 1995)* p 12. DOI: 10.2118/28975-MS
[13] Batalov S A, Andreev V E, Lobankov V M and Mukhametshin V Sh 2019 Numerical simulation of oil formation with regulated disturbances. Oil recovery quality simulation *J. of Phys.: Conf. Ser.* **1333(3)** 032006. DOI: 10.1088/1742-6596/1333/3/032006

IOP Conf. Series: Materials Science and Engineering 1064 (2021) 012070 doi:10.1088/1757-899X/1064/1/012070

Microwave separation of persistent oil emulsions

R A Nafikova[1], N N Soloviev[2], Sh G Mingulov[1] and A R Safiullina[1]

[1]Ufa State Petroleum Technological University, Branch of the University in the City of Oktyabrsky, 54a, Devonskaya St., Oktyabrsky, Republic of Bashkortostan, 452607, Russian Federation

[2]LLC Gazprom VNIIGAZ, Staraya Basmannaya St., 20, b. 8, Moscow, Russian Federation

E-mail: rnafikova@gmail.com, N_Soloviev@vniigaz.gazprom.ru, info@of.ugntu.ru

Abstract. At present, the development of more effective methods for preparing such oils requires special attention due to the growing share of hard-to-extract high-viscous hydrocarbon reserves. The study examined the effectiveness of the thermochemical method and the method combining microwave radiation intended to separate high-viscous emulsions resulting from the thermal methods to increase oil recovery. The results of the research justify the need to use a complex impact aimed at separating persistent oil emulsions.

1. Introduction

Modern oil production has the increasing share of hard-to-recover hydrocarbon reserves [1–3], in particular, heavy and high-viscous oil with a viscosity of 30 mPa · s and higher. Such hydrocarbon resources are several times higher than the reserves of traditional oils and therefore are considered as a reserve for oil production. In Russia, hard-to-recover oil reserves are estimated at more than 7 billion tons. In this regard, it is necessary to pay attention not only to efficient production methods [4–7] which are currently quite complex and require significant costs but also to the issues of collecting and preparing such oils.

The work aimed to study the effects of microwaves on persistent oil emulsions. Meanwhile, the authors stood following tasks: to determine the initial content of mechanical impurities and the water content; to select empirically the time and power of exposure to eliminate the negative effects of microwaves; in a centrifuge, to treat hydrocarbons with a saturated aqueous solution of sodium chloride in different proportions to create favourable conditions conducive to more efficient separation of hydrocarbon mixtures.

2. Methods and materials

The study used a comparative analysis of the effectiveness of the separation of persistent emulsions by the thermochemical method followed by centrifugation and a method combining microwave irradiation (in different operation modes of the plant).

3. Results and discussion

For the preparation of heavy high-viscosity oils, in some cases, technologies include the use of hydrocarbon solvents (light oils, petroleum naphtha fractions, condensate, etc. with long-term sludge) at a temperature of at least 60 ° C. Solvents do not exclude the use of other known processing methods. The most common method is the treatment of high-viscous oil with a diluent based on the oil fraction and a demulsifier, and the oil is heated before the introduction of the demulsifier.

Another option is to introduce a hydrocarbon solvent containing oil-soluble demulsifier and ballast water into the oil. Oil processing occurs at intensive mixing with the introduced reagents [8].

According to the authors of [9], the combined method, including mixing with light oil in a vast range of ratios (from 1: 1 to 1: 100), long-term sludge (7–24 h), elevated processing temperature (up to 80 ° C), increased consumption of effective demulsifier (in particular, dissolvan 4411–250 g / t) helps to successfully destabilize and destroy high-viscous emulsions resulting from the thermal methods to increase oil recovery, the content of solids in which does not exceed 1%. The authors consider it effective to dose a large amount of demulsifier into the intermediate layer followed by feeding the intermediate layer at the beginning of the process, into the crude oil.

Methods using electric, centrifugal and magnetic fields are suitable for the preparation of heavy oils containing a significant number of solids [10]. However, these methods did not receive proper development and did not find wide application in the practice of preparing slop oils. There are combined methods for processing slop oils including treatment with specially developed reagent compositions, ballast (drainage) waters at temperatures of 60–80 °C. The compositions include demulsifiers, wetting agents and other chemicals in various ratios. Combined methods can successfully destroy slop and pit emulsions containing up to 12% of mechanical impurities by weight.

The methods used today for breaking oil-water emulsions are based on the contact heating methods and chemical additives (demulsifiers). Contact thermal methods have excessively high-energy costs, and chemical additives can pollute wastewater and processed products. An alternative is contactless microwave demulsification technology patented more than two decades ago. Microwaves have high penetrating power and act on water, oil and interfacial films selectively providing a separate supply of energy. It results in a significant increase in the phase separation rate and a decrease in power consumption compared to conventional heating methods [11, 12].

Studies conducted in various countries have established the high efficiency of microwave energy in various fields of economic activity: industry, agriculture, medicine, etc. The search for new applications of microwave processing methods aimed at increasing efficiency and expansion of the applications of ultra-high frequency fields occupying the frequency range from 300 MHz to 300 GHz continues successfully.

An electromagnetic wave penetrates a material and interacts with charged particles. The combination of such microscopic processes leads to the absorption of field energy in the object. The quantum theory helps to obtain a full description of this effect. However, the successful design of microwave devices requires only to the macroscopic properties of the material medium described by classical physics.

Depending on the location of the charges, the molecules of the dielectric medium can be polar and non-polar. In some molecules, the location of charges is so symmetrical that their electric dipole moment is zero without an external electric field. Polar molecules have some electric dipole moment even without an external field. Applying external electric field results in polarization of nonpolar molecules and breaking the symmetry of the arrangement of their charges, and the molecule acquires some electric moment.

Under the influence of microwaves on oil sludge, the temperature of the latter increases two times faster than the temperature of water and 10–20 times faster than the temperature of solid rocks. In this case, the rate of thermal action process on the system is much higher than with classical heating methods. The microwave action excites the dipole rotation of medium molecules in the presence of powerful intermolecular bonds resulting in hysteresis between the applied field and the induced response, and the stored energy is released during relaxation in the form of heat. Theoretically, such treatment should lead to more efficient separation of high-viscous mixtures into their constituent components: hydrocarbons, water, solids.

The studies of the microwave effect on various samples of persistent oil emulsions showed that the main energy is absorbed by the aqueous phase concentrating in globules coated with an armour shell.

It results in arising volumetric heat sources in water globules and their intense heating leading to the destruction of the armour shell [11, 12].

However, the burst temperature of the shell varies depending on the strength and thickness of the armour. Low temperatures are sufficient to rupture thin shells. As a result of the destruction (melting) of the shells, the droplets merge, and the emulsion exfoliates [12].

With significant strength and thickness of the shell, the burst temperature and pressure inside the shell can be so large that the shell ruptures and a single drop of water produces many small droplets. It leads to the formation of a finely dispersed even more stable system, i.e. there is a negative effect of microwave exposure.

For laboratory studies of the microwave effect on hydrocarbon mixtures, we used a Panasonic Microwave Oven (2.45 GHz, 900 W) household microwave oven with an adjustable output power range. The object of the study was samples of persistent oil emulsions. A preliminary analysis of the samples showed that close (almost identical) densities of the dispersion medium and the dispersed phase are a significant factor in the stability of the studied objects. We determined the initial content of mechanical impurities by GOST 6370-83 "Petroleum products and additives. Method for determination of mechanical impurities" and water content by GOST 2477-65 "Oil and oil products. Methods for determining the water content": sample No. 1 the water content is 28% of the mass, mechanical impurities are 3.1% of the mass; sample No. 2 the water content is 27% of the mass, mechanical impurities are 3.7% of the mass.

Studies on the destruction of the samples used the same conditions to obtain comparable results. We took and dispersed samples as quickly as possible to prevent them from cooling. The initial temperature of all experimental samples was 75 ºC. Then we experimentally calibrated the exposure time and power so that after processing the temperature of the samples did not exceed 90 °C to avoid negative effects of the microwaves. The exposure time was 50 seconds. The heating power was 50% of the nominal (Figure 1).

To create favourable conditions for the separation of hydrocarbon mixtures in a centrifuge, we performed the processing of hydrocarbons with a saturated aqueous solution of sodium chloride (brine) in different proportions. It required the saline because the density of the aqueous phase in the process of demulsification increases, and thereby contributes to a greater separation of water from oil. Centrifugation was carried out at sample temperatures of 90 °C and speeds of 2600 rpm. The average value for the two samples allowed to determine the water content, solids (Figs. 2, 3).

Figure 1. Photograph of separation by centrifugal distillation for 10 minutes of the sample after microwave exposure

Figure 2. Photograph of separation by centrifugal distillation for 10 minutes of the sample and the addition of 15% saltwater after microwave exposure

Figure 3. Photograph of separation by centrifugal distillation for 10 minutes of the sample and the addition of 20% saltwater after microwave exposure

The study used a comparative analysis of the effectiveness of the separation of persistent emulsions by the thermochemical method followed by centrifugation and a method combining microwave irradiation (in different operation modes of the plant).

Laboratory studies have shown that the sample heating to 90 °C by the usual method and centrifuging for 10 minutes only results in the initial separation of water. At the same time, processing the sample with microwaves allows reaching a temperature of 90 °C and leads to a complete separation of the sample phases.

The centrifugation time ranged from 2 to 10 minutes in the experiments.

The viscosity of the diluted hydrocarbon was 28 MPa•s, and the viscosity of the initial hydrocarbon was 75 MPa•s at temperatures of 90 °C. It was found that the weighting of the aqueous phase leads to an intensification of the separation of hydrocarbon mixture with water (Figure 3).

As a result of the experiments, we recorded the separation of samples into hydrocarbon, water and solid phases.

Experimental data confirmed that microwave exposure promotes a deeper separation of obsolete emulsions in addition to other methods (mixing to separate the densities of different phases, heating, centrifugation).

4. Conclusion

Analysis of the results showed that the use of microwaves reduces the cost of reducing the viscosity of resistant emulsions to effectively separate them. The increase in the water phase with using microwave action increases the demulsification and reduces the separation time in a centrifuge.

Comprehensive preparation with microwave exposure allows the separation of persistent emulsions, including slop oils, intermediate layers and other oil sludges since the separation rate of the hydrocarbon mixture increases due to the destructive effect of electromagnetic fields on the armour shells of water droplets of emulsions.

The main advantages of today's microwave installations compared to traditional technologies are the ecological purity of the technological process, its simple mechanization and automation, a significant reduction of technological time for the creation of the final product, reduction of energy and economic.

References

[1] Yakupov R F, Mukhametshin V Sh, Khakimzyanov I N and Trofimov V E 2019 Optimization of Reserve Production From Water Oil Xones of D3ps Horizon of Shkapovsky Oil Field Using Horizontal Wells *Georesursy* **21(3)** 55–61. DOI: 10.18599/grs.2019.3.55-61

[2] Stenkin A V, Kotenev Yu A, Mukhametshin V Sh and Sultanov Sh Kh 2019 Use of Low-Mineralized Water for Displacing Oil from Clay Productive Field Formations *IOP Conf. Ser.: Mater. Sci. Eng.* **560(1)** 012202. DOI: 10.1088/1757-899X/560/1/012202

[3] Valeev A S, Kotenev Yu A, Mukhametshin V Sh and Sultanov Sh Kh 2019 Substantiation of the Recovery of Residual Oil from Low-Productive and Heterogeneous Formations in Western Siberia by Improving the Waterflood System Using Gas and Water-Gas Impacts *IOP Conf. Ser.: Mater. Sci. Eng.* **560(1)** 012204. DOI: 10.1088/1757-899X/560/1/012204

[4] Mukhametshin V V and Kuleshova L S 2019 Justification of Low-Productive Oil Deposits Flooding Systems in the Conditions of Limited Information Amount *SOCAR Proceedings* **2** 16–22. DOI: 10.5510/OGP20190200384

[5] Akhmetov R T, Kuleshova L S and Mukhametshin V V 2019 Application of the Brooks-Corey model in the conditions of lower cretaceous deposits in terrigenous reservoirs of Western Siberia *IOP Conf. Ser.: Mater. Sci. Eng.* **560(1)** 012004. DOI: 10.1088/1757-899X/560/1/012004

[6] Kuleshova L S, Kadyrov R R, Mukhametshin V V and Safiullina A R 2019 Design changes of injection and supply wellhead fittings operating in winter conditions *IOP Conf. Ser.: Mater. Sci. Eng.* **560(1)** 012072. DOI: 10.1088/1757-899X/560/1/012072

[7] Kuleshova L S, Kadyrov R R, Mukhametshin V V and Akhmetov R T 2019 Auxiliary equipment for downhole fittings of injection wells and water supply lines used to improve their performance in winter *IOP Conf. Ser.: Mater. Sci. Eng.* **560(1)** 012071. DOI: 10.1088/1757-899X/560/1/012071

[8] Garifullin M F, Ismagilov I Kh, Sahabutdinov R Z, Tronov V P and Shireev A I, Patent A. S. 1397473 USSR MPK C 10 G 33/04 Method of dehydration of high viscosity oil (23 May 1988)

[9] Pozdnyshev G N, Fattakhova R Sh and Bril D M 1983 Formation of stable oil emulsions in the application of thermal methods of influence on the formation and ways of their destruction *Oil industry* **16(65)** 44

[10] Schwuger M J 1982 Effects of adsorption on detergency phenomena: II *J. of American Oil Chemist's Society* **59(6)** 258–272

[11] Minnigalimov R Z and Nafikova R A 2008 Improvement of oil-slimes processing technology *Oil Industry* **4** 105–107

[12] Nafikova R A 2012 Improvement of oil-field sludge treatment technology for centrifuging using super-high frequency current *Problems of Gathering, Treatment and Transportation of oil and oil products* **4(90)** 169–174

Enhancement of efficiency of the hydraulic distributor of the hydraulic drive of the bottom-hole pump

R I Suleimanov and M Ya Khabibullin

Ufa State Petroleum Technological University, Branch of the University in the City of Oktyabrsky, 54a, Devonskaya St., Oktyabrsky, Republic of Bashkortostan, 452607, Russian Federation

E-mail: rustamsul@rambler.ru, info@of.ugntu.ru

Abstract. The article deals with the hydraulic drive of the bottom-hole pump and analyzes the hydraulic drive operation. The authors outline peculiar failures of the hydraulic distributor of oil system control and suggest a technical solution for efficient operation of a hydraulic distributor designed to redirect the flow of liquids circulating in the hydraulic drive of the bottom-hole pump. The solution allows increasing the inter-repair period and reducing the energy consumption of the hydraulic drive.

1. Introduction

Years-long experience in operating wells equipped with bottom-hole pumps shows that a conventional rocking machine has a number of drawbacks. The inconvenience of mounting on the well consists in the need of arranging a bulky foundation. Large rotating masses create a potential risk of injury to service personnel. Failure statistics leaves much to be desired: constantly torn drive belts, jammed gearboxes and bearings, frayed or broken rope suspension, unbalanced crank counterweights [1, 2].

The hydraulic drive of bottom-hole pump is devoid of all these disadvantages. Design of the hydraulic drive is reliable and trouble-free, the number of shutdowns for technical reasons is reduced, and it is possible to control the operation of the well drive and smoothly change the operating mode in real time [2-5].

2. Materials and methods

The hydraulic drive is designed to give a reciprocating motion to the plunger of a bottom-hole pump when pumping liquid out of oil wells. The hydraulic drive operates at a load on the wellhead rod up to 80 kN and the ambient temperature from minus 400 to plus 400 °C.

The hydraulic drive (Figure 1) provides a wide range of stepless speed control with good smoothness of movement, the ability to work in dynamic modes with the required quality of transients, protection of the system from overload and precise control of operating forces. The use of hydraulic cylinders makes it possible to create a rectilinear motion without kinematic transformations, as well as to provide a certain ratio of forward and reverse speeds [6-7].

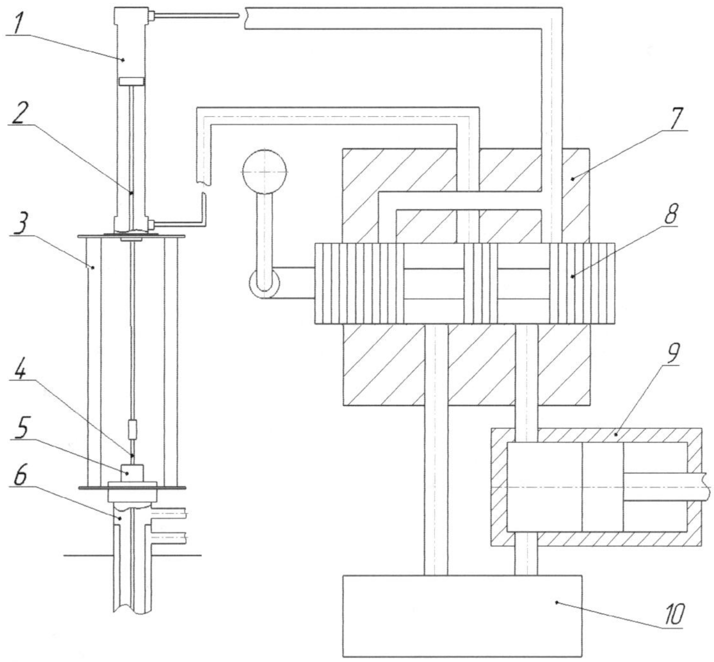

Figure 1. Diagram of the hydraulic drive of bottom-hole pump: 1 – hydraulic cylinder; 2 – rod of hydraulic cylinder ; 3 – pedestal; 4 – polished rod; 5 – wellhead equipment; 6 – well; 7 – hydraulic distributor; 8 – spool; 9 – hydraulic motor; 10 – oil station

Main advantages of the hydraulic drive: 1) sufficiently high efficiency; 2) increased rigidity; 3) required durability; 4) reduced metal consumption and weight as compared to other drives of bottom-hole pumps; 5) reduced terms of installation, dismantling and commissioning [8].

The hydraulic drive is used to perform the following actions:

1) commissioning of a periodically operating well stock that has been put out of action;

2) development of wells after drilling and major repairs;

3) operation on wells equipped with single-lift installations of dual operation.

The hydraulic drive of the bottom-hole pump contains two modules connected by pipelines:

1) the first module contains a base frame with a working hydraulic cylinder installed on it,

2) the second module contains an additional frame with an oil station installed on it.

Practice shows that the growth of individual capacities in hydraulic cylinders increases the pressure of liquid in hydraulic systems, the speed of rod movement, and the intensity of their use. As a result, there is a sharp increase in operating loads, deterioration of the conditions and the mode of hydraulic cylinders operation, and then an inevitable decrease in their reliability, eventually leading to failure of the hydraulic cylinder [9].

When analyzing the operation of hydraulic cylinders for installation of bottom-hole pumps, a number of reasons that reduce the performance of individual elements of the hydraulic system have been identified, namely:

1) the cylinder body has uneven wear and a residual deformation caused by the pressure of internal sealing surface with the formation of marks, scratches, scuffs and cracks on it, which eventually leads to the rupture of the body;

2) the guide sleeves and pistons have uneven wear on the perimeter and width of the rubbing surfaces, the formation of scratches and scuffing, destruction of the piston;

3) the seals for all purposes are subject to aging of the material, uneven wear of sealing surfaces along the perimeter and width of the seal, the appearance of marks and scratches, and rupture of the

seal;

4) the support surfaces have wear of sliding surfaces, partial or complete destruction of components;

5) failure of threaded connections;

6) the rod has wear of the compacted surface, the formation of marks, scratches, scuffs, dents and corrosion centers on it, thread failure and curvature of the rod.

The performance of hydraulic systems is sharply reduced when the oil viscosity changes during heating, which leads to a change in the speed of movement of the working bodies. Due to the presence of internal leaks, it is difficult to accurately coordinate the movements of the hydraulic drive. Wear of the seals and the working surface of the piston increases leakage. The most prone to wear surface of the sleeve mirror of the hydraulic cylinder is in the area near the cover with the guide sleeve. When the rod is fully extended, due to the friction forces in the hydraulic cylinder trunnions, it is loaded with a transverse force, which leads to wear of the guide sleeve and the surface of the sleeve mirror. Wear of surfaces, as a rule, does not exceed 1 mm [10].

The hydraulic system is controlled by a spool-type distributor.

The precision pair (body – spool) is manufactured with small gaps. Increasing the gap by 0.06 mm is considered to be unacceptable and requires replacing the hydraulic distributor. However, only the use of thin anti-friction coatings makes it possible to extend the service life of spool distributors.

Details of hydraulic drive units are labor-intensive in manufacturing. Improving the design is almost impossible. With the advent of new materials and technological techniques for manufacturing parts, this task for hydraulic cylinders can be successfully solved. The contact surface of the guide sleeve is covered with a thin layer of anti-friction coating made of modified polyamide. The outer surface of the hydraulic cylinder piston is also covered with the same anti-friction coating. There are two options for applying an anti-friction coating: 1) contact surfaces of the spool; 2) contact surfaces of the distributor housing.

External leakage from the working cylinder occurs when the guide sleeve of the hydraulic cylinder has wear and for this reason the seals can not cope with sealing.

Shortcomings in the operation of the hydraulic distributor have been identified:

1) the possibility of hydraulic shocks at the moment when the spool belts overlap the channels of the hydraulic distributor, since the time of blocking channels may be quite short, and the termination of liquid flow may be too fast, which may lead to the occurrence of a hydraulic shock;

2) high energy consumption. Practice shows that the basic design of the spool predetermines the range of power consumption of the hydraulic drive within 13-14 kW, while the traditional rocking machine consumes 9.7–9.9 kW, an average excess being equal to 29.4 %. High power consumption is associated with artificial resistance in the hydraulic distributor. This resistance is due to the need of creating a back pressure to ensure the control pressure. This pressure is 0.7 MPa. This leads to additional energy losses of about 20-24 kW per day.

When the spool moves in the hydraulic distributor, almost the entire flow of the working fluid is immediately directed to the hydraulic cylinder, which immediately gains maximum speed.

When the spool returns to the neutral position, the fluid flow into the hydraulic cylinder is abruptly blocked. As a result, at the beginning of movement and stop of the operating mechanism, hydraulic shocks and significant pressure surges take place and negatively affect the operation of the hydraulic system.

Under proportional control, the movement of the spool ensures smooth and precise control of the flow rate entering the hydraulic motor. This allows for changing the speed of the hydraulic motor in a wide range – from zero to maximum. The pressure is also changed smoothly, without significant jumps. The operation of the hydraulic system is noticeably improved, its durability increases. Proportional control increases the safety of the system, reduces the amount of internal leaks, improves the operation of all components of the hydraulic drive, increases the accuracy and ease of operation of the machine [11, 12].

3. Results and discussion

To avoid negative phenomena, a spool design with high-precision slots or belts is proposed (Figure 2). These slots serve as throttles.

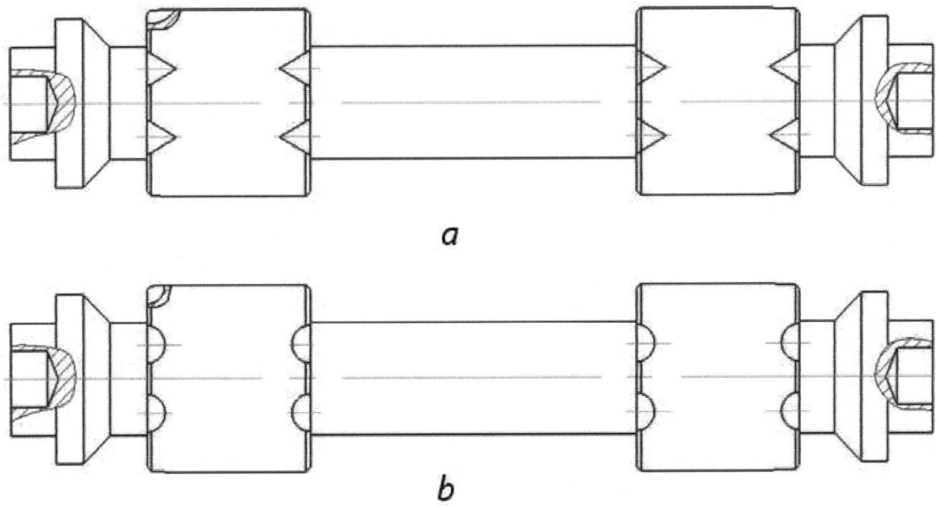

Figure 2. Spool design: a – with triangular slots; b – with semicircular slots

Movement of the spool is carried out by an external force. With the gradual movement of the spool, the area of operation windows between its slots and the bores in the housing changes proportionally, which allows controlling the flow rate. The controlled part of the flow enters the hydraulic cylinder, and the remainder of the liquid is drained into the hydraulic tank. In the hydraulic distributor with the center closed, the remaining power fluid enters the hydraulic tank at a pressure greater than the opening of the safety valve. In this case, we use the principle of regulating the flow of power fluid with a throttle installed at the entrance of the hydraulic motor and a safety valve.

In the hydraulic distributor with the center open, the remaining power fluid enters the hydraulic tank at a lower pressure. They apply the principle of regulating the flow of working fluid with a throttle installed parallel to the hydraulic motor. The safety valve is closed, and the control valve with an open center is presented in the form of throttles.

When the spool moves from the neutral position towards its opening, the area of the operation windows in the discharge and flow lines increases, and in the parallel line – decreases. Flow control in this way is more economical than using throttle control with a safety valve.

This is because part of the power fluid entering the drain through a throttle in a parallel line converts significantly less flow energy into heat than when passing through a safety valve.

This control method provides the required accelerated, uniform or slow motion of the hydraulic motor. Hydraulic shocks are excluded in this case.

Different shape and size of the slots in the spool provide the required flow characteristics of the power fluid.

Figures 3 and 4 show the change in the flow rate directed to the hydraulic motor, depending on the movement of the spool of the hydraulic distributor without slots and with slots. Different shape and size of the slots in the spool provide the required flow characteristics of the power fluid. These characteristics are ideal. The opening area of the throttle windows of the hydraulic distributor changes linearly when the spool moves.

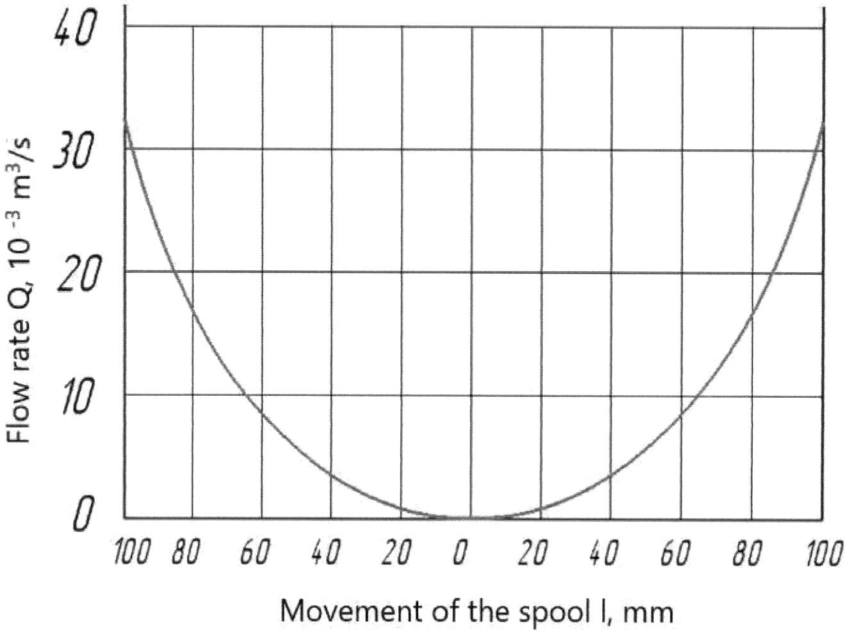

Figure 3. Graph of dependence of the flow of the hydraulic distributor on the movement of the spool with triangular slots

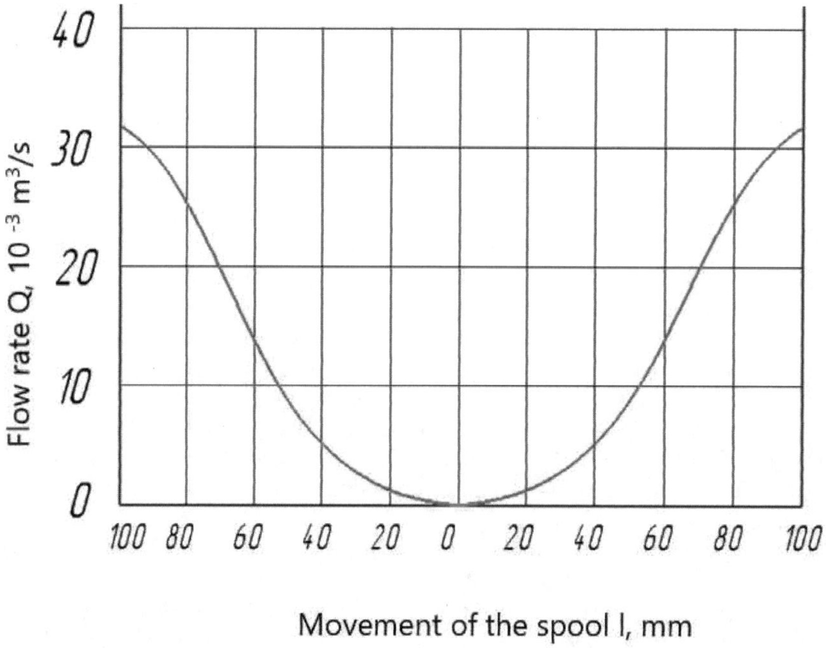

Figure 4. Graph of dependence of the flow of the hydraulic distributor on the movement of the spool with semicircular slots

Figures 3 and 4 show the change in the flow rate directed to the hydraulic motor, depending on the movement of the spool of the hydraulic distributor for different designs.

These characteristics are ideal. The opening area of the hydraulic distributor windows changes linearly when the spool moves.

Geometric shape of the high-precision slots of the spool determines the configuration of the regulating characteristics of such a hydraulic distributor, and therefore, the law of speed regulation in the hydraulic cylinder.

Figure 2 shows two spools of a proportional hydraulic distributor. One of them has triangular throttling slits, while the other has semicircular slits.

In both cases, the opening area of the throttling windows changes non-linearly, but in different proportions. When moving the spool, the nature of the flow rate change directed to the hydraulic cylinder is different. The speed of the hydraulic motor will also be different.

When moving the spool from the neutral position to the working position, the flow from the pump enters one or another cavity of the hydraulic cylinder. The hydraulic cylinder rod is extended or retracted; the motor shaft rotates to the left or right side, i.e. the movement is reversed. Thus, the hydraulic distributor controls the flow direction. At the same time, hydraulic distributors can also regulate the flow rate, i.e. change the flow rate of the power fluid entering the hydraulic motor. For this purpose, throttling slots are made on the working surfaces of the spool. They allow for smooth opening of the area of the operation windows when moving the spool from the neutral to the working position and thereby control the amount of flow.

Let us consider the change in the flow rate passing through the spool, depending on the area of the operation window and the pressure drop in it. The maximum movement of the spool is 6 mm, the 100% flow rate at its extreme position is $32\cdot10^{-3}$ m^3/min, and the value of the pressure difference between the discharge channel of the distributor and the hydraulic motor is $\Delta P = 1.2$ MPa (Figure 5).

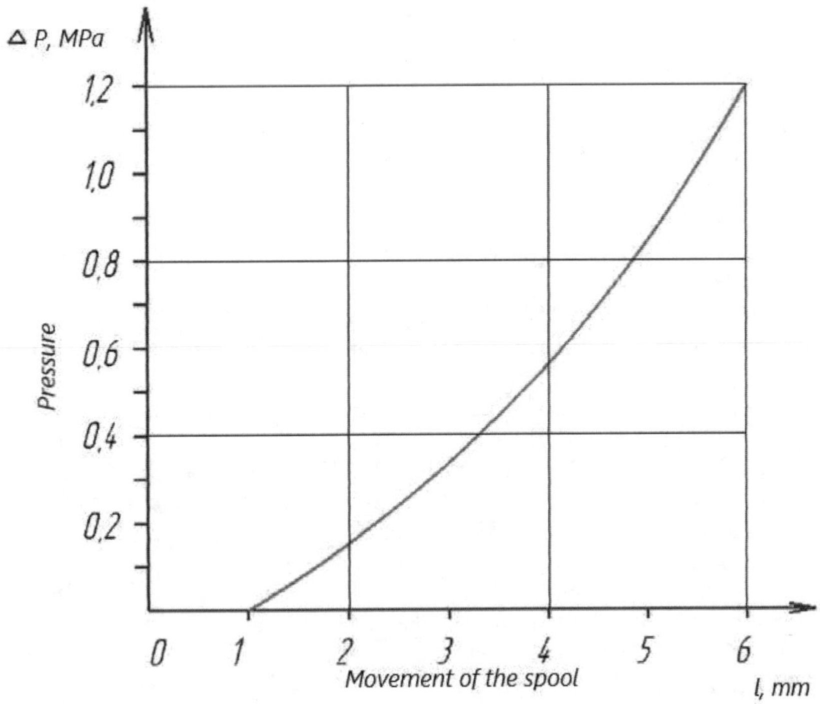

Figure 5. Dependence of pressure change on a spool position

The pressure drop and flow rate begin to increase after overcoming the insensitivity zone of the spool equal to 1.0 mm. This overlap is necessary to prevent the flow of power fluid into the hydraulic motor when the spool is in a neutral position.

The flow rate changes due to an increase in the pressure drop on the spool. The value of pressure drop at a constant external load regulates the movement of the spool, changing the cross-section area for the flow of power fluid.

The flow rate entering the hydraulic motor will affect the working pressure in it. Thus, depending on the position of the throttling spool, the flow rate of the working fluid changes proportionally.

When the spool reaches its extreme position in the working position (6 mm), the pressure drop on the distributor will increase to the set value (1.2 MPa), the flow rate will be maximum ($32 \cdot 10^{-3}$ m^3/min).

4. Conclusion

The considered technical solution makes it possible to improve the performance of hydraulic cylinder parts that have anti-friction polymer coatings. At the same time, the rod and the sleeve mirror of the hydraulic cylinder operate in the best conditions and maintain high surface quality.

This technical solution allows for repeated use of expensive parts of complex shapes, reduces the risk of destruction of the hydraulic distributor body parts due to hydraulic shocks, increases the repair period by 35-40 %, and also reduces the power consumption of the hydraulic drive by 6.7%, which is an average of 10.5 kW.

References

[1] Ketelsen S, Padovani D, Andersen T O, Ebbesen M K and Schmidt L 2019 Classification and review of pump-controlled differential cylinder drives *Energies* **12(7)** 1293

[2] Malrait F, Jebai A K and Ejjabraoui K 2019 Power conversion optimization for hydraulic systems controlled by variable speed drives *J. of Process Control* **74** 133-146

[3] Velusamy K, Sundararajan S, Thukkaiyamoopar Anbarasu R R C and Sellamuthu M 2019 Optimum Sizing of Hydraulic Pump in Vibratory Roller Vehicle Propulsion System *SAE Technical Papers*

[4] Khabibullin M Ya and Suleimanov R I 2019 Automatic packer reliability prediction under pulsed transient flooding of hydrocarbon reservoirs *IOP Conf Ser.: Mater. Sci. Eng.* **560(1)** 012024. DOI: 10.1088/1757-899X/560/1/012024

[5] Gaspar J F, Kamarlouei M, Sinha A, Robles E and Soares C G 2016 Speed control of oil-hydraulic power take-off system for oscillating body type wave energy converters *Renewable Energy* **97** 769-783

[6] Suleimanov R I, Gabdrakhimov M S, Khabibullin M Y, Zaripova L M and Vasilyeva E R 2018 The study of hydraulic hammer device in drilling tool assembly in hydraulical rotary drilling *Int. J. of Engineering and Technology* **7(2)** 28-30

[7] Rogachev M K, Mukhametshin V V and Kuleshova L S 2019 Improving the Efficiency of Using Resource Base of Liquid Hydrocarbons in Jurassic Deposits of Western Siberia *J. of Mining Institute* **240** 711-715. DOI: 10.31897/PMI.2019.6.711

[8] Zaripova L M and Gabdrakhimov M S 2019 Restoration of intake capacity of injection well by vibrations *IOP Conf. Ser.: Earth Env.* **378(1)** 012120. DOI: 10.1088/1755-1315/378/1/012120.

[9] Batalov S A, Andreev V E, Lobankov V M and Mukhametshin V Sh 2019 Numerical simulation of oil formation with regulated disturbances. Oil recovery quality simulation *J. of Phys.: Conf. Ser.* **1333(3)** 032006. DOI: 10.1088/1742-6596/1333/3/032006

[10] Abdyukova R Ya 2019 Studies on operation and types of drilling pump valves *IOP Conf. Ser.: Mater. Sci. Eng.* **560(1)** 012050. DOI: 10.1088/1757-899X/560/1/012050

[11] Batalov S A, Andreev V E, Lobankov V M and Mukhametshin V Sh 2019 Numerical Simulation of the Oil Reservoir with Regulated Disturbances. Oil recovery stability simulation *J. of Phys.: Conf. Ser.* **1333(3)** 032007. DOI:10.1088/1742-6596/1333/3/032007

[12] Pobedza J and Sobczyk A 2014 Properties of high pressure water hydraulic components with modern coatings *Advanced Materials Research* **849** 100-107

Selection and justification of efficiency parameters of a noise-resistant downhole telemetry system

K T Tyncherov, V Sh Mukhametshin and N Rakhimov

Ufa State Petroleum Technological University, Branch of the University in the City of Oktyabrsky, 54a Devonskaya St., Oktyabrsky, Republic of Bashkortostan, 452607, Russian Federation

E-mail: academic-mvd@mail.ru, vsh@of.ugntu.ru

Abstract. The article examines the issue of selecting and justifying efficiency parameters of a noise-resistant downhole telemetry system that measures and transmits borehole information to the surface via an electromagnetic communication channel. The work is devoted to the continuation of research on the synthesis of systems capable of transmitting messages from oil wells equipped with sucker rod pumps from a depth of up to two thousand meters. To ensure noise immunity, coding is proposed in the residual class system with the possibility of error correction in the neural network basis.

1. Introduction

One of the main tasks of increasing the degree of oil reserves recovery is the availability of reliable information about deposits [1–5]. When conducting hydrodynamic studies of oil and gas wells using submersible telemetry systems, one of the most important issues is the transmission of well information to the so-called "day surface" [6]. In this case, the transmission of information through an electromagnetic communication channel under the influence of various interference is particularly difficult. Interference is any disturbance in the information transmission channel that causes random deviations of the received message from the transmitted one and makes it difficult to receive it. Interference (or noise) immunity is understood as the ability of an information system to withstand the harmful effects of interference. As a result of interference, the received message will differ to some extent from the transmitted one. Therefore, noise immunity can be characterized as the degree of correspondence of the received message to the transmitted one for a given interference. When comparing several systems, the most noise-resistant will be the one that, with the same interference, will provide a smaller difference between the received and transmitted messages.

Telemetry information is data on temperature, pressure, flow rate, water content in the mixture, etc. To ensure the noise immunity of the communication channel, the research uses transmitters with increased power and supersensitive receivers. They reduce the range to low and ultra-low frequencies and duplicate broadband signal transmission. In order to detect, localize and correct computational errors during digital signal processing, the research uses modular number systems with this ability [7–9].

In its most general form, the solution to the problem of increasing noise immunity of telemetric information transmission encounters certain difficulties and, to obtain practical results, the problem can be divided into two subtasks:

1. Optimization of the “digital channel” subsystem, implying the choice of the best signal type, type of carrier, modulation method and the corresponding best reception method.

2. Optimization of the subsystem “encoder-decoder” of a digital channel.

The results obtained make it possible to effectively solve the problem of increasing the efficiency of the development of deposits with hard-to-recover backflows [10–12].

2. Materials and methods

General statement of the problem of synthesis of a noise resistant telemetric system

In [6], the conditions for solving the first subproblem are described in sufficient detail, which implies the selection and justification of devices for an electromagnetic channel, primary and secondary sealing, an autonomous power supply system, etc.

In this work, special attention is paid to the formulation of the second problem associated with increasing the noise immunity of information transmission.

The multilevel structure of the telemetry system and the complex functional interconnection of the individual elements of the system with each other make it necessary to classify it as a complex system. The quality of the system is usually characterized by a large set of indicators written in the form of a vector.

$$E = (E_1, E_2, \dots, E_n). \quad (1)$$

The optimal system is considered to be the one to which the largest (smallest) value of some function

$$E = \varphi(E_1, E_2, \dots, E_n), \quad (2)$$

from private quality indicators $E_1, E_2, \dots, E_n$,where E is the system efficiency, and $\varphi(E_1, E_2, \dots, E_n)$ is the target function of the system.

In the simplest cases, the efficiency of the system is assessed by the most essential parameters

$$\varphi(E_1, E_2, \dots, E_n) = E_\vartheta, \quad (3)$$

where E_ϑ is the main quality indicator that needs to be maximized. Generally, a systematic approach is necessary, which assesses the efficiency of the system as a whole by a set of parameters. In this case, it is necessary to take into account the most essential parameters of the system being developed.

The desire to take into account all parameters leads to the complication of the assessment results. However, over-limiting the number of parameters may lead to the estimate being too rough. In this regard, it is fashionable to propose an estimate in the form of a linear function

$$E = \sum_{1=1}^{m} \lambda_i \zeta_i, \quad (4)$$

where m is the number of parameters taken into account. The relative values of the weight λ_i and the relative values of the parameters ζ_i are determined as follows. Let the main parameters of the system be set by the values $C_i (C_i > 0)$by the upper and lower limits C_{imax} and C_{imin}, respectively. Then the relative value of the parameter having the upper bound will be determined by the ratio:

$$\lambda_i = C_i / C_{imax}, \quad (5)$$

and the relative value of the parameter having the lower bound is by the ratio:

$$\lambda_i = C_i / C_{imin}. \quad (6)$$

Each parameter is characterized by some weight$\rho_i$or relative weight:

$$\lambda_i = \rho_i / \sum_{1=1}^{k} \rho_i. \quad (7)$$

With this definition of the parameters included in (4), it is obvious that $0 \leq E \leq 1$. The best system is one for which E is larger. The choice of the weighting coefficients ρ_i is determined relatively arbitrarily, which equally applies to the number of parameters n.

Evaluation of the effectiveness of a telemetry system with an electromagnetic communication channel by one indicator is attractive in that it is expressed by the same number and takes into account all the selected parameters of the system. However, modern systems cannot always be fully characterized by one indicator. The score for several indicators may be more specific. Evaluations of the telemetry system for several indicators are also justified by the fact that they are only the basis for analysis in order to make a decision. The final decision, as a rule, is based not only on quantitative calculation data, but also on experience, intuition, as well as additional considerations that were not taken into account when constructing a mathematical model.

Selection of the set of parameters for estimating the efficiency of an immunity telemetric system
In order to ensure high noise immunity when transmitting telemetric information from the well to the day surface, the following characteristics can be selected:
- temporary;
- reliable;
- operational
- probabilistic.

From each subset it is necessary to select one defining parameter, which most characterizes its subset.

Such parameters that determine the quality of the system functioning include:

- the probability of non-detection of an error by the P_{ND}, system, which is an assessment of the reliability of the transfer of information such as pressure, flow rate, water content in the mixture and temperature;
- noise immunity P_n, defined as the probability of correct message reception in the form of a noise-like signal with modular coding;
- probability of loss (error detection) P_{ED}– probability of transformation of the transmitted message into an erasure signal;
- transmission rate R, characterized by the ratio of the number of received messages in time t_i to time t;
- time of delivery of messages T.

In systems with a non-deterministic number of repetitions, the transmission time of each message depends on the specific conditions of passage through the communication channel and is a random variable. The average transmission time of one message for characteristics of system properties is determined during the research.

One of the urgent tasks of improving the system is also to reduce hardware costs. This problem can be solved not only on the basis of simplifying algorithms, processing, but also on the main reduction in memory. The complexity of the latter can be estimated using known indicators.

3. Conclusion

The considered first set of aspects is essential for choosing a set of parameters. Parameters P_{ND}, P_n, R and $T$are usually determined on the assumption of using absolutely reliable equipment operating in conditions of real interference caused by the natural background of the earth, interference from nearby high-power power cables, etc. Therefore, the parameter values calculated under these conditions determine the quality transmission algorithms and methods for protecting information from errors.

Some of the listed characteristics will be used to assess the increase in noise immunity of information transmission in the telemetry system of wells equipped with sucker rod pumps, which ultimately will improve the reliability of receiving telemetry information transmitted via the electromagnetic communication channel, and thus, qualitatively conduct hydrodynamic studies of an oil well.

The need to transfer protective coding to a neural network basis should be noted as a second set of aspects [6]. The most attractive in this regard are systems based on neural networks using coding by means of modular information representation. This aspect is becoming especially relevant in the period of the emergence of a relatively new organization of ultra-high-speed computing - neural networks.

Neural networks (NN) begin to play an increasingly important and noticeable role at all stages of the life cycle of computing process control in digital signal processing. Currently, there are three main scientific directions in the field of neuromathematics:

• theory of neural networks;

• neuroinformatics (using the theory of neural networks to develop algorithms for solving problems);

• neurons and neurocomputers.

Now there is every reason to believe that any tasks in the field of digital signal processing should be more efficiently solved on neurocomputers, since the algorithm of any task can be represented in a logical basis with a controlled number of layers of neurons. This means that the neural network algorithm for solving any problem at the logical level is more parallel than any conceivable physical implementation of it. Consequently, the potential for high-speed information processing is initially incorporated in NN.

The special value of neural networks, in relation to the issues of ensuring high noise immunity, lies in the fact that they are self-organizing. This means that the system is capable of learning, and, consequently, of independent detection and correction of errors that arise within it.

At the same time, the creation of this theory gave rise to many questions related to its implementation. As it turned out, the models of modern artificial neural networks are for the most part a purely mathematical interpretation of the construction of biological neural networks, which leads to the loss of a significant part of the capabilities of their biological prototype.

In this regard, the most appropriate is the use of modular number systems, and in particular, the system of residues.

The experience of designing the first computing devices based on modular information coding has shown that, in terms of increasing the speed and reliability of processing numerical information, residue systems have more capabilities than conventional positional number systems. As is known, the noise immunity of processors is ensured by introducing various forms of redundancy: hardware, software and time redundancy. From the point of view of providing hardware redundancy, one can use a special coding that has not only parallelism properties, but also natural correcting capabilities. Studies have shown that in the numbers theory, there is a modular number system in residues based on the Chinese Sun Tzu remainder theorem. This non-positional modular system not only allows parallelizing the execution of machine operations, but also has the ability to provide information redundancy. The uniqueness of the residue system lies in the fact that the information processing hardware in this system is easily controlled and diagnosed. This is due to the specific features of the representation and processing of code structures in the system of residues and the manifestation of failures in them [9].

References

[1] Rogachev M K and Mukhametshin V V 2018 Control and regulation of the hydrochloric acid treatment of the bottomhole zone based on field-geological data *J. of Mining Institute* **231** 275-280. DOI: 10.25515/PMI.2018.3.275.

[2] Mukhametshin V V and Kuleshova L S 2019 Justification of Low-Productive Oil Deposits Flooding Systems in the Conditions of Limited Information Amount *SOCAR Proceedings* **2** 16–22. DOI: 10.5510/OGP20190200384

[3] Rogachev M K, Mukhametshin V V and Kuleshova L S 2019 Improving the efficiency of using resource base of liquid hydrocarbons in Jurassic deposits of Western Siberia *J. of Mining Institute* **240** 711-715. DOI: 10.31897/PMI.2019.6.711

[4] Akhmetov R T, Kuleshova L S and Mukhametshin V V 2019 Application of the Brooks-Corey model in the conditions of lower cretaceous deposits in terrigenous reservoirs of Western Siberia *IOP Conf. Ser.: Mater. Sci. Eng.* **560(1)** 012004. DOI: 10.1088/1757-899X/560/1/012004

[5] Kuleshova L S, Kadyrov R R, Mukhametshin V V and Safiullina A R 2019 Design changes of injection and supply wellhead fittings operating in winter conditions *IOP Conf. Ser.: Mater. Sci. Eng.* **560(1)** 012072. DOI: 10.1088/1757-899X/560/1/012072

[6] Tyncherov K T, Mukhametshin V Sh and Khuzina L B 2017 Method to control and correct telemtry well information in the basis of residue number system *J. of Fundamental and Applied*

Sciences **9(2S)** 1370–1374. DOI: 10.4314/jfas.v9i2s.848

[7] Yakupov R F, Mukhametshin V Sh and Tyncherov K T 2018 Filtration model of oil coning in a bottom water-drive reservoir *Periodico Tche Quimica* **15(30)** 725-733

[8] Akushskii I Y and Yuditskii D I 1968 *Machine arithmetic in residual classes* (Moscow: Soviet radio) 440 p

[9] Chervyakov N I, Veligosha A V, Tyncherov K T and Velikikh S A 1998 Use of modular coding for high-speed digital filter design *Cybernetics and Systems Analysis* **34** 254

[10] Kuleshova L S, Kadyrov R R, Mukhametshin V V and Akhmetov R T 2019 Auxiliary equipment for downhole fittings of injection wells and water supply lines used to improve their performance in winter *IOP Conf. Ser.: Mater. Sci. Eng.* **560(1)** 012071. DOI: 10.1088/1757-899X/560/1/012071

[11] Mukhametshin V V 2017 Eliminating uncertainties in solving bottom hole zone stimulation tasks *Bull. of the Tomsk Polytechnic University. Geo Assets Engineering* **328(7)** 40–50

[12] Akhmetov R T, Mukhametshin V V, Andreev A V and Sultanov Sh Kh 2017 Some Testing Results of Productive Strata Wettability Index Forecasting Technique *SOCAR Proceedings* **4** 83–87. DOI: 10.5510/OGP20170400334

IOP Conf. Series: Materials Science and Engineering 1064 (2021) 012073 doi:10.1088/1757-899X/1064/1/012073

Geological and technological justification of the bottom-hole zone treatment of wells and formations of the Langepas group of fields

S A Yaskin[1], V V Mukhametshin[2] and L S Kuleshova[3]

[1] TPE "Urayneftegaz" LLC "LUKOIL–Western Siberia", 116a, Lenin St., Uray, Tyumen region, Khanty-Mansi Autonomous Area – Yugra, 628285, Russian Federation
[2] Ufa State Petroleum Technological University, 1, Kosmonavtov st., Ufa, Republic of Bashkortostan, 450062, Russian Federation
[3] Ufa State Petroleum Technological University, Branch of the University in the City of Oktyabrsky, 54a, Devonskaya St., Oktyabrsky, Republic of Bashkortostan, 452607, Russian Federation

E-mail: vv@of.ugntu.ru, markl212@mail.ru

Abstract. Based on the geological and routine analysis are represented the effective application areas of methods of simulating production and enhanced oil recovery of fields developed by the TPP "Langepasneftegaz". Based on laboratory experiments, the composition of composite systems of recovery methods on productive formations has been established. The technologies adaptation of the bottom-hole zone treatment of wells and formations of the Langepas group of fields was carried out.

1. Introduction

The fields developed by the TPP "Langepasneftegaz" are located in the Nizhnevartovsk district of the Khanty-Mansi national district in the central part of the Tyumen region.

The main share of the annual oil production of TPP "Langepasneftegaz" is obtained from facilities that are in the production decline stage, characterized by the high water cut of the produced fluid, active reserves depletion and an efficiency decrease of water flooding. The problem solution of production level stabilization of hydrocarbon crude is unthinkable without the use of new progressive methods and technologies that improving the effectiveness of traditional methods of stimulation formations, which is confirmed by the actual performance indicators of the TPP "Langepasneftegaz".

The emerging positive trend is particularly enhanced by the large-scale targeted use of a wide range of technologies, such as the development of the Uryevskoye field. It accounts for about 40 % of oil production in the Langepas region. The methods, that have a high technological effect and are provided with special installations for preparing and injecting large composition volumes into the formation have found widespread industrial introduction in this field. Such technologies are the use of SPL, thermotropic gellant "Metka", as well as the integrated technology of SPL+STEL.

Technologies currently being implemented involve the use of relatively small volumes of polymer solutions, to which some cross-linking agents are added to create a bond between individual polymer molecules. The bottomhole zone of the injection well is treated using this scheme. As crosslinkers, salts of polyvalent metals (chromium, aluminum, iron) are used, which are capable of forming multiple bonds with polymer molecules in the form of branched structures with high rheological properties. The cross-linking agent turns the polymer solution into a gel (sometimes the polymer is crosslinked so tightly that the resulting compound resembles rubber). In some cases, pre-crosslinked dry polymer compositions are used in waterproofing processes. These compositions swell, increasing in free volume by 5-6 times when in contact with fresh water and by 2-3 times when in contact with mineralized water.

The essence of the method with the using of crosslinked polymer systems is the addition of a cross-linking agent to neighboring reactive groups of a polymer molecule. In this case, network polymer structures can be formed. In the course of the implementation of such technologies, it is possible to provide for using of slowly cross-linking polymer-crosslinker compositions, as a result of which they can be advanced deep into the highly permeable zones of the formation over considerable distances and, therefore, the distribution of flows in the formations can be effectively regulated.

One type of crosslinked polymer systems technology is polymer flooding technology combined with viscoelastic compositions, the essence of which is that during the injection of a polymer slug, the bottomhole zone of injection wells, as often as required, is treated with small (30-100 m^3) volumes of crosslinked polymer systems with small (4–24 h) gelation time. In this case, the cross-linking agent is attached to reactive groups belonging to different polymer molecules. The resulting hydrogels have very low mobility, a high initial shear gradient, and pronounced viscoelastic properties.

Depending on the geological and physical parameters of the reservoirs, the state of field development and economic constraints, various technologies for using crosslinked polymer systems can be implemented to increase oil recovery and treat the bottomhole zone.

2. Materials and methods

Based on the generalization of the implementation results of physical-chemical methods on the fields Langepas region formulated ways of improving their effectiveness: the selection of the most efficient technologies; finding cheaper and technological agents; adaptation of technologies to the field situations Langepas region; scientific and technical support for technology implementation.

The analysis and synthesis of the laboratory results on the technologies adaptation to enhance oil recovery, oil production intensification and water cut reduction to the field situation of TPP "Langepasneftegaz" optimized technological parameters and compositions of composite systems of stimulation methods.

3. Results and Discussion

To efficiency improvement of mud-acid treatment of wells acid dispersions of the following compositions are proposed:

- acid dispersion № 1: neonol AF-6-1.75 %; sulfonol-1.25 %; hydrochloric acid (12 %) - 50 % (HCI-6 %); hydrofluoric acid (5 %) - 47 % (HF-2.4 %);
- acid dispersion # 2: OPT-BHA-8.0 %; sulfonol-3.0 %; hydrochloric acid (12 %) – 50 % (HCI – 6%); hydrofluoric acid (5%) – 39 % (HF-2%).

These acid compositions were recommended for BHT production and injection wells. In addition, an acidic micro-emulsion of the following composition can be used for BHT injection wells: neonol AF-6 – 5 %: SNOB – 2.5 %; hydrochloric acid (12 %) – 50 % (HCI – 6 %); hydrofluoric acid (5 %) – 40 % (HF – 2 %).

Gel compositions based on waste from the WMS-5 catalyst production and water solutions of hydrochloric acid are recommended as rigid water-insulating compositions 7.5/7.5 – 8/8 % by weight.

To simultaneously increase the displacement efficiency and the sweep efficiency, the following compositions are recommended:

- synterol AFM-12-volgonat-PAA with a SAS concentration of 5-10 g/dm^3 at a ratio of 7: 3 and 6 : 4 and a polymer concentration of 0.5-1, .0 g/dm^3;
- synterol AFM-12-sulfonol-PAA with a ASA concentration of 5-10 g/dm^3 at a ratio of 5 : 5 and a

polymer concentration of 0.5-1.0 g/dm^3;

- neonol Af9-6-volgonat with a SAS concentration of 10 g / dm^3 at a ratio of 7: 3 and 6: 4 and subsequent polymer injection with a concentration of 0.5–1.0 g/dm^3.

For high-temperature formations have been optimized the compositions of thermotropic gel-forming compositions "Metka", "Galka", PB-3P-1, which can be used both at the early and late stages of field development.

By changing the ratio of components and introducing additional components into the composition, their surface-active, rheological and kinetic parameters can be adjusted to adapt to specific geological and physical conditions. The use of inorganic gelling based on the aluminum salt-urea-SAS-water system leads to an increase in the heterogeneous stratum coverage of water flooding, conformance control and enhanced oil recovery of the heterogeneous stratum.

The following gelling compositions are recommended for cooled layers: "GALKA-THERMOGEL" (40-70 °C)," GALKA-THERMOGEL-NT "(20-40 °C) based on urea and aluminum chloride, as well as a composition consisting of the reagent "Metka" and Cenomanian water (based on methylcellulose and urea).

A method and computer program for calculating the thermal field of bottom-hole formation zones have been developed. Studies of thermal profiles of injection wells when cold water is injected into them allowed us to formulate the main technological requirements for thermotropic gelling, clarify the processes mechanisms occurring in the bottom-hole zone, and rank the target objects by the types of recommended thermotropic systems.

To calculate the parameters of high conductivity zones of productive formations has developed a method and were conducted indicator studies at the Pokamasovskoe and Lokosovskoe fields. Studies on the tracer fluid injection at the Pokamasovskoe and Lokosovskoe fields of the TPP "Langepasneftegaz" have revealed the presence of zones of low filtration resistance in these fields and the general patterns inherent in these zones. Based on the literature data analysis, it can be argued that the conducted field tests are one of the largest among such studies conducted in Russia and abroad. Studies have shown that the channel volumes of low filtration resistance are very low, the cumulative volume of these channels in the influence zone of injection wells is measured in units or tens of m^3 (the maximum value is 55 m^3). The relative share of these channels from the gross pore volume of the productive reservoir is hundredths or thousandths of a percent. Due to the higher channel properties, the relative share of water flow filtered through high productivity zones is significantly higher and reaches tenths of a percent (maximum value of 2.4 %).

The problem of predicting the application of physical-chemical methods to enhanced oil recovery in specific geological and physical conditions of productive formations has been successfully solved. The justification for the use of physical and chemical EOR in a group of fields should be based on the positive experience of using these methods in close geological and physical conditions [1-9]. To determine the proximity or similarity of the geological and physical conditions of different formations of fields, a procedure for grouping the productive formations under consideration is necessary. For this purpose, the grouping of fields in the Langepas region by the method of pattern recognition theory (main components, cluster analysis) according to a set of controlled geological and geophysical parameters allowed us to identify four fairly homogeneous groups, and in each group are typical objects for each group.

As a result of geological-routine analysis of the selected objects with the most efficient system development (formation SFP8 Urevskoye field), and the objects projected recovery rates which do not reach the design level (formations SE1 Las-Eganskoye field, BW6 Potochnoe, and ACB1/3 South-Pokachevskoye fields). As a result of the performed geological and mathematical modeling and the conducted technical and economic analysis, the optimal technologies set enhanced oil recovery, oil production intensification and reducing the water cut is established differentially for the selected field groups. The first methods group is aimed at conformance control of injection wells, as well as at stimulating the fluid influx to the producing wells and reducing the water cut of the extracted products, and it is recommended to perform the processing systemically, i.e. simultaneous processing of injection (sedimentation-forming compositions) and production (methods of stimulating production). These methods include cyclic water flooding, water solutions of CP, MAT. The second and third groups of technologies are aimed at the pre-washing of remaining oil. They include the following methods: SPDS, ESS, ESS+STEL, RT-3P-1 (OPS), RT-3P-1 (OPS), aluminosilicate, SAS+ alkali+polymer. These methods are used in monolithic relatively homogeneous formations with low clays content. The fourth group of technologies is aimed at stimulating inflows in development wells, filter cake removal of formations, and reducing the water cut of extracted products. The group includes STEL, MAT, SPL (injection wells), SPL (development wells), SPL+STEL, the aluminosilicate.

4. Conclusion

A complex of laboratory, theoretical, hydrodynamic and geological field researches on the adaptation of various methods of exposure to specific geological and physical conditions and develop hardware components set for their implementation allowed to significantly increase technical and economic efficiency, targeting and success of the technology, reasonably go from the stage of industrial experiment works to large-scale commercial introduction of physical-chemical methods of enhanced oil recovery, oil production intensification and water cut reduction at the region fields.

References

[1] Beaudette-Hodsman C, MacLeod B and Venkatadri R 2008 Production of High Quality Water for Oil Sands Application *Int. Thermal Operations and Heavy Oil Symp. (Calgary, Alberta, Canada, 20-23 October 2008)* p 8. DOI: 10.2118/117840-MS

[2] Mukhametshin V V and Kuleshova L S 2019 Justification of Low-Productive Oil Deposits Flooding Systems in the Conditions of Limited Information Amount *SOCAR Proceedings* **2** 16–22. DOI: 10.5510/OGP20190200384

[3] Bortolotti V, Macini P and Srisuriyachai F 2009 Laboratory Evaluation of Alkali and Alkali-Surfactant-Polymer Flooding Combined with Intermittent Flow in Carbonatic Rocks *Asia Pacific Oil and Gas Conf. & Exhibition (Jakarta, Indonesia, 4-6 August 2009)* p 13. DOI: 10.2118/122499-MS

[4] Yakupov R F, Mukhametshin V Sh, Khakimzyanov I N and Trofimov V E 2019 Optimization of reserve production from water oil zones of D3ps horizon of Shkapovsky oil field by means of horizontal wells *Georesources* **21(3)** 55-61. DOI: 10.18599/grs.2019.3.55-61

[5] Stenkin A V, Kotenev Yu A, Mukhametshin V Sh and Sultanov Sh Kh 2019 Use of low-mineralized water for displacing oil from clay productive field formations *IOP Conf. Ser.: Mater. Sci. Eng.* **560(1)** 012202. DOI: 10.1088/1757-899X/560/1/012202

[6] Hodgin J E and Harrell D R 2006 The Selection, Application, and Misapplication of Reservoir Analogs for the Estimation of Petroleum Reserves *SPE Annual Technical Conf. and Exhibition (San Antonio, Texas, USA, 24-27 September 2006)* p 15. DOI: 10.2118/102505-MS

[7] Rogachev M K, Mukhametshin V V and Kuleshova L S 2019 Improving the efficiency of using resource base of liquid hydrocarbons in Jurassic deposits of Western Siberia *J. of Mining Institute* **240** 711-715. DOI: 10.31897/PMI.2019.6.711

[8] Valeev A S, Kotenev Yu A, Mukhametshin V Sh and Sultanov Sh Kh 2019 Substantiation of the recovery of residual oil from low-productive and heterogeneous formations in Western Siberia by improving the waterflood system using gas and water-gas impacts *IOP Conf. Ser.: Mater. Sci. Eng.* **560(1)** 012204. DOI: 10.1088/1757-899X/560/1/012204

[9] Yakupov R F, Mukhametshin V Sh and Tyncherov K T 2018 Filtration model of oil coning in a bottom water-drive reservoir *Periodico Tche Quimica* **15(30)** 725-733

[10] Akhmetov R T, Kuleshova L S and Mukhametshin V V 2019 Application of the Brooks-Corey model in the conditions of lower cretaceous deposits in terrigenous reservoirs of Western Siberia *IOP Conf. Ser.: Mater. Sci. Eng.* **560(1)** 012004. DOI: 10.1088/1757-899X/560/1/012004

Appropriate use of dissolved gas energy in fields at the final stage of development

L F Yusupova, E M Almukhametova, D S Nosirov and A R Safiullina

Ufa State Petroleum Technological University, Branch of the University in Oktyabrsky, 54a Devonskaya St., Oktyabrsky, Republic of Bashkortostan, 452607, Russian Federation

E-mail: shalilya@yandex.ru, info@of.ugntu.ru

Abstract. In order to improve the development of fields at a later stage, this article considered the use of water-gas repression together with the dissolved gas energy. The theoretical basis of the study is based on the application of the method of physical modeling of oil displacement by water. As a result, it was found that waterflooding would be more effective when such an activity is carried out in the early stages of development. Waterflooding can also be effective in old, depleted fields with shallow depths of oil-bearing horizons. The article states that in flooded deposits, where the reservoir pressure did not drop below the bubble point pressure, it is advisable to reduce the reservoir pressure to the bubble point pressure (to carry out forced withdrawal). After that, water injection should be resumed with 30-50% compensation. It has been established that in order to increase the efficiency of the development of fields developed in the dissolved gas mode, waterflooding must be started at a reservoir pressure that is 30–70% less than the bubble point pressure. The best option for the development of such deposits is the presence of active edge waters. If they are absent, it is advisable to end water in the aquifer with compensation of withdrawals from 20 to 40% after the start of gas extraction from oil.

1. Introduction

Water-gas repression on a productive stratum is considered one of the most effective methods to enhance its oil recovery. Thus, the studies of Hnatiuk G. A. and Liskiewicz I. conducted in70-80 years proved that water and gas injection is an effective method to increase the final oil recovery factor [1]. This method was thoroughly learned for the conditions of the Samotlor and Urengoi oil fields. Prior to the study of the early stage of field development, the introduction of this method at the Samotlo field showed its high efficiency. In this case, gas is by-passed from the gas reservoir under pressure higher than the reservoir pressure of the oil horizon.

Currently, many oil fields in the Volga-Ural province have entered a late stage of development, and since the main fields (Timan-Pechora, West Siberian, Volgo-Ural) have high gas factors and are developed with waterflooding [2-5], we considered the possibility application of water-gas repression using the energy of gas dissolved in oil.

2. Materials and methods

In order to identify a rational way of using the energy of gas dissolved in oil, based on the analysis of literature data of experimental studies carried out before, two possible options for the development of fields were modeled, which consisted of a combination of operating modes of water-driven and

IOP Conf. Series: Materials Science and Engineering 1064 (2021) 012074 doi:10.1088/1757-899X/1064/1/012074

dissolved gas:

- development of the natural regime of dissolved gas to a drop in pressure in the reservoir followed by water injection (the conditions of the reservoir operation in the depletion mode with subsequent waterflooding were simulated;
- waterflooding regime with the maintenance of reservoir pressure higher than the saturation pressure to full waterflooding of the formation followed by oil recovery in the depletion mode and injection using water.

In both the first and second cases, the main driving force in the processes of oil displacement is the combination of gas energy, which is separated from oil after the pressure in the reservoir decreases, and water energy, injected following (or simultaneous) gas evolution [6]. Gas released from oil can fill part of the pores previously occupied by oil and contribute to its additional displacement from the formation during flooding (replacement effect). On the other hand, the presence of free gas causes a decrease in the phase permeability of the porous medium.

The dissolved gas mode is driven by the driving force - the energy of the gas that is released from the oil at a pressure below the bubble point pressure. In essence, this mode is similar to the elastic-water pressure mode, because the pressure in the reservoir does not decrease instantly, but gradually. The difference between this mode and the elastic-water-pressure mode is that gas is released from oil under the action of the gas expansion energy. Gas bubbles are elastic force carriers that induce carbonated oil to move. As in the elastic-water pressure, and in the dissolved gas mode, there are two phases distinguished.

During the development of the first phase, the pressure drop in the well is transmitted throughout the entire formation gradually: the radius of the conditional feed loop, on which the initial pressure is maintained, begins to increase. When the pressure starts to decrease at the natural boundary of the reservoir (oil-bearing contour), and the drainage area remains unchanged, he second phase begins [7]. A characteristic feature of the dissolved gas regime is the constancy of the oil-bearing contour, during the production of oil and gas, the reserves of the deposit are depleted (oil saturation decreases within the limits of a constant initial volume). The rate of saturation drop is much lower than the rate of pressure drop, and this is the second characteristic feature of the regime and the main reason for its low efficiency. The gas content of reservoir oil is constantly changing: first it increases, reaches its maximum value, and then, as the reservoir is developed, decreases.

Since, with a decrease in reservoir pressure below the bubble point pressure, the oil viscosity increases, and the phase permeability for oil changes, the oil recovery factor also decreases. At about 50% water saturation, water breaks through to the production wells very quickly. The oil displacement front practically does not arise, and is not formed from oil globules. Therefore, oil recovery does not increase until the displacement front breaks through. It is in this sense that the combination of the energy of dissolved gas and waterflooding is important, which can be both natural (active edge waters) and artificial, when water is forcibly injected into injection wells [8].

The second direction of research - the option of developing a field with a combination of modes of water pressure and dissolved gas, discusses the following. The field is developed by waterflooding and with the subsequent achievement of a higher saturation pressure, and ultimately leading to a complete flooding of production wells. After that, the reservoir pressure in the field is reduced to the bubble point pressure and below, by means of forced withdrawal and limiting the rate of water injection.

3. Results and discussion

The article discusses the results of experimental modeling of these development options in the geological and industrial conditions of the above fields. The study was carried out by physical modeling of fluid filtration processes in the pore space of the reservoirs of the indicated fields. Experiments were performed on reservoir models with natural sampling of coring material. Recombined samples of reservoir oils and waters were used as reservoir fluids.

The study modelled the corresponding values of reservoir pressure and temperature. To study the processes of filtration of inhomogeneous liquids in laboratory studies, it used methods of approximate modeling and the similarity theory, which at one time was developed in detail by Efros D. A. and is a system of seven equations [9]. Subsequently, many researchers have comprehensively analyzed these equations and made conclusions regarding the filtration parameters of inhomogeneous fluids.

First of all, this concerns those parameters that are determined in advance, even before the start of development processes: physical properties of a porous medium, liquids and gases, surface tension at the liquid-gas interface, etc. In addition, some parameters can be determined not only by similarity conditions, but also by the very program of experience, when under given conditions it is possible to minimize a number of factors that affect the process of oil displacement. For example, in experiments in which gas dissolution should be avoided, the pressure drop should be small in relation to the total pressure level [10, 11]. Based on the conditions of similarity, performed in the experiments, we used models that consisted of natural core material, which ensured the geometric similarity of the pore space, also ensured the similarity of the capillary pressure and related ratios through equality of contact angles. For the processes of oil displacement, the external conditions of similarity are the criteria Π_1 and Π_2, which are calculated according to the following dependencies:

$$\Pi_1 = \frac{\sigma}{\sqrt{\frac{K_a}{m}}}\Delta P;$$

$$\Pi_2 = \frac{\sigma}{K_a gradP},$$

where Π_1 and Π_2 are similarity criteria;σ –interfacial tension at the fluid boundary, N / m; K_a – absolute permeability, m^2; m – porosity; ΔP – pressure drop, Pa; grad$P$ –pressure gradient, Pa / m.

A characteristic feature of these criteria is that, starting from some values of Π_1 and Π_2, they cease to affect the characteristics of filtration processes [12]. Due to this feature, the existence of a self-similarity region was established, when the fluid movement does not change when one or several parameters that determine this movement are changed, namely $\Pi_1 < 0.5$ and $\Pi_2 > 0.3 \cdot 10^6$. Based on these criteria, the minimum length Lmin of the reservoir model will be determined by the formula:

$$L_{min} = 6 \times 10^5 \sqrt{K_a \cdot m}$$

If the porosity of the rock is 10-16%, and the permeability is 40-170 · 10-3 μm^2, then the length of the model should be at least 110-240 mm. Since in the experiments carried out the length of the models was 420–450 mm, the conditions of the laboratory studies performed with the proper completeness ensured the self-similarity of the processes according to the criteria Π_1 and Π_2.

For an experimental study of the processes of filtration and displacement of oil by water in accordance with similarity criteria, in relation to the parameters and conditions of occurrence of productive horizons of the Timan-Pechora, West Siberian, Volga-Ural fields, reservoir models were built as follows. Samples of sandstones were taken with a diameter of 28 mm, porosity from 0.10 to 0.16 and permeability from 1.0 · 10-3 to 11.0 · 10-3 μm^2. The compiled model from core material with approximately the same reservoir properties was placed in core holders, and thus its total absolute permeability was determined. After that, the samples were saturated with formation water, and the open porosity of the model was determined by weighing.

Residual water saturation was reproduced by the displacement method. First, from a water-saturated model, the water was mixed with transformer oil, after which the oil was replaced with kerosene. The model prepared in this way was injected with oil recombined in reservoir conditions [13]. At each stage of fluid injection, the phase permeability of the model was measured for each of them. To model the residual naphtha water saturation, reservoir water and oil from the above fields with the parameters shown in Table 1 were used.

For the recombination of oil samples, the study used the associated gas of the above fields, and for the displacement of oil it was fresh (river) water. The injection rate during the displacement process was ~ 0.17 · 10-9 m3 / sec. Experimental studies were carried out using the setup (Figure 1). Modeling of reservoir pressure and fluid injection was carried out using presses 1 and 2 of the installation, which are connected with containers 3, 4, 9, 13 using appropriate transmission channels. The working volumes of the containers were filled with fluids diverting for the experiments.

In the experiment, a core separator 5 was used, which makes it possible to create cylindrical models of a formation with a diameter of 28 mm and a length of 500 mm. The overburden pressure was modeled by lateral compression of the rock with an elastic cuff, the pressure on which was created using a hand press 6. Reservoir pressure was simulated using a pump 8 and a container with a piston separator 9 and was maintained by a pressure regulator 7. At the exit from the model, the fluids were separated in the separator 10, after which the liquid phase got to the measuring burette 11, and the gas volume was measured by the gas meter 12 or the so-called gas measuring burette.

Table 1. Geological and physical data for modeling

Parameters	Deposits of the Timan-Pechora horizons	Deposits of the West Siberian horizons	Deposits of the Volga-Ural horizons
Depth of occurrence, m	2240	2575	2550
Porosity	0.084	0.106	0.077
Permeability, 10-3 μm^2	4.9	4.0	1.5
Oil saturation	0.68	79	0.77
Reservoir temperature, ° C	70	70	52
Initial reservoir pressure, MPa	30.7	34.3	20.0
Oil viscosity in reservoir conditions, mPa · s	0.94	0.77	0.81
Volume factor	1.475	1.67	1.315
Paraffin content,%	12.4	9.3	10.0
Saturation pressure, MPa	26.0	34.3	17.4
Density of oil, kg / m^3	842	830	810
Gas content, m^3 / t	165.0	611	115.6
Formation water salinity, g / dm^3	180.0	118.0	115.2
Density of water, kg / m^3	1120	1074	1035

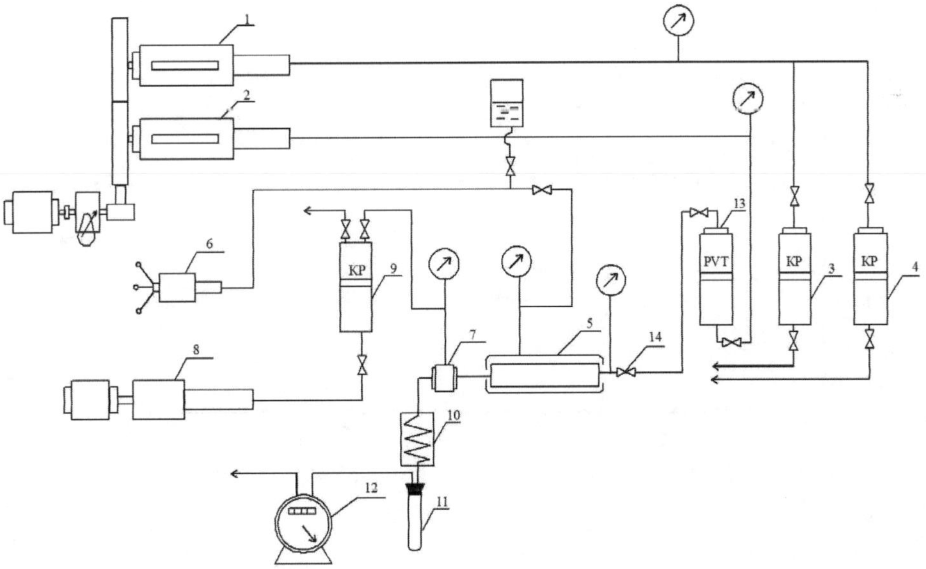

Figure 1. Schematic diagram of the installation: 1, 2 - installation presses; 3, 4, 9 - containers; 5 - reservoir model; 6 - press; 7 - reducer; 8 - pump; 10 - separator; 11 - measuring burette; 12 - gas meter; 13 - PVT bomb; 14 - valve

The results of the experimental work reproduced the development of the natural regime of dissolved gas followed by water injection (the conditions of reservoir operation in the depletion mode followed by waterflooding were simulated), shown in Figures 2 and 3. By using the obtained data for the conditions of the fields of the West Siberian horizons (Fig. 2) it established that in the case of a decrease in reservoir pressure to 0.75 Psat, about 45% of oil is squeezed out from the model; and when the reservoir pressure decreases to 0.5 Psat and 0.25 Psat, 58% and 61% of oil is squeezed out, respectively.

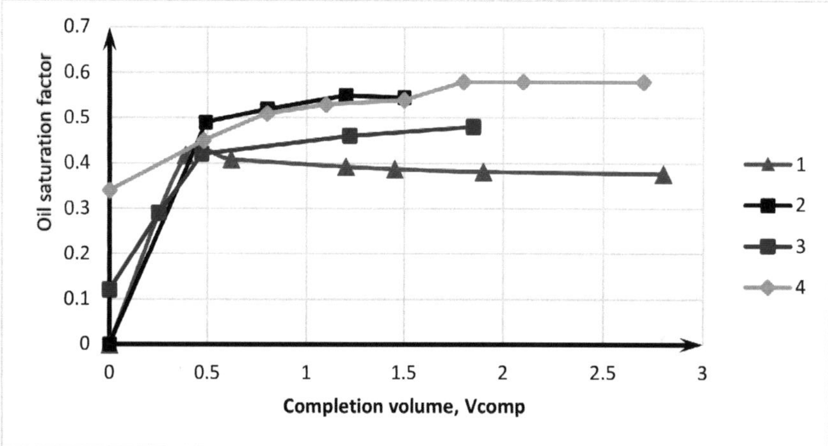

Figure 2. Experimental dependences of oil recovery on reservoir pressure and injection volume, obtained on the model of menilite deposits of the Timan-Pechora horizons: 1 - curve of oil displacement beyond reservoir pressure (P_{dis}), higher oil saturation pressure (P_{sat}); 2 - curve of oil displacement beyond reservoir pressure $P_{dis} = 0.75\ P_{sat}$; 3 - curve of oil displacement at $P_{dis} = 0.5\ P_{sat}$; 4 – oil displacement curve at $P_{dis} = 0.25\ P_{sat}$

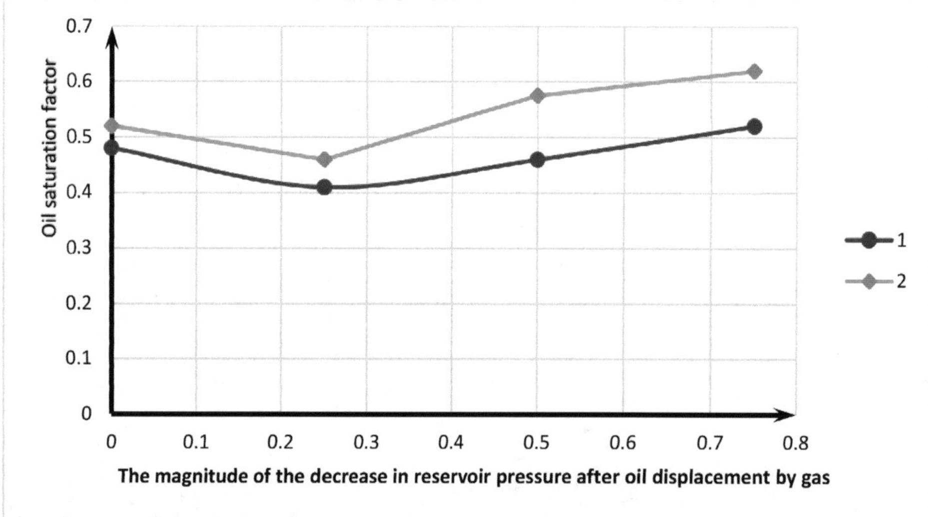

Figure 3. Experimental dependence of oil recovery on reservoir pressure and volume of gas injection, obtained on the model of fields in the West Siberian horizons. 1 - at the time of water injection, 2 - at the time of gas injection.

Thus, with a decrease in reservoir pressure to 0.75 P_{sat} the oil displacement coefficient is slightly lower than the pressure above the bubble point pressure (0.46 versus 0.52). However, when the reservoir pressure is reduced to 0.50 P_{sat} and 0.25 P_{sat} (curves 3, 4, Figure 2), oil recovery is achieved at 0.58 and 0.61, respectively, i.e. more than in the “clean” water pressure mode (curve 1, Figure 2). It should also be noted that the oil recovery factor during the anhydrous period is lower than when exposed to gas. These results are shown in Fig. 3 in the form of the dependence of the change in the oil recovery factor on the magnitude of the decrease in reservoir pressure in the dissolved gas mode with subsequent gas injection [14].

Thus, our results indicate that waterflooding has the greatest effect at the early stage of field development, and in depleted formations, i.e. already at the final stage, as a result of natural development, where reservoir pressures are very low (much lower than the oil saturation pressure with water), further gas injection can be effective. Gas injection can be applied to old fields at a late stage of development and with already depleted reservoir pressure.

4. Conclusion

Thus, having conducted the analysis of the feasibility of using gas energy in fields at the final stage, we can highlight the following main points:

- waterflooding of deposits will be most effective if it is carried out at the initial stage of field development;
- in depleted reservoirs that were developed in the dissolved gas mode, when the reservoir pressure is low (close to atmospheric), so further waterflooding may be less effective. It is advisable to introduce such flooding in old depleted fields with a shallow depth of occurrence of oil and gas reservoirs;
- in flooded reservoirs, where the reservoir pressure does not decrease in relation to the bubble point pressure, it is advisable to reduce the reservoir pressure to a value low from the bubble point pressure. After that, water injection should be restored with 30–50% compensation;
- for the development of fields developed in the mode of dissolved gas, it is effective to start waterflooding at reservoir pressure below 30–70% of the oil saturation pressure. The best option for the development of such deposits is the presence of active marginal waters, and in the absence of them, it is advisable, after the start of gas extraction from oil, to pump water into the marginal area with compensation for withdrawals from 20 to 40%.

References

*[1]*Wu B, Bahri C, Tan C, Li Q, Rahim A, Helmi M and Friedel T 2010 Sand Production Prediction for a Mature Oil Field-A Case Study *SPE Asia Pacific Oil and Gas Conf. and Exhibition. Society of Petroleum Engineers*

[2] Rogachev M K and Mukhametshin V V 2018 Control and regulation of the hydrochloric acid treatment of the bottomhole zone based on field-geological data *J. of Mining Institute* **231** 275-280. DOI: 10.25515/PMI.2018.3.275

[3] Yakupov R F, Mukhametshin V Sh, Khakimzyanov I N and Trofimov V E 2019 Optimization of reserve production from water oil zones of D3ps horizon of Shkapovsky oil field by means of horizontal wells *Georesursy* **21(3)** 55-61. DOI: 10.18599/grs.2019.3.55-

[4] Kuleshova L S, Kadyrov R R, Mukhametshin V V and Safiullina A R 2019 Design changes of injection and supply wellhead fittings operating in winter conditions *IOP Conf. Ser.: Mater. Sci. Eng.* **560(1)** 012072. DOI: 10.1088/1757-899X/560/1/012072

[5] Stenkin A V, Kotenev Yu A, Mukhametshin V Sh and Sultanov Sh Kh 2019 Use of low-mineralized water for displacing oil from clay productive field formations *IOP Conf. Ser.: Mater. Sci. Eng.* **560(1)** 012202. DOI: 10.1088/1757-899X/560/1/012202

[6] Almukhametova E M 2018 Developing the technology of non-stationary treatment through the example of the deposit of Severnye Buzachi *Advances in Engineering Research (AER)* **157** 23-26

[7] Khamehchi E and Reisi E 2015 Sand production prediction using ratio of shear modulus to bulk compressibility (case study) *Egyptian J. of Petroleum* **24(2)** 113-118

[8] Shakirov A A 2018 Informativity of transient processes accompanying hydrodynamic well survey *Advances in Engineering Research (AER)* **157** 558-561

[9] Ikporo B and Sylvester O 2015 Effect of sand invasion on oil well production: a case study of Garon field in the Niger Delta *The Int. J. of Engineering and Science* **4(5)** 64-72

[10] Valeev A I, Almukhametova E M, Kuzmina V V and Anisimov V V 2018 Geology of minerals of the Urengoy deposit *Advances in Engineering Research (AER)* **177** 521-524

[11] Akhmetov R T, Mukhametshin V V, Andreev A V and Sultanov Sh Kh 2017 Some Testing Results of Productive Strata Wettability Index Forecasting Technique *SOCAR Proceedings* **4** 83–87. DOI: 10.5510/OGP20170400334

[12] Selivanova M V, Tyncherov K T, Ikhsanova F A, Kalmykov I A and Olenev A A 2019 A device for paired zeroing of numbers in a residue system *J. of Phys.: Conf. Ser.* **1333(2)** 022017. DOI: 10.1088/1742-6596/1333/2/022017

[13] Tananykhin D S 2015 Development of energy-efficient technologies to increase the recovery factor of unconsolidated formations *Science of SCIENCE* **7** 3

[14] Almukhametova E M, Karimova A A, Almukhametov F F, Gabdrakhmanov N K, Gabzalilova A Kh and Garifullina Z A 2019 Study of water-oil mixture separation in bottomhole area via drift flow model *IOP Conf. Ser.: Mater. Sci. Eng.* **560(1)** 012005. DOI: 10.1088/1757-899X/560/1/012005

[15] Selivanova M V, Tyncherov K T, Ikhsanova F A, Kalmykov I A and Olenev A A 2019 A method of paired zeroing of numbers in a residue system *J. of Phys.: Conf. Ser.* **1333(2)** 022015. DOI: 10.1088/1742-6596/1333/2/022015

Thermomanometric system for diagnosing the technical condition of the ESP by the level of its vibration

L Z Zainagalina[1] and E B Dumler[2]

[1]Ufa State Petroleum Technological University, Branch of the University in Oktyabrsky, 54a Devonskaya St., Oktyabrsky, Republic of Bashkortostan, 452607, Russian Federation
[2]Ufa State Petroleum Technological University, 1 Kosmonavtov St., Ufa, Republic of Bashkortostan, 450062, Russian Federation

E-mail: zlz11@mail.ru, info@of.ugntu.ru

Abstract. The article presents the results of the analysis of the main vibration causes of a submersible electric centrifugal pump and its consequences. It gives recommendations for minimizing the vibration with the installation of an electric centrifugal pump (ESP). The article analyzes the problem of breaks' occurrence in the submersible unit assemblies is a result of the influence of various factors during the operation of the ESP. This type of failure is the most dangerous, since the fall of equipment to the bottom of the well is accompanied by damage to pump unit assemblies, cable rupture, jamming of equipment in the casing, and blockage of the filtration zone. Elimination of such accidents is characterized by significant costs for lifting the fallen equipment and restoring the bottom of the well, while the economic losses associated with an increase in the volume of work and the time of the forced downtime of the well increases significantly. The research identifies a number of reasons causing the accident, carries out the analysis of existing techniques and methods of eliminating "flights", and proposes a telemetry device to solve this problem.

1. Introduction

The overwhelming majority of high and medium production wells are equipped with electric centrifugal pump (ESP) installations, which allow lifting fluids in a wide variation in productivity and production conditions with a sufficiently high value of the turnaround time. Electric centrifugal pumps are dynamic vane machines. High rotational speed, multistage, complex design with a large number of stages and difficult working conditions lead to various types of loads, including vibration.

Along with this, oil production by the ESP unit is significantly complicated due to vibrations arising in the working bodies of the installation, which greatly increases the risk of sudden premature failures in underground equipment. At the same time, the most critical complication caused by vibration is the "flight" of the ESP - spontaneous disintegration of downhole equipment into parts (tubing string, submersible pump, and submersible motor, hydraulic protection). The analysis of emergency failures caused by the "flight" showed that the main (in terms of the number of failures) places of dismemberment are: tubing in the immediate vicinity of the pump unit suspension point and the pump section casing due to the destruction of studs in the flange joints [1-3]. From the analysis of the diagram of the causes of failure of wells equipped with installations of electric centrifugal pumps (Figure 1), it follows that one third of equipment failures occurs due to increased vibration.

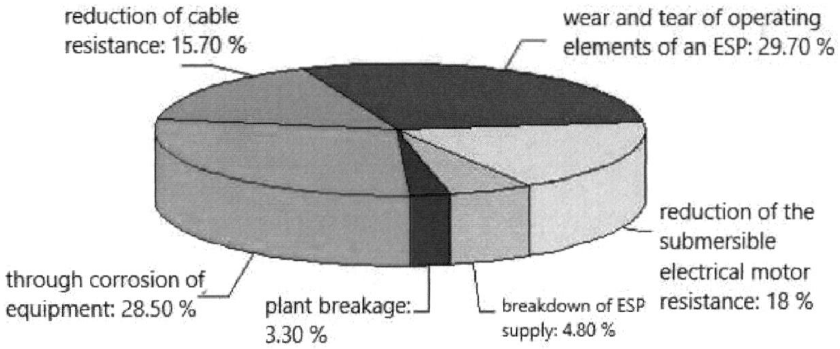

Figure 1. Diagram of the causes of failures in the oil fund of wells equipped with installations of electric centrifugal pumps.

Typical causes of vibration during ESP operation include the following:

- design features of the ESP: (the length of the installation with a small diameter, which increases its flexibility and reduces the dynamic rigidity);
- presence of factory defects in equipment: the main defects of the submersible electric motor (SEM) (imbalance) of the rotor, misalignment of the rotor shafts, defects in bearing assemblies (decrease in the bearing capacity of the lubricating layer, production of friction material, etc.);
- natural aging of the pumping unit, which is associated with the gradual wear of the working parts of the pump and violation of the centering of rotor shafts;
- poor quality manufacturing of units and parts of the installation;
- high percentage of errors when assembling the installation;
- large curvature of the well in the ESP suspension zone. Uneven distribution of the load on the ESP supports leads to their uneven radial wear, which turns into one-sided wear. The result of this action is an increase in the vibration level of the ESP. Vibration is transmitted from the ESP shaft to the hydraulic protection shaft. Mechanical seals of hydraulic protection against vibration begin to pass formation fluid into the SEM cavity, which leads to breakdown of the SEM winding;
- pump operation under conditions of increased free gas content of the liquid. During the operation of the ESP, a large layer of foam always accumulates above the pump, which does not allow to accurately determine the position of the dynamic level of the liquid, and, consequently, the bottomhole pressure. As a result, there are inconsistencies between operating modes of the reservoir and the pump, a decrease in pressure at the pump intake, an increase in the amount of free gas, ingress of large volumes of free gas into the pump, which can lead to increased vibration or disruption of supply;
- in the event of a failure of the wellhead bypass valve, increased pressure in the annulus artificially squeezes out the dynamic level, which creates an additional load on the submersible motor and increases vibration and can lead to premature failure of the ESP.

Thus, the vibration of the ESP is the root cause of almost all premature failures of its elements or a factor that multiplies these failures.

2. Materials and methods

Vibration analysis provides important information about the processes in various mechanical devices related to rotation, rolling, sliding, movement of liquids, gases, etc. The article presents the results of the analysis of the main causes of vibration and their consequences and gives recommendations on minimizing the effect of vibration on the reliability of the installation.

3. Results and discussion

A special type of failure appeared in the 1990s, called the "ESP flight". ESP flight is a failure that manifests itself in the spontaneous disintegration of downhole equipment into parts (tubing strings,

IOP Conf. Series: Materials Science and Engineering 1064 (2021) 012075 doi:10.1088/1757-899X/1064/1/012075

parts of a submersible pump, electric motor and hydraulic protection) and falling to the bottom of the well. This type of failure is the most dangerous type of failure, since the fall of downhole equipment to the bottom of the well is accompanied by damage to the units, assemblies, jamming of the fallen equipment in the casing, which makes it unrepairable. Elimination of such accidents is characterized by significant costs for lifting the fallen equipment and restoring the bottom of the well, while the economic losses associated with an increase in the volume of work and time of the forced downtime of the well increase significantly.

According to the given regularities between defects of the electric centrifugal installation and the main frequencies of the vibration spectrum, a classification of the sources of vibration was compiled in accordance with Figure 2, which made it possible to facilitate the identification of defects [4].

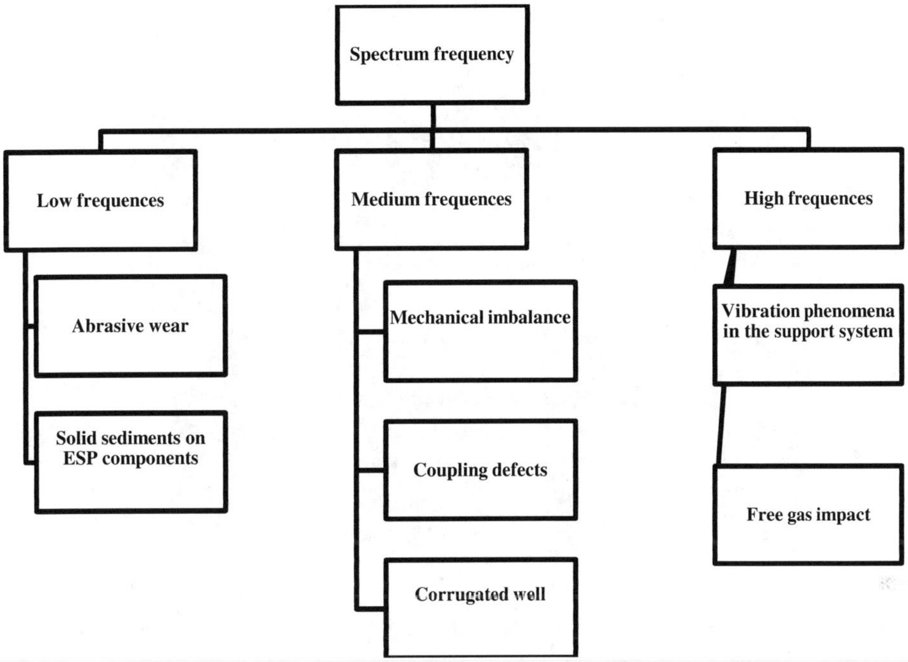

Figure 2. Classification of sources of vibration in the ESP

The impact of vibrations on parts of electric centrifugal pumps leads to pump malfunctions, exit from the operating mode, wear of seals, supporting elements, bearings, couplings and adapters, unscrewing of the ESP installation elements, as well as destruction of pump elements. The most difficult to eliminate accidents of submersible electric centrifugal installations for oil production are complications associated with the dismemberment of downhole equipment and its fall to the bottom ("flight"). Such consequences entail significant expenditures of time and material resources for their elimination, which indicates the relevance of creating means of damping vibration loads in the ESP [5, 6].

The impact of vibration on parts of electric centrifugal pumps leads to pump malfunctions, exit from the operating mode, wear of seals, supporting elements, bearings, couplings and adapters, unfolding of the ESP installation elements, as well as destruction of pump elements (Figure 3) [7]. The unbalanced forces arising from vibration provoke the initial closure of the contacting rubbing surfaces, between which there may be mechanical impurities (abrasive) with a hardness of 6-8 on the Mohs scale.

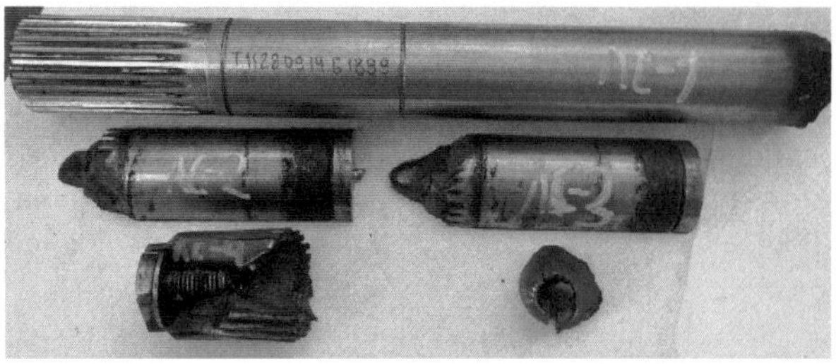

Figure 3. Fatigue failure of the ESP shaft

The presence of an abrasive in the pumped liquid in combination with a significant contact force of the friction surfaces leads to catastrophic abrasive and vibro-abrasive wear of the shafts (a), impellers (b), guide vanes (c) and other pump parts (d) (Figure 4).

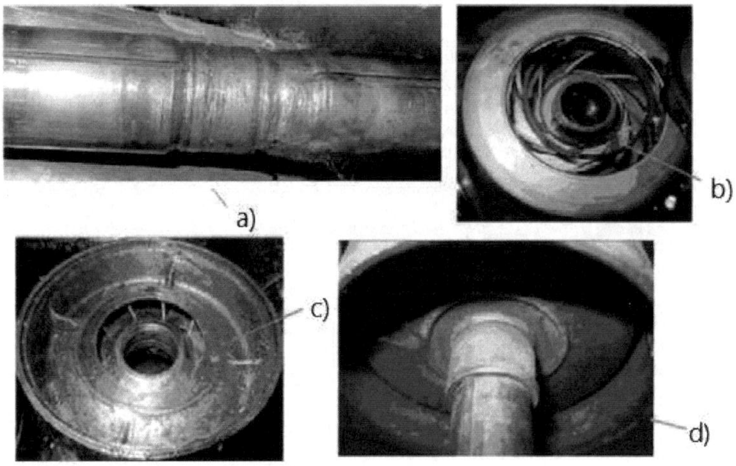

Figure 4. Wear of working elements of the ESP: a - vibration abrasive wear of the shaft; b - through hole in the seats of textolite washers; c - vibration abrasive wear of the guide vane collar; d - abrasive wear of the diffuser guide sleeve

During the extraction of an abrasive medium, an unacceptable vibro-abrasive local wear of the rubbing surfaces of the working bodies of the ESP occurs from the point of view of the theory of fatigue strength.

Complications associated with the dismemberment of submersible equipment and its flight to the bottom are the most difficult to eliminate accidents of electric centrifugal installations for oil production (Figure 5) [6].

Figure 5. ESP flight, tubing break

In this case, the main reason for the flight of the pump unit assemblies is the high vibration level of the pump unit, which leads to fatigue failure of the metal.

As you can see, the effect of vibration loads on a pumping unit can lead to serious consequences, entailing a significant investment of time and material resources.

The development of diagnostic methods to combat the influence of vibration will prevent serious accidents, as well as improve the pump performance and the stability of its units.

There are many ways to deal with the sources of vibration:

- use of special vibration dampers [8, 9];
- means for eliminating the imbalance of the ESP rotating elements;
- design and composition of ESP units to prevent abrasive wear;
- devices for combating the presence of free gas, such as dispensers and gas separators;
- special couplings for ESP operation in deviated wells.

The analysis of emergency failures of the "flight" type showed that the main (in terms of the number of failures) places of dismemberment are: tubing in the immediate vicinity of the pump unit suspension point and the pump section casing due to the destruction of studs in the flange joints. Investigation of the sites of dismemberment indicates fatigue failure of the metal under the influence of variable vibration loads.

The relevance of the problem of ESP vibration control for diagnostics of failures is evidenced by the research results set forth in the works [6, 7, 9, 10]. In this regard, at oil-producing enterprises, there recently has been an increase of interest in measures that allow diagnosing failures, increasing the service life and predicting further dynamics of the ESP unit during operation.

If there is a timely and accurate diagnosis of the vibration level of the ESP, it is possible to prevent or at least delay the failure of the installation. In particular, if it is known that vibration is caused by clogging of the pump, then the situation could be corrected by periodic flushing or one-time acid treatments.

When gas enters the working parts of the pump, gas cavities are formed, the size of which is commensurate with the dimensions of the stage channel. In this case, there is a deterioration in the energy exchange between the impeller and the liquid. The particles of the liquid surrounding the bubble are under the action of the ever increasing difference between the pressure of the liquid and the pressure inside the bubble and move to its center at an accelerated rate. With complete condensation of the bubble, a collision of liquid particles occurs, accompanied by an instantaneous local increase in pressure, reaching hundreds of megapascals. This leads to destruction of the pump running surface as a result of increased vibration. If the cause of vibration is the formation of gas cavities in the pump, it is possible to recommend changing the operating mode of the pump, for example, transferring it to intermittent operation. In addition, the availability of information about the current vibration level will allow to adequately assess the prospects for ESP failure-free operation during production stimulation: wells with a high vibration level should not be recommended to increase the frequency [11, 12].

Theoretical and experimental studies show that vibration analysis provides important information about the processes in various mechanical devices associated with rotation, rolling, sliding, movement of liquids, gases, etc. The location of discrete frequencies of the vibrogram and their amplitudes make it possible to determine the state of health of the system, to detect incipient failures associated, for example, with wearing. Functioning disturbances tend to increase vibration. Certain discrete frequencies correspond to many mechanical units and devices on the vibrogram. For a rotary mechanism, such as an engine, compressor, pump, fan, turbine, there is a series of discrete frequencies that are multiples of the rotor speed. A number of criteria have been developed to establish permissible vibration levels for machines with rotating mechanisms. Vibration magnitude peaks and peak displacements are used as parameters in the criteria. The main sources of vibration in this case are deformation or damage to rotor parts, deformation of the housing and base, eccentricity in bearings, damage to bearings, and others [7, 10].

In the mid-80s, Aliev I. M. conducted experiments to study the informative signs of diagnostics, which included determination of the amplitude-frequency characteristics of submersible electric motors and submersible units (SEM with water protection and a pump) of various power, manufacturing accuracy and degree of wear. It was found that as the ESP wears out, the energy of the oscillatory process is redistributed: the amplitude of the first harmonic (50 Hz) decreases, and the second (100 Hz) increases. In particular, it was found that for the new ESP, the ratio of the amplitudes of these harmonics satisfies the condition $A_2 / A_1 < 1$, and for the worn one - the condition $A_2 / A_1 > 1$. This ratio was accepted as the main diagnostic feature [8].

Until recently, direct measurements of ESP vibration were not carried out due to the absence or high cost of submersible telemetry sensors. In this case, indirect information about the presence of a risk of failure of the installation due to vibration could be obtained based on the analysis of changes in wellhead pressure.

The graph shows a section of wellhead pressure change in a well where a failure occurred within two weeks after commissioning due to tubing breakage due to a crack in the sleeve, where one can observe systematic pressure changes with a frequency of 5 seconds, apparently caused by vibration of the installation (Figure 6) [10].

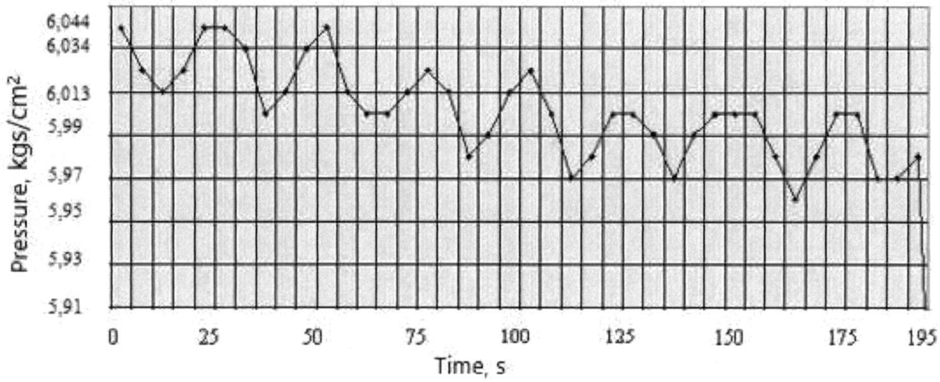

Figure 6. Wellhead pressure fluctuations

Thermomanometric systems (TMS) are used to check the technical condition of the ESP during operation in the production well.

A thermomanometric (telemetric) system is a set of technical means intended for measuring and transmitting to the controller of the ESP control station of current parameters of the well (oil production or injection). Depending on the equipment and configuration, the TMS can measure the following parameters:

- temperature of the surrounding (reservoir) fluid;
- engine oil temperature;
- temperature of the motor windings;
- formation fluid pressure at the ESP suspension point;
- formation fluid pressure at the ESP outlet;
- vibration of the submersible motor in radial and axial directions;
- insulation resistance of the transformer-submersible cable-motor stator winding system;
- fluid consumption;
- other parameters with special options.

The list of measured parameters may slightly differ depending on the manufacturer and customer requirements. The most famous manufacturers of these systems are OJSC "Alnas", CJSC "Electon" and LLC "Borets". Information from the submersible unit to the ground unit is fed through a power cable. In the ground part of the system, the received data is processed.

The TMS consists of a submersible (often abbreviated as TMSP) and a surface unit (TMSN). In this case, the submersible unit is a primary transducer for measuring the parameters of formation fluid and SEM, and the ground unit is used to receive and transform information transmitted by the TMSP, with subsequent transmission of data to the controller of the control station (CSC), as well as to measure the insulation resistance.

The transfer of information from the submersible to the ground TMS unit can occur both in digital and in analog form. As an example, a typical graph of vibration changes over time is presented for a well equipped with an ESP (Figure 7).

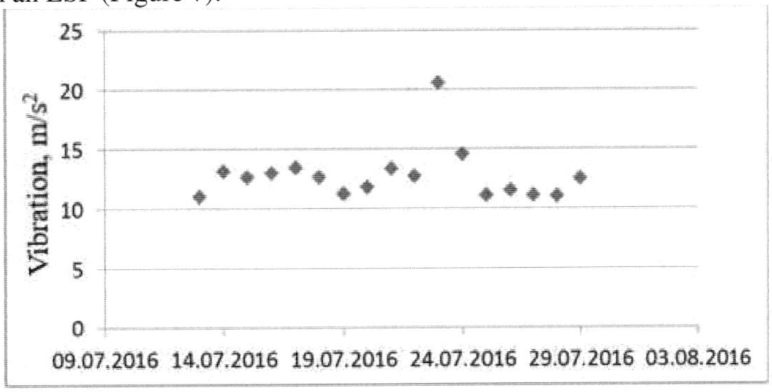

Figure 7. Changes in vibration over time

Vibration in TMS has the dimension of vibration acceleration (m / s^2). It is generally accepted that when the value of 20 m / s2 is exceeded, the installation switches to a critical mode of operation with a high risk of failure.

Vibration measurements allow not only to determine the risk of ESP failure, but also to diagnose the operation of its components.

4. Conclusion

Analysis of the state of operation of submersible electric centrifugal plants showed that many geological and technical factors and technological parameters cause the appearance of vibration loads in the ESP.

Thus, vibration is not only a source of problems during ESP operation, but also a valuable source of diagnostics of the condition of a submersible installation. There are many factors affecting the manifestation of vibration loads in the ESP, which indicates the vastness and significance of this area of scientific activity. To eliminate the consequences of ESP failures due to vibrations, expensive repairs and shutdown of the well may be required. Therefore, there is an economic feasibility of studying these issues.

In this regard, the direction of ESP diagnostics based on measured vibration indicators is one of the most promising in the context of automated control of oil production by a mechanized method.

References

[1] Mukhametshin V V and Kuleshova L S 2019 Justification of Low-Productive Oil Deposits Flooding Systems in the Conditions of Limited Information Amount *SOCAR Proceedings* **2** 16–22. DOI: 10.5510/OGP20190200384

[2] Suleimanov R I, Khabibullin M Ya and Suleimanov R I 2019 Analysis of the reliability of the power cable of an electric-centrifugal pump unit *IOP Conf. Ser.: Earth Env.* **378(1)** 012054. DOI: 10.1088/1755-1315/378/1/012054.

[3] Suleimanov R I, Zainagalina L Z, Khabibullin M Ya, Zaripova L M and Kovalev N O 2018 Studying heat-affected zone deformations of electric arc welding *IOP Conf. Ser.: Mater. Sci. Eng.* **327(3)** 032053. DOI: 10.1088/1757-899X/327/3/032053.

[4] Rivin E I 2001 *Passive Vibration Isolation* (New York: ASME Press)

[5] Polyakov V N, Chizhov A P, Kotenev Yu A and Mukhametshin V Sh 2019 Results of System Drilling Techniques and Completion of Oil and Gas Wells *IOP Conf. Ser.: Earth Env.* **378(1)**

012119. DOI: 10.1088/1755-1315/378/1/012119

[6] Zainagalina L Z, Petrova L V and Petrov V A 2019 Analysis of the efficiency of telemetric systems for drilling wells *IOP Conf. Ser.: Mater. Sci. Eng.* **560(1)** 012096. DOI: 10.1088/1757-899X/560/1/012096.

[7] Andreev A V, Mukhametshin V Sh and Kotenev Yu A 2016 Deposit Productivity Forecast in Carbonate Reservoirs with Hard to Recover Reserves *SOCAR Proceedings* **3** 40-45. DOI: 10.5510/OGP20160300287

[8] Zotov A N, Valeev A R and Tichonov A Y 2014 Application of oscillatory system with quasi-zero stiffness for creation artificial gravity *8th European Nonlinear Dynamics Conf. ENOC* (6-11 July 2014, Vienna, Austria)

[9] Khabibullin M Ya, Suleimanov R I, Sidorkin D I and Arslanov I G 2017 Parameters of damping of vibrations of tubing string in the operation of bottomhole pulse devices *Chemical and Petroleum Engineering* **53(5-6)** 378-384

[10] Wang Y C and Lakes R S 2004 Extreme stiffness systems due to negative stiffness elements *American J. of Physics* **72(1)** 40-50

[11] Rogachev M K, Mukhametshin V V and Kuleshova L S 2019 Improving the efficiency of using resource base of liquid hydrocarbons in Jurassic deposits of Western Siberia *J. of Mining Institute* **240** 711-715. DOI: 10.31897/PMI.2019.6.711

[12] Mukhametshin V V and Andreev V E 2018 Increasing the efficiency of assessing the performance of techniques aimed at expanding the use of resource potential of oilfields with hard-to-recover reserves *Bull. of the Tomsk Polytechnic University. Geo Assets Engineering* **329(8)** 30–36

IOP Conf. Series: Materials Science and Engineering 1064 (2021) 012076 doi:10.1088/1757-899X/1064/1/012076

Analytical studies of bottom hole assembly dynamics in directional well while drilling with coiled-tubing system

L Z Zainagalina[1], E Ya Zinatullina[2] and A R Safiullina[1]

[1] Ufa State Petroleum Technological University, Branch of the University in the City of Oktyabrsky, 54a, Devonskaya St., Oktyabrsky, Republic of Bashkortostan, 452607, Russian Federation

[2] Ufa State Petroleum Technological University, 1, Kosmonavtov st., Ufa, Republic of Bashkortostan, 450062, Russian Federation

E-mail: safiullina.a.r@yandex.ru, info@of.ugntu.ru

Abstract. The article is devoted to the relevant topic of more accurate selection of the design of the bottom-hole assembly (BHA) with a hydraulic test stand to create the necessary axial stress on the bit during drilling with coiled tubing rigs. A mathematical model of the BHA as a dynamic system with linear coefficients during longitudinal vibrations while drilling wells with a coiled-tubing system has been developed and analytically investigated. Analytical dependencies for determining the dynamic parameters of coiled-tubing BHA by a hydraulic test stand have been refined when operating in a directional well.

1. Introduction

At present, the most relevant method of production stimulation is the construction of sidetracks from the old operation well stock to ensure effective hydraulic communication in the well-reservoir system and involvement in the development of weakly drained stagnant zones [1–5].

There are several options for wiring sidetracks both with traditional and coiled tubing installations. These options are as follows:

- pilot installations, which clarify the geological structure of the deposit in poorly drilled areas or in the boundary part of an oil field, where the conditions of bedding are variable;
- undirected installations are drilled by telemetry systems without navigation assistance;
- obliquely directed installations with the zenith angle less than 50°, drilled using telemetry systems along a given trajectory (orientable);
- gentle installations with zenith angle within 50–65°;
- horizontal installations with the maximum zenith angle exceeding 65°, including multi-barrel ones [6, 11].

The development of special downhole assemblies with flexible drill pipes of augmented reliability aimed to drill wells with coiled tubing units with regards to the dynamic processes occurring during their construction, not only increases the quality of the wellbore being formed but also predetermines further trouble-free operation of production equipment [7].

The disadvantage of using heavy-weight drill pipes to create an axial load on a bit is the large amount of time and physical effort required for their assembly and disassembly during pulling-and-running operations. In addition, extraheavy drill rods cannot be used when drilling horizontal wells [8].

2. Materials and methods

To meet the objectives, the following methods were used: modern calculation and experimental methods, which showed good convergence of laboratory and calculated data, including special software packages for mathematical modeling.

Part of the research was carried out using the modern specialized program Ansys Workbench, computer technology, and computer modeling methods.

3. Results and Discussion

The elastic-stressed state of the bit-screw motor assembly in a directional well has been studied.

During the study, the distances from the bit to the point of contact of the string with the borehole walls were determined. Next, the deflection of the assembly was determined with x changing from 0 to l_k, and then the bending moment acting from the bit to the point of contact of the screw motor with the borehole wall was determined.

Similar calculations were made at $\alpha = 45°$, $\alpha = 60°$, $\alpha = 90°$ and the influence curves of the distance from the bit to the point of contact with the borehole wall from the zenith angle of the well were plotted (Figure 1).

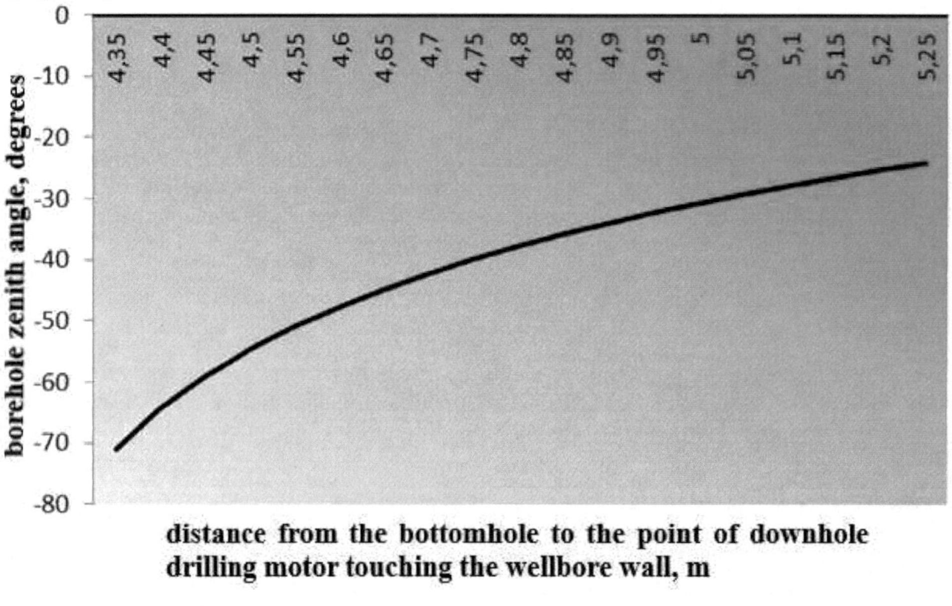

Figure 1. Dependence of distance from the bit to the point of contact with the borehole wall on the zenith angle of the borehole

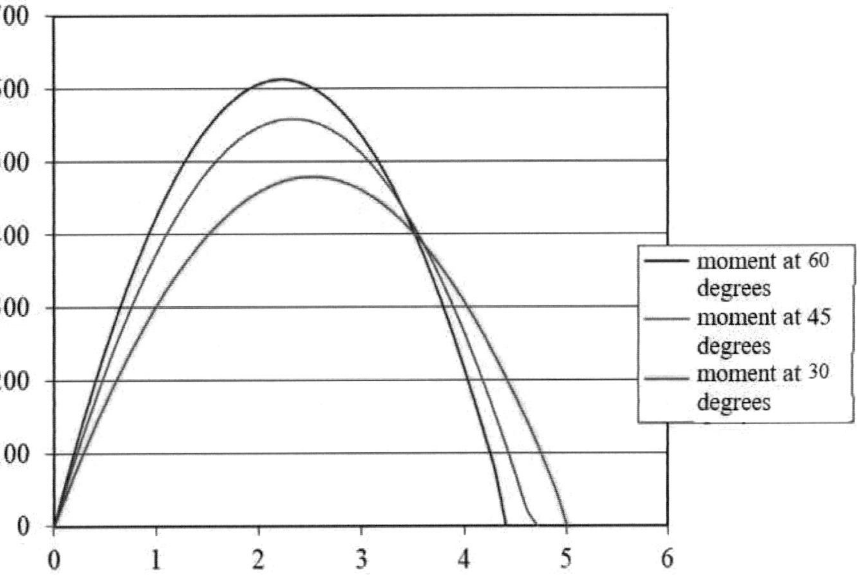

Figure 2. Values of bending moments influencing a screw motor

Calculations have shown that the DR3-95M.5/6.50 engine is characterized by the minimum values of the elastic-stressed state of the engine body and provides better permeability in the wellbore compared to other standard sizes of downhole motors.

Similarly, calculations were made for downhole motors of standard sizes D-76.2000.45, D-63. The corresponding curves of the dependence of the distance from the bit to the point of contact with the borehole wall from the zenith angle of the borehole and curves of bending moments acting on the drilling assembly were constructed [9].

The calculation results are shown in Table 1.

Table 1. Design parameters of screw motors

Parameters	Screw motor model		
	D-63	D-76.2000.45	DR3-95M.5/6.50
Axial load P, kN	20	20	60
Borehole apparent radius r, m	6.5	3.5	1.4
Reduced moment of inertia, m4	$5.5 \cdot 10^6$	$11.9 \cdot 10^6$	$25.9 \cdot 10^6$
Weight per unit length of the screw motor qB, N / m	19.6	24.838	44.559
Stiffness coefficient k	0.211	0.282	0.3315
Bit diameter D□, mm	76.0	83	112.0
Internal diameter of the engine housing D1, mm	56.7	68.4	84.5
Stator diameter d, mm	41.5	50.2	62.7
Weight, kg	65	115	220
Length, mm	3300	4630	5 345

Analytical studies of the coil-tubing BHA performance in a directional well were carried out.

Longitudinal vibrations of a drilling tool being a part of a coil-tubing BHA with a hydraulic loader were considered when drilling lateral horizontal wellbores [10–13].

Let us write down the equations of the considered assembly motion:

$$\begin{cases} m_2\ddot{X}_3 + K_2X_3 + K_1(X_3 - X_2) = 0, \\ m_1\ddot{X}_1 + C(\dot{X}_2 - \dot{X}_1) = P \cdot \cos \omega t, \\ K_1(X_3 - X_2) = C(\dot{X}_2 - \dot{X}_1). \end{cases}$$

This system was solved within the Mathcad package. Figures 3–5 show the obtained dependences $X_i(t)$ under the condition of admissibility of the stress-strain state and dynamic equilibrium.

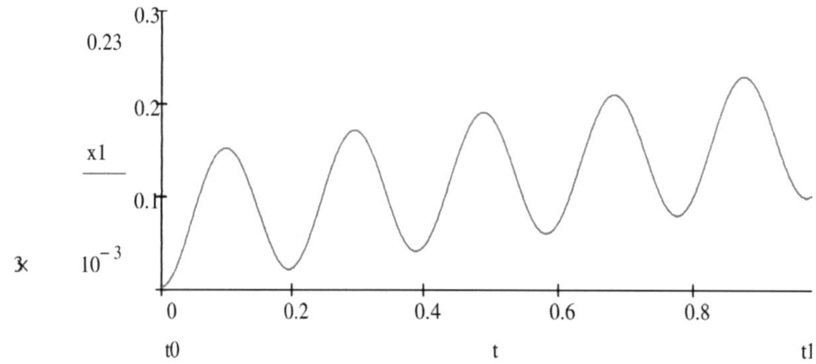

Figure 3. Bottom-hole assembly displacement dependence

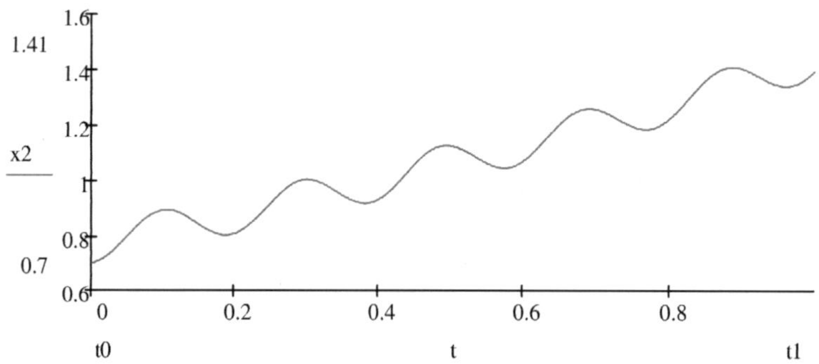

Figure 4. Dependence of hydraulic test stand movement

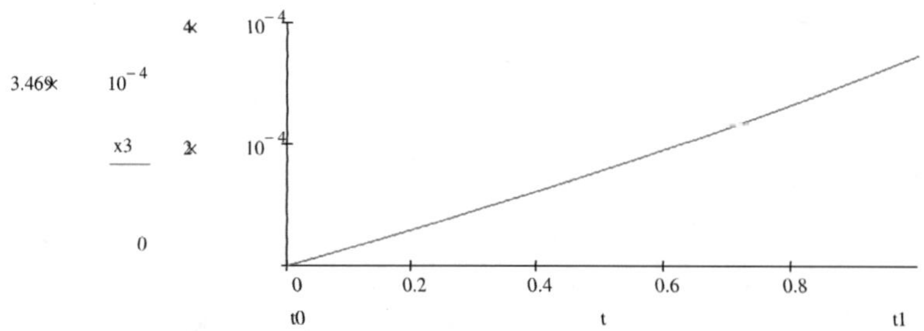

Figure 5. Smoothed characteristic of dynamically disturbed

As can be seen from Figures 3–5, with correctly selected parameters of the coil-tubing BHA, in addition to creating an axial load, the hydraulic test stand operate as a damper that dampens, ground vibrations of the layout in addition to other things.

The stress-strain state of the hydraulic test stand was assessed on a computer using the Workbench 16.0 program. The calculation results are shown in Figures 6 and 7.

Figure 6 shows that the maximum deformation occurs in the upper part of the case and is equal to 0.0004 mm. Figure 7 shows that the maximum stresses occurring in the body of the hydraulic test stand are 394.25 MPa.

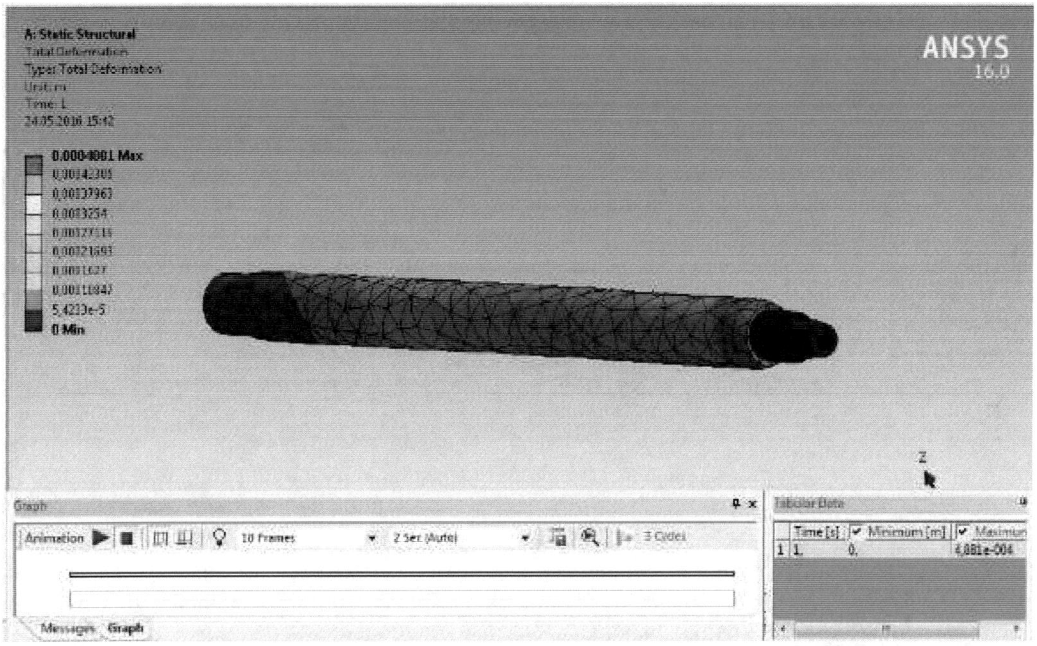

Figure 6. Full deformation of hydraulic test stand

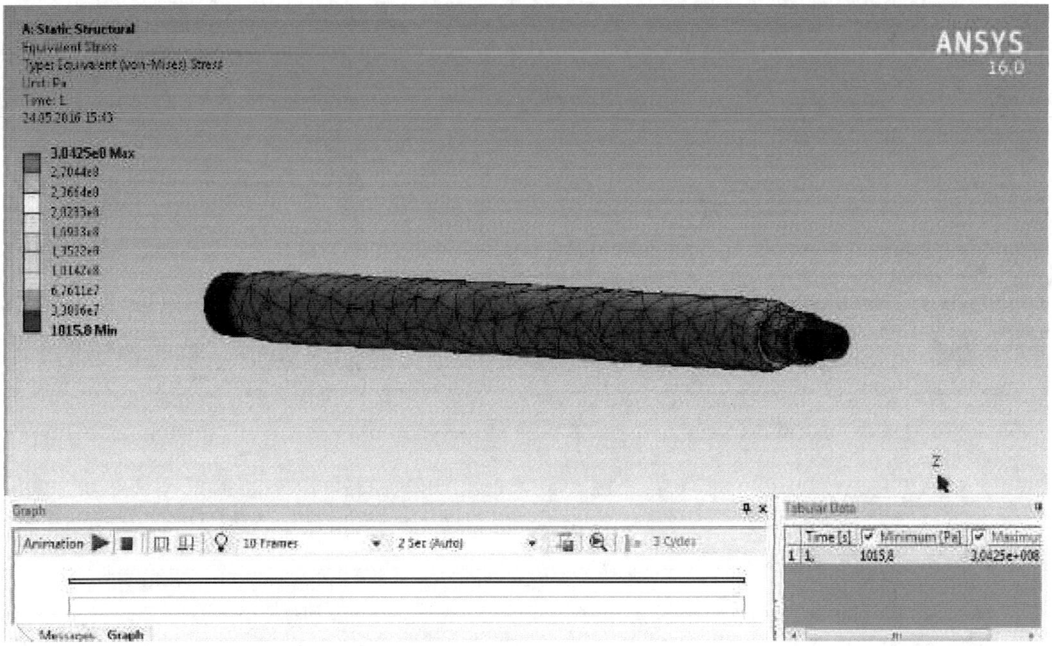

Figure 7. Equivalent stresses of hydraulic test stand

4. Conclusion

The effect of the bending moment on the coil-tubing BHA in a directional well has been analyzed analytically. It has been found that with the proposed coil-tubing BHA design implying the hydraulic test stand being an engine, the engine torque (M_{eng}) should be greater than the coil-tubing BHA bending moment.

The stress-strain state of the hydraulic loader has been assessed on a computer using the Workbench 16.0 program. The results have showed that the maximum deformation occurs at the top of the case and are equal to 0.0004 millimeters. The maximum stresses arising in the body of the hydraulic test stand are 394.25 MPa. These indicators are permissible during the operation of the hydraulic test stand.

A mathematical model describing the operation of the coil-tubing BHA, represented by a system of second-order stationary differential equations with linearized coefficients, the solution of which enabled to calculate the operating parameters of the layout and its basic units (axial load, mass, frequency, dynamic stiffness of modules and regularities of movement of the coil-tubing BHA with a hydraulic loader) has been developed. It has been found that the hydraulic test stand suppresses vibrations arising on the bit in addition to creating the required axial load.

References

[1] Yakupov R F, Mukhametshin V Sh, Khakimzyanov I N and Trofimov V E 2019 Optimization of reserve production from water oil zones of D3ps horizon of Shkapovsky oil field by means of horizontal wells *Georesursy* **21(3)** 55-61. DOI: 10.18599/grs.2019.3.55-61

[2] Rogachev M K and Mukhametshin V V 2018 Control and regulation of the hydrochloric acid treatment of the bottomhole zone based on field-geological data *J. of Mining Institute* **231** 275-280. DOI: 10.25515/PMI.2018.3.275

[3] Stenkin A V, Kotenev Yu A, Mukhametshin V Sh and Sultanov Sh Kh 2019 Use of low-mineralized water for displacing oil from clay productive field formations *IOP Conf. Ser.: Mater. Sci. Eng.* **560(1)** 012202. DOI: 10.1088/1757-899X/560/1/012202

[4] Kuleshova L S, Kadyrov R R, Mukhametshin V V and Akhmetov R T 2019 Auxiliary equipment for downhole fittings of injection wells and water supply lines used to improve their performance in winter *IOP Conf. Ser.: Mater. Sci. Eng.* **560(1)** 012071. DOI: 10.1088/1757-899X/560/1/012071

[5] Valeev A S, Kotenev Yu A, Mukhametshin V Sh and Sultanov Sh Kh 2019 Substantiation of the recovery of residual oil from low-productive and heterogeneous formations in Western Siberia by improving the waterflood system using gas and water-gas impacts *IOP Conf. Ser.: Mater. Sci. Eng.* **560(1)** 012204. DOI: 10.1088/1757-899X/560/1/012204

[6] Suleimanov R I, Zainagalina L Z, Khabibullin M Ya, Zaripova L M and Kovalev N O 2018 Studying heat-affected zone deformations of electric arc welding *IOP Conf. Ser.: Mater. Sci. Eng.* **327(3)** 032053. DOI: 10.1088/1757-899X/327/3/032053.

[7] Nurgaliev R Z, Urazakov K R and Khabibullin M Ya 2019 Improving efficiency of packers subjected to alternate axial loads *Oil Industry* **4** 94-97. DOI: 10.24887/0028-2448-2019-4-94-97.

[8] Yagafarova Kh N, Arslanov I G and Teregulov R A 2018 Performance analysis of surface reducing gear of rod driven screw pump with involute gearing and Novikov gearing *Advances in Engineering Research (AER)* **157** 627-630

[9] Zainagalina L Z, Petrova L V and Petrov V A 2019 Analysis of the efficiency of telemetric systems for drilling wells *IOP Conf. Ser.: Mater. Sci. Eng.* **560(1)** 012096. DOI: 10.1088/1757-899X/560/1/012096.

[10] Muhammed A and Childs D W 2016 Vibration modeling and experimental results of two-phase twin-screw pump *J. of Engineering for Gas Turbines and Power* **38(9)** 092601

[11] Khabibullin M Ya and Suleimanov R I 2019 Automatic packer reliability prediction under pulsed transient flooding of hydrocarbon reservoirs *IOP Conf. Ser.: Mater. Sci. Eng.* **560(1)**

012024. DOI: 10.1088/1757-899X/560/1/012024.
[12] Hussein M F M, François S, Schevenels M, Hunt H E M, Talbot J P and Degrande G 2014 The fictitious force method for efficient calculation of vibration from a tunnel embedded in a multi-layered half-space *J. Sound Vib.* **333(25)** 6996-7018
[13] Selivanova M V, Tyncherov K T, Ikhsanova F A, Kalmykov I A and Olenev A A 2019 A device for paired zeroing of numbers in a residue system *J. of Phys.: Conf. Ser.* **1333(2)** 022017. DOI: 10.1088/1742-6596/1333/2/022017

IOP Conf. Series: Materials Science and Engineering 1064 (2021) 012077 doi:10.1088/1757-899X/1064/1/012077

Mathematical model of propagation of liquid vibrations in a tubing string

L M Zaripova

Ufa State Petroleum Technological University, Branch of the University in the City of Oktyabrsky, 54a, Devonskaya St., Oktyabrsky, Republic of Bashkortostan, 452607, Russian Federation

E-mail: Lilyabert31@mail.ru, info@of.ugntu.ru

Abstract. Effectiveness of wave processes in tubing strings, wells and long pipelines may be limited by damping of hydrodynamic waves along the length of the channel. Violation of thermodynamic equilibrium in reservoir fluids during their rise to the surface and in ground pipelines may be accompanied by the release of asphalt-resin-paraffin components in the flow of crystals due to a decrease in temperature and pressure. Subsequent depositing of crystals takes place as they grow, due to adhesion to the wall of pipes, the surface of equipment. Deposits are most intensified in areas of local expansion and narrowing of the cross-section area of the flow (occurrence of vortices).

1. Introduction

Paraffin is a mixture of higher methane hydrocarbons. Refined paraffin is a colorless or white crystalline mass, odorless and tasteless, slightly greasy to the touch, the specific gravity of which ranges from 0.907 to 0.915 at 15 ° C. The specific gravity of the crude paraffin is 0.881–0.905. The melting point of paraffin depends on the molecular weight and ranges from 49 to 60 ° C. At lower temperatures, it is solid.

The depth of the onset of paraffin deposition during oil production from deposits varies widely. For example, during the exploitation of oil fields in the Perm region, the depth of the onset of paraffin deposition is 1000 m, in the Republic of Tatarstan - 1200-1250 m and lower, in the Republic of Bashkortostan - 400-1700 m, in the fields of South Mangyshlak - 1000-1100 m. At the Nizhnesortymsky oil field in West Siberia, paraffin deposits were observed in the depth interval of 300–1400 m, and at the Urengoyskoye gas condensate field - at a depth of 800–1600 m. These results were obtained by studying the deposits when lifting the tubing string from the wells. The significant depth of the onset of paraffin deposition is due to the crystallization of more refractory paraffins primarily. In this regard, during oil production in the tubing string, the most refractory paraffins accumulate in asphalt-resin-paraffin deposits (ARPD).

At a high resin content, solid paraffinic hydrocarbons crystallize in aggregate form in the form of small crystals. Being polar substances, resins are adsorbed on small crystals and cause paraffin aggregation. If at the initial moment many small crystalline nuclei are formed, then upon further cooling, solid paraffinic hydrocarbons released from the solution crystallize on these nuclei. This ultimately leads to the formation of small paraffin crystals in solution.

For oil with a high content of resins and asphaltenes, close to the temperature of its saturation with paraffin, the process of aggregate crystallization is more characteristic - a process in which crystals grown independently of each other from different crystallization centers turn out to be randomly cemented into a single system of high-molecular compounds.

With a deeper cooling of highly paraffinic oil, the process of aggregate-dendritic crystallization is possible, where the aggregate is formed not by single crystals of paraffin, but by individual dendrites.

Thus, the ratio of paraffins and natural depressants has a significant effect on the crystallization process. With a small content of natural oil depressants, dendritic crystallization occurs with an external phase from molecular groups of paraffins, and with an excess of natural depressants, aggregate crystallization with an external phase from molecular groups of resins and asphaltenes occurs.

Decreases in temperature and pressure may be accompanied by the release of asphalt-resin-paraffin components in the flow of crystals when the thermodynamic equilibrium in layered fluids is disturbed. The problem of attenuation of the amplitude of hydrodynamic pressure waves in the tubing string and ground pipeline, taking into account the friction of the liquid against the pipe walls is covered in [1-11].

When using wave processes, reducing the intensity of asphalt-resin-paraffin deposits (ARPD) on the inner surface of pipe systems, as well as increasing the speed of their erosion, for example, by hot liquid (washing liquid), is provided by the following additional processes:

– increase in the total friction path, respectively, the intensity (speed) of destruction of the surface layer of the ARPD due to additional wave-like movement of the washout liquid relative to the average flow rate (increasing the total contact time);

– reduction of adhesion forces in the inner surface of pipes, due to the difference in densities of metal and the ARPD and, as a result, the radial components of inertial forces (in the limit, when the breaking forces from the flow of the extracted reservoir fluid exceed the adhesion forces) [12-14].

2. Materials and methods

When flushing, the efficiency of the process of dewaxing with a hot liquid (for example, water) of a tubing string in a well or a pipeline is limited by the distribution of hydrodynamic waves (HDW) along its length. Without external damping, the propagation of HDW is limited, mainly by the forces of internal friction in the fluid itself. When the fluid moves in the column or in the part of it narrowed by the ARPD, due to the friction forces on the inner surface of the channel, external damping forces arise, the influence of which increases with narrowing of the channel cross-sectional area and with an increase in the HDW power.

As a result, for real oilfield technological problems, characterized by a high intensity of HDW emitted by the hydrogenerators and their relatively low frequency (for the case under consideration, in the lower intervals of the sound frequency range), the effect of internal friction becomes negligible compared to the effect of flow friction against the inner surface of the column.

The effectiveness of the influence of HDW on the dewaxing process is directly related to the intensity of their decay. When moving in pipes, the main reason for the damping of waves is the absorption of their energy by the forces of friction against the channel walls and, to a much lesser extent, the forces of internal friction between the particles of the medium itself.

Therefore, we will initially consider the effect of friction on the damping of longitudinal waves of the medium against the walls of the waveguide. The influence on the damping of internal friction, in general, is considered in numerous educational literature and therefore is not presented in this article. Let us assess the influence of external friction on the damping of the HDW in more detail.

The propagation of flat longitudinal HDW in a cylindrical channel, taking into account external damping, i.e. due to the friction of the liquid against the walls, is characterized by a well-known nonlinear expression

$$\frac{\partial^2 u}{\partial t^2}+\frac{\partial}{\partial t}(au^2)=c^2\frac{\partial^2 u}{\partial x^2}, \tag{1}$$

reduced to a linearized one for harmonic waves, i.e. telegraphic equation

$$\frac{\partial^2 u}{\partial t^2}+\frac{8\omega a}{3\pi}x\frac{\partial u}{\partial t}=c^2\frac{\partial^2 u}{\partial x^2}, \tag{2}$$

where $u = \Delta P / (\rho c)$ – velocity in the considered cross section with the amplitude of vibration of the fluid U; $\Delta P = \rho a u^2$ – hydrodynamic pressure; F – cross-sectional area of fluid flow; χ – wetted perimeter; $\omega = 2\pi \nu$ – circular frequency of the HDW; $X = U/\omega$ – amplitude of vibration of the particles of the liquid; $a = \lambda/8\delta$; $\delta = F/\chi$ – hydraulic radius of the flow cross-section; $\delta = F/\chi$ – hydraulic radius of the flow cross- section (in the pipe flow) $4\delta = d_т$, where $d_т$ – internal diameter of pipe); λ – coefficient of hydraulic resistance.

To determine the depth of the HDW propagation in the flow, we will reveal the coefficient of hydraulic resistance λ for turbulent flow in a cylindrical channel of Newtonian liquids.

3. Results and Discussion

At the turbulent flow of Newtonian liquids in cylindrical channels (corresponding to the Reynolds number Re > Re_{kr} = 2100–2320 – upper limit of Re_{kr} = 2320 – for technical water; the presence of a transient flow mode in the interval Re_{kr} = 2100–4000 is ignored in the given task), the coefficients of hydraulic resistance may decrease up to a multiple of a smaller value. This decrease is accompanied by a reduced degree of damping of liquid pressure HDW and, accordingly, an increase in the effectiveness of their influence on a particular technological process along the length of the tubing string in a well or a long pipeline.

In order to simplify the wave equation (1) as much as possible, the coefficient of hydraulic resistance λ in a cylindrical channel will be determined using the Blasius formula

$$\lambda = 0{,}3164\,\mathrm{Re}^{-0{,}25}. \tag{3}$$

In this case, the wave equation (1) will have the form

$$\frac{\partial^2 u}{\partial t^2}+0.3164A\sqrt[4]{\frac{\eta}{4\delta\rho}}\frac{\partial u^{1.75}}{\partial t}=c^2\frac{\partial^2 u}{\partial x^2};\quad A=1. \tag{4}$$

Thus, a strongly nonlinear wave equation is obtained even in the simplified version.

According to the work of B. I. Mitelman, in the turbulent regime in the area above the transition, i.e. in the area of quadratic friction with the lower boundary

$$u/\nu = 2{,}4\cdot 10^6,\ \mathrm{m}^{-1}, \tag{5}$$

where the velocity u and kinematic viscosity ν are measured in m/s and m^2/s, respectively, the coefficient of hydraulic resistance in pipes with a diameter d_t is equal to

$$\lambda = 0{,}0121\, d_t^{-0{,}226}, \tag{6}$$

and instead of (4) by analogy with (2) we have a linearized equation

$$\frac{\partial^2 u}{\partial t^2}+X\frac{\omega\lambda}{3\pi\delta}\frac{\partial u}{\partial t}=c^2\frac{\partial^2 u}{\partial x^2} \tag{7}$$

or, for harmonic HDW, the expression

$$\omega^2 U+i\frac{\omega\lambda}{3\pi\delta}U^2=c^2\frac{\partial^2 u}{\partial x^2}, \tag{8}$$

the solution of which leads to Painleve transcendental function and after integration has the form

$$\left(\frac{dU}{dx}\right)^2-imU^3+nU^2+C_1=0, \tag{9}$$

where

$$n = \omega^2 / c^2 ; \quad m = 2\lambda\, \omega / (9\, \pi\, \delta\, c^2) ; \quad \delta = d_t / 4 . \tag{10}$$

Extracting the square root from the resulting equation, we obtain the desired solution via incomplete elliptic integrals of the 1st kind, which is not given due to the bulkiness.

Under boundary conditions: at x = 0

$$U = U_0 ; \quad \frac{dU}{dx} = 0; \tag{11}$$

in equation (9)

$$C_1 = i\, m\, U_0^3 - n\, U^2 . \tag{12}$$

Since we are primarily interested in a qualitative picture of the relative losses of hydrodynamic velocity or pressure, we will limit ourselves to a particular solution (9) at $C_1 = 0$. For a harmonic function, this physically corresponds to the case of placing the center of coordinates on the ridge of the HDW. In this case, instead of (9), we will consider the equation

$$\left(\frac{dU}{dx}\right)^2 - imU^3 + nU^2 = 0 \tag{13}$$

or in a transcendent form:

$$\frac{dU}{U\sqrt{n - imU}} = dx, \tag{14}$$

the solution of which corresponds to the formula:

$$\left.\begin{array}{l} when\ n > 0 \\ x = \dfrac{1}{\sqrt{n}} \ln \dfrac{\sqrt{n - imU} - \sqrt{n}}{\sqrt{n - imU} + \sqrt{n}}; \\ when\ n < 0 \\ x = \dfrac{1}{\sqrt{-n}} arctg \dfrac{\sqrt{n - imU}}{\sqrt{-n}}. \end{array}\right\} \tag{15}$$

The real part of the obtained complex expression corresponds to the desired hydrodynamic displacement velocity. In this case, after the transformations, we get: at n > 0

$$x = n^{-0.5} \ln \sqrt{\left(\left|\sqrt{1 + m^2 U^2 n^{-2}}\right| - 1\right)\left(3 + 2\sqrt{2}\right)^{-1}} . \tag{16}$$

4. **Conclusion**

Stationary ground-based or borehole hydrogenerators should be used only (or primarily) in subinfra- or infrasound frequency ranges. With a pipeline length being small, the use of HDW in the lower areas of the sound range may also be satisfactory in special cases. High-frequency HDW should only be used when overlaying low-frequency ones.

At the lower end of the tubing string in the well, it is better to use a low-frequency pulsator.

References

[1] Mukhametshin V V and Kuleshova L S 2019 Justification of Low-Productive Oil Deposits Flooding Systems in the Conditions of Limited Information Amount *SOCAR Proceedings* **2** 16–22. DOI: 10.5510/OGP20190200384

[2] Yilbas B S, Patel F and Karatas C 2015 *Laser Surface Engineering* (Waugh J.L.G. Woodhead Publishing) pp 97–105

[3] Stepanova N V and Razumakov A A 2013 *The 8 Int. forum on strategic technologies (IFOST*

2013) vol 1 pp 240–242
[4] Yakupov R F, Mukhametshin V Sh, Khakimzyanov I N and Trofimov V E 2019 Optimization of reserve production from water oil zones of D3ps horizon of Shkapovsky oil field by means of horizontal wells *Georesourses* **21(3)** 55–61. DOI: 10.18599/grs.2019.3.55-61
[5] Ribeiro P M V and Pedroso L J 2016 A mode superposition approach for fluid-structure interaction problems *Revista Internacional de Metodos Numericos para Calculo y Diseno en Ingenieria* **32(2)** 79–90
[6] Rogachev A S, Shkodich N F, Vadchenko S G, Baras F, Chassagnon R, Sachkova N V and Boyarchenko O D 2013 *Int. J. Self-Propag. High-Temp. Synth* **22(4)** 210–216
[7] Bataev I A, Bataev A A, Mali V I and Pavliukova D V 2012 *Materials and Design* **35** 225–23
[8] Suleimanov R I, Khabibullin M Ya and Suleimanov Re I 2019 Analysis of the reliability of the power cable of an electric-centrifugal pump unit *IOP Conf. Ser.: Earth Env. Sci.* **378(1)**
[9] Davydov A Yu 2018 Utilization of Expert Systems for Screw Pump Sets with Surface Drive Management *Advances in Engineering Research (AER)* **157** 129–133
[10] Galiullina I F and Kadyrov R R 2018 Technical and economic background for siting production of well-killing liquid at oil fields *IOP Conf. Ser.: Earth Env.* **194(8)**
[11] Gilmanova A M 2019 Consideration of various factors influencing acoustic logging equipment production accuracy *IOP Conf. Ser.: Mater. Sci. Eng.* **560(1)** 012063. DOI: 10.1088/1757-899X/560/1/012063
[12] Zaripova L M and Gabdrakhimov M S 2019 Layout for drilling seismic wells by vibration method *IOP Conf. Ser.: Earth Env.* **350(1)**
[13] Zaripova L M and Gabdrakhimov M S 2019 A mathematical model of the pulsator for cleaning paraffin deposits of pipelines and downhole equipment *J. of Phys.: Conf. Ser.* **1333(3)**
[14] Zaripova L M and Gabdrakhimov M S 2019 Restoration of intake capacity of injection well by vibrations *IOP Conf. Ser.: Earth Env.* **378(1)**

IOP Conf. Series: Materials Science and Engineering 1064 (2021) 012078 doi:10.1088/1757-899X/1064/1/012078

A mathematical model of a vibrator with an axial striker mechanism

L B Khuzina, L M Zaripova and M S Gabdrakhimov

Ufa State Petroleum Technological University, Branch of the University in the City of Oktyabrsky, 54a, Devonskaya St., Oktyabrsky, Republic of Bashkortostan, 452607, Russian Federation

E-mail: lhyzina@yandex.ru, info@of.ugntu.ru

Abstract. To be operated effectively by the production and injection wells, the nature of oil reservoir production is largely determined. The efficient operation of wells depends on geological and technological factors. This refers to their exploitation with oil output equal to the potential capabilities of the reservoir when it is fully served by the filtration process. The bottom-hole zone (BHZ) is in a non-equilibrium thermodynamic state of energy and mass exchange with the well and the reservoir. The contamination of the bottom-hole zone has a significant impact on the effectiveness of wells.

1. Introduction

The deterioration in properties of the bottom-hole zone is due to: the permeability of the well-killing and washing liquids in the process of underground repair of the wells; the penetration of mechanical impurities and corrosion products of metals during well-killing or washing; the deformation of rocks at the bottom hole during drilling; the reduction of penetrating and poriness with the increasing of effective stress; the reduction of the phase permeability for the liquid (oil) while the bottom-hole pressure reducing below the level of the saturation pressure of the oil reservoir by the gas; the reduction of the phase permeability for the oil from the water saturation of the reservoir during mining (using water-flooding operation, in the case of formation of water cones, etc.); the hydration of the particles of the argillaceous cement of the terrigenous reservoir when saturated it with fresh water; the precipitation and formation of asphaltene-resin-paraffin compounds of oil or salts from the produced water when temperature-pressure conditions are changing [1-5].

The analysis of the reasons affecting the permeability of geological rock in the bottom-hole zones of wells educed that the choking of the filtration rock channels with the solid particles of the argillaceous solution, the particles of drilled rock, sand, silt, etc. in the process of various technological operations reduce the relative permeability for oil by 5-6 times [6-10].

A downhole vibrator has been created for vibration-wave cleaning of the BHZ, and the design scheme of the vibrator is described [11-12].

2. Materials and methods

The downhole vibrator consists of a housing assembly, upper and lower adapters, a spring-loaded seat with a through-hole, a spring, a calibrated sleeve, inside which a balancer is installed, made at the same time with a disk, which is attached through an axis. The housing assembly is provided with a shoulder.

The tool operates as follows. The downhole vibrator is lowered into the well, and after the circulation restoring, the jet of service fluid acts on the blade of the balancer, takes it out of balance, due to hydraulic forces, the balancer begins to sway from one extreme position to another. When the passage section of the seat is blocked by a disk, a water hammer occurs.

The spring stiffness is selected taking into account that during normal operation of the downhole vibrator, the seat is to be in its original position. If the seat channel is clogged with mechanical impurities, then under the influence of excessive pressure of the service fluid, the seat is pressed out and mechanical impurities are washed out.

3. Results and discussion

The mechanic process of the downhole vibrator.

To ensure the most stable forced vibrations for the vibrator, if possible, the following conditions should be fulfilled:

- the load on the seat-striker is equal to the pressure of the high-speed flow of the liquid on the counter surface of the seat plus the pressure drop in the inner hole of the seat-striker;
- the load on the seat-striker is greater than the spring's elastic force when it is pre-deformed.

The hydraulic losses in the inner hole of the seat-striker can be determined by a well-known formula:

$$(P_1 - P_2) = V_1^2(2g)^{-1}(\xi_{lhr} + \xi_{jc}D_2^2 d^{-2}\varepsilon^{-2}), \quad (1)$$

where$\varepsilon = \frac{d_{jc}}{d}$ is the coefficient of jet contraction; $d_{jc}$is the diameter of the contraction flow at the entrance of the chamber of the pendulum apparatus in the cavity of the seat-striker with a diameter *d*; $\xi_{lhr}$and ξ_{jc} are the coefficients of the local hydraulic resistance at the compression of strings and the sudden expansion of it at the exit from the cavity diameter *d*, respectively; V_1 is the flow rate of the fluid in the cavity of the pendulum oscillations when applying the water hammer; *d* is the diameter of the inner seat-striker; *D* is the outer diameter of the seat-striker.

The operation scheme of the seat-striker is shown in figure 1.

We can rewrite the conditions for ensuring stable forced vibrations for the vibrator taking into account (1):

$$P = \frac{V_1^2}{2g}F_k + (P_1 - P_2)F_k + \Delta P\frac{\pi d^2}{4} \geq k(h_{fr} - h), \quad (2)$$

where F_k is the area of the end surface of the seat-striker,

$$F_k = \frac{\pi(D^2 - d^2)}{4}; \quad (3)$$

k is the spring stiffness; h_{fr} is the height of the spring in a free state; *h* is the height of the springs after deformation during the vibrator assembling (figure 1) and under the additional influence of the dynamic pressure of the fluid flow; ΔR is the variable load on the seat during periodic overlapping the holes *d* of the seat-striker of the pendulum valve and changing from zero to the value of ΔP_{wh}; ΔR_{wh} is the amplitude of water hammer occurring at the time of overlapping the inner hole *d* of the seat-striker of the pendulum valve; *D* and *d* are the outer and inner diameters of the seat-striker, respectively.

When the pendulum periodically closing to momentary overlaps the flow passage section with a diameter of *d* in the seat-striker, a hydraulic hammer occurs, which affects the seat with a variable load of amplitude ΔP. The value of the water hammer of ΔP_{wh} can be determined by the formula of N.E. Zhukovsky: the disturbing variable load (on the seat) at $t < \tau = \frac{2L}{c}$ (the case of installing vibrators in a well located hundreds of meters away from the bottom-hole or at the surface):

$$\Delta P_{wh} = \rho c V_1; \quad (4)$$

when $t < \tau = \frac{2L}{c}$

$$\Delta P_{wh} = \rho c(V_1 - V_0), \quad (5)$$

where c is the speed of the strike wave (for practical calculations, without taking into account the compressibility of fluid and elasticity of pipe walls, equal to the speed of sound); L is the length of the section of the column from the placement of the vibrator to the drilling bit jet nozzles (the source of flow rate); t is the overlap time of flow (in one cycle of the pendulum oscillation, the flow is blocked twice), $t = (2v)^{-1}$; p is the fluid density; V_{fh} is the average rate of the fluid hole of the seat; V_1 is the flow rate of the fluid in the cavity of the pendulum oscillations when applying the water hammer.

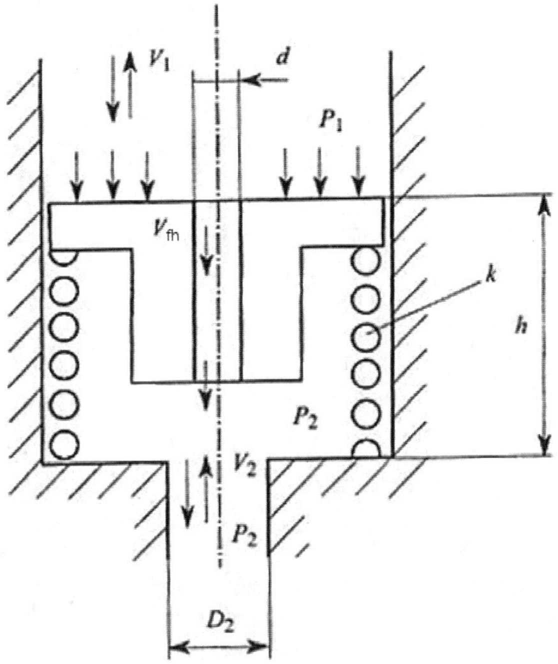

Figure 1. The scheme of the seat-striker: P_1 is the hydraulic load on the seat during water hammer; P_2 is the hydraulic load during water hammer under the seat; V_{fh} is the average rate of the fluid hole of the seat; V_1 is the flow rate of the fluid in the cavity of the pendulum oscillations when applying the water hammer; V_2 is the flow rate of fluid when applying the water hammer in the cavity under the seat; h is the height of the springs after deformation; k is the stiffness of the spring; d is the diameter of the inner seat-striker; D is the outer diameter of the seat-striker.

With a smooth change in the flow rate in the column (the high non-reactivity with a large length of it) during the time τ of the loss rate.

$$(V_1 - V) = \frac{V_1 \tau}{t}. \tag{6}$$

where V is the rate of the fluid flow.

Since in this case, $t > \tau$, i.e. the hydraulic strike is not immediate and, properly, incomplete, its value will be slightly less and equal to

$$\Delta P = \frac{F_k \Delta P_{wh} \tau}{t}. \tag{7}$$

In this case, formula (7) for determining the value of the disturbing variable load that is acting on the seat-striker when the pendulum valve periodically overlaps the hole can be converted into an approximate relationship

$$\Delta P = \frac{2\rho L V_1}{t} F_k = \frac{4\rho L V_1 v}{g} F_k. \tag{8}$$

Figure 2 shows the results obtained by the formula (8) for calculating the dependence of the amplitude of the disturbing variable load ΔP and the frequency of forced vibrations acting on the seat-striker when the pendulum of the downhole vibrator moves from the flow of the washing liquid Q. The assumed placement of the vibrator from the bottom of the well is $L = 2.5$ m.

The end surface area of the valve seat is

$$F_k = \frac{\pi(0{,}1^2 - 0{,}03^2)}{4} = 0{,}0071\ m^2. \quad (9)$$

The rate of the liquid flow in the cavity of the pendulum oscillations when applying the water hammer is equal to the rate of the liquid flow (V_1=V). The density of the liquid $p = 10^3$ kg/m^3.

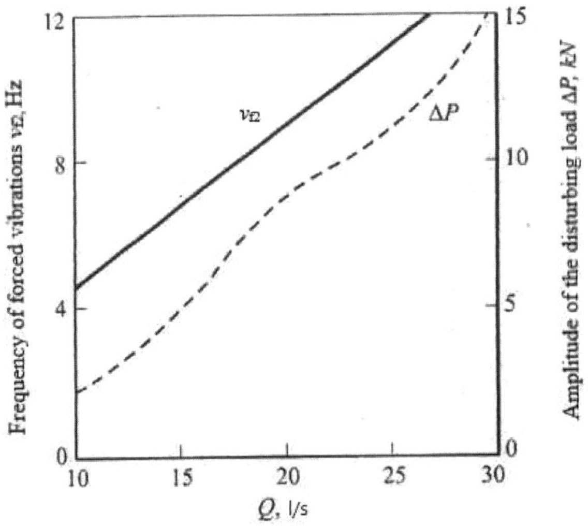

Figure 2. The influence of the flow rate Q on the frequency of the forced vibrations v_{f2} and the amplitude of the disturbing load ΔP of the pendulum vibrator, m: $l_1 = 0.052$; $l_2 = 0.064$; $l_3 = 0.032$.

4. Conclusion

Thus, the obtained values of the variable load ΔP are comparable with the results of bench studies of downhole vibrators.

References

[1] Yilbas B S, Patel F and Karatas C 2015 *Laser Surface Engineering* (Waugh J.L.G. Woodhead Publishing) pp 97-105

[2] Khabibullin M Ya and Suleimanov R I 2018 Choice of Bottom-Hole Filter Construction for Viscous Fluid Filtration *Advances in Engineering Research (AER)* **157** 272-275. DOI: 10.2991/aime-18.2018.52

[3] Kuleshova L S, Kadyrov R R, Mukhametshin V V and Akhmetov R T 2019 Auxiliary Equipment for Downhole Fittings of Injection Wells and Water Supply Lines Used to Improve Their Performance in Winter *IOP Conf. Ser.: Mater. Sci. Eng.* **560(1)** 012071. DOI: 10.1088/1757-899X/560/1/012071

[4] Rogachev A S, Shkodich N F, Vadchenko S G, Baras F, Chassagnon R, Sachkova N V and Boyarchenko O D 2013 *Int. J. Self-Propag. High-Temp. Synth.* **22(4)** 210–216

[5] Bataev I A, Bataev A A, Mali V I and Pavliukova D V 2012 *Materials and Design* **35** 225-234

[6] Akhmetov R T, Kuleshova L S and Mukhametshin V V 2019 Application of the Brooks-Corey Model in the Conditions of Lower Cretaceous Deposits in Terrigenous Reservoirs of Western Siberia *IOP Conf. Ser.: Mater. Sci. Eng.* **560(1)** 012004. DOI: 10.1088/1757-899X/560/1/012004

[7] Suleimanov R I, Khabibullin M Ya and Suleimanov Re I 2019 Analysis of the Reliability of the Power Cable of an Electric-Centrifugal Pump Unit *IOP Conf. Ser.: Earth Env.* **378(1)**

[8] Davydov A Yu 2018 Utilization of Expert Systems for Screw Pump Sets with Surface Drive Management *Advances in Engineering Research (AER)* **157** 129–133
[9] Galiullina I F and Kadyrov R R 2018 Technical and Economic Background for Sitting Production of Well-Killing Liquid at Oil Fields *IOP Conf. Ser.: Earth Env.* **194(8)**
[10] Gilmanova A M 2019 Consideration of Various Factors Influencing Acoustic Logging Equipment Production Accuracy *IOP Conf. Ser.: Mater. Sci. Eng.* **560(1)** 012063. DOI: 10.1088/1757-899X/560/1/012063
[11] Zaripova L M and Gabdrakhimov M S 2019 A Mathematical Model of the Pulsator for Cleaning Paraffin Deposits of Pipelines and Downhole Equipment *J. of Phys.: Conf. Ser.* **1333(3)**
[12] Zaripova L M and Gabdrakhimov M S 2019 Restoration of Intake Capacity of Injection Well by Vibrations *IOP Conf. Ser.: Earth Env.* **378(1)**

IOP Conf. Series: Materials Science and Engineering 1064 (2021) 012079 doi:10.1088/1757-899X/1064/1/012079

Influence of oscillatory phenomena in vertical well on volume of outflow from production string

L M Zaripova and M S Gabdrakhimov

Ufa State Petroleum Technological University, Branch of the University in the City of Oktyabrsky, 54a, Devonskaya St., Oktyabrsky, Republic of Bashkortostan, 452607, Russian Federation

E-mail: Lilyabert31@mail.ru, info@of.ugntu.ru

Abstract. During well operation, there is a leaky cement ring, which is due to various reasons, the main of which are the technology of casing production strings and the used grouting materials that do not regard the actual geological and technical conditions of wells. Thus, the most important task is to ensure high-quality annulus insulation during well cementing. Theoretical studies which can be used to calculate the volume of fluid outflowing from a cemented and uncemented casing regarding the pressure distribution in a vertical well on conditions that data on the characteristics of the casing and working fluid is known are an urgent issue.

1. Introduction

When the vertical wellhead is closed by a valve preventing the liquid from flowing out, a pressure p_0 exceeding the atmospheric pressure p_a is applied to the surface of the liquid from the side of the valve. After the instantaneous valve opening, part of the liquid will pour out due to an increase in its volume caused by a decrease in density, and to the production string narrowing caused by a decrease in pressure in the liquid [1-11]. Let us calculate the spilled volume under the assumption that the fluid is homogeneous and its compressibility does not depend on pressure.

2. Materials and methods

Let us introduce the following notation: H is the height of the column with liquid in a stress-free state; r is the inner radius of the column in a stress-free state; R is the outer radius of the column in a stress-free state; δ is a column wall thickness; β is a fluid compressibility ratio; E is elastic modulus of the column material; ρ is liquid density at an arbitrary pressure p; ρa is liquid density at atmospheric pressure pa; z is a coordinate axis descending the borehole axis.

3. Results

Let us use the obtained equation relating pressure and density:

$$\beta\rho dp = d\rho. \tag{1}$$

Taking into account the condition $p=p_a$ with $\rho=\rho_a$ according to (1) we obtain the following:

$$\rho = \rho_a e^{\beta(p-p_a)}. \tag{2}$$

Knowing the pressure $p = p_0$ at the wellhead with $z = 0$, according to (2) we obtain the liquid density at the level z=0:

$$\rho_0 = \rho_a e^{\beta(p_0 - p_a)}.$$

At an arbitrary depth z we select an infinitely small area dz. Along this interval, the pressure of the liquid column changes by the value $dp(z) = \rho(z)gdz$. Upon substituting this value in (1) and integrating, we obtain the dependence of the following fluid density on the depth:

$$\rho(z) = \frac{\rho_0}{1 - \beta\rho_0 gz},$$

where ρ_0 is the density of the liquid at a valve level, i.e. at z=0.

Let us find the dependence of the fluid pressure on the depth. Equality (1) can be rewritten as $\beta dp = d(\ln\rho)$. Having integrated regarding the conditions $p = p_0$, $\rho = \rho_0$ with $z = 0$, we obtain the following:

$$p(z) = p_0 - \frac{1}{\beta}\ln(1 - \beta\rho_0 gz).$$

This internal pressure expands the pipe, causing its radius to increase. Let us find the inner radius under the assumption that there is no external influence on the pipe. Increasing the radius means increasing the circumference. In turn, the increase in circumference is due to circumferential tensile stress in the pipe material. At depth, the stress in a thick-walled pipe caused by fluid pressure p (z) is as follows:

$$\sigma(z) = \frac{R^2 + r^2}{(R + r)\delta} p(z).$$

Let us designate the initial circumference $l = 2\pi r$, herewith, $\Delta l(z)$ is the increase in this length. Let us derive an equation relating specific elongation with the stress and we get the following:

$$\frac{\Delta l(z)}{l} = \frac{\sigma(z)}{E}.$$

Hence, we obtain the following equation:

$$\Delta l(z) = \frac{l}{E}\sigma(z).$$

Therefore, the new circumference is equal to the following:

$$L(z) = l + \Delta l(z) = l\left[1 + \frac{\sigma(z)}{E}\right].$$

Pipe radius at depth z is as follows:

$$r(z) = \frac{L(z)}{2\pi} = r\left[1 + \frac{\sigma(z)}{E}\right],$$

cross-sectional area is equal to:

$$S(z) = \pi r^2(z) = \pi r^2\left[1 + \frac{\sigma(z)}{E}\right]^2.$$

The mass of the compressed fluid is determined by the formula given below:

$$m = \int_{(V)} \rho(z)dV = \int_0^H \rho(z)S(z)dz.$$

IOP Conf. Series: Materials Science and Engineering 1064 (2021) 012079 doi:10.1088/1757-899X/1064/1/012079

Let us apply subintegral functions:

$$m = \pi r^2 \rho_0 \int_0^H \frac{1}{1-\beta\rho_0 gz} \left\{ 1 + \frac{R^2 + r^2}{(R+r)E\delta} \left[p_0 - \frac{1}{\beta} \ln(1-\beta\rho_0 gz) \right] \right\}^2 dz.$$

Calculation of the integral leads to the following expression:

$$m = \frac{\pi r^2}{3Ag} \left\{ \left[1 + Ap_0 - \frac{A}{\beta} \ln(1-\beta\rho_0 gH) \right]^3 - (1+Ap_0)^3 \right\},$$

which designates the following:

$$A = \frac{R^2 + r^2}{(R+r)E\delta}.$$

We have found a mass of liquid compressed in a pipe. Now let us open the latch. The pressure p_0 will decrease to atmospheric p_a, part of the liquid will pour out and the following mass will remain in the pipe:

$$m_a = \frac{\pi r^2}{3Ag} \left\{ \left[1 + Ap_a - \frac{A}{\beta} \ln(1-\beta\rho_a gH) \right]^3 - (1+Ap_a)^3 \right\}.$$

Consequently, the mass of the poured liquid $\Delta m = m - m_a$ is equal to the following:

$$\Delta m = \frac{\pi r^2}{3Ag} \left\{ \left[1 + Ap_0 - \frac{A}{\beta} \ln(1-\beta\rho_0 gH) \right]^3 - (1+Ap_0)^3 - \right.$$

$$\left. - \left[1 + Ap_a - \frac{A}{\beta} \ln(1-\beta\rho_a gH) \right]^3 + (1+Ap_a)^3 \right\}.$$

At atmospheric pressure, this mass outside the pipe will have a density ρ_a. Therefore, the spilled volume $\Delta V = \Delta m / \rho_a$ is as follows:

$$\Delta V = \frac{\pi r^2}{3Ag\rho_a} \left\{ \left[1 + Ap_0 - \frac{A}{\beta} \ln(1-\beta\rho_0 gH) \right]^3 - (1+Ap_0)^3 - \right.$$

$$\left. - \left[1 + Ap_a - \frac{A}{\beta} \ln(1-\beta\rho_a gH) \right]^3 + (1+Ap_a)^3 \right\}.$$

$E \to \infty$.corresponds to the case when there is a cement stone behind the production casing which does not allow the pipe to expand. Then $A \to 0$, and, using the approximations we obtain the following:

$$(1+Ap)^3 \to 1 + 3Ap,$$

$$\left[1 + Ap - \frac{A}{\beta} \ln(1-\beta\rho gH) \right]^3 \to 1 + 3Ap - \frac{3A}{\beta} \ln(1-\beta\rho gH),$$

Out of the last equality we find the following:

$$\Delta V_c = \frac{\pi r^2}{\beta \rho_a g} \ln \frac{1-\beta \rho_a g H}{1-\beta \rho_0 g H}.$$

As you can see, the volume ΔVc pouring out of the cemented column does not depend on the characteristics of the column E and δ.

The table for pipes of various diameters shows the calculated values of the outflowing volume obtained with the following data:

$$E = 2{,}1 \cdot 10^{11} \text{ Pa}, \ \delta = 7 \cdot 10^{-3} \text{ m}, \ H = 2000 \text{ m}, \ g = 9{,}81 \text{m/s}^2, \ \beta = 4 \cdot 10^{-10} \text{ Pa}^{-1},$$
$$p_a = 10^5 \ Pa \ \ p_0 = 3 \cdot 10^7 \ Pa \ \ \rho_a = 1250 \text{ kg/m}^3.$$

Table 1. Values of outflowing volume are calculated for different diameters

Column outer diameter, mm	Outflowing volume, l	
	from an uncemented column	from cemented column
60.3	77	69
73	116	102
89	176	151

4. Conclusion

Formulas enabling to determine the density of the liquid and the pressure in it with regards to the depth and the elastic properties of the liquid are obtained. Formulas enabling to calculate the volume of outflowing fluid from cemented and uncemented casings under the situation of known data on the casing and working fluid characteristics are obtained.

References

[1] Yilbas B S, Patel F and Karatas C 2015 *Laser Surface Engineering* (Waugh J.L.G. Woodhead Publishing) pp 97-105
[2] Stepanova N V and Razumakov A A 2013 *The 8 Int. forum on strategic technologies (IFOST 2013)* vol 1 pp 240–242
[3] Yakupov R F, Mukhametshin V Sh, Khakimzyanov I N and Trofimov V E 2019 Optimization of reserve production from water oil zones of D3ps horizon of Shkapovsky oil field by means of horizontal wells *Georesursy* **21(3)** 55-61. DOI: 10.18599/grs.2019.3.55-61
[4] Rogachev A S, Shkodich N F, Vadchenko S G, Baras F, Chassagnon R, Sachkova N V and Boyarchenko O D 2013 *Int. J. Self-Propag. High-Temp. Synth.* **22(4)** 210–216
[5] Bataev I A, Bataev A A, Mali V I and Pavliukova D V 2012 *Materials and Design* **35** 225–234
[6] Mukhametshin V V and Kuleshova L S 2019 Justification of Low-Productive Oil Deposits Flooding Systems in the Conditions of Limited Information Amount *SOCAR Proceedings* **2** 16–22. DOI: 10.5510/OGP20190200384
[7] Suleimanov R I, Khabibullin M Ya and Suleimanov Re I 2019 Analysis of the reliability of the power cable of an electric-centrifugal pump unit *IOP Conf. Ser.: Earth Env.* **378(1)**
[8] Davydov A Yu 2018 Utilization of expert systems for screw pump sets with surface drive management *Advances in Engineering Research (AER)* **157** 129-133
[9] Galiullina I F and Kadyrov R R 2018 Technical and economic background for siting production of well-killing liquid at oil fields *IOP Conf. Ser.: Earth Env.* **194(8)** 082013
[10] Zaripova L M and Gabdrakhimov M S 2019 A mathematical model of the pulsator for cleaning paraffin deposits of pipelines and downhole equipment *J. of Phys.: Conf. Ser.* **1333(3)**
[11] Zaripova L M and Gabdrakhimov M S 2019 Restoration of intake capacity of injection well by vibrations *IOP Conf. Ser.: Earth Env.* **378(1)**

AUTHOR INDEX